Wavelets:
A Tutorial in Theory and Applications

Wavelet Analysis and Its Applications

The subject of wavelet analysis has recently drawn a great deal of attention from mathematical scientists in various disciplines. It is creating a common link between mathematicians, physicists, and electrical engineers. This book series will consist of both monographs and edited volumes on the theory and applications of this rapidly developing subject. Its objective is to meet the needs of academic, industrial, and governmental researchers, as well as to provide instructional material for teaching at both the undergraduate and graduate levels.

This is the second volume of the series. It is a compilation of twenty-two chapters covering the basic theory, analysis, algorithms, and several of the most important applications of wavelets. The series editor is very grateful to the authors of these chapters for their very fine contributions to this volume.

This is a volume in
WAVELET ANALYSIS AND ITS APPLICATIONS

CHARLES K. CHUI, SERIES EDITOR
Texas A&M University, College Station, Texas

A list of titles in this series appears at the end of this volume.

Wavelets:
A Tutorial in Theory and Applications

EDITED BY

CHARLES K. CHUI

Department of Mathematics
Texas A&M University
College Station, Texas

ACADEMIC PRESS, INC.

Harcourt Brace Jovanovich, Publishers

Boston San Diego New York
London Sydney Tokyo Toronto

This book is printed on acid-free paper. ∞

ACADEMIC PRESS, INC.
1250 Sixth Avenue, San Diego, CA 92101-4311

United Kingdom Edition published by
ACADEMIC PRESS LIMITED
24–28 Oval Road, London NW1 7DX

LIBRARY OF CONGRESS CATALOG CARD NUMBER: 91-58833
ISBN 0-12-174590-2

Printed in the United States of America

93 94 95 BB 9 8 7 6 5 4

Contents

Preface

Wavelet analysis is a rapidly developing area in the mathematical sciences which is emerging as an independent field of investigation. Moreover, it has already created a common link between mathematicians and electrical engineers, and has even drawn a great deal of attention from scientists and engineers in other disciplines.

This volume consists of twenty-two chapters contributed by specialists in various areas in this field. The material compiled is fairly wide in scope and ranges from the development of the basic theory such as orthonormal wavelet bases to applied problems such as second generation compact image coding. Most of the chapters are tutorial in nature as suggested by the title of this book. The classification into seven parts is not intended to separate these chapters into different areas, since there is obviously no distinct boundary between them. Rather, the objective is to serve as a reader's guide and to convey a general idea of what this volume is about. Hence, even within each part of this volume, the chapters are arranged according to subject matter rather than the alphabetical order of their authors.

Pollen's discussion of Daubechies' first scaling function constitutes the first chapter of this volume not only because the construction of compactly supported orthonormal wavelets by Daubechies made a very great impact on the development of this field, but Pollen's presentation starts from the very first principles. The final two chapters, on the other hand, discuss the state-of-the-art directions in applications of wavelets to image coding engineered by Mallat and to acoustic signal compressions using wavelet packets introduced by Meyer, Coifman, Quake, and Wickerhauser.

Part I of this volume consists of three chapters that are directly or indirectly concerned with orthogonal wavelets. Immediately following Pollen's chapter, Heller, Resnikoff, and Wells discuss wavelet matrices and their fundamental invariants. These arise as the defining coefficients of compactly supported wavelet systems. Walter compares the development of orthogonal wavelets with that of distributions and discusses the interaction between these two theories.

In order to obtain compactly supported wavelets that are continuous as well as symmetric or antisymmetric, one has to abandon orthogonality. Part II of this volume is devoted to a study of semi-orthogonal and nonorthogonal wavelets. A wavelet is said to be semi-orthogonal if any two different dyadic scales of it are orthogonal, and it is called nonorthogonal if no such orthogonality is required. Of course, being a wavelet, it must have a dual; and hence nonorthogonal wavelets are also called biorthogonal wavelets in the wavelet literature. Battle expounds the block spin approach to constructing semi-orthogonal wavelets that are cardinal spline functions, while Unser and Aldroubi present such wavelets from a point of view of signal analysis. Since symmetric or antisymmetric compactly supported continuous semi-orthogonal wavelets cannot have compactly supported duals, nonorthogonal wavelets are also important. Cohen not only details the construction procedure of such wavelets, but also describes the corresponding class of subband coding schemes and frame bounds that are essential to stability in wavelet decomposition and reconstruction schemes. The chapter by Feauveau is another tutorial on this topic from a different point of view. It should be pointed out that Cohen and Feauveau, along with Daubechies, are the main contributors to the theory of nonorthogonal wavelets.

In addition to wavelet bases generated by integer translations of dyadic dilations, there are other local bases that are also important. Part III consists of three chapters on different wavelet-like local bases. Alpert introduces the ideas of wavelets within the context of mathematical physics and constructs several types of wavelet-like bases with illustrations on algorithmic solutions of a variety of integral and differential equations. Auscher describes a procedure for constructing wavelets with preassigned boundary conditions on a bounded interval, by applying the approximation theoretic notion of Hermite interpolation. The following chapter by Auscher, Weiss, and Wickerhauser, gives a detailed account of the local sine and cosine bases of Coifman and Meyer. These authors also discuss some of the applications of these local bases and particularly show that they can be used to yield arbitrarily smooth wavelets.

In the first three parts described above, only the univariate theory of wavelets is considered. Part IV of this volume is devoted to a study of the multivariate theory. To start with, Madych studies with great care the elementary properties of multiresolution analysis in the multivariate setting. The chapter by Berger and Wang discusses multiresolution analyses from the point of view of subdivision schemes. In particular, applications to surface generation are presented in this chapter. In Stöckler's chapter on the construction of non-tensor product multivariate wavelets, a universal tool to identify special properties of a wavelet basis is developed. Important properties such as compact supports and generalized linear phases are considered. A highlight of this chapter is the explicit formulation of a symbol matrix that yields compactly supported wavelets with generalized linear phase in any dimension. Multivariate box splines are used to construct examples of such wavelets.

Wavelet series representations are important because the coefficients contain important local information on time-frequency or phase-space. Other important windowed integral transforms are short-time Fourier and windowed Radon transforms. Part V of this volume consists of two chapters on these transforms. Feichtinger and Gröchenig analyze series expansions of signals with respect to Gabor wavelets. Using Heisenberg group techniques, they are able to design stable iterative algorithms for signal analysis and synthesis, yielding information on time-frequency localization. In the chapter on windowed Radon transforms by Kaiser and Streater, a reconstruction formula is derived which inverts the Radon transform. The ideas in this chapter originated in relativistic quantum theory in mathematical physics.

In order to apply decomposition and reconstruction algorithms to signal analysis or to analyze solutions of certain mathematical modeling problems, one has to work with a sample space. Part VI of this volume is devoted to a study of sampling and interpolatory theory and related problems. The first chapter written by Benedetto is a comprehensive tutorial article on this important subject. It starts with the classical sampling theorem and the Paley-Wiener Theorem, leading up to a detailed exposition of frames and irregular sampling. The second chapter authored by Aldroubi and Unser discusses Shannon's sampling theory and the Gabor transform using the framework of wavelet transforms. In the chapter by Seip, on the other hand, a complete description of sampling and interpolation in the Bargmann-Fock space is given using Beurling's density concept. A brief discussion on continuous wavelet transforms, Bergman spaces, and a von Neumann type lattice is also included in this chapter.

Part VII of this volume is a collection of four chapters on different important applications of wavelets. The chapter by Jaffard and Laurençot provides a review of the algorithms for the construction of orthonormal wavelets. It details the main properties of wavelet decompositions, shows how to use these bases in the study of a large class of operators, and gives both theoretical and numerical applications to partial differential equations. One of the most important applications of wavelets is to separate an analog signal into octaves, yielding local time-frequency information. The chapter by Gopinath and Burrus presents the filter bank theory of such applications. As already mentioned earlier in this preface, Mallat has been very successful in applying the wavelet transform to image coding. In the chapter by Froment and Mallat, a compact image coding algorithm that separates the edges from texture information is introduced. It allows the user to adapt the coding precision to the properties of the human visual perception. As in Mallat's earlier work, multiscale edges are detected from local maxima of the wavelet transform modulus. Orthogonal wavelet packets introduced by Coifman, Meyer, Quake, and Wickerhauser are discussed by Wickerhauser in the last chapter of the volume. This chapter also discusses some results in acoustic signal compression with wavelet packets using a simple counting cost function.

This volume is a compilation of various important topics in wavelet analysis and its applications. In addition to these twenty-two chapters, a fairly extensive bibliography is included at the end of this volume. It is hoped that the reader will find at least one topic relevant to his or her own research or can learn about the subject of wavelets in general. Although the TeX files of most of the chapters were prepared by the authors themselves, Robin Campbell has been very helpful in preparing the remaining TeX files. My wife, Margaret, was of great help in unifying the format in all the chapters, and both she and my assistant Stephanie Sellers spent long hours in preparing the manuscript in camera-ready form. I am greatly indebted to all of them. Finally, to the editorial office of Academic Press, and particularly to Charles Glaser, I am grateful for their efficient assistance.

College Station, Texas
October, 1991

Charles K. Chui

Part I

Orthogonal Wavelets

Daubechies' Scaling Function on $[0,3]$

David Pollen

Abstract. We construct Daubechies' scaling function supported on $[0,3]$ from first principles and prove that it is continuous everywhere, left-differentiable at dyadic rationals and nowhere right-differentiable on dyadic rationals on $[0,3)$. Furthermore, we prove that its integer translates are orthonormal and that its definite integral equals one. This scaling function is one of an infinite class of scaling functions introduced by Daubechies in [1] for the purpose of constructing orthonormal bases of compactly supported wavelets. The particular scaling function studied in this paper is distinguished from the others by the property that it is the simplest scaling function which can be used to construct a complete orthonormal wavelet basis of $L^2(\mathbf{R})$ whose primary wavelet is continuous.

§1. Preliminaries

A function f is said to have *support* in the interval $[0,3]$, if $f(x)=0$ when $x \notin [0,3]$, and $[0,3]$ is the smallest closed interval for which this holds. For each $j \in \mathbf{Z}$, the set of *dyadic rationals of level* j is defined to be

$$\mathbf{D}_j = \{\frac{k}{2^j} : k \in \mathbf{Z}\} .$$

Hence, we have $\mathbf{D}_{-1} = 2\mathbf{Z}$, $\mathbf{D}_0 = \mathbf{Z}$, and $\mathbf{D}_1 = \frac{1}{2}\mathbf{Z}$. Define the ring of dyadic rationals ("the *dyadics*") by $\mathbf{D} = \cup_{k\in\mathbf{Z}}\mathbf{D}_k$. Note that $\mathbf{D}$ is not a field. In this paper, many of our calculations will be in the quadratic number field

$$\mathbf{Q}(\sqrt{3}) = \{\alpha + \beta\sqrt{3} : \alpha, \beta \in \mathbf{Q}\} .$$

This number field has a conjugation operation denoted by overscore, namely:

$$\overline{(\alpha + \beta\sqrt{3})} = (\alpha - \beta\sqrt{3}) .$$

Throughout this paper, we let

$$a = \frac{1+\sqrt{3}}{4} .$$

Wavelets–A Tutorial in Theory and Applications
C. K. Chui (ed.), pp. 3–13.

ISBN 0-12-174590-2

Then a and $\overline{a}$ are the two roots of $8x^2 - 4x - 1 = 0$. The following upper and lower bounds will be important to many of our calculations:

$$0.5 < a = 0.68301... < 1.0$$
$$-0.25 < \overline{a} = -0.183012... < 0 .$$

§2. The scaling function on D

In this section we show that there exists a unique function φ defined on $\mathbf{D}$ that satisfies the following conditions:

1. If $x \in \mathbf{D}$ then φ satisfies the *scaling relation*

$$\begin{aligned}\varphi(x) =& a\varphi(2x) + (\overline{1-a})\varphi(2x-1) \\ &+ (1-a)\varphi(2x-2) + \overline{a}\varphi(2x-3) ;\end{aligned}$$

2. φ satisfies the *normalization* constraint

$$\sum_{k \in \mathbf{Z}} \varphi(k) = 1 ;$$

and

3. φ vanishes outside of $[0, 3]$.

Our first step toward the proof will be to investigate functions $x \mapsto \varphi(x)$ that satisfy these three conditions for x restricted to $\mathbf{Z}$.

First assume that at least one solution φ exists for x restricted to $\mathbf{Z}$. Then, by successively substituting $x = 0, 1, 2,$ and 3 into the scaling relation and requiring that φ be zero outside of the interval $[0, 3]$, we obtain the following relation

$$\begin{pmatrix} \varphi(0) \\ \varphi(1) \\ \varphi(2) \\ \varphi(3) \end{pmatrix} = \begin{pmatrix} a & 0 & 0 & 0 \\ (1-a) & \overline{1-a} & a & 0 \\ 0 & \overline{a} & (1-a) & \overline{1-a} \\ 0 & 0 & 0 & \overline{a} \end{pmatrix} \begin{pmatrix} \varphi(0) \\ \varphi(1) \\ \varphi(2) \\ \varphi(3) \end{pmatrix} .$$

This matrix equation and the normalization constraint have exactly one common solution, namely:

$$\begin{aligned} \varphi(0) &= 0 \quad , & \varphi(1) &= \tfrac{1+\sqrt{3}}{2} \quad , \\ \varphi(2) &= \tfrac{1-\sqrt{3}}{2} \quad , & \varphi(3) &= 0 \quad . \end{aligned}$$

These are necessary conditions that must be satisfied by every solution φ.

If a solution exists, then the values of φ on $\mathbf{D} \backslash \mathbf{Z}$ are determined by the values of φ on $\mathbf{Z}$ and can be recursively computed from the scaling relation.

The scaling relation expresses the value of φ at $x \in \mathbf{D}_j$ as a weighted linear sum of $\varphi(2x)$, $\varphi(2x-1)$, $\varphi(2x-2)$, $\varphi(2x-3)$, all of which are values of φ at dyadic points belonging to the previous level. This recursion can be iterated until $\varphi(x)$ is expressed in terms of values of φ at dyadics at level 0, which are precisely the values of φ on $\mathbf{Z}$. If $x \notin [0,3]$, then for an integer x not in $\{0,1,2,3\}$, $\varphi(x)$ must be zero. For a dyadic x, an iteration of the scaling relation expresses $\varphi(x)$ in terms of φ evaluated on integers other than $\{0,1,2,3\}$; hence $\varphi(x)=0$. Thus, if there exists a function φ on $\mathbf{D}$ satisfying the required properties, then it must be unique.

Furthermore, it follows from this analysis that a solution φ exists and takes on the values described above on the integers. This function is (the restriction to $\mathbf{D}$) of *Daubechies' four coefficient continuous scaling function on* $[0,3]$ which was introduced in [1] and is further investigated in [2]. The values of φ on the half-integers are:

$$\varphi(1/2) = \frac{2+\sqrt{3}}{4} \; ;$$
$$\varphi(3/2) = 0 \; ;$$
$$\varphi(5/2) = \frac{2-\sqrt{3}}{4} \; .$$

§3. Basic results about φ on D

Examining the values of φ at the integers and half-integers reveals two important facts, namely:

Theorem 1. $x \in \mathbf{D}$ *implies* $\varphi(x) \in \mathbf{D}[\sqrt{3}]$.

Theorem 2. *The values of* φ *are conjugate symmetric about* $3/2$; *i.e., if* $x \in \mathbf{D}$, *then* $3-x \in \mathbf{D}$ *and* $\varphi(3-x) = \overline{\varphi(x)}$.

These elementary theorems can both be proved by induction on the level of the dyadic argument. First verify that the result holds for the dyadic points of level 0; *i.e.*, the integers. For the induction step, use the scaling relation and the induction hypothesis to prove the result for the next level of points.

This scheme of proof is elementary but extremely useful. It can, for instance, be used to derive the following two properties of φ for every dyadic x:

Theorem 3. φ *satisfies the property of partition of unity*

$$\sum_{k\in\mathbf{Z}} \varphi(x-k) = 1 \; .$$

Theorem 4. φ *satisfies the property of partition of* x

$$\sum_{k\in\mathbf{Z}} \left(\frac{3-\sqrt{3}}{2} + k\right) \varphi(x-k) = x \; .$$

From these properties of partition of unity and the partition of x, and using the fact that the support of φ is contained in $[0,3]$ we have the following result.

Theorem 5. *For $0 \le x \le 1$ and $x \in \mathbf{D}$, φ satisfies the properties of interval translation:*

$$2\varphi(x) + \varphi(x+1) = x + \frac{1+\sqrt{3}}{2}\ ;$$

$$2\varphi(2+x) + \varphi(x+1) = -x + \frac{3-\sqrt{3}}{2}\ ;$$

$$\varphi(x) - \varphi(x+2) = x + \frac{-1+\sqrt{3}}{2}\ .$$

Combining the properties of partition of φ with the above properties and with the scaling relation yields the *half-interval recursions*:

Theorem 6. *For $0 \le x \le 1$ and $x \in \mathbf{D}$,*

$$\varphi(\frac{0+x}{2}) = a\varphi(x)\ ;$$

$$\varphi(\frac{1+x}{2}) = \overline{a}\varphi(x) + ax + \frac{2+\sqrt{3}}{4}\ ;$$

$$\varphi(\frac{2+x}{2}) = a\varphi(1+x) + \overline{a}x + \frac{\sqrt{3}}{4}\ ;$$

$$\varphi(\frac{3+x}{2}) = \overline{a}\varphi(1+x) - ax + \frac{1}{4}\ ;$$

$$\varphi(\frac{4+x}{2}) = a\varphi(2+x) - \overline{a}x + \frac{3-2\sqrt{3}}{4}\ ;$$

$$\varphi(\frac{5+x}{2}) = \overline{a}\varphi(2+x)\ .$$

The above six formulae relate the values of φ in the six half-integer intervals in $[0,3]$ to the three unit intervals in $[0,3]$. These formulae also provide a recursive method for computing the values of φ on $\mathbf{D}$ which is similar to the formulae provided by the scaling relation. However, the new recursion method is more efficient. The value of φ at one level can be computed from one value of φ on the previous level. In contrast, using the scaling relation requires four values of φ from the previous level.

§4. Extension of the scaling function to R

Definition 7. *For* $0 \leq x \leq 1$, *define an operator* K *that satisfies:*

$$K(f)(\frac{x}{2}) = af(x) \ ;$$

$$K(f)(\frac{x+1}{2}) = \overline{a}f(x) + ax + \frac{2+\sqrt{3}}{4} \ ;$$

$$K(f)(\frac{x+2}{2}) = af(1+x) + \overline{a}x + \frac{\sqrt{3}}{4} \ ;$$

$$K(f)(\frac{x+3}{2}) = \overline{a}f(1+x) - ax + \frac{1}{4} \ ;$$

$$K(f)(\frac{x+4}{2}) = af(2+x) - \overline{a}x + \frac{3-2\sqrt{3}}{4} \ ;$$

$$K(f)(\frac{x+5}{2}) = \overline{a}f(2+x) \ .$$

For $x \notin [0,3]$, *define* $K(f)(x) = 0$.

Lemma 8. $K(\varphi) = \varphi$ on $\mathbf{D}$.

Let $g_0 : \mathbf{R} \to \mathbf{R}$ denote the function that is piecewise linear function with break-points at $\mathbf{Z}$ and assumes the values of φ on the integers. Then g_0 is continuous with support $[0,3]$. For $n \in \mathbf{N}$, define $g_n = K(g_{n-1}) = K^n(g_0)$. Outside the interval $[0,3]$, let g_n be zero.

For an arbitrary continuous function f, $K(f)$ may be multiply defined on $\{0, 1/2, 1, 3/2, 2, 5/2, 3\}$ possibly with essential discontinuities. However, each g_n must be continuous everywhere and must have support $[0,3]$. Furthermore, each g_n is linear on every dyadic interval $[\frac{m}{2^n}, \frac{m+1}{2^n}]$ for $m \in \mathbf{Z}$.

Denote by $||f||_\infty$ the maximum of the absolute value of f. It follows by investigating each of the six half-intervals that if $||f||_\infty \leq 3$ then $||K(f)||_\infty \leq 3$. Since $||g_0||_\infty \leq 3$, then for all k, $||g_n||_\infty \leq 3$. Fix $j, k \in \mathbf{Z}_+$. From the definition of K and the fact that $0 \leq |\overline{a}| \leq a$, we have

$$||g_k - g_{k+j}||_\infty \leq a^k ||g_0 - g_j||_\infty \leq a^k(||g_0||_\infty + ||g_j||_\infty) \leq 6a^k.$$

The first inequality follows from the definition of K and can be proved using the recursive technique on each of the six half-intervals; the second follows from the triangle inequality; and the third is a consequence of the preceeding paragraph. Hence, we have

$$||g_k - g_{k+j}||_\infty \leq 6a^k.$$

Recall that $a < 1$. Thus, the sequence $\{g_k\}$ is a Cauchy sequence of functions with respect to the norm $||\ ||_\infty$; and for each x, the sequence $g_k(x)$ is a Cauchy sequence in $\mathbf{R}$. Define the function g by

$$g(x) = \lim_{n \to \infty} g_n(x) \ .$$

Then g is continuous, being the limit of a uniformly convergent sequence of continuous functions. Note that the $\{g_n\}$ converges to g uniformly with geometrically decreasing error whose ratio is given by the value of a.

Since g_k and φ agree on $\mathbf{D}_k$, the dyadic points of level k, the restriction of g to $\mathbf{D}$ is equal to φ. Hence, g is a continuous extension of φ, and we extend φ to $\mathbf{R}$ by defining $\varphi(x) = g(x)$ for all $x \in \mathbf{R}$.

The conjugate symmetry property of φ about 3/2 cannot be extended – indeed, it cannot even be directly stated – for the extension of φ to a continuous function on $\mathbf{R}$. However, because $\mathbf{D}$ is dense in $\mathbf{R}$, *every other previously stated property* of φ, such as the scaling relation, or the partition of unity, extends to $\mathbf{R}$ by continuity.

§5. Differentiability of the scaling function

We will now show that φ is not right-differentiable at zero. First, observe that since $\varphi(0) = 0$, we have

$$\begin{aligned}\lim_{j\to\infty} \frac{\varphi(1/2^j) - \varphi(0)}{1/2^j} &= \lim_{j\to\infty} \frac{\varphi(1/2^j)}{1/2^j} \\ &= \lim_{j\to\infty} \frac{a^j \varphi(1)}{1/2^j} \\ &= \lim_{j\to\infty} (2a)^j \varphi(1) \ ,\end{aligned}$$

where the recursion on the first half-interval is used. But this limit diverges, since $\varphi(1) \neq 0$ and $1 < 2a$. This establishes the following result.

Theorem 9. *φ is not right-differentiable at zero.*

Now consider any $k/2^j \in \mathbf{D}$. Using the half-integral recursions again, we can derive the *local dyadic similarity property* of φ, namely:

Theorem 10. *There exist $\alpha, \beta, \gamma \in \mathbf{Q}(\sqrt{3})$, depending on k and j, such that for any x, $0 \leq x \leq 1$,*

$$\varphi((k+x)/2^j) = \alpha\varphi(x) + \beta x + \gamma \ .$$

This theorem asserts that, with the exception of an additive linear term and a multiplicative constant, φ "looks the same" on every dyadic interval.

If we combine the local similarity property with the fact that φ is not right-differentiable at zero, we immediately arrive at the following.

Theorem 11. *φ is not right-differentiable at any dyadic in $[0,3)$.*

This theorem was first proved in [2] by other means. (Obviously, φ is right-differentiable outside of $[0,3)$, since it is zero there.)

We now investigate the left-differentiability of φ at $x=3$, the right edge of the support of φ. We will show that φ is right-differentiable at 3. Let x_k be an arbitrary sequence whose limit is 3, such that $x_k \leq 3$. Without loss of generality, by taking a subsequence, we can assume that $x_j \leq x_{j+1}$ for all j, and we can also assume that there is at most one element of the sequence in each interval $[3-1/2^j, 3-1/2^{j+1}]$. By re-indexing the sequence we can denote this element by x_j. Hence, x_j is indexed by some infinite subsequence of $\mathbf{N}$ which will not affect the limit.

Next, observe that

$$\varphi'(3^-) = -\lim_{j\to\infty}\left(\frac{\varphi(3)-\varphi(x_j)}{3-x_j}\right) .$$

By the last half-integral recursion, we see that

$$|\varphi(3)-\varphi(x_j)| \leq |\overline{a}|^j ||\varphi||_\infty .$$

Furthermore, by adjusting our above sequence, we arranged that

$$1/2^j \leq 3-x_j \leq 1/2^{j+1} .$$

Hence, since $|2\overline{a}|<1$, we have

$$|\varphi'(3^-)| \leq \lim_{j\to\infty} ||\varphi||_\infty |2\overline{a}|^j = 0 .$$

Thus, φ is left-differentiable at 3 with left-derivative equal to 0. By the local similarity property of φ on dyadic intervals, the following result is an immediate consequence.

Theorem 12. *φ is left-differentiable at every dyadic.*

We denote the left-derivative by φ', and proceed to investigate its values. By left-differentiating the scaling relation for φ, we obtain the *scaling relation* for φ', namely:

$$\frac{1}{2}\varphi'(x) = a\varphi'(2x) + (\overline{1-a})\varphi'(2x-1) + (1-a)\varphi'(2x-2) + \overline{a}\varphi'(2x-3) ,$$

which is analogous to the scaling relation for φ but has a factor of $1/2$ on the left side. This scaling relation for φ' will allow us to recursively compute the values of φ' on $\mathbf{D}$, once the values of φ' on the integers are known. The scaling relation for φ' yields

$$\frac{1}{2}\begin{pmatrix} \varphi'(0) \\ \varphi'(1) \\ \varphi'(2) \\ \varphi'(3) \end{pmatrix} = \begin{pmatrix} a & 0 & 0 & 0 \\ (1-a) & \overline{1-a} & a & 0 \\ 0 & \overline{a} & (1-a) & \overline{1-a} \\ 0 & 0 & 0 & \overline{a} \end{pmatrix} \begin{pmatrix} \varphi'(0) \\ \varphi'(1) \\ \varphi'(2) \\ \varphi'(3) \end{pmatrix} .$$

This equation is analogous to the matrix equation relating the values of φ on the integers, except that the left-side factor of $1/2$. In addition to the matrix equation, the values of φ' are related by the following *normalization constraint* for φ' which is obtained by left-differentiating the partition of x property of φ and then substituting $x = 0$:

$$\sum_{k \in \mathbf{Z}} \left(\frac{3-\sqrt{3}}{2} + k \right) \varphi'(-k) = 1 .$$

The matrix equation with the normalization constraint has the unique solution $\varphi'(0) = 0$, $\varphi'(1) = 1$, $\varphi'(2) = -1$, and $\varphi'(3) = 0$. By using the scaling relation for φ', we have the following result by induction.

Theorem 13. *If $x \in \mathbf{D}$, then $\varphi'(x) \in \mathbf{D}[\sqrt{3}]$ and*

$$\varphi'(x) = -\overline{\varphi'(3-x)} .$$

§6. Orthogonality properties of the scaling function

Theorem 14. *(Definite Integral of the Scaling Function)*

$$\int_{\mathbf{R}} \varphi(x)dx = 1 .$$

Proof: This result is a direct consequence of the property of partition of unity, by utilizing a Minkowski-type argument. For $K \in \mathbf{N}$ with $K \geq 10$, consider the function

$$f_K(x) = \sum_{k=-K}^{K} \varphi(x-k) .$$

Since φ has support $[0, 3]$, the property of partition of unity shows that f_K is one on $[-K+3, K-3]$, zero on $(-\infty, -K-3]$, and zero on $[K+3, \infty)$. The value of this function is non-constant on $[-K-3, -K+3]$ and on $[K-3, K+3]$. We do not care what values it attains on these intervals, but simply note that the absolute value of the functions f_K is bounded on these intervals by a bound C independent of K because the scaling function is bounded in absolute value by 3 and there are at most 4 translates of φ that contribute to every value of $f_K(x)$. Therefore, we have

$$2K+1-C \leq \int_{\mathbf{R}} f_K(x)dx \leq 2K+1+C ,$$

where C is independent of K. Hence ,

$$1 = \lim_{K\to\infty}\left(\frac{1}{2K+1}\right)\int_{\mathbf{R}}\sum_{k=-K}^{K}\varphi(x-k)dx = \int_{\mathbf{R}}\varphi(x)dx$$

from which the theorem follows immediately.

We now prove that the integer translates of the scaling function are orthonormal.

Theorem 15. *(Orthonormality Property of the Scaling Function)*

$$\int_{\mathbf{R}} \varphi(x)\varphi(x-k)dx = \delta_{k,0} ,$$

for each $k \in \mathbf{Z}$.

Proof: Define

$$\Lambda_k = \int_{\mathbf{R}} \varphi(x)\varphi(x-k)dx .$$

We wish to show that $\Lambda_k = \delta_{k,0}$. By a simple change of variables, we see that $\Lambda_k = \Lambda_{-k}$. Additionally, for $k \geq 3$, the supports of $\varphi(x)$ and $\varphi(x-k)$ are disjoint; so $\Lambda_k = 0$. Therefore, we need only prove the theorem for Λ_0, Λ_1, and Λ_2. For each $k = 0, 1$, and 2, take the definition of Λ_k and substitute the scaling relation for $\varphi(x)$ and for $\varphi(x-k)$ to obtain an expression for Λ_k as a finite sum of terms of the form:

$$\int_{\mathbf{R}} \varphi(2x-r)\varphi(2x-2k-s)dx ,$$

where $r, s \in \mathbf{Z}$. By another simple change of variables, this expression is in fact seen to be $\frac{1}{2}\Lambda_{s+2k-r}$, which must be either zero or half of Λ_0, Λ_1, or Λ_2. Thus, we obtain three linear constraints on Λ_0, Λ_1, and Λ_2. Since these are homogeneous, we do not have a unique solution. However, for every solution, we necessarily have $\Lambda_1 = 0$ and $\Lambda_2 = 0$. This establishes the theorem for all $k \neq 0$. Using the property of partition of unity, our result on the definite integral of φ, and the fact that we have proven that $\Lambda_k = 0$ for $k \neq 0$, we may conclude that

$$\begin{aligned} 1 &= \int_{\mathbf{R}} \varphi(x)dx \\ &= \int_{\mathbf{R}} \varphi(x) \left(\sum_{k \in \mathbf{Z}} \varphi(x-k) \right) dx \\ &= \sum_{k \in \mathbf{Z}} \int_{\mathbf{R}} \varphi(x)\varphi(x-k)dx \\ &= \sum_{k \in \mathbf{Z}} \Lambda_k \\ &= \Lambda_0 \ . \end{aligned}$$

The interchange of summation and integration is justified by the fact that φ has compact support.

§7. The corresponding wavelet

As usual, we define the *corresponding wavelet* function $\psi(x)$ by

$$\begin{aligned} \psi(x) = &- \overline{a}\varphi(2x) + (1-a)\varphi(2x-1) \\ &- \overline{(1-a)}\varphi(2x-2) + a\varphi(2x-3) \ . \end{aligned}$$

From this definition, it immediately follows that ψ is a continuous function which is supported on $[0,3]$ and has zero mean. The latter property should be emphasized.

Theorem 16. *(Definite Integral of the Wavelet)*

$$\int_{\mathbf{R}} \psi(x) = 0 \ .$$

Using the scaling relation and the definition of the wavelet and the orthonormality of the scaling function, we also have the following results.

Theorem 17. *(Orthogonality Properties of the Wavelet)*

$$\int_{\mathbf{R}} \psi(x)\psi(x-k)dx = \delta_{k,0} \ ,$$

for all $k \in \mathbf{Z}$.

Theorem 18. *(Orthogonality Properties of the Scaling Function and the Wavelet)*

$$\int_{\mathbf{R}} \varphi(x)\psi(x-k)dx = 0 \ ,$$

for all $k \in \mathbf{Z}$.

From the above theorem, and the fact that the definite integral of the wavelet is zero, a substitution of the partition of x of the scaling function yields the following result.

Theorem 19. *(Vanishing of the First Moment of the Wavelet)*

$$\int_{\mathbf{R}} x\psi(x)dx = 0 \ .$$

Finally we conclude by stating that the set of function $\{2^{j/2}\psi(2^j - k)\}$, $j, k \in \mathbf{Z}$, is a complete orthonormal basis of $L^2(\mathbf{R})$. The proof can be completed easily by applying the results we already derived, but we refer the interested reader to a more general treatment in [3].

Acknowledgements

The author is pleased to acknowledge fruitful discussions with Andy Latto, David Plummer, David Linden, and Howard Resnikoff.

References

1. Daubechies, I., Orthonormal bases of compactly supported wavelets, *Comm. Pure and Appl. Math.* **41** (1988), 909–996.
2. Daubechies, I. and J. C. Lagarias, Two-scale difference equations, Parts I and II, AT&T Bell Labs., 1988, preprint.
3. Lawton, W., Tight frames of compactly supported affine wavelets, *J. Math. Phys.* **31** (8) (1990), 1898–1901.

This research was supported in part by DARPA and monitored by AFOSR under contract F 49620-89-C-0125.

David Pollen
Department of Mathematics
Princeton University
Princeton, NJ 08544
pollen@math.princeton.edu

Wavelet Matrices and the Representation of Discrete Functions

Peter N. Heller, Howard L. Resnikoff, and Raymond O. Wells, Jr.

Abstract. This paper includes a formulation of the notion of wavelet matrices and their fundamental invariants. These arise as coefficient systems which define compactly supported wavelet systems which satisfy m-scale scaling equations, for $m \geq 2$. A Haar wavelet matrix is a square wavelet matrix, and there is a characteristic Haar mapping from the class of all wavelet matrices to Haar wavelet matrices. The Haar wavelet matrices are equivalent to a unitary group of a specific dimension, while the wavelet matrices admit a unitary action and the characteristic Haar mapping is an equivariant mapping under this action. Moreover, on the fibres of this mapping there is an infinite dimensional Lie group defined, and in specific cases, this allows for a complete classification of all wavelet matrices. In particular, a given wavelet matrix consists of a first row (the scaling vector or "low-pass" filter), and the remaining rows (the wavelet vectors or "high-pass" filters). An algorithm for determining the wavelet vectors in terms of the scaling vectors is given. In analogy with the representation of L^2 functions by means of a wavelet system (an orthonormal expansion), there is a representation of arbitrary discrete functions in terms of a wavelet matrix used as a discrete orthogonal system.

§1. Introduction

In this paper we introduce and present a number of properties of *wavelet matrices*. These notions were inspired by the work of Daubechies [5] and Mallat [13], among others (see the survey by Meyer [16]). In these papers one finds compactly supported wavelet systems which are defined by a finite number of coefficients, $a_0, \ldots, a_{2g-1}$, the scaling coefficients, which satisfy specific orthogonality conditions. In particular, one defines a scaling function $\varphi(x)$ as a compactly supported solution of

$$\varphi(x) = \sum_{k=0}^{2g-1} a_k \varphi(2x - k).$$

Wavelets–A Tutorial in Theory and Applications
C. K. Chui (ed.), pp. 15–50.

ISBN 0-12-174590-2

Moreover, there is an associated set of coefficients $b_k = (-1)^k a_{2g-1-k}$, which defines the wavelet function associated to the scaling function of the wavelet system. The wavelet function is denoted by $\psi(x)$ and is defined by

$$\psi(x) = \sum_{k=0}^{2g-1} b_k \varphi(2x - k).$$

The wavelet system is defined in terms of rescalings by powers of 2 of these two functions $\phi(x)$ and $\psi(x)$, and these provide, in most cases, an orthonormal basis for $L^2(\mathbb{R})$ (see [8]). The coefficients $\{a_k\}$ and $\{b_k\}$ form the basis of the Mallat algorithm [13] which allows one to compute approximations to the scaling and wavelet expansion coefficients at any level from the scaling expansion coefficients at a higher level, and this allows very fast computation of the wavelet expansion coefficients in practice.

The wavelet matrices we introduce below are generalizations of the $2 \times 2g$ matrix of the form

$$\begin{pmatrix} a_0 & \dots & a_{2g-1} \\ b_0 & \dots & b_{2g-1} \end{pmatrix}, \tag{1.1}$$

where the a's and b's are as above. We generalize these matrices to $m \times mg$ matrices, where m is the *rank* of the matrix, satisfying specific orthogonality conditions given in Section 2, and we call the integer g the *genus* of the wavelet matrix, *i.e.*, the number of $m \times m$ blocks in the matrix. We associate with these an algebra of $g \times g$ matrices with Laurent polynomials as coefficients, which allows one to generate higher genus matrices from lower genus ones. These Laurent matrices are closely related to the concepts of polyphase factorizations for multirate filtering, as developed by Vaidyanathan [23] and Vetterli [24]. In Section 2 we present the definitions of wavelet matrices, their fundamental invariants, some elementary properties, and some examples. We note that the wavelet matrices of rank m will correspond to wavelet systems with multiplier m, replacing the multiplier 2 used in both the Daubechies and Mallat papers ([5] and [13]). In this paper we do not pursue the study of the orthogonal systems of functions in $L^2(\mathbb{R})$ associated with multiplier m greater than 2, but concentrate on the algebraic and geometric properties of the coefficient systems for such orthogonal systems (the set of wavelet matrices of rank m). We note that in a wavelet system defined by a scaling equation of the form,

$$\varphi(x) = \sum_{\lambda \in \Lambda} a_\lambda \varphi(Mx - \lambda), \quad x \in \mathbb{R}^n$$

for Λ some lattice in $\mathbb{R}^n$, the *multiplier* M is in general a *linear transformation* (see, *e.g.*, [9]), and there will usually be integer or rational invariants associated with such a multiplier which will correspond to the rank we use in this paper. In the case studied here, the rank and the multiplier coincide, but, in general, they do not, and hence the additional terminology.

The first wavelet systems were the Haar functions, and these have been generalized to higher dimensions and higher rank [9]. This leads naturally to the notion of a *Haar wavelet matrix* and their classification which is described in Section 3. A Haar wavelet matrix is essentially a square wavelet matrix, and the set of all Haar wavelet matrices of a given rank is equivalent to a specific orthogonal or unitary group (depending on which field the matrices are defined over) of a given rank. There is a *canonical Haar wavelet matrix* which corresponds to the identity element of the corresponding matrix group. Moreover, there is a characteristic mapping from wavelet matrices to Haar wavelet matrices of the same rank, and the classification of the wavelet matrices can be reduced to the classification of those wavelet matrices whose characteristic Haar wavelet matrix is the canonical Haar wavelet matrix. In addition, a number of classical examples of specific matrices which have different origins in mathematics and signal processing can all be seen to be Haar wavelet matrices of specific types, and these include: the finite Fourier transform matrices, the discrete cosine transform matrix, Hadamard and Walsh matrices which play a role in coding theory, Rademacher matrices, and Chebyshev matrices. All of these matrices have rows and columns with specific orthogonality properties and other defining conditions. The notion of Haar wavelet matrix offers a unifying point of view for these disparate notions.

In Section 4 we discuss in some detail the algebraic and geometric structure of the space of compact wavelet matrices of a given rank and genus, a space which we call a *wavelet space* corresponding to a given field, rank and genus. We first show that there is an infinite-dimensional group structure on each fibre of the characteristic Haar mapping described earlier (which is a Lie group for the case of real or complex fields), and all such fibres are isomorphic. The product is called the Pollen product and was introduced in Pollen's fundamental paper [17].

In this section we also address the question: given a scaling vector, can one find wavelet vectors so that the combined scaling and wavelet vectors form a wavelet matrix? If this is the case, how unique is the choice of wavelet vectors? The general answer seems to have the form: given a scaling vector and a choice of the Haar matrix there is a set of wavelet vectors which forms with the given scaling vector a wavelet matrix whose characteristic Haar matrix is the given Haar matrix. For rank 2 this is true, and under a nondegeneracy hypothesis on the scaling vector, there is also uniqueness. For arbitrary rank and for genus ≤ 3 we find that, under a nondegeneracy hypothesis on the scaling vector, there is an algorithm which gives the desired wavelet vectors by solving a specific square linear system. For rank ≥ 3 uniqueness fails; however, we conjecture that there is uniqueness up to the action of a finite group. The algorithm is defined in general, but the invertibility of the linear system has only been verified for small genus; it may well be true in general. Numerical experiments support this hypothesis for higher genus as well. The scaling vectors in rank 2 have been classified by Pollen ([17] and [18]) and Wells [26] in two different manners, which give, with the above results a classification of the corresponding wavelet matrices. This classification is described briefly in Section 4.

Finally, in Section 5, we show that any discrete function admits a locally finite orthogonal series representation in terms of the rows of a given wavelet matrix. This representation can be related to the orthonormal series representation of an L^2 interpolation of the given discrete function. In practice, when one uses a computer system to represent continuous functions, one must use some form of discretization, and the analysis we present here shows how to obtain an orthogonal series representation of a discrete function in terms of a wavelet matrix without going through the L^2 theory and then discretizing. A special case is described which shows that a discrete function has a Fourier-wavelet matrix expansion, where the wavelet matrix is a Haar wavelet matrix of the finite-Fourier-transform type. The expansion becomes an infinite series in powers of the m-th root of unity, with coefficients corresponding to finite Fourier transform of shifted distinct and nonoverlapping blocks of m values of the given discrete function.

§2. Wavelet matrices

Let $\mathbb{F}$ be a subfield of the field $\mathbb{C}$ of complex numbers. Typically, $\mathbb{F}$ will be the rational numbers $\mathbb{Q}$, the real numbers $\mathbb{R}$, the field $\mathbb{C}$ itself, or some algebraic number field. Throughout this paper $\mathbb{F}$ will denote such a generic field unless specified explicitly. Many of the algebraic results do not depend on which field is chosen, while some of the geometric results do.

Consider a system of m formal Laurent series in an indeterminate Z with coefficients in a subfield $\mathbb{F}$ of $\mathbb{C}$

$$a^s(Z) := \sum_{k=-\infty}^{\infty} a_k^s Z^k, \quad 0 \le s < m \tag{2.1}$$

and form the m-row vector $a(Z)$ whose elements are the Laurent series $a^s(Z)$,

$$a(Z) := (a^s(Z)). \tag{2.2}$$

Construct the partial Laurent series

$$P[a]_r^s(Z) := \sum_{k=-\infty}^{\infty} a_{mk+r}^s Z^{mk}, \quad 0 \le r < m \tag{2.3}$$

and form the $m \times m$ matrix whose elements are the partial Laurent series, *i.e.*,

$$P[a](Z) := \begin{pmatrix} P[a]_0^0(Z) & & \cdots & & P[a]_{m-1}^0(Z) \\ \vdots & \ddots & & & \vdots \\ & \cdots & P[a]_r^s(Z) & \cdots & \\ \vdots & & & \ddots & \vdots \\ P[a]_0^{m-1}(Z) & & \cdots & & P[a]_{m-1}^{m-1}(Z) \end{pmatrix}. \tag{2.4}$$

We call $P[a](Z)$ the *Laurent matrix* associated with the array $a = (a_k^s)$ in Equation (2.1).

We note that the square $m \times m$ Laurent matrix $P[a](Z)$ has the same information as the array (a_k^s) in (2.1), and given any such Laurent matrix $M(Z)$, there is an associated $a(Z)$ of the form (2.2) such that $M(Z) = P[a](Z)$. This is most easily illustrated in the special case where $m = 2$, where we see that

$$P[a]_0^0(Z) = \cdots + a_{-2}^0 Z^{-2} + a_0^0 + a_2^0 Z^2 + \cdots$$
$$P[a]_1^0(Z) = \cdots + a_{-1}^0 Z^{-2} + a_1^0 + a_3^0 Z^2 + \cdots$$

and similarly for the second row. Thus the left-hand column of the Laurent matrix contains the even coefficients and the right-hand column contains the odd coefficients of the array (a_k^s).

Consider an array $a = (a_k^s)$ as in (2.1) where $a_k^s \in \mathbb{F}$. There are m rows in this array, and possibly an infinite number of columns. The array of coefficients a is said to be a *wavelet* of *rank* m if

$$P[a](Z)P[a]^*(Z^{-1}) = mI, \tag{2.5}$$

$$\sum_{r=0}^{m-1} P[a]_r^s(1) = m\delta^{s,0}, \tag{2.6}$$

where "$*$" denotes the transposed conjugate and I denotes the identity matrix of the appropriate rank. If a wavelet matrix a of rank m has only a finite number of columns, then it follows from (2.5) and (2.6) that the number of columns is of the form mg, for some positive integer g. We call this integer g the *genus* of the wavelet matrix a. The genus is the number of $m \times m$ block matrices constituting a. In this case the wavelet matrix a is said to be *compact*, which will correspond to compactly supported wavelet systems.

The set of all wavelet matrices of rank m and genus g over the field $\mathbb{F}$ will be denoted $\mathrm{WM}(m, g; \mathbb{F})$. If the field is fixed but does not play a special role, the abbreviated notation $\mathrm{WM}(m, g)$ will be used. Most often we will consider $\mathbb{F}$ to be the field of real or complex numbers. If $a \in \mathrm{WM}(m, 1; \mathbb{F})$ then a is said to be a *Haar wavelet matrix of rank* m.

Comparison of coefficients of corresponding powers of Z in (2.5) yields the (quadratic) orthogonality relations for the rows of a:

$$\sum_k \bar{a}_{k+mr'}^{s'} a_{k+mr}^s = m\delta^{s',s}\delta_{r',r}. \tag{2.7}$$

Thus, the vectors

$$a^s := (a_0^s, \ldots, a_{mg-1}^s) \tag{2.8}$$

have length equal to $\sqrt{m}$ and are pairwise orthogonal. The linear conditions (2.6) are equivalent to the simpler formula

$$\sum_k a_k^s = m\delta^{s,0}. \tag{2.9}$$

The vector a^0 is called the *scaling vector* and each of the a^s for $0 < s < m$ is called a *wavelet vector*. The linear condition (2.9) says simply that the sum of the components of the scaling vector is the rank m, while the sum of the components of each of the wavelet vectors is 0. Because of the interpretation of the rows of a wavelet matrix as taps of a digital filter, a^0 will sometimes be referred to as the "low-pass" vector, and each of the vectors a^s, $0 < s \leq m$, as a "high-pass" vector. We will refer to (2.5) and (2.6) or equivalently (2.7) and (2.9) as the *quadratic* and *linear* conditions defining a wavelet matrix, respectively.

It will sometimes be helpful to employ an alternative notation for the elements of compact wavelet matrices. Assume that $a_k^s = 0$ unless $0 \leq k < mg$. Write

$$a_k := a_k^0, \quad b_k^s := a_k^s \tag{2.10}$$

where $0 < s < m$ and $0 \leq k < mg$. In addition we shall denote the scaling and wavelet vectors

$$\alpha = (a_0, \ldots, a_{mg-1}) \tag{2.11}$$

and

$$\beta^s = (b_0^s, \ldots, b_{mg-1}^s), \quad s = 1, \ldots, m-1, \tag{2.12}$$

respectively.

Finally we define $\mathrm{SV}(m, g; \mathbb{F})$ to be the set of all scaling vectors of rank m and genus g for the coefficient field $\mathbb{F}$; that is, we let

$$\mathrm{SV}(m, g; \mathbb{F}) := \left\{ \alpha = (a_0, \ldots, a_{mg-1}) \in \mathbb{F}^{mg}: \right. \tag{2.13}$$

$$\left. \sum_k \bar{a}_{k+ml} a_{k+ml'} = m\delta_{l,l'}, \sum_{k=0}^{mg-1} a_k = m \right\}.$$

Thus there is a natural mapping

$$\mathrm{WM}(m, g; \mathbb{F}) \xrightarrow{\sigma} \mathrm{SV}(m, g; \mathbb{F}) \tag{2.14}$$

given by

$$\sigma(a) = a^0.$$

One question we will investigate in Section 4 is when this mapping is onto, and, when it is onto, what is the structure of the fibres. In other words, given a scaling vector α, 1) find corresponding wavelet vectors $\beta^1, \ldots, \beta^{m-1}$ such that

$$\begin{pmatrix} \alpha \\ \beta^1 \\ \vdots \\ \beta^{m-1} \end{pmatrix}$$

is a wavelet matrix, if this is possible, and 2) how unique are the β's for a given α. The answer to this question will, in principle, reduce the classification problem for wavelet matrices to the classification of the corresponding set of scaling vectors.

We shall give now some simple examples of wavelet matrices. At this point we shall merely check that they satisfy the linear and quadratic conditions for wavelet matrices, and we shall have much more to say about some of them later in the chapter.

Example 1. (Haar Wavelet Matrices of Rank 2) The matrices

$$\begin{pmatrix} 1 & 1 \\ 1 & -1 \end{pmatrix} \quad \begin{pmatrix} 1 & 1 \\ -1 & 1 \end{pmatrix} \tag{2.15}$$

are both wavelet matrices of rank 2, and as one can check they are the only square wavelet matrices of rank 2 with real coefficients. More generally, we can see easily that every Haar wavelet matrix of rank 2 has the form

$$\begin{pmatrix} 1 & 1 \\ -e^{i\theta} & e^{i\theta} \end{pmatrix}, \tag{2.16}$$

where $\theta \in \mathbb{R}$.

Example 2. (Daubechies' Wavelet Matrix of rank 2 and genus 2) Let

$$a_{D2} = \frac{1}{4}\begin{pmatrix} 1+\sqrt{3} & 3+\sqrt{3} & 3-\sqrt{3} & 1-\sqrt{3} \\ -1+\sqrt{3} & 3-\sqrt{3} & -3-\sqrt{3} & 1+\sqrt{3} \end{pmatrix}. \tag{2.17}$$

This is a wavelet matrix discovered by Daubechies [5], and is part of a more general pattern. It corresponds to a wavelet system which has a compactly supported continuous scaling function.

Example 3. (Wavelet matrices of rank 2 and genus 2) Define a periodic one-parameter family of scaling vectors by:

$$a_0(\theta) = \frac{1}{2}(1 + \cos\theta - \sin\theta) \tag{2.18}$$

$$a_1(\theta) = \frac{1}{2}(1 + \cos\theta + \sin\theta) \tag{2.19}$$

$$a_2(\theta) = \frac{1}{2}(1 - \cos\theta + \sin\theta) \tag{2.20}$$

$$a_3(\theta) = \frac{1}{2}(1 - \cos\theta - \sin\theta), \tag{2.21}$$

and define an associated wavelet vector $b_k(\theta) := (-1)^{k+1}a_{3-k}(\theta)$. Then the matrix

$$a(\theta) := \begin{pmatrix} a_0(\theta) & a_1(\theta) & a_2(\theta) & a_3(\theta) \\ b_0(\theta) & b_1(\theta) & b_2(\theta) & b_3\theta) \end{pmatrix} \tag{2.22}$$

is a one-parameter family of wavelet matrices of rank 2 and genus 2. If we consider this example and its associated family where we reverse the sign on the second row, we obtain all real-valued wavelet matrices of rank 2 and genus 2. Moreover, the example a_{D2} is a special case of this example for $\theta = \pi/6$.

Example 4. (Flat Wavelet Matrices) A *flat wavelet matrix* is a wavelet matrix with the property that all of the coefficients in the matrix have the same absolute value. We see that the Haar matrices above have this property. We now give some additional ones, which at the same time provide examples of higher rank and genus. First we have the following example which has rank 2 and genus 4:

$$\frac{1}{2}\begin{pmatrix} 1 & 1 & 1 & -1 & 1 & 1 & -1 & 1 \\ 1 & 1 & 1 & -1 & -1 & -1 & 1 & -1 \end{pmatrix}. \tag{2.23}$$

The next example is rank 4 and genus 8:

$$\frac{1}{2}\begin{pmatrix} 1 & 1 & 1 & 1 & 1 & -1 & 1 & -1 & 1 & 1 & 1 & 1 & -1 & 1 & -1 & 1 \\ 1 & 1 & 1 & 1 & 1 & -1 & 1 & -1 & -1 & 1 & -1 & 1 & 1 & 1 & 1 & 1 \\ -1 & -1 & 1 & 1 & -1 & -1 & 1 & 1 & -1 & 1 & 1 & -1 & -1 & -1 & -1 & 1 \\ -1 & -1 & 1 & 1 & -1 & -1 & 1 & 1 & -1 & -1 & -1 & 1 & -1 & 1 & 1 & -1 \end{pmatrix} \tag{2.24}$$

Real flat wavelet matrices are generalizations of Hadamard matrices, and are useful in the context of channel coding [3].

Next we present an important example of a wavelet matrix which is not compact. There are also many examples of such matrices implicit in the earlier extensive work on noncompact wavelet systems (see [14], [15], [10], [2], and [13], all of which precede the fundamental paper by Daubechies [5] on compactly supported wavelet systems. The book by Meyer [16] gives a good overview of this subject).

Example 5. (The Wavelet Matrix a_{sinc}) For each $m > 1$ there is a wavelet matrix a_{sinc} which is not compact. This example is important not only as a concrete illustration of the general theory but also because of its special relation to signal processing applications. Moreover, the scaling function associated with the wavelet matrix a_{sinc} is universal in the sense that it is independent of the rank m.

The sequence a_k defined by the coefficients of the formal Laurent series

$$1 + \frac{2}{\pi}\sum_{k=0}^{\infty}\frac{(-1)^k}{2k+1}\{Z^{2k+1} + Z^{-(2k+1)}\} \tag{2.25}$$

is the first row α of a noncompact system of wavelet coefficients of rank 2. The second row β is defined by $b_0 = 1$, $b_{2k} = 0$, and $b_{2k+1} = -a_{2k+1}$. The wavelet matrix a_{sinc} is defined by

$$a_{\text{sinc}} := \begin{pmatrix} \alpha \\ \beta \end{pmatrix}.$$

More generally, we can define a wavelet matrix of rank m and infinite genus by the following formula which will depend on a choice of a Haar wavelet matrix $h = (h_k^s)$ of rank m (see Section 3):

$$a_{\text{sinc}} = (a_k^s), \quad 0 \leq s < m, \quad -\infty \leq k \leq \infty, \tag{2.26}$$

where

$$a_k^s = e^{-\pi i(\frac{m-1}{m})\frac{k}{m}} \left\{ \frac{1}{m} \sum_{l=0}^{m-1} h_l^s e^{\frac{2\pi ikl}{m^2}} \right\} \frac{\sin\left(\frac{\pi k}{m^2}\right)}{\left(\frac{\pi k}{m^2}\right)} \tag{2.27}$$

for $k \neq 0$, and for $k = 0$ we have

$$a_0^s = \begin{cases} 1 & s = 0 \\ 0 & s \neq 0. \end{cases} \tag{2.28}$$

For a further discussion of this example see [22].

§3. Haar wavelet matrices

The set of wavelet matrices with genus equal to one plays a special role in the theory of wavelets. We shall let

$$H(m; \mathbb{F}) := \mathrm{WM}(m, 1; \mathbb{F}) \tag{3.1}$$

and we shall call the elements of $H(m; \mathbb{F})$ *Haar wavelet matrices*[1] *of rank* m. We shall see in the next subsections that the set of Haar wavelet matrices of rank m over the complex numbers is a homogeneous space which is isomorphic to the Lie group $U(m-1)$ of unitary $(m-1) \times (m-1)$ matrices, and that there is a distinguished Haar wavelet matrix which corresponds to the identity element of the group $U(m-1)$; we will call it the canonical Haar wavelet matrix. Moreover, there is a natural mapping from wavelet matrices to Haar wavelet matrices of the same rank.

3.1. The canonical Haar wavelet matrix

We will now provide a characterization of Haar wavelet matrices. Recall that the unitary group $U(m)$ of order m is the group of $m \times m$ complex matrices U such that $U^*U = I$.

Theorem 6. *An $m \times m$ complex matrix h is a Haar wavelet matrix if and only if*

$$h = \begin{pmatrix} 1 & 0 \\ 0 & U \end{pmatrix} \mathbf{h} \tag{3.2}$$

[1] We use the term "Haar wavelet matrix" instead of simply "Haar matrix" as this latter expression has been used for a special type of matrix arising in the Haar transform using the Haar basis for $L^2(\mathbb{R})$ (see, *e.g.*, Pratt [19]). The classical Haar matrix is a special case of what we are calling a Haar wavelet matrix.

where $U \in U(m-1)$ is a unitary matrix and $\mathbf{h}$ *is the canonical Haar matrix of rank m, which is defined by*

$$\mathbf{h} := \begin{pmatrix} 1 & 1 & \cdots & \cdots & \cdots & \cdots & 1 \\ -(m-1)\sqrt{\frac{1}{m-1}} & \sqrt{\frac{1}{m-1}} & \cdots & \cdots & \cdots & \cdots & \sqrt{\frac{1}{m-1}} \\ \vdots & \ddots & \ddots & \cdots & \cdots & \cdots & \vdots \\ 0 & \cdots & 0 & -s\sqrt{\frac{m}{s^2+s}} & \sqrt{\frac{m}{s^2+s}} & \cdots & \sqrt{\frac{m}{s^2+s}} \\ \vdots & \cdots & \cdots & \cdots & \ddots & \ddots & \vdots \\ 0 & \cdots & \cdots & \cdots & 0 & -\sqrt{\frac{m}{2}} & \sqrt{\frac{m}{2}} \end{pmatrix} \tag{3.3}$$

where $s = (m-k)$ and $k = 0, 1, \ldots, m-1$ are the row numbers of the matrix.

Proof: The proof of this theorem will use the following lemma.

Lemma 7. *If a is a Haar wavelet matrix then*

$$a_r = a_r^0 = 1 \quad \textit{for} \quad 0 \leq r < m. \tag{3.4}$$

Proof: From Equations (2.7) and (2.9) we have that $\Sigma \bar{a}_k a_k = m$ and $\Sigma a_k = m$. It follows that $\Sigma \bar{a}_k = m$ also. Now we compute that

$$\sum_k |a_k - 1|^2 = \sum_k (\bar{a}_k a_k - a_k - \bar{a}_k - 1) = 0, \tag{3.5}$$

which implies that $a_k = 1$, for $k = 0, \ldots, m-1$. ■

Continuing with the proof of Theorem 6, Lemma 7 implies that the elements of the first row of a Haar wavelet matrix are all equal to 1. Regarding the remaining $m-1$ rows, it is clear that $\begin{pmatrix} 1 & 0 \\ 0 & U \end{pmatrix}$ is a Haar matrix whenever h is a Haar matrix and $U \in U(m-1)$. Hence the action of U can be employed to develop a canonical form for h. The first step "rotates" the last row of h so that its first $m-2$ entries are zero. Since the rows of a Haar matrix are pairwise orthogonal and of length equal to $\sqrt{m}$, the orthogonality of the first and last rows implies that the last row can be normalized to have the form

$$\left(0, \ldots, 0, -\sqrt{\frac{m}{2}}, \sqrt{\frac{m}{2}}\right).$$

Repetition of this argument for the preceding rows yields the result. ■

We have the following simple corollaries of this theorem.

Corollary 8. *If $h', h'' \in H(m; \mathbb{C})$ are two Haar matrices, then there exists a unitary matrix $U \in U(m-1)$ such that*

$$h' = \begin{pmatrix} 1 & 0 \\ 0 & U \end{pmatrix} h''.$$

Corollary 9. *If a is a real wavelet matrix, i.e., if $a_k^s \in \mathbb{R}$, then a is a Haar matrix if and only if*

$$a = \begin{pmatrix} 1 & 0 \\ 0 & O \end{pmatrix} \mathbf{h} \tag{3.6}$$

where $O \in \mathrm{O}(m-1)$ is an orthogonal matrix and $\mathbf{h}$ is the canonical Haar matrix of rank m.

3.2. The characteristic Haar wavelet matrix of a wavelet matrix

Recall the Laurent matrix $P[a](Z)$ given in (2.4) and associated with an array $a = (a_k^s)$ as in (2.1). Define

$$\chi(a) := P[a](1). \tag{3.7}$$

We now have the following theorem which relates wavelet matrices to Haar matrices.

Theorem 10. *If $a \in \mathrm{WM}(m, g; \mathbb{F})$ then $\chi(a) \in H(m; \mathbb{F})$, and thus there is a well-defined mapping*

$$\mathrm{WM}(m, g; \mathbb{F}) \xrightarrow{\chi} H(m; \mathbb{F}). \tag{3.8}$$

This theorem justifies the designation of $\chi(a)$ as being the *characteristic Haar wavelet matrix* associated with the given wavelet matrix a.

Proof: The elements of the matrix $h = \chi(a)$ are

$$h_r^s = \sum_{k=-\infty}^{\infty} a_{mk+r}^s. \tag{3.9}$$

Substituting $Z = 1$ in (2.5) yields

$$hh^* = mI; \tag{3.10}$$

thus, $\frac{1}{\sqrt{m}} h$ is a unitary matrix.

We must show that h satisfies the conditions for a wavelet matrix of genus 1. The linear condition is

$$\begin{aligned} \sum_{r=0}^{m-1} h_r^s &= \sum_{r=0}^{m-1} \sum_{k \in \mathbb{Z}} a_{mk+r}^s \\ &= \sum_l a_l \\ &= m\delta^{s,0}. \end{aligned}$$

Regarding the quadratic constrains, we see by (3.10) that they are satisfied, but they also follow directly from (2.7). Namely, we have

$$\begin{aligned}\sum_{r=0}^{m-1} \bar{h}_r^{s'} h_r^s &= \sum_{r=0}^{m-1} \sum_{k',k\in\mathbb{Z}} \bar{a}_{mk'+r}^{s'} a_{mk+r}^s \\ &= \sum_l \sum_n \bar{a}_n^{s'} a_{n+ml}^s \\ &= m\delta^{s',s}. \quad \blacksquare\end{aligned}$$

Corollary 11. *If a is a complex wavelet matrix of rank m and $\chi(a)$ is the characteristic Haar wavelet matrix of a, then there exists a unitary matrix $U \in \mathrm{U}(m-1)$ such that*

$$\tilde{a} := \begin{pmatrix} 1 & 0 \\ 0 & U \end{pmatrix} a \tag{3.11}$$

is a wavelet matrix whose characteristic Haar matrix is the canonical Haar matrix.

The above corollary is an example of the fact that there is a left-action of the unitary group $\mathrm{U}(m-1)$ on the wavelet matrices $\mathrm{WM}(m,g;\mathbb{C})$ given by

$$U \in \mathrm{U}(m-1), a \in \mathrm{WM}(m,g;\mathbb{C}) \mapsto \begin{pmatrix} 1 & 0 \\ 0 & U \end{pmatrix} a. \tag{3.12}$$

Corollary 12. *The mapping χ in (3.8) is equivariant[2] with respect to the left actions of $\mathrm{U}(m-1)$ on $\mathrm{WM}(m,g;\mathbb{C})$ and $\mathrm{H}(m;\mathbb{C})$.*

3.3. Tensor products of Haar wavelets matrices

Theorem 13. *If $a \in \mathrm{H}(m')$ and $b \in \mathrm{H}(m'')$, then*

$$a \otimes b \in \mathrm{H}(m'm'').$$

Proof: Let $h = a \otimes b$ and set $m = m'm''$. Then

$$h_{km+r}^{lm+s} = a_k^l b_r^s$$

so

$$\sum_q h_q^p = \sum_k \sum_r a_k^l b_r^s = m\delta^{l,0}\delta^{s,0},$$

[2] If A and B are two spaces with a left-group action by a group G, then a mapping $f\colon A \to B$ is *equivariant* if f is compatible with the actions, *i.e.*, $f(g\cdot x) = g\cdot f(x)$, where $g\cdot x$ denotes the left action in A and $g\cdot y$ denotes the left action in B for $g\in G$, $x\in A$, $y \in B$.

which are the linear wavelet matrix conditions (2.9). The quadratic norm conditions for the wavelet vectors are

$$\begin{aligned}\sum_q |h_q^p|^2 &= \sum_k \sum_r |a_r^l b_r^s|^2 \\ &= \left(\sum_k |a_k^l|^2\right)\left(\sum_r |b_r^s|^2\right) \\ &= m'm'' = m.\end{aligned}$$

The vanishing of the quadratic cross-correlation terms is shown similarly. ■

Application of this construction to the canonical Haar wavelet matrix

$$\mathbf{h} = \begin{pmatrix} 1 & 1 \\ -1 & 1 \end{pmatrix}$$

yields Haar wavelet matrices of rank 2^n. For $n = 2$ the result is

$$\begin{pmatrix} 1 & 1 & 1 & 1 \\ -1 & 1 & -1 & 1 \\ -1 & -1 & 1 & 1 \\ 1 & -1 & -1 & 1 \end{pmatrix}.$$

This is the classical method of Sylvester for constructing Hadamard matrices (see Example 17 below). The following corollary of the above theorem is easy to verify, where we recall that all of the entries of a flat wavelet matrix have the same absolute value.

Corollary 14. *The tensor products of two flat Haar wavelets matrices is a flat Haar wavelet matrix.*

3.4. Examples of Haar wavelet matrices

In this subsection we want to illustrate the wide variety of Haar wavelet matrices which have already appeared in the literature in a number of special cases. We will expand on some of these examples as they will be useful in applications, but for the moment will concentrate on special cases of the general concept of a Haar wavelet matrix.

Example 15. (The Finite Fourier Transform Matrix) Let $m > 1$ be an integer and

$$\omega := e^{2\pi i/m} \tag{3.13}$$

be a primitive m-th root of unity. The *finite Fourier transform*[3] *matrix* of rank

[3] The finite Fourier transform is referred to in the signal processing literature as the *discrete Fourier transform* (often abbreviated as DFT). We prefer the word "finite" as it refers to a finite cyclic group of order m.

m is (see [4] and [12])

$$\Omega_m := \begin{pmatrix} 1 & & \dots & & 1 \\ & \ddots & & & \\ & & \omega^{sk} & & \\ & & & \ddots & \\ 1 & & \dots & & \omega^{(m-1)^2} \end{pmatrix}, \tag{3.14}$$

where $0 \leq k, s < m$. One sees that Ω_m is a Haar wavelet matrix over $\mathbb{C}$; if $m = 2$ then Ω_m is defined over $\mathbb{R}$ and is identical to the canonical Haar matrix $\mathbf{h}$.

Example 16. (Discrete Cosine Transform) Consider the following $m \times m$ matrix d defined by

$$d = (d_k^s) \quad \text{where} \quad d_k^s = \begin{cases} 1 & \text{if} \quad s = 0 \\ \sqrt{2} \cos\left[\frac{s(2k+1)\pi}{2m}\right] & \text{if} \quad s \neq 0. \end{cases} \tag{3.15}$$

We see that this matrix satisfies

$$dd^t = mI,$$

and hence is a Haar wavelet matrix, since the first row satisfies the condition $\Sigma_k d_k^0 = m$. This matrix is the matrix of the *discrete cosine transform* which is used in image compression (see [20]).

Example 17. (Hadamard Matrices) A *Hadamard matrix* is a square matrix h whose entries are ± 1 such that [11]

$$h^t h = hh^t = I. \tag{3.16}$$

The theorem of Hadamard [6] is the simple observation that the square of the determinant of an $m \times m$ Hadamard matrix is equal to m^m. It is easy to see that an $m \times m$ Hadamard matrix whose first row has entries all equal to 1 is simply a Haar wavelet matrix of rank m with integer entries (*i.e.*, defined over the integers $\mathbb{Z}$).

We can easily extend the result of Hadamard to the case of Haar matrices whose entries are not necessarily integers.

Theorem 18. *Let h be a Haar matrix of rank m. Then*

$$(\det h)^2 = m^m. \tag{3.17}$$

Proof: The rows of a Haar matrix are pairwise orthogonal and, considered as vectors, each has squared length equal to m. The determinant of the Haar matrix is just the volume spanned by these vectors. ■

It is known that a Hadamard matrix has rank equal to 1, 2 or $4n$. While there are constructions that produce infinitely many Hadamard matrices, they have not yet been fully classified. The least value of $4n$ for which examples of Hadamard matrices are not known in 268 (Lidl and Pilz [11]). Walsh matrices [25] are collections of Hadamard matrices that have been ordered in a specific way. Thus there is no fundamental difference between the Hadamard and Walsh categories. In particular, a Walsh matrix is a particular kind of Haar wavelet matrix. Rademacher has studied another special case of Haar matrices [21].

Example 19. (Chebyshev matrices) These are a family of examples of Haar wavelet matrices defined by the formula

$$h^s_{\text{Chebr}} := C^s(m) \sum_{k=0}^{s} (-1)^k \binom{s}{k} \binom{s+k}{k} \frac{r!(m+1-k)!}{(r-k)!(m+1)!}, \tag{3.18}$$

where

$$C^s(m) = \sqrt{\frac{(2s+1)((m+1)!)^2}{m(m+s)!(m-s-1)!}}. \tag{3.19}$$

It is known (cp. [1]) that the rows of this matrix are orthogonal and otherwise satisfy the wavelet matrix constraints for a Haar matrix.

§4. The algebraic and geometric structure of wavelet matrix spaces

4.1. The wavelet group

Let h be a Haar wavelet matrix of rank m and let $\mathrm{WM}_h(m, g; \mathbb{F}) := \chi^{-1}(h)$, where χ is given by (3.7). Then

$$\mathrm{WM}(m, g; \mathbb{F}) = \bigcup_{h \in \mathbf{H}(m,g;\mathbb{F})} \mathrm{WM}_h(m, g; \mathbb{F}) \tag{4.1}$$

is a disjoint union.

We define the set

$$\mathrm{WG}_h(m; \mathbb{F}) := \bigcup_{1 \le g} \mathrm{WM}_h(m, g; \mathbb{F}), \tag{4.2}$$

and we shall see that this can be equipped with a noncommutative product (called the Pollen product) and which, moreover, will turn out to be an infinite dimensional Lie group equipped with this product when the underlying field is the real or complex numbers. We will call $\mathrm{WG}_h(m; \mathbb{F})$ the *wavelet group of rank m at the Haar wavelet matrix h*. The unit element of this Lie group is the Haar matrix h. If h' and h'' are two Haar wavelet matrices then the corresponding wavelet groups $\mathrm{WG}_{h'}(m; \mathbb{F})$ and $\mathrm{WG}_{h''}(m; \mathbb{F})$ are isomorphic.

This product will provide a practical method for constructing wavelet matrices with large genus as products of wavelet matrices with small genus. Moreover, Pollen proved [17] that every element in $\mathrm{WG}_h(2;\mathbb{F})$ has a unique expression as a product of wavelet matrices belonging to $\mathrm{WM}_h(2,2;\mathbb{F})$, where $\mathbb{F}$ is an arbitrary infinite subfield of $\mathbb{C}$. In this sense, the wavelet matrices of genus 2 are the "prime" elements of $\mathrm{WG}_h(2;\mathbb{F})$. In particular, a rational wavelet matrix of rank 2 is the Pollen product of rational wavelet matrices of rank 2 and genus $g = 2$.

Let $a', a'' \in \mathrm{WG}_h(m;\mathbb{F})$ where $h \in H(m,\mathbb{F})$. We have seen that if $a \in \mathrm{WM}(m, mg : \mathbb{C})$ then $P[a](Z)$ is a unitary matrix of rank m, and $P[a](1) = \chi(a)$. Hence

$$P[a'](Z)h^{-1}P[a''](Z)$$

is unitary and, moreover,

$$P[a'](1)h^{-1}P[a''](1) = hh^{-1}h = h.$$

Therefore there exists a wavelet matrix $a \in \mathrm{WG}_h(m;\mathbb{F})$ and a Laurent polynomial $P[a](Z)$ such that

$$P[a](Z) = P[a'](Z)h^{-1}P[a''](Z),$$

and we define the *Pollen product* by the formula

$$a' \diamondsuit_h a'' := a. \tag{4.3}$$

If the Haar matrix h is clear from the context, then we drop the h subscript and write simply

$$a' \diamondsuit a'' := a, \tag{4.4}$$

for $a', a'' \in \mathrm{WG}_h(m;\mathbb{F})$.

Theorem 20. *$G = \mathrm{WG}_h(m;\mathbb{F})$ is a group with the noncommutative product*

$$(a', a'') \mapsto a' \diamondsuit_h a''.$$

The unit element of G is h. If $\mathbb{F} = \mathbb{R}$ (resp. $\mathbb{C}$), then G is an infinite dimensional real (resp. complex) Lie group. If h and h'' are Haar matrices of the same rank, then

$$P[a'](Z) \mapsto P[a''] := P[a'](Z)h'^{-1}h'' \tag{4.5}$$

induces an isomorphism

$$\mathrm{WG}_h(m;\mathbb{F}) \overset{\cong}{\to} \mathrm{WG}_h(m;\mathbb{F}).$$

The proof of this theorem is elementary and is omitted. The difficulty lay in *formulating* a suitable product for wavelet matrices.

We now give an example of the Pollen product in a special case.

Example 21. (The Pollen Product for Rank 2 and Genus 3) Consider $m = 2, g = 3$. It is convenient to adopt an indexing for the entries of the wavelet matrix a that "centers" the matrix entries around the zero index. Therefore let

$$a = \begin{pmatrix} a_{-2} & a_{-1} & a_0 & a_1 & a_2 & a_3 \\ b_{-2} & b_{-1} & b_0 & b_1 & b_2 & b_3 \end{pmatrix}. \tag{4.6}$$

The corresponding Laurent matrix has the form

$$P[a](Z) = \begin{pmatrix} a_{-2}Z^{-2} + a_0 + a_2Z^2 & a_{-1}Z^{-2} + a_1 + a_3Z^2 \\ b_{-2}Z^{-2} + b_0 + b_2Z^2 & b_{-1}Z^{-2} + b_1 + b_3Z^2 \end{pmatrix}. \tag{4.7}$$

For simplicity we shall assume that the entries of a are real numbers. Then the characteristic Haar wavelet matrix $\chi(a)$ associated with a is one of two possible Haar matrices as in Example 1. Now let a' and a'' be two given elements of $\mathrm{WM}(2, 3; \mathbb{R})$ in the form (4.6). Writing $a' \diamondsuit a''$ for $a' \diamondsuit_h a''$, we want to calculate $a = a' \diamondsuit a''$.

Calculation shows that the Laurent polynomial $P[a']\mathbf{h}^{-1}P[a''](Z)$ will have nonzero coefficients for the powers $Z^{-4}, \ldots, Z^6$ which implies that the genus of a is 5. A tedious but straightforward calculation yields the formula for the coefficients of the product wavelet matrix. For instance,

$$a_4 = \frac{1}{2}\{a'_{-2}(a''_{-2} - b''_{-2}) + a'_{-1}(a''_{-2} + b''_{-2})\},$$

and the formulas for $a_{-3}, \ldots, a_6$, as well as the b_k's, will all be of a similar, but often more complex, nature.

4.2. The wavelet space for rank 2 and genus g

Let $\mathrm{SV}(2, g; \mathbb{F})$ be the set of *scaling vectors of rank 2 and genus* g as given by (2.13). From (3.7) and (2.14) we have the diagram,

$$\begin{array}{ccc} \mathrm{WM}(2, g; \mathbb{F}) & \xrightarrow{\sigma} & \mathrm{SV}(2, g; \mathbb{F}) \\ \downarrow \chi & & \\ H(2; \mathbb{F}) & & \end{array} \tag{4.8}$$

We define the *fibre* of $\mathrm{WM}(2, g; \mathbb{F})$ over the point $h \in \mathbf{H}(2, \mathbb{F})$ to be the inverse image of a point h under the mapping χ, and we denote this by

$$\mathrm{WM}_h(2, g; \mathbb{F}) := \chi^{-1}(h), \quad \text{for} \quad h \in \mathrm{H}(2, \mathbb{F}). \tag{4.9}$$

Note that in this case $\mathrm{H}(2, \mathbb{F})$ is a group consisting of two elements.

Let us introduce a useful reformulation of the orthogonality conditions which define a wavelet matrix. For a fixed rank m define the operators

$$L\colon \mathbb{F}^{mg} \to \mathbb{F}^{mg} \quad R\colon \mathbb{F}^{mg} \to \mathbb{F}^{mg} \tag{4.10}$$

by

$$L(a_0, a_1, \ldots, a_{mg-1}) := (a_m, a_{m+1}, \ldots, a_{2g-1}, 0, \ldots, 0), \tag{4.11}$$

and

$$R(a_0, a_1, \ldots, a_{mg-1}) := (0, \ldots, 0, a_0, a_1, \ldots, a_{mg-m-1}), \tag{4.12}$$

the *left-shift* and *right-shift* operators by m units[4]. We have the usual inner product on $\mathbb{F}^{mg}$, and we see that the adjoint L^* of L is precisely R, and conversely. Moreover, both L and R are nilpotent of order g, *i.e.*, $L^g = R^g = 0$.

Now let $\alpha \in \mathbb{F}^{mg}$, $\beta^s \in \mathbb{R}^{mg}$, then

$$a = \begin{pmatrix} \alpha \\ \beta^1 \\ \vdots \\ \beta^{m-1} \end{pmatrix}$$

is a wavelet matrix of rank m and genus g if and only if the linear conditions (2.5) are satisfied and

$$\begin{array}{ll} \langle L^j\alpha, \alpha\rangle = m\delta^{j,0}, & j = 0, \ldots, g-1, \\ \langle L^j\alpha, \beta^s\rangle = 0, & j = 0, \ldots, g-1, s = 0, \ldots, m-1, \\ \langle L^j\beta^s, \alpha\rangle = 0, & j = 0, \ldots, g-1, s = 0, \ldots, m-1, \\ \langle L^j\beta^s, \beta^{s'}\rangle = m\delta^{s,s'}, & j = 0, \ldots, g-1, s = 0, \ldots, m-1. \end{array} \tag{4.13}$$

Similarly, a vector $\alpha \in \mathbb{F}^{mg}$ is a scaling vector in $\mathrm{SV}(m, g; \mathbb{F})$ if and only if the linear condition (2.9) is satisfied and

$$\langle L^j, \alpha\rangle = m\delta^{j,0}, \quad j = 0, \ldots, g-1.$$

Note that there are natural inclusions

$$\mathrm{SV}(m, g-1; \mathbb{F}) \stackrel{i_1, i_2}{\longrightarrow} \mathrm{SV}(m, g; \mathbb{F}) \tag{4.14}$$

given by

$$(a_0, \ldots, a_{mg-g-1}) \stackrel{i_1}{\longmapsto} (a_0, \ldots, a_{mg-g-1}, 0, \ldots, 0)$$

or by

$$(a_0, \ldots, a_{mg-g-1}) \stackrel{i_2}{\longmapsto} (0, \ldots, 0, a_0, \ldots, a_{mg-g-1}).$$

We will say that a scaling vector is *nondegenerate* if it is not in the image of i_1 or i_2 in (4.14). This is the same as saying that the vector consisting of all of the last m entries or the first m entries of the scaling vector α is not zero.

[4] The operator $L = L(m)$ and its adjoint depends on the integer m, but we will be using these operators for a fixed m in any given context of this paper (the rank of the wavelet matrices being considered), and it is less cumbersome to omit the m dependence.

Thus α is nondegenerate if and only if $L^{g-1}\alpha \neq 0$ and $R^{g-1}\alpha \neq 0$, which we will use extensively below.

Lemma 22. *Let $\alpha \in \mathrm{SV}(m, g; \mathbb{F})$ be a nondegenerate scaling vector, then the vectors*

$$\{\alpha, L\alpha, L^2_\alpha, \ldots, L^{g-1}\alpha\}$$

and the vectors

$$\{\alpha, R_\alpha, \ldots, R^{g-1}_\alpha\}$$

are linearly independent in $\mathbb{F}^{mg}$.

Proof: Since $L^{g-1}\alpha \neq 0$, it follows that $L^k\alpha \neq 0$, $k = 0, \ldots, g-2$. Suppose that

$$\sum_{k=0}^{g-1} c_k L^k \alpha = 0, \tag{4.15}$$

then, applying L^{g-1} to (4.15), we obtain, using the nilpotence of L,

$$c_0 L^{g-1}\alpha = 0,$$

which implies that $c_0 = 0$. Applying L^{g-2}, L^{g-3}, etc., successively, we see that $c_1 = \cdots = c_{g-1} = 0$. The vector of right-shifts of α is treated in the same manner. ■

Now consider a wavelet matrix of the form

$$\begin{pmatrix} \alpha \\ \beta^1 \\ \vdots \\ \beta^{g-1} \end{pmatrix},$$

where α is nondegenerate. We will say that the *wavelet vectors are nondegenerate* if

$$R^{g-1}\beta^j \neq 0, \quad j = 1, \ldots, g-1. \tag{4.16}$$

This is the "transpose" of part of the nondegeneracy condition for a scaling vector. By the same arguments as in the proof of Lemma 22, we have the following lemma.

Lemma 23. *Let β^j be a nondegenerate wavelet vector, then the vectors*

$$\{\beta^j, R\beta^j, \ldots, R^{g-1}\beta^j\}$$

are linearly independent.

We will say that a wavelet matrix is *nondegenerate* if its scaling vector and wavelet vectors are all nondegenerate.

We have the following theorem which shows that the scaling vector of a wavelet matrix of rank 2 determines the wavelet vector, provided that the characteristic Haar wavelet matrix of the wavelet matrix has been specified.

Theorem 24. *The mapping*

$$\mathrm{WM}(2,g;\mathbb{F})\xrightarrow{\sigma}\mathrm{SV}(2,g;\mathbb{F})$$

is onto. Moreover, for a given Haar matrix $h \in \mathrm{H}(2;\mathbb{F})$ and a given nondegenerate scaling vector $a \in \mathrm{SV}(2,g;\mathbb{F})$, there is a unique nondegenerate wavelet vector β such that

$$a = \begin{pmatrix}\alpha\\ \beta\end{pmatrix} \in \mathrm{WM}(w,g;\mathbb{F})$$

and $\chi(a) = h$.

Proof: Suppose that we are given $\alpha \in \mathrm{SV}(2,g;\mathbb{R})$, then it is easy to verify that if we let

$$\beta_k := (-1)^{k+1}\alpha_{2g-k},$$

then $a = \binom{\alpha}{\beta} \in \mathrm{WM}(2,g;\mathbb{F})$. Thus the mapping σ is trivialy onto. Moreover, it is easy to see that if α is nondegenerate, then β is also.

Suppose that β and $\tilde{\beta}$ are two nondegenerate wavelet vectors such that

$$\begin{pmatrix}\alpha\\ \beta\end{pmatrix} \text{ and } \begin{pmatrix}\alpha\\ \tilde{\beta}\end{pmatrix}$$

are both wavelet matrices, where α is a nondegenerate scaling vector. If

$$V_\alpha := \mathrm{span}\{\alpha, L\alpha, \ldots, L^{g-1}\alpha\},$$

then V_α is g-dimensional by Lemma 22, and, moreover, by Lemma 23 we see that

$$\{\beta, R\beta, \ldots, R^{g-1}\beta\}$$

and

$$\{\tilde{\beta}, R\tilde{\beta}, \ldots, R^{g-1}\tilde{\beta}\}$$

are both bases for $V_\alpha^\perp$. Thus we see that, for some $c_j \in \mathbb{F}$, $j = 0, \ldots, g-1$,

$$\tilde{\beta} = c_0\beta + c_1 R\beta + \cdots + c_{g-1}R^{g-1}\beta. \tag{4.17}$$

Now, operating on (4.17) by R^{g-1}, we obtain

$$R^{g-1}\tilde{\beta} = c_0 R^{g-1}\beta,$$

which implies that $c_0 \neq 0$, since $R^{g-1}\tilde{\beta}$ is not the zero vector, by hypothesis.

Now we want to show that $c_1 = \cdots = c_{g-1} = 0$, but this follows inductively starting at the top. Namely,

$$\begin{aligned}
0 &= \langle \tilde{\beta}, R^{g-1}\tilde{\beta}\rangle, \\
&= \langle c_0\beta + c_1 R\beta + \cdots + c_{g-1}R^{g-1}\beta, R^{g-1}(c_0\beta + \cdots + cg - 1R^{g-1}\beta)\rangle, \\
&= \langle c_0\beta + \cdots + c_{g-1}R^{g-1}\beta, c_0 R^{g-1}\beta\rangle, \\
&= c_{g-1}c_0\langle R^{g-1}\beta, R^{g-1}\beta\rangle,
\end{aligned}$$

which implies that $c_{g-1} = 0$, since $c_0 \neq 0$, and $R^{g-1}\beta \neq 0$. In a similar manner one sees that $c_k = 0$ for $1 \leq k < g-1$. Thus we obtain that $\tilde{\beta} = c_0\beta$. Now, using this fact and our hypothesis that our two wavelet matrices have the same Haar characteristic matrices, we obtain

$$\chi\begin{pmatrix}\alpha\\ \beta\end{pmatrix} = \chi\begin{pmatrix}\alpha\\ \tilde{\beta}\end{pmatrix} = \chi\begin{pmatrix}\alpha\\ c_0\beta\end{pmatrix}, \tag{4.18}$$

and since χ maps rows of a wavelet matrix linearly to rows of a Haar matrix, it follows from the relation (4.18) that $\beta = c_0\beta$, and hence $c_0 = 1$, and $\beta = \tilde{\beta}$. ■

We now want to discuss the geometry of the space $\mathrm{SV}(2, g; \mathbb{R})$ itself, restricting our attention to the special case where $\mathbb{F} = \mathbb{R}$. Let T^l denote the l-torus, i.e., T^l is the Cartesian product of l copies of the unit circle, $T^l = S^1 \times \cdots \times S^1$.

Theorem 25. (a) $\mathrm{SV}(2, g; \mathbb{R})$ *is a compact real-algebraic variety*[5] *of dimension* $g-1$.

(b) *There is an algebraic mapping*

$$T^{g-1} \xrightarrow{\pi} \mathrm{SV}(2, g; \mathbb{R}), \tag{4.19}$$

which is onto. Moreover π is one-to-one on the complement of a subvariety of T^{g-1} of lower dimension.

(c) *There is a compact convex set $K \subset \mathbb{R}^{g-1}$, such that K has nonempty interior K^0, and K is the closure of K^0, and there is an algebraic mapping*

$$\mathrm{SV}(2, g; \mathbb{R}) \xrightarrow{\rho} K, \tag{4.20}$$

such that π is onto, and is finite-to-one at all points.

Let us denote by $\mathrm{SV}_{\mathrm{rd}}(2, g; \mathbb{R}) := K$, where K is given as in Theorem 25 above. We call this space the *reduced scaling vector space*, and we see that we have the diagram, letting $d = g-1$ be the dimension of the scaling vector space,

$$T^d \xrightarrow{\pi} \mathrm{SV}(2, g; \mathbb{R}) \xrightarrow{\rho} \mathrm{SF}_{\mathrm{rd}}(2, g; \mathbb{R}) \subset \mathbb{R}^d. \tag{4.21}$$

[5] A real-algebraic variety (or subvariety) is the solution of polynomial equations in $\mathbb{R}^N$ for some N.

The torus T^d and the coordinates of $\mathbb{R}^d \supset \mathrm{SV}_{\mathrm{rd}}(2, g; \mathbb{R})$ provide two different parametrizations of the scaling vector space. The mapping π is one-to-one almost everywhere, but is degenerate along tori of smaller dimensions embedded in T^d, while ρ has a generic degree equal to 2^d.

Remark. The proof depends on results developed in other papers, primarily [17], [18] and [26], but we wanted to formulate the basic results to give the reader an overview of the parametrization of the basic wavelet space for rank 2, which, as we see by Theorem 24, is the same as the parametrization of the scaling vectors given in Theorem 25 (modulo the choice of the characteristic Haar wavelet matrix).

First of all, we note that $\mathrm{SV}(2, g; \mathbb{R})$ is a compact real-algebraic variety. This follows since, by definition, it is the solution of a set of real-algebraic equations in $\mathbb{R}^{2g}$, one of which is the equation of a sphere (which forces the compactness). Part (b) implies that dim $\mathrm{SV}(2, g; \mathbb{R}) \leq d$, while part (c) implies that dim $\mathrm{SV}(2, g; \mathbb{R}) \geq d$, and this yields (a). Part (b) is a consequence of the Pollen factorization theorem ([17] and [18]). This theorem asserts that any wavelet matrix of rank 2 and genus g is the unique product of wavelet matrices of genus 2. In fact Pollen's results yield the dimension of the variety in a complete manner, including a description of the subvariety where π fails to be one-to-one. Part (c) is a parametrization given in [26] and is an alternate approach to the parametrization problem, allowing some identification of wavelet matrices in a finite-to-one manner, but obtaining a global coordinate system. This parametrization uses a Fourier transform description of the defining equations of a scaling vector.

4.3. The wavelet space for higher rank

In this section we want to generalize some of the constructions of wavelet matrices from rank 2 to higher rank. To begin with we recall the following diagram combining the mappings (3.7) and (2.14):

$$\begin{array}{ccc} \mathrm{WM}(m, g; \mathbb{F}) & \xrightarrow{\sigma} & \mathrm{SV}(m, g; \mathbb{F}) \\ \downarrow \chi & & \\ \mathrm{H}(g; \mathbb{F}) & & \end{array} \tag{4.22}$$

We saw in Subsection 4.2 that for $r = 2$, the mapping σ is onto, and moreover, up to the choice of a Haar matrix, scaling vector uniquely determines a wavelet matrix. Moreover, we have a rather complete characterization of the scaling vectors in $\mathrm{SV}(2, g; \mathbb{F})$ given by the parametrizations in [17], [18], and [26].

For higher rank, the situation is far less complete, although there are specific results which are of interest in themselves, some of which we shall present here. For rank $r = 2$, the scaling vectors of Daubechies [5] have played an important role in the applications, and they are specific distinguished subsets of the parameter space (see, in particular, Wells [26]). For higher rank, the complete parametrization of $\mathrm{SV}(r, g; \mathbb{F})$ is unknown at this time, but the distinguished subset of the space which corresponds to maximal vanishing of moments (the Daubechies condition) has been constructed for arbitrary rank and genus in Heller [7]. This provides many examples of scaling vectors for higher rank wavelet matrices. We now ask the question: given any scaling vector of higher rank, does there exist wavelet vectors corresponding to the given scaling vectors? The following theorem shows by construction that the answer is affirmative in certain cases.

Theorem 26. *The mapping*

$$\mathrm{WM}(m, g; \mathbb{F}) \xrightarrow{\sigma} \mathrm{SV}(m, g; \mathbb{F})$$

is onto for $m \geq 2, 1 \leq g \leq 3$.

The proof will be carried out in the remainder of this section. As we will see the construction is quite general, and reduces to proving that a certain system of linear equations (defined for any genus) is invertible. We conjecture that this will always be the case, although it is only verified for the case where the genus $g \leq 3$.

Suppose now that we are given a nondegenerate scaling vector α for a rank m genus g wavelet matrix;

$$\alpha = (a_0, a_1, \ldots, a_{gm-1}).$$

Recall that in (4.10) we defined the left-shift operator L and the right-shift operator R acting on vectors in $\mathbb{F}^{mg}$. Thus, since α is nondegenerate, the vectors $L^k\alpha$, $k = 0, 1, \ldots, g-1$ are linearly independent (Lemma 22), and we may consider the vector space they span:

$$V_\alpha = \left\{ \sum_{k=0}^{g-1} c_k L^k \alpha \right\}.$$

We denote the orthogonal complement of V_α by $V_\alpha^\perp$; that is

$$V_\alpha^\perp := \{ y \in \mathbb{F}^{mg} \colon \ \langle y, x \rangle = 0 \quad \forall \, x \in V_\alpha \}.$$

In particular,

$$y \in V_\alpha^\perp \Rightarrow y \perp \alpha.$$

Observe that

$$L\colon\ V_\alpha \longrightarrow V_\alpha, \text{ and } R\colon\ V_\alpha^\perp \longrightarrow V_\alpha^\perp.$$

Furthermore, since

$$\langle L^k\alpha, R^{k'}\alpha \rangle = \langle L^{k+k'}\alpha, \alpha \rangle = 0 \text{ for } k' > 0, \tag{4.23}$$

we see that the right-shifted vectors $R^k\alpha$ belong to $V_\alpha^\perp$ for $k > 0$.

It is often useful to examine each subblock of length m of the scaling vector; we denote the k-th such subblock by α_k;

$$\alpha_k = (a_{m(k-1)}, a_{m(k-1)+1}, \ldots, a_{m(k-1)+m-1}).$$

We can express the α_k using the shift operator formulation:

$$\alpha_k = R^{g-1}L^k\alpha$$

displays each α_k as the rightmost m components of an mg-vector, while

$$\alpha_k = L^{g-1}R^{g-1-k}\alpha$$

displays each α_k as the leftmost m components of an mg-vector.

We will now construct a wavelet matrix

$$a = \begin{pmatrix} \alpha \\ \beta^1 \\ \vdots \\ \beta^r \end{pmatrix}$$

such that $\sigma(a) = \alpha$ and $\chi(a) = h$, where

$$h = (h_k^s) = \begin{pmatrix} h^0 \\ \vdots \\ h^{m-1} \end{pmatrix}.$$

We must construct wavelet vectors b^j satisfying the orthogonality conditions

$$\langle L^k\alpha, \beta^j\rangle = 0 \quad \forall\, k, \tag{4.24}$$

$$\langle L^k\beta^k, \alpha\rangle = 0 \quad \forall\, k, \tag{4.25}$$

$$\langle L^k\beta^j, \alpha^{j'}\rangle = m\delta_{j,j'}\delta_{k,0}, \tag{4.26}$$

and the Haar condition that $\chi(a) = h$, where h is the given Haar matrix, *i.e.*,

$$\sum_{l=0}^{g-1} b^j_{lm+r} = h^j_r. \tag{4.27}$$

From (4.24) the β^j must lie in the space $V_\alpha^\perp$, and we already have on hand $g-1$ vectors $R^k\alpha$ in $V_\alpha^\perp$. We will use these right-shifted versions of α to build β^j, adding in one "error" vector E^j to fulfill the Haar condition (4.27). The correct *Ansatz* is to define β^j to be of the form

$$\beta^j = \sum_{l=0}^{g-2} c^j_l R^l\alpha + E^j, \tag{4.28}$$

where $E^j = (0, 0, \ldots, 0, E_0^j, E_1^j, \ldots, E_{m-1}^j)$ is an mg-vector which is nonzero only in its rightmost m entries. We can now find linear equations which determine the constants c_l^j and E_n^j.

We define the length m subblocks of the wavelet vectors

$$\beta_k^j = (b_{(k-1)m}^j, b_{(k-1)m+1}^j, \ldots, b_{(k-1)m+m-1}^j).$$

Thus

$$\beta^j = (\beta_0^j, \beta_1^j, \ldots, \beta_{m-1}^j).$$

In analogy with the α_k, we have

$$\beta_k^j = R^{g-1} L^k \beta,$$

where the right hand side is an mg-vector all of whose entries are zero, except for the rightmost m, which form β_k^j. The Haar condition (4.27) becomes:

$$\sum_{k=0}^{g-1} \beta_k^j = h^j \Rightarrow$$

$$\sum_{k=0}^{g-1} R^{g-1} L^k \beta^j = h^j.$$

Substituting (4.28) into this equation, we find

$$\sum_{l=0}^{g-2} c_l^j \left\{ \sum_{k=0}^{g-1} R^{g-1} L^k R^l \alpha \right\} + E^j = h^j. \tag{4.29}$$

This yields m inhomogeneous linear equations in the variables c_k^j and E_n^j. Equation (4.26) requires

$$0 = \langle L^k \beta^j, \alpha \rangle = \langle \beta^j, R^k \alpha \rangle \Rightarrow$$

$$\sum_l c_l^j \langle R^l \alpha, R^k \alpha \rangle + \langle E^j, R^k \alpha \rangle = 0 \text{ when } 0 \leq k \leq g-1. \tag{4.30}$$

This gives us $g-1$ more linear equations among the variables c_k^j and E_n^j. In fact, we will show that Equation (4.30) for $l = g-1$ is a consequence of the previous $g-2$ equations and the Haar condition. Thus we take the m Equation (4.29) and the Equation (4.30) for $l = 0, 1, \ldots, g-2$ to form an $(m+g-1) \times (m+g-1)$ linear system.

Lemma 27. *For genus $g \leq 3$, the linear system (4.29) and (4.30) is invertible.*

Proof: The case $g = 1$ is trivial.

Suppose that $g = 2$. We use I to denote the $m \times m$ identity matrix and $\mathbf{1}$ to denote the m-vector of ones:

$$\mathbf{1} = (1, 1, \ldots, 1).$$

In a slight abuse of notation, we use E^j to denote the m-vector $(E_0^j, E_1^j, \ldots, E_{m-1}^j)$. Suppose $g = 2$. Then we have

$$\beta^j = c_0^j \alpha + E^j.$$

$$\begin{pmatrix} \alpha_0 \cdot \alpha_0 + \alpha_1 \cdot \alpha_1 & \alpha_1 \\ \mathbf{1}^{\mathbf{t}} & I \end{pmatrix} \begin{pmatrix} c_0^j \\ (E^j)^{\mathbf{t}} \end{pmatrix} = \begin{pmatrix} 0 \\ (h^j)^{\mathbf{t}} \end{pmatrix}.$$

Now if we denote the matrix on the left by A,

$$\det(A) = \begin{vmatrix} \alpha_0 \cdot \alpha_0 + \alpha_1 \cdot \alpha_1 & \alpha_1 \\ \mathbf{1}^{\mathbf{t}} & I \end{vmatrix} = \begin{vmatrix} \alpha_0 \cdot \alpha_0 + \alpha_1 \cdot \alpha_1 - \alpha_1 \cdot \mathbf{1} & 0 \\ \mathbf{1}^{\mathbf{t}} & I \end{vmatrix}.$$

Thus

$$\det(A) = \alpha_0 \cdot \alpha_0 + \alpha_1 \cdot \alpha_1 - \alpha_1 \cdot \mathbf{1} = \alpha_0 \cdot \alpha_0 > 0,$$

since $|\alpha_0|^2 \neq 0$ if and only if $L^{g-1}\alpha \neq 0$, and α is nondegenerate.

Suppose that $g = 3$. Now we have that

$$\beta^j = c_0^j \alpha + c_1^j R\alpha + E^j.$$

$$\begin{pmatrix} \alpha_0 \cdot \alpha_0 + \alpha_1 \cdot \alpha_1 + \alpha_2 \cdot \alpha_2 & 0 & \alpha_2 \\ 0 & \alpha_0 \cdot \alpha_0 + \alpha_1 \cdot \alpha_1 & \alpha_1 \\ \mathbf{1}^{\mathbf{t}} & (\mathbf{1} - \boldsymbol{a_2})^{\mathbf{t}} & I \end{pmatrix} \begin{pmatrix} c_0^j \\ c_1^j \\ (E^j)^{\mathbf{t}} \end{pmatrix} = \begin{pmatrix} 0 \\ 0 \\ (h^j)^{\mathbf{t}} \end{pmatrix}.$$

Again denoting the matrix on the left by A,

$$\begin{aligned} \det(A) &= \begin{vmatrix} \alpha_0 \cdot \alpha_0 + \alpha_1 \cdot \alpha_1 + \alpha_2 \cdot \alpha_2 & 0 & \alpha_2 \\ 0 & \alpha_0 \cdot \alpha_0 + \alpha_1 \cdot \alpha_1 & \alpha_1 \\ \mathbf{1}^{\mathbf{t}} & (\mathbf{1} - \alpha_2)^{\mathbf{t}} & I \end{vmatrix} \\ &= \begin{vmatrix} \alpha_0 \cdot \alpha_0 + \alpha_1 \cdot \alpha_1 + \alpha_2 \cdot \alpha_2 - \alpha_2 \cdot \mathbf{1} & \alpha_2 \cdot (\alpha_2 - \mathbf{1}) & 0 \\ -\alpha_1 \cdot \mathbf{1} & \alpha_0 \cdot \alpha_0 + \alpha_1 \cdot \alpha_1 + \alpha_1 \cdot (\alpha_2 - \mathbf{1}) & 0 \\ \mathbf{1}^{\mathbf{t}} & (\mathbf{1} - \alpha_2)^{t} & I \end{vmatrix}. \end{aligned}$$

Thus

$$\det(A) = \begin{vmatrix} \alpha_0 \cdot \alpha_0 + \alpha_1 \cdot \alpha_1 + \alpha_2 \cdot \alpha_2 - \alpha_2 \cdot \mathbf{1} & \alpha_2 \cdot (\alpha_2 - \mathbf{1}) \\ -\alpha_1 \cdot \mathbf{1} & \alpha_0 \cdot \alpha_0 + \alpha_1 \cdot \alpha_1 + \alpha_1 \cdot (\alpha_2 - \mathbf{1}) \end{vmatrix}$$

$$\begin{aligned} \det(A) &= \begin{vmatrix} \alpha_0 \cdot \alpha_0 + \alpha_1 \cdot \alpha_1 + \alpha_0 \cdot \alpha_1 & \alpha_1 \cdot \alpha_0 \\ -\alpha_1 \cdot \alpha_1 & \alpha_0 \cdot \alpha_0 - \alpha_1 \cdot \alpha_0 \end{vmatrix} \\ &= (\alpha_0 \cdot \alpha_0)^2 + (\alpha_0 \cdot \alpha_0)(\alpha_1 \cdot \alpha_1) - (\alpha_1 \cdot \alpha_0)^2. \end{aligned}$$

However, Schwarz' inequality tells us that

$$(\alpha_1 \cdot \alpha_0)^2 \leq (\alpha_0 \cdot \alpha_0)(\alpha_1 \cdot \alpha_1)$$

and so

$$\det(A) \geq (\alpha_0 \cdot \alpha_0)^2 > 0,$$

again using the fact that α is nondegenerate. ■

Remark. We conjecture that the determinant of the linear system is nonzero for any positive g, ensuring uniqueness of the solution $\{c_0^j, \ldots, c_{g-2}^j, E_0^j, \ldots, E_{m-1}^j\}$.

We now have the following theorem.

Theorem 28. *Let the genus $g \geq 1$. If $\{c_0^j, \ldots, c_{g-2}^j, E_0^j, \ldots, E_{m-1}^j\}$ is a solution to the equations (4.30) and (4.29), then the wavelet system given by*

$$\beta^j = \sum_{l=0}^{g-2} c_l^j R^l \alpha + E^j \tag{4.31}$$

satisfies the orthogonality conditions (4.24)-(4.26), as well as the Haar condition (4.27). Moreover, each wavelet vector β^j is nondegenerate if and only if $c_0^j \neq 0$.

Proof:

(i) First observe that the Haar condition (4.27) holds by construction of the b^j:

$$\sum_{l=0}^{g-2} c_l^j \left\{ \sum_{k=0}^{g-1} R^{g-1} L^k R^l \alpha \right\} + E^j = h^j,$$

as does Equation (4.25) for $0 \leq l \leq g-2$:

$$\sum_k c_k^j \langle R^k \alpha, R^l \alpha \rangle + \langle E^j, R^l \alpha \rangle = 0.$$

(ii) Next let us show that Equation (4.25) holds for $l = g-1$:

$$\langle L^{g-1} \beta^j, \alpha \rangle = \langle \beta^j, R^{g-1} \alpha \rangle = \sum_{l=0}^{g-2} c_l^j \langle R^l \alpha, R^{g-1} \alpha \rangle + \langle E^j, R^{g-1} \alpha \rangle,$$

and we must show this quantity to be zero. From the Haar condition (4.27) we know that

$$h^j = \sum_{k=0}^{g-1} L^{g-1} R^k \left\{ \sum_{l-0}^{g-2} c_l^j R^l \alpha + E^j \right\} = L^{g-1} E^j + \sum_{k=0}^{g-1} \sum_{l=0}^{g-2} c_l^j L^{g-1} R^{k+l} \alpha.$$

so that the orthogonality of the Haar matrix rows implies

$$0 = \langle \mathbf{1}, h^j \rangle \Rightarrow$$

$$0 = \left\langle \sum_{n=0}^{g-1} L^{g-1} R^n \alpha, L^{g-1} E^j + \sum_{k=0}^{g-1} \sum_{l=0}^{g-2} c_l^j L^{g-1} R^{k+l} \alpha \right\rangle.$$

The first term here is

$$\left\langle \sum_{n}^{g-1} L^{g-1} R^n \alpha, L^{g-1} E^j \right\rangle = \sum_{n}^{g-1} \langle R^n \alpha, E^j \rangle$$

$$= \langle R^{g-1} \alpha, E^j \rangle + \sum_{n}^{g-2} \langle R^n \alpha, E^j \rangle$$

$$= \langle R^{g-1} \alpha, E^j \rangle - \sum_{n}^{g-2} \sum_{l}^{g-2} c_l^j \langle R^n \alpha, R^l \alpha \rangle.$$

Thus it suffices to show that the second term satisfies

$$\left\langle \sum_{n}^{g-1} L^{g-1} R^n \alpha, \sum_{k}^{g-1} \sum_{l}^{g-2} c_l^j L^{g-1} R^{k+l} \alpha \right\rangle = \sum_{n}^{g-1} \sum_{l}^{g-2} c_l^j \langle R^n \alpha, R^l \alpha \rangle. \tag{4.32}$$

The left-hand side can be rewritten

$$\sum_{l}^{g-2} c_l^j \left\langle \sum_{n}^{g-2} L^{g-1} R^n \alpha, \sum_{k}^{g-1} L^{g-1} R^{k+l} \alpha \right\rangle. \tag{4.33}$$

However, for any integer S,

$$\langle R^l \alpha, R^{l+s} \alpha \rangle = \sum_{k=0}^{g-1-l-s} \alpha_k \cdot \alpha_{k+s}$$

$$= \sum_{k=0}^{g-1} \langle L^{g-1} R^{k+l+s} \alpha, L^{g-1} R^{k+1} \alpha \rangle. \tag{4.34}$$

The sums inside the inner product of (4.33) are taken over a rectangle, $0 \leq n \leq g-1$, and $l \leq k+l \leq g-1$. Keeping in mind that $R^k \alpha = 0$ for $k > g-1$, we can rewrite this double sum in two parts along diagonals of the rectangle:

$$\left\langle \sum_{n}^{g-2} L^{g-1} R^n \alpha, \sum_{k}^{g-1} L^{g-1} R^{k+l} \alpha \right\rangle = \sum_{n=0}^{g-1} \langle L^{g-1} R^{n+k} \alpha, L^{g-1} R^{l+k} \alpha \rangle$$

$$+ \sum_{k=1}^{g-1} \sum_{k'=0}^{g-1} \langle L^{g-1} R^{k'} \alpha, L^{g-1} R^{l+k+k'} \alpha \rangle$$

and by (4.34) this is

$$= \sum_{n=0}^{g-1} \langle R^n \alpha, R^l \alpha \rangle + \sum_{k=1}^{g-1} \langle \alpha, R^{l+k} \alpha \rangle$$

$$= \sum_{n=0}^{g-1} \langle R^n \alpha, R^l \alpha \rangle.$$

Thus (4.33) becomes

$$\sum_{n=0}^{g-1} \sum_{l}^{g-2} c_l^j \langle R^n \alpha, R^l \alpha \rangle$$

and we are done.

(iii) Next consider Equation (4.24). If $k = 0$, we want

$$\langle \alpha, \beta \rangle = 0,$$

which is also equation (4.25) with $k = 0$, satisfied by construction. If $k > 0$ in (4.24) then we want

$$0 = \langle L^k \alpha, \beta^j \rangle = \langle L^{k-1} \alpha, R\beta^j \rangle;$$

however,

$$R\beta^j = \sum_{l=0}^{g-2} c_l^j R^{l+1} \alpha$$

and so

$$\langle L^{k-1} \alpha, R\beta^j \rangle = \sum_{l=0}^{g-2} c_l^j \langle L^{k-1} \alpha, R^{l+1} \alpha \rangle = 0.$$

(iv) Similarly, Equation (4.26) can be rewritten

$$\langle \beta^j, R^k \beta^{j'} \rangle = m\delta_{j,j'} \delta_{k,0},$$

and the fact that h_r^j is a Haar matrix, with

$$h^j \cdot h^{j'} = m\delta j, j'$$

ensures that this holds; the proof is similar to that of (ii).

Finally, we want to show that each wavelet vector β^j constructed as above is nondegenerate if and only if $c_0^j \neq 0$. We have from (4.31)

$$\beta^j = \sum_l c_l^j R^l \alpha + E^j \tag{4.35}$$

and applying R^{g-1} to this equation we find that

$$R^{g-1} \beta^j = c_0^j R^{g-1} \alpha$$

since $RE^j = 0$, and $R^g = 0$. Since α is nondegenerate, we have that $R^{g-1}\alpha \neq 0$, and so β^j is nondegenerate if and only if $c_0^j \neq 0$. ■

§5. Wavelet matrix series

5.1. A discrete orthonormal expansion

The main theorem of this section is one of the key links between the mathematical theory of wavelets and the practical applications of wavelets. The theorem exhibits a locally finite compact wavelet matrix series expansion for an arbitrary discrete function.

If

$$f(x) = \sum_n f_n(x)$$

where f_n is a sequence of functions $x \mapsto f_n(x)$ defined on some infinite set, (*e.g.*, $\mathbb{Z}$ or $\mathbb{R}$) then $f(x)$ will have a meaning that a prescribed by the type of convergence that is assumed. Let us suppose that for each x only finitely many of the numbers $f_n(x)$ are not zero. For such values of x the infinite series reduces to a finite sum. If the sum is finite for all x in the common domain of the f_n, then the series is said to be *locally finite*, and for it, no additional criteria of convergence are required. We recall that in (2.10) we introduced the notation

$$a_k := a_k^0, \quad b_k^s := a_k^s, \quad s \geq 1.$$

Theorem 29. *Let $f: \mathbb{Z} \to \mathbb{C}$ be an arbitrary function defined on the integers and let a be a compact wavelet matrix of rank m and genus g. Then f has a unique wavelet matrix expansion*

$$f(n) = \sum_{l=-\infty}^{\infty} c_l a_{ml+n} + \sum_{s=1}^{m} \sum_{k=-\infty}^{\infty} c_k^s b_{mk+n}^s \tag{5.1}$$

where

$$c_l = \frac{1}{m} \sum_n f(n) \bar{a}_{ml+n} \tag{5.2}$$

$$c_k^s = \frac{1}{m} \sum_n f(n) \bar{b}_{mk+n}^s. \tag{5.3}$$

The wavelet matrix expansion is locally finite, i.e., for given n only finitely many terms of the series are different from zero.

Remark. Using the language of signal processing, the first term in (5.2) is the "low-pass" part of the expansion, and the second term is the "high-pass" part of the expansion

Proof: The proof has two parts. First we will show that an arbitrary function f: $\mathbb{Z} \to \mathbb{C}$ has an orthogonal expansion of the form (5.1). Then we will prove that if an expansion of the form (5.1) exists, its coefficients are given by (5.2) and (5.3).

The construction of a wavelet matrix expansion depends on the expansion for the discrete analogue of the Dirac distribution. Consider the Kronecker delta, which we will now write in the form

$$\delta(n', n) := \begin{cases} 1 & \text{if } n' = n \\ 0 & \text{else.} \end{cases} \tag{5.4}$$

We want to show that $\delta(n', n)$ has the wavelet matrix expansion

$$\delta(n', n) = \frac{1}{m} \sum_l \bar{a}_{ml+n'} a_{ml+n} + \frac{1}{m} \sum_{0<s<m} \sum_l \bar{b}^s_{ml+n'} b^s_{ml+n}, \tag{5.5}$$

$$= \frac{1}{m} \sum_{0 \le s<m} \sum_l \bar{a}^s_{ml+n'} a^s_{ml+n}, \tag{5.6}$$

but this follows immediately from (2.7).

If f: $\mathbb{Z} \to \mathbb{C}$ is an arbitrary function, then f can be written as

$$f(n) = \sum_{n'} f(n') \delta(n', n)$$

whence

$$f(n) = \frac{1}{m} \sum_{n'} f(n') \sum_{0 \le s<m} \sum_L \bar{a}^s_{ml+n'} a^s_{ml+n} \tag{5.7}$$

$$= \sum_l c_l a_{ml+n} + \sum_{0<s<m} \sum_l c^s_l a^s_{ml+n}. \quad \blacksquare \tag{5.8}$$

The L^2 norm of the discrete function f: $\mathbb{Z} \to \mathbb{C}$ is defined by

$$\|f\|^2 := \sum_{n \in \mathbb{Z}} |f(n)|^2. \tag{5.9}$$

The norm of f is expressed in terms of the coefficients by *Parseval's formula* for wavelet matrix expansions:

Theorem 30. *If*

$$f(x) = \sum_l c_l a_{ml+n} + \sum_{0<s<m} \sum_l c^s_l a^s_{ml+n}$$

and if $\|f\| \le \infty$, *then*

$$\|f\|^2 = m \left\{ \sum_k |c_n|^2 + \sum_{0<s<m} \sum_n |c^s_n|^2 \right\}. \tag{5.10}$$

Suppose that $p(x)$ is a polynomial of degree d.

Theorem 31. *A necessary and sufficient condition that*

$$p(n) = \sum_k c_k a_{mk+n} \tag{5.11}$$

is that

$$\sum_l l^r b_l^s = 0 \quad \textit{for} \quad 0 \leq r \leq d \quad \textit{and} \quad 0 < s < m. \tag{5.12}$$

Proof: Calculate the coefficients of b_k^s in the expansion of $p(n)$. ■

Suppose the condition (5.12) of Theorem 31 is satisfied. Let

$$p(x) = \sum_{l=0}^{d} p_l x^l.$$

The coefficient c_k is

$$c_k = \sum_{l=0}^{d} p_l \sum_n n^l \bar{a}_{mk+n}.$$

Hence the coefficients c_k are linear combinations of the moments of a_k. In particular,

$$1 = \frac{1}{m} \sum_k a_{mk+n} \tag{5.13}$$

and

$$n = \sum_l l a_l - m \sum_k k a_{mk+n} \tag{5.14}$$

Example 32. (Fourier-Wavelet Matrix Expansion) If the characteristic Haar matrix of a wavelet matrix $a \in \mathrm{WM}(m, g; \mathbb{C})$ is the finite Fourier transform matrix Ω_m (see Example 15), then the wavelet matrix expansion for a function $f \colon \mathbb{Z} \to \mathbb{C}$ is called a *Fourier-wavelet matrix expansion.*

If $g = 1$ then $a = \Omega_m$, and we recall that we let $\omega = \exp(2\pi i/m)$ be a primitive m-th root of unity. In this case, the Fourier-wavelet matrix expansion for f has the form

$$f(ml + r) = \sum_{k=-\infty}^{\infty} \left\{ \sum_{s=0}^{m-1} \hat{f}_k(s) \omega^{rs} \right\}, \quad r = 0, \ldots, m-1. \tag{5.15}$$

The inner sum is the finite Fourier transform for the restriction of f to the interval $(mk, mk+m-1)$. That is to say,

$$\hat{f}_k(s) := \frac{1}{m}\sum_{p=0}^{m-1} f(mk+p)\bar{\omega}^{ps}.$$

In this case, *i.e.*, for $g = 1$, the Fourier-wavelet series represents f as a short time Fourier series, *i.e.*, a sequence of finite Fourier series for nonoverlapping "windows" of length m. Notice that the right-hand side of (5.15) is independent of l.

Define

$$D_Z := Z\frac{d}{dZ}. \tag{5.16}$$

where $\frac{d}{dZ}$ is the operator of formal differentiation with respect to the indeterminate Z.

The action of D_Z on the formal Laurent series $a(Z)$ is

$$D_Z a(Z) = Z\frac{d}{dZ}a(Z) = \sum_k n a_n^s Z^n. \tag{5.17}$$

This formula motivates the following definitions:

The l-th *discrete moment* of a is

$$\mathrm{Mom}_l(a) := (D_Z^l a)(1) = \sum_n n^l a_n^s. \tag{5.18}$$

The l-th *partial discrete moment* of a is

$$\mathrm{Mom}_l(a:r) := (D_Z^l p[a])(1) = \sum_k (mk+r)^l a_{mk+r}^s, \tag{5.19}$$

where $0 \le r < m$.

A wavelet matrix a is said to be *low-pass of degree* d if

$$\sum_k k^e b_{mk+n}^s = 0, \quad 0 < s < m \tag{5.20}$$

for all integers e such that $0 \le e \le d$.

Remark. This is analogous to having vanishing moments for functions integrable on the real axis.

Lemma 33. *Every wavelet matrix is low-pass of degree 0.*

This follows immediately from the definitions.

Theorem 34. *If a is a low-pass wavelet matrix of rank m and of degree d and if $p(x)$ is a polynomial of degree $\leq d$, then the wavelet matrix expansion of the restriction of $p(x)$ to $\mathbb{Z}$ is*

$$p(n) = \sum_k c_k a_{mk+n} \tag{5.21}$$

where

$$c_k = \frac{1}{m} \sum_l p(l) \bar{a}_{mk+l} \tag{5.22}$$

The proof of this follows simply from the definitions and will be omitted.

We have the following two special cases of this result, the first being a discrete version of a partition of unity, and the second being a "partition of" a linear function.

Corollary 35. *For any wavelet matrix a of rank m*

$$\sum_k a_{mk+n} = 1. \tag{5.23}$$

Corollary 36. *If a is a wavelet matrix of rank m which is low-pass of degree 1, then*

$$n = \sum_k c_k a_{mk+n} \tag{5.24}$$

where

$$c_K = \frac{1}{m} \sum_l l \bar{a}_{mk+l}. \tag{5.25}$$

Remark. It can be shown that a Haar wavelet matrix of rank m can be rotated into a Haar wavelet matrix of rank m for which all but one of the wavelet moments vanishes.

References

1. Abramowitz, M. and I. Stegun, *Handbook of Mathematical Functions*, National Bureau of Standards, Washington D. C., 1972.
2. Battle, G., A block spin construction of ondelettes, Part I: Lemarié functions, *Comm. Math. Phys.* **110** (1987), 601–615
3. Blahut, R. E., *Digital Transmission of Information*, Addition-Wesley, Reading, MA, 1990.
4. Burrus, C. S. and T. W. Parks, *DFT/FFT and Convolution Algorithms: Theory and Implementation*, Wiley Interscience, New York, 1985.
5. Daubechies, I., Orthonormal bases of compactly supported wavelets, *Comm. Pure Appl. Math.* **41** (1988), 906–966.
6. Hadamard, J., Resolution d'une question relative aux determinants, *Bull. Sci. Math.* **17** (2) pt. 1 (1893), 240–246.
7. Heller, P. N., Higher rank Daubechies scaling coefficients, Aware Technical Report, Aware, Inc., Cambridge, MA, to appear.
8. Lawton, W. M., Necessary and sufficient conditions for constructing orthonormal wavelets bases, *J. Math. Phys.* **32** (1) (1991), 57–61.
9. Lawton, W. M. and H. L. Resnikoff, Multidimensional wavelet base, Aware Technical Report, Aware, Inc., Cambridge, MA, *SIAM J. Math. Anal.*, submitted.
10. Lemarié, P.H., Ondelettes à localisation exponentielle, *J. Math. Pure et Appl.* **67** (1988), 227–236.
11. Lidl, R. and G. Pilz, *Applied Abstract Algebra*, Springer-Verlag, New York, 1990.
12. Lipson, John D., *Elements of Algebra and Algebraic Computing*, Addison-Wesley, Reading, MA, 1981.
13. Mallat, S., Multiresolution approximation and wavelet orthonormal bases of $L^2(\mathbb{R})$, *Trans. Amer. Math. Soc.* **315** (1989), 69–87.
14. Meyer, Y., Principe d'incertitude, bases Hilbertiennes et algèbres d'opérateurs, *Séminaire Bourbaki* **662**, (1985–86).
15. Meyer, Y., Ondelettes et functions splines, *Séminaire EDP*, Ecole Polytechnique, Paris, December 1986.
16. Meyer, Y., *Ondelettes et Opérateurs*, in two volumes, Hermann, Paris,1990.
17. Pollen, D., $SU_I(F[z, 1/z])$ for F a subfield of $\mathbb{C}$, *J. Amer. Math. Soc.* (1990).
18. Pollen, D., Parametrization of compactly supported wavelets, Aware Technical Report, Aware, Inc., Cambridge, MA, to appear.
19. Pratt, W. K., *Digital Image Processing*, Wiley, New York, 1978.
20. Press, W. H., B. F. Flannery, S. Teukolsky, and W. Vetterline, *Numerical Recipes in C*, Cambridge University Press, New York, 1988.
21. Rademacher, H., Einige sätze von allgemeinen orthogonalfunktionen, *Math. Ann.* **87** (1922), 122–138.
22. Resnikoff, H. L. and R. O. Wells, Jr., *Wavelet Analysis*, in preparation.

23. Vaidyanathan, P. P., Quadrature mirror filter banks, M-band extensions, and perfect-reconstruction techniques, *IEEE ASSP Magazine* **4** (3) (1987), 4–20.
24. Vetterli, M., A theory of multirate filter banks, *IEEE Trans. ASSP* **35** (3) (1987), 356–372.
25. Walsh, J. L., A closed set of orthogonal functions, *Amer. J. Math.* **45** (1923), 5–24.
26. Wells, Jr., R. O., Parametrizing smooth compactly supported wavelets, *Trans. Amer. Math. Soc.*, 1991, to appear.

This material is based upon work supported by AFOSR under grant number 90-0334 which was funded by DARPA, and by Aware, Inc.

Peter N. Heller
Aware, Inc.
University Place
124 Mt. Auburn Street
Cambridge, MA 02138

Howard L. Resnikoff
Aware, Inc.
University Place
124 Mt. Auburn Street
Cambridge, MA 02138

Raymond O. Wells, Jr.
Department of Mathematics
Rice University
Houston, TX 77251-1892
wells@rice.edu

Wavelets and Generalized Functions

Gilbert G. Walter

Abstract. Both wavelets and generalized functions or distributions have a similar history. They both were applied to problems before their theory was developed or even before they were completely understood. The two theories that were subsequently developed interact considerably. Some of these interactions are presented in this work. They include the expansions of distributions in terms of wavelets of functions, the introduction of wavelets of distributions with point support, and the distribution solutions of dilation equations. The theory is used to obtain sampling theorems for subspaces such as quadratic splines.

§1. Introduction

The development of the subject of wavelets is similar to the earlier development of generalized functions. Both fields began with an attempt to solve problems arising in technology by using new techniques whose theory was not completely understood.

Generalized functions (or distributions) began with the Heaviside calculus and an informal use of the delta function in engineering and physics. This was put into a rigorous mathematical form by Schwartz [18], Korevaar [9], and others. As a result the subject became even more widely used and is now an integral part of partial differential equations [8], Fourier transforms [5], and signal analysis [22].

Wavelets began with a technique that seemed to work better for seismic analysis than previous techniques [19]. The original theory involved representation of functions by means of integral transform [6], but was soon extended to non-orthogonal series representation (see [3] and [7]), and eventually to orthogonal series (see [4] and [10]). These theories were put to use in a number of new areas such as image processing, signal analysis, and data compression (see [1], [20], and [10]).

However the two fields have more in common than their historical similarity. Many of the arguments in wavelet theory involve convergence in the sense

Wavelets–A Tutorial in Theory and Applications
C. K. Chui (ed.), pp. 51–70.

ISBN 0-12-174590-2

of distributions [4]; wavelet expansions hold for certain classes of distributions [12]; these are wavelets of point support which must be tempered distributions [24]; the solutions to the dilation equations arising in wavelet theory are, in general, also tempered distributions [21]. In addition there is a natural relation between wavelet subspaces and delta sequences; that is, sequences of functions converging to the delta function.

In this work we shall present a few aspects of the relation between the two subjects. We first give quick surveys of the two subjects mainly in order to fix our notation. Then we discuss the expansion of distributions in terms of wavelets of ordinary functions. This will be followed by a section on sampling theorems which will use the previous section but is also related to distributions in a natural way. The next section will consider distribution solutions of dilation equations and wavelets of distributions with point support.

§2. Background on wavelets and distributions

2.1. The theory of orthonormal bases of wavelets in $L^2(\mathbb{R})$ may be found in a number of places. A detailed introduction may be found in [4] while a more general complete theory is in [12]. Here, we merely sketch a few elements of the theory in one dimension.

We begin with a "scaling function" φ, a real-valued function on $\mathbb{R}$ which is r times differentiable and whose derivatives are continuous and rapidly decreasing. That is, φ satisfies

$$|\varphi^{(k)}(t)| \leq C_{pk}(1+|t|)^{-p}, \qquad k=0,1,\ldots,r; \quad p \in \mathbb{Z}, \quad t \in \mathbb{R}. \tag{2.1}$$

Associated with φ is a multiresolution analysis of $L^2(\mathbb{R})$; that is, a nested sequence of closed subspaces $\{V_m\}_{m\in\mathbb{Z}}$ such that

$$\begin{aligned} &\text{(i)} && \{\varphi(t-n)\} \text{ is an orthonormal basis of } V_0;\\ &\text{(ii)} && \cdots \subset V_{-1} \subset V_0 \subset V_1 \subset \cdots \subset L^2(\mathbb{R});\\ &\text{(iii)} && f \in V_m \Leftrightarrow f(2\cdot) \in V_{m+1};\\ &\text{(iv)} && \bigcap_m V_m = \{0\}, \overline{\bigcup_m V_m} = L^2(\mathbb{R}). \end{aligned} \tag{2.2}$$

Clearly $\{\sqrt{2}\,\varphi(2t-n)\}$ is an orthonormal basis for V_1 since the map $f \mapsto \sqrt{2}\,f(2\cdot)$ is an isometry from V_0 onto V_1. Since $\varphi \in V_1$, it must have an expansion of the form

$$\varphi(t) = \sum_k c_k \sqrt{2}\,\varphi(2t-k), \qquad \{c_k\} \in \ell^2, \qquad t \in \mathbb{R}. \tag{2.3}$$

Example 1. A simple example which however does not satisfy (2.1) is given by $\varphi(t) = \chi_{[0,1)}(t)$, the characteristic function of the unit interval. Its translates $\{\varphi(t-n)\}$ are orthonormal and V_0 is composed of piecewise constant functions.

It is not obvious that a multiresolution analysis exists for φ other than this example. In order to find other φ's, two approaches have been used. One begins with another definition of V_0 in terms of a (non-orthogonal) Riesz basis $\{\theta(t-n)\}$ of translates of a fixed function $\theta(t)$. The φ's are then found by means of an orthogonalization procedure as considered by Meyer [11].

The other approach begins with the "dilation equation" (2.3) and tries to choose $\{c_k\}$ such that (2.2) is satisfied.

Example 2. The simplest example of the orthogonalization approach leads to the *Franklin wavelets*. The initial function is the "hat function" (or second order B-spline) $\theta(t) = (1-|t-1|)\chi_{[0,2)}(t)$. This corresponds to a V_0 composed of continuous piecewise linear functions with breaks at the integers. Its Fourier transform is

$$\hat{\theta}(w) = \int_{-\infty}^{\infty} e^{-iwt}\theta(t)dt = \left[\frac{1-e^{-iw}}{iw}\right]^2 = \frac{e^{-iw}\sin^2 w/2}{(w/2)^2}. \tag{2.4}$$

To find the orthogonal φ we use the isometry property of the Fourier transform:

$$\begin{aligned}\delta_{0n} &= \int_{-\infty}^{\infty} \varphi(t-n)\varphi(t)dt = \frac{1}{2\pi}\int_{-\infty}^{\infty} \widehat{\varphi}(w)e^{-iwn}\overline{\widehat{\varphi}(w)}dw \qquad (2.5)\\ &= \frac{1}{2\pi}\sum_{k=-\infty}^{\infty}\int_0^{2\pi} |\widehat{\varphi}(w+2\pi k)|^2 e^{-iwn}dw \\ &= \frac{1}{2\pi}\int_0^{2\pi}\sum_k |\widehat{\varphi}(w+2\pi k)|^2 e^{-iwn}dw.\end{aligned}$$

This is the integral which gives the Fourier coefficient of the periodic function $|\widehat{\varphi}^*(w)|^2 = \sum|\widehat{\varphi}(w+2\pi k)|^2$, which by orthogonality, must be δ_{0n}. Hence, $|\widehat{\varphi}^*(w)|^2$ which is equal to its Fourier series, is identically equal to 1.

Since $\varphi \in V_0$, we have $\varphi(t) = \sum a_n\theta(t-n)$ for some $\{a_n\} \in \ell^2$. Again by taking Fourier transforms we find

$$\widehat{\varphi}(w) = \sum_n a_n e^{-iwn}\hat{\theta}(w) = \alpha(w)\hat{\theta}(w) \tag{2.6}$$

and hence,

$$|\widehat{\varphi}^*(w)|^2 = |\alpha(w)|^2|\hat{\theta}^*(w)|^2 = 1. \tag{2.7}$$

$|\hat{\theta}^*(w)|^2$ can be found in closed form by summing the series

$$\sum_k \left|\frac{\sin{(w+2\pi k)/2}}{(w+2\pi k)/2}\right|^4.$$

By substituting the result in (2.7), and solving for $\alpha(w)$, we find φ to be

$$\widehat{\varphi}(w) = \frac{\sin^2 w/2}{(w/2)^2}\left(1-(2/3)\sin^2\frac{w}{2}\right)^{-\frac{1}{2}}. \tag{2.8}$$

Example 3. The **Daubechies wavelets** are based on the other approach. We look for $\{c_k\}$ in (2.3) such that the orthogonality condition is satisfied. This is again given in terms of the Fourier transform

$$\widehat{\varphi}(w) = \frac{1}{\sqrt{2}} \sum c_k e^{-ikw/2} \widehat{\varphi}\left(\frac{w}{2}\right) = m_0\left(\frac{w}{2}\right) \widehat{\varphi}\left(\frac{w}{2}\right) \tag{2.9}$$

which translates into the orthogonality condition

$$\left|m_0\left(\frac{w}{2}\right)\right|^2 + \left|m_0\left(\frac{w}{2} + \pi\right)\right|^2 = 1 \tag{2.10}$$

on the 2π periodic function m_0 from (2.7), and the condition $m(0) = 1$. If c_k is chosen by

$$\begin{aligned} c_0 &= \nu(\nu - 1)/(\nu^2 + 1)\sqrt{2}, \\ c_1 &= (1 - \nu)/(\nu^2 + 1)\sqrt{2}, \\ c_2 &= (\nu + 1)/(\nu^2 + 1)\sqrt{2}, \\ c_3 &= \nu(\nu + 1)/(\nu^2 + 1)\sqrt{2}, \end{aligned}$$

where $\nu \in \mathbb{R}$ then (2.10) will be satisfied [4, p. 946].

This does not directly give $\widehat{\varphi}(w)$ or $\varphi(t)$ which rather must be found recursively by starting with, say, $\varphi_0(t) = \chi_{[0,1)}(t)$ and then defining

$$\varphi_n(t) = \sqrt{2} \sum_k c_k \quad \varphi_{n-1}(2t - k). \tag{2.11}$$

This sequence converges to a continuous function when $\nu = \pm 1/\sqrt{3}$, but in general, does not, although it does converge weakly [21].

Once we have the scaling function $\varphi(t)$, we may use it to construct the "mother wavelet" $\psi(t)$. For orthonormal wavelets, it must be chosen such that $\{\psi(t - n)\}$ is an orthonormal basis of the space W_0, given by the orthogonal complement of V_0 in V_1. Then

$$V_1 = V_0 \oplus W_0.$$

If such a $\psi(t)$ can be found, then $2^{m/2}\psi(2^{m/2}t - n) = \psi_{nm}(t)$ is an orthonormal basis of W_m, the dilation space of W_0. Indeed from (2.2) it follows that

$$\bigoplus_{m \in Z} W_m = L^2(\mathbb{R})$$

and hence $\{\psi_{n,m}\}_{n,m \in Z}$ is an orthonormal basis of $L^2(\mathbb{R})$.

Again there are two methods of finding $\psi(t)$. One uses the orthogonalization procedure similar to that illustrated in Example 2. But $\psi(t)$ must also be orthogonal to $\varphi(t - n)$. Hence the two conditions

$$\begin{aligned} &\text{a)} \quad \sum_k \widehat{\psi}(t + n\pi k)\overline{\widehat{\varphi}(t + 2\pi k)} = 0 \\ &\text{b)} \quad \sum_k |\widehat{\psi}(t + 2\pi k)|^2 = 1 \end{aligned} \tag{2.12}$$

must be satisfied. Since $\psi(t) \in V_1$, it has an expansion similar to (2.3), namely:

$$\psi(t) = \sum_k d_k \sqrt{2}\, \varphi(2t - k), \tag{2.13}$$

and hence,

$$\widehat{\psi}(w) = m_1\left(\frac{w}{2}\right) \widehat{\varphi}\left(\frac{w}{2}\right). \tag{2.14}$$

This may be substituted into (2.12) and solved for $m_1\left(\frac{w}{2}\right)$ which is then used in (2.14).

The other approach involving the dilation equation is straightforward; $\psi(t)$ is defined as

$$\psi(t) := \sqrt{2} \sum_k c_{1-k}(-1)^k \varphi(2t - k). \tag{2.15}$$

Then it is merely a matter of checking that the two orthogonality conditions are satisfied. [4, p. 944].

2.2. The background needed for generalized functions is much more limited since we shall only consider tempered distributions. An easy to read reference is [26], while a more complete source is [18].

To define tempered distributions, we begin with the space $\mathcal{S}$ of rapidly decreasing C^∞ functions on $\mathbb{R}$; that is, functions which satisfy (2.1) for all integers $k \geq 0$. An element $\theta \in \mathcal{S}$ is usually called a *test function*. Rather than defining a topology on $\mathcal{S}$, we define convergence in $\mathcal{S}$ of the sequence $\{\theta_\nu\}$ to θ to be the uniform convergence for $t \in \mathbb{R}$ of

$$|t|^p |\theta_\nu^{(k)}(t) - \theta^{(k)}(t)|$$

to zero as $\nu \to \infty$. Here, p and k are nonnegative integers.

We denote by $\mathcal{S}_r$ the space containing $\mathcal{S}$ and for which (2.1) is satisfied only for $k \leq r$ but for all p. The convergence is the same as that of $\mathcal{S}$ with the same restriction on k.

The space $\mathcal{S}$ contains all the Hermite function $\{h_n\}$ [18], and hence is dense in $L^2(\mathbb{R})$ in its norm. It is a linear space which is complete in terms of its own convergence, and is closed under multiplication by t and differentiation [26].

Since the Fourier transform converts differentiation into multiplication by a multiple of t and vice-versa, $\mathcal{S}$ is closed under the Fourier transform. A *tempered distribution* is an element of the dual space $\mathcal{S}'$ of $\mathcal{S}$. This space is composed of all continuous linear functionals on $\mathcal{S}$; functions from $\mathcal{S}$ to C which are linear

$$\langle f, \alpha_1\theta_1 + \alpha_2\theta_2 \rangle = \alpha_2 \langle f, \theta_1 \rangle + \alpha_2 \langle f, \theta_2 \rangle$$

and continuous; that is,

$$\theta_n \to 0 \Rightarrow \langle f, \theta_n \rangle \to 0.$$

Each locally integrable function f of polynomial growth belongs to $\mathcal{S}'$ in the sense that

$$\langle f, \theta \rangle = \int_{-\infty}^{\infty} f(t)\theta(t)dt.$$

Operations on $\mathcal{S}'$ are defined in terms of the corresponding operations on $\mathcal{S}$. For example, the derivative is defined by

$$\langle f', \theta \rangle = -\langle f, \theta' \rangle.$$

Similar definitions give us addition, multiplication by a scalar, dilation, translation, and multiplication by a polynomial. $\mathcal{S}'$ is a linear space under the first two operations, and if convergence is defined as weak convergence (*i.e.*, $f_n \to f$ in $\mathcal{S}'$ if $\langle f_n, \theta \rangle \to \langle f, \theta \rangle$ for each $\theta \in \mathcal{S}$), it is a complete linear space.

An element $f \in \mathcal{S}'$ is of order r if there is a continuous measure μ of polynomial growth $\left(\int \frac{|d\mu|}{(1+|x|)^k} < \infty \right)$ such that $f = D^r \mu$. Such an f belongs to $\mathcal{S}'_r$, the dual space of $\mathcal{S}_r$, and in fact characterizes it. Some examples of elements of $\mathcal{S}'$ are the "delta function", $\langle \delta_\alpha, \varphi \rangle = \varphi(\alpha)$; the Dirac comb $\sum\limits_{n=-\infty}^{\infty} \delta(t-n)$, the pseudo function $pv\frac{1}{t}$ given by

$$\left\langle pv\frac{1}{t}, \varphi(t) \right\rangle = pv \int_{-\infty}^{\infty} \frac{\varphi(t)}{t} dt,$$

as well as all derivatives of locally integral functions of polynomial growth. In particular $\delta = Dh$, where h is the Heaviside function $h(x) = \begin{cases} 1 & x > 0 \\ 0 & x \leq 0. \end{cases}$

The definition of Fourier transform may be extended from $\mathcal{S}$ to $\mathcal{S}'$ by the same device as the other operations. For $f \in \mathcal{S}'$ we define $\hat{f} = \mathcal{F}f$ to be that element of $\mathcal{S}'$ given by

$$\langle \hat{f}, \varphi \rangle = \langle f, \widehat{\varphi} \rangle.$$

It is easy to check that this is a linear functional and that it is continuous. Moreover, the map $f \mapsto \hat{f}$ is a continuous linear transformation from $\mathcal{S}'$ onto $\mathcal{S}'$ whose inverse exists and is continuous as well.

Some examples of Fourier transforms on $\mathcal{S}'$ are

$$\begin{aligned}
&\mathcal{F}\delta(t-T) = e^{-iw\tau};\\
&\mathcal{F}\sum_k \delta(t-2\pi k) = \sum_n \delta(w-n);\\
&\mathcal{F}(t^k e^{-it\tau}) = i^k 2\pi \delta^{(k)}(w+\tau);\\
&\mathcal{F}\left(pv\frac{1}{t}\right) = -\pi i \text{ sgn } w.
\end{aligned}$$

§3. Expansion of distributions in wavelets

We assume that we have a multiresolution analysis (2.2) with scaling function $\varphi \in \mathcal{S}_r$. Then ψ, the mother wavelet, is also in $\mathcal{S}_r$ [12, p. 70]. The expansions of a tempered distribution of order r with respect to $\{\varphi(t-n)\}$ and $\{\psi(t-n)\}$ therefore exists and the coefficients satisfy

$$\begin{aligned} a_n &= \langle f, \varphi(-n)\rangle = \langle D^r[(t^2+1)^k d\mu], \varphi(\cdot - n)\rangle \\ &= \langle d\mu, (t^2+1)^k(-1)^r \varphi^{(r)}(t-n)\rangle = 0(|n|^k) \end{aligned} \tag{3.1}$$

for some $k \in \mathbb{Z}$, where μ is a bounded (regular Borel) measure. The same is true for $\psi(t-n)$.

The coefficients of $\theta \in \mathcal{S}_r$ can similarly be calculated as

$$\int_{-\infty}^{\infty} \theta(t)\varphi(t-n)dt = 0(|n|^{-p}), \quad \text{for all} \quad p \in \mathbb{Z}.$$

Thus the expansion of f converges in the sense of $\mathcal{S}'_r$ to some $f_0 \in \mathcal{S}'_r$; that is,

$$\sum_n a_n \int \varphi(t-n)\theta(t)dt = \langle f_0, \theta\rangle.$$

But the convergence of the expansion of f is much stronger than this since $\sum a_n\varphi(t-n)$ converges uniformly on bounded sets as do its first r derivatives. This follows from the inequality

$$\begin{aligned} \sum_n |a_n \varphi^{(j)}(t-n)| &\le C \sum_n \frac{(|n|+1)^k}{(|t-n|+1)^{k+2}} \\ &\le C \sum_n \sum_m \binom{k}{m} \frac{(|n-t|+1)^{k-m}|t|^m}{(|n-t|+1)^{k+2}} \\ &\le C \sum_n \frac{1}{(|n-t|+1)^2}(1+|t|)^k \\ &\le C'(1+|t|)^k. \end{aligned} \tag{3.2}$$

In fact we have shown that the limiting function and its r derivatives are continuous functions of polynomial growth on $\mathbb{R}$. We denote by T_0 the space of all functions given by such series; more precisely, we have the following.

Definition 4. *Let $\{a_n\}$ be a sequence of complex numbers such that $a_n = 0(|n|^k)$ for some $k \in \mathbb{Z}$; then $T_0 = \{f | f(t) = \sum_n a_n\varphi(t-n)\}$ and $U_0 = \{g | g(t) = \sum_n a_n\psi(t-n)\}$.*

Hence both T_0 and U_0 are composed of functions in C^r of polynomial growth. We denote by T_m and U_m their corresponding dilation spaces

($f \in T_0 \Leftrightarrow f(2^m t) \in T_m$). Then we may expect that a multiresolution analysis of $\mathcal{S}'_r$ exists, namely:

$$\cdots \subset T_{-1} \subset T_0 \subset T_1 \subset \cdots \subset T_m \subset \cdots \mathcal{S}'_r \tag{3.3}$$

and

$$\bigcap_m T_m = \{0\}, \qquad \overline{\bigcup_m T_m} = \mathcal{S}'_r,$$

where the closure is in the topology of $\mathcal{S}'_r$, but we must check a few things. Since $\varphi(t) \in V_1 \subset T_1$ where V_1 is part of the $L^2(\mathbb{R})$ multiresolution analysis, it follows that all partial sums $\sum_{|n|\le N} a_n \varphi(t-n) \in T_1$.

However, the corresponding series may not be in T_1 unless $\{c_k\}$ in (2.3) is sufficiently rapidly decreasing. We therefore assume $c_k = 0(|k|^{-m})$ all $m \in \mathbb{Z}$. Then the series becomes

$$\begin{aligned}\sum_n a_n \varphi(t-n) &= \sum_n a_n \sum_k \sqrt{2}\, c_k \varphi(2t-2n-k) \\ &= \sum_n a_n \sum_j c_{j-2n} \sqrt{2}\, \varphi(2t-j) \\ &= \sum_j \left(\sum_n a_n c_{j-2n}\right) \sqrt{2}\, \varphi(2t-j) \\ &= \sum_j a'_j \sqrt{2}\, \varphi(2t-j).\end{aligned} \tag{3.4}$$

Clearly if a_n is of polynomial growth say, $O(|n|^k)$, so is a'_n.

It can also be shown that T_0 is closed in the sense of $\mathcal{S}'_r$. This follows from the fact that $f_m \to 0$ in $\mathcal{S}'_r$ is equivalent to

$$f_m = D^r[(1+t^2)^k d\mu_m] \quad \text{and} \quad d|\mu_m| \to 0,$$

(see [26]). Hence by (3.1) we deduce that $a_{mn} = \langle f_m, \varphi(\cdot - n)\rangle \to 0$ and

$$|a_{mn}| \le C(n^2+1)^k.$$

From this it follows that if $f_m \to f$ in $\mathcal{S}'_r$, the coefficients of f are of polynomial growth and hence its series $\sum_n a_n \varphi(t-n)$ is in T_0.

Just as in the L^2 case, we can express each $f_1 \in T_1$ uniquely as $f_1 = f_0 + g_0$ where $f_0 \in T_0$ and $g_0 \in U_0$, in spite of the fact that we do not have orthogonality. Indeed since

$$a_n^0 = \langle f_1, \varphi(t-n)\rangle = \sum_k c_k a'_{2n+k} = \sum_j a'_j c_{j-2n}, \tag{3.5a}$$

and

$$b_n^0 = \langle f_1, \psi(t-n)\rangle = \sum_k (-1)^k c_{1-k} a'_{2n+k} = \sum_j a'_j c_{2n+1-j}(-1)^j, \quad (3.5\text{b})$$

we have the same decomposition algorithm as in the L^2 case [4, p. 937]. Similarly the reconstruction algorithm is

$$a'_n = \sum_k c_{n-2k} a_k^o + \sum_k c_{1-n+2k}(-1)^n b_k^o, \tag{3.6}$$

(see [4, p. 938]). Thus we use (3.5a) and (3.5b) to define f_0 and g_0 respectively, and (3.6) to deduce that $f_1 = f_0 + g_0$. We summarize in the following.

Theorem 5. *Let the scaling function $\varphi \in \mathcal{S}_r$ with a dilation equation (2.3) satisfying $c_k = O(|k|^p)$ for all $p \in \mathbb{Z}$ with an associated multiresolution analysis in $L^2(\mathbb{R})$; let $\psi \in \mathcal{S}_r$ be the mother wavelet given in (2.15). Then there exists a multiresolution analysis (3.3) of closed dilation subspaces $\{T_m\}$ whose union is dense in $\mathcal{S}'_r$ and whose intersection is $\{0\}$; the closed subspaces U_m are the complementary subspaces of T_m in T_{m+1} and*

$$T_m = U_0 \oplus U_1 \oplus \cdots \oplus U_m \oplus T_0,$$

where $\oplus$ denotes the non-orthogonal direct sum.

The only thing we have not shown is that $\bigcap_m T_m = \{0\}$ and $\bigcup_m T_m$ is dense in $\mathcal{S}'_r$. The latter follows from the fact that $L^2(\mathbb{R})$ is dense in $\mathcal{S}'_r$ and $\bigcup_m V_m \subset \bigcup_m T_m$, while the former follows from an argument involving the partial sums.

As a consequence of Theorem 5, we have the usual wavelet expansion theorems; that is, for $f \in \mathcal{S}'_r$

$$f(t) = \sum_{m,n} b_{mn}\psi_{mn}(t), \tag{3.7}$$

where $\psi_{mn}(t) = 2^{m/2}\psi(2^n t - n)$, with convergence in $\mathcal{S}'_r$. We also form the mixed expansion

$$f(t) = \sum_n a_n \varphi(t-n) + \sum_{m=0}^{\infty}\sum_n b_{mn}\psi_{mn}(t). \tag{3.8}$$

The map P_m, which takes $\mathcal{S}'_r$ into T_m, is a continuous projector taking tempered distributions into C^r functions of polynomial growth. In particular,

$$P_m(\delta(t-\alpha)) = q_m(t,\alpha), \tag{3.9}$$

where $q_m(t,w)$ is the reproducing kernel of the Hilbert space V_m. It also has the reproducing property for T_m; that is, for $f \in T_m$ we have

$$\langle f, q_m(\cdot, w)\rangle = f(w).$$

§4. An application to sampling theorems

The classical Shannon sampling theorem [19] says that a π-band limited function $f(t)$, $t \in \mathbb{R}$, may be recovered from its value on the integers by the formula

$$f(t) = \sum_n f(n) S_n(t) \tag{4.1}$$

where $S_n(t) = \frac{\sin \pi(t-n)}{\pi(t-n)}$ and the convergence is uniform on $\mathbb{R}$. (A π-band limited function is one whose Fourier transform has support on $[-\pi, \pi]$ and belongs on $L^2(\mathbb{R})$.) This has been extended to functions of polynomial growth [22] and to many other spaces of functions with different associated sampling functions $S_n(t)$ [2]. A general theory for f belonging to reproducing kernel Hilbert spaces (RKHS) has been developed in [14]. Since the subspaces V_m of $L^2(\mathbb{R})$ are such RKHS, it seems that a sampling theorem for many such spaces can be proved. Indeed, it is easy to derive a sampling function and not much harder to show uniform convergence [23].

4.1. Hilbert space sampling in V_0

We suppose that there is a multiresolution analysis of $L^2(\mathbb{R})$ as in (2.2) with the added condition that

$$\widehat{\varphi}^*(w) = \sum_k \widehat{\varphi}(w + 2\pi k) \neq 0, \quad w \in \mathbb{R}. \tag{4.2}$$

The sampling function $S_n(t) = S(t-n)$ for V_0 is defined as

$$\widehat{S}(w) := \widehat{\varphi}(w)/\widehat{\varphi}^*(w). \tag{4.3}$$

Then $S(t) \in \mathcal{S}_r \cap V_0$, $S(k) = \delta_{0k}$, and $\{S(t-n)\}$ is a Riesz basis of V_0 [23]. Hence

$$f(t) = \sum_n f(n) S(t-n) \tag{4.4}$$

since $f(k) = \sum a_n S(k-n) = a_k$, with convergence in $L^2(\mathbb{R})$. To show the uniform convergence of (4.4), we use the reproducing kernel

$$q(x,y) = \sum \varphi(x-n)\varphi(y-n)$$

and integrate the difference between $f(t)$ and the partial sums of (4.4) with it, yielding

$$\begin{aligned} |f(t) - \sum_{|n| \le N} f(t-n)| &\le \left| \int_{-\infty}^{\infty} q(t,s) \left[f(s) - \sum_{|n| \le N} f(n) S(s-n) \right] ds \right| \\ &\le q(t,t) \left\| f - \sum_{|n| \le N} f(n) S(\cdot - n) \right\| . \end{aligned}$$

Since $q(t,t)$ is bounded the series converges uniformly.

We note that we can replace the φ in the definition in (4.3) by any $\theta \in V_0$ such that $\hat{\theta}^*(w) \neq 0$. This is clear since $\hat{\theta}(w) = h(w)\widehat{\varphi}(w)$ where $h(w)$ is periodic. Hence $\hat{\theta}^*(w) = h(w)\widehat{\varphi}^*(w)$ and

$$\widehat{S}(w) = \hat{\theta}(w)/\hat{\theta}^*(w) = \widehat{\varphi}(w)/\widehat{\varphi}^*(w).$$

We also note that $\widehat{\varphi}^*(w) = \sum_n \varphi(n)e^{-iwn}$, since the Fourier series coefficients of $\widehat{\varphi}^*$ are exactly $\varphi(n)$, and that (4.4) is valid for $f \in T_0$.

4.2. Distribution space sampling in T_0

We begin with the same definition of the sampling function $S_n(t)$ in (4.3), but without the added condition that $\widehat{\varphi}^*(w) \neq 0$. Rather we require only that its zeros be isolated and of finite multiplicity $\leq r$ in $[-\pi,\pi]$. Then the function

$$\widehat{\varphi}^*(w)/\prod_i \sin(w-\alpha_i), \tag{4.5}$$

where the α_i's are the zeros of $\widehat{\varphi}^*(w)$ (counted according to multiplicity), has no zeros and is periodic. Since $\widehat{\varphi}^*(w) \in C^\infty$, the reciprocal of (4.5) is also in C^∞ and periodic and its Fourier coefficients are rapidly decreasing.

The Fourier series of $\frac{1}{\sin(w-\alpha)}$ may be found from the Laurent series of $h(z) = \frac{2i}{ze^{-i\alpha}+z^{-1}e^{i\alpha}}$ to be

$$\frac{1}{\sin(w-\alpha)} \sim 2i\sum_{n=0}^{\infty}(-1)^n e^{(2\alpha+1)ni}e^{-iw(2n+1)}.$$

It converges in the sense of $\widehat{S'_r}$ (*i.e.*, its inverse Fourier transform converges in S'_r.) Its product with a C^∞ periodic function is again in $\widehat{S'_r}$. The same is true of the repeated factors $1/\prod_i \sin(w-\alpha_i)$. Hence

$$\widehat{\varphi}^{*-1}(w) = \sum_n \alpha_n e^{-iwn}, \qquad \alpha_n = O(|n|^r), \tag{4.6}$$

and the sampling function $S(t)$ as the inverse Fourier transform of $\widehat{\varphi}^{*-1}(w)\widehat{\varphi}(w)$ is given by

$$S(t) = \sum_n \alpha_n \varphi(t-n). \tag{4.7}$$

Hence $S(t) \in T_0$. Clearly $S(k) = \delta_{0k}$. To show that the sampling series converges, we observe that for $f \in T_0$ we have

$$\begin{aligned}\hat{f}(w) &= \sum a_n \widehat{\varphi}(w)e^{-iwn} = \hat{a}^*(w)\widehat{\varphi}(w)\\ &= \hat{a}^*(w)\widehat{\varphi}^*(w)\widehat{S}(w)\end{aligned}$$

the product of a periodic distribution with $\widehat{S}(w)$. The inverse Fourier transform gives

$$f(t) = \sum_n b_n S(t-n),$$

with convergence in $\mathcal{S}'_r$. For $f(t) \in \mathcal{S}_r$, the convergence is also uniform, and we have

$$f(k) = \sum b_n S(k-n) = b_k.$$

Hence the sampling theorem (4.4) holds for T_0 under greater generality.

Example 6. Cubic splines. Let $\theta_3(t)$ be the basic cubic spline function, with

$$\hat{\theta}_3(w) = \left[\frac{1-e^{-iw}}{iw}\right]^4, \qquad w \in \mathbb{R}.$$

Then $\theta_3(t)$ is a cubic polynomial in each interval $(n, n+1)$, $n \in \mathbb{Z}$, has support on $[0,4]$ and belongs to $\mathcal{S}_2$. It may be orthogonalized [12, p. 61] to yield

$$\widehat{\varphi}(w) = \left[\frac{\sin w/2}{w/2}\right]^4 [P_8(\cos w/2)]^{-1/2}$$

where P_8 is a polynomial of degree 6 which is positive on [0,1]. The sampling function [23] has Fourier transform

$$\widehat{S}(w) = \hat{\theta}_3(w)/\hat{\theta}^*_3(w)$$

where

$$\hat{\theta}^*_3(w) = \frac{1}{6}e^{-iw}[1+4e^{-iw}+e^{-2iw}]$$

and hence

$$S_3(t) = \sqrt{3}\left[\sum_{n=0}^{\infty}(\sqrt{3}-2)^{n+1}\theta_3(t-n+1) + \sum_{n=1}^{\infty}(\sqrt{3}-2)^{n-1}\theta_3(t+n+1)\right]. \tag{4.8}$$

Since the coefficients are rapidly decreasing, $S_3(t) \in V_0 \cap \mathcal{S}_2$. Hence the only advantage of this theory is that each $f \in T_0$ has a sampling expansion.

Example 7. Quadratic splines. Here, $\theta_2(t)$ is the basic spline of order 2,

$$\hat{\theta}_2(w) = \left[\frac{1-e^{-iw}}{iw}\right]^3, \qquad w \in \mathbb{R}.$$

It is a quadratic polynomial in $(n, n+1)$, $n \in \mathbb{Z}$, has support in [0,3] and belongs to $\mathcal{S}_1$. The periodic extension is $\hat{\theta}^*_2(w) = \frac{1}{2}e^{-iw}[1+e^{-iw}]$ which has a zero at $w = \pm\pi$. Its inverse converges in the sense of $\mathcal{S}'_1$ and is used to calculate

$$S_2(t) = 2\sum_{n=0}^{\infty}(-1)^n \theta_2(t-n+1). \tag{4.9}$$

Since $S_2(t) \notin V_0$, this space has no sampling theorems. But $S_2(t) \in T_0$ and hence for $f \in V_0 \cap \mathcal{S}_1$ we have

$$f(t) = \sum f(n) S_n(t)$$

convergent in $\mathcal{S}'_r$, and also uniformly convergent for $t \in \mathbb{R}$.

This setting enables us to treat sampling by means of a traditional engineering approach. A signal $f(t) \in T_0$ is sampled at the integers and an impulse train

$$f^*(t) = \sum_n f(n)\delta(t-n) \tag{4.10}$$

is formed. This series converges in the sense of $\mathcal{S}'_r$. The projection of $f^*(t)$ onto T_0 (actually a type of low pass filter) gives

$$P_0 f^*(t) = \sum_n f(n) q(t-n, 0),$$

where $q(t,u)$ is the reproducing kernel of V_0. Since $\hat{q}(w,0) = \widehat{\varphi}(w)\overline{\widehat{\varphi}^*(w)}$ and $\widehat{S}(w) = \widehat{\varphi}(w)/\widehat{\varphi}^*(w)$, it follows that $\hat{q}(w,0) = |\widehat{\varphi}^*(w)|^2 \widehat{S}(w)$. Hence by multiplying $\widehat{P_0 f}^*(w)$ by $|\widehat{\varphi}^*(w)|^{-2}$ and taking the inverse Fourier transform, we recover our original function from the sampling series (4.4). This again requires $\widehat{\varphi}^*(w) \neq 0$.

§5. Wavelets of distributions

We have previously required that the scaling function $\varphi \in \mathcal{S}_r$. We now remove this restriction to allow elements of $\mathcal{S}'$ as possible choices. However, we need a Hilbert space structure for which we take the Sobolev spaces H^{-s} defined as

$$H^{-s} := \left\{ f \in \mathcal{S}' \colon \int_{-\infty}^{\infty} |\hat{f}(w)|^2 (w^2+1)^{-s} dw < \infty \right\}$$

with inner product given by

$$\langle f, g \rangle_{-s} = \frac{1}{2\pi} \int_{-\infty}^{\infty} \hat{f}(w)\overline{\hat{g}(w)}(w^2+1)^{-s} dw \tag{5.1}$$

(see [17]). If $s > \frac{1}{2}$, $\delta \in H^{-s}$, and we may consider "scaling functionals" with support on discrete sets. This has been done in [24]; a short version is given here.

5.1. Distribution with point support

In this section we take $s = 1$ and introduce a multiresolution analysis of H^{-1} composed of dilation subspaces $\{V_m\}$ where

$$V_0 = \{ f \in H^{-1} \colon \text{supp } f \in \mathbb{Z} \}.$$

Then clearly

$$\cdots \subset V_{-1} \subset V_0 \subset V_1 \subset \cdots \subset V_m \subset \cdots \subset H^{-1}.$$

It can also be shown that

$$f(t) = \sum_k a_k \delta(t-k), \qquad \{a_k\} \in \ell^2, \tag{5.2}$$

belongs to V_0, and in fact [24] characterizes V_0. This also shows that

$$\overline{\bigcup_m V_m} = H^{-1},$$

since partial sums of dilations of (5.2) are dense in H^{-1}. This comes from the fact that their Fourier transforms are trigonometric polynomials dense in $L^2(w^2+1)^{-1}$. However,

$$\bigcap_{m=-\infty} V_m = \{C\delta\} \neq \{0\},$$

since $\delta \in V_m$ for all m.

If we try to define the spaces W_m as in the usual approach in Section 2, we are confronted with the fact that H^{-s} is not homogeneous; that is, dilations are not isometries on H^{-s}. However, they are homeomorphisms which map H^{-s} into itself with a modified norm. Hence we introduce the following inner product on H^{-1}:

$$\langle f, g\rangle_{-1,\alpha} = \frac{1}{2\pi}\int_{-\infty}^{\infty} \hat{f}(w)\overline{\hat{g}(w)}(w^2+\alpha^2)^{-1}dw, \quad \alpha > 0. \tag{5.3}$$

We note that the V_m are the same whatever α is, but the orthogonal complement of V_0 in V_1 depends on α. We denote by $W_{0,\alpha}$ this orthogonal complement with respect to (5.3) with the corresponding dilation spaces $W_{m,\alpha}$.

We now need to construct an orthonormal basis of V_0 composed of translation of some φ. A modification of the standard orthogonalization procedure does the trick. Let $\varphi \in V_0$ of the form (5.2); then

$$\begin{aligned}\delta_{0k} = \langle \varphi, \varphi(\cdot - k)\rangle_{-1,\alpha} &= \int_{-\infty}^{\infty} |\widehat{\varphi}(w)|^2 e^{-iwk}(w^2+\alpha^2)^{-1}dw \\ &= \frac{1}{2\pi}\int_0^{2\pi}\left\{\sum_n |\widehat{\varphi}(w+2\pi n)|^2[(w+2\pi n)^2+\alpha^2]^{-1}\right\} e^{-iwk}dw,\end{aligned}$$

and δ_{0k} is the Fourier coefficient of the function in { }, which belongs to $L^2(0,2\pi)$. Hence it must be equal to its Fourier series (in the L^2 sense),

$$\sum_n |\widehat{\varphi}(w+2\pi n)|^2[(w+2\pi n)^2+\alpha^2]^{-1} = \sum_k \delta_{0k}e^{iwk} = 1. \tag{5.4}$$

But $|\widehat{\varphi}|^2$ is periodic (2π) and hence we may solve (5.4) for $|\widehat{\varphi}|^2$ which is

$$|\widehat{\varphi}(w)|^2 = \left[\sum_n ((w + 2\pi n)^2 + \alpha^2)^{-1}\right]^{-1}. \tag{5.5}$$

The positive square root of (5.5) is one solution; the general solution is obtained by multiplying it by a unimodular periodic function. We denote the positive solution by $\widehat{\varphi}(w, \alpha)$,

$$\widehat{\varphi}(w, \alpha) := \left[\sum_n ((w + 2\pi n)^2 + \alpha^2)^{-1}\right]^{-1/2} \tag{5.6}$$

and shall concentrate on it. We have

Proposition 8. *Let $\varphi(t, \alpha)$ be given by (5.6); then $\{\varphi(t - n, \alpha)\}$ is an orthonormal basis of V_0 with the inner product (5.3).*

The fact that it is complete in V_0 follows easily from the definition.

The next step is to find an orthonormal basis of $W_{0,\alpha}$; that is, the wavelets themselves. We look for an element $\psi \in W_{0,\alpha}$ whose translates are orthonormal and complete in $W_{0,\alpha}$. This may be translated into conditions similar to (5.4) as follows:

$$\begin{aligned} &\text{(i)} \quad \sum_n \widehat{\psi}(w + 2\pi n)((w + 2\pi n)^2 + \alpha^{-2})^{-1} = 0; \\ &\text{(ii)} \quad \sum_n |\widehat{\psi}(w + 2\pi n)|^2((w + 2\pi n)^2 + \alpha^2)^{-1} = 1. \end{aligned} \tag{5.7}$$

The first condition is a restatement of the fact that $\psi(t)$ is orthogonal to $\delta(t-k)$ for all $k \in \mathbb{Z}$.

Now however $\widehat{\psi}(w)$ is periodic (π) since $\psi \in V_1$; thus both series may be simplified by separating the even and odd terms which gives us

$$\begin{aligned} &\text{(i)} \quad \widehat{\psi}(w)a(w, \alpha) + \widehat{\psi}(w + \pi)a(w + 2\pi, \alpha) = 0; \\ &\text{(ii)} \quad |\widehat{\psi}(w)|^2 a(w, \alpha) + |\widehat{\psi}(w + \pi)|^2 a(w + 2\pi, \alpha) = 1, \end{aligned} \tag{5.8}$$

where

$$a(w, \alpha) := \sum_n ((w + 4\pi n)^2 + \alpha^2)^{-1}. \tag{5.9}$$

These two equations are easily solved for $|\widehat{\psi}|^2$, yielding

$$|\widehat{\psi}(w)|^2 = \frac{a(w + 2\pi, \alpha)}{a(w, \alpha)(a(w + 2\pi, \alpha) + a(w, \alpha))}.$$

However, the positive square root does not satisfy (5.8)(i), but rather must be multiplied by $e^{-iw/2}$. We define

$$\widehat{\psi}(w,\alpha) := e^{-iw/2}[a(w+2\pi,\alpha)/a(w,\alpha)(a(w+2\pi,\alpha)+a(w,\alpha))]^{1/2}. \quad (5.10)$$

Proposition 9. *Let $\psi(t,\alpha)$ be given by (5.10); then $\{\psi(t-n,\alpha)\}$ is an orthonormal basis of $W_{0,\alpha}$.*

Again the completeness is straightforward. To find an orthonormal basis of the dilation space V_1, we choose α' such that the map $D_2 f = \sqrt{2}\, f(2\cdot)$ is a multiple of an isometry from V_0 to V_1. The proper choice is $\alpha' = 2\alpha$, since

$$\|f\|_{-1,\alpha} = \frac{1}{2}\|D_2 f\|_{-1,2\alpha},$$

and hence $\{2^{3/2}\varphi(2t-n,\alpha'/2)\}$ is an orthonormal basis of V_1.

We now start with $\alpha' = 1$ in each V_m, and find the corresponding α in V_0, so that we can get an orthonormal basis of V_m. This gives us

$$\varphi_{m,n}(t) = 2^{3m/2}\varphi(2^m t - n, 2^{-m}), \qquad n = 0, \pm 1, \ldots\,. \quad (5.11)$$

For the orthogonal complement, we begin with $W_0 = W_{0,1}, \ldots, W_m = W_{m,1}, \ldots$ and thereby obtain, similarly,

$$\psi_{m,n}(t) = 2^{3m/2}\psi(2^m t - n, 2^{-m}), \qquad n = 0, \pm 1, \ldots\,. \quad (5.12)$$

Corollary 10. *Let $\varphi_{m,n}(t)$ and $\psi_{m,n}(t)$ be given by (5.11) and (5.12) respectively; then $\{\varphi_{0,n}(t)\}_{n=-\infty}^{\infty}$ and $\{\psi_{m,n}(t)\}_{m=0,n=-\infty}^{\infty\ \ \infty}$ form an orthonormal basis of H^{-1} with the usual inner product (5.1).*

Hence we have the usual multiresolution analysis

$$H^{-1} = V_0 \oplus \bigoplus_{m=0}^{\infty} W_m = \{\delta\} \oplus \bigoplus_{m=-\infty}^{\infty} W_m \quad (5.13)$$

where $\{\delta\}$ is the space spanned by the δ function.

Our theory is almost complete. We still need to get simpler expressions for $\psi(t,\alpha)$ by summing the series for $a(w,\alpha)$ of (5.9). We begin with the fact that

$$\sum_{n=-\infty}^{\infty} \frac{1}{x+2\pi n} = \frac{1}{2}\frac{\cos\frac{x}{2}}{\sin\frac{x}{2}}$$

from the Fourier series of e^{-ixt} on (0,1) evaluated at $t = 0$. Then

$$\begin{aligned} a(w,\alpha) &= \frac{i}{4\alpha}\sum_n \frac{1}{\frac{w+\alpha i}{2}+2\pi n} - \sum_n \frac{1}{\frac{w-\alpha i}{2}+2\pi n} \\ &= \frac{i}{8\alpha}\left\{\frac{\cos(w+i\alpha)/4}{\sin(w+i\alpha)/4} - \frac{\cos(w-\alpha i)/4}{\sin(w-\alpha i)/4}\right\} \\ &= \frac{i}{8\alpha}\frac{\sin(-i\alpha/2)}{\sin^2 w/4\cos^2\alpha i/4 - \cos^2 w/4\sin^2\alpha i/4} \\ &= \frac{1}{4\pi}\frac{\sinh\alpha/2}{\cos\alpha/2 - \cos w/2}. \end{aligned}$$

This is substituted into (5.10) to obtain

$$\widehat{\psi}(w,\alpha) = e^{-iw/2}(\cosh \alpha/2 - \cos w/2)\sqrt{\frac{4\alpha}{\sinh \alpha}} \tag{5.14}$$

after a little manipulation. The inverse Fourier transform is easily found to be, for $\alpha = 2^{-m}$,

$$\psi(t, 2^{-m}) = C\left\{\cosh\,(2^{-m-1})\delta\left(t-\frac{1}{2}\right) - \frac{1}{2}\delta(t) - \frac{1}{2}\delta(t-1)\right\}, \tag{5.15}$$

where C is a positive constant.

We can also find $\varphi(t,\alpha)$ to be given by

$$\widehat{\varphi}(w,\alpha) = \left[\frac{2\alpha(\cosh\,\alpha - \cos w)}{\sinh\,\alpha}\right]^{1/2}. \tag{5.16}$$

However this leads to an infinite Fourier series whose sum is not as simple [24]. This really does not matter that much since the important result is (5.15) which gives an orthonormal basis of H^{-1} each of whose elements has support on 3 points (or fewer in the case of $\delta(t)$).

5.2. Dilation equations with distributional solutions

An alternative starting point for orthogonal wavelets is the dilation equation (2.3)

$$\varphi(t) = \sqrt{2}\sum_k c_k \varphi(2t-k)$$

rather than the multiresolution analysis. Usually only the dilation equations with continuous solutions are considered but in many cases distribution solutions exist when these are no continuous solutions [21].

In this section we shall assume that only a finite number of non-zero terms exist in (2.3). If $\varphi \in \mathcal{S}'$ has compact support then $\widehat{\varphi}$ is an entire function and

$$\widehat{\varphi}(w) = (1/\sqrt{2})\sum c_k e^{-iwk/2}\widehat{\varphi}(w/2).$$

If we assume further that $\widehat{\varphi}(0) = 1$, then

$$\sum c_k = \sqrt{2}. \tag{5.17}$$

Under these conditions it was shown in [21] that the dilation equation (2.3) has a unique nontrivial distributional solution with compact support. If (2.3) has only one term, say $c_0 = \sqrt{2}$, then the solution may be found explicitly. In fact, it is $\varphi(t) = \delta(t)$ since $2\delta(2t) = \delta(t)$.

If two terms are non-zero, say c_0 and c_1, then there is never a distribution solution of the form

$$\varphi(t) = \sum_{k=0}^{n} a_k \delta(t - t_k). \tag{5.18}$$

We first observe that the support of φ must be in $[-1, 1]$, and hence we may take $0 \le t_0 < t_1 < t_2 < \cdots < t_n \le 1$. From the dilation equation we have

$$\sum_{k=0}^{n} a_k \delta(t - t_k) = \sqrt{2} \sum_{k=0}^{n} \left\{ c_0 \frac{a_k}{2} \delta\left(t - \frac{t_k}{2}\right) + c_1 \frac{a_k}{2} \delta\left(t - \frac{t_k + 1}{2}\right)\right\}. \tag{5.19}$$

The coefficients of $\delta(t - t_k)$ on both sides must be equal. If $t_0 = 0$ then $a_0 = \sqrt{2}\, c_0 \frac{a_0}{2}$ and $a_0 = 0$ since $c_0 \neq \sqrt{2}$. If $t_0 > 0$ then $\sqrt{2}\, \frac{c_0}{2} a_0 = 0$ and $a_0 = 0$ again. Since $\frac{t_k}{2} < t_k < \frac{t_k+1}{2}$, the same argument can be repeated to conclude that $a_1 = a_2 = \cdots = a_n = 0$. We can also extend the argument to more than two nonzero c_k's. In fact we have used only the first sum on the right-hand side of (5.19).

Proposition 11. *The distributional solution to the dilation equation (2.3) has no solution of the form (5.18) if $c_n = 0$ for $|n| \ge \mathbb{N}$ but at least two c_n's are non-zero.*

One might expect that derivatives of L^2 solutions to the dilation equation are solutions to a similar dilation equation, but this is not the case since

$$\frac{d}{dt}\varphi(2t - k) = 2\varphi'(2t - k)$$

and condition (5.17) is violated.

Since the distributional solutions have compact support they are finite order and indeed by the Paley-Wiener-Schwartz theorem [18] their Fourier transforms satisfy

$$|\widehat{\varphi}(w)| \le C(1 + |w|)^M, \qquad w \in \mathbb{R}$$

and are entire functions of exponential type. Thus they belong to the Sobolev spaces H^{-s} for $s > M + \frac{1}{2}$.

Proposition 12. *Let $\varphi(t)$ be the distributional solution to the dilation equation (2.3) satisfying $\widehat{\varphi}(0) = 1$. Then $\varphi \in H^{-s}$ for $s > \log_2 \sum |c_n|$.*

We adapt the proof in [21], and again denote the Fourier transform of (2.3) to be as in (2.9)

$$\widehat{\varphi}(w) = m_0\left(\frac{w}{2}\right) \widehat{\varphi}\left(\frac{w}{2}\right)$$

where $m_0(w) = \frac{1}{\sqrt{2}} \sum c_n e^{-iwn}$. It can be shown by iterating this equality that

$$\widehat{\varphi}(w) = \lim_{k \to \infty} \prod_{j=1}^{k} m_0(w/2^j) = \lim_{k \to \infty} G_k(w)$$

(see [4] and [21]), where the convergence is almost uniform on $\mathbb{C}$.

Hence a uniform bound on $G_k(w)$ should give us our result. Since

$$|m_0(w)| \leq \frac{1}{\sqrt{2}} \sum |c_n| = C \geq 1,$$

it follows that for $2^{p-1} \leq |w| < 2^p$, $k > p$

$$\begin{aligned} |G_k(w)| &= |m_0(w/2)| \ldots |m_0(w/2^p)||G_{k-p}(w/2p)| \\ &\leq C^p \sup_{|w| \leq 1} |G_{k-p}(w)| \leq C^p C_1; \end{aligned}$$

similarly for $k \leq p$

$$|G_k(w)| \leq C^k \leq C^p C_1.$$

The inequality $2^{p-1} \leq |w|$ is now converted to one using C; that is,

$$C^{p-1} = 2^{(p-1)M} \leq |w|^M$$

where $M = \log_2 C$. Thus

$$|G_k(w)| \leq CC_1 |w|^{\log_2 C}, \qquad 1 \leq |w|$$

from which the conclusion follows since

$$\log_2 C = \log_2 \sum |c_n| - \frac{1}{2}.$$

It should be noted that if all the c_n's are positive then $\log_2 C = 0$ and $\varphi \in H^{-1}$. In this case as well as the others, one can then use the orthogonalization procedure of the last section.

References

1. Beylkin, G., R. Coifman, and V. Rokhlin, Fast wavelet transform and numerical algorithms I, *Comm. Pure and Appl. Math.* **44** (1991), 141–183.
2. Butzer, P. L., W. Splettstößer, and R. L. Stens, The sampling theorem and linear predictions in signal analysis, *Jahresber. Deutsch. Math.-Verein.* **90** (1988), 1–60.
3. Daubechies, I., The wavelet transform, time-frequency localization and singal analysis, *IEEE Trans. Inform. Theory* **36** (1990), 961–1005.
4. Daubechies, I., l Orthonormal bases of compactly supported wavelets, *Comm. Pure and Appl. Math.* **41** (1988), 909–996.
5. Gelfand, I. M. and G. E. Shilov, *Generalized Functions*, Vol. 1, 3, Academic Press, New York, 1964, 1967.
6. Grossmann, A. and J. Morlet, Decomposition of Hardy functions into square integrable wavelets of constant shape, *SIAM J. Math. Anal.* **15** (1984), 723–736.

7. Heil, C. E. and D. F. Walnut, Continuous and discrete wavelet transforms, *SIAM Review* **31** (1989), 628–666.
8. Hörmander, L., *Linear Partial Differential Equations Operators*, Springer, Berlin, 1963.
9. Korevaar, J., Distributions defined by fundamental sequences, *Nederi Akad. Wetensch. Proc. Ser* **A 58** (1955), 368–389, 483–503, 663–674.
10. Mallat, S., Multiresolution approximations and wavelet orthonormal bases of $L^2(\mathbb{R})$, *Trans. Amer. Math. Soc.* **315** (1989), 69–87.
11. Meyer, Y., The Franklin Wavelets, 1988, preprint.
12. Meyer, Y., *Ondeletts et Opérateurs I*, Hermann, Paris, 1990.
13. Morlet, J., G. Arens, I. Fourgeau, and D. Giard, Wave propagation and sampling theory, Part II, *Geophysics* **47** (1982), 203–236.
14. Nashed, M. Z. and G. G. Walter, General sampling theorems for functions in reproducing kernel Hilbert spaces, *Math. Control Signals Systems* **4** (1991), 373–412.
15. Paley, R. E. A. C. and N. Wiener, *Fourier Transforms in the Complex Domain*, Colloq. Publ. Vol. 29, Amer. Math. Soc., Providence, RI, 1934.
16. Papoulis, A., *Signal Analysis*, McGraw-Hill, New York, 1977.
17. Rudin, W., *Functional Analysis*, McGraw-Hill, New York, 1973.
18. Schwartz, L., *Théore des Distributions*, in two volumes, Hermann, Paris, 1951.
19. Shannon, C. E., Communication in the presence of noise, *Proc. IRE* **37** (1949), 10–21.
20. Strang, G., Wavelets and dilation equations, *SIAM Review* **31** (1989), 614–627.
21. Volkmer, H., Distributional and square summable solutions of dilation equations, 1990, preprint.
22. Walter, G., Sampling bandlimited functions of polynomial growth, *SIAM J. Math. Anal.* **19** (1988), 1198–1203.
23. Walter, G., A sampling theorem for wavelet subspaces, 1990, preprint.
24. Walter, G., Discrete discrete wavelets, 1989, preprint.
25. Young, R. M. *An Introduction to Non-Harmonic Fourier Series*, Academic Press, New York, 1980.
26. Zemanian, A. H., *Distribution Theory and Transform Analysis*, McGraw-Hill New York, 1965.

This research was supported in part by NSF Grant DMS–901526.

Gilbert G. Walter
Department of Mathematical Sciences
University of Wisconsin-Milwaukee
P.O. Box 413
Milwaukee, WI 53201

ggw@archimedes.math.uwm.edu

Part II

Semi-orthogonal and Nonorthogonal Wavelets

Cardinal Spline Interpolation and the Block Spin Construction of Wavelets

Guy Battle

Abstract. The construction of wavelets that are inter-scale orthogonal (pre-ondelettes) but not necessarily intra-scale orthogonal is a major part of the block spin approach to wavelets. We review this in the case of Lemarié wavelets and relate it to the cardinal spline approach of Chui and Wang. We advertize the computational value of the pre-ondelette basis as opposed to that of the orthonormal basis of wavelets.

§1. Introduction

An interesting variety of ondelettes (orthogonal wavelets) have been constructed over the past few years, and the multi-scale resolution scheme of Meyer and Mallat seems to be the most popular framework in which to construct them. Even the ingenious construction by Daubechies [6] of compactly supported class C^N ondelettes is based on it. By contrast, the block spin construction machine of this author [1] has a mathematical physics flavor. Another contrast can be found in the order in which the orthogonalization is done. The major part of the latter construction produces functions which we dub *pre-ondelettes.*

Definition 1. *A pre-ondelette is a square-integrable function* φ *such that* $\varphi(2^r x + m)$ *is orthogonal to* $\varphi(2^s x + n)$ *for all* $m, n \in \mathbb{Z}^d$ *and all integers* $r \neq s$.

Translates of a pre-ondelette on the same scale are not necessarily orthogonal. Our *final* step is to orthogonalize on each scale by applying the inverse square root of the overlap matrix associated with the functions living on that scale and so the derivation of an ondelette from a pre-ondelette is routine as long as the matrix is positive and invertible. This same procedure is the *initial* step in the Meyer-Mallat construction, and it is applied to translates of a function η with the property

Wavelets–A Tutorial in Theory and Applications
C. K. Chui (ed.), pp. 73–90.

ISBN 0-12-174590-2

$$\eta(x) = \sum_n c_n \eta(2x - n). \tag{1.1}$$

The derivation of a function whose copies on different scales are orthogonal is the latter step in their construction.

Last year it was recognized independently by Charles Chui and this author that for the class of Lemarié functions [8] the pre-ondelette expansion of an arbitrary function has greater computational value than the ondelette expansion. Chui was able to carry this idea much further, and his development is described in [3], from the point of view of approximation theory. In the case of the Lemarié functions (ondelettes with exponential decay) Lemarié was aware of the correspondence with certain trigonometric polynomials arising in the fundamental work of Schoenberg [9] on cardinal spline interpolation, but these special polynomials are useful in deriving an explicit position space representation of the Lemarié pre-ondelettes rather than the Lemarié ondelettes. Chui and his co-workers have greatly expanded this program to obtain constructions of new pre-ondelettes having more convenient properties as well as the explicit position space representations needed for numerical work.

Remark. The term "ondelette" – the French word for "wavelet" – is fading from the literature, and we have gradually succumbed to the more popular term "wavelet". We will often refer to a pre-ondelette as a "wavelet" – but not a "pre-wavelet". Many workers in the field use the term "wavelet" whether the basis is completely orthogonal or not.

The relevance of wavelets to mathematical physics can be traced back to renormalization group ideas of Gawedzki and Kupiainen [7], and we have already described the connection between wavelets and Gaussian fixed points of the renormalization group [2]. Our aim here is to give a brief review of the block spin construction of wavelets given there and then relate it to the approximation-theoretic approach of Chui and Wang [4]. We should note that Lemarié has recently given a nice account [9] of the connection between the multi-resolution analysis of Meyer and Mallat and cardinal spline interpolation. Finally, we should mention that Chui and Wang [5] have constructed spline wavelets of any order smoothness with compact support. They are pre-ondelettes. (The Daubechies wavelet is not a spline, but of course her basis is intra-scale as well as inter-scale orthogonal.)

In this pedagogical exposition we concentrate almost exclusively on the class of Lemarié pre-ondelettes and discuss this case to death. A Lemarié pre-ondelette is a $C^{N-\varepsilon}$ pre-ondelette also having exponential decay and vanishing moments of order $\leq N$, where N can be made as large as desired by adjusting a construction parameter. Thus we are discussing a class of pre-ondelette bases with each basis arising from a polynomial in a manner yet to be described. The corresponding class of polynomials is known in approximation theory as the *Euler-Frobenius* class of polynomials, but our analysis here will be more naive and pedestrian than that of an approximation theorist. We will also consider only the *one*-dimensional case because in the case of scalar L^2-wavelets there is no interesting dependence on dimension. Let χ be the characteristic function of the unit interval [0,1]. In our case the function η of Equation (1.1) is just the N-fold convolution of χ with itself – known in approximation theory as the *Nth-order B-spline*. The Lemarié pre-ondelettes are themselves splines – indeed, the "mother wavelet" generating the basis turns out to be an infinite series of integer translates of the Nth-order B-spline. The values of the coefficients depend only on the complex roots of the Euler-Frobenius polynomial appearing as $Q_M(z)$ in Section 4 with $N = M + 1$ – specifically, they are computed as residues of poles corresponding to zeros of $Q_M(z)$. This result can be derived from the explicit *momentum* space representation of Lemarié pre-ondelettes, which has been in the literature for several years [1]. This computational success cannot be duplicated for the corresponding ondelettes because the momentum expression for the latter contains a radical.

In Section 2 we briefly review our block spin construction of wavelets with the averaging constraints defined by arbitrary η. To avoid confusion we recall here that our "unit scale" pre-ondelettes [1,2] are chosen by averaging constraints graphically given by

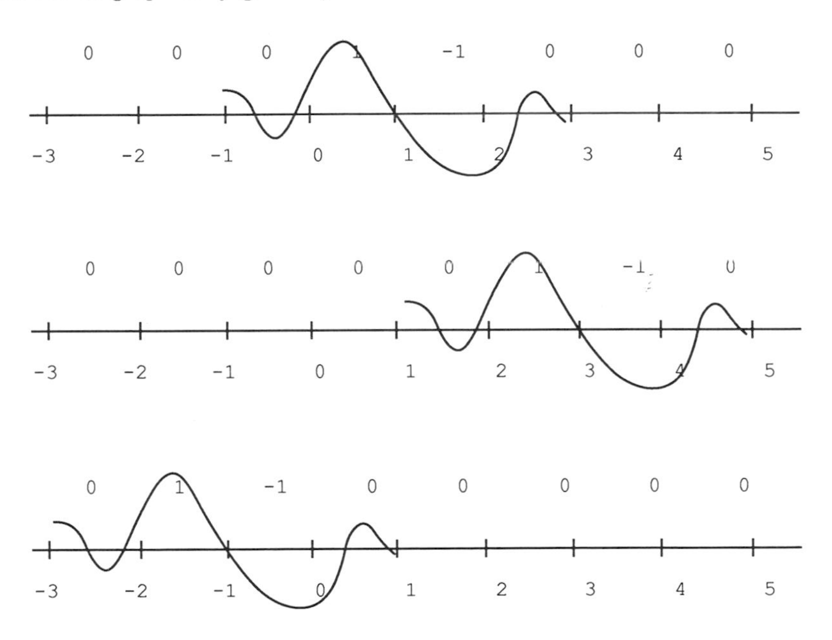

in the case where η happens to be the Nth-order B-spline. Thus all of our "unit-scale" pre-ondelettes are 2-unit translates of the "mother wavelet" φ described by, say, the first figure. Logically, the "unit-scale" should be the next scale down, but we wish to keep our notation and terminology consistent with (see [2] and [7]), and with our choice, the spline wavelets have integer knots. As far as the figure above is concerned, we are thinking of the averaging constraints

$$\int \frac{1}{p^M} \hat{\varphi}(p) \overline{\hat{\chi}(p)} e^{-imp} dp = \delta_{m0} - \delta_{m1}, \tag{1.2}$$

which easily imply the conditions (2.4) in Section 2 by the formula

$$\widehat{\chi}(p) = \frac{e^{ip} - 1}{ip}. \tag{1.3}$$

In Section 3 we recall the explicit momentum expression for Lemarié pre-ondelettes (including the important phase factor omitted in [1]). We verify directly that copies on different scales are orthogonal, incidentally correcting the flawed calculation in the Introduction of [1] and showing that the integer parameter M does not have to be even.

In Section 4 we show how the explicit position space representation of the Lemarié pre-ondelettes can be derived. The expression enables us to compute exact rates of exponential fall-off. Although the set of functions is only partially orthogonal, bear in mind that it is still a basis in the following sense:

Theorem 2. *Let $\{\varphi_k\}$ be the set of Lemarié pre-ondelettes for a given value of the construction parameter M. For every square-integrable function f there is a unique sequence $\{a_k\}$ indexed by the pre-ondelettes such that*

$$f = \sum_k a_k \varphi_k \tag{1.4}$$

in the L^2-convergent sense.

Actually, the computation of the coefficients is more interesting than the theorem itself (which immediately follows from the invertibility of the overlap matrix). When an arbitrary configuration is expanded in these functions, the calculation of coefficients is convenient because *the dual basis functions have an explicit position space representation as well.* We derive it in Section 6, and the residue calculation of coefficients involves yet another class of special polynomials $z^{M+1}K_M(z)$. In Section 7 we show that actually

$$z^M K_M(z) = 2^{2M+2} Q_M(z^2), \tag{1.5}$$

from which we will be able to conclude:

Theorem 3. *The exponential decay rate for a dual basis pre-ondelette is exactly one-half the rate for a pre-ondelette on the same scale, regardless of the value of M.*

Chui and Wang also prove (1.5), and we relate their derivation [4] of the Lemarié pre-ondelettes to ours in Section 5.

§2. Construction of the mother wavelet

Let η be a real-valued square-integrable function and $\{c_m\}$ a real-valued summable sequence such that

$$\eta(x) = \sum_m c_m \eta(2x - m). \tag{2.1}$$

For example,

$$\hat{\eta}(p) = \widehat{\chi}(p)^N \tag{2.2}$$

defines such a function with

$$c_m = \begin{cases} 2^{-N+1}\binom{N}{m}, & 0 \leq m \leq N, \\ 0, & \text{otherwise.} \end{cases} \tag{2.3}$$

Indeed, this is the class of functions involved in the construction of the Lemarié functions, but in this section we consider the general $\eta(x)$.

Our approach to the construction of wavelets is already described in [2] for an arbitrary number of dimensions. In our one-dimensional context, we construct the "mother wavelet" by minimizing $\|\varphi\|_2$ with respect to the constraints

$$\int \varphi(x - m)\eta(x)dx = (-1)^m c_{1-m}, \tag{2.4}$$

where m ranges over the integers. Let φ_0 be the solution and recall that

$$\int \varphi_0(2^r x \quad 2n)\eta(x)dx = 0 \tag{2.5}$$

for all integers n and all positive integers r. The details are in [2], but this condition follows from the identity

$$\sum_m c_m (-1)^m c_{1-m} = 0. \tag{2.6}$$

On the other hand, (2.5) implies (with $n = 2^{r-1}m + k$)

$$\varphi_0(x) + \lambda\varphi_0(2^r x - 2k), \qquad k \in \mathbb{Z},$$

satisfies (2.4) for all real λ and integers $r > 0$. Hence,

$$\begin{aligned} &\int \varphi_0(x)\varphi_0(2^r x - 2k)dx \\ =&\frac{1}{2}\frac{d}{d\lambda}\int (\varphi_0(x) + \lambda\varphi_0(2^r x - 2k))^2 dx|_{\lambda=0} \\ =&\ 0. \end{aligned} \tag{2.7}$$

and so inter-scale orthogonality is established.

Remark. This orthogonality argument differs from the one given in Section 2 of [2]. The latter reasoning is also quite standard, but depends on the linear functionals in (2.4) being bounded. By contrast, only the existence of φ_0 is needed here – a milder condition when one solves this problem in a massless Sobolev space instead of in L^2.

The solution of this constrained minimization problem is routine by now. We get

$$\widehat{\varphi}_0(p) = c\frac{\hat{\eta}(p)}{\sum_\ell |\hat{\eta}(p+2\pi\ell)|^2}\sum_m (-1)^m c_{1-m} e^{imp} \tag{2.8}$$

in momentum space. The basis $\{\varphi_0(2^r x + 2m)\}$ generated from this "mother wavelet" is inter-scale orthogonal but not intra-scale orthogonal. However, an orthonormal basis can be derived from this basis by the standard translation-invariant orthogonalization on each scale, provided that the overlap matrix

$$S_{mn} = \int \varphi_0(x+2m)\overline{\varphi_0(x+2n)}dx. \tag{2.9}$$

is invertible. This condition reduces to the requirement

$$\frac{\left|\sum_m (-1)^m c_{1-m} e^{imp}\right|^2}{\sum_\ell |\hat{\eta}(p+2\pi\ell)|^2} + \frac{\left|\sum_m c_{1-m} e^{imp}\right|^2}{\sum_\ell |\hat{\eta}(p+2\pi\ell+\pi)|^2} > 0, \tag{2.10}$$

which is satisfied for the class of cases given by (2.2). We will have occasion to consider the overlap matrix again, but we mention the orthogonalization

$$\psi_0(x+2m) = \sum_n (S^{-1/2})_{mn}\varphi_0(x+2n) \tag{2.11}$$

only to dismiss it. It is the basis $\{\varphi_0(2^r x + 2m)\}$ that we truly wish to study, as we have already explained in Section 1. In position space, we have

$$\varphi_0(x) = \sum_n a_n \eta(x-n), \tag{2.12}$$

$$a_n = \frac{c}{2\pi}\int_0^{2\pi} \frac{e^{-inp}}{\sum_\ell |\hat{\eta}(p+2\pi\ell)|^2}\sum_m (-1)^m c_{1-m} e^{imp} dp, \tag{2.13}$$

and we will be concerned with the calculation of a_n for our specific class (2.2) of cases.

§3. The Lemarié case

In this section we examine the explicit momentum representation of φ_0 in the case given by (2.2). Actually, we derived this expression in [1] several years ago, but an important phase factor was omitted there. This changes the orthogonality argument given in the Introduction of [1], making it applicable to *all* nonnegative integer values of the construction parameter M – not just even values.

Remark. This issue does not affect the essential correctness of [1]. Inter-scale orthogonality is guaranteed by the principle discussed in the preceding section.

We give here an independent proof of the orthogonality based on the momentum representation of the solution. The (correct) solution is given by (with $N = M + 1$)

$$\widehat{\varphi}_0(p) = ce^{ip}\frac{(1-e^{-ip})^{M+1}\widehat{\chi}(p)^{M+1}}{\sum\limits_n |\widehat{\chi}(p+2\pi n)|^{2M+2}}. \tag{3.1}$$

The phase factor e^{ip} appears because the factor

$$\sum_\ell [P_M^-(-\ell) - P_M^-(-\ell+1)]e^{i\ell p} \tag{3.2}$$

arises in the derivation recalled from [1], where

$$P_M^\pm(n) = \begin{cases} (\pm 1)^n \binom{M}{n}, & 0 \le n \le M, \\ 0, \text{ otherwise.} \end{cases} \tag{3.3}$$

The point is that

$$\sum_n P_M^\pm(n)e^{inp} = (1 \pm e^{ip})^M, \tag{3.4}$$

and so (3.2) reduces to

$$(1-e^{-ip})^M - e^{ip}(1-e^{-ip})^M = -e^{ip}(1-e^{-ip})^{M+1}. \tag{3.5}$$

From the point of view of the last section, we are simply stating that

$$\sum_m (-1)^m c_{1-m} e^{imp} = -e^{ip}(1-e^{-ip})^N, \tag{3.6}$$

if c_m is given by (2.3). We now demonstrate the inter-scale orthogonality directly. Without loss of generality, we compare the unit scale to the 2^r scale for an arbitrary integer $r > 0$, namely:

$$\int \varphi_0(x+2m)2^{-r/2}\varphi_0(2^{-r}x+2\ell)dx = 2^{r/2}\int e^{i2mp}e^{-i2^{r+1}\ell p}\hat{\varphi}_0(p)\overline{\hat{\varphi}_0(2^r p)}dp$$
$$= c2^{r/2}\int e^{i2mp}e^{-i2^{r+1}\ell p}e^{ip}e^{-i2^r p}(1-e^{-ip})^{M+1}(1-e^{i2^r p})^{M+1}\times$$
$$\frac{\hat{\chi}(p)^{M+1}}{\sum_n |\hat{\chi}(p+2\pi n)|^{2M+2}}\frac{\overline{\hat{\chi}(2^r p)^{M+1}}}{\sum_n |\hat{\chi}(2^r p+2\pi n)|^{2M+2}}dp. \tag{3.7}$$

On the other hand, we have the identities

$$\hat{\chi}(2^r p) = (1+e^{i2^{r-1}p})\cdots(1+e^{i2p})(1+e^{ip})\hat{\chi}(p), \tag{3.8}$$
$$(1+e^{-i2^{r-1}p})\cdots(1+e^{-i2p})(1+e^{-ip})(1-e^{-ip}) = 1-e^{-i2^r p}, \tag{3.9}$$

from which we obtain

$$\int \varphi_0(x+2m)2^{-r/2}\varphi_0(2^{-r}x+2\ell)dx$$
$$= c2^{r/2}\int e^{i2mp}e^{-i2^{r+1}\ell p}e^{ip}e^{-i2^r p}(1-e^{i2^r p})^{M+1}\times$$
$$\frac{|\hat{\chi}(p)|^{2M+2}}{\sum_n |\hat{\chi}(p+2\pi n)|^{2M+2}}\frac{(1-e^{-i2^r p})^{M+1}}{\sum_n |\hat{\chi}(2^r p+2\pi n)|^{2M+2}}dp. \tag{3.10}$$

Since $|\hat{\chi}(p)|^{2M+2}$ is the only factor in the integrand that does not have period 2π, the decomposition

$$\int F(p)dp = \sum_n \int_0^{2\pi} F(p+2\pi n)dp \tag{3.11}$$

collapses the integral to

$$\int \varphi_0(x+2m)2^{-r/2}\varphi_0(2^{-r}x+2\ell)dx$$
$$= c2^{r/2}\int_0^{2\pi} e^{i2mp}e^{-i2^{r+1}p}e^{ip}e^{-i2^r p}\frac{|1-e^{i2^r p}|^{2M+2}}{\sum_n |\hat{\chi}(2^r p+2\pi n)|^{2M+2}}dp. \tag{3.12}$$

Every factor in this integral in π-periodic except e^{ip}, and so the integral is zero.

§4. Lemarié pre-ondelettes in position space

In this section we derive an explicit expression for φ_0 in position space,and our starting point is a standard set of identities that Lemarié uses in his own analysis [8] of his ondelettes. The periodic function

$$\sum_n |\hat{\chi}(p+2\pi n)|^{2M+2} \tag{4.1}$$

certainly arises in that context and Schoenberg encounters this expression as well [10]. It is actually a trigonometric polynomial because the Poisson summation formula in this case evaluates a compactly supported spline over all integers. But one can be more explicit. Since

$$4\sin^2\left(\frac{1}{2}p\right)\sum_n \frac{1}{(p+2\pi n)^2} = \sum_n |\hat{\chi}(p+2\pi n)|^2 \equiv 1, \tag{4.2}$$

we may differentiate the identity

$$\sum_n \frac{1}{(p+2\pi n)^2} = \frac{1}{4}\csc^2\left(\frac{1}{2}p\right) \tag{4.3}$$

$2M$ times to obtain

$$(2M-1)!\sum_n \frac{1}{(p+2\pi n)^{2M+2}} = \frac{1}{4}\frac{d^{2M}}{dp^{2M}}\csc^2\left(\frac{1}{2}p\right), \tag{4.4}$$

so (4.1) is just

$$4^M\frac{\sin^{2M+2}\left(\frac{1}{2}p\right)}{(2M-1)!}\frac{d^{2M}}{dp^{2M}}\csc^2\left(\frac{1}{2}p\right). \tag{4.5}$$

Now *even*-order derivatives of $\csc^2(\frac{1}{2}p)$ can be shown (inductively) to be polynomials in $\csc^2(\frac{1}{2}p)$ and $\cot^2(\frac{1}{2}p)$, homogeneous of degree equal to two plus the order of the derivative, and with at least one factor of $\csc^2(\frac{1}{2}p)$ in each term. Thus (4.5) is a polynomial in $\cos^2(\frac{1}{2}p)$ only, and with no power higher than $\cos^{2M}(\frac{1}{2}p)$. Lemarié discusses this trigonometric polynomial in [7]. We may write it as an Mth-degree polynomial in $\cos p$, and we denote this last polynomial by R_M. In momentum space (3.1) becomes

$$\widehat{\varphi}_0(p) = ce^{ip}\frac{(1-e^{-ip})^{M+1}\hat{\chi}(p)^{M+1}}{R_M(\cos p)}, \tag{4.6}$$

and so in position space we have a spline with integer knots realized as an expansion in integer translates of the $(M+1)$-fold convolution of χ with itself.

$$\varphi(x) = \sum_n a_n(\chi *)^{M+1}(x+n), \tag{4.7}$$

$$a_n = c\int_0^{2\pi} e^{-inp}\frac{e^{ip}(1-e^{-ip})^{M+1}}{R_M(\cos p)}dp. \tag{4.8}$$

We have the alternate contour expressions

$$a_n = -ic \oint_{|z|=1} z^{-n} \frac{(1-z^{-1})^{M+1}}{R_M(\frac{1}{2}z + \frac{1}{2}z^{-1})} dz, \tag{4.9}$$

$$a_n = ic \oint_{|z|=1} z^{n-2} \frac{(1-z)^{M+1}}{R_M(\frac{1}{2}z + \frac{1}{2}z^{-1})} dz, \tag{4.10}$$

and we can avoid computing residues at $z = 0$ for almost all n by using (4.9) for $n \leq -1$ and (4.10) for $n \geq 2 - M$. The point is that the function

$$Q_M(z) = z^M R_M\left(\frac{1}{2}z + \frac{1}{2}z^{-1}\right) \tag{4.11}$$

is a $2M$th degree polynomial with no zero at $z = 0$ and

$$z^{-n} \frac{(1-z^{-1})^{M+1}}{R_M(\frac{1}{2}z + \frac{1}{2}z^{-1})} = z^{-n-1} \frac{(z-1)^{M+1}}{Q_M(z)}, \tag{4.12}$$

$$z^{n-2} \frac{(1-z)^{M+1}}{R_M(\frac{1}{2}z + \frac{1}{2}z^{-1})} = z^{n+M-2} \frac{(1-z)^{M+1}}{Q_M(z)}. \tag{4.13}$$

Let $\alpha_1, \ldots, \alpha_k$ be the roots of $Q_M(z)$ in the unit disk. (There are none on the boundary, since R_M $(\cos p)$ never vanishes.) Thus, we have

$$a_n = 2\pi c \sum_{\iota=1}^{k} \operatorname{Res}\left(z^{-n-1} \frac{(z-1)^{M+1}}{Q_M(z)}, \alpha_\iota\right) \tag{4.14}$$

for $n \leq -1$ and

$$a_n = -2\pi c \sum_{\iota=1}^{k} \operatorname{Res}\left(z^{n+M-2} \frac{(1-z)^{M+1}}{Q_M(z)}, \alpha_\iota\right) \tag{4.15}$$

for $n \geq 2 - M$. This covers all values of n unless $M = 0, 1$. The case where $M = 0$ is trivial because the pre-ondelettes are just Haar functions in that case, while the $M = 1$ regularity is a case interesting enough to examine completely, anyway.

Since

$$\frac{d^2}{dp^2} \csc^2\left(\frac{1}{2}p\right) = \frac{1}{2}\csc^4\left(\frac{1}{2}p\right) + \csc^2\left(\frac{1}{2}p\right)\cot^2\left(\frac{1}{2}p\right), \tag{4.16}$$

the trigonometric polynomial is just

$$\frac{1}{2} + \cos^2\left(\frac{1}{2}p\right) = 1 + \frac{1}{2}\cos p, \tag{4.17}$$

and so by (4.11), we have

$$Q_1(z) = \frac{1}{4}z^2 + z + \frac{1}{4}. \tag{4.18}$$

The roots are $-2 \pm \sqrt{3}$, so that the only root in the unit disk is $-2 + \sqrt{3}$. Hence,

$$a_n = 2\pi c \frac{1}{\frac{1}{2}\sqrt{3}}(-2+\sqrt{3})^{-n-1}(-3+\sqrt{3})^2 \tag{4.19}$$

for $n \leq -1$ and

$$a_n = 2\pi c \frac{1}{\frac{1}{2}\sqrt{3}}(-2+\sqrt{3})^{n-1}(3-\sqrt{3})^2 \tag{4.20}$$

for $n \geq 1$. When $n = 0$, there is a pole at $z = 0$ as well. This yields

$$a_0 = 4\pi c \frac{1}{\sqrt{3}} \frac{(-3+\sqrt{3})^2}{-2+\sqrt{3}} + 8\pi c. \tag{4.21}$$

Finally, since $\chi * \chi$ has compact support, it follows from (4.7) that the rate ρ of exponential decay for $\varphi_0(x)$ is precisely given by

$$(2-\sqrt{3})^n = e^{-\rho n}, \tag{4.22}$$

and so

$$\rho = \ln\left(\frac{1}{2-\sqrt{3}}\right) = \ln(2+\sqrt{3}) \approx 1.316. \tag{4.23}$$

For arbitrary M,

$$\rho = \ln\left(\frac{1}{|\alpha_0|}\right), \tag{4.24}$$

where α_0 is the root of $Q_M(z)$ in the unit disk with the largest modulus.

§5. Cardinal spline interpolation

We now compare our approach to the cardinal spline approach of Chui and Wang [4]. In approximation theory the *fundamental Nth-order cardinal interpolating spline* is the solution to minimizing $\|f\|_2$ in $\mathcal{H}_N$ with respect to the constraints

$$f(m) = \delta_{m0}, \qquad m \in \mathbb{Z}, \tag{5.1}$$

where $\mathcal{H}_N$ is the subspace of $L^2(\mathbb{R})$ generated by $\{\eta_N(x+m)\}$ with $\eta = \eta_N$ given by (2.2). This choice of η is the *Nth-order B-spline* [9], although we are not using standard notation for it. On the other hand, notation for the solution of this constrained minimization problem is usually L_N, and we use that here.

The point of departure in [4] is the observation that the function

$$\frac{d^N}{dx^N} L_{2N}(2x-1) \tag{5.2}$$

is the "mother wavelet" of an inter-scale orthogonal basis. Here we adapt their statement to our choice of "unit scale", so our "mother wavelet" is actually

$$\frac{d^N}{dx^N} L_{2N}(x-1). \tag{5.3}$$

The basis is the same, of course, but described with a shift in the index r – namely $\{L_{2N}^{(N)}(2^r x - 2m - 1)\}$. The orthogonality property follows from the linear expansions

$$\begin{aligned} \frac{d^N}{dx^N} L_{2N}(x-1) &= \sum_n b_n \frac{d^N}{dx^N} \eta_{2N}(x-n) \\ &= \sum_m \nu_m \eta_N(x-m) \end{aligned} \tag{5.4}$$

together with the integration by parts

$$\begin{aligned} &\int L_{2N}^{(N)}(x-1)\eta_N(2^r x - m)dx \\ &= (-1)^N 2^{rN} \int L_{2N}(x-1)\eta_N^{(N)}(2^r x - m)dx \\ &= \sum_n \mu_n L_{2N}(2^{-r}(m+n)-1) \\ &= 0, \qquad r < 0. \end{aligned} \tag{5.5}$$

The point is that $\eta_N^{(N)}$ is only a linear combination of Dirac delta functions $\delta(\cdot - n)$, while for $r < 0$, $2^{-r}(m+n)-1$ is an integer other than zero.

How is this cardinal spline interpolation related to the constrained minimization described in Section 2? First observe that

$$\begin{aligned} &\int L_{2N}^{(N)}(x-1)\eta_N(x-m)dx \\ &= (-1)^N \int L_{2N}(x-1)\eta_N^{(N)}(x-m)dx \\ &= (-1)^N \sum_{n=0}^{N} (-1)^n \binom{N}{n} L_{2N}(m+n-1) \\ &= (-1)^{N-m+1} c_{1-m}, \end{aligned} \tag{5.6}$$

so that

$$(-1)^{N+1} \frac{d^N}{dx^N} L_{2N}(x-1) \tag{5.7}$$

satisfies the linear constraints (2.4). Next observe that the L^2-norm of this function is automatically minimal with respect to these constraints. This is not to be confused with the minimizing L^2-norm of L_{2N} in the cardinal spline interpolation problem above. Instead, it follows from the fact that for *any* function satisfying (2.4), the L^2-norm is *determined* if the function lies in $\mathcal{H}_N$ – and the function (5.7) certainly does, as can be seen from (5.4). Therefore

$$\varphi_0(x) = (-1)^{N+1} \frac{d^N}{dx^N} L_{2N}(x-1), \tag{5.8}$$

and so the connection is established.

Finally, the polynomial $Q_M(z)$ that we derived in the last section is quite familiar in spline interpolation [10] as we have mentioned already in Section 1. It is the Euler-Frobenius polynomial associated with the $2N$th-order B-spline (with $N = M + 1$). This polynomial is central to the analysis in [4] as well.

§6. The dual basis

In this section we derive the position space representation for the dual basis. Since our basis of pre-ondelettes is scale-orthogonal, we may concentrate on the unit-scale pre-ondelettes without loss of generality. Bear in mind that our "unit-scale" functions are 2-unit translates of φ, so the overlap matrix is given by (2.9) and is index-translation-invariant. Clearly S is diagonalized by

$$S_{mn} = \int_0^{2\pi} e^{i2(m-n)t} w(t)\, dt, \tag{6.1}$$

$$\begin{aligned} w(t) &= \sum_\ell |\widehat{\varphi}_0(t+\pi\ell)|^2 \\ &= \frac{c|1-e^{it}|^{2M+2}}{\sum_n |\hat{\chi}(t+2\pi n)|^{2M+2}} + \frac{c|1+e^{it}|^{2M+2}}{\sum_n |\hat{\chi}(t+\pi+2\pi n)|^{2M+2}} \\ &= c\frac{|1-e^{it}|^{2M+2}}{R_M(\cos t)} + c\frac{|1+e^{it}|^{2M+2}}{R_M(-\cos t)}, \end{aligned} \tag{6.2}$$

while the dual basis $\{u_m\}$ is given by

$$u_m(x) = \sum_n (S^{-1})_{mn} \varphi_0(x+2n). \tag{6.3}$$

Hence,

$$(S^{-1})_{mn} = \int_0^{2\pi} e^{i2(m-n)t} \frac{1}{w(t)}\, dt \tag{6.4}$$

and so in particular $u_m(x) = u_0(x+2m)$. Now an obvious Poisson summation yields

$$\begin{aligned}\hat{u}_0(p) &= \frac{c}{w(p)}\widehat{\varphi}_0(p) \\ &= ce^{ip}\frac{(1-e^{-ip})^{M+1}\hat{\chi}(p)^{M+1}}{w(p)R_M(\cos p)} \\ &= ce^{ip}\frac{(1-e^{-ip})^{M+1}R_M(-\cos p)\hat{\chi}(p)^{M+1}}{R_M(-\cos p)|1-e^{ip}|^{2M+2}+R_M(\cos p)|1+e^{ip}|^{2M+2}}. \end{aligned} \tag{6.5}$$

Thus, we have

$$u_0(x) = \sum_n q_n(\chi*)^{M+1}(x+n), \tag{6.6}$$

$$\begin{aligned} q_n = c\int_0^{2\pi} & e^{-inp}e^{ip}(1-e^{-ip})^{M+1}R_M(-\cos p) \\ & \cdot [R_M(-\cos p)(1-e^{ip})^{M+1}(1-e^{-ip})^{M+1} \\ & + R_M(\cos p)(1+e^{ip})^{M+1}(1+e^{-ip})^{M+1}]^{-1}dp. \end{aligned} \tag{6.7}$$

The alternate contour expressions for q_n are

$$q_n = -ic\oint_{|z|=1} z^{-n}(1-z^{-1})^{M+1}(-1)^M Q_M(-z)K_M(z)^{-1}dz, \tag{6.8}$$

$$\begin{aligned} K_M(z) = & (-1)^M Q_M(-z)(1-z)^{M+1}(1-z^{-1})^{M+1} \\ & + Q_M(z)(1+z)^{M+1}(1+z^{-1})^{M+1}, \end{aligned} \tag{6.9}$$

$$q_n = ic\oint_{|z|=1} z^{n-2}(1-z)^{M+1}(-1)^M Q_M(-z)K_M(z)^{-1}dz, \tag{6.10}$$

where we have multiplied the numerator and denominator of each integrand by z^M to obtain these expressions. Now

$$z^{M+1}K_M(z) = Q_M(z)(1+z)^{2M+2} - Q_M(-z)(1-z)^{2M+2} \tag{6.11}$$

is a polynomial for which $z = 0$ is a zero, but the residue of the integrand at $z = 0$ will be zero for almost all n if we avoid negative powers of z in our choice of integrands. Since

$$z^{-n}\frac{(1-z^{-1})^{M+1}}{K_M(z)}(-1)^M Q_M(-z) = -z^{-n}\frac{(1-z)^{M+1}}{z^{M+1}K_M(z)}Q_m(-z), \tag{6.12}$$

$$\begin{aligned} & z^{n-2}\frac{(1-z)^{M+1}}{K_M(z)}(-1)^M Q_M(-z) \\ & \quad = z^{n+M-1}\frac{(1-z)^{M+1}}{z^{M+1}K_M(z)}(-1)^M Q_M(-z), \end{aligned} \tag{6.13}$$

we accomplish this by using (6.8) when $n \leq 0$ and turning to (6.10) when $n \geq 1$. Therefore, if we let $\beta_1, \ldots, \beta_\ell$ denote the roots of $z^{M+1}K_M(z)$ in the unit disk, including 0, we have

$$q_n = -2\pi c \sum_{\iota=1}^{\ell} \operatorname{Res}\left(z^{-n} \frac{(1-z)^{M+1}}{z^{M+1}K_M(z)} Q_m(-z), \beta_\iota \right) \tag{6.14}$$

for $n \leq 0$, and

$$q_n = -2\pi c(-1)^M \sum_{\iota=1}^{\ell} \operatorname{Res}\left(z^{n+M-1} \frac{(1-z)^{M+1}}{z^{M+1}K_M(z)} Q_M(-z), \beta_\iota \right) \tag{6.15}$$

for $n \geq 1$.

Consider the case $M = 1$ and recall (4.18) to obtain

$$\begin{aligned} z^2K_1(z) &= \left(\frac{1}{4}z^2 + z + \frac{1}{4}\right)(1+z)^4 - \left(\frac{1}{4}z^2 - z + \frac{1}{4}\right)(1-z)^4 \\ &= \left(\frac{1}{2}z^2 + \frac{1}{2}\right)(4z + 4z^3) + 2z(z^4 + 6z^2 + 1) \\ &= 4z^5 + 16z^3 + 4z. \end{aligned} \tag{6.16}$$

The roots are

$$z = 0, \pm\sqrt{-2 \pm \sqrt{3}}, \tag{6.17}$$

and those in the unit disk are $0, \pm i\sqrt{2-\sqrt{3}}$. Thus the poles of the analytic functions are simple and the residue calculation is easy.

Only in the case $n = 0$ does the residue at $z = 0$ come up. We have

$$\begin{aligned} q_0 = &-2\pi c \operatorname{Res}\left(\frac{(1-z)^2}{z^2K_1(z)} Q_1(-z), 0 \right) \\ &- 4\pi c \operatorname{Re}\left[\operatorname{Res}\left(\frac{(1-z)^2}{z^2K_1(z)} Q_1(-z), i\sqrt{2-\sqrt{3}} \right) \right] \end{aligned} \tag{6.18}$$

because our polynomials have real coefficients, and so

$$\begin{aligned} q_0 = -\frac{\pi c}{8} - 4\pi c \operatorname{Re}\Bigg[&\left(1 - i\sqrt{2-\sqrt{3}}\right)^2 \left(-\frac{1}{2} + \frac{1}{4}\sqrt{3} \right. \\ &\left. - i\sqrt{2-\sqrt{3}} + \frac{1}{4} \right) \frac{1}{i\sqrt{2-\sqrt{3}}} \frac{1}{2\sqrt{3}} \frac{1}{2i\sqrt{2-\sqrt{3}}} \Bigg]. \end{aligned} \tag{6.19}$$

For $n > 0$ (resp. $n < 0$) q_n is given by the same expression with the term $-\frac{1}{10}\pi c$ absent and the factor $(i2-\sqrt{3})^n$ (resp. $(i2-\sqrt{3})^{-n}$) included in the brackets. The exact rate of exponential decay for $u_0(x)$ is

$$\ln\left(\frac{1}{\sqrt{2-\sqrt{3}}}\right) = \frac{1}{2}\ln\left(\frac{1}{2-\sqrt{3}}\right) = \frac{1}{2}\ln\left(2+\sqrt{3}\right). \tag{6.20}$$

This number is approximately .658, but more to the point, it is exactly one-half the exponential decay rate for φ because the roots of $z^2K_1(z)$ are 0 together with the square roots of the roots of $Q_1(z)$. This is no accident, as we show in the next section.

§7. The polynomial relation

We have just seen from our calculations in the $M = 1$ case that

$$zK_1(z) = 16Q_1(z^2). \tag{7.1}$$

We now extend this to (1.5) for arbitrary M. The change of variable $z = e^{ip}$ yields

$$\csc^2\left(\frac{1}{2}p\right) = -\frac{4z}{(z-1)^2}, \tag{7.2}$$

$$\frac{d}{dp}f(e^{ip}) = iz\frac{df}{dz}\Big|_{z=e^{ip}}, \tag{7.3}$$

from which it follows that

$$\begin{aligned} 4^M \frac{\sin^{2M+2}\left(\frac{1}{2}p\right)}{(2M-1)!}\frac{d^{2M}}{dp^{2M}}\csc^2\left(\frac{1}{2}p\right) &= -\frac{4^{M+1}}{(2M-1)!}\frac{(z-1)^{2M+2}}{(-4z)^{M+1}}\left(iz\frac{d}{dz}\right)^{2M}\frac{z}{(z-1)^2} \\ &= z^{-M-1}\frac{(z-1)^{2M+2}}{(2M-1)!}\left(z\frac{d}{dz}\right)^{2M}\frac{z}{(z-1)^2}. \end{aligned} \tag{7.4}$$

Thus

$$Q_M(z) = z^{-1}\frac{(z-1)^{2M+2}}{(2M-1)!}\left(z\frac{d}{dz}\right)^{2M}\frac{z}{(z-1)^2}, \tag{7.5}$$

and so by (6.11) we have

$$
\begin{aligned}
z^{M+1}K_M(z) &= z^{-1}\frac{(z^2-1)^{2M+2}}{(2M-1)!}\Big(z\frac{d}{dz}\Big)^{2M}\frac{z}{(z-1)^2} \\
&\quad - z^{-1}\frac{(z^2-1)^{2M+2}}{(2M-1)!}\Big(z\frac{d}{dz}\Big)^{2M}\frac{z}{(z+1)^2} \\
&= 4z^{-1}\frac{(z^2-1)^{2M+2}}{(2M-1)!}\Big(z\frac{d}{dz}\Big)^{2m}\frac{z^2}{(z^2-1)^2}. \qquad (7.6)
\end{aligned}
$$

because $z[(z+1)^2-(z-1)^2]=4z^2$. On the other hand, if we make the change of variable $w=z^2$, then

$$
z\frac{d}{dz}f(z^2) = 2w\frac{df}{dw}\Big|_{w=z^2} \qquad (7.7)
$$

shows that

$$
\begin{aligned}
z^M K_M(z) &= 2^{2M+2}w^{-1}\frac{(w-1)^{2M+2}}{(2M-1)!}\Big(w\frac{d}{dw}\Big)^{2M}\frac{w}{(w-1)^2} \\
&= 2^{2M+2}Q_M(w). \qquad (7.8)
\end{aligned}
$$

Chui and Wang proved (1.5) in their own work [4] as well.

References

1. Battle, G., A block spin construction of ondelettes Part I: Lemarié functions, *Comm. Math. Phys.* **110** (1987), 601–615.
2. Battle, G., Wavelets: a renormalization group point of view, in *Wavelets and Their Applications*, G. Beylkin, R. Coifman, I. Daubechies, S. Mallat, Y. Meyer, L. Raphael, and B. Ruskai (eds.), Jones and Bartlett, Cambridge, MA, 1992, to appear.
3. Chui, C. K., *An Introduction to Wavelets*, Academic Press, Boston, MA, 1992.
4. Chui, C. K. and J. Z. Wang, A cardinal spline approach to wavelets, *Proc. Amer. Math. Soc.*, 1991, to appear.
5. Chui, C. K. and J. Z. Wang, On compactly supported spline wavelets and a duality principle, *Trans. Amer. Math. Soc.*, 1991, to appear.
6. Daubechies, I., Orthonormal bases of compactly supported wavelets, *Comm. Pure and Appl. Math.* **41** (1988), 909–996.
7. Gawedzki, K. and A. Kupiainen, A rigorous block spin approach to massless lattice theories, *Comm. Math. Phys.* **77** (1980), 31–64.
8. Lemarié, P., Ondelettes à localization exponentielle, *J. Math. Pures et Appl.* **67** (1988), 227–236.
9. Lemarié, P., Some remarks on wavelet theory and interpolation, Université de Paris-Sud, preprint.
10. Schoenberg, I., Cardinal interpolation and spline functions, *J. Approx. Theory* **2** (1969), 167—206.

This research was supported in part by the National Science Foundation under Grant No. DMS-9024867.

Guy Battle
Department of Mathematics
Texas A&M University
College Station, TX 77843–3368
FAX #: (409) 845 6028

Polynomial Splines and Wavelets–
A Signal Processing Perspective

Michael Unser and Akram Aldroubi

Abstract. We discuss a unified framework of representation of signals using polynomial spline basis functions and wavelets. The emphasis is on signal processing interpretation of these techniques and fast implementation using digital filters. First, we consider the derivation of filtering algorithms for polynomial spline interpolation and approximation. We indicate how this latter process can be viewed as an extension of Shannon's classical sampling theory. Second, we consider the construction of an extended family of polynomial spline wavelet transforms in relation to a multiresolution analysis of L_2. Particular examples include the Battle/Lemarié orthogonal wavelets, the B-spline wavelets of compact support of Chui and Wang, and the cardinal (or fundamental) spline wavelets. Two equivalent ways of understanding these wavelet transforms are the biorthogonal formulation and the filter bank interpretation. We give explicit wavelet and filter formulas for splines of any order odd n and discuss the implementation of fast algorithms. The theory is illustrated with image processing examples. The connection between multiresolution pyramids and wavelet representations is also made explicit.

§1. Introduction

Polynomial splines play a central role in approximation theory and numerical analysis. These functions have a number of desirable properties that make them useful in a variety of applications ([17], [32], [53], and [54]). Not surprisingly, polynomial splines have had a significant impact on the early development of the theory of the wavelet transform ([7], [11], [12], [26], [28], [31], and [42]). The features that make them particularly attractive in this context are as follows:

(1) Polynomial spline basis (or scaling) functions have a simple explicit analytic form. The standard basis functions are the B-splines which are

Wavelets–A Tutorial in Theory and Applications
C. K. Chui (ed.), pp. 91–122.

ISBN 0-12-174590-2

compactly supported and can be generated by repeated convolutions of a rectangular pulse [36].

(2) Polynomial splines have good smoothness properties: a polynomial spline of order n is a function of class C^{n-1}. Practically, this means that it is possible to achieve any degree of regularity by simply increasing the order of the splines. Precise convergence rates for the approximation of smooth functions and their derivatives by splines are also available [40].

(3) In the simplest case $n = 0$, polynomial splines are piecewise constant functions. At the other extreme when $n \to +\infty$, they converge to bandlimited function ([4], [30], and [39]). The corresponding wavelet expansions are the classical Haar and modulated-sinc (ideal bandpass) transforms. Until the mid 80s, these two examples were the only known representatives of what is now called the wavelet transform.

(4) Because of their simple analytic form, splines are easy to manipulate. Operations such as differentiation and integration can be performed in a straightforward manner. Polynomial splines therefore constitute the method of choice for designing finite element methods for the numerical solution of differential equations [32]. These properties should also simplify the development of numerical algorithms for image processing (edge detection, etc.).

Several polynomial spline wavelets have been described in the literature; these include some of the earliest examples of orthogonal wavelet transforms. The oldest one is the well known Haar transform [21]. The next example was constructed by Strömberg using polynomial splines of order n [42]. Later, Battle and Lemarié independently constructed orthogonal spline wavelet transforms using symmetrical basis functions with an exponential decay ([7] and [26]). These latter representations are particular cases of the general class of orthogonal wavelet transforms defined by S. Mallat and can therefore be computed using his fast algorithm ([27] and [28]). Chui and Wang proposed an alternative representation using nonorthogonal spline wavelets obtained from the $(n+1)$th derivative of a fundamental (or cardinal) spline of order $2n+1$ [12]. They also constructed B-spline wavelets with compact support that are the natural counterparts of the classical B-spline functions [11]. The B-spline wavelets were also discovered independently by Unser *et al.* and used to construct an extended family of polynomial spline wavelets related to the former by reversible filtering [48]. It has also been demonstrated that the B-spline wavelets converge to a modulated Gaussian as the order of the spline goes to infinity [51]. This provides a link between the wavelet and Gabor transforms ([6], [15], and [20]). It also suggests that it is possible to construct wavelets for which the product of the time-frequency localizations is arbitrarily close to the optimal bound specified by the uncertainty principle.

The purpose of this chapter is to present a unified framework for this whole class of polynomial spline wavelet transforms together with some general design principles. We will emphasize a signal processing formulation and relate these approaches to some recent filter-based algorithms for polynomial spline interpolation and approximation ([46], [49], and [52]). To simplify the presentation, we will cast it in the recently proposed framework of biorthogonal wavelet transforms ([14], [24], [34], and [60]). In this respect, the present formulation is also an extension of the earlier approaches of Unser *et al.* and Chui and Wang who constructed dual basis functions in order to obtain certain polynomial spline expansions ([11] and [52]).

This chapter is organized as follows. Section 2 provides a review of the main properties of polynomial splines. It also presents filtering algorithms for the classical problems of polynomial spline interpolation and approximation. Section 3 considers the construction of an extended family of polynomial spline wavelet transforms that encompasses all the examples mentioned above. The emphasis is deliberately put on a formulation using discrete signal processing operators. This allows us to construct basis functions with certain specific properties based on the solution of discrete convolution equations. We also show how to extend Mallat's fast algorithm for the present class of nonorthogonal spline wavelets. In Section 4, we consider certain aspects of these wavelet transforms that are more directly related to signal processing. In particular, we look at their multiband filter characteristics and discuss some practical implementation issues. We also show how to extend these techniques for higher-dimensional signals and present some image processing examples. A special application is the generation of polynomial spline pyramids for multiresolution image analysis and processing.

1.1. Notations and operators

L_2 is the vector space of measurable, square-integrable, one-dimensional functions $f(x), x \in \mathbb{R}$. The inner product of $f(x) \in L_2$ with $g(x) \in L_2$ is written as $\langle f(x), g(x) \rangle$. The central B-spline of order n is denoted by $\beta^n(x)$; this function, which is piecewise polynomial of degree n, is generated by repeated convolution of a B-spline of order 0

$$\beta^n(x) = \beta^0 * \beta^{n-1}(x), \tag{1}$$

where $\beta^0(x)$ is the characteristic function in the interval $[-\frac{1}{2}, \frac{1}{2})$, and where the symbol $*$ denotes the convolution operation.

l_2 is the space of square summable sequences (or discrete signals) $a(k), k \in \mathbb{Z}$. A sequence $a(k)$ is formally characterized by its z-transform, which we denote by a capital letter

$$A(z) = \sum_{k=-\infty}^{+\infty} a(k) z^{-k}. \tag{2}$$

This correspondence is also expressed as: $a(k) \stackrel{z}{\longleftrightarrow} A(z)$. The discrete Fourier transform of $a(k)$ is obtained by replacing z by $e^{j2\pi f}$ in $A(z)$. The convolution between two discrete sequences $a \in l_2$ and $b \in l_2$ is

$$b * a(k) = \sum_{l=-\infty}^{+\infty} b(l)a(k-l) \overset{z}{\longleftrightarrow} B(z)A(z). \tag{3}$$

It is sometimes useful to view b as the impulse response of a shift invariant filter that is applied to a. This filter is entirely described by its transfer function $B(z)$. If $B(z)$ is a polynomial with complex roots that are not on the unit circle, the inverse operator $(b)^{-1}$ exists and is uniquely defined by the equation

$$(b)^{-1} * b(k) = \delta_0(k), \tag{4}$$

where $\delta_0(l) := \delta_{0,l}$ denotes the unit pulse at the origin; $\delta_{k,l}$ is the Kronecker symbol. The up-sampling by a factor of two is defined as

$$[b]_{\uparrow 2}(k) = \begin{cases} b(k/2), & k \text{ even} \\ 0, & k \text{ odd} \end{cases} \overset{z}{\longleftrightarrow} B(z^2). \tag{5}$$

The dual operation is the decimation by a factor of two

$$[b]_{\downarrow 2}(k) = b(2k) \overset{z}{\longleftrightarrow} \frac{1}{2}\left(B(z^{1/2}) + B(-z^{1/2})\right). \tag{6}$$

Another operation is the modulation operator which acts by changing the sign of every other sample

$$\tilde{a}(k) = (-1)^k a(k) \overset{z}{\longleftrightarrow} A(-z). \tag{7}$$

The symbol $'$ is used to represent the time reversal of a continuous-time function or a discrete sequence; *i.e.*, $g'(x) = g(-x)$ and $a'(k) = a(-k)$.

We also introduce a special notation for the discrete B-spline of order n obtained by sampling $\beta^n(x)$ at the integers

$$b^n(k) = \left.\beta^n(x)\right|_{x=k}. \tag{8}$$

The sequence b^n plays a crucial role in many of the derivations. An example where it arises is the calculation of inner products between shifted B-spline basis functions

$$\langle \beta^n(x), \beta^n(x-k) \rangle = b^{2n+1}(k); \tag{9}$$

a result that follows directly from the convolution property (1).

§2. Polynomial splines

In this section, we review some of the fundamental properties of polynomial splines. We also derive algorithms for spline interpolation and approximation. We have tried to relate these problems to well known signal processing concepts. Such a formulation will be beneficial in two important respects: (i) it will provide us with efficient filter-based algorithms, and (ii) it will give us new insights for understanding some of these problems.

The basic function space S^n that is considered here is the space of polynomial splines of order n with knot points at the integers. These functions are polynomials of degree n in each interval $[k, k+1)$ with the additional smoothness constraint that the polynomial segments are connected in a way that guarantees the continuity of the function and its $(n-1)$ first derivatives.

2.1. B-splines and scaling functions

A fundamental result due to Schoenberg is that any polynomial spline function of degree n can be represented by a linear combination of shifted B-splines ([36] and [38])

$$g^n(x) = \sum_{k=-\infty}^{+\infty} c(k)\beta^n(x-k), \tag{10}$$

where the basis functions are defined recursively by Equation (1); the superscript denotes the spline order and also the degree of the piecewise polynomial segments. The B-spline representation (10) is unique; the polynomial spline $g^n(x)$ is therefore entirely characterized by the sequence of its B-spline coefficients $\{c(k) : k \in \mathbb{Z}\}$. The principle of this expansion for $n = 3$ (cubic splines) is illustrated in Figure 1.

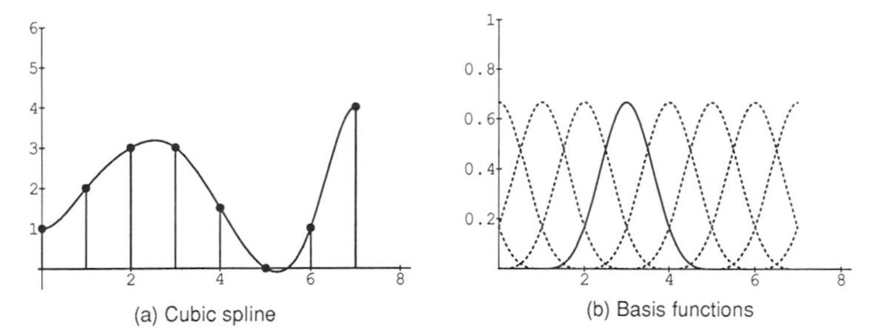

Figure 1. B-spline representation of a cubic spline polynomial.

The B-splines defined by Equation (1) are symmetric bell-shaped functions; they have a number of attractive properties. First, and most importantly, they are compactly supported; in fact, the support of these functions is minimal among all polynomial splines of order n [38]. Second, they have a simple analytical form in both the time and frequency domain. An explicit B-spline formula using piecewise polynomials of degree n is

$$\beta^n(x) = \sum_{j=0}^{n+1} \frac{(-1)^j}{n!} \binom{n+1}{j} \left[x + \tfrac{n+1}{2} - j\right]_+^n, \tag{11}$$

where $[x]_+^n = \max\{0, x\}^n$ is the one-sided power function of degree n. The Fourier transform of a B-spline of order n can be found from Equation (1) and is given by

$$\beta^n(x) \overset{FT}{\longleftrightarrow} B^n(f) = \text{sinc}^{n+1}(f) \tag{12}$$

where $\text{sinc}(x) = \frac{\sin \pi x}{\pi x}$ and where f is the frequency variable.

It is also possible to construct alternative sets of shift invariant basis functions by taking linear combinations of B-splines. This result can be stated as follows:

Proposition 1 [50]. *The set of functions* $\{\varphi^n(x-k) : k \in \mathbb{Z}\}$ *with*

$$\varphi^n(x) = \sum_{k=-\infty}^{+\infty} p(k)\beta^n(x-k), \tag{13}$$

is a basis of S^n *provided that* p *is an invertible convolution operator from* l_2 *into itself.*

Hence, the basic space of polynomial spline functions can also be defined as

$$S^n = \left\{ g : g(x) = \sum_{k \in \mathbb{Z}} \check{c}(k)\varphi^n(x-k), \check{c} \in l_2 \right\}, \tag{14}$$

where φ^n is given by Equation (13). In accordance with the terminology used for the wavelet transform, we will call φ^n a scaling function. The reason for considering basis functions that are different from B-splines is that this will allow us to define polynomial spline representations with other useful properties (*e.g.*, orthogonality, canonical representation, etc.).

2.2. Spline interpolation

In practice, it is often of interest to determine a polynomial spline that precisely interpolates a given sequence of data points $\{g(k)\}$. This problem is commonly referred to as the cardinal spline interpolation problem [37]. In [49], we have described a simple procedure for obtaining the solution by digital filtering. The key idea is that the B-spline coefficients in Equation (10) can be determined by convolution

$$c(k) = \left(b^n\right)^{-1} * g(k), \tag{15}$$

where $\left(b^n\right)^{-1}$ is the convolution inverse of the discrete B-spline kernel of order n. The stability of this inverse filter is guaranteed from a result in [4]. The interpolation algorithm described by Equation (15) can be implemented quite efficiently using recursive digital filters [49]. A detailed comparison with the standard matrix algorithm that uses LU decomposition is provided in [54].

Likewise, the generalized spline coefficients in Equation (14) are obtained by a change of variable, which in the present case, can be performed by simple convolution

$$\breve{c}(k) = (p)^{-1} * c(k) = (p * b^n)^{-1} * g(k). \tag{16}$$

Note that the kernel $p * b^n$ is precisely the sampled version of the scaling function $\varphi^n(x)$; *i.e.*, $\varphi^n(k) = p * b^n(k)$.

A situation of special interest occurs when $p = (b^n)^{-1}$, in which case the coefficients are the sampled values themselves. This latter representation is referred to as the cardinal spline (C-spline) representation. The corresponding basis functions $\eta^n(x)$ are the cardinal (or fundamental) splines. These functions provide the impulse response of a polynomial spline interpolator; they have been shown to converge to sinc(x) (the impulse response of an ideal interpolator) as the order of the spline goes to infinity [4]. The cardinal splines for $n = 3$ and $n \to +\infty$ are represented in Figure 2. The interpolation property of these functions stems from the fact that they are precisely equal to one at the origin and that they vanish at all other integer values.

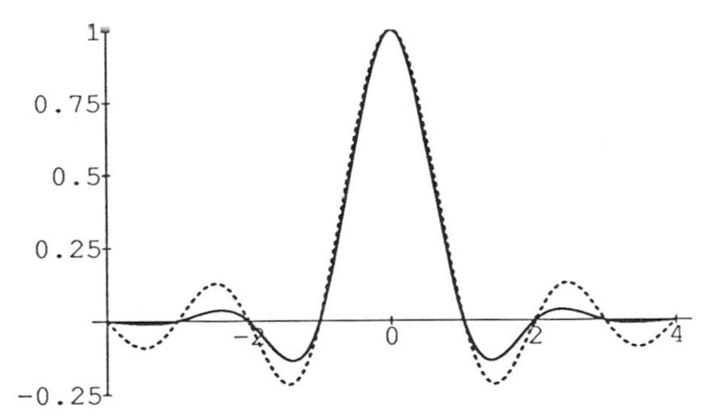

Figure 2. Cardinal (or fundamental) splines for $n = 3$ (solid line) and $n \to +\infty$ (dashed line). These functions are the impulse responses of a cubic spline interpolator and an ideal lowpass filter, respectively.

Based on this observation, we can interpret Equation (16) as a global change of coordinate system from cardinal to generalized spline representation. The key result is that this process can be accomplished by digital filtering. This point of view is to be contrasted with the standard approach to polynomial spline interpolation which uses a matrix formulation. Although both approaches are essentially equivalent, the present formulation may have the advantage of simplicity. It also offers a new signal processing interpretation. Further, it can result in more efficient algorithms, especially when the filters are implemented recursively.

2.3. Spline approximation

Another related problem is to find the minimum error polynomial spline approximation of a function $g(x)$ in the L_2-norm. This process is equivalent to determining the orthogonal projection of $g(x)$ on S^n. We have shown in [49] that the B-spline coefficients of this approximation can be determined by simple inner product

$$c(k) = \langle g(x),\ \mathring{\beta}^n(x-k)\rangle, \tag{17}$$

where $\mathring{\beta}^n(x)$ is the dual spline function which is given by Equation (13) with $p = (b^{2n+1})^{-1}$.

This result is a particular case of a generalized sampling theorem that can be stated as follows.

Theorem 2. *Let $V = \text{span}\{\varphi(x-k), k \in \mathbb{Z}\}$ be a closed subspace of L_2 such that $b(k) = \langle \varphi(x), \varphi(x+k)\rangle$ is an invertible convolution operator. Then, there exists a biorthogonal function $\mathring{\varphi}(x) \in V$ such that the orthogonal projection of a function $g(x) \in L_2$ on V is given by*

$$g_{(0)}(x) = \sum_{k=-\infty}^{+\infty} \langle g(x), \mathring{\varphi}(x-k)\rangle\ \varphi(x-k) \tag{18}$$

and

$$\langle \varphi(x-k),\ \mathring{\varphi}(x-l)\rangle = \delta_{k,l}. \tag{19}$$

A proof that uses the explicit derivation of $\mathring{\varphi}(x)$ is given in [48]. We will make use of this result later on to find explicit formulas for biorthogonal spline basis functions and wavelets.

Here again, the filtering interpretation can provide us with some new insights. The first step is to observe that this procedure has many similarities with the standard signal processing approach to discretization dictated by Shannon's Sampling Theorem ([25] and [41]). To make this link explicit, we consider the generalized basis functions specified by Proposition 1 and construct the projection of a function g on S^n as follows. First, we convolve the function $g(x)$ with the continuous-time filter whose impulse response is $\mathring{\varphi}^n(-x)$. The expansion coefficients in Equation (18) can be obtained by sampling this prefiltered signal

$$\check{c}(k) = \langle g(x), \mathring{\varphi}^n(x-k)\rangle = \mathring{\varphi}^{n\prime} * g(x)\Big|_{x=k}. \tag{20}$$

The polynomial spline approximation is then determined from the reconstruction formula

$$g^n(x) = \sum_{k=-\infty}^{+\infty} \check{c}(k)\varphi^n(x-k), \tag{21}$$

which is also equivalent to convolving the delta weighted sequence $\sum_{k\in\mathbb{Z}} \check{c}(k)\delta(x-k)$ with the interpolation filter $\varphi^n(x)$. This three step paradigm is schematically outlined by the block diagram in Figure 3. The prefilter has a role that is analogous to the anti-aliasing lowpass filter required in conventional sampling theory. In the case of the cardinal representation $\left(p = (b^n)^{-1}\right)$, the postfilter $\eta^n(x)$ is a true interpolation filter (see Figure 2). The corresponding optimal prefilter $\mathring{\eta}^n(x)$ is a polynomial spline of order n with B-spline coefficients $\mathring{p}(k) = \left(b^{2n+1}\right)^{-1} * b^n(k)$ [52]. The frequency responses of these operators are also displayed in Figure 3. Interestingly, both filters converge to the ideal lowpass filter as the order of the spline tends to infinity. The convergence has been proven to occur pointwise and in all L_p-norms with $1 \leq p < +\infty$ in the frequency domain ([4] and [52]).

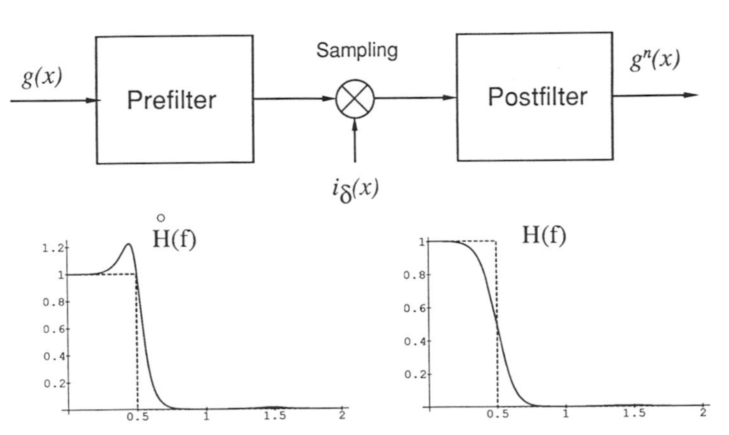

Figure 3. Equivalent block diagram for the polynomial spline approximation of signals and frequency responses of the corresponding cardinal filters for $n = 3$ (solid line) and $n \to +\infty$ (dashed line).

The polynomial spline approximation procedure described by Equation (18) (or equivalently by Equations (20) and (21)) is therefore asymptotically equivalent to bandlimiting a function to the frequency interval $[-\frac{1}{2}, \frac{1}{2}]$. In this sense, it can be seen as an extension of classical sampling theory.

2.4. Dilation property and binomial kernels

A result that turns out to be crucial for the construction of wavelets is that a B-spline expanded by a factor of two is itself a function of S^n when n is odd. This property is expressed by the dilatation (or two scale difference) equation

$$\beta^n(x/2) = \sum_{k=-\infty}^{+\infty} u_2^n(k)\beta^n(x-k) \tag{22}$$

where u_2^n is the binomial kernel of order odd n

$$u_2^n(k) = \begin{cases} \frac{1}{2^n}\binom{n+1}{k+\frac{n+1}{2}} \\ 0, \text{ otherwise} \end{cases}, \ |k| \leq \tfrac{n+1}{2} \quad \stackrel{DFT}{\longleftrightarrow} \quad U_2^n(f) = 2\cos^{n+1}(\pi f). \tag{23}$$

A simple proof of this result can be obtained in the frequency domain. Specifically, we use the trigonometric identity $\sin 2\theta = 2\cos\theta \sin\theta$ to rewrite the Fourier transform of $\beta^n(x/2)$ as

$$2\,\mathrm{sinc}^{n+1}(2f) = \frac{2\sin^{n+1}(2\pi f)}{(2\pi f)^{n+1}} = 2\cos^{n+1}(\pi f)\frac{\sin^{n+1}(\pi f)}{(\pi f)^{n+1}},$$

where the right-most term is precisely $B^n(f)$, as defined by Equation (12). For n odd, the function $2\cos^{n+1}(\pi f)$ is a trigonometric polynomial with periodicity one; it is the discrete Fourier transform of the binomial kernel u_2^n given by Equation (23).

Interestingly enough, the kernel u_2^n also comes into play when computing derivatives of B-spline functions. For instance, it is well known that derivatives of polynomial splines can be obtained by taking finite differences of lower order splines [36]; that is

$$\frac{d^m\beta^n(x)}{dx^m} = \Delta_m \beta^{n-m}(x), \tag{24}$$

where Δ_m is the central mth order difference operator. Since the weighting coefficients associated with Δ_m are the binomial coefficients with an alternating change of sign, this formula for m even can be rewritten as

$$\frac{d^m\beta^n(x)}{dx^m} = 2^{m-1} \sum_{k=-\infty}^{+\infty} \tilde{u}_2^{m-1}(k)\beta^{n-m}(x-k), \tag{25}$$

where the symbol "$\sim$" denotes the modulation operator (see Equation (7)).

§3. Polynomial spline wavelets

One of the standard ways of constructing a wavelet transform is to consider a sequence of embedded subspaces that forms a multiresolution analysis. For this purpose, we consider the fine-to-coarse sequence of dyadic contractions and dilatations of our basic spline space S^n with the restriction that n is now odd. Specifically, $V_{(i)}$, the function space at resolution (i), is defined as

$$V_{(i)} = \left\{ g_{(i)} : g_{(i)}(x) = \sum_{k\in\mathbb{Z}} c_{(i)}(k)\varphi_{i,k}(x), c_{(i)} \in l_2 \right\} \tag{26}$$

where the normalized basis functions $\varphi_{i,k} = 2^{-i/2}\varphi^n(2^{-i}x - k)$ are obtained by translation (index k) of the generalized scaling function specified by Proposition 1 expanded by a factor of 2^i ; the coefficient $2^{-i/2}$ is an inner product normalization across scale. Note that the factor 2^i also represents the spacing between the knot points. The dilation property (22) states that the basis functions of $V_{(1)}$ are included in $V_{(0)} = S^n$ provided that n is odd. Hence, we have a sequence of nested subspaces

$$\cdots \supset V_{(-1)} \supset V_{(0)} \supset V_{(1)} \cdots \supset V_{(i)} \cdots$$

that can be shown to define a multiresolution analysis of L_2 in the sense defined by S. Mallat [28].

We can now apply this multiresolution analysis to any function $g \in L_2$ and compute a sequence of fine-to-coarse approximations $\{g_{(i)}, i \in \mathbb{Z}\}$ by repeated orthogonal projection of g into $V_{(i)}$ with $i \in \mathbb{Z}$. Based on Theorem 2, we have an explicit expression for the polynomial spline approximation of g at resolution level (i)

$$g_{(i)} = \sum_{k \in \mathbb{Z}} \langle g, \mathring{\varphi}_{i,k} \rangle \varphi_{i,k}, \tag{27}$$

where the functions $\varphi_{i,k} \in V_{(i)}$ and $\mathring{\varphi}_{i,k} \in V_{(i)}$ form a biorthogonal pair.

For each resolution level (i), we define the residual signal $\underline{g}_{(i)} = g_{(i-1)} - g_{(i)}$, which is included in $W_{(i)}$, the orthogonal complementary subspace of $V_{(i)}$ with respect to $V_{(i-1)}$. If we now assume that a wavelet function ψ can be defined such that $W_{(0)} = \text{span}\{\psi(x-k), k \in \mathbb{Z}\}$, we can use Theorem 2 to write a formula for the residual signal that is in all points analogous to Equation (27). It is the expansion

$$\underline{g}_{(i)} = \sum_{k \in \mathbb{Z}} \langle g, \mathring{\psi}_{i,k} \rangle \psi_{i,k}, \tag{28}$$

where the normalized basis functions are given by $\psi_{i,k} = 2^{-i/2}\psi(2^{-i}x - k)$ and where $\mathring{\psi}_{i,k} \in W_{(i)}$ is the biorthogonal function specified in Theorem 2. By summing up the residual contributions over all resolution levels, we obtain the full wavelet decomposition of a function $g \in L_2$

$$g = \sum_{(i,k) \in \mathbb{Z}^2} \langle g, \mathring{\psi}_{i,k} \rangle \psi_{i,k}. \tag{29}$$

Since the residual spaces $W_{(i)}$ are orthogonal to each other by construction, the wavelet and its dual are biorthogonal not only within the same resolution level (index k) but also across all scales (index i)

$$\langle \psi_{i,k}, \mathring{\psi}_{j,l} \rangle = \delta_{i,j}\delta_{k,l}. \tag{30}$$

This last equation is consistent with the full expansion (29). We also note that the role of ψ (the synthesis wavelet) and $\mathring{\psi}$ (the analysis wavelet) can readily be interchanged.

The crucial point in this analysis will be to construct polynomial spline wavelets that effectively span the residual space $W_{(0)}$. We will show that this can be done in a systematic way.

3.1. B-spline wavelets of compact support

An admissible wavelet function ψ must satisfy two essential properties. First, it must be included in the finer resolution approximation space. Specifically, this means that the function $\psi(x/2)$ is a polynomial spline of order n

and that it can be represented by its B-spline expansion

$$\psi(x/2) = \sum_{k\in\mathbb{Z}} s(k)\beta^n(x-k).$$

Second, the wavelet must be orthogonal to the polynomial spline subspace at the same resolution level. This leads to the following constraint

$$\forall\, k \in \mathbb{Z},\ \langle \psi(x/2), \beta^n(x/2-k)\rangle = \left[s * u_2^n * b^{2n+1}\right]_{\downarrow 2}(k) = 0,$$

where the explicit representation of this inner product in term of discrete convolution operators is obtained by making use of Equations (22) and (9). Upsampling by a factor of two and taking the z-transform, we get

$$\frac{1}{2}\left(S(z)U_2^n(z)B_1^{2n+1}(z) + S(-z)U_2^n(-z)B_1^{2n+1}(-z)\right) = 0$$

where $U_2^n(z)$ and $B_1^{2n+1}(z)$ are the z-transforms of u_2^n and b^{2n+1}, respectively. It can be verified by substitution that a particular solution of this equation is

$$S(z) = zU_2^n(-z)B_1^{2n+1}(-z).$$

This solution also corresponds to the sequence s of minimal length. Taking the inverse z-transform (c.f. Equation (7), in particular), we obtain an explicit formula for the B-spline wavelet of order n

$$\underline{\beta}^n(x/2) = \sum_{k=-\infty}^{+\infty} \tilde{u}_2^n * \tilde{b}^{2n+1}(k+1)\beta^n(x-k). \tag{31}$$

The wavelet $\underline{\beta}^n$ is represented by an underlined symbol. An alternative but equivalent characterization of B-spline wavelets using the $(n+1)$th derivative of a B-spline of order $2n+1$ has been proposed by Chui and Wang [11]; the link between these two formulations is provided by Equation (25). One of the differences, however, is that this latter construction uses causal B-splines instead of central ones which makes it also applicable for n even.

The B-spline wavelet generates $W_{(0)}$, and consequently all the wavelet spaces $W_{(i)}$ ([11] and [48]). Hence, the set of functions $\{\underline{\beta}^n(2^{-i}x-k), (i,k) \in \mathbb{Z}^2\}$ is an unconditional basis of L_2.

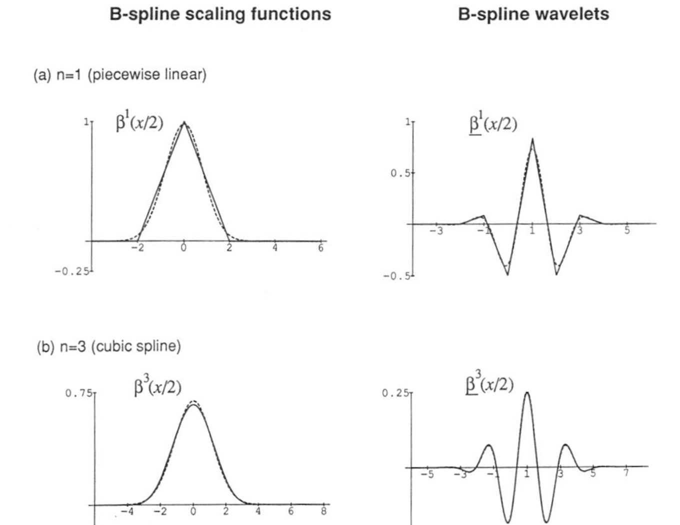

Figure 4. Examples of B-spline scaling functions and wavelets with their corresponding Gabor approximation (dashed lines) at resolution level (1).

Since β^n has a compact support and both u_2^n and b^{2n+1} are finite impulse response (FIR) operators, the function $\underline{\beta}^n$ defined by Equation (31) is compactly supported as well. The B-spline functions at resolution level (1) and their corresponding wavelets, as determined from Equatin (31), are shown in Figure 4 for $n = 1$ and 3. The wavelets are symmetrical functions that are shifted by one unit with respect to the standard B-spline functions. These functions may be thought of as modulated versions of the basic B-spline kernels. In fact, we have demonstrated elsewhere that the functions β^n and $\underline{\beta}^n$ tend to Gaussians and modulated Gaussians as the order of the spline tends to infinity [51]. Specifically, we have the approximation formulas

$$\beta^n(x) \cong \sqrt{\frac{6}{\pi(n+1)}} \exp\left(-\frac{6x^2}{(n+1)}\right) \tag{32}$$

$$\underline{\beta}^n(x) \cong \frac{4a^{n+1}}{\sqrt{2\pi(n+1)}\sigma_0} \cos\left(2\pi\ f_0(2x-1)\right) \exp\left(-\frac{(2x-1)^2}{2\sigma_0^2(n+1)}\right) \tag{33}$$

with $a = 0.697066$, $f_0 = 0.409177$ and $\sigma_0^2 = 0.561145$. These functions are superimposed in dashed line in Figure 4. It can be seen from this graph that the quality of the approximation is already surprisingly good for $n = 1$ and $n = 3$. The B-spline basis functions and wavelets are therefore good approximations of Gabor functions which are known to be maximally localized in both time and frequency. This result suggests that B-spline wavelet representations should be close-to-optimal with respect to the uncertainty principle. We have verified numerically that the product of the time and frequency uncertainties for the cubic spline wavelet is already within 2% of the limit specified by the uncertainty principle.

3.2. Generalized spline wavelets

Once we have constructed one wavelet that satisfies the required conditions, we can generate many others by reversible linear transformation. These generalized spline wavelets can be characterized as follows

$$\psi^n(x) = \sum_{k=-\infty}^{+\infty} q(k)\underline{\beta}^n(x-k) \tag{34}$$

where q is an invertible convolution operator from l_2 into itself. It is not difficult to show that the functions $\{\psi^n(x-k), k \in \mathbb{Z}\}$ generate $W_{(0)}$. Therefore, they can be used to represent any function according to the wavelet expansion (29).

At this stage, the parameters p and q in Equations (13) and (34) have been left unspecified. Our later goal will be to select these parameters in a judicious way in order to construct scaling functions and wavelets with certain specific properties. A principle that will be used over and over again is to first write an equation for the property that the functions should satisfy, express it in terms of discrete convolution operators, and finally solve the corresponding difference equation. For this purpose, we will need an explicit expression for the inner product of B-spline wavelets which is given by

$$\begin{aligned} \langle \underline{\beta}^n(x), \underline{\beta}^n(x-k)\rangle &= \frac{1}{2}\left[\tilde{u}_2^n * \tilde{u}_2^n * \tilde{b}^{2n+1} * \tilde{b}^{2n+1} * b^{2n+1}\right]_{\downarrow 2}(k) \\ &= b^{2n+1} * \left[\tilde{b}^{2n+1} * b^{2n+1}\right]_{\downarrow 2}(k). \end{aligned} \tag{35}$$

3.3. Biorthogonal basis functions

To fully characterize the wavelet transform, we need to determine the biorthogonal functions that appear in Equations (27) and (28). We know from Theorem 2 that these functions must themselves be included in the approximation subspace. Hence, they can be represented by the expansions

$$\mathring{\varphi}^n(x) = \sum_{k=-\infty}^{+\infty} r(k)\beta^n(x-k) \tag{36}$$

$$\mathring{\psi}^n(x) = \sum_{k=-\infty}^{+\infty} s(k)\underline{\beta}^n(x-k) \tag{37}$$

where the sequences r and s still need to be specified. The function $\mathring{\varphi}^n$ must satisfy the biorthogonality condition (19), that is

$$\langle \mathring{\varphi}^n(x), \varphi^n(x-k) \rangle = r * p' * b^{2n+1}(k) = \delta_0(k)$$

where δ_0 is the identity sequence. Solving this equation, we get

$$r = \left(p' * b^{2n+1}\right)^{-1}. \tag{38}$$

Similarly, we rewrite the biorthogonality condition for $\mathring{\psi}^n(x)$

$$\langle \mathring{\psi}^n(x), \psi^n(x-k) \rangle = s * q' * b^{2n+1} * \left[\tilde{b}^{2n+1} * b^{2n+1}\right]_{\downarrow 2}(k) = \delta_0(k)$$

and solve for s, which yields

$$s = \left(q' * b^{2n+1} * \left[\tilde{b}^{2n+1} * b^{2n+1}\right]_{\downarrow 2}\right)^{-1}. \tag{39}$$

The direct inner product calculation suggested by Equation (29) for obtaining the coefficients in the wavelet expansion is simple conceptually but not very efficient computationally. Stéphane Mallat has proposed a fast algorithm that uses quadrature mirror filters for computing orthogonal wavelet transforms [28]. We will now show how to extend this approach for nonorthogonal basis functions.

3.4. The fast wavelet algorithm

The fast wavelet algorithm uses the fact that the approximation spaces $V_{(i)}$ are nested and that the computations at coarser resolutions can entirely be based on the approximations at the next finest level. In other words, we have

$$c_{(i+1)}(k) = \langle g, \mathring{\varphi}_{i+1,k} \rangle = \langle g_{(i)}, \mathring{\varphi}_{i+1,k} \rangle \tag{40}$$

$$d_{(i+1)}(k) = \langle g, \mathring{\psi}_{i+1,k} \rangle = \langle g_{(i)}, \mathring{\psi}_{i+1,k} \rangle \tag{41}$$

where $g_{(i)}$, the projection of g on the finer approximation space $V_{(i)}$, is itself represented by the generalized spline expansion

$$g_{(i)} = \sum_{k=-\infty}^{+\infty} c_{(i)}(k) \varphi_{i,k}. \tag{42}$$

The fast pyramid algorithm can be derived by performing an appropriate change of variable. For this purpose, we express the dual basis functions at the coarser level in term of the dual scaling functions at the finer resolution level

$$\mathring{\varphi}^n(x/2) = \sum_{k=-\infty}^{+\infty} v'(k) \mathring{\varphi}^n(x-k) \tag{43}$$

$$\mathring{\psi}^n(x/2) = \sum_{k=-\infty}^{+\infty} \underline{v}'(k)\mathring{\varphi}^n(x-k), \tag{44}$$

where the sequences v and $\underline{v}$ are given by

$$v(k) = \left[\left(p * b^{2n+1}\right)^{-1}\right]_{\uparrow 2} * p * b^{2n+1} * u_2^n(k) \tag{45}$$

$$\underline{v}(k+1) = \left[\left(q * b^{2n+1}\right)^{-1}\right]_{\uparrow 2} * p * \tilde{u}_2^n(k). \tag{46}$$

Rewriting the functions $\mathring{\varphi}_{i+1,k}$ and $\mathring{\psi}_{i+1,k}$ in terms of their expansions (43) and (44), and using the biorthogonality of φ^n and $\mathring{\varphi}^n$ to evaluate the inner product, we get the following wavelet decomposition algorithm

$$c_{(i+1)}(k) = \left\langle \sum_{l=-\infty}^{+\infty} c_{(i)}(l)\varphi_{i,l},\ \mathring{\varphi}_{i+1,k} \right\rangle = \frac{1}{\sqrt{2}} \left[v * c_{(i)}\right]_{\downarrow 2}(k) \tag{47}$$

$$d_{(i+1)}(k) = \left\langle \sum_{l=-\infty}^{+\infty} c_{(i)}(l)\varphi_{i,l},\ \mathring{\psi}_{i+1,k} \right\rangle = \frac{1}{\sqrt{2}} \left[\underline{v} * c_{(i)}\right]_{\downarrow 2}(k) \tag{48}$$

where $\sqrt{2}$ is a scale normalization factor. This process involves two basic operations: filtering and down-sampling. The full wavelet decomposition can be obtained by applying this algorithm iteratively starting at the finest resolution level.

For reconstructing the coefficients $c_{(i)}$, we use the fact that the basis functions at the coarser level are included in the finer approximation space. We can therefore write the expansions

$$\varphi^n(x/2) = \sum_{k=-\infty}^{+\infty} w(k)\varphi^n(x-k), \tag{49}$$

$$\psi^n(x/2) = \sum_{k=-\infty}^{+\infty} \underline{w}(k)\varphi^n(x-k), \tag{50}$$

where the weighting sequences w and $\underline{w}$ are determined from Equations (22) and (31) by simple change of variable

$$w(k) = [p]_{\uparrow 2} * (p)^{-1} * u_2^n(k) \tag{51}$$

$$\underline{w}(k-1) = [q]_{\uparrow 2} * (p)^{-1} * \tilde{u}_2^n * \tilde{b}^{2n+1}(k). \tag{52}$$

We then use the fact that the function at the finer level is equal to the sum of its coarser resolution approximation plus the residue

$$g_{(i)} = \sum_{k=-\infty}^{+\infty} c_{(i+1)}(k)\varphi_{i+1,k} + \sum_{k=-\infty}^{+\infty} d_{(i+1)}(k)\psi_{i+1,k}.$$

Replacing $\psi_{i+1,k}$ and $\varphi_{i+1,k}$ by their expressions given by Equations (49) and (50), we find by identification with Equation (42) that

$$c_{(i)}(k) = \frac{1}{\sqrt{2}}\left(w * \left[c_{(i+1)}\right]_{\uparrow 2}(k) + \underline{w} * \left[d_{(i+1)}\right]_{\uparrow 2}(k)\right) \tag{53}$$

which yields the reconstruction algorithm. This procedure is also applied iteratively beginning with the coarsest level of the pyramid.

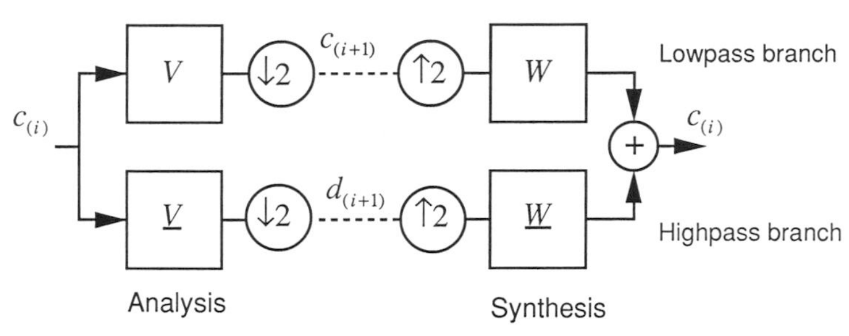

Figure 5. Perfect reconstruction filter bank for the fast wavelet algorithm.

The convolution operators v, $\underline{v}$, w, and $\underline{w}$ define a perfect reconstruction filter bank which is schematically represented in Figure 5. The fact that the decomposition is reversible introduces strong interdependencies among the various filters. A well known result in multirate filter bank theory ([57] and [59]), is that the transfer functions of the filters must satisfy the constraints

$$\begin{cases} V(z)W(z) + \underline{V}(z)\underline{W}(z) = 2 \\ V(z)W(-z) + \underline{V}(z)\underline{W}(-z) = 0, \end{cases} \tag{54}$$

which are necessary and sufficient for a perfect reconstruction. In other words, they guarantee the reversibility of the filter bank algorithm. By extending the results of Rioul for perfect reconstruction FIR/FIR filter banks [34], we can prove that Equation (54) is equivalent to the following set of conditions

$$\left[v * w\right]_{\downarrow 2}(k) = \langle v'(l), w(l+2k)\rangle_{l_2} = \delta_0(k), \tag{55a}$$

$$\left[\underline{v} * \underline{w}\right]_{\downarrow 2}(k) = \langle \underline{v}'(l), \underline{w}(l+2k)\rangle_{l_2} = \delta_0(k), \tag{55b}$$

$$\left[\underline{v} * w\right]_{\downarrow 2}(k) = \langle \underline{v}'(l), w(l+2k)\rangle_{l_2} = 0, \tag{55c}$$

$$\left[v * \underline{w}\right]_{\downarrow 2}(k) = \langle v'(l), \underline{w}(l+2k)\rangle_{l_2} = 0. \tag{55d}$$

These relations are quite general and must be satisfied for any orthogonal or biorthogonal wavelet transform algorithm. By rewriting these conditions as l_2-inner products, we can interpret them as biorthogonality ((a) and (b)) and orthogonality ((c) and (d)) conditions on the corresponding discrete sequences. Another implication of these results is that the specification of either w and $\underline{w}$ (or v and $\underline{v}$) alone is sufficient to characterize the whole system; the complementary filters in each branch can be determined from the perfect reconstruction constraints.

3.5. Specific wavelet transforms

A suitable selection of the convolution operators p and q in Equations (13) and (34) allows the construction of a whole variety of polynomial spline wavelets with certain specific properties. A motivation in this process may also be to simplify the structure of the filters given by Equations (45), (46), (51), and (52). In this section, we will consider several such examples and also indicate how the present formulation relates to previous approaches. Some of these wavelets for $n = 3$ (cubic spline) are shown in Figure 6.

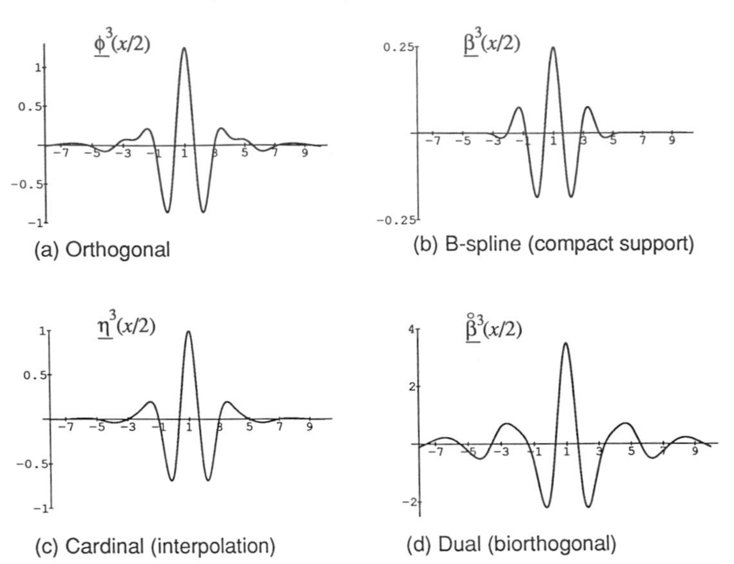

(a) Orthogonal (b) B-spline (compact support)

(c) Cardinal (interpolation) (d) Dual (biorthogonal)

Figure 6. Examples of cubic spline wavelets.

3.5.1. Orthogonal representation (O-splines)

The Battle and Lemarié polynomial spline scaling functions and wavelets are orthogonal ([7] and [26]). These functions will be denoted by $\phi^n(x)$ and $\underline{\phi}^n(x)$, respectively. Within our present framework, imposing such an orthogonality constraint leads to the conditions

$$\langle \phi^n(x), \phi^n(x-k) \rangle = p * p' * b^{2n+1}(k) = \delta_0(k),$$

$$\langle \underline{\phi}^n(x), \underline{\phi}^n(x-k) \rangle = q * q' * \left[\tilde{b}^{2n+1} * b^{2n+1}\right]_{\downarrow 2} * b^{2n+1}(k) = \delta_0(k).$$

By solving these difference equations, we find that

$$\begin{cases} p_a = \left(b^{2n+1}\right)^{-1/2} \\ q_a = \left(\left[\tilde{b}^{2n+1} * b^{2n+1}\right]_{\downarrow 2} * b^{2n+1}\right)^{-1/2}, \end{cases} \tag{56}$$

where the notation $(b)^{-1/2}$ denotes the symmetrical square root inverse of the convolution operator b; *i.e.*, $(b)^{-1/2} * b * (b)^{-1/2} = \delta_0$. Another equivalent way of finding these weighting sequences is to search for the basis functions that are identical to their biorthogonal complement given by Equations (38) or (39). It can be verified that the corresponding filters v, $\underline{v}$, w and $\underline{w}$, given by Equations (45), (46), (49), and (50) define a quadrature mirror filter bank, that is, $v = w'$ and $\underline{v} = \underline{w}' = \tilde{v} * \delta_1$, a result that is in agreement with Mallat's general theory of orthogonal wavelet transforms.

This approach of constructing orthogonal basis functions is not the only one. Another solution is to decompose the symmetrical convolution operator b^{2n+1} into a cascade of two complementary causal and anti-causal filters

$$b^{2n+1} = b_+^{2n+1} * b_-^{2n+1}, \tag{57}$$

where $b_+^{2n+1}(k) = 0$ for $k < 0$, and $b_-^{2n+1} = (b_+^{2n+1})'$. This decomposition is best performed in the z-transform domain by determining the roots of B_1^{2n+1} and factoring this polynomial into a product of simple terms. The causal weighting sequences

$$\begin{cases} p_+ = \left(b_+^{2n+1}\right)^{-1} \\ q_+ = \left(\left[\tilde{b}_+^{2n+1} * b_+^{2n+1}\right]_{\downarrow 2} * b_+^{2n+1}\right)^{-1} \end{cases} \tag{58}$$

will generate one-sided orthogonal basis functions (ϕ_+^n) and wavelets $(\underline{\phi}_+^n)$. Note that the scaling functions ϕ_+^n are equivalent to the ρ-functions first described by Strömberg [42]. The one-sided wavelets $\underline{\phi}_+^n$, on the other hand, are new. There is also a dual representation that uses time-reversed basis functions; *i.e.*, $\phi_-^n(x) = \phi_+^n(-x)$ and $\underline{\phi}_-^n(x) = \underline{\phi}_+^n(-x)$. The advantage of this formulation is that the z- transform of the resulting quadrature mirror filter is now a ratio of two polynomials in z; the wavelet decomposition algorithm can therefore be implemented recursively using standard digital filtering techniques [33]. This representation may be better suited than the previous one for certain real time signal processing applications such as speech analysis. A potential limitation is that the corresponding filters do not have a linear phase. Their frequency responses, however, are the same as in the previous case.

3.5.2. Basic representation (B-splines)

For $p = q = \delta_0$, we have a representation that uses B-spline scaling functions and wavelets (c.f. Subsections 2.1 and 3.1). The choice of this representation tends to simplify the reconstruction process (indirect wavelet transform) since the corresponding filters are binomial or modulated binomial FIR kernels. Moreover, the decomposition filters have a simple recursive structure and can be implemented efficiently using the fast algorithms described in [49].

3.5.3. Cardinal or fundamental representation (C-splines)

Another possibility is to construct wavelets for which the coefficients are precisely the samples of the residual signal. The scaling functions $\eta^n(x)$ that satisfy this condition have already been encountered in Subsection 2.2. The corresponding interpolation conditions can be expressed as follows

$$\eta^n(k) = p * b^n(k) = \delta_0(k)$$

$$\underline{\eta}^n(k+\tfrac{1}{2}) = q * \left[\tilde{u}_2^n * \tilde{b}^{2n+1} * b^n\right]_{\downarrow 2}(k) = \delta_0(k).$$

Solving these equations, we get the cardinal weighting sequences

$$\begin{cases} p_c = (b^n)^{-1} \\ q_c = \left(\left[b^n * \tilde{u}_2^n * \tilde{b}^{2n+1}\right]_{\downarrow 2}\right)^{-1} \end{cases}. \tag{59}$$

The corresponding cardinal wavelet decomposition has the advantage of allowing a good visualization of the underlying continuous signals since the expansion coefficients are the sample values of the spline and wavelet signal components. Furthermore, the computational complexity of the reconstruction algorithm can be reduced by a factor of two since it is only necessary to compute the finer resolution coefficients that are between knot points (interpolation). It can be seen from Figure 6 that the cardinal and orthogonal cubic spline wavelet functions are similar. The essential difference, however, is that the former basis functions are precisely equal to one for $x = 1$ (center of symmetry) and vanish for all other odd-valued indices.

3.5.4. Dual representation (D-splines)

Another interesting case is to try to simplify as much as possible the structure of the analysis filters $\underline{v}$ and v given by Equations (45) and (46). This leads to the dual spline representation ([48] and [52]) with

$$\begin{cases} p_d = \left(b^{2n+1}\right)^{-1} \\ q_d = \left(\left[\tilde{b}^{2n+1} * b^{2n+1}\right]_{\downarrow 2} * b^{2n+1}\right)^{-1} \end{cases}, \tag{60}$$

which uses precisely the same filters as the B-spline wavelet transform but in reverse order. This situation corresponds to a flow graph transpose of the B-spline case. These wavelets are denoted by $\underline{\mathring{\beta}}^n(x)$ and are called dual because they are indeed the dual of the B-spline wavelets as specified by Equations (37) and (39). A different – but equivalent – definition of these functions is also given in [11].

3.5.5. The spline derivative wavelet

A last interesting case is the spline wavelet that was obtained by Chui and Wang from the $(n+1)$th derivative of a cardinal spline of order $2n+1$ [12]. The centered version of this particular wavelet, which we denote by $\chi^n(x)$, is determined as follows

$$\chi^n(x) = \frac{d^{n+1}\eta^{2n+1}(2x-1)}{dx^{n+1}} = 2^n \sum_{k=-\infty}^{+\infty} \left(b^{2n+1}\right)^{-1} * \tilde{u}_2^n(k+1)\beta^n(2x-k), \tag{61}$$

and can be written in terms of the modulated binomial kernel using (25). It is another particular instance of the generalized spline wavelet (34) with

$$q_e = 2^n \left(\left[b^{2n+1} * \tilde{b}^{2n+1} \right]_{\downarrow 2} \right)^{-1} . \tag{62}$$

This equivalence is not difficult to establish if one uses the property that $\left[\tilde{b} * b\right]_{\downarrow 2 \uparrow 2} = \tilde{b} * b$ for any sequence b.

3.5.6. Comments

The digital filters for the fast wavelet algorithm for all these cases can be obtained from Equations (45), (46), (51), and (52) by substituting the corresponding expressions for p and q. Most of these wavelets are symmetrical which also implies that the corresponding analysis and synthesis filters have linear phase. This symmetry property is especially relevant in image processing applications where it is very important to preserve spatial relationships in the transformed domain. It is also essential if one wants a good localization in time (or space).

It is also worth mentioning that one of the distinctive features of wavelets considered here is that they are constrained to be orthogonal between resolution levels while they are not necessarily orthogonal within the same level. We have seen in Subsection 3.3 that one of the implications of this requirement is that the wavelet ψ and its dual $\mathring{\psi}$ span the same residual space $W_{(0)}$. This condition is usually not implicit in general biorthogonal schemes ([14], [24], [34], and [60]); in fact, other examples of polynomial spline wavelets not satisfying this constraint have been proposed ([14] and [18]). A limitation of these approaches is that the biorthogonal wavelet $\mathring{\psi}$, which does not span the same subspace as the wavelet ψ, usually lacks regularity. On the other hand, both the analysis and synthesis filters can have a finite impulse response which may have some practical advantages [14].

§4. Signal and image processing applications

4.1. Filter bank interpretation

In Subsection 2.3, we have presented a filtering interpretation of polynomial spline approximation and provided the connection with classical sampling theory. The same type of interpretation can also be made for the wavelet decomposition [60]. For instance, the wavelet coefficients in (29) can be obtained, at least conceptually, by sampling the output of a multiband analysis filter bank. Each branch corresponds to a particular resolution level (i); it is a bandpass filter with an impulse response $2^{-i/2}\mathring{\varphi}^n\left(-2^{-i}x\right)$. The bandwidth of these filters is inversely proportional to 2^i but the quality factor Q remains constant for all resolutions. The parameter Q classically refers to the ratio of the time and frequency resolutions of a filter [33]. The multiband characteristics of the analysis filter bank for the dual and cardinal cubic spline wavelet transforms are shown in Figure 7. In the first case, the impulse responses are precisely the B-spline wavelets which explains why the frequency responses are nearly Gaussian (see [51]). In the second case (cardinal representation), the spectral characteristics are closer to that of an ideal bandpass filter bank. In fact, it can be proven that the corresponding analysis and synthesis wavelet filters both tend to ideal bandpass filters as the order of the spline goes to infinity ([4] and [48]).

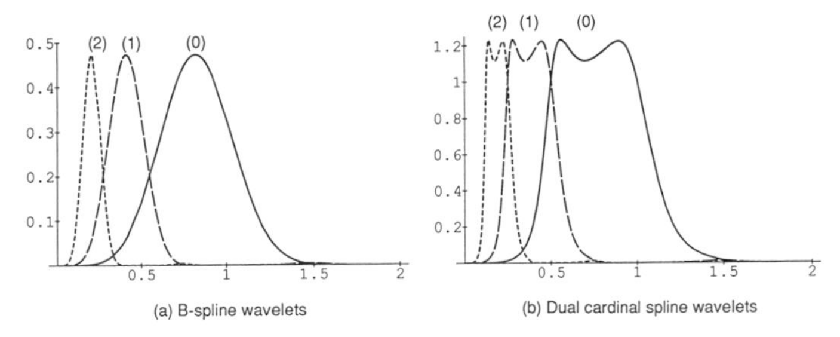

Figure 7. Multiband characteristics of the analysis wavelets for the dual and cardinal cubic spline wavelet expansions.

Similarly, we can define a complementary reconstruction filter bank with comparable properties using the wavelet functions $2^{-i/2}\varphi^n\left(2^{-i}x\right), i \in \mathbb{Z}$. The fact that the wavelet transform is reversible is also expressed as a general condition on the frequency responses of these filters

$$\sum_{i=-\infty}^{+\infty} \Psi(2^i f)\mathring{\Psi}(2^i f) = 1, \tag{63}$$

where $\Psi(f)$ and $\mathring{\Psi}(f)$ denote the frequency responses of $\psi(x)$ and $\mathring{\psi}(x)$, respectively. This relation, which is also valid in the general context of biorthogonal wavelet transforms, means that the whole system acts like an all-pass filter.

4.2. Implementation

When it comes to computing these wavelet transforms, the most efficient approach is certainly the fast wavelet algorithm described in Subsection 3.4. Although the filter bank implementation appears to be relatively straightforward, there are several practical issues that need to be dealt with. These are briefly discussed below.

4.2.1. Choice of the finer resolution signal model

In signal and image processing applications, our initial data representation is a sequence (or array) of sample values $g(k)$. A consistent way of choosing the initial signal model is to determine the polynomial spline that interpolates $g(k)$. By convention, the initial step size is assumed to be one. Based on the results in Subsection 2.2, the initial spline coefficients $c_{(0)}$ at level (0) are obtained by convolution (prefilter) using Equation (16). The advantage of using such an initialization procedure is that $g_{(0)}$ is uniquely defined and is independent of our choice of basis functions. Likewise, we can apply the inverse procedure (postfilter) to reconstruct the digital signal from its finer resolution spline coefficients. Clearly, the cardinal representation with $p = (b^n)^{-1}$ avoids the use of such filters.

4.2.2. Specification of boundary conditions

Since the signals or sequences encountered in practice have a finite extent, we have to introduce some boundary conditions. For practical convenience and to avoid discontinuities, the signals are usually extended on both sides by using their mirror image, a standard practice in image processing. A consistent implementation of these boundary conditions across scale is important to guarantee that the wavelet transform is fully reversible. This issue is further discussed in [48, Appendix C].

4.2.3. Digital filter implementation

The digital filters associated with the various wavelets transforms can be determined from Equations (45), (46), (51), and (52). Explicit formulas for the corresponding impulse and frequency responses are given in [48] (Tables 3 and 4). When the operators are FIR (*e.g.*, B-spline reconstruction and dual analysis filters), the implementation is straightforward. For all other infinite impulse response (IIR) filters, there are basically two implementation strategies.

The first is to use a truncated FIR approximation. This technique has the advantage of simplicity but introduces approximation errors. The simplest design method is to evaluate the frequency response of a given filter and to perform an inverse FFT to obtain the coefficients of the impulse response. The impulse response is then truncated to an appropriate length to satisfy a prescribed tolerance error. We note that this technique is the only one applicable for the implementation of the symmetric orthogonal transformation.

The second approach is a recursive implementation which has the advantage of providing an exact algorithm. To illustrate this principle, we consider the evaluation of the wavelet coefficients in the basic representation

$$\sqrt{2}d_{(i+1)}(k) = \left[\left[\left(b^{2n+1}\right)^{-1}\right]_{\downarrow 2} * \tilde{u}_2^n * c_{(i)}\right]_{\downarrow 2}(k) = \left(b^{2n+1}\right)^{-1} * \left[\tilde{u}_2^n * c_{(i)}\right]_{\downarrow 2}(k),$$

which can be evaluated in two steps. The first is to prefilter with the FIR kernel $\tilde{u}_2^n$, while down-sampling by a factor of two. The second is to apply the postfilter $(b^{2n+1})^{-1}$ which corresponds to a direct B-spline transform of order $2n+1$ [49]. This filter can be implemented very efficiently by decomposing it into a cascade of first order causal and anti-causal recursive filters (similar to Equation (57)) with a total complexity of approximately $2n$ additions and $2n$ multiplications per sample point. A similar procedure is applicable for the implementation of all other IIR filters in the basic, cardinal, and dual representations.

4.3. Two dimensional extensions

All techniques discussed so far can be readily extended to higher dimensions through the use of tensor product splines [40]. The corresponding basis functions and filters are separable, implying that higher dimensional problems can be solved by successive one-dimensional processing along the various dimensions of the data. The application of this principle for polynomial spline interpolation and approximation is straightforward. Similarly, higher dimensional wavelet transforms can be computed by successive one-dimensional transformations. For two dimensional images, there are in fact four types of basis functions corresponding to the different cross-products. A theoretical justification for this procedure is given in [28].

Most of the interpolation and approximation techniques in Section 2 are also applicable with some modifications for multivariate non-tensor product splines, but the computations tend to be more involved. Chui *et al.* describe a general procedure for the construction of compactly supported box-spline wavelets [10].

4.4. Image processing examples

An example of application of the polynomial spline approximation techniques discussed in Section 2.3 is the generation of polynomial spline pyramids [50]. Image pyramids are data structures for representing image copies at multiple resolutions [35]; their use in image processing preceded the development of the wavelet transform. These representations are particularly useful for improving the efficiency of many image processing tasks such as edge detection [43], object recognition, and image segmentation ([22] and [35]). Their main advantage is to provide a simple mechanism for adjusting the pixel size to optimize the performance of a given image processing algorithm. Furthermore, the use of a coarse-to-fine strategy can dramatically improve the execution speed and convergence properties of iterative algorithms which proceed by successive refinement. Such pyramids have also been widely used for image compression, especially in progressive transmission schemes ([9], [44], and [45]).

Image pyramids are usually generated by repeated application of a REDUCE operation (lowpass filtering and down-sampling by a factor of two). A complementary EXPAND operation (up-sampling by a factor of two and lowpass filtering) provides the mapping of a coarser representation onto the next finer level. The filters used for both of these operations are usually identical and typically Gaussian ([8] and [9]). This procedure is in many ways analogous to what happens in the lowpass branch of the fast wavelet transform algorithm described in Subsection 3.4 (c.f. Figure 5). The main difference, however, is that the conventional pyramid filters are usually not designed to minimize the loss of information from one level to the next. The multiresolution approximation procedure described in Subsections 2.4 and 3.1, on the other hand, is optimum in the sense that it minimizes the error measured in the L_2-norm. To illustrate the effect that this property may have on performance, we present a comparison between the commonly used Gaussian pyramid [8] and the cubic spline pyramid obtained from our multiresolution analysis.

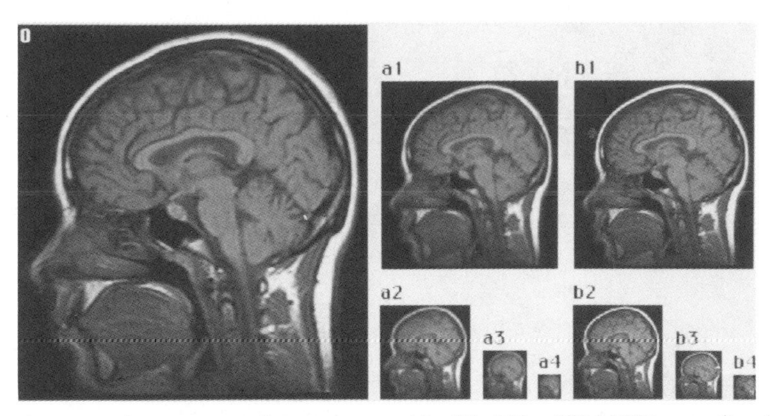

Figure 8. Comparison of image pyramids. (0): 238 × 253 MRI image (level 0), (a1-a4): levels 1 to 4 of the Gaussian pyramid with $a = 3/8$, (b1-b4): levels 1 to 4 of the cardinal cubic spline pyramid.

For our experiments, we have chosen the cardinal spline representation which provides a precise rendition of the underlying continuous functions in terms of their samples. Its use also avoids the need for an initial prefilter, otherwise necessary for a consistent polynomial spline representation (c.f. Subsection 4.2). The corresponding decimation filter, which is successively applied along the rows and columns of the image, is given by Equation (45) with $p = (b^3)^{-1}$ and $n = 3$. The Gaussian and cubic spline pyramids for a typical magnetic resonance imaging (MRI) slice are shown in Figure 8. Note that the sharpness of the cubic spline pyramid is preserved at all resolution levels while the corresponding images in the Gaussian pyramid seem increasingly blurred by comparison.

A visual display of the loss of information that occurs during resolution conversion is provided by the difference (or Laplacian) pyramid; it consists of

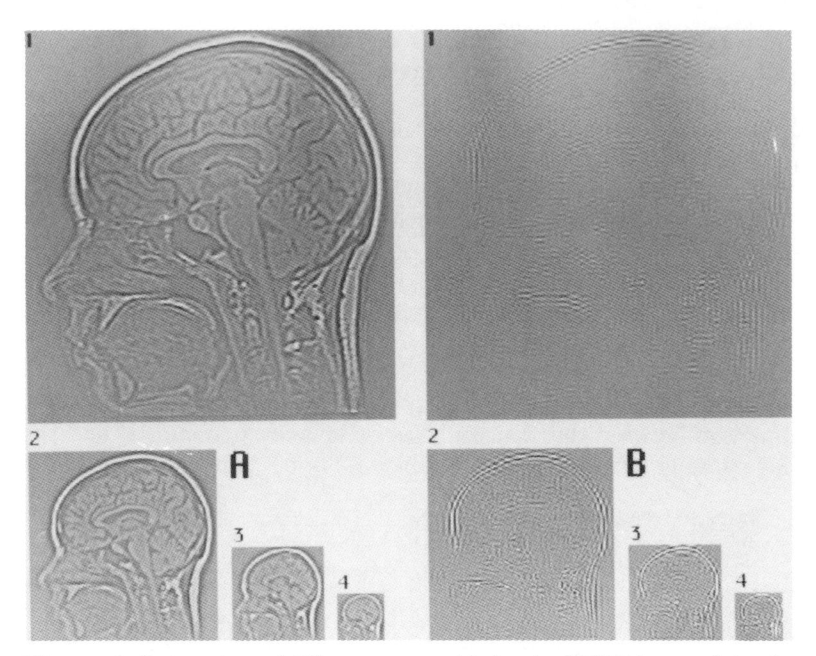

Figure 9. Comparison of difference pyramids for the "MRI" image. (a1-a4): first four levels of the basic Laplacian pyramid, (b1-b4): first four difference images for the cardinal cubic spline pyramid.

the sequence of difference images between two consecutive levels. In the case of the cubic spline pyramid, the coarser resolution level is interpolated to the finer grid using a cubic spline interpolator with an expansion factor of two. The corresponding filter is given by Equation (51) with $n = 3$ and $p = (b^3)^{-1}$. Figure 9 provides a comparison between this data structure for $n = 3$ and Burt's Laplacian pyramid (LP) with a parameter value of $a = \frac{3}{8}$ [9]. The same intensity scaling factors were applied to all images to facilitate the comparison. For LP, the amount of information displayed at each resolution level is quite significant and the main subject is still recognizable. In the case of the cubic spline pyramid, the energy of the difference is reduced drastically and only very high frequency details are visible. At resolution level (1), there is an improvement by as much as 8dB. The polynomial spline pyramid should therefore provide an attractive alternative to standard multiresolution techniques. Our preliminary results indicate that this type of approach allows improved image coding according to the lossy scheme developed by Burt and Adelson ([3] and [47]). There are also many other potential applications including feature extraction ([8] and [23]), image segmentation [55], and edge detection [29].

The residual information displayed at the various levels of the difference pyramid in Figure 9 is precisely what is coded in a more compact way by the wavelet coefficients. The cubic C-spline wavelet transform with a depth of two

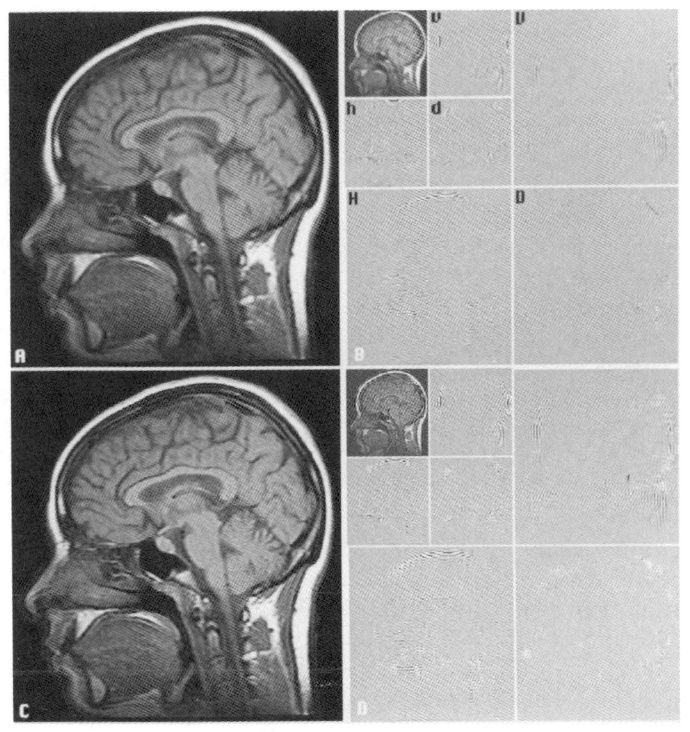

Figure 10. (a) original "MRI" image; (b) cardinal cubic spline wavelet transform; (c) cubic B-spline coefficients ; (d) cubic B-spline wavelet transform.

is shown in Figure 10b; each wavelet sub-image had its gray scale linearly expanded for maximum contrast display. This example illustrates the fact that the wavelet components in the upper right (V), lower left (H), and lower right (D) quadrants tend to amplify high resolution vertical, horizontal and diagonal edges, respectively. The same quantitative behavior also applies for the mid-range frequency components (v, h, and d). The lower resolution image in the upper left corner is equivalent to the level (2) image in the cubic spline pyramid (Figure 9-b2). The finer (resp., mid range) resolution channels V, H, and D (resp., v, h, and d) provide a compressed – but equivalent – representation of the difference image in Figure 9-b1 (resp., Fig. 9-b2). It is precisely this last property that makes the wavelet transform particularly attractive for image coding ([5], [18], and [28]). It should be noted that this type of approach falls into the more general class of subband coding techniques that have been developed over the last decade in signal processing ([1], [58], and [61]). In principle, one should expect the wavelet algorithms to give better compression rates than predictive pyramidal approaches of the type described by Burt and Adelson or some least squares extension ([9] and [47]). However, pyramid algorithms are slightly more robust to quantization errors because of the intrinsic redundancy of the representation. Moreover, they are better suited for progressive data transmission in the sense that the quantization errors introduced at coarser resolutions can still be corrected at a finer level. This is obviously not possible for the wavelet transform because all scales are decoupled by construction. The choice between either one of these techniques is therefore not a clear cut issue and strongly depends on the type of application.

Our last example is an illustration of the cubic B-spline wavelet transform. The first step is to compute the B-spline coefficients at level (0) which are displayed in Figure 10c. The corresponding prefilter $(b^3)^{-1}$ is implemented recursively using the procedure described in [49]. Note that this process is a special form of image sharpening. The coefficients of the B-spline wavelet transforms with a depth of two are displayed in Figure 10d with approximately the same constrast as in Figure 10b. The B-spline wavelet channels appear to be visually sharper than their cardinal counterparts. This representation is fully reversible and related to the one displayed in Figure 10b by a simple change of coordinate system (reversible digital filter).

An essential characteristic of the B-spline wavelet transform is the excellent time-frequency (or space-frequency) localization of its basis functions. This representation is in many respects similar to a hierarchical or wavelet-like Gabor transform ([16] and [20]). It should therefore be useful for the time-frequency or time-scale analysis of non-stationary signals [19]. Another related application is the segmentation of textured images which could be based on a multiresolution extension of the method described in [56].

§5. Conclusion

The main purpose of this chapter has been to present an overview of polynomial spline approximation techniques and wavelet transforms in a signal processing context. In particular, we have emphasized the link between these recent techniques and classical signal representation theories (Shannon's Sampling Theory and the Gabor transform). We have also shown that the wavelet and pyramid techniques are closely related. In fact, they can be linked together through the concept of a multiresolution analysis first formalized by Mallat. Most of the algorithms that have been described are computationally very efficient and also quite appropriate for signal processing. These properties should make polynomial splines useful in a variety of applications including data acquisition and display, multiresolution image processing, non-stationary signal analysis, and data compression.

A feature that makes polynomial splines stand apart from other representations is that they provide a simple explicit mapping between the discrete and continuous signal domains. This correspondence may be useful for developing new processing algorithms for problems that are better formulated in a continuous framework (*e.g.*, differentiation, edge detection, etc...) (see [54]).

Most of the principles presented here are quite general and not restricted to polynomial splines. In this respect, we should cite some recent work of

Chui and Wang, which presents a general framework for compactly supported wavelets [13]. Aldroubi *et al.* describe general families of wavelet transforms with a regularity index n that share many of the features of polynomial spline wavelets including their convergence properties as n goes to infinity [2].

References

1. Adelson, E. H., E. Simoncelli, and R. Hingorani, Orthogonal pyramid transforms for image coding, in *Proc. SPIE Conf. Visual Communication and Image Processing*, Cambridge, MA, 1987, 50–58.
2. Aldroubi, A. and M. Unser, Families of wavelet transforms in connection with Shannon sampling theory and the Gabor transform, in *Wavelets– A Tutorial in Theory and Applications*, C. K. Chui (ed.), Academic Press, Boston, 1992.
3. Aldroubi, A., M. Unser, and M. Eden, Asymptotic properties of least squares spline filters and application to multi-scale decomposition of signals, in *Proc. International Symposium on Information Theory and its Applications*, Waikiki, Hawaii, 1990, 271–274.
4. Aldroubi, A., M. Unser, and M. Eden, Cardinal spline filters : stability and convergence to the ideal sinc interpolator, NCRR Report 20/90, National Institutes of Health, 1990.
5. Antonini, M., M. Barlaud, P. Mathieu, and I. Daubechies, Image coding using vector quantization in the wavelet domain, in *Proc. Int. Conf. Acoust., Speech and Signal Processing*, Albuquerque, NM, 1990, 2297–2300.
6. Bastiaans, M. J., Gabor's expansion of a signal into Gaussian elementary signals, *Proc. IEEE* **68** (1980), 538–539.
7. Battle, G., A block spin construction of ondelettes. Part I: Lemarié functions, *Comm. Math. Phys.* **110** (1987), 601–615.
8. Burt, P. J., Fast algorithms for estimating local image properties, *Comput. Graphics and Image Processing* **21** (1983), 368–382.
9. Burt, P. J. and E. H. Adelson, The Laplacian pyramid as a compact code, *IEEE Trans. Comm.* **31** (1983), 337–345.
10. Chui, C. K., J. Stöckler, and J. D. Ward, Compactly supported box-spline wavelets, CAT Report #230, Texas A&M University, 1990.
11. Chui, C. K. and J. Z. Wang, On compactly supported spline wavelets and a duality principle, *Trans. Amer. Math. Soc.*, 1991, to appear.
12. Chui, C. K. and J. Z. Wang, A cardinal spline approach to wavelets, *Proc. Amer. Math. Soc.*, 1991, to appear.
13. Chui, C. K. and J. Z. Wang, A general framework of compactly supported splines and wavelets, CAT Report #219, Texas A&M University, 1990.
14. Cohen, A., I. Daubechies, and J. C. Feauveau, Biorthogonal bases of compactly supported wavelets, *Comm. Pure and Appl. Math.*, 1991, to appear.
15. Daubechies, I., The wavelet transform, time-frequency localization and signal analysis, *IEEE Trans. Inform. Theory* **36** (1990), 961–1005.

16. Daugman, J. G., Complete discrete 2-D Gabor transforms by neural networks for image analysis and compression, *IEEE ASSP* **36** (1988), 1169–1179.
17. de Boor, C., *A Practical Guide to Splines*, Springer-Verlag, New York, 1978.
18. Feauveau, J. C., P. Mathieu, M. Barlaud, and M. Antonini, Recursive biorthogonal wavelet transform for image coding, in *Proc. Int. Conf. Acoust., Speech and Signal Processing*, Toronto, Canada, 1991, 2649–2652.
19. Flandrin, P., Some aspects of non-stationary signal processing with emphasis on time-frequency and time-scale methods, in *Wavelets: Time-frequency Methods and Phase Space*, J. M. Combes, A. Grossmann and P. Tchamitchian (eds.), Springer-Verlag, New York, 1989, 68–98.
20. Gabor, D., Theory of communication, *J. IEE (London)* **93** (1946), 429–457.
21. Haar, A., Zur Theorie der orthogonalen funktionensysteme, *Math. Ann.* **69** (1910), 331–371.
22. Harlow, C. A. and S. A. Eisenbeis, The analysis of radiographic images, *IEEE Trans. Computers* **22** (1973), 678–688.
23. Hashimoto, M. and J. Sklansky, Multiple-order derivatives for detecting local image characteristics, *Comput. Vision, Graph., Image Processing* **39** (1987), 28–55.
24. Herley, C. and M. Vetterli, Linear phase wavelets: theory and design, in *Proc. Int. Conf. Acoust., Speech and Signal Processing*, Toronto, Canada, 1991.
25. Jerri, A. J., The Shannon sampling theorem-its various extensions and applications: A tutorial review, *Proc. IEEE* **65** (1977), 1565–1596.
26. Lemarié, P. G., Ondelettes à localisation exponentielles, *J. Math. Pures et Appl.* **67** (1988), 227–236.
27. Mallat, S. G., Multiresolution approximations and wavelet orthogonal bases of $L^2(R)$, *Trans. Amer. Math. Soc.* **315** (1989), 69–87.
28. Mallat, S. G., A theory of multiresolution signal decomposition: the wavelet representation, *IEEE Pattern Anal. Machine Intell.* **11** (1989), 674–693.
29. Marr, D. and E. Hildreth, Theory of edge detection, *Proc. Roy. Soc. London* **B 207** (1980), 187–217.
30. Marsden, M. J., F. B. Richards, and S. D. Riemenschneider, Cardinal spline interpolation operators on l^p data, *Indiana Univ. Math. J.* **24** (1975), 677–689.
31. Meyer, Y., *Ondelettes et Opérateurs I: Ondelettes*, Hermann, Paris, France, 1990.
32. Prenter, P. M., *Splines and Variational Methods*, Wiley, New York, 1975.
33. Proakis, J. G. and D. G. Manolakis, *Introduction to Digital Signal Processing*, Macmillan, New York, 1990.
34. Rioul, O., A discrete-time multiresolution theory unifying octave band filter banks, Pyramid and wavelet transforms, *IEEE Trans. Signal Proc-*

cessing, to appear.

35. Rosenfeld, A., *Multiresolution Image Processing*, Springer-Verlag, New, York, 1984.
36. Schoenberg, I. J., Contribution to the problem of approximation of equidistant data by analytic functions, *Quart. Appl. Math.* **4** (1946), 45-99, 112–141.
37. Schoenberg, I. J., Cardinal interpolation and spline functions, *J. Approx. Theory* **2** (1969), 167–206.
38. Schoenberg, I. J., *Cardinal Spline Interpolation*, CBMS-NSF series in Applied Math. # 12, SIAM Publ., Philadelphia, PA, 1973.
39. Schoenberg, I. J., Notes on spline functions III: on the convergence of the interpolating cardinal splines as their degree tends to infinity, *Israel J. Math.* **16** (1973), 87–92.
40. Schumaker, L. L., *Spline Functions : Basic Theory*, Wiley, New York, 1981.
41. Shannon, C. E., Communication on the presence of noise, *Proc. I.R.E.* **37** (1949), 10–21.
42. Strömberg, J. O., A modified Franklin system and higher-order spline system of R^n as unconditional bases for Hardy spaces, in *Proc. Conf. in Harmonic Analysis in honor of Antoni Zygmund*, Vol. II, W. Beckner *et al.* (ed.), Wadsworth Math. Series, 1983, 475–493.
43. Tanimoto, S. and T. Pavlidis, A hierarchical data structure for picture processing, *Comput. Graphics and Image Processing* **4** (1975), 104–119.
44. Tanimoto, S. L., Image transmission with gross information first,*Comput. Graphics and Image Processing* **9** (1979), 72–76.
45. Tzou, K. H., Progressive image transmission: a review and comparison of techniques, *Optical Eng.* **26** (1987), 581–589.
46. Unser, M., Recursive filters for fast B-spline interpolation and compression of digital images, *Proc. SPIE* **1232**, *Medical Imaging IV: Image Capture and Display*, 1990, 337-347.
47. Unser, M., An improved least squares Laplacian pyramid for image compression, *Signal Processing*, to appear.
48. Unser, M., A. Aldroubi, and M. Eden, A family of polynomial spline wavelet transforms, NCRR Report 153/90, National Institutes of Health, 1990.
49. Unser, M., A. Aldroubi, and M. Eden, Fast B-spline transforms for continuous image representation and interpolation, *IEEE Pattern Anal. Machine Intell.* **13** (1991), 277–285.
50. Unser, M., A. Aldroubi, and M. Eden, The L_2 polynomial spline pyramid : a discrete multiresolution representation of continuous signals, *IEEE Pattern Anal. Mach. Intell.*, to appear.
51. Unser, M., A. Aldroubi, and M. Eden, On the asymptotic convergence of B-spline wavelets to Gabor functions, *IEEE Trans. Inform. Theory*, (special issue on Wavelet transform), to appear.
52. Unser, M., A. Aldroubi, and M. Eden, Polynomial spline signal approximations : filter design and asymptotic equivalence with Shannon's sampling

theorem, *IEEE Trans. Inform. Theory*, to appear.

53. Unser, M., A. Aldroubi, and M. Eden, B-spline signal processing. Part I : theory, *IEEE Trans. Signal Processing*, to appear.
54. Unser, M., A. Aldroubi, and M. Eden, B-spline signal processing. Part II : efficient design and applications, *IEEE Trans. Signal Processing*, to appear.
55. Unser, M. and M. Eden, Multiresolution feature extraction and selection for texture segmentation, *IEEE Pattern Anal. Mach. Intell.* **11** (1989), 717–728.
56. Unser, M. and M. Eden, Non-linear operators for improving texture segmentation based on features extracted by spatial filtering, *IEEE Trans. Syst. Man Cybern.* **20** (1990), 804–815.
57. Vaidyanathan, P. P., Quadrature mirror filter banks, M-band extensions and perfect-reconstruction technique, *IEEE ASSP Mag.* **4** (1987), 4–20.
58. Vetterli, M., Multi-dimensional sub-band coding: some theory and algorithms, *Signal Processing* **6** (1984), 97–112.
59. Vetterli, M., A theory of multirate filter banks, *IEEE ASSP* **35** (1987), 356–372.
60. Vetterli, M. and C. Herley, Wavelets and filter banks: theory and design, *IEEE ASSP*, 1992, to appear.
61. Woods, J. W. and S. D. O'Neil, Sub-band coding of images, *IEEE ASSP* **34** (1986), 1278–1288.

Michael Unser
Mathematics and Image Processing Group/BEIP
National Center for Research Resources
Building 13, Room 3W13
National Institutes of Health
Bethesda, MD 20892
unser@helix.nih.gov

Akram Aldroubi
Mathematics and Image Processing Group/BEIP
National Center for Research Resources
Building 13, Room 3W13
National Institutes of Health
Bethesda, MD 20892
aldroubi%aamac.dnet@dxi.nih.gov

Biorthogonal Wavelets

Albert Cohen

Abstract. In this chapter, we study the construction of biorthogonal bases of wavelets which generalize the orthonormal bases and have interesting properties in signal processing. We describe the class of subband coding schemes associated with these wavelets and we give necessary and sufficient conditions for frame bounds which ensure the stability of the decomposition-reconstruction algorithm. We finally present the example of compactly supported spline wavelets which can be generated by this approach. The results presented in this chapter are mainly joint work with I. Daubechies and J. C. Feauveau.

§1. Introduction

In recent years, orthonormal wavelet bases have revealed to be a powerful tool in applied mathematics and digital signal processing. The possibility of data compression offered by a multiscale decomposition leads to some very good results in speech [8] or image [1] coding or fast numerical analysis of operators [2].

One of the main reasons for this success is the existence of a Fast Wavelet Transform algorithm (FWT) which only requires a number of operations proportional to the size of the initial discrete data. This algorithm relates the orthonormal wavelet bases with more classic tools of digital signal processing such as subband coding schemes and discrete filters.

We can describe in four steps the connections between these different domains:

a) Wavelets bases are usually defined from the data of a multiresolution analysis; *i.e.*, a ladder of approximation subspaces of $L^2(\mathbb{R})$

$$\{0\} \to \cdots V_1 \subset V_0 \subset V_{-1} \cdots \to L^2(\mathbb{R}) \tag{1}$$

which satisfy the following properties

$$f(x) \in V_j \Leftrightarrow f(2x) \in V_{j-1} \Leftrightarrow f(2^j x) \in V_0. \tag{2}$$

Wavelets–A Tutorial in Theory and Applications
C. K. Chui (ed.), pp. 123–152.

ISBN 0-12-174590-2

There exists a *scaling function* $\phi(x)$ in V_0 such that

$$\{\phi_k^j\}_{k\in\mathbb{Z}} = \{2^{-j/2}\phi(2^{-j}x - k)\}_{k\in\mathbb{Z}} \tag{3}$$

is an orthonormal basis for V_j. The function ϕ has to satisfy a two-scale difference equation which expresses the embedded structure of the V_j-spaces

$$\phi(x) = 2\sum_{n\in\mathbb{Z}} h_n\phi(2x - n). \tag{4}$$

The wavelet ψ is then defined by

$$\psi(x) = 2\sum_{n\in\mathbb{Z}}(-1)^n h_{1-n}\phi(2x - n) = 2\sum_{n\in\mathbb{Z}} g_n\phi(2x - n) \tag{5}$$

and its integer translates $\{\psi(x-k)\}_{k\in\mathbb{Z}}$ form an orthonormal basis for the orthogonal complement W_0 of V_0 in V_{-1}. The functions

$$\{\psi_k^j\}_{k\in\mathbb{Z}} = \{2^{-j/2}\psi(2^{-j}x - k)\}_{k\in\mathbb{Z}},$$

thus, characterize the additional details between two levels of approximation (V_j and V_{j-1}). By a *telescoping argument* using the ladder structure (1), the whole set $\{\psi_k^j\}_{j,k\in\mathbb{Z}}$ is an orthonormal basis of $L^2(\mathbb{R})$. More details on multiresolution analysis can be found in [14] and [15].

b) In the Fourier domain, Equations (4) and (5) can be rewritten as

$$\hat{\phi}(2\omega) = m_0(\omega)\hat{\phi}(\omega) \text{ with } m_0(\omega) = \sum_{n\in\mathbb{Z}} h_n e^{-in\omega} \tag{6}$$

$$\widehat{\psi}(2\omega) = m_1(\omega)\hat{\phi}(\omega) = e^{-i\omega}\overline{m_0(\omega+\pi)}\hat{\phi}(\omega) \tag{7}$$

where $m_0(\omega)$ is a 2π periodic function that satisfies the following two properties (due to the multiresolution analysis axioms)

$$|m_0(\omega)|^2 + |m_0(\omega+\pi)|^2 = 1; \tag{8}$$

$$m_0(0) = 1 \quad \text{and} \quad m_0(\pi) = 0. \tag{9}$$

Here, m_0 and m_1 are the transfer functions of a pair of low-pass and high-pass filters known in signal processing as Conjugate Quadrature Filters (CQF, see [16]). These discrete filters are the key of the FWT algorithm: To analyze a discrete signal s_n, one identifies it with the coordinates of a function in V_0; *i.e.*, $f_0 = \sum_{n\in\mathbb{Z}} s_n\phi(x-n)$. The coordinates of the signal in V_1 (resp. W_1) are then obtained by applying the discrete filter $\overline{m_0(\omega)}$ (resp. $\overline{m_1(\omega)}$) followed by a decimation of one sample out of two to keep the same total amount of information. It is then possible to iterate this decomposition process on the coarser approximation in the following way: $V_1 \to V_2 \overset{\perp}{\oplus} W_2$, $V_2 \to V_2 \overset{\perp}{\oplus} W_3, \ldots$.

The reconstruction stage consists of refining the decimated sequence of approximation (resp. detail) coefficients by using $m_0(\omega)$ (resp. $m_1(\omega)$) as interpolating filters and adding these two components multiplied by two to get the finer approximation. This sequence of operations — filtering, downsampling, interpolation, reconstruction — is known in signal processing as a (two-channel) subband coding scheme, as illustrated in Figure 1.

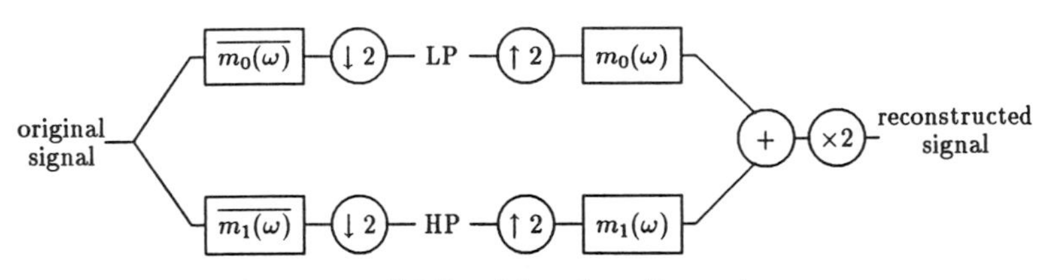

Figure 1. CQF subband coding scheme.

$\downarrow 2$: removes one sample out of two

$\uparrow 2$: insert a zero between each two samples.

c) Apparently the functions ϕ and ψ do not play any role in this algorithm involving only the CQF pair. However, in many application, it is interesting to have these filters associated with a smooth wavelet and scaling function. Indeed, iterating (6) at each scale leads to

$$\hat{\phi}(\omega) = \prod_{k=1}^{+\infty} m_0(2^{-k}\omega). \tag{10}$$

This formula corresponds to an infinite number of iterations of the refinement process used in the reconstruction algorithm starting from a single low scale approximation coefficient. In the time domain, the limit of this process is the scaling function ϕ and if one starts the reconstruction on a detail coefficient, the limit is the wavelet ψ.

The smoothness of ϕ and ψ will thus appear in the aspect of the low scale components which play an important role in data compression since many high scale details are thrown away.

d) In practice, the starting point to a multiscale analysis is a 2π periodic function $m_0(\omega)$ which satisfies Equations (8) and (9).

The scaling function is then defined by the infinite product (10). It has been shown in [7] and [12] that for a generic choice of $m_0(\omega)$, the scaling function is in $L^2(\mathbb{R})$ and satisfies

$$\langle \phi(x) | \phi(x-k) \rangle = \delta_{0,k}. \tag{11}$$

The wavelet is then derived from Equation (7). A particularly interesting class of CQF is the set of trigonometric polynomials $m_0(\omega)$ satisfying Equations (8) and (9) since they correspond to finite impulse response (FIR) filters. They lead to compactly supported wavelets which have been constructed by I. Daubechies in [10] with the possibility of arbitrarily high regularity for ϕ and ψ by choosing $m_0(\omega)$ in a smart way. ■

Unfortunately the CQF present some serious disadvantages for signal processing:

(1) They cannot be both FIR and linear phase (*i.e.*, with real and symmetrical coefficients) except the Haar filter which has no great interest since the associated wavelet $\psi = \psi_{[0,\frac{1}{2}]} - \chi_{[0,1]}$ is not continuous.
(2) Since they are the solution of the quadratic Equation (8), their coefficients are usually algebraic numbers with no simple expressions.
(3) Their design uses the Féjer-Riesz lemma (for FIR filters): $|m_0|^2$ is constructed at first and m_0 is derived using this lemma. However, this technique does not generalize to the multidimensional case.
(4) In the case of FIR filters, the subspaces V_j have no simple and direct definition other than Span $\{\phi_k^j\}_{k\in\mathbb{Z}}$. For example, they cannot be composed of spline functions except for the Haar case.

For all these reasons, these filters are often rejected by the engineers for some specific applications. However, these disadvantages are not related to the structure of the subband coding scheme itself and they can be removed by using a more general class of filters. More precisely, we shall allow the decomposition and the reconstruction filters to be different. The result is a pair of *dual filters* $\{m_0, \widetilde{m}_0\}$ which have to satisfy

$$\overline{m_0(\omega)}\widetilde{m}_0(\omega) + \overline{m_0(\omega+\pi)}\widetilde{m}_0(\omega+\pi) = 1. \tag{12}$$

These filters have been introduced in signal processing by M. Vetterli (see [17]).

Is it possible to mimic, in this more general setting, the construction of orthonormal wavelets from discrete filter that we describe previously? The answer is yes, but the orthonormality has been lost and the result is a pair of biorthogonal wavelet bases $\{\psi_k^j, \widetilde{\psi}_k^j\}_{j,k\in\mathbb{Z}}$ which allow the following decomposition of any function f in $L^2(\mathbb{R})$

$$\hat{f} = \sum_{j,k\in\mathbb{Z}} \langle f|\widetilde{\psi}_k^j\rangle \psi_k^j = \sum_{j,k\in\mathbb{Z}} \langle f|\psi_k^j\rangle \widetilde{\psi}_k^j. \tag{13}$$

In the next section of this chapter, we shall introduce the class of dual filters and their relation to biorthogonal wavelets.

In the third section, we shall discuss the additional conditions that must be filled by the filters to obtain biorthogonal wavelets bases. An important problem that does not occur in the orthonormal case is the frame bounds which relate the L^2 norm of a function and the ℓ^2 norm of its coordinates in the expansion of Equation (13). These bounds are indeed crucial for the stability of the decomposition-reconstruction algorithm. Two different strategies will be presented to check that these new wavelets form stable (or unconditional, or Riesz) bases.

Finally, we shall show in the last section that it is possible to build a (nonorthonormal) wavelet basis generated by a compactly supported spline function. Our approach is different from the technique developed by C. K. Chui in [3].

§2. Dual filters and dual wavelets

2.1. Dual filters

Let us consider, in the most general sense, the subband coding scheme described in Figure 2. The decomposition is performed by the pair $\{\overline{\tilde{m}_0}, \overline{\tilde{m}_1}\}$, whereas $\{m_0, m_1\}$ are used for the reconstruction. A discrete signal s_n can be represented by its discrete Fourier transform; *i.e.*,

$$s(\omega) = \sum_{n \in \mathbb{Z}} s_n e^{-in\omega}. \tag{14}$$

The decomposition stage transforms s_n into an approximation sequence a_n and a detail sequence d_n defined by

$$a(2\omega) = \frac{1}{2}(\overline{\tilde{m}_0(\omega)}s(\omega) + \overline{\tilde{m}_0(\omega+\pi)}s(\omega+\pi)), \tag{15}$$

and

$$d(2\omega) = \frac{1}{2}(\overline{\tilde{m}_1(\omega)}s(\omega) + \overline{\tilde{m}_1(\omega+\pi)}s(\omega+\pi)). \tag{16}$$

And thus, the reconstructed signal r_n can be written

$$r(\omega) = \alpha(\omega)s(\omega) + \beta(\omega)s(\omega+\pi) \tag{17}$$

with

$$\alpha(\omega) = m_0(\omega)\overline{\tilde{m}_0(\omega)} + m_1(\omega)\overline{\tilde{m}_1(\omega)}, \tag{18}$$

and

$$\beta(\omega) = m_0(\omega)\overline{\tilde{m}_0(\omega+\pi)} + m_1(\omega)\overline{\tilde{m}_1(\omega+\pi)}. \tag{19}$$

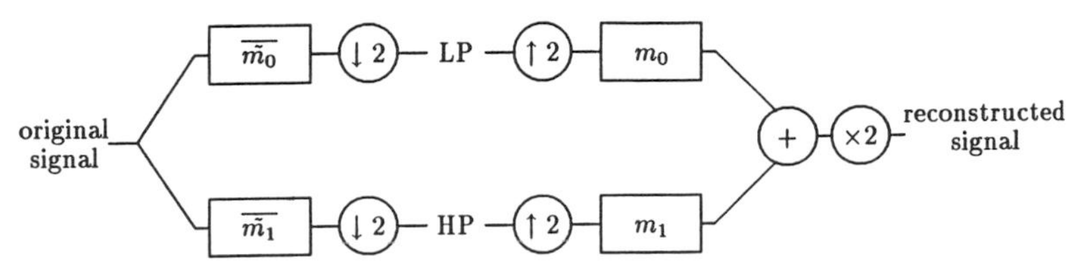

Figure 2. A general 2 channels subband coding scheme.

Perfect reconstruction is achieved for any signal if and only if $\alpha(\omega) = 1$ and $\beta(\omega) = 0$ for all ω in $[-\pi, \pi]$. This leads to the following system in which m_0 and m_1 can be considered as solutions and $\widetilde{m}_0$ and $\widetilde{m}_1$ as parameters namely:

$$\begin{cases} m_0(\omega)\overline{\widetilde{m}_0(\omega)} + m_1(\omega)\overline{\widetilde{m}_1(\omega)} = 1 \\ m_0(\omega)\overline{\widetilde{m}_0(\omega+\pi)} + m_1(\omega)\overline{\widetilde{m}_1(\omega+\pi)} = 0. \end{cases} \tag{20}$$

If we want to avoid the infinite impulse response solutions, we need to impose that the determinant is a monomial $\alpha e^{ik\omega}$, $\alpha \neq 0$, $k \in \mathbb{Z}$. Up to a shift and scalar multiplication on the filters, we choose, for sake of convenience, $\alpha = -1$ and $k = 1$. This leads to

$$\overline{m_0(\omega)}\widetilde{m}_0(\omega) + \overline{m_0(\omega+\pi)}\widetilde{m}_0(\omega+\pi) = 1, \tag{21}$$

and

$$m_1(\omega) = e^{-i\omega}\overline{\widetilde{m}_0(\omega+\pi)}, \qquad \widetilde{m}_1(\omega) = e^{-i\omega}\overline{m_0(\omega+\pi)}. \tag{22}$$

Equations (21) and (22) are thus the most general for finite impulse response subband coders with exact reconstruction (in the two channels case). We call m_0 and $\widetilde{m}_0$ *dual filters.*

Clearly the special case $m_0 = \widetilde{m}_0$ corresponds to the CQF. However, the disadvantages of the CQF can be avoided:

(1) If m_0 is fixed, $\widetilde{m}_0$ can be found as the solution of a Bezout problem, which is equivalent to a linear system on the coefficients. The Féjer-Riesz lemma is no more needed.
(2) The coefficients can be very simple numerically and, in particular, they can have finite binary expansions, which are very useful for the implementation. They can also be real and symmetrical.

We now want to mimic the construction of wavelets in this more general setting. For this, we shall assume that the dual trigonometric polynomials m_0 and $\widetilde{m}_0$ satisfy

$$m_0(0) = \widetilde{m}_0(0) = 1 \quad \text{and} \quad m_0(\pi) = \widetilde{m}_0(\pi) = 0. \tag{23}$$

2.2. Dual wavelets and scaling functions

Let us define, first in the sense of tempered distribution, the dual scaling functions and wavelets by

$$\hat{\phi}(\omega) = \prod_{k=1}^{+\infty} m_0(2^{-k}\omega), \tag{24}$$

$$\hat{\tilde{\phi}}(\omega) = \prod_{k=1}^{+\infty} \widetilde{m}_0(2^{-k}\omega), \tag{25}$$

$$\widehat{\psi}(\omega) = m_1\left(\frac{\omega}{2}\right)\hat{\phi}\left(\frac{\omega}{2}\right), \tag{26}$$

$$\widehat{\tilde{\psi}}(\omega) = \widetilde{m}_1\left(\frac{\omega}{2}\right)\hat{\tilde{\phi}}\left(\frac{\omega}{2}\right). \tag{27}$$

Since we want to use these functions to analyze $L^2(\mathbb{R})$ it is necessary that ϕ and $\tilde{\phi}$ are both square integrable. Note that in the CQF-orthonormal case, this is always satisfied as a consequence of Equation (8) (see [7]). For dual filters this is false in general.

In this section we shall assume that ϕ and $\tilde{\phi}$ (and thus ψ and $\widetilde{\psi}$) are in $L^2(\mathbb{R})$. Some precise conditions on m_0 and $\widetilde{m}_0$ for this square integrability to hold will be stated in the next section.

Starting from this first assumption, we are going to prove the following result:

Theorem 1. *For any function f in $L^2(\mathbb{R})$, we have in the L^2 sense,*

$$f = \lim_{J \to +\infty} \sum_{j=-J}^{J} \sum_{k \in \mathbb{Z}} \langle f | \psi_k^j \rangle \widetilde{\psi}_k^j = \lim_{J \to +\infty} \sum_{j=-J}^{J} \sum_{k \in \mathbb{Z}} \langle f | \widetilde{\psi}_k^j \rangle \psi_k^j. \tag{28}$$

We remark that this does not mean that $\{\psi_k^j, \widetilde{\psi}_k^j\}_{j,k \in \mathbb{Z}}$ are biorthogonal bases, or even frames, since the summations are made in a precise way. However, this is a first step toward the construction of biorthogonal Riesz bases.

The proof of Theorem 1 is based on several lemmas that we shall comment here and prove in the appendix. We first introduce formally two approximation operators,

$$P_j(f) = \sum_{k \in \mathbb{Z}} \langle f | \tilde{\phi}_k^j \rangle \phi_k^j, \tag{29}$$

and

$$\widetilde{P}_j(f) = \sum_{k \in \mathbb{Z}} \langle f | \phi_k^j \rangle \tilde{\phi}_k^j, \tag{30}$$

and two detail operators,

$$\Delta_j(f) = \sum_{k \in \mathbb{Z}} \langle f | \widetilde{\psi}_k^j \rangle \psi_k^j, \tag{31}$$

and

$$\tilde{\Delta}_j(f) = \sum_{k \in \mathbb{Z}} \langle f | \psi_k^j \rangle \widetilde{\psi}_k^j. \tag{32}$$

Lemma 2. *The operators P_j, $\widetilde{P}_j$, Δ_j, and $\tilde{\Delta}_j$ are bounded on $L^2(\mathbb{R})$. Moreover, their norm is independent of j.*

This result gives a rigorous meaning to the definition of these operators on $L^2(\mathbb{R})$. We then havethe following.

Lemma 3. *For all j in $\mathbb{Z}$,*

$$P_{j-1} = P_j + \Delta_j \quad \text{and} \quad \widetilde{P}_{j-1} = \widetilde{P}_j + \tilde{\Delta}_j. \tag{33}$$

These two identities are essentially equivalent to the perfect reconstruction property of the subband coding scheme. By a *telescoping argument* we immediately obtain, for $J > 0$,

$$P_{-J-1} = P_J + \sum_{j=-J}^{J} \Delta_j, \tag{34}$$

and

$$\widetilde{P}_{-J-1} = \widetilde{P}_J + \sum_{j=-J}^{J} \tilde{\Delta}_j. \tag{35}$$

The next stage consists of letting J tend to $+\infty$ and using:

Lemma 4. *For all f in $L^2(\mathbb{R})$,*

$$\lim_{j\to+\infty} \|P_j(f)\|_{L^2} = \lim_{j\to+\infty} \|\widetilde{P}_j(f)\|_{L^2} = 0, \tag{36}$$

and

$$\lim_{j\to-\infty} \|P_j(f) - f\|_{L^2} = \lim_{j\to-\infty} \|\widetilde{P}_j(f) - f\|_{L^2} = 0. \tag{37}$$

The first limit is just a consequence of $\phi, \tilde{\phi} \in L^2(\mathbb{R})$. The second limit is due to some specific identities satisfied by ϕ and $\tilde{\phi}$ because of the hypotheses on the dual filters m_0 and $\widetilde{m}_0$.

Combining Equations (34), (35), (36), and (37), we clearly obtain the result of Theorem 1; *i.e.*, for any f in $L^2(\mathbb{R})$, we have

$$\lim_{J\to+\infty} \left\| f - \sum_{j=-J}^{J} \sum_{k\in\mathbb{Z}} \langle f|\widetilde{\psi}_k^j\rangle \psi_k^j \right\|_{L^2} = \lim_{J\to+\infty} \left\| f - \sum_{j=-J}^{J} \sum_{k\in\mathbb{Z}} \langle f|\psi_k^j\rangle \widetilde{\psi}_k^j \right\|_{L^2} = 0. \tag{38}$$

We now examine *the gap* existing between this system of *dual wavelets* and a pair of biorthogonal Riesz bases which will be obtained in Section 3.

2.3. From dual wavelets to biorthogonal Riesz bases

Recall that a Riesz basis of a Hilbert space H is a family of vectors $\{e_\lambda\}_{\lambda\in\Lambda}$ such that

(1) The finite linear combinations of e_λ are dense in H.
(2) There exist two strictly positive constants C_1 and C_2 such that, for any finite family of coefficients $\{\alpha_\lambda\}_{\lambda\in\Lambda_f}$ ($\Lambda_f \subset \Lambda$),

$$C_1 \sum_{\lambda\in\Lambda_f} |\alpha_\lambda|^2 \le \left\| \sum_{\lambda\in\Lambda_f} \alpha_\lambda e_\lambda \right\|_H^2 \le C_2 \sum_{\lambda\in\Lambda_f} |\alpha_\lambda|^2. \tag{39}$$

An equivalent definition for a Riesz basis (see [19]) is given by the following.:

(1) The vectors $\{e_\lambda\}_{\lambda\in\Lambda}$ are linearly independent, and
(2) $\{e_\lambda\}_{\lambda\in\Lambda}$ is a frame; *i.e.*, there exist two strictly positive constants D_1 and D_2 such that for any f in H

$$D_1\|f\|_H^2 \le \sum_{\lambda\in\Lambda} |\langle f|e_\lambda\rangle|^2 \le D_2\|f\|_H^2. \tag{40}$$

We shall rather use this second definition for ψ_k^j and $\widetilde{\psi}_k^j$. For these dual wavelets, the following holds.

Theorem 5. *$\{\psi_k^j, \widetilde{\psi}_k^j\}_{j,k\in\mathbb{Z}}$ are a pair of biorthogonal Riesz bases if and only if*

(1) *for all j, j', k, k' in $\mathbb{Z}$,*

$$\langle \psi_k^j | \widetilde{\psi}_{k'}^{j'} \rangle = \delta_{j,j'}\delta_{k,k'}. \tag{41}$$

(2) *There exist two strictly positive constants C and $\widetilde{C}$ such that, for all f in $L^2(\mathbb{R})$*

$$\sum_{j,k\in\mathbb{Z}} |\langle f|\psi_k^j\rangle|^2 \le C\|f\|^2 \tag{42}$$

and

$$\sum_{j,k\in\mathbb{Z}} |\langle f|\widetilde{\psi}_k^j\rangle|^2 \le \widetilde{C}\|f\|^2. \tag{43}$$

Proof:

a) If Equation (41) is satisfied, then any f in the closed linear span of the ψ_k^j with $(j,k) \ne (j_0,k_0)$ satisfies $\langle f|\widetilde{\psi}_{k_0}^{j_0}\rangle = 0$ and thus $\psi_{k_0}^{j_0}$ cannot be in this closed linear span. Thus, the ψ_k^j are linearly independent and the proof is similar for $\widetilde{\psi}_k^j$.

b) By Equation (28), we have

$$\psi_{k_0}^{j_0} = \lim_{J\to+\infty} \sum_{j=-J}^{J} \sum_{k\in\mathbb{Z}} \langle \psi_{k_0}^{j_0} | \widetilde{\psi}_k^j \rangle \psi_k^j, \tag{44}$$

and, thus, $\psi_{k_0}^{j_0}(1 - \langle \psi_{k_0}^{j_0}|\widetilde{\psi}_{k_0}^{j_0}\rangle)$ lies in the closure of the ψ_k^j for $(j,k) \ne (j_0,k_0)$. By linear independence, this implies

$$\langle \psi_{k_0}^{j_0} | \widetilde{\psi}_{k_0}^{j_0} \rangle = 1. \tag{45}$$

Isolating any $\langle \psi_{k_0}^{j_0}|\widetilde{\psi}_k^j\rangle\psi_k^j$ in (44) we also obtain for any $(j,k) \ne (j_0,k_0)$, by lincar independence

$$\langle \psi_{k_0}^{j_0} | \psi_k^j \rangle = 0, \tag{46}$$

and thus (41).

c) Combining Equations (28), (42), and (43), we obtain for any f in $L^2(\mathbb{R})$,

$$\begin{aligned}\|f\|^2 &= \lim_{J\to+\infty} \sum_{j=-J}^{J} \sum_{k\in\mathbb{Z}} \langle f|\widetilde{\psi}_k^j\rangle\langle\psi_k^j|f\rangle \\ &\leq \left(\sum_{j,k\in\mathbb{Z}} |\langle f|\psi_k^j\rangle|^2\right)^{1/2} \left(\sum_{j,k\in\mathbb{Z}} |\langle f|\widetilde{\psi}_k^j\rangle|^2\right)^{1/2} \\ &\leq C^{1/2}\|f\| \left(\sum_{j,k\in\mathbb{Z}} |\langle f|\widetilde{\psi}_k^j\rangle|^2\right)^{1/2}.\end{aligned}$$

This leads to

$$C^{-1}\|f\|^2 \leq \sum_{j,k\in\mathbb{Z}} |\langle f|\widetilde{\psi}_k^j\rangle|^2, \tag{47}$$

and similarly

$$\widetilde{C}^{-1}\|f\|^2 \leq \sum_{j,k\in\mathbb{Z}} |\langle f|\psi_k^j\rangle|^2. \tag{48}$$

The lower frame bounds are thus directly deduced from the upper bounds. According to the second definition and to the identity (41), $\{\psi_k^j, \widetilde{\psi}_k^j\}_{j,k\in\mathbb{Z}}$ are a pair of biorthogonal Riesz bases. Any f in $L^2(\mathbb{R})$ can be written

$$f = \sum_{j,k\in\mathbb{Z}} \langle f|\widetilde{\psi}_k^j\rangle\psi_k^j = \sum_{j,k\in\mathbb{Z}} \langle f|\psi_k^j\rangle\widetilde{\psi}_k^j \tag{49}$$

where these expansions are unique and converge unconditionally. ■

Before examining the type of dual filters leading to such bases, we shall prove two technical results that we shall use in the next section to check the hypothesis of Lemma 5. The first one deals with Equation (41).

Lemma 6. *Let ϕ_n and $\tilde{\phi}_n$ be defined for $n > 0$ by*

$$\hat{\phi}_n(\omega) = \prod_{k=1}^{n} m_0(2^{-k}\omega)\chi_{[-2^n\pi,2^n\pi]}(\omega), \tag{50}$$

and

$$\hat{\tilde{\phi}}_n(\omega) = \prod_{k=1}^{n} \widetilde{m}_0(2^{-k}\omega)\chi_{[-2^n\pi,2^n\pi]}(\omega). \tag{51}$$

Then if ϕ_n and $\tilde{\phi}_n$ converge in $L^2(\mathbb{R})$ to ϕ and $\tilde{\phi}$,

$$\langle\phi(x-p)|\tilde{\phi}(x-\ell)\rangle = \delta_{p,\ell}, \tag{52}$$

and

$$\langle \psi_k^j | \widetilde{\psi}_{k'}^{j'} \rangle = \delta_{j,j'} \delta_{k,k'}. \tag{53}$$

Proof: By recursion we can establish

$$\langle \phi_n(x-p) | \tilde{\phi}_n(x-\ell) \rangle = \delta_{p,\ell}. \tag{54}$$

Indeed,

$$\langle \phi_n(x-p) | \tilde{\phi}_n(x-\ell) \rangle = \frac{1}{2\pi} \int_{-2^n\pi}^{2^n\pi} \left(\prod_{k=1}^{n} m_0(2^{-k}\omega) \overline{\widetilde{m}_0(2^{-k}\omega)} \right) e^{i(\ell-p)\omega} d\omega$$

$$= \frac{2^n}{2\pi} \int_{-\pi}^{\pi} \left(\prod_{k=0}^{n-1} m_0(2^k\omega) \overline{\widetilde{m}_0(2^k\omega)} \right) e^{i2^n(\ell-p)\omega} d\omega$$

$$= \frac{2^n}{2\pi} \int_{-\frac{\pi}{2}}^{\frac{\pi}{2}} \left(\prod_{k=2}^{n-1} m_0(2^k\omega) \overline{\widetilde{m}_0(2^k\omega)} \right)$$

$$[m_0(\omega) \overline{\widetilde{m}_0(\omega)} + m_0(\omega+\pi) \overline{\widetilde{m}_0(\omega+\pi)}] e^{i2^n(\ell-p)\omega} d\omega$$

$$= \langle \phi_{n-1}(x-p) | \tilde{\phi}_{n-1}(x-\ell) \rangle$$

$$= \cdots = \frac{1}{2\pi} \int_{-\pi}^{\pi} e^{i(\ell-p)\omega} d\omega = \delta_{p,\ell}.$$

Consequently the L^2 convergence of ϕ_n and $\tilde{\phi}_n$ implies

$$\langle \phi(x-p) | \tilde{\phi}(x-\ell) \rangle = \delta_{p,\ell} \tag{55}$$

which can also be written as

$$\sum_{k \in \mathbb{Z}} (\bar{\hat{\phi}} \hat{\tilde{\phi}})(\omega + 2k\pi) = 1. \tag{56}$$

We now have

$$\sum_{k \in \mathbb{Z}} (\overline{\widehat{\psi}} \widehat{\widetilde{\psi}})(\omega + 2k\pi) = \sum_{k \in \mathbb{Z}} (\overline{m}_1 \widetilde{m}_1 \bar{\hat{\phi}} \hat{\tilde{\phi}}) \left(\frac{\omega}{2} + k\pi \right)$$

$$= \left[\overline{m}_1 \widetilde{m}_1 \left(\frac{\omega}{2} \right) + \overline{m_1} \widetilde{m}_1 \left(\frac{\omega}{2} + \pi \right) \right] \sum_{k \in \mathbb{Z}} (\bar{\hat{\phi}} \hat{\tilde{\phi}}) \left(\frac{\omega}{2} + 2k\pi \right)$$

$$= 1,$$

and

$$\sum_{k \in \mathbb{Z}} (\overline{\widehat{\psi}} \widehat{\widetilde{\phi}})(\omega + 2k\pi) = \sum_{k \in \mathbb{Z}} (\overline{m}_1 \widetilde{m}_0 \bar{\hat{\phi}} \hat{\tilde{\phi}}) \left(\frac{\omega}{2} + k\pi \right)$$

$$= \left[\overline{m}_1 \widetilde{m}_0 \left(\frac{\omega}{2} \right) + \overline{m_1} \widetilde{m}_0 \left(\frac{\omega}{2} + \pi \right) \right] \sum_{k \in \mathbb{Z}} (\bar{\hat{\phi}} \hat{\tilde{\phi}}) \left(\frac{\omega}{2} + 2k\pi \right)$$

$$= 0.$$

Consequently,

$$\langle \psi(x-k) | \widetilde{\psi}(x-\ell) \rangle = \delta_{k,\ell} \tag{57}$$

and

$$\langle \psi(x-k) | \tilde{\phi}(x-\ell) \rangle = 0. \tag{58}$$

If we now define $V_j = \mathrm{Span}\{\phi_k^j\}_{k\in\mathbb{Z}}$, $W_j = \mathrm{Span}\{\psi_k^j\}_{k\in\mathbb{Z}}$, and $\widetilde{V}_j = \mathrm{Span}\{\tilde{\phi}_k^j\}_{k\in\mathbb{Z}}$, $\widetilde{W}_j = \mathrm{Span}\{\widetilde{\psi}_k^j\}_{k\in\mathbb{Z}}$, it follows from Equation (56) that for any j, $\widetilde{V}_j$ is orthogonal to W_j. Since, for all $j' < j$, we have $\widetilde{W}_{j'} \subset \widetilde{V}_j$, it follows that W_j and $\widetilde{W}_{j'}$ are orthonormal when $j' < j$ and, by a symmetrical argument, when $j < j'$. Consequently, we obtain by Equation (55),

$$\langle \psi_k^j | \widetilde{\psi}_{k'}^{j'} \rangle = \delta_{j,j'} \delta_{k,k'}, \tag{59}$$

and the lemma is proved. ■

The last lemma of this section deals with the upper frame bounds (42) and (43).

Lemma 7. *Suppose that the function ϕ satisfies*

$$\sup_{\omega\in\mathbb{R}} \sum_{k\in\mathbb{Z}} |\hat{\phi}(\omega + 2k\pi)|^{2-\sigma} < +\infty, \tag{60}$$

for some $\sigma > 0$, and

$$\sup_{\omega\in\mathbb{R}} (1+|\omega|)^\sigma |\hat{\phi}(\omega)| < +\infty. \tag{61}$$

Then there exists a constant C such that, for all f in $L^2(\mathbb{R})$,

$$\sum_{j,k\in\mathbb{Z}} |\langle f | \psi_k^j \rangle|^2 \leq C \|f\|^2. \tag{62}$$

The same holds for $\widetilde{\psi}$ and $\tilde{\phi}$.

Proof: Since $\widehat{\psi}(\omega) = m_1(\frac{\omega}{2})\hat{\phi}(\frac{\omega}{2})$ has at least a first order zero at the origin, using Equations (58) and (57), we may conclude

$$\sum_{j\in\mathbb{Z}} |\widehat{\psi}(2^j\omega)|^\sigma \leq C_1, \tag{63}$$

and

$$\sum_{k\in\mathbb{Z}} |\widehat{\psi}(\omega + 2k\pi)|^{2-\sigma} \leq C_2 \tag{64}$$

uniformly in ω.

Using the Plancherel and the Poisson formulas, we can derive for all f in $L^2(\mathbb{R})$,

$$\begin{aligned}
\sum_{k\in\mathbb{Z}} |\langle f|\psi_k^j\rangle|^2 &= \frac{1}{4\pi^2}\sum_{k\in\mathbb{Z}} 2^j \left|\int_{\mathbb{R}} \hat{f}(\omega)\overline{\widehat{\psi}(2^j\omega)e^{-i2^jk\omega}}d\omega\right|^2 \\
&= \frac{1}{4\pi^2}\sum_{k\in\mathbb{Z}} 2^{-j}\left|\int_{\mathbb{R}} \hat{f}(2^{-j}\omega)\overline{\widehat{\psi}(\omega)e^{-ik\omega}}d\omega\right|^2 \\
&= \frac{2^{-j}}{2\pi}\int_0^{2\pi}\left|\sum_{\ell\in\mathbb{Z}} \hat{f}(2^{-j}(\omega+2\ell\pi))\overline{\widehat{\psi}(\omega+2\ell\pi)}\right|^2 d\omega \\
&\le \frac{2^{-j}}{2\pi}\int_0^{2\pi}\left(\sum_{\ell\in\mathbb{Z}} |\hat{f}(2^{-j}(\omega+2\ell\pi))|\,|\widehat{\psi}(\omega+2\ell\pi)|^{\frac{\sigma}{2}}|\widehat{\psi}(\omega+2\ell\pi)|^{1-\frac{\sigma}{2}}\right)^2 d\omega \\
&\le \frac{2^{-j}}{2\pi}\int_0^{2\pi}\left(\sum_{\ell\in\mathbb{Z}} |\hat{f}(2^{-j}(\omega+2\ell\pi))|^2|\widehat{\psi}(\omega+2\ell\pi)|^{\sigma}\right)\left(\sum_{\ell\in\mathbb{Z}}|\widehat{\psi}(\omega+2\ell\pi)|^{2-\sigma}\right)d\omega \\
&\le C_2\frac{2^{-j}}{2\pi}\int_{\mathbb{R}} |\hat{f}(2^{-j}\omega)|^2|\widehat{\psi}(\omega)|^{\sigma}d\omega \\
&\le \frac{1}{2\pi}C_2\int_{\mathbb{R}} |\hat{f}(\omega)|^2|\widehat{\psi}(2^j\omega)|^{\sigma}d\omega.
\end{aligned}$$

Summing over all the scales $j \in \mathbb{Z}$ and using Equation (59), we obtain

$$\begin{aligned}
\sum_{j,k\in\mathbb{Z}} |\langle f|\psi_k^j\rangle|^2 &\le \frac{C_1C_2}{2\pi}\int_{\mathbb{R}} |\hat{f}(\omega)|^2 d\omega \\
&= C_1C_2\|f\|^2,
\end{aligned}$$

and this concludes the proof. ■

We are now ready to characterize the dual filters which lead to biorthogonal wavelet bases, such that the functions ϕ and $\tilde{\phi}$ satisfy the hypotheses of Lemmas 6 and 7.

§3. Biorthogonal wavelet bases and stable subband coding schemes

In this section we shall present two strategies to design the dual filters so that the associated dual wavelets ψ and $\widetilde{\psi}$ generate a pair of biorthogonal Riesz bases. In other words, we shall establish some conditions for the stability of the FWT algorithm since this is equivalent to the frame bounds inequalities that we require on our multiscale bases.

The first strategy uses estimates on the decay at infinity of the functions $\hat{\phi}(\omega)$ and $\hat{\tilde{\phi}}(\omega)$ that can also be found in [6]. It furnishes a sufficient condition for biorthogonality and stability.

The second strategy is based on the study of two operators associated with the dual filters. We show here that it leads to sufficient conditions and it was proved in [4] that this criterion is also necessary. It is thus a sharper strategy but it is only tractable for filters of reasonable size.

3.1. A Fourier criterion

Since we have assumed that m_0 and $\widetilde{m}_0$ vanish at $\omega = \pi$, it is possible to express these filters in a factorized form

$$m_0(\omega) = \left(\frac{1+e^{i\omega}}{2}\right)^N p(\omega), \tag{65}$$

and

$$\widetilde{m}_0(\omega) = \left(\frac{1+e^{i\omega}}{2}\right)^{\widetilde{N}} \tilde{p}(\omega). \tag{66}$$

The following results give a sufficient criterion based on the properties of the trigonometric polynomials p and $\tilde{p}$.

Theorem 8. *Suppose that the function $p(\omega)$ satisfies*

$$\inf_{j>0}\left[\max_{\omega\in\mathbb{R}}\left|\prod_{k=1}^{j} p(2^{-k}\omega)\right|^{\frac{1}{j}}\right] < 2^{N-\frac{1}{2}}. \tag{67}$$

Then

(1) $\hat{\phi}_n(\omega) = \prod_{k=1}^{n} m_0(2^{-k}\omega)\chi_{[-2^n\pi,2^n\pi]}(\omega)$ *converges to* $\hat{\phi}(\omega)$ *in* $L^2(\mathbb{R})$*, and*

(2) *The conditions in (60) and (61) of Lemma 7 are satisfied.*

If $\tilde{p}(\omega)$ satisfies a similar hypothesis, then the dual filters m_0 and $\widetilde{m}_0$ generate biorthogonal Riesz bases of wavelets.

Proof: The hypothesis (67) implies that for some $j > 0$,

$$\max_{\omega\in\mathbb{Z}}\left|\prod_{k=1}^{j} p(2^{-j}\omega)\right| < 2^{j(N-\frac{1}{2}-\epsilon)}, \quad (\epsilon > 0). \tag{68}$$

We now write

$$|\hat{\phi}_n(\omega)| = \left|\sum_{k=1}^{n}\left(\frac{1+e^{i2^{-k}\omega}}{2}\right)\right|^N \left|\prod_{k=1}^{n} p(2^{-k}\omega)\right| \chi_{[-2^n\pi,2^n\pi]}(\omega)$$

$$= \left|\frac{\sin\left(\frac{\omega}{2}\right)}{2^n\sin(2^{-n-1}\omega)}\right|^N \left|\prod_{k=1}^{n} p(2^{-k}\omega)\right| \chi_{[-2^n\pi,2^n\pi]}(\omega).$$

Since $|\sin(2^{-n-1}\omega)| > C2^{-n-1}|\omega|$ when ω is in $[-2^n\pi, 2^n\pi]$, we have

$$|\hat{\phi}_n(\omega)| \leq C(1+|\omega|)^{-N} \left| \prod_{k=1}^{n} p(2^{-k}\omega) \right| \chi_{[-2^n\pi, 2^n\pi]}(\omega). \tag{69}$$

We treat the second product in the following way: The contribution of the factors $p(2^{-k}\omega)$ for $k \geq \frac{\log(1+|\omega|)}{\log 2}$ can be globally majorated by a constant since the infinite product converges to a smooth function. We divide the other factors in packets of size j that we can majorate using Equation (68). This leads to

$$\begin{aligned} |\hat{\phi}_n(\omega)| &\leq C(1+|\omega|)^{-N} \left(2^{j(N-\frac{1}{2}-\epsilon)}\right)^{\frac{\log(1+|\omega|)}{j\log 2}} \\ &\leq C(1+|\omega|)^{-\frac{1}{2}-\epsilon}. \end{aligned}$$

Since this bound is independent of n and $\hat{\phi}_n(\omega)$ converges pointwise to $\hat{\phi}(\omega)$, we can apply Lebesgue's dominated convergence theorem to conclude that ϕ_n converges in $L^2(\mathbb{R})$ to ϕ. At the limit, we also have

$$|\hat{\phi}(\omega)| \leq C(1+|\omega|)^{-1/2-\epsilon} \tag{70}$$

which immediately implies the conditions (60) and (61) of Lemma 7 by choosing $0 < \sigma < \min\left(\frac{1}{2}, \frac{4\epsilon}{2\epsilon+1}\right)$. If $\tilde{p}(\omega)$ satisfies a similar condition, then the results of Lemmas 6 and 7 can be applied to ψ and $\widetilde{\psi}$ and by Theorem 5, the families $\{\psi^j_k, \widetilde{\psi}^j_k\}_{j,k\in\mathbb{Z}}$ are biorthogonal Riesz bases for $L^2(\mathbb{R})$. ■

This criterion was used in [6] to construct many biorthogonal wavelet bases but it is not sharp. In particular, Equation (70) is not strictly necessary to have ϕ and ψ in $L^2(\mathbb{R})$. These functions can be very lacunary; *i.e.*, their Fourier transform can have a bad decay but only at some points which occur less and less frequently at infinity so that they are still square integrable. We now present a sharper criterion based on a different approach.

3.2. A matrix base criterion

Let us first introduce the basic tool which will be used in this approach.

Definition 9. *Let $m_0(\omega)$ be a trigonometric polynomial such that $m_0(0) = 1$ and $m_0(\pi) = 0$. The transition operator T_0 associated with this filter acts on 2π periodic functions in the following way*

$$T_0 f(\omega) = \left|m_0\left(\frac{\omega}{2}\right)\right|^2 f\left(\frac{\omega}{2}\right) + \left|m_0\left(\frac{\omega}{2}+\pi\right)\right|^2 f\left(\frac{\omega}{2}+\pi\right). \tag{71}$$

This operator appears in the works of W. Lawton [12] and J.P. Conze and A. Raugi [9] for the study of the orthonormal case. It can also be useful to estimate the Sobolev regularity of the scaling function associated to m_0

(see [11], [18], and [5]). Here, we will need two lemmas which give some basic properties of T_0.

Lemma 10. *Let* $m_0(\omega) = \sum_{k=0}^{N} c_k e^{ik\omega}$ *and* T_0 *the associated transition operator, and define the* $2N+1$ *dimensional space*

$$E_N = \left\{ \sum_{k=-N}^{N} c_k e^{ik\omega} \,\middle|\, (c_{-N}, \ldots, c_N) \in \mathbb{C}^{2N+1} \right\}, \tag{72}$$

and its subspace

$$F_N = \left\{ \sum_{k=-N}^{N} c_k e^{ik\omega} \,\middle|\, \sum_{k=-N}^{N} c_k = 0 \right\}. \tag{73}$$

Then E_N *and* F_N *are stable under the action of* T_0.

Proof: From the definition of T_0, it is clear that if $|m_0|^2$ and f are two elements of E_N, then $T_0 f$ is also in E_N.

A trigonometric polynomial f is in F_N if and only if $f \in E_N$ and $f(0) = 0$. Consequently, if f is in F_N, we have

$$T_0 f(0) = |m_0(0)|^2 f(0) + |m_0(\pi)|^2 f(\pi) = 0, \tag{74}$$

which proves that F_N is stable under T_0. ∎

Remarks. If we consider the Fourier expansion

$$|m_0(\omega)|^2 = \sum_{k=-N}^{N} H_k e^{ik\omega}, \tag{75}$$

then the matrix of T_0 restricted to E_N is given by

$$M_0 = (2H_{i-2j})_{i,j=-N\ldots N} = 2 \begin{bmatrix} H_N & 0 & \cdots & \cdots & 0 \\ H_{N-2} & H_{N-1} & H_N & & \vdots \\ \vdots & \vdots & \vdots & & 0 \\ H_{-N} & H_{-N+1} & \vdots & & H_N \\ 0 & 0 & H_{-N} & & H_{N-2} \\ \vdots & \vdots & & & \vdots \\ 0 & 0 & \cdots & \cdots & H_{-N} \end{bmatrix}. \tag{76}$$

Since $|m_0(0)|^2 = \sum_{k=-N}^{N} H_k = 1$ and $|m_0(\pi)|^2 = \sum_{k=-N}^{N} (-1)^k H_k = 0$, it follows that

$$\sum_k H_{2k} = \sum_k H_{2k+1} = \frac{1}{2} \tag{77}$$

and thus the row vector $\mu = (1, 1, \ldots, 1)$ satisfies

$$\mu T_0 = \mu. \tag{78}$$

This is another way to show that F_N is stable since $F_N = (\mathbb{C}\mu)^\perp$ but it also shows that 1 is an eigenvalue of T_0 and if this eigenvalue is not degenerated then it is not in the spectrum of T_0 restricted to F_N.

The second lemma makes a connection between the iteration of the operator T_0 and the sequence ϕ_n which must converge to ϕ in $L^2(\mathbb{R})$.

Lemma 11. *For any 2π periodic function $f(\omega)$, it follows that*

$$\int_{-\pi}^{\pi} T_0^n f(\omega)d\omega = \int_{-2^n\pi}^{2^n\pi} f(2^{-n}\omega)\prod_{k=1}^{n}|m_0(2^{-k}\omega)|^2 d\omega = \int f(2^{-n}\omega)|\hat{\phi}_n(\omega)|^2 d\omega. \tag{79}$$

Proof: We prove it by induction. Equation (79) is trivial for $n = 0$ and if it is satisfied for some $n \geq 0$, then

$$\int_{-\pi}^{\pi} T_0^{n+1} f(\omega)d\omega = \int_{-\pi}^{\pi} T_0^n (T_0 f)(\omega)d\omega$$

$$= \int_{-2^n\pi}^{2^n\pi} T_0 f(2^{-n}\omega)\prod_{k=1}^{n}|m_0(2^{-k}\omega)|^2 d\omega$$

$$= \int_{-2^n\pi}^{2^n\pi} [f(2^{-n-1}\omega)|m_0(2^{-n-1}\omega)|^2 + f(2^{-n-1}\omega+\pi)|m_0(2^{-n-1}\omega+\pi)|^2]$$

$$\prod_{k=1}^{n}|m_0(2^{-k}\omega)|^2 d\omega$$

$$= \int_{-2^{n+1}\pi}^{2^{n+1}\pi} f(2^{-n-1}\omega)\prod_{k=1}^{n+1}|m_0(2^{-k}\omega)|^2 d\omega. \quad \blacksquare$$

We are now ready to state a criterion based on transition operators.

Theorem 12. *Let λ be the largest eigenvalue of T_0 restricted to F_N. If $|\lambda| < 1$, then*

(1) *ϕ_n converges to ϕ in $L^2(\mathbb{R})$, and*

(2) *the conditions (60) and (61) of Lemma 7 are satisfied.*

If the same holds for the operator $\tilde{T}_0$ associated to $\tilde{m}_0$, then the dual filters m_0 and $\tilde{m}_0$ generate biorthogonal Riesz bases of wavelets.

Proof: Let us define the trigonometric polynomial c by

$$c(\omega) = 1 - \cos\omega. \tag{80}$$

It is clear that $c(\omega)$ is in F_N. Applying Lemma 11 and using the hypothesis $|\lambda| < 1$, we obtain

$$\begin{aligned}\int |\hat{\phi}_n(\omega)|^2 c(2^{-n}\omega)d\omega &= \int_{-\pi}^{\pi} T_0^n c(\omega)d\omega \\ &\leq \sqrt{2\pi}\|T_0^n c\|_{L^2} \\ &\leq C\left(\frac{1+|\lambda|}{2}\right)^n = C2^{-n\epsilon},\end{aligned}$$

with $\epsilon = \frac{1}{\log 2}[\log 2 - \log(1+|\lambda|)] > 0$ because $|\lambda| < 1$. This leads to a Littlewood-Paley type of estimate. Indeed, since $c(\omega)$ is positive and $c(\omega) \geq 1$ when $\frac{\pi}{2} \leq |\omega| \leq \pi$, we have

$$\int_{2^{n-1}\pi \leq |\omega| \leq 2^n\pi} |\hat{\phi}_n(\omega)|^2 d\omega \leq C2^{-n\epsilon}. \tag{81}$$

This estimate is also valid if we replace ϕ_n by ϕ because we have $|\hat{\phi}(\omega)| = |\hat{\phi}_n(\omega)\hat{\phi}(2^{-n}\omega)| \leq |\hat{\phi}_n(\omega)| \max_{|\omega|\leq\pi}(|\hat{\phi}(\omega)|)$ when $|\omega| \leq 2^n\pi$. Consequently

$$\int_{2^{n-1}\pi \leq |\omega| \leq 2^n\pi} |\hat{\phi}(\omega)|^2 d\omega \leq C2^{-n\epsilon}, \tag{82}$$

which means that ϕ is not only in $L^2(\mathbb{R})$ but also in the Besov space $B_2^{\epsilon,\infty}(\mathbb{R})$.

Let us now prove the L^2 convergence of ϕ_n to ϕ. Since $m_0(0) = \hat{\phi}(0) = 1$, there exists an α in $(0;\pi]$ such that

$$|\omega| \leq \alpha \Rightarrow |\hat{\phi}(\omega)| \geq C > 0. \tag{83}$$

We now divide ϕ_n in two parts: $\phi_n = \phi_n^1 + \phi_n^2$ with

$$\hat{\phi}_n^1(\omega) = \hat{\phi}_n(\omega)\chi_{[-2^n\alpha,2^n\alpha]}(\omega), \tag{84}$$

and

$$\hat{\phi}_n^2(\omega) = \hat{\phi}_n(\omega)\left[\chi_{[-2^n\pi,2^n\pi]}(\omega) - \chi_{[-2^n\alpha,2^n\alpha]}(\omega)\right]. \tag{85}$$

Clearly $\hat{\phi}_n^1(\omega)$ converges pointwise to $\hat{\phi}(\omega)$ and by (83), we have

$$|\hat{\phi}_n^1(\omega)| \leq \frac{|\hat{\phi}(\omega)|}{C}. \tag{86}$$

By Lebesgue's theorem, ϕ_n^1 converges to ϕ in L^2. We also have

$$\begin{aligned}\int |\hat{\phi}_n^2(\omega)|^2 d\omega &= \int_{2^n\alpha \leq |\omega| \leq 2^n\pi} |\hat{\phi}_n(\omega)|^2 d\omega \\ &\leq \frac{1}{c(\alpha)} \int |\hat{\phi}_n(\omega)|^2 c(2^{-n}\omega)d\omega \\ &\leq C2^{-n\epsilon},\end{aligned}$$

and thus ϕ_n^2 tends to zero in L^2 and consequently ϕ is also the L^2 limit of ϕ_n.

To prove Equations (60) and (61), we shall use the estimate (82). We first remark that since $m_0(\pi) = m_0(-\pi) = 0$, the scaling function satisfies

$$\hat{\phi}(2k\pi) = 0 \qquad \text{if} \quad k \in \mathbb{Z}/\{0\}. \tag{87}$$

Using a first order Taylor development, we can write

$$\begin{aligned}
\sum_{k\in\mathbb{Z}} |\hat{\phi}(\omega + 2k\pi)|^{2-\sigma} &\leq \int_{\mathbb{R}} \left| \frac{d}{d\omega}[|\hat{\phi}|^{2-\sigma}] \right| d\omega \\
&= \int \left| \frac{d}{d\omega}[(|\hat{\phi}|^2)^{1-\frac{\sigma}{2}}] \right| d\omega \\
&= \int \left| \left(1 - \frac{\sigma}{2}\right) \frac{d}{d\omega}[|\hat{\phi}|^2] \right| |\hat{\phi}|^{-\frac{\sigma}{2}} d\omega \\
&\leq \int |2-\sigma| \left| \frac{d\hat{\phi}}{d\omega} \right| |\hat{\phi}(\omega)|^{1-\frac{\sigma}{2}} d\omega \\
&\leq C \left(\int \left| \frac{d\hat{\phi}}{d\omega} \right|^2 d\omega \right)^{1/2} \left(\int |\hat{\phi}(\omega)|^{2-\sigma} d\omega \right)^{1/2}.
\end{aligned}$$

The first factor proportional to the L^2 norm of $x\phi(x)$ which is finite since ϕ is square integrable and compactly supported.

To evaluate the second factor we compute the integral of $|\hat{\phi}(\omega)|^{2-\sigma}$ on a dyadic ring $2^{n-1}\pi \leq |\omega| \leq 2^n\pi$. By the Hölder inequality and (82) we obtain

$$\begin{aligned}
\int_{2^{n-1}\pi \leq |\omega| \leq 2^n\pi} |\hat{\phi}(\omega)|^{2-\sigma} d\omega &\leq \left[\int_{2^{n-1}\pi \leq |\omega| \leq 2^n\pi} |\hat{\phi}(\omega)|^2 d\omega \right]^{1-\frac{\sigma}{2}} (2^n\pi)^{\frac{\sigma}{2}} \\
&\leq C 2^{-n\epsilon(1-\frac{\sigma}{2}) + n\frac{\sigma}{2}}.
\end{aligned}$$

The second factor will thus be finite and (57) will be satisfied if we choose σ such that $\epsilon(1 - \frac{\sigma}{2}) - \frac{\sigma}{2} > 0$; *i.e.*, $\sigma < \frac{2\epsilon}{1+\epsilon}$. We apply the same method to show (58). If $|\omega|$ is in $[2^{n-1}\pi, 2^n\pi]$ for $n \geq 1$, we can write

$$\begin{aligned}
|\hat{\phi}(\omega)|^2 &\leq \int_{2^{n-1}\pi \leq |\omega| \leq 2^n\pi} \left| \frac{d}{d\omega}[|\hat{\phi}|^2] \right| d\omega \\
&\leq C \left(\int_{\mathbb{R}} \left| \frac{d\hat{\phi}}{d\omega} \right|^2 d\omega \right)^{1/2} \left(\int_{2^{n-1}\pi \leq |\omega| \leq 2^n\pi} |\hat{\phi}(\omega)|^2 d\omega \right)^{1/2} \\
&\leq C 2^{-n\frac{\epsilon}{2}} \leq C(1 + |\omega|)^{-\frac{\epsilon}{2}},
\end{aligned}$$

and thus (61) is satisfied if we choose $\sigma \leq \frac{\epsilon}{4}$.

This concludes the proof of the theorem. ■

Remarks.

(1) In the proof we use estimates of $|\hat{\phi}|^2$ or its integral on dyadic rings $2^{n-1}\pi \leq |\omega| \leq 2^n\pi$, for $n \geq 1$. The case $|\omega| \leq \pi$ does not cause any problem since $\hat{\phi}(\omega)$ is a smooth function.

(2) The criterion that we have established is in fact sharp; *i.e.*, it is necessary and sufficient to have biorthogonal Riesz bases. The complete proof of this fact, which is very technical, can be found in [4].

Using our theoretical results, we shall now build an important family of biorthogonal wavelets where the functions ϕ and ψ are piecewise polynomial.

§4. Biorthogonal wavelets and splines

The connections between wavelet theory and spline theory come out naturally from the general framework of multiresolution analysis. Indeed, for a given $N \geq 1$, we can define a multiresolution analysis $\{V_j\}_{j\in\mathbb{Z}}$ by

$$V_j = \left\{ f \in L^2(\mathbb{R}) \cap C^{N-1}(\mathbb{R}) | \left(\frac{d}{dx}\right)^{N+1} f = \sum_{k\in\mathbb{Z}} \alpha_k \delta_{2^j k} \right\}, \tag{88}$$

where $\delta_{2^j k}$ is the delta function localized at the point $2^j k$. In other words, V_j is the space of square integrable functions which are piecewise polynomial of degree N on the dyadic intervals $[2^j k, 2^j(k+1)]$ with C^{N-1} continuity at the nod points $2^j k$.

A natural generator for these spaces is the Nth degree B-spline function defined by

$$\phi^N(x) = (*)^{N+1}\chi_{[0;1]}. \tag{89}$$

It is then well known that the integer translates of ϕ generate V_0. For example, in the case of the linear splines ($N = 1$), ϕ is the *hat function*; *i.e.*, $\phi^1(x) = \max(0, 1 - |x-1|\}$.

However, it is clear that the translates of ϕ are not orthonormal. Still $\{\phi^N(x-k)\}_{k\in\mathbb{Z}}$ forms a Riesz bases for V_0; *i.e.*, there exists $C_2 \geq C_1 > 0$, such that for any sequence α_k in $\ell^2(\mathbb{Z})$,

$$C_1 \sum_{k\in\mathbb{Z}} |\alpha_k|^2 \leq \left\| \sum_{k\in\mathbb{Z}} \alpha_k \phi^N(x-k) \right\|_{L^2}^2 \leq C_2 \sum_{k\in\mathbb{Z}} |\alpha_k|^2 \tag{90}$$

which can also be expressed by

$$0 < C_1 \leq \sum_{k\in\mathbb{Z}} |\hat{\phi}^N(\omega + 2k\pi)|^2 \leq C_2. \tag{91}$$

This allows an orthonormalization process that preserves the structure of a family generated by the translates of a single function. The new scaling function is defined by

$$\hat{\phi}_o^N(\omega) = \hat{\phi}^N(\omega) \left(\sum_{k\in\mathbb{Z}} |\hat{\phi}^N(\omega + 2k\pi)|^2 \right)^{-1/2}. \tag{92}$$

Its translates are orthonormal by construction but it is not compactly supported (it has exponential decay). The same holds for the associated wavelet. In this construction, due to Battle and Lemarié [13], the CQF filters do not have a finite impulse response, and this fact constitutes a major disadvantage for implementation.

Using the results of the two previous sections, we shall now construct biorthogonal bases of compactly supported wavelets where the functions ϕ^N and ψ^N are spline functions of order N.

We simply choose for ϕ^N the B-spline function defined by (89) since we do not require anymore the orthonormality of its integer translates. The low pass filter associated with this scaling function is given by

$$m_0^N(\omega) = \left(\frac{1+e^{-i\omega}}{2}\right)^{N+1}. \tag{93}$$

To find a dual filter, it is useful to define the following polynomial

$$P_L(y) = \sum_{j=0}^{L-1} \binom{L-1+j}{j} y^j, \tag{94}$$

which is the solution of the Bezout equation

$$(1-y)^L P_L(y) + y^L P_L(1-y) = 1. \tag{95}$$

By a change of variable, we obtain

$$\left[\cos\left(\frac{\omega}{2}\right)\right]^{2L} P_L\left(\sin^2\left(\frac{\omega}{2}\right)\right) + \left[\sin\left(\frac{\omega}{2}\right)\right]^{2L} P_L\left(\cos^2\left(\frac{\omega}{2}\right)\right) = 1, \tag{96}$$

and by a shift

$$\begin{aligned} &\left[\frac{1+e^{i\omega}}{2}\right]^{2L} \left[e^{-iL\omega} P_L\left(\sin^2\left(\frac{\omega}{2}\right)\right)\right] \\ &\quad + \left[\frac{1-e^{i\omega}}{2}\right]^{2L} \left[(-1)^L e^{-iL\omega} P_L\left(\cos^2\left(\frac{\omega}{2}\right)\right)\right] = 1. \end{aligned} \tag{97}$$

This formula gives a solution for all the values of N smaller than $2L-1$. Indeed, if $N+1 \le 2L$, we can take as a dual filter

$$\widetilde{m}_0^{N,L}(\omega) = \left[\frac{1+e^{i\omega}}{2}\right]^{2L-N-1} P_L\left(\sin^2\left(\frac{\omega}{2}\right)\right) e^{-iL\omega}. \tag{98}$$

In other words, for a fixed $N, \widetilde{m}_0^{N,L}$ is a dual filter for m_0^N if $2L \ge N+1$.

How can we choose the parameter L in an optimal way? At first, it seems natural to choose the smallest value of L such that $2L \ge N+1$. Unfortunately, this choice does not lead, except in the Haar case ($N = 0$), to a dual scaling function $\tilde{\phi}_{N,L}$ which is square integrable. More precisely the filter $\widetilde{m}_0^{N,L}$ does not satisfy in this case the conditions of Theorems 8 and 12.

A very simple example to illustrate this problem is the case of the linear splines $N = 1$. If we choose $L = 1$, then the dual filter is given by

$$\widetilde{m}_0^{1,1}(\omega) = e^{-i\omega}. \tag{99}$$

By formula (25), we see that

$$\hat{\tilde{\phi}}^{1,1}(\omega) = e^{-i\omega}, \tag{100}$$

which is not square integrable. Observe that $\phi^{1,1}$ is a delta function centered on 1 and that in the distribution sense we still have

$$\langle \phi^1(x-k) | \tilde{\phi}^{1,1}(x-\ell) \rangle = \delta_{k,\ell}, \tag{101}$$

but the subband coding scheme has no chances to be numerically stable (in the ℓ^2 sense). For this reason, we need to choose a larger L, such that $\widetilde{m}_0^{N,L}(\omega)$ satisfies the conditions of Theorem 8 or 12. In [6], we prove that $\tilde{\phi}^{N,L}$ can be made arbitrarily regular if we choose a large enough L for a fixed N. This means that the Fourier transform of $\hat{\tilde{\phi}}^{N,L}$ has an arbitrarily high rate of decay at infinity and that $\widetilde{m}_0^{N,L}$ will satisfy the conditions of Theorem 8. Note that the regularity of the functions $\tilde{\phi}$ and $\tilde{\psi}$ is not very important in applications since we only use them in the decomposition stage by inner product with the function to be analyzed. As explained in the introduction, smoothness is mostly important in the reconstruction process and we shall rather take ϕ^N and $\psi^{N,L}$ as synthesis functions since they are piecewise polynomials. For a given value of N, the best choice for L is thus the smallest value such that $\widetilde{m}_0^{N,L}$ satisfies the conditions of Theorem 12 (which are sharp).

For $N = 1$, this value is $L = 2$; *i.e.*,

$$\widetilde{m}_0^{1,2}(\omega) = \left(\frac{1+e^{i\omega}}{2}\right)^2 e^{-2i\omega}\left(1 + 2\sin^2\left(\frac{\omega}{2}\right)\right). \tag{102}$$

For $N = 2$, this value is also $L = 2$; *i.e.*,

$$\widetilde{m}_0^{2,2}(\omega) = \left(\frac{1+e^{i\omega}}{2}\right) e^{-2i\omega}\left(1 + 2\sin^2\left(\frac{\omega}{2}\right)\right). \tag{103}$$

For $N = 3$, this value is $L = 4$; *i.e.*,

$$\widetilde{m}_0^{3,3}(\omega) = \left(\frac{1+e^{i\omega}}{2}\right)^4 e^{-4i\omega}\left(1 + 4\sin^2\left(\frac{\omega}{2}\right) + 10\sin^4\left(\frac{\omega}{2}\right) + 20\sin^6\left(\frac{\omega}{2}\right)\right). \tag{104}$$

We illustrate in Figures 3 and 4 the first two cases (linear and quadratic). Note that, for even N, the wavelets are shifted odd functions, whereas for odd N, they are shifted even functions. We show in these figures the results for the minimal value of L (in the first column) and for the next value (in the second column).

Observe the chaotic aspect of the functions $\tilde{\phi}$ and $\tilde{\psi}$ in the quadratic case. One can check that these functions, although square integrable, do not satisfy the decay condition $(1+|\omega|)^{-1/2-\epsilon}(|\hat{\tilde{\phi}}| + |\hat{\tilde{\psi}}|) < +\infty$. This means that only Theorem 12 can be applied here. This also explains the fractal aspect of these graphs: To be square integrable, $\tilde{\phi}$ and $\tilde{\psi}$ must have a lacunary structure, typical of these fractal figures.

Finally, let us mention a slightly different construction of compactly supported spline wavelet which is due to C. K. Chui and J. Z. Wang [9]. In this construction, the spaces W_j are kept orthogonal and the wavelet are nonorthogonal only inside a given scale. The advantages of this framework is that the wavelet decomposition is still orthogonal and ϕ and ψ are still symmetric or antisymmetric. Furthermore, the discussion on frame bounds is much simpler and that the dual wavelet is also a spline function. However the dual wavelet and filter are no longer compactly supported.

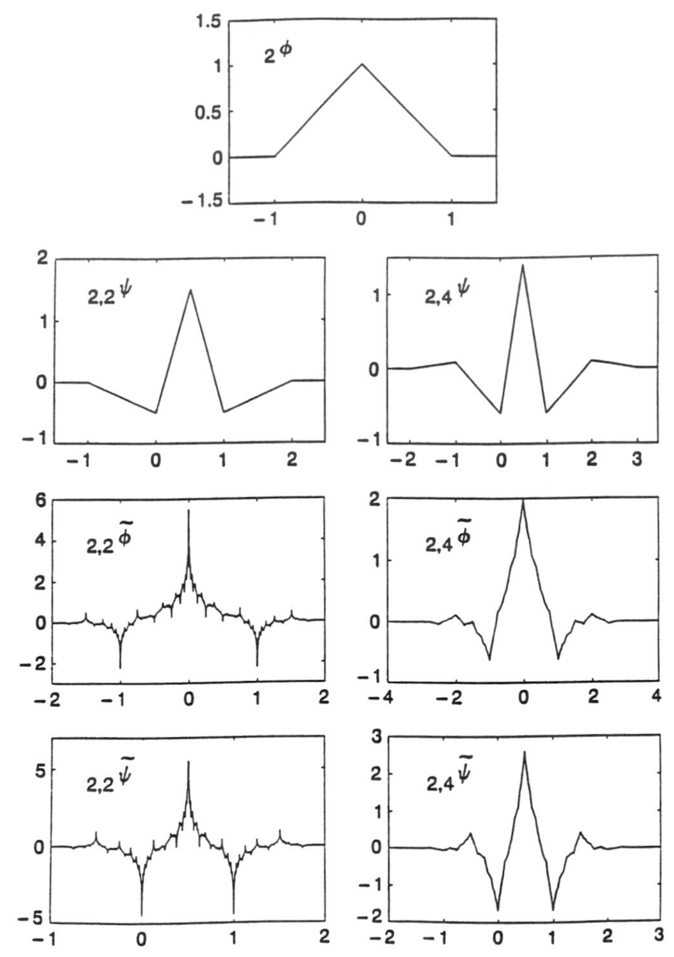

Figure 3. Graphs of the functions $\phi, \psi, \tilde{\phi}$, and $\widetilde{\psi}$ in the spline case.

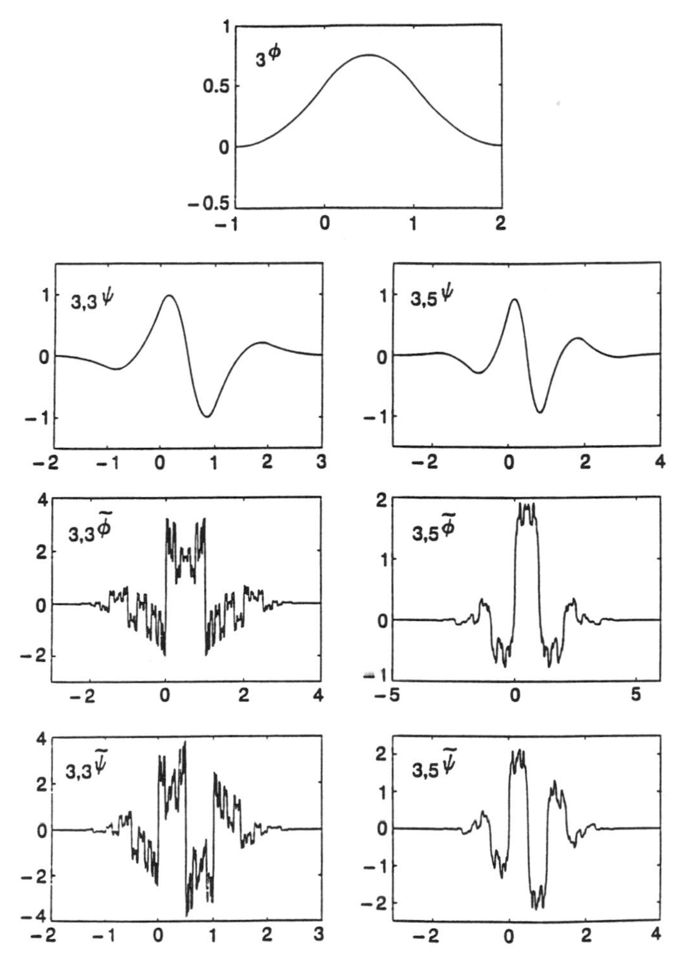

Figure 4. Graphs of the functions $\phi, \psi, \tilde{\phi}$, and $\tilde{\psi}$ in the quadratic case.

Appendix: Proof of some lemmas

Proof of Lemma 2

We shall prove this result for P_j, the argument being the same for $\widetilde{P}_j$, Δ_j, and $\tilde{\Delta}_j$.

First, from the definition (29) of P_j, we have

$$P_j(f(y))(x) = P_0(f(2^j y))(2^{-j}x). \tag{105}$$

It is thus sufficient to prove that P_0 is a continuity operator on $L^2(\mathbb{R})$. We will then have $\|P_0\| = \|P_j\|$ for all j. Since both ϕ and $\tilde{\phi}$ are in L^2 and compactly supported, we can define the two trigonometric polynomials

$$t(\omega) = \sum_{k\in\mathbb{Z}} |\hat{\phi}(\omega + 2k\pi)|^2 = \sum_{n\in\mathbb{Z}} \langle \phi(x)|\phi(x-n)\rangle e^{in\omega}, \tag{106}$$

and

$$\tilde{t}(\omega) = \sum_{k\in\mathbb{Z}} |\hat{\tilde{\phi}}(\omega + 2k\pi)|^2 = \sum_{n\in\mathbb{Z}} \langle \tilde{\phi}(x)|\tilde{\phi}(x-n)\rangle e^{in\omega}. \tag{107}$$

For any sequence $\{e_n\}$ in $\ell^2(\mathbb{Z})$ we have

$$\begin{aligned}
\int_{\mathbb{R}} \left|\sum_{n\in\mathbb{Z}} e_n \phi(x-n)\right|^2 dx &= \frac{1}{2\pi}\int_{\mathbb{R}} \left|\sum_{n\in\mathbb{Z}} e_n e^{-in\omega}\right|^2 |\hat{\phi}(\omega)|^2 d\omega \\
&= \frac{1}{2\pi}\int_{-\pi}^{\pi} \left|\sum_{n\in\mathbb{Z}} e_n e^{-in\omega}\right|^2 \sum_{k\in\mathbb{Z}} |\hat{\phi}(\omega+2k\pi)|^2 d\omega \\
&\leq \max\left(t(\omega)\right) \sum_{n\in\mathbb{Z}} |e_n|^2.
\end{aligned}$$

For all f in $L^2(\mathbb{R})$, we can estimate the ℓ^2 norm of the sequence $\langle f|\tilde{\phi}(x-n)\rangle$ in the following way.

$$\begin{aligned}
\sum_{n\in\mathbb{Z}} |\langle f|\tilde{\phi}(x-n)\rangle|^2 &= \frac{1}{4\pi^2}\sum_{n\in\mathbb{Z}} \left|\int \hat{f}(\omega)\overline{\hat{\tilde{\phi}}(\omega)} e^{in\omega} d\omega\right|^2 \\
&= \frac{1}{2\pi}\int_{-\pi}^{\pi} \left|\sum_{k\in\mathbb{Z}} \hat{f}(\omega+2k\pi)\overline{\hat{\tilde{\phi}}(\omega+2k\pi)}\right|^2 d\omega \\
&\leq \frac{1}{2\pi}\int_{-\pi}^{\pi} \left(\sum_{k\in\mathbb{Z}} |\hat{f}(\omega+2k\pi)|^2\right)\left(\sum_{k\in\mathbb{Z}} |\hat{\tilde{\phi}}(\omega+2k\pi)|^2\right) d\omega \\
&\leq \max\left(\tilde{t}(\omega)\right)\|f\|^2.
\end{aligned}$$

It follows from these two estimates that for all f in $L^2(\mathbb{R})$

$$\|P_0 f\|_{L^2}^2 \leq \max\left(\tilde{t}(\omega)\right)\max\left(t(\omega)\right)\|f\|_{L^2}^2, \tag{108}$$

and thus $\|P_0\| \leq [\max(t)\max(\tilde{t})]^{1/2}$. ■

Proof of Lemma 3

We shall use the Fourier coefficients $h_n, g_n, \tilde{h}_n, \tilde{g}_n$ of the filters $m_0, m_1, \widetilde{m}_0$, and $\widetilde{m}_1$. For any f in $L^2(\mathbb{R})$, we have

$$\begin{aligned}
P_0(f) + \Delta_0(f) &= \sum_{k\in\mathbb{Z}} \langle f|\tilde{\phi}(x-k)\rangle\phi(x-k) + \sum_{k\in\mathbb{Z}} \langle f|\widetilde{\psi}(x-k)\rangle\psi(x-k) \\
&= \sum_{n,m\in\mathbb{Z}} \left(\sum_{k\in\mathbb{Z}} \bar{\tilde{h}}_{n-2k}h_{m-2k} + \sum_{k\in\mathbb{Z}} \bar{\tilde{g}}_{n-2k}g_{m-2k}\right) \\
&\quad 4\langle f|\tilde{\phi}(2x-h)\rangle\phi(2x-m) \\
&= \sum_{n,m\in\mathbb{Z}} \left(\sum_{k\in\mathbb{Z}} [\bar{\tilde{h}}_{n-2k}h_{m-2k} + (-1)^{n+m}\bar{\tilde{h}}_{1-m+2k}h_{1-n+2k}]\right) \\
&\quad 4\langle f|\tilde{\phi}(2x-n)\rangle\phi(2x-m) \\
&= \sum_{m,n\in\mathbb{Z}} 4c_{m,n}\langle f|\tilde{\phi}(2x-n)\rangle\phi(2x-m).
\end{aligned}$$

When $m - n = 2p$, we have

$$c_{m,n} = \sum_{k\in\mathbb{Z}} \bar{\tilde{h}}_k h_{k+m-n} = \sum_{k\in\mathbb{Z}} \bar{\tilde{h}}_k h_{k+2p} = \frac{1}{2}\delta_{0,p} \tag{109}$$

because of the duality relation (21). When $m - n = 2p + 1$, then

$$\begin{aligned}
c_{n,m} &= \sum_{k\in\mathbb{Z}} \bar{\tilde{h}}_{n-2k}h_{m-2k} - \sum_{k\in\mathbb{Z}} \bar{\tilde{h}}_{1-m+2k}h_{1-n+2k} \\
&= 0
\end{aligned}$$

because the two sums contain exactly the same terms. We thus have

$$c_{m,n} = \frac{1}{2}\delta_{n,m} \tag{110}$$

and

$$P_0(f) + \Delta_0(f) = 2\sum_{k\in\mathbb{Z}} \langle f|\tilde{\phi}(2x-k)\rangle\phi(2x-k) = P_{-1}(f). \tag{111}$$

By Equation (105) we can rescale this identity to obtain for all j

$$P_j + \Delta_j = P_{j-1}; \tag{112}$$

the same holds for $\widetilde{P}_j$, $\tilde{\Delta}_j$, and $\widetilde{P}_{j-1}$. ■

Proof of Lemma 4

We will use the following well known approximation result: If f is in $L^2(\mathbb{R})$, for all $\epsilon > 0$, there exists a function g which is a finite linear combination of intervals characteristic functions $\left(g = \sum_{k=1}^{n} \alpha_k \chi_{I_k}\right)$ such that $\|f-g\| \leq \epsilon$. Using Lemma 2, this also gives

$$\|P_j(f)\| = \|P_j(f-g+g)\| \leq \|P_j(g)\| + C\epsilon, \tag{113}$$

and

$$\|P_j(f) - f\| \leq \|P_j(g) - g\| + (C+1)\epsilon. \tag{114}$$

It is thus sufficient to prove the limits (36) and (37) in the case where $f = \chi_{[a,b]}$.

We first prove Equation (36). Using the same argument as in the proof of Lemma 2, we obtain

$$\begin{aligned}
\|P_j(f)\|^2 &\leq C \sum_{k\in\mathbb{Z}} |\langle f|\tilde{\phi}_k^j\rangle|^2 \\
&= C \sum_{k\in\mathbb{Z}} \left| \int_a^b \overline{\tilde{\phi}_k^j(x)} dx \right|^2 \\
&\leq C \sum_{k\in\mathbb{Z}} \int_a^b 2^{-j} |\tilde{\phi}(2^{-j}x - k)|^2 dx \\
&= C \sum_{k\in\mathbb{Z}} \int_{2^{-j}a-k}^{2^{-j}b-k} |\tilde{\phi}(x)|^2 dx
\end{aligned}$$

when j goes to $+\infty$. It is clear that this expression tends to zero if $\tilde{\phi}$ is in $L^2(\mathbb{R})$.

To prove (37), we shall directly evaluate $P_j(f) = P_j(\chi_{[a,b]})$. We first remark that, because of the hypothesis (23), we have

$$\hat{\phi}(2k\pi) = \hat{\tilde{\phi}}(2k\pi) = \delta_{0,k}. \tag{115}$$

This leads to the following summation formulas which are valid for almost all x:

$$\sum_{k\in\mathbb{Z}} \phi(x-k) = \sum_{k\in\mathbb{Z}} \tilde{\phi}(x-k) = 1, \tag{116}$$

and

$$\int \phi(x) = \int \tilde{\phi}(x) = 1. \tag{117}$$

We know that ϕ and $\tilde{\phi}$ are compactly supported; *i.e.*, both vanish out of an interval $[-s, s]$. For $j < 0$ and $|j|$ large enough, we can derive a pointwise estimate of

$$P_j(f)(x) = \sum_{k\in\mathbb{Z}} \langle f|\tilde{\phi}_k^j\rangle \phi_k^j(x). \tag{118}$$

(1) If x is out of $[a-2^{j+1}s, b+2^{j+1}s]$, then either $\phi_k^j(x) = 0$ or $\langle f|\tilde{\phi}_k^j\rangle = 0$ and thus $P_j(f)(x) = 0$.

(2) If x is inside $[a+2^{j+1}s, b-2^{j+1}s]$, then either $\phi_k^j(x) = 0$, or $\langle f|\tilde{\phi}_k^j\rangle = \int \overline{\tilde{\phi}_k^j} dx = 2^{j/2}$ and thus

$$P_j(f)(x) = \sum_{k\in\mathbb{Z}} 2^{j/2}\phi_k^j(x) = \sum_{k\in\mathbb{Z}} \phi(2^{-j}x - k) = 1. \tag{119}$$

In these two cases, $P_j(f)(x)$ is exactly $f(x)$. We still have to evaluate $\int_{R_j} |P_j(f) - f|^2$ where R_j is the residual domain $R_j = [a - 2^{j+1}s, a + 2^{j+1}s] \cup [b - 2^{j+1}s, b + 2^{j+1}s]$. Now note that when x is in R_j, we have

$$f(x) = \chi_{R_{j+1}}(x)f(x) = f_j(x) \quad (R_j \subset R_{j+1}), \tag{120}$$

and

$$P_j(f)(x) = P_j(f_j)(x). \tag{121}$$

Consequently

$$\begin{aligned} \int_{R_j} |P_j(f)(x) - f(x)|^2 dx &= \int_{R_j} |P_j(f_j)(x) - f_j(x)|^2 dx \\ &\le \int_{\mathbb{R}} |P_j(f_j) - f_j|^2 dx \\ &\le C\|f_j\|^2, \end{aligned}$$

and $\|f_j\|^2$ tends to zero when j goes to $-\infty$. This concludes the proof of the lemma. ∎

References

1. Antonini, M., M. Barlaud, P. Mathieu, and I. Daubechies, Image coding using wavelet transform, *IEEE Trans. ASSP*, to appear.
2. Beylkin, G., R. Coifman, and V. Rokhlin, Fast wavelet transform, and numerical algorithms I, *Comm. Pure and Appl. Math.* **44** (1991), 141–183.
3. Chui, C. K. and J. Z. Wang, A general frame work of compactly supported splines and wavelets, CAT Report #219, Texas A&M University, College Station, TX, 1990.
4. Cohen, A. and I. Daubechies, A stability criterion for biorthogonal wavelet bases and their related subband coding schemes, AT&T Bell Laboratories, 1991, preprint.
5. Cohen, A. and I. Daubechies, Nonseparable bidimensional wavelet bases, AT&T Bell Laboratories, 1991, preprint.
6. Cohen, A., I. Daubechies, and J. C. Feauveau, Biorthogonal bases of compactly supported wavelets, *Comm. Pure and Appl. Math.*, 1991, to appear.

7. Cohen, A., Ondelettes, analyses multirésolutions et filtres miroir en quadrature, *Annales de l'IHP, analyse non linéaire* **7** (5) (1990).
8. Coifman, R., Y. Meyer, S. Quacke, and M. V. Wickerhauser, Signal processing and compression with wave packets, *Proceedings of the Conference on Wavelets*, Marseilles, Spring 1989.
9. Conze, J. P. and A. Raugi, Fonctions harmoniques pour un opérateur de transition et applications, Dept. de Math., Université de Rennes, France, 1990, preprint.
10. Daubechies, I., Orthonormal bases of compactly supported wavelets, *Comm. Pure and Appl. Math.* **41** (1988), 909–996.
11. Eirola, T., Sobolev characterization of solutions of dilation equations, Helsinki University of Technology, submitted to *SIAM J. Math. Anal.*, 1991.
12. Lawton, W., Necessary and sufficient conditions for constructing orthonormal wavelets bases, *J. Math. Phys.* **32** (1) (1991), 57–61.
13. Lemarié, P. G., Ondelettes à localisation exponentialle, *J. Math. Pure et Appl.* **67** (1988), 227–236.
14. Mallat, S., A theory for multiresolution signal decomposition: the wavelet representation, *IEEE Pattern Anal. and Machine Intell.* **11** (7) (1989), 674–693.
15. Meyer, Y., *Ondelettes et Opérateurs*, in two volumes, Hermann, Paris, 1990.
16. Smith, M. J. T. and T. P. Barnwell, Exact reconstruction techniques for tree structured subband coders, *IEEE ASSP* **34** (1986), 434–441.
17. Vetterili, M., Filter banks allowing perfect reconstruction, *Signal Processing* **10** (1986), 219–244.
18. Villemoes, L., Energy moments in time and frequency for two scale difference equation solutions and wavelets, Math. Institute, Technical Univ. of Denmark, DK2800 Lyngby, Denmark, 1991, preprint.
19. Young, R. M., *An Introduction to Nonharmonic Fourier Series*, Academic Press, New York, 1980.

Albert Cohen
CEREMADE
Université Paris IX Dauphine
Place du Maréchal de Lattre de Tassigny
75016 Paris, France

Nonorthogonal Multiresolution Analysis Using Wavelets

Jean-Christophe Feauveau

Abstract. Orthogonal multiresolution analysis theory and algorithms are well understood by now. However, for many applications such as image compression or edge detection, the orthogonal framework is not satisfactory since certain natural constraints cannot be achieved. For instance, it is not possible to use finite impulse response linear phase filters. To remove these drawbacks, we introduce here nonorthogonal multiresolution analyses which may be considered as a generalization of the orthogonal setting and provide more freedom for the design of multiresolution analyses.

§1. Introduction

Orthogonal multiresolution analysis (OMRA) defined by S. Mallat [10] and Y. Meyer ([12] and [13]) is a powerful tool for studying signals with features at various scales. In applications, the practical implementation of this transformation is performed by using a basic filter bank $(m_0(\omega), \overline{m}_0(\omega+\pi)e^{-i\omega})$, where $m_0(\omega)$ is a low-pass transfer function and $\overline{m}_0(\omega+\pi)e^{-i\omega}$ a high-pass one.

The OMRA theory imposes the reconstruction condition $|m_0(\omega)|^2 + |m_0(\omega+\pi)|^2 = 1$. This relation is a very strict constraint for filter design as we want to optimize filters for a particular task. For instance, image segmentation generally uses an optimal high-pass detection filter $f(\omega)$ which does not satisfy this orthogonal reconstruction condition. Moreover, it is known that it is not possible to find such a filter $m_0(\omega)$ with finite impulse response (FIR) and linear phase (e.g. symmetric or antisymmetric functions), which are natural requirements for many applications, such as image compression [1]. This paper is aimed at producing a framework for possibly nonorthogonal multiresolution analyses which can naturally use the usual optimized detection or smoothing filters.

We will recall in Section 2 some of the OMRA theory limitations. To generalize this theory we relax the orthogonality condition in Section 3: this will lead to nonorthogonal multiresolution analyses. The mathematical details

Wavelets–A Tutorial in Theory and Applications
C. K. Chui (ed.), pp. 153–178.

ISBN 0-12-174590-2

for nonorthogonal multiresolution analysis are much more complicated, but we still have all the interpretative and algorithmic properties developed by S. Mallat [10]. To conclude this section, we will develop decomposition and reconstruction algorithms which will be used in practice. Finally, some multiresolution analyses examples related to linear, quadratic and cubic cardinal spline interpolation spaces are studied in Section 4.

Before closing the introduction, we want to link this work with the existing literature. Basically, two multiresolution analysis approaches using wavelets have been developed using the orthogonal framework. In a pure wavelet context, before the introduction of the multiresolution approximation spaces by S. Mallat, Ph. Tchamitchian [17] has already defined some families of biorthogonal wavelets (possibly with compact support). The drawback of this approach is the lack of a pyramidal algorithm for wavelet coefficient computation, due to the lack of multiscale approximation spaces. In the orthogonal multiresolution framework (*i.e.*, with the use of an orthogonal projector family), P. G. Lemarié [9] defined some nonorthogonal bases $(\psi_{j,k})_{k\in\mathbb{Z}}$ for inter-scale spline spaces with compactly supported functions. Nevertheless, this work does not consider the computation of wavelet coefficients (in the sequel, this task will be performed using the dual analysis).

Several sources have contributed in the spirit of this paper. For instance, D. LeGall and A. Tabatabai [7] have considered some nonorthogonal perfect reconstruction filter banks. From this result, it was logical to look for multiresolution analyses which use these filter banks. To provide a general framework, we keep the global approach developed by S. Mallat [10] in the orthogonal case. Lastly, a very fruitful joint work with I. Daubechies and A. Cohen [3] enables us to derive accurate conditions for perfect reconstruction filter banks to generate nonorthogonal multiresolution analyses, namely: Theorem 16.

§2. Limitations of orthogonal multiresolution analysis

For many applications of the multiresolution analysis, it is natural to analyze the signal or image using a symmetrical scale function ϕ. Except for the Haar function, we cannot find an FIR function $m_0(\omega)$ leading to OMRA satisfying this requirement. It is possible to use an FIR approximation of a symmetrical filter $m_0(\omega)$; nevertheless, it is often necessary to analyze a signal through many scales and the induced approximation error increases at each decomposition step.

Let us recall that the usual criteria for optimal multiresolution analysis are expressed in terms of regularity, space localization, spectral localization or oscillation of the wavelet ψ. Let $\widehat{\psi}$ be the Fourier transform of ψ. Following I. Daubechies [6], the wavelet regularity can be expressed by:

$$\int_{\mathbb{R}} (1+|\omega|)^N |\widehat{\psi}(\omega)| d\omega < +\infty$$

with N as large as possible. Localization and oscillating characteristics are application dependent. For instance, it is often useful to look for ϕ and ψ such that:

$$\langle f, \phi \rangle \approx f(0) \quad \text{and} \quad \langle f, \psi \rangle \approx 0$$

for all regular functions. To satisfy these approximative relations, the following formulae are sufficient:

$$\int_{\mathbb{R}} x^n \phi(x) dx = 0 \quad \text{for} \quad 0 < n \leq M_0, \qquad \phi(0) = 0$$

and

$$\int_{\mathbb{R}} x^n \psi(x) dx = 0 \quad \text{for} \quad 0 \leq n \leq M_1.$$

This is a simple consequence of a Taylor development. A more accurate study of these constraints for OMRA can be found in [4].

In practice, the design of a filter $m_0(\omega)$ verifying the perfect reconstruction relation is performed via a Riesz algorithm ([15] and [16]), and these constraints are difficult to take into account. To deal with these limitations, we are going to introduce in the next section the nonorthogonal multiresolution analysis framework.

§3. Nonorthogonal multiresolution analysis

Let us recall that the adjective "orthogonal" applied to multiresolution analyses means orthogonal projections. In this case, the projectors are completely characterized by their projection spaces. In the following, we remove the *a priori* orthogonality condition and this will provide a more flexible framework which allow for the removal of the limitations in the previous section.

3.1. The model and its mathematical characterization

In this subsection, the model of the nonorthogonal multiresolution analysis (NOMRA) is presented in Subsection 3.1.1; the dual multiresolution analysis is discussed in Subsection 3.1.2; and the mathematical characterization of NOMRA is described in Subsection 3.1.3.

3.1.1. The model

To see what a nonorthogonal multiresolution analysis should be, we use a weaker formal model introduced by S. Mallat in the orthogonal case [10]. Let $f(x)$ be a measurable finite energy signal and A_j the projection used to approximate the signal at the resolution 2^j. Using the orthogonal case, we now develop the natural properties of the projector A_j in a nonorthogonal multiresolution analysis framework.

(1) A_j is a continuous linear operator. If $A_j f(x)$ is the approximation of $f(x)$ at resolution 2^j, then $A_j f(x)$ is not changed by an approximation at the same resolution, namely: $A_j \circ A_j = A_j$. Thus, the operator A_j is a projector, characterized by its projection space V_j and its projection direction which is denoted Z_j. The space V_j can be interpreted as the set of the approximated functions of $L^2(\mathbb{R})$ at resolution 2^j.

(2) All the informations useful to compute the approximate signal at resolution 2^{j+1} are contained in the approximate signal at resolution 2^j, so:

$$V_j \subset V_{j+1}, \qquad \forall j \in \mathbb{Z}.$$

Moreover, a signal which has no more information at resolution 2^j must have no information at lower resolutions:

$$Z_{j+1} \subset Z_j, \qquad \forall j \in \mathbb{Z}.$$

(3) The approximation operation must be similar at every resolution. Thus, these approximation spaces must be deduced from each other by a simple dilation of functions:

$$f(x) \in V_j \Leftrightarrow f(2x) \in V_{j+1}, \qquad \forall j \in \mathbb{Z},$$

and

$$f(x) \in Z_j \Leftrightarrow f(2x) \in Z_{j+1}, \qquad \forall j \in \mathbb{Z}.$$

(4) The translation of $k2^{-j}$ of the approximate function at scale 2^j is equal to the approximate function at scale 2^j translated by $k2^{-j}$. Considering property (3), we can express this condition with $j = 0$:

$$A_0(f(x))(y) = g(y) \Leftrightarrow A_0(f(x-k))(y) = g(y-k), \qquad \forall f \in L^2(\mathbb{R}), \forall k \in \mathbb{Z}.$$

(5) The approximate signal at resolution 2^j can be characterized by 2^j equidistant coefficients per unit length. Moreover, Property (4) must be translated on these coefficients. Mathematically, this is expressed by an isomorphism (*i.e.*, a bijective linear continuous operator with continuous inverse) ξ_0 for V_0 onto $\ell^2(\mathbb{Z})$ which commute under the group action $(\mathbb{Z}, +)$: there exist two strictly positive real numbers σ_0 and σ_1 such that:

$$\sigma_0 \|f\|_{L^2} \leq \|\xi_0(f)\|_{\ell^2} \leq \sigma_1 \|f\|_{L^2},$$

and

$$\xi_0(A_0(f(x))) = (u_n)_{n \in \mathbb{Z}} \Leftrightarrow \xi_0(A_0(f(x-k))) = (u_{n-k})_{n \in \mathbb{Z}}.$$

(6) The system provides a complete decomposition of $L^2(\mathbb{R})$:

$$\overline{\lim_{j \to +\infty} V_j} = \overline{\bigcup_{j \in \mathbb{Z}} V_j} = L^2(\mathbb{R}),$$

and

$$\lim_{j \to -\infty} V_j = \bigcap_{j \in \mathbb{Z}} V_j = \{0\}.$$

Definition 1 . *A sequence of projectors $(A_j)_{j\in\mathbb{Z}}$ satisfying the six properties described above is called a nonorthogonal multiresolution analysis (NOMRA) of $L^2(\mathbb{R})$.*

To conclude this subsection, let us make some remarks:

(a) Orthogonal multiresolution analyses are included in the set of NOMRA. Although a better terminology should be merely "multiresolution analyses" without the adjective "nonorthogonal", we do not do so here because there exist several other multiresolution theories, such as [14], [18], [20], or [19], for instance.

(b) In the orthogonal case, $Z_j = (V_j)^\perp$ and the conditions in Property (2) are equivalent; the same remark can be applied to Property (3). In the general case, it is no longer possible to describe the projector A_j with V_j and we must split the condition into two parts.

(c) According to Property (1), since the projectors A_j are continuous, it follows that the subspaces V_j and Z_j are necessarily closed.

(d) Let D_y: $L^2(\mathbb{R}) \to L^2(\mathbb{R})$ defined by $D_y(f)(x) = f(yx)$, be the dilation operator. Property (3) is equivalent to:

$$D_2 \circ A_j = A_{j+1} \circ D_2, \qquad \forall j \in \mathbb{Z}. \tag{1}$$

(e) Let T_k: $L^2(\mathbb{R}) \to L^2(\mathbb{R})$ defined by $T_k(f)(x) = f(x-k)$, be the translation operator. Property (4) can be written:

$$T_k \circ A_0 = A_0 \circ T_k, \qquad \forall k \in \mathbb{Z}. \tag{2}$$

3.1.2. The dual multiresolution analysis

Before giving a mathematical description of the properties defining a NOMRA, we show that the previous properties define, in fact, two NOMRA.

Let us first recall the dual morphism notion in Hilbert spaces. Let $(E, \langle,\rangle_E)$ and $(F, \langle,\rangle_F)$ be two Hilbert spaces and A a continuous operator from $(E, \langle,\rangle_E)$ to $(F, \langle,\rangle_F)$. Then, there exists a unique continuous operator A^* from $(F, \langle,\rangle_F)$ to $(E, \langle,\rangle_E)$ such that:

$$\langle A(f), g\rangle_F = \langle f, A^*(g)\rangle_E, \qquad \forall f \in E, \ \forall g \in F.$$

By definition, A^* is the adjoint morphism of A.

Theorem 2. *Let $(A_j)_{j\in\mathbb{Z}}$ be a NOMRA with projectors defined by the spaces $(V_j)_{j\in\mathbb{Z}}$ and the projection directions $(Z_j)_{j\in\mathbb{Z}}$. Then, the dual family $(A_j^*)_{j\in\mathbb{Z}}$ is a NOMRA with projectors on spaces $V_j^* = (Z_j)^\perp$ and projection directions $Z_j^* = (V_j)^\perp$. In such a case, $(A_j^*)_{j\in\mathbb{Z}}$ will be called the dual NOMRA associated with $(A_j)_{j\in\mathbb{Z}}$. Moreover, if ξ_0 is an isomorphism from V_0 onto $\ell^2(\mathbb{Z})$ and $\xi_0' = \xi_0 \circ A_0$, then the adjoint operator of ξ_0', denoted by ${\xi'}_0^*$, is an operator from $\ell^2(\mathbb{Z})$ onto V_0^*. In this case, the inverse ξ_0^{adj} of ξ_0^*, defined on its image, is an isomorphism from V_0^* onto $\ell^2(\mathbb{Z})$.*

Proof: Let $(A_j)_{j\in\mathbb{Z}}$ be a NOMRA, so that A_j is a continuous projector on V_j in the direction Z_j. These two spaces are necessarily closed and A_j^* is the continuous projector on $V_j^* = (Z_j)^\perp$ in the direction $Z_j^* = (V_j)^\perp$. Indeed, if $(f,g) \in (L^2(\mathbb{R}))^2$, we can find four functions $(f_1, f_2, g_1, g_2) \in V_j \times Z_j \times V_j^* \times Z_j^*$ such that:

$$f = f_1 + f_2, \quad \text{and} \quad g = g_1 + g_2,$$

and we have

$$\langle A_j(f), g\rangle = \langle f, A_j^*(g)\rangle = \langle f_1, g_1\rangle.$$

We have to verify that $(A_j^*)_{j\in\mathbb{Z}}$ is a NOMRA. Properties (1) and (6) are obviously satisfied. For Property (2), we simply apply the orthogonality to the inclusion provided by Property (2) and satisfied by $(A_j)_{j\in\mathbb{Z}}$. Property (3) for $(A_j)_{j\in\mathbb{Z}}$ can be represented by $D_2 \circ A_j = A_{j+1} \circ D_2$, and we can write

$$A_j^* \circ D_2^* = D_2^* \circ A_{j+1}^*.$$

But, for f and g in $L^2(\mathbb{R})$, we have

$$\langle D_2(f), g\rangle = \int_{\mathbb{R}} f(2x)\bar{g}(x)dx = \frac{1}{2}\int_{\mathbb{R}} f(x)\bar{g}\left(\frac{x}{2}\right)dx = \frac{1}{2}\langle f, D_{1/2}(g)\rangle,$$

and $D_2^* = \frac{1}{2}D_{1/2}$. Hence, $A_j^* \circ D_{1/2} = D_{1/2} \circ A_{j+1}^*$, which can be written as $D_2 \circ A_j^* = A_{j+1}^* \circ D_2$ (this is Property (3)).

The proof of Property (4) is very similar: by the hypothesis, we have $T_k \circ A_0 = A_0 \circ T_k$ and clearly $T_k^* = T_{-k}$. Thus, we have $T_k \circ A_0^* = A_0^* \circ T_k$ for every integer k. Finally, we have to verify Property (5) for $(A_j^*)_{j\in\mathbb{Z}}$. Let ξ_0 be the isomorphism from V_0 onto $\ell^2(\mathbb{Z})$ introduced by $(A_j^*)_{j\in\mathbb{Z}}$. Let $\xi_0' = \xi_0 \circ A_0$ be the operator from $L^2(\mathbb{R})$ to $\ell^2(\mathbb{Z})$ and ${\xi'}_0^* = A_0^* \circ \xi_0^*$ the adjoint of ξ_0' (which is an operator from $\ell^2(\mathbb{Z})$ on $L^2(\mathbb{R})$). We have the relation:

$$\langle \xi_0'(f), u\rangle_{\ell^2} = \langle f, {\xi'}_0^*(u)\rangle_{L^2}, \qquad \forall f \in \ell^2(\mathbb{Z}), \ \forall u \in \ell^2(\mathbb{Z}).$$

Let us point out the following properties:

(i) ${\xi'}_0^*$ is continuous;
(ii) ${\xi'}_0^*$ is injective;
(iii) The space V_0^* is the range of ${\xi'}_0^*$.

To justify (ii), note that if ${\xi'}_0^*(u) = 0$, then the duality equality using $f = \xi_0^{-1}(u)$ leads to $\|u\|_{\ell^2} = 0$. To verify (iii) we have ${\xi'}_0^* = A_0^* \circ \xi_0^*$ and the range is clearly included in V_0^*. The operator ξ_0^* is an isomorphism from $\ell^2(\mathbb{Z})$ onto V_0. If for every g in V_0^* we can find an h in V_0, such that $A_0^*(h) = g$, then the property is proved. But for every g in V_0^* we can find f in $L^2(\mathbb{R})$ such that $A_0^*(f) = g$. Moreover, we can find f_1 in $(Z_0^*)^\perp = V_0$ and f_2 in $Z_0^* = (V_0)^\perp$, such that $f = f_1 + f_2$. So we have $A_0^*(f_1) = g$ with f_1 in V_0.

Let ξ_0^{adj} be the inverse of ${\xi'}_0^*$ on its range V_0^*. We can summarize our knowledge about ξ_0^{adj} as follows: $(\xi_0^{adj})^{-1}$ is a continuous linear bijection between two Banach spaces. The open mapping theorem allows us to conclude that ξ_0^{adj} is continuous. So ξ_0^{adj} is an isomorphism from V_0^* onto $\ell^2(\mathbb{Z})$. The invariance of ξ_0^{adj} under the action of $\mathbb{Z}$ is easy to prove by using duality brackets. We can conclude that $(A_j^*)_{j\in\mathbb{Z}}$ satisfies Property (5).

This competes the proof of Theorem 2.

To conclude this subsection, let us observe two points.

(i) When $(A_j)_{j\in\mathbb{Z}}$ is an orthogonal multiresolution analysis, the families $(A_j^*)_{j\in\mathbb{Z}}$ and $(A_j)_{j\in\mathbb{Z}}$ are clearly identical.

(ii) We easily verify that the dual NOMRA $(A_j^{**})_{j\in\mathbb{Z}}$ of a dual NOMRA $(A_j^*)_{j\in\mathbb{Z}}$ is exactly $(A_j)_{j\in\mathbb{Z}}$: the bi-dual multiresolution analysis is identical to itself.

3.1.3. The mathematical characterization

Let $(A_j)_{j\in\mathbb{Z}}$ be a NOMRA. Property (6) can be written equivalently as:

$$\lim_{j\to+\infty} Z_j = \bigcap_{j\in\mathbb{Z}} Z_j = \{0\}$$

and

$$\overline{\lim_{j\to-\infty} Z_j} = \overline{\bigcup_{j\in\mathbb{Z}} Z_j} = L^2(\mathbb{R}).$$

For the purpose of a later interpretation in Section 3.4, we define an inter-scale projection direction $W_j \subset Z_j$ by:

$$V_{j+1} = V_j \oplus W_j \quad \text{and} \quad A_j(f) = 0, \qquad \forall\, f \in W_j. \tag{3}$$

So we have $Z_j = \overline{\bigoplus_{p\geq j} W_p}$, and a NOMRA $(A_j)_{j\in\mathbb{Z}}$ is completely characterized by the sequences $(V_j)_{j\in\mathbb{Z}}$ and $(W_j)_{j\in\mathbb{Z}}$. Let us observe that, being the kernel of the continuous projector A_j restricted to the Hilbert space V_{j+1}, W_j is closed.

The following theorem, which is easy to prove, allows us to give a mathematical definition of NOMRA.

Theorem 3. *A NOMRA $(A_j)_{j\in\mathbb{Z}}$ defines a unique sequence of projection spaces $(V_j)_{j\in\mathbb{Z}}$ and a unique sequence of inter-scale spaces $(W_j)_{j\in\mathbb{Z}}$. These spaces are necessarily closed. Conversely, let $(V_j)_{j\in\mathbb{Z}}$ and $(W_j)_{j\in\mathbb{Z}}$ be two sequences of closed subspaces of $L^2(\mathbb{R})$ such that:*

(i) $\forall j \in \mathbb{Z}, V_j \subset V_{j+1}$;

(ii) $\overline{\lim_{j\to+\infty} V_j} = \overline{\bigcup_{j\in\mathbb{Z}} V_j} = L^2(\mathbb{R})$;

(iii) $\lim_{j\to-\infty} V_j = \bigcap_{j\in\mathbb{Z}} V_j = \{0\}$;

(iv) $\forall j \in \mathbb{Z}$,

$$f(x) \in V_j \Leftrightarrow f(2x) \in V_{j+1} \quad \textit{and}$$
$$f(x) \in W_j \Leftrightarrow f(2x) \in W_{j+1};$$

(v) $\forall k \in \mathbb{Z}$,

$$f(x) \in V_0 \Rightarrow f(x-k) \in V_0 \quad \text{and}$$
$$f(x) \in W_0 \Rightarrow f(x-k) \in W_0;$$

(vi) *there exists an isomorphism ξ_0 from V_0 onto $\ell^2(\mathbb{Z})$ which commutes with integer translations.*

Then the sequence $(A_j)_{j\in\mathbb{Z}}$, such that A_j is the projector on V_j in the direction $\overline{\bigoplus_{p\geq j} W_p}$, defines a NOMRA.

3.2. Nonorthogonal multiresolution approximation theory

The purpose of this subsection is to study the structure of the projection spaces that define a NOMRA. We first recall some mathematical tools in Subsection 3.2.1. The projection spaces and the explicit computation of the signal approximation at scale 2^j will be discussed in Subsection 3.2.2.

3.2.1. Some mathematical tools

In the following, we use the usual notations $f_j(x)$ and $f_{j,k}(x)$ introduced in the orthogonal theory, namely:

$$f_j(x) = 2^{j/2} f(2^j x), \qquad \forall x \in \mathbb{R};$$
$$f_{jk}(x) = 2^{j/2} f(2^j x - k), \qquad \forall x \in \mathbb{R}.$$

Later in Subsection 3.2.2, we will also need the following notation: $U_p = \{u_{k-p}\}_{k\in\mathbb{Z}}$ for any $U = \{u_k\}_{k\in\mathbb{Z}} \in \ell^2(\mathbb{Z})$. Lastly we will use the Kronecker symbol $\delta_{p,q}$ defined by

$$\delta_{p,q} = \begin{cases} 1 & \text{for} \quad p = q \\ 0 & \text{for} \quad p \neq q \end{cases}$$

and the Kronecker sequence Δ is denoted by

$$\Delta = \Delta_0 = (\delta_{0,k})_{k\in\mathbb{Z}}.$$

The study of functional spaces associated with NOMRA is closely related to the concept of unconditional basis. Let us recall the definition and a characterization of such bases.

Definition 4. *A sequence $(f(x;k))_{k\in\mathbb{Z}}$ of functions in $L^2(\mathbb{R})$ is an unconditional family if we can find two strictly positive real numbers ρ_0 and ρ_1 such that for all sequence $(x_k)_{k\in\mathbb{Z}}$ in $\ell^2(\mathbb{Z})$, the function $g(x) = \sum_{n=-\infty}^{+\infty} x_n f(x;n)$ satisfies*

$$\rho_0 \|\{x_k\}\|_{\ell^2} \leq \|g\|_{L^2} \leq \rho_1 \|\{x_k\}\|_{\ell^2}.$$

Lemma 5. *If $f(x;k) = f_{j,k}(x)$, then $(f_{j,k}(x))_{k\in\mathbb{Z}}$ is an unconditional family if and only if there exist two strictly positive real numbers σ_0 and σ_1 such that for every ω*

$$\sigma_0 \leq \sum_{k=-\infty}^{+\infty} |\hat{f}(\omega + 2k\pi)|^2 \leq \sigma_1.$$

The proof of this lemma is a simple application of Poisson's summation formula (see [11]).

Let us now recall the definition of bi-orthogonality. This notion was previously introduced in the wavelet theory framework by Ph. Tchamitchian [17].

Definition 6. *Two sequences* $((f(x;k))_{k\in\mathbb{Z}}, (g(x;k)_{k\in\mathbb{Z}})$ *of functions in* $L^2(\mathbb{R})$ *are said to be bi-orthogonal if:*

$$\langle f(x;p), g(x;q)\rangle = \delta_{p,q}.$$

In the wavelet context, we have the following result:

Lemma 7. *The two sequences* $((\phi_{j,k})_{k\in\mathbb{Z}}$ *and* $(\phi'_{j,k})_{k\in\mathbb{Z}})$ *are bi-orthogonal if and only if:*

$$\forall\,\omega\in\mathbb{R},\qquad \sum_{k=-\infty}^{+\infty}\hat{\phi}(\omega+2k\pi)\bar{\hat{\phi}}'(\omega+2k\pi)=1. \tag{4}$$

Let us notice that the bi-orthogonality property for $j = 0$ implies the bi-orthogonality for every j. Equation (4) is a straightforward application of Poisson's formula for $j = 0$. In the orthogonal case (*i.e.*, $\phi = \phi'$), we find the usual relation $\sum_{k=-\infty}^{+\infty} |\hat{\phi}(\omega+2k\pi)|^2 = 1$.

3.2.2. Projection of a signal into scale spaces

To understand the structure of the space defined by a NOMRA, we are going to study the consequences of Properties (1) to (6). First, in the general non-orthogonal framework, we deduce from Property (5) that V_j has the same structure as the orthogonal case.

Theorem 8. *Let* $(A_j)_{j\in\mathbb{Z}}$ *be a NOMRA on* $L^2(\mathbb{R})$ *and* ξ_0 *be the isomorphism induced by Property (4). Let* $U \in \ell^2(\mathbb{Z})$ *and* $\phi \in L^2(\mathbb{R})$ *be such that* $U = \xi_0(\phi)$. *Then the two following properties are equivalent:*

(i) *the family* $(U_p)_{p\in\mathbb{Z}}$ *is an unconditional basis of* $\ell^2(\mathbb{R})$;

(ii) *for every integer* j, *the family* $(\phi_{j,k})_{k\in\mathbb{Z}}$ *is an unconditional basis of* V_j.

Such a function ϕ *will be called a scaling function of the analysis.*

The Kronecker sequence Δ is the canonical unconditional basis of $\ell^2(\mathbb{R})$, and this is why we will call $\phi = \xi_0^{-1}(\Delta)$ the scaling function of the analysis (for a fixed isomorphism ξ_0).

The proof is straightforward: the isomorphism commutes under the action of $\mathbb{Z}$; hence $U = \xi_0(\phi)$ implies $U_p = \xi_0(\phi_{0,p})$. Moreover, ξ_0 is a topological isomorphism, and it must keep the unconditional characteristics of the bases.

Let us note that the set of possible scaling functions is defined by $\xi_0^{-1} \circ \mu(U_0)$, where μ is any topological isomorphism of $\ell^2(\mathbb{Z})$. Accordingly, many choices are possible, and we should impose the orthogonality of the $(\phi_{j,k})_{k\in\mathbb{Z}}$. It turns out to be a bad idea for a practical implementation of these NOMRA.

Usually, the orthogonality hypothesis on projectors leads to explicit formula to calculate the coordinates of $A_j(f)$ in the orthonormal basis of V_j using scalar products of f with the basis elements. This property leads to inter-scale and degraded signals coordinates interpretation in term of filtering and subsampling of the original signal. We are going to show that these properties are preserved in the general case.

Proposition 9. *Let $(A_j)_{j\in\mathbb{Z}}$ be a NOMRA and $\phi \in L^2(\mathbb{R})$ the scaling function. There exists a function $\phi^* \in L^2(\mathbb{R})$ such that:*

$$A_j(f) = \sum_{-\infty}^{+\infty} \langle f, \phi^*_{j,k}\rangle \phi_{j,k}, \qquad \forall f \in L^2(\mathbb{R}). \tag{5}$$

Moreover, if ϕ is induced by the isomorphism ξ_0, and if ξ_0^{adj} is the isomorphism associated to the dual NOMRA using Theorem 2, then

$$\phi^* = (\xi_0^{adj})^{-1} \circ \xi_0(\phi).$$

The proof of Proposition 9 is delayed to Appendix A.

This result is very important because it provides the projections with an explicit means of calculation. Thus, the approximation $A_j(f)$ of f at scale 2^j is completely characterized by its coordinates on the basis $(\phi_{j,k})_{k\in\mathbb{Z}}$ which are calculated using scalar products. Moreover, these coordinates can be interpreted as a sampling of the convolution $f(x)$ with $\bar{\phi}^*_j(-x)$ at the dyadic points $2^{-j}k$. Let us remark that in the orthogonal case, we have the equality $\phi^* = \phi$, and Propositoin 9 is easily verified.

Equation (5) applied to $f(x) = \phi_{j,k}(x)$ yields the bi-orthogonality of the families $((\phi_{j,k})_{k\in\mathbb{Z}}, (\phi^*_{j,k})_{k\in\mathbb{Z}})$. Using Lemma 7, we also have

$$\sum_{k=-\infty}^{+\infty} \hat{\phi}(\omega + 2k\pi)\bar{\hat{\phi}}^*(\omega + 2k\pi) = 1, \qquad \forall x \in \mathbb{R}. \tag{6}$$

For a future pyramidal implementation purpose, we now study the relations between coordinates at two consecutive scales 2^{j+1} and 2^j.

Proposition 10. *There exist two sequences $(a_k)_{k\in\mathbb{Z}}$ and $(b_k)_{k\in\mathbb{Z}}$ in $\ell^2(\mathbb{Z})$ such that:*

$$\phi\left(\frac{x}{2}\right) = 2\sum_{n=-\infty}^{+\infty} a_n \phi(x+n) \tag{7}$$

and

$$\phi^*\left(\frac{x}{2}\right) = 2\sum_{n=-\infty}^{+\infty} b_n \phi^*(x+n). \tag{8}$$

Equivalently, these two sequences are the impulse responses of two functions α and β in $L^2[0, 2\pi)$ such that:

$$\hat{\phi}(2\omega) = \alpha(\omega)\hat{\phi}(\omega) \tag{9}$$

and

$$\hat{\phi}^*(2\omega) = \beta(\omega)\hat{\phi}^*(\omega). \tag{10}$$

The proof of Proposition 10 can be found in Appendix B.

We have introduced two 2π-periodic funtions $\alpha(\omega)$ and $\beta(\omega)$. Let us observe that $\alpha(\omega) = \beta(\omega) = m_0(\omega)$ in the orthogonal case. As we will prove in Subsection 3.4, Equation (8) will enable us to compute the coordinates of the projection at scale 2^j from the projection at scale 2^{j+1}. This is why we are going to precisely study those two filter functions.

3.3. The two-scale functions

For implementation and NOMRA design, we are going to apply to $\alpha(\omega)$ and $\beta(\omega)$ results in the previous section. First, from Equation (4), we have the identity:

$$\sum_{k=-\infty}^{+\infty} \hat{\phi}(\omega + 2k\pi)\bar{\hat{\phi}}^*(\omega + 2k\pi) = 1,$$

and using Equations (9) and (10), we find:

$$\sum_{k=-\infty}^{+\infty} \alpha(\omega + k\pi)\bar{\beta}(\omega + k\pi)\hat{\phi}(x + k\pi)\bar{\hat{\phi}}^*(x + k\pi) = 1.$$

Splitting this sum according to k is even or odd, we derive the following result.

Proposition 11. *The functions $\alpha(\omega)$ and $\beta(\omega)$ satisfy the relation:*

$$\alpha(\omega)\bar{\beta}(\omega) + \alpha(\omega + \pi)\bar{\beta}(\omega + \pi) = 1. \tag{11}$$

Let us remark that in the orthogonal case, we find the usual relation $|\alpha(\omega)|^2 + |\alpha(\omega + \pi)|^2 = 1$. We now prove a weaker version of this relation in the general case.

Proposition 12. *We can find four strictly positive real numbers ρ_0, ρ_1, σ_0 and σ_1 such that:*

$$\rho_0 \leq |\alpha(\omega + \pi)|^2 + |\alpha(\omega)|^2 \leq \rho_1, \tag{12}$$

and

$$\sigma_0 \leq |\beta(\omega + \pi)|^2 + |\beta(\omega)|^2 \leq \sigma_1, \tag{13}$$

for all $\omega \in [0, 2\pi)$.

Proof: By the hypothesis, $(\phi_{j,k})_{k\in\mathbb{Z}}$ is an unconditional family. In view of Lemma 5, there exist two strictly positive real numbers τ_0 and τ_1 such that:

$$\tau_0 \leq G(\omega) = \sum_{k=-\infty}^{+\infty} |\hat{\phi}(\omega + 2k\pi)|^2 \leq \tau_1, \qquad \omega \in \mathbb{R}.$$

Since we clearly have the identity

$$G(2\omega) = |\alpha(\omega+\pi)|^2 G(\omega) + |\alpha(\omega)|^2 G(\omega+\pi),$$

it follows that $\frac{\tau_0}{\tau_1} \leq |\alpha(\omega+\pi)|^2 + |\alpha(\omega)|^2 \leq \frac{\tau_1}{\tau_0}$, which proves the first part of the result. To obtain Equation (13), we apply Equation (12) to the dual analysis.

To conclude this subsection, we will precise the numerical values of α and β in 0 and π. Let us recall that, in practice, we want to use some regular NORMA. Let us assume that ϕ and ϕ^* are in $L^1(\mathbb{R}) \cap L^2(\mathbb{R})$ which is a very weak hypothesis. Then, $\hat{\phi}$ and $\hat{\phi}^*$ are necessarily continuous functions and Equations (9) and (10) lead to:

$$\alpha(0) = 1 \tag{14}$$

and

$$\beta(0) = 1. \tag{15}$$

We now prove that under the assumption ϕ and ϕ^* in $L^1(\mathbb{R}) \cap L^2(\mathbb{R})$, we also have:

$$\alpha(\pi) = 0, \tag{16}$$

and

$$\beta(\pi) = 0. \tag{17}$$

Indeed, the unconditional basis $(\phi_{j,k})_{k\in\mathbb{Z}}$ of V_j allows us to build an orthonormal basis $(\phi'_{j,k})_{k\in\mathbb{Z}}$ using the function ϕ' defined by:

$$\hat{\phi}'(\omega) = \frac{\hat{\phi}(\omega)}{\sqrt{\sum_{k=-\infty}^{+\infty} |\hat{\phi}(\omega+2k\pi)|^2}}.$$

The associated numerical filter is:

$$m_0(\omega) = \sqrt{\frac{G(2\omega)}{G(\omega)}}\alpha(\omega), \quad \text{with} \quad G(\omega) = \sum_{k=-\infty}^{+\infty} |\hat{\phi}(\omega+2k\pi)|^2.$$

We have:

$$|m_0(\omega+\pi)|^2 + |m_0(\omega)|^2 = 1,$$

and we just proved that $m_0(0) = 1$ because $\hat{\phi}'$ is also continuous. We conclude that $m_0(\pi) = 0$, which implies Equation (16). Equation (17) is obtained by duality.

3.4. Wavelet representation of a signal

In order to describe a signal using non-redundant information, it is natural to characterize the details which disappear between scales 2^{j+1} and 2^j. Using the ideas developed in the orthogonal case, we now introduce a (nonorthogonal) wavelet to describe the inter-scale spaces $(W_j)_{j\in\mathbb{Z}}$ introduced in Subsection 3.1.

Let $(A_j)_{j\in\mathbb{Z}}$ be a NORMA and $(V_j)_{j\in\mathbb{Z}}$ the associated projection spaces. We define the inter-scale projection direction $(W_j)_{j\in\mathbb{Z}}$ as in Equation (3). They are closed subspaces such that:

$$V_{j+1} = V_j \oplus W_j \tag{18}$$

and

$$A_j(f) = 0, \qquad \forall f \in W_j. \tag{19}$$

Once more, we will show that W_j has a distinctive structure.

Theorem 13. *Let $(A_j)_{j\in\mathbb{Z}}$ be a NOMRA, $\phi(x)$ the scaling function and $\phi^*(x)$, $\alpha(\omega)$ and $\beta(\omega)$ the associated functions defined previously. Then, there exists a function $\psi(x)$ in $L^2(\mathbb{R})$ defined by:*

$$\widehat{\psi}(2\omega) = \bar{\beta}(\omega+\pi)e^{-i\omega}\hat{\phi}(\omega), \tag{20}$$

such that $(\psi_{j,k})_{k\in\mathbb{Z}}$ is an unconditional basis of W_j; that is,

$$W_j = \overline{Span(\psi_{j,k})_{k\in\mathbb{Z}}}.$$

The proof of Theorem 13 is delayed to Appendix C.

Definition 14. *A function ψ will be called a wavelet if the family $(\psi_{j,k})_{(j,k)\in\mathbb{Z}^2}$ is an unconditional basis of $L^2(\mathbb{R})$.*

Theorem 13 is important from the theoretical point of view. Nevertheless, in an algorithmic perspective, we must find a means to calculate the information lost between two consecutive scales. It is precisely the role of the following function ψ^* defined here:

Proposition 15. *Under the hypotheses in Theorem 13, there exists a function $\psi^*(x)$ in $L^2(\mathbb{R})$ defined by:*

$$\widehat{\psi}^*(2\omega) = \bar{\alpha}(\omega+\pi)e^{-i\omega}\hat{\phi}^*(\omega), \tag{21}$$

such that for every f in $L^2(\mathbb{R})$ we have:

$$A_{j+1}(f) - A_j(f) = \sum_{k=-\infty}^{+\infty} \langle f, \psi^*_{j,k}\rangle \psi_{j,k}. \tag{22}$$

In particular, for every function f in W_j,

$$f = \sum_{k=-\infty}^{+\infty} \langle f, \psi^*_{j,k}\rangle \psi_{j,k}. \tag{23}$$

The proof of this proposition is delayed to Appendix D.

To conclude this section, let us note that Equation (23) expresses the fact that $(\psi_{j,k})_{k\in\mathbb{Z}}$ and $(\psi^*_{j,k})_{k\in\mathbb{Z}}$ are bi-orthogonal families.

3.5. Algorithms from the nonorthogonal multiresolution analysis

The aim of this Subsection is to derive simple algorithms to decompose and reconstruct the signal to be analyzed.

Let $(A_j)_{j\in\mathbb{Z}}$ be a NOMRA, $(V_j)_{j\in\mathbb{Z}}$ and $(W_j)_{j\in\mathbb{Z}}$ be the corresponding scale and inter-scale spaces. Also, let $\phi(x)$, $\phi^*(x)$, $\psi(x)$, $\psi^*(x)$, $\alpha(\omega)$, and $\beta(\omega)$ be the functions previously associated with the analysis. We want to analyze a measurable finite energy signal $\mathcal{J} \in L^2(\mathbb{R})$. Let us denote by $\mathcal{J}_j$ the projection onto V_j and by $\mathcal{I}_j$ the projection of the lost detail onto W_j; that is,

$$\mathcal{J}_j = \sum_k C_{j,k}\phi_{j,k} \quad \text{with} \quad C_{j,k} = \langle \mathcal{J}, \phi^*_{j,k}\rangle, \quad \text{and} \tag{24}$$

$$\mathcal{I}_j = \sum_k S_{j,k}\psi_{j,k} \quad \text{with} \quad S_{j,k} = \langle \mathcal{J}, \psi^*_{j,k}\rangle. \tag{25}$$

We set $I_j = (C_{j,k})_{k\in\mathbb{Z}}$ and $J_j = (S_{j,k})_{k\in\mathbb{Z}}$. The sequence I_j is a representation of $\mathcal{J}$ at scale 2^j, and J_j is a representation of the details lost between scales 2^{j+1} and 2^j. We will assume that the input of the algorithm is I_0: the sampling of $\mathcal{J}^*_{(x)} * \bar{\phi}^*_{j,k(-x)}$ at rate 1.

3.5.1. The decomposition algorithm

The aim of the basic step of the NOMRA decomposition algorithm is to split I_0 into two parts I_{-1} and J_{-1}. We can iterate this decomposition by splitting I_{-1} again, etc.. A decomposition up to a scale 2^j, $j \in -\mathbb{N}$, will provide the representation $J_{-1} \cup J_{-2} \cup J_{-3} \cup \cdots \cup J_j$ and I_j.

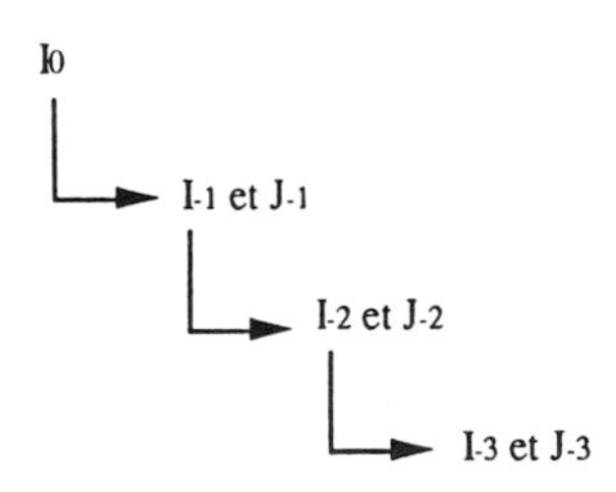

Figure 1. Decomposition from scale 2^0 up to scale 2^{-3}.

We now wish to find a means for computating I_{j-1} and J_{j-1} from I_j. Equation (7) can be written:

$$\phi^*_{j,k} = \sqrt{2}\sum_{n=-\infty}^{+\infty} b_n \phi^*_{j+1,2k-n}$$

and for every integer j and k we have:

$$C_{j,k} = \langle \mathcal{J}, \phi^*_{j,k} \rangle = \sqrt{2} \sum_{n=-\infty}^{+\infty} \bar{b}_n \langle \mathcal{J}, \phi^*_{j+1,2k-n} \rangle.$$

If we set $u(n) = \sqrt{2}\,\bar{b}_n$, then

$$C_{j,k} = \sum_p C_{j+1,2k-p} u(p). \tag{26}$$

In the same way, the definition of the dual wavelet (21) gives:

$$\psi^*_{j,k} = \sqrt{2} \sum_{n=-\infty}^{+\infty} (-1)^n \bar{a}_{-n} \phi^*_{j+1,2k-n+1}.$$

If we set $v(n) = \sqrt{2}(-1)^n a_{-n}$, then

$$S_{j,k} = \sum_p C_{j+1,2k+1-p} v(p). \tag{27}$$

The numerical filters used here are defined by:

$$\bar{\beta}(-\omega) = \frac{1}{\sqrt{2}} \sum_{n=-\infty}^{+\infty} u(n) e^{in\omega}, \tag{29}$$

and

$$\alpha(-\omega + \pi) = \frac{1}{\sqrt{2}} \sum_{n=-\infty}^{+\infty} v(n) e^{in\omega}. \tag{30}$$

From Equations (15) and (16), the function $\bar{\beta}(-\omega)$ provides a low-pass filter which leads to a smoothing of the information from one scale to the other. The function $\alpha(-\omega + \pi)$ gives a high-pass filter which detects the events between the scales 2^{j+1} and 2^j.

Let us observe that for real impulse responses. We have:

$$\bar{\beta}(-\omega) = \beta(\omega) \quad \text{and} \quad \alpha(-\omega + \pi) = \bar{\alpha}(\omega + \pi).$$

3.5.2. The reconstruction algorithm

We want to reconstruct the coordinates I_{j+1} from I_j and J_j. For this purpose, we write the relation $V_{j+1} = V_j \oplus W_j$ applied to a signal of $L^2(\mathbb{R})$.

$$\sum_{p=-\infty}^{+\infty} C_{j+1,p} \phi_{j+1,p} = \sum_{p=-\infty}^{+\infty} C_{j,p} \phi_{j,p} + \sum_{p=-\infty}^{+\infty} C_{j,p} \psi_{j,p}.$$

The Fourier transform of this equality leads to:

$$C_{j+1,p} = \sum_{k=-\infty}^{+\infty} C_{j,k}\bar{u}(2k-p) + \sum_{k=-\infty}^{+\infty} S_{j,k}\bar{v}(2k+1-p), \tag{31}$$

where the filters impulse responses are given by (29) and (30). Once more, the structure of the reconstruction algorithm in the general case is the same as the orthogonal case.

3.6. Design of nonorthogonal multiresolution analysis

Let α and β be two filter functions in $L^\infty[0, 2\pi)$ satisfying relations (14) to (17). Just as the orthogonal case, we want to build nonorthogonal multiresolution analyses from these filters.

We first note that if we have a NOMRA using two filters (α, β), then the associated scaling function ϕ and the dual scaling function ϕ^* satisfy:

$$\hat{\phi}(\omega) = \prod_{n=1}^{+\infty} \alpha\left(\frac{\omega}{2^n}\right)\hat{\phi}(0), \tag{32}$$

and

$$\hat{\phi}^*(\omega) = \prod_{n=1}^{+\infty} \beta\left(\frac{\omega}{2^n}\right)\hat{\phi}^*(0). \tag{33}$$

Moreover $\hat{\phi}(0)\bar{\hat{\phi}}^*(0) = 1$ [9]. Indeed, Equations (15) and (20) lead to $\hat{\phi}(2k\pi) = 0$ for every integer $k \neq 0$.

We now give an important theorem which provides sufficient conditions for the filter functions (α, β) to define a NOMRA. This theorem is an adaptation of a famous one stated by I. Daubechies [5] in the orthogonal case. Let γ be a filter function and assume that we have:

$$\gamma(\omega) = \left(\frac{1+e^{i\omega}}{2}\right)^{N_\gamma} \mathcal{F}_\gamma(\omega) \quad \text{with} \quad \mathcal{F}_\gamma(\omega) \in C^0(\mathbb{R}). \tag{34}$$

Furthermore, define the real numbers:

$$b_\gamma(j) = \sup_{\omega\in\mathbb{R}}(|\mathcal{F}_\gamma(\omega)\mathcal{F}_\gamma(2\omega)\ldots\mathcal{F}_\gamma(2^{j-1}\omega)|^{1/j}) \quad \text{and} \quad b_\gamma = \inf_{j>0}(b_\gamma(j)).$$

Theorem 16. *Let $(\alpha, \beta) \in C^0(\mathbb{R})$ be two continuous filter functions satisfying the following hypotheses:*

(i)

$$\begin{cases} \alpha(0) = 1, \beta(0) = 1, \alpha(\pi) = 0, \beta(\pi) = 0; \\ \alpha(\omega)\bar{\beta}(\omega) + \alpha(\omega+\pi)\bar{\beta}(\omega+\pi) = 1. \end{cases}$$

(ii)

$$b_\alpha < 2^{N_\alpha - 1/2} \quad \text{and} \quad b_\beta < 2^{N_\beta - 1/2}. \tag{35}$$

Then, the functions ϕ and ϕ^* obtained from Equations (32) and (33) define a unique nonorthogonal multiresolution analysis. Moreover, we have the inequalities:

$$|\hat{\phi}(\omega)| \leq c(1+|\omega|)^{-N_\alpha+\nu_\alpha},$$

and

$$|\hat{\phi}^*(\omega)| \leq c^*(1+|\omega|)^{-N_\beta+\nu_\beta},$$

where c and c^* are strictly positive numbers and $\nu_\gamma = \frac{\ln(b_\gamma)}{\ln(2)}$ for $\gamma = \alpha$ or β.

The inequalities in (35) seem to be difficult to satisfy. Yet, we can replace the lower bounds on every j by $j = 1$ or 2 in order to use them more easily. This technique will be used in the sequel. We do not give a proof of Theorem 16 because it is just a tedious adaptation of the Daubechies's theorem. The interested reader is referred to a paper by Cohen, Daubechies and Feauveau [3] on the mathematical development, where similar results are proved.

§4. Examples of nonorthogonal multiresolution analyses

We have developed a theoretical framework for generalizing the orthogonal multiresolution analysis defined by S. Mallat. For the design of these nonorthogonal multiresolution analyses, Theorem 16 provides a practical approach by using a couple of digital filters represented by two functions α and β. The purpose of this section is to put the theory into practical use. We will explicitly compute some pairs (α, β) that generate a NOMRA defined by projectors on the cardinal splines spaces. The function α defining the scaling function ϕ (namely: the B-spline), also defines the projection spaces V_j. So, this function α will be chosen. To completely design a NOMRA, it remains to define the directions of projection. Or equivalently, we may define a filter function β in order to get the dual scaling function ϕ^*.

In the following, α is chosen such that ϕ be a B-spline of degrees 1, 2 or 3. These cases are interesting because spline interpolations, generally a cubic spline, are often used in signal processing. In this case, we have the coordinates $I_0 = (C_{0,k})_{k\in\mathbb{Z}}$, which are exactly the input of the multiresolution decomposition algorithm (or the result of the reconstruction). Before we give a short review of spline functions, let us observe that for a fixed α, we generally have an infinitely many possible choices for the function β for designing a NOMRA. Indeed, projection spaces will always be the same, but the projection directions will change from one scale to another.

P. G. Lemarié [8] and G. Battle [2] were the first to note that interpolation spaces of cardinal splines lead naturally to multiresolution analyses. These spaces are defined as follows: For every nonnegative integer n, there exists a unique real valued function $\phi(x;n)$, called the B-spline such that

(i) $\phi(x;n)$ is compactly supported in $\left[-\frac{n}{2}-1, \frac{n}{2}\right]$ if n is even, and in $\left[-\frac{n-1}{2}, \frac{n+1}{2}\right]$ if n is odd,

(ii) for every integer p, $\phi(x;n)$ restricted to $(p,p+1)$ is a polynomial of degree at most n,
(iii) $\phi(x;n)$ is in $C^{n-1}(\mathbb{R})$,
(iv) $\|\phi(x;n)\|_{L^2} = 1$.

The Fourier transform of $\phi(x;n)$ is easy to compute. Indeed, there exists a constant a_n such that

$$\hat{\phi}(\omega;n) = a_n \left(\frac{\sin(\omega/2)}{\omega/2}\right)^{n+1} g(\omega),$$

with $g(\omega) = e^{i\omega/2}$ if n is even, and 1 if n is odd.

Let us define the spaces $V_j(n) = \overline{\mathrm{Span}(\phi_{j,k}(x;n))_{k\in\mathbb{Z}}}$. Elements in $V_0(n)$ are functions in $L^2(\mathbb{R}) \cap C^{n-1}(\mathbb{R})$ which are piecewise polynomial of degree at most n between consecutive integers. The spaces $(V_j(n))_{j\in\mathbb{Z}}$ define an orthogonal multiresolution analysis associated to the following scaling function:

$$\hat{\phi}(\omega;n;orth) = \frac{\hat{\phi}(\omega;n)}{\sqrt{\sum_{k=-\infty}^{+\infty} |\hat{\phi}(\omega+2k\pi;n)|^2}}$$

which provides orthonormal bases $(\phi_{j,k}(x;n;orth))_{k\in\mathbb{Z}}$ of $V_j(n)$.

However, with the exception of $n = 0$, the associated digital filters are not finite impulse response (FIR) filters. Removing the orthogonality, we will define NOMRA on the same projection spaces associated to nonorthogonal unconditional bases $(\phi_{j,k}(x;n))_{k\in\mathbb{Z}}$ leading to FIR digital filters. After a preliminary study of the reconstruction relation, we will develop the particular cases $n = 1$ for the linear interpolation, $n = 2$ for the quadratic interpolation and $n = 3$ for the cubic spline interpolation.

4.1. Preliminary study

The spline spaces $V_j(n)$ provide a good candidate for NOMRA projection spaces. It remains to define some possible projection directions (e.g. spaces $Z_j(n)$ or $W_j(n)$) and for implementation purpose, we want to use FIR filters.

For every n, there exists a filter function $\alpha(\omega)$ such that:

$$\hat{\phi}(2\omega;n) = \alpha(\omega)\hat{\phi}(\omega;n) \Leftrightarrow \alpha(\omega) = (\cos(\omega/2))^{n+1} g(\omega),$$

with $g(\omega) = e^{i\omega/2}$ if n is even and $g(\omega) = 1$ if n is odd. To find a function $\beta(\omega)$ which, together with $\alpha(\omega)$, satisfies the conditions of Theorem 16, we will use the following useful relation:

For every integer N, we have

$$(\cos^2(\omega/2))^N \left(\sum_{k=0}^{N-1} C_{N-1+k}^k (\sin^2(\omega/2))^k\right) + (\sin^2(\omega/2))^N$$
$$\left(\sum_{k=0}^{N-1} C_{N-1+k}^k (\cos^2(\omega/2))^k\right) = 1.$$

Thus, for $n = 2p - 1$ and $n + m = N$, we find:

$$\alpha(\omega)(\cos^2(\omega/2))^m \left(\sum_{k=0}^{p+m-1} C_{N-1+k}^k (\sin^2(\omega/2))^k \right) \\ + \alpha(\omega+\pi)(\sin^2(\omega/2))^m \left(\sum_{k=0}^{p+m-1} C_{N-1+k}^k (\cos^2(\omega/2))^k \right) = 1,$$

and the function

$$\beta(\omega) = (\cos(\omega/2))^{2m} \left(\sum_{k=0}^{p+m-1} C_{N-1+k}^k (\sin^2(\omega/2))^k \right) \tag{36}$$

satisfies condition (i) of Theorem 16 for all $m \geq 1$. For $n = 2p$, we have a similar result with

$$\beta(\omega) = (\cos(\omega/2))^{2m+1} \left(\sum_{k=0}^{p+m-1} C_{N-1+k}^k (\sin^2(\omega/2))^k \right) e^{i\omega/2}. \tag{37}$$

We now apply these results to $n = 1, 2,$ and 3.

4.2. Linear interpolation

In this subsection we set $n = 1$ and write $\phi(x)$ instead of $\phi(x; 1)$, and V_j instead of $V_j(1)$. A short computation leads to:

$$\alpha(\omega) = \frac{1 + \cos(\omega)}{2}.$$

Equation (36) used with $N = 2$ and $m = 1$ provides:

$$\beta(\omega) = \cos^2(\omega/2) \left(\sum_{k=0}^{1} C_{1+k}^k (\sin^2(\omega/2))^j \right) = \frac{1}{4}(3 + 2\cos(\omega) - \cos(2\omega)).$$

Let us remark that the high-pass filter of the decomposition algorithm $\alpha(\omega+\pi)$, which detects the inter-scale singularities, is a kind of second derivative.

The associated wavelet is:

$$\psi(x) = \frac{1}{8}(-\phi(2x-3) - 2\phi(2x-2) + 6\phi(2x-1) - 2\phi(2x) - \phi(2x+1)).$$

Condition (i) in Theorem 16 is satisfied, and a computation gives:

$$N_\alpha = 2, N_\beta = 2, b_\alpha = 1, \quad \text{and} \quad b_\beta \leq b_\beta(2) < 2.6$$

which proves that these digital filters define a multiresolution analysis.

4.3. Quadratic interpolation

In this subsection we consider $n = 2$ and again use $\phi(x)$ instead of $\phi(x; 2)$, and V_j instead of $V_j(2)$. First, from the definition of $\hat{\phi}(\omega)$ we have

$$\alpha(\omega) = (\cos(\omega/2))^3 e^{i\omega/2} = \frac{1}{8}(e^{-i\omega} + 3 + 3e^{i\omega} + e^{2i\omega}).$$

Let us note that the digital filter associated with $\alpha(\omega + \pi)$ detects the first variation of the analyzed signal.

Let $\beta(\omega)$ be defined by:

$$\beta(\omega) = \frac{1}{4}(-e^{i\omega} + 3 + 3e^{i\omega} - e^{2i\omega}).$$

Condition (i) of Theorem 16 is obviously satisfied. But there is a major difficulty: this function $\beta(\omega)$ does not fulfill Condition (ii) in Theorem 16, and computer simulation shows that the associated dual function ϕ^* is not square integrable. So, we use another function $\beta(\omega)$ in order to stay in the $L^2(\mathbb{R})$ framework. Using Equation (37) and $m = 2$ and $N = 4$, we find:

$$\beta(\omega) = (\cos(\omega/2))^5 e^{i\omega/2} \frac{1}{2}(16 - 29\cos(\omega) + 20\cos^2(\omega) - 5\cos^3(\omega)).$$

Conditions (i) and (ii) of Theorem 16 are fulfilled. Indeed, we have

$$b_\beta < b_\beta(2) \leq 20 < 2^{5-1/2}.$$

4.4. Cubic interpolation

In this subsection we will again write $\phi(x)$ instead of $\phi(x; 3)$. The associated function $\alpha(\omega)$ is:

$$\alpha(\omega) = (\cos(\omega/2))^4 = \frac{1}{8}(3 + 4\cos(\omega) + \cos(2\omega)),$$

and Equation (36) allows us to select a filter function $\beta(\omega)$ for $m = 4$ and $N = 6$, namely:

$$\begin{aligned}\beta(\omega) = \cos^8(\omega/2)(1 &+ 6\sin^2(\omega/2) + 21\sin^4(\omega/2) + 56\sin^6(\omega/2) \\ &+ 126\sin^8(\omega/2) + 252\sin^{10}(\omega/2)),\end{aligned}$$

which is the first to fulfill all the conditions in Theorem 16. In particular, we have

$$b_\beta < b_\beta(2) \leq 160 < 2^{8-1/2}.$$

In this case, the detection filter $\alpha(\omega+\pi)$ which extracts the wavelet coefficients is used to compute the second derivative of a signal.

Appendix A. (Proof of Proposition 9)

For every $f \in L^2(\mathbb{R})$, we can write

$$A_j(f) = \sum_{k=-\infty}^{+\infty} C_{j,k}(f)\phi_{j,k}.$$

The linear projection $C_{j,k}$: $L^2(\mathbb{R}) \to \mathbb{R}$ is continuous. Thus, it is an element of the topological dual of $L^2(\mathbb{R})$. But $L^2(\mathbb{R})$ can be identified by its dual space. Indeed, using the Riesz-Fischer's theorem [15], there exists a unique function $\phi^*(x; j, k)$ such that

$$\forall f \in L^2(\mathbb{R}), \qquad C_{j,k}(f) = \int_{\mathbb{R}} f(x)\bar{\phi}^*(x; j, k)dx = \langle f(x), \phi^*(x; j, k)\rangle.$$

Moreover, Equation (1) can be written as

$$\sum_{k=-\infty}^{+\infty} C_{j,k}(f)\phi_{j,k} = \sum_{k=-\infty}^{+\infty} C_{j+1,k}(D_2 \circ f)\sqrt{2}\phi_{j,k}.$$

The uniqueness of the decomposition implies:

$$\langle f(x), \phi^*(x; j, k)\rangle = \sqrt{2}\langle f(2x), \phi^*(x; j+1, k)\rangle = \frac{1}{\sqrt{2}}\left\langle f(x), \phi^*\left(\frac{x}{2}; j+1, k\right)\right\rangle.$$

Since the scalar product defines an isomorphism from $L^2(\mathbb{R})$ onto its dual space, we can conclude that

$$\phi^*(x; j+1, k) = \sqrt{2}\phi^*(2x; j, k). \tag{38}$$

For $j = 0$, the second commutation relation as in Equation (2) implies:

$$\sum_{k=-\infty}^{+\infty} C_{0,k+p}(f)\phi_{0,k} = \sum_{k=-\infty}^{+\infty} C_{0,k}(T_{-p} \circ f)\phi_{0,k}, \qquad \forall p \in \mathbb{Z}.$$

Thus, $\langle f(x), \phi^*(x; 0, k+p)\rangle = \langle f(x+p, \phi^*(x; j, k)\rangle = \langle f(x), \phi^*(x-p; j, k)\rangle$, and we have

$$\phi^*(x; 0, k) = \phi^*(x-k; 0, k). \tag{39}$$

Using (38) and (39), we can write, successively,

$$\phi^*(x; j, k) = 2^{j/2}\phi^*(2^j x; 0, k) = 2^{j/2}\phi^*(2^j x - k; 0, 0) = \phi^*_{j,k}(x),$$

where we set $\phi^*(x) = \phi^*(x; 0, 0)$. This completes the proof of Equation (5). Let us now define $\phi' - (\xi_0^{adj})^{-1} \circ \xi_0(\phi)$. We have $\xi_0(\phi) = \Delta_0 = (\delta_{0,k})_{k\in\mathbb{Z}}$, and we want to prove $\phi' = \phi^*$.

For every function $f(x)$ in $L^2(\mathbb{R})$, we have:

$$\langle f, \phi' \rangle = \langle \xi_0 \circ A_0(f), \Delta_0 \rangle = \left\langle \xi_0 \left(\sum_{k=-\infty}^{+\infty} \langle f, \phi_{0,k}^* \rangle \right), \Delta_0 \right\rangle$$
$$= \sum_{k=-\infty}^{+\infty} \langle f, \phi_{0,k}^* \rangle \langle \xi_0(\phi_{0,k}), \Delta_0 \rangle,$$

and thus,

$$\langle f, \phi' \rangle = \sum_{k=-\infty}^{+\infty} \langle f, \phi_{0,k}^* \rangle \langle \Delta_k, \Delta_0 \rangle = \langle f, \phi_{0,0}^* \rangle$$

and the last relation is proved. This completes the proof of Proposition 9.

Appendix B. (Proof of Proposition 10)

The relations about ϕ are obtained from $\phi_{-1,0} \in V_0$. Using the basis of V_0, we have:

$$\phi\left(\frac{x}{2}\right) = 2 \sum_{n=-\infty}^{+\infty} a_n \phi(x+n)$$

with $\sum_{n=-\infty}^{+\infty} |a_n|^2 < \infty$, because $(\phi_{0,k})_{k \in \mathbb{Z}}$ is an unconditional basis. Let us define

$$\alpha(\omega) = \sum_{n=-\infty}^{+\infty} a_n e^{in\omega},$$

which is a function in $L^2[0, 2\pi)$. We have $\hat{\phi}(2\omega) = \alpha(\omega)\hat{\phi}(\omega)$ which yields ϕ. From Proposition 9, we also have $\phi^* = (\xi_0^{adj})^{-1} \circ \xi_0(\phi) = (\xi_0^{adj})^{-1}(\{\delta_{0,k}\}_{k \in \mathbb{Z}})$. Thus, ϕ^* is the scaling function of a NOMRA, and the previous result on ϕ can be applied to ϕ^*. Let us point out that the relation:

$$\phi^*\left(\frac{x}{2}\right) = 2 \sum_{n=-\infty}^{+\infty} b_n \phi^*(x+n)$$

necessarily implies that $\bar{b}_n = \frac{1}{2}\langle \phi_{0,-n}, \phi_{-1,0}^* \rangle$, by using the bi-orthogonailty. This completes the proof of Proposition 10.

Appendix C. (Proof of Theorem 13)

Let us first note that $\psi(x)$ is in $L^2(\mathbb{R})$. Indeed, since $\beta(\omega)$ is in $L^2[0, 2\pi)$, $\psi(x)$ is in V_1. We have to prove the theorem for $j = 0$, and the general case can be deduced from this case. To prove that $\{\psi_{0,k}\}_{k \in \mathbb{Z}}$ is an unconditional basis of its span, we use the following lemma.

Lemma 17. *The sequence* $\{\psi_{0,k}\}_{k\in\mathbb{Z}}$ *is an unconditional family.*

Proof: Let us introduce the functions:

$$G(\omega) = \sum_{k=-\infty}^{+\infty} |\hat{\phi}(\omega + 2k\pi)|^2 \quad \text{and} \quad L(\omega) = \sum_{k=-\infty}^{+\infty} |\widehat{\psi}(\omega + 2k\pi)|^2.$$

Since the sequence $\{\phi_{0,k}\}_{k\in\mathbb{Z}}$ defines an unconditional family, there exist two strictly positive real numbers σ_0 and σ_1 such that

$$\sigma_0 \leq G(\omega) \leq \sigma_1 \quad \text{for all} \quad \omega \in \mathbb{R}.$$

The function L is defined almost everywhere because ψ is in $L^2(\mathbb{R})$. We easily show that

$$L(2\omega) = |\beta(\omega + \pi)|^2 G(\omega) + |\beta(\omega)|^2 G(\omega + \pi).$$

This completes the proof of Lemma 17.

We remark that if f is in $W_0' = \overline{\text{Span}(\psi_{0,k})_{k\in\mathbb{Z}}}$, then $f(x)$ is in $L^2(\mathbb{R})$ and the sequence of its coordinates on the basis $\{\psi_{0,k}\}_{k\in\mathbb{Z}}$ is in $\ell^2(\mathbb{Z})$. We now prove that $W_0' = W_0$. We first prove the inclusion $W_0' \subset W_0$. Clearly $W_0' \subset V_1$, and we have to prove that the restriction to A_0 to W_0' is the null operator. To do so, it is necessary and sufficient to show that $A_0(\psi_{0,0}) = 0$, or equivalently:

$$\langle \psi_{0,0}, \phi_{0,k}^* \rangle = 0, \quad \text{for all} \quad k \in \mathbb{Z}.$$

Consider

$$\langle \psi_{0,0}, \phi_{0,k}^* \rangle = \int_{\mathbb{R}} \psi(x)\phi^*(x - k)dx = 2\int_{\mathbb{R}} \hat{\psi}(2\omega)\bar{\hat{\phi}}^*(2\omega)e^{2ik\omega}d\omega,$$

and

$$\langle \psi_{0,0}, \phi_{0,k}^* \rangle = 2\int_{\mathbb{R}} \bar{\beta}(\omega + \pi)\bar{\beta}(\omega)\hat{\phi}(\omega)\beta(\omega)\bar{\hat{\phi}}^*(\omega)e^{-i(2k-1)\omega}d\omega,$$

which is equal to $2\int_0^{2\pi} \bar{\beta}(\omega + \pi)\bar{\beta}(\omega)e^{i(2k-1)\omega}d\omega$, by using Lemma 7, applied to the bi-orthogonality of $(\phi_{j,k})_{k\in\mathbb{Z}}$ and $\{\phi_{j,k}^*\}_{k\in\mathbb{Z}}$. This last integral is always zero, and this proves the first inclusion.

To prove the converse inclusion $W_0' \supset W_0$, it is sufficient to show that $V_1 \subset V_0 \oplus W_0'$. That is, it remains to show that $\phi_{1,0}$ and $\phi_{1,1}$ are both in $V_0 \oplus W_0'$, since this implies $\phi_{1,k} \in V_0 \oplus W_0'$ for every integer k. Now, using Equation (11), we can write:

$$\hat{\phi}(\omega) = [\bar{\beta}(\omega) + \bar{\beta}(\omega + \pi)]\alpha(\omega)\hat{\phi}(\omega) + [e^{i\omega}(\alpha(\omega + \pi) - \alpha(\omega))]e^{-i\omega}\bar{\beta}(\omega + \pi)\hat{\phi}(\omega),$$

or equivalently:

$$\hat{\phi}\left(\frac{\omega}{2}\right) = \left[\bar{\beta}\left(\frac{\omega}{2}\right) + \bar{\beta}\left(\frac{\omega}{2} + \pi\right)\right]\hat{\phi}(\omega) + \left[e^{i\omega/2}\left(\alpha\left(\frac{\omega}{2} + \pi\right) - \alpha\left(\frac{\omega}{2}\right)\right)\right]\widehat{\psi}(\omega).$$

The terms between the brackets are both 2π-periodic funtions in $L^2[0, 2\pi)$, and this gives $\phi_{1,0} \in V_0 \oplus W_0'$. In the same manner, since

$$\hat{\phi}(\omega)e^{-i\omega} = [e^{-i\omega}(\bar{\beta}(\omega) - \bar{\beta}(\omega+\pi))]\alpha(\omega)\hat{\phi}(\omega) + [\alpha(\omega+\pi)+\alpha(\omega)]e^{-i\omega}\bar{\beta}(\omega+\pi)\hat{\phi}(\omega)$$

we have $\phi_{1,1} \in V_0 \oplus W_0'$.

This completes the proof of Theorem 13.

Appendix D. (Proof of Proposition 15)

Clearly, $\psi^*(x)$ is in $L^2(\mathbb{R})$ because $\alpha(\omega)$ is bounded and ϕ^* is in $L^2(\mathbb{R})$. As usual, the case $j = 0$ implies the general case. To prove Equation (23), we note that by using the fact that $\{\psi_{0,k}\}_{k\in\mathbb{Z}}$ is a basis of W_0, it is sufficient to verify the case $f = \psi_{0,k}$; that is, $\langle \psi_{0,0}, \psi_{0,k}^* \rangle = \delta_{0,k}$, where $\delta_{0,k}$ is the Kronecker symbol. This is simple, since

$$\begin{aligned}
\langle \psi_{0,0}, \psi_{0,k}^* \rangle &= \frac{1}{2\pi} \int_{\mathbb{R}} e^{ik\omega} \widehat{\psi}(\omega) \overline{\widehat{\psi}}^*(\omega) d\omega \\
&= \frac{1}{\pi} \int_{\mathbb{R}} e^{i2k\omega} \alpha(\omega+\pi) \bar{\beta}(\omega+\pi) \hat{\phi}(\omega) \bar{\hat{\phi}}^*(\omega) d\omega \\
&= \frac{1}{\pi} \int_0^{2\pi} e^{i2k\omega} \alpha(\omega+\pi) \bar{\beta}(\omega+\pi) d\omega = \frac{1}{\pi} \int_0^{\pi} e^{i2k\omega} d\omega = \delta_{0,k}
\end{aligned}$$

by using Equation (11).

To prove (22), let us notice that:

(i) $\langle \ell, \psi_{0,k}^* \rangle = 0$ for all $\ell \in \mathbb{Z}_1$. Indeed, $\psi_{0,k}^*$ is in $V_1^* = (Z_1)^\perp$.
(ii) $\langle g, \psi_{0,k}^* \rangle = 0$ for all $g \in V_0$. We have to prove $\langle \phi_{0,n}, \psi_{0,k}^* \rangle = 0$.

Using a change of variable, we verify $\langle \phi_{0,0}, \psi_{0,k}^* \rangle = 0$, as follows:

$$\begin{aligned}
\langle \phi_{0,0}, \psi_{0,k}^* \rangle &= \frac{1}{2\pi} \int_{\mathbb{R}} e^{ik\omega} \hat{\phi}(\omega) \overline{\widehat{\psi}}^*(\omega) d\omega = \frac{1}{\pi} \int_{\mathbb{R}} e^{i(2k+1)\omega} \alpha(\omega+\pi) \alpha(\omega) \hat{\phi}(\omega) \bar{\hat{\phi}}^*(\omega) d\omega \\
&= \frac{1}{\pi} \int_0^{2\pi} e^{i(2k+1)\omega} \alpha(\omega+\pi) \alpha(\omega) d\omega = 0.
\end{aligned}$$

We are now able to prove Equation (22). For every f in $L^2(\mathbb{R})$, we can find g, h and ℓ, in V_0, W_0, and Z_1, respectively, such that $f = g + h + \ell$. Using (23) we have:

$$A_1(f) - A_0(f) = h = \sum_{k=-\infty}^{+\infty} \langle h, \psi_{0,k}^* \rangle \psi_{0,k}.$$

The two previous remarks allow us to write:

$$A_1(f) - A_0(f) = \sum_{k=-\infty}^{+\infty} \langle f, \psi_{0,k}^* \rangle \psi_{0,k}.$$

This completes the proof of Proposition 15.

References

1. Antonini, M., M. Barlaud, and P. Mathieu, Image coding using vector quantization in the wavelet transform domain, in *Proc. Int. Conf. Acoust., Speech and Signal Processing*, Albuquerque, NM, 1990, 2297–2300.
2. Battle, G., A block spin construction of ondelettes, Part 1: Lemarié functions, *Comm. Math. Phys.* **110** (1987), 601–615.
3. Cohen, A., I. Daubechies, and J. C. Feauveau, Bi-orthogonal bases of compactly supported wavelets, *Comm. Pure and Appl. Math.*, 1991, to appear.
4. Cohen, A., Construction de bases d'ondelettes a-Hölderiennes, *CEREMADE*, Univ. Paris-Dauphine, preprint.
5. Daubechies, I., Orthogonal bases of compactly supported wavelets, *Comm. Pure and Appl. Math.* **41** (1988), 909–996.
6. Daubechies, I., The wavelet transform, time-frequency localization and signal analysis, *IEEE Trans. Inform. Theory* **36** (1990), 961–1005.
7. Le Gall, D., and A. Tabatabai, Sub-band coding of digital images using symmetric short kernel filters and arithmetic coding techniques, *Proc. ICASSP*, 1988, 761–764.
8. Lemarié, P. G., Ondelettes à localisation exponentielle, *J. Math. Pure et Appl.* **67** (1988), 227–236.
9. Lemarié, P. G., Constructions d'ondelettes-splines, 1987, unpublished.
10. Mallat, S., A theory for multiresolution signal decomposition: the wavelet representation, *IEEE Pattern Anal. and Machine Intell.* **11** (7) (1989), 674–693.
11. Mallat, S., Multiresolution approximation and wavelet orthonormal bases of $L^2(\mathbf{R})$, *Trans. Amer. Math. Soc.* **315** (1989), 69–87.
12. Meyer, Y., Principe d'incertitude, bases Hilbertiennes et algèbre d'opérateurs, *Séminaire Bourbaki* **662**, 1985–1986.
13. Meyer, Y., Ondelettes, fonctions splines et analyses graduées, Rapport CEREMADE **8703**, 1987.
14. Rosenfeld, A. and M. Thurson, Edge and curve detection for visual scene analysis, *IEEE Trans. Comput.* **C-20** (1971).
15. Rudin, W., *Real and Complex Analysis*, McGraw-Hill, 3rd edition, 1987.
16. Smith, M. J. T. and T. P. Barnwell III, Exact reconstruction techniques for tree structured subband coders, *IEEE ASSP* **34** (1986), 434–441.
17. Tchamitchian, Ph., Biorthogonalité et théorie des opérateurs, *Rev. Mat. Iberoamericana* **3** (2) (1987).
18. Torre, V. and T. Poggio, On edge detection, *IEEE Pattern Anal. and Machine Intell.* **8** (2) (1986), 146–163.
19. Witkin, A., Scale space filtering, *Proc. Int. Joint Conf. Artif. Intell.*, Karlsruhe, West Germany, 1980, 1019–1021.
20. Yuille, A. L. and T. Poggio, Scaling theorems for zero crossings, *IEEE Pattern Anal. and Machine Intell.* **8** (1) (1986).

Jean-Christophe Feauveau
MATRA-MS2I
38 Bd Paul Cexanne
78052 Saint Quentin en Yvelines
France

Part III

Wavelet-like Local Bases

Wavelets and Other Bases for Fast Numerical Linear Algebra

Bradley K. Alpert

Abstract. The fundamental ideas of wavelets are introduced within the context of mathematical physics. We present essential background notions of mathematical bases, and discuss Fourier, polynomial, and wavelets bases in this light. We construct several types of wavelets and wavelet-like bases and illustrate their use in algorithms for the solution of a variety of integral and differential equations.

§1. Introduction

Problems of physics, requiring the numerical solution of differential and integral equations, rank among the most compute-intensive applications currently feasible. The field of scientific computation is concerned with both hardware and algorithmic improvement for the modelling of increasingly complex problems. Recently, the development of new mathematical bases for scientific computation has enabled the construction of algorithms which are dramatically more efficient than earlier algorithms. Wavelets permit the accurate representation of a variety of functions and operators without redundancy. Through the ability to represent local, high-frequency information with localized basis elements, wavelets allow adaption in a straightforward, consistent fashion.

For a variety of applications, sparse matrix representations of differential operators have been the key to efficient algorithms. Integral operators, by contrast, are represented in classical bases as dense matrices. These representations lead to algorithms which, for large-scale problems, are often prohibitively slow. The most notable exception to this rule is for convolutional operators, which are represented in the Fourier basis as diagonal matrices, and which have correspondingly fast algorithms. Wavelets can be viewed as a "diagonalizing" basis for a wider class of integral operators. The quotes are necessary here, because the statement is only approximately true. Wavelet expansions of integral operators are not exactly diagonal; rather, they have a peculiar band

Wavelets–A Tutorial in Theory and Applications
C. K. Chui (ed.), pp. 181–216.

ISBN 0-12-174590-2

structure. Furthermore, the sparse band structure represents an approximation (to arbitrary precision) of the original integral operator.

Despite these caveats, the solution of a wide range of integral equations is transformed using wavelets from a direct procedure requiring order $O(n^3)$ operations to one requiring only order $O(n)$. Here n is the number of points in the discretization of the domain.

Many time-dependent problems formulated as partial differential equations require adaptive representations for carrying out the time evolution. Typically, a small part of the domain has most of the activity, and the representation must have high resolution there. In the rest of the domain, such high resolution is wasted (and costly). Various adaptive mesh techniques have been developed to address this issue, but they often suffer accuracy problems in the application of operators, multiplication of functions, and so forth. Wavelets offer promise in providing a consistent, simple adaptive framework.

In Section 2 we provide background to the study of wavelets by reviewing the definition and some examples of mathematical bases. In Section 3 we present several constructions of wavelets and wavelet-like bases and discuss their fundamental properties. In Section 4 we introduce the application of wavelets to the solution of integral equations and to the representation of differential operators. Finally, we summarize in Section 5.

§2. Function representations in mathematical physics

2.1. Mathematical bases

The *norm* of a sequence $\alpha = \langle \alpha_1, \alpha_2, \ldots \rangle$, a measure of its size, will be defined by the formula

$$\|\alpha\|_{l^2} = \Big(\sum_n {\alpha_n}^2 \Big)^{1/2}.$$

The space l^2 consists of the square summable sequences α: $\|\alpha\|_{l^2} < \infty$.

The norm of a function $f : \Omega \to \mathbb{R}$, will be defined as the L^2-norm

$$\|f\|_{L^2} = \left(\int_\Omega f(x)^2 \, dx \right)^{1/2},$$

where Ω is the domain of f. In this chapter we restrict ourselves primarily to $\Omega = \mathbb{R}$ and to functions f with $\|f\|_{L^2} < \infty$ (*i.e.*, $f \in L^2$); the function norm $\|f\|_{L^2}$ will be abbreviated $\|f\|$. A sequence of functions $f_1, f_2, \ldots$ *converges* to a function f if the difference $f_n - f$ becomes arbitrarily small, *i.e.*, if $\|f_n - f\| \to 0$ as $n \to \infty$. A series of functions $\sum_n f_n$ converges to f if the sequence of partial sums $S_1, S_2, \ldots$ converges to f. The partial sum S_n is defined by the formula

$$S_n(x) = \sum_{j=1}^{n} f_j(x).$$

The *linear span*, or closure $\overline{F}$, of a set of functions $F = \{f_1, f_2, \ldots\} \subset L^2$ is the set of linear combinations

$$f(x) = \sum_{n=1}^{\infty} \alpha_n f_n(x)$$

that are contained in L^2. The restriction that the sum be square integrable is equivalent to the requirement that the sequence of coefficients $\langle \alpha_1 \|f_1\|, \alpha_2 \|f_2\|, \ldots \rangle$ be square summable, or in l^2. A set of functions F is *linearly independent* if any proper subset F' of F has linear span $\overline{F'}$ which is a proper subset of the linear span $\overline{F}$ of F:

$$F' \subset F \text{ and } F' \neq F \quad \Rightarrow \quad \overline{F'} \neq \overline{F}.$$

A set of functions F is a *basis* for a space S if $\overline{F} = S$ and F is linearly independent. We will concern ourselves with bases for $L^2(\mathbb{R})$.

2.1.1. Stability and orthogonality

If a function is represented in terms of functions that are not linearly independent, the representation is not unique. For example, if

$$\text{linear span}\{f_1, f_2, f_3, \ldots\} = \text{linear span}\{f_2, f_3, \ldots\},$$

and

$$f(x) = \sum_n \alpha_n f_n(x),$$

then the coefficient α_1 can be chosen arbitrarily. On the other hand, linear independence is enough to guarantee representational uniqueness in theory, but is not sufficient when numerical computations are carried out to finite precision. The functions $\langle f_0, f_1, \ldots \rangle$, defined by the formula

$$f_n(x) = \begin{cases} x^n & \text{if } x \in [0,1]; \\ 0 & \text{otherwise;} \end{cases}$$

are linearly independent but f_{11} can be represented as a linear combination of $f_0, \ldots, f_{10}$ to 6 digit accuracy. Perturbation of a function

$$f(x) = \sum_{n=0}^{11} \alpha_n f_n(x)$$

by one part in a million can correspond to variations of $\|\alpha\|_{l^2}$ by several percent (even if each f_j were normalized so $\|f_j\| = 1$). The monomials are therefore seriously deficient for numerical use as a basis. In computations, they are replaced by the orthogonal polynomials.

The *inner product* of two functions $f, g \in L^2(\mathbb{R})$ is the integral

$$\langle f, g \rangle = \int_{-\infty}^{\infty} f(x)\, g(x)\, dx.$$

Two functions f and g are *orthogonal* if $\langle f, g \rangle = 0$. An orthogonal basis is a basis in which the functions $f_1, f_2, \ldots$ are pairwise orthogonal: $\langle f_i, f_j \rangle = 0$ for $i \neq j$. The basis is orthonormal if $\langle f_i, f_j \rangle = \delta_{ij}$.

An orthonormal basis is, in a sense, a natural representation language for functions. Perturbation of the coefficients in the representation of a function produces a commensurate perturbation in the function. Similarly, if the function is perturbed the coefficients change nearly the same amount: the representation is stable. Another important characteristic of orthonormal bases is that it is simple to determine the expansion coefficients. Given an orthonormal basis $\{f_1, f_2, \ldots\}$ for $L^2(\mathbb{R})$ and a function $f \in L^2(\mathbb{R})$, the coefficients in the expansion

$$f(x) = \sum_{n=1}^{\infty} \alpha_n\, f_n(x)$$

are given by the inner product

$$\alpha_n = \langle f_n, f \rangle = \int_{-\infty}^{\infty} f_n(x)\, f(x)\, dx \qquad n \in \mathbb{N}.$$

Without orthogonality, the coefficients must be obtained by the (often expensive) solution of a system of equations. We mention at this point that recent results suggest that *frames* may offer the simplicity of orthogonal bases without their rigidity see [10] and [11]).

2.1.2. Truncated expansions

Though a function f may be specified by an infinite expansion, actual computations require finite representations. Generally, an infinite basis is abbreviated to a finite basis, which corresponds to truncating the expansion. If the first n terms are retained, we have

$$f(x) = \sum_{j=1}^{n} \alpha_j\, f_j(x) + E_n(x),$$

where the truncation error $E_n(x)$ is given by

$$E_n(x) = \sum_{j=n+1}^{\infty} \alpha_j\, f_j(x).$$

The computation cost generally increases with n, so it is desirable for the error E_n to decay with increasing n as rapidly as possible. If k is the largest integer such that the quantity

$$n^k \|E_n\|$$

is bounded as $n \to \infty$, then we say the expansion for f is kth-order convergent. If, instead,

$$\sup_n n^k \|E_n\| < \infty$$

for all k, the expansion is super-algebraicly convergent.

2.2. Classical bases

We now give a few concrete examples of bases in use for a variety of computational tasks.

The fundamental representation for periodic functions is the Fourier basis for $L^2[-\pi, \pi]$,

$$f_n(x) = e^{inx}, \qquad n \in \mathbb{Z},$$

which is an orthogonal basis under the (complex) inner product

$$\langle f, g \rangle = \int_{-\pi}^{\pi} f(x)\, \tilde{g}(x)\, dx,$$

where $\tilde{g}$ is the complex conjugate of g. Many properties of the Fourier basis make it well-suited to computation. Differentiation and integration of functions represented in the Fourier basis is simple:

$$\frac{df_n}{dx}(x) = n\, f_n(x), \qquad \int f_n(x)\, dx = \frac{1}{n}\, f_n(x).$$

The transformation of a function tabulated at n equispaced points on the interval $[-\pi, \pi]$ into an n-term Fourier expansion, or its inverse transformation, is accomplished rapidly by the Fast Fourier Transform. In addition, convolutions are diagonal operators in the Fourier basis. Suppose we are given an integral operator

$$(\mathcal{K}f)(x) = \int_{-\infty}^{\infty} K(x, y)\, f(y)\, dy,$$

where the kernel $K(x, y)$ is convolutional, *i.e.*,

$$K(x, y) = k(x - y).$$

The element K_{mn} of the matrix representing the operator $\mathcal{K}$ in the Fourier basis, given by the formula

$$K_{mn} = \int_{-\pi}^{\pi} \int_{-\pi}^{\pi} k(x - y)\, e^{imx}\, e^{-iny}\, dx\, dy, \qquad m, n \in \mathbb{Z},$$

satisfies $K_{mn} = 0$ if $m \neq n$, as can be seen by integrating. The ability in the Fourier basis to manipulate convolutions efficiently as diagonal operators leads to much of the strength of the basis as a computational tool.

Another class of bases are the orthogonal polynomials. Given an interval $I \subset \mathbb{R}$ and a positive weight function $\omega : I \to \mathbb{R}$, we can define an inner product

$$\langle f, g \rangle_\omega = \int_I f(x)\, g(x)\, \omega(x)\, dx.$$

There is a sequence of polynomials, $p_0, p_1, \ldots$, with p_j of degree j, which is orthogonal with respect to the weight ω, *i.e.*, $\langle p_m, p_n \rangle_\omega = 0$ if $m \neq n$. The sequence forms a basis for the functions defined on the interval I which are square integrable with weight ω. The sequence is uniquely determined by ω up to leading coefficients, and can be computed by the Gram-Schmidt orthogonalization process. For $I = [-1, 1]$ and $\omega(x) = 1$, the sequence is the Legendre polynomials; they form an orthogonal basis for $L^2[-1, 1]$. The transformation of tabulated functions to expansions of orthogonal polynomials is inexpensive in certain cases (including Legendre [5]). Differentiation and integration is generally simple and fast; the formulae for these operations depend on the weight ω.

2.3. Time-frequency localization

One issue which arises commonly in physical problems, and to which classical bases are not well suited, is the representation of very short, high-frequency, signals. An example in music is the attack, or beginning, of a played note, which introduces high-frequency components that decay rapidly as the note is sustained. In image processing, edges (localized high freqencies) are encountered at irregular spacings. The same is true for seismic data. Generally, there is a need for bases for which the elements representing the highest frequencies are most localized in time (or space).

The short-time Fourier transform, in which the exponential e^{inx} is multiplied by a localized window function such as $e^{-(x-a)^2/b}$, for various values of a, is an attempt to localize the Fourier basis elements. It does not however, localize different frequencies separately. The window width parameter b must be chosen for the degree of localization desired. The short-time Fourier transform also possesses the complication that no choice of window function leads to an orthogonal basis.

Wavelet bases have been developed to handle time-frequency localization cleanly.

§3. Construction of bases of wavelets

The term *wavelets* denotes a family of functions of the form

$$w_{a,b}(x) = |a|^{-1/2}\, w\left(\frac{x-b}{a}\right), \qquad a, b \in \mathbb{R},\ a \neq 0, \tag{1}$$

obtained from a single function w by the operations of dilation and translation. Such families, while named rather recently (Grossman and Morlet [13]), were used much earlier in the study of certain integral operators. Recent developments, however, have created widespread interest in wavelets among mathematicians and engineers. Explicit constructions of functions w leading to bases for $L^2(\mathbb{R})$ have propelled wavelets into applications in signal processing and the numerical solution of integral equations and partial differential equations.

In this section we present constructions of wavelets and wavelet-like bases; in Section 4 we give a sampling of their numerical applications.

3.1. Haar basis

The simplest example of a basis of wavelets, the Haar basis, consists of piecewise constant functions. The "pieces," or intervals on which the functions are constant, are of arbitrarily small size, and the basis is complete for $L^2(\mathbb{R})$. We start by defining the function $\phi : \mathbb{R} \to \mathbb{R}$ to be the characteristic function of the interval $[0, 1)$,

$$\phi(x) = \begin{cases} 1, & \text{if } x \in [0,1); \\ 0, & \text{otherwise.} \end{cases}$$

The integer translates of ϕ span the space V_0 of functions constant on unit intervals,

$$V_0 = \left\{ f \in L^2(\mathbb{R}) \;\middle|\; f(x) = \sum \alpha_k \, \phi(x-k) \right\}.$$

The sum here is taken over all $k \in \mathbb{Z}$; the requirement that $f \in L^2(\mathbb{R})$ is equivalent to $\sum_k \alpha_k{}^2 < \infty$. From the space V_0 we define for each $n \in \mathbb{Z}$ the space V_n by taking dilates of functions in V_0,

$$f(x) \in V_n \quad \Leftrightarrow \quad f(2^n x) \in V_0.$$

V_1 consists of functions constant on intervals of length 2, V_{-1} consists of functions constant on intervals of length $\frac{1}{2}$, and so forth. We therefore have the containment hierarchy

$$\cdots \subset V_2 \subset V_1 \subset V_0 \subset V_{-1} \subset V_{-2} \cdots \tag{2}$$

(note the decreasing subscripts). The only function contained in all subspaces is constant on $\mathbb{R}$ (and contained in L^2), hence, is identically zero; the union of the subspaces contains functions arbitrarily close to any function in L^2 so its closure coincides with L^2,

$$\bigcap_n V_n = \{\mathbf{0}\}, \qquad \overline{\bigcup_n V_n} = L^2(\mathbb{R}).$$

The closure $\overline{\cup V_n}$ is spanned by dilates and translates $\phi(2^n x - k)$ of ϕ, but the dilates and translates are not linearly independent (so, do not form a basis). This fact is evident from the containment hierarchy (2). To construct a basis, one can exploit the hierarchy and construct *difference* spaces: for each $n \in \mathbb{Z}$ we define the space W_n to be the orthogonal complement of V_n in V_{n-1},

$$W_n \oplus V_n = V_{n-1}, \qquad W_n \perp V_n. \tag{3}$$

It is easily verified that W_n is a dilate of W_0,

$$f(x) \in W_n \quad \Leftrightarrow \quad f(2^n x) \in W_0,$$

and that, analogous to V_0, the space W_0 is spanned by integer translates of a single function,

$$W_0 = \left\{ f \in L^2(\mathbb{R}) \;\middle|\; f(x) = \sum \alpha_k \, w(x-k) \right\}.$$

Here the translated function $w : \mathbb{R} \to \mathbb{R}$ is

$$w(x) = \begin{cases} 1, & \text{if } x \in [0, 1/2); \\ -1, & \text{if } x \in [1/2, 1); \\ 0, & \text{otherwise.} \end{cases}$$

Through (3) L^2 can be decomposed into a direct sum of the W_n,

$$\bigoplus_n W_n = \overline{\bigcup_n V_n} = L^2(\mathbb{R}),$$

so L^2 is spanned by dilates and translates of w,

$$L^2(\mathbb{R}) = \left\{ f \;\middle|\; f(x) = \sum \alpha_{nk} \, w_{nk}(x) \text{ with } \sum {\alpha_{nk}}^2 < \infty \right\}.$$

The normalized dilates and translates $w_{nk}(x) = 2^{-n/2} w(2^{-n}x - k)$, $n, k \in \mathbb{Z}$ form an orthonormal basis for $L^2(\mathbb{R})$. They are wavelets, according to the definition (1), with coefficients a, b taking the discrete values $a = 2^n$ and $b = 2^n k$ for $n, k \in \mathbb{Z}$.

One last point about the Haar basis. Toward the goal of accurate, practical representation of functions the Haar basis provides little help. Given a function f with several continuous derivatives, the expansion of f in the Haar basis,

$$f(x) = \sum_{n,k \in \mathbb{Z}} \alpha_{nk} \, w_{nk}(x),$$

converges only slowly. In general, as the number of terms in a truncated expansion for f doubles, the error from neglecting the discarded terms is cut in half, so the expansion is first-order convergent. For typical functions, high accuracy is achieved only with a very large number of terms. This slow convergence limits the practical value of the Haar basis for numerical applications.

3.2. Multiresolution analysis

The Haar basis is not new, but until recently there was no known orthogonal basis of wavelets in which the wavelet function w was even continuous, much less differentiable. In 1985 Y. Meyer [17] constructed such a basis with $w \in C^\infty(\mathbb{R})$. This was a surprise which seemed very improbable, and Meyer [16] and S. Mallat [15] developed a framework, the *multiresolution analysis,* in which to understand these bases. The Haar basis was presented in Subsection 3.1 within this framework; we now make the framework explicit.

A multiresolution analysis [9] consists of a sequence of closed subspaces V_n, $n \in \mathbb{Z}$, in $L^2(\mathbb{R})$ such that they lie in a containment hierarchy,

$$\cdots \subset V_2 \subset V_1 \subset V_0 \subset V_{-1} \subset V_{-2} \cdots, \tag{4}$$

they have intersection that is trivial and union that is dense in $L^2(\mathbb{R})$,

$$\bigcap_n V_n = \{\mathbf{0}\}, \qquad \overline{\bigcup_n V_n} = L^2(\mathbb{R}), \tag{5}$$

they are dilates of one another,

$$f(x) \in V_n \quad \Leftrightarrow \quad f(2^n x) \in V_0, \tag{6}$$

and there exists a *scaling* function $\phi \in V_0$ whose integer translates span V_0,

$$V_0 = \left\{ f \in L^2(\mathbb{R}) \;\middle|\; f(x) = \sum \alpha_k \, \phi(x-k) \right\}. \tag{7}$$

The spaces V_n are all alike when the scale is ignored; nevertheless, a journey up the containment hierarchy can be thought of (roughly) as adding ever higher frequencies while retaining low frequencies.

The multiresolution analysis leads directly to a scalewise, orthogonal decomposition of $L^2(\mathbb{R})$. The orthogonal complement of V_n in V_{n-1}, denoted by W_n, is the building block:

$$W_n \oplus V_n = V_{n-1}, \qquad W_n \perp V_n. \tag{8}$$

The spaces W_n, $n \in \mathbb{Z}$, are dilates of W_0 and their direct sum is L^2,

$$\bigoplus_n W_n = \overline{\bigcup_n V_n} = L^2(\mathbb{R}). \tag{9}$$

The space W_0 is spanned by integer translates of a function w. Intuitively, the space V_{-1}, spanned by integer translates of two functions ($\phi(2x)$ and $\phi(2x - 1)$), is twice the size of V_0, spanned by integer translates of ϕ; W_0 is the difference between these two spaces. The argument is made rigorous using group representations (omitted here). Using this characterization of W_0, and equation (9), we write L^2 as the space spanned by dilates and translates of w,

$$L^2(\mathbb{R}) = \left\{ f \;\middle|\; f(x) = \sum \alpha_{nk}\, w_{nk}(x) \text{ with } \sum {\alpha_{nk}}^2 < \infty \right\}, \tag{10}$$

where $w_{nk}(x) = 2^{-n/2} w(2^{-n}x - k)$.

What can be said about the wavelet function w? We have $w \in W_0 \subset V_{-1}$, so

$$w(x) = \sum_{k \in \mathbb{Z}} b_k\, \phi(2x - k),$$

for some coefficients $\ldots b_{-1}, b_0, b_1, \ldots$. Furthermore, a similar expansion holds for the scaling function ϕ (since $\phi \in V_0 \subset V_{-1}$):

$$\phi(x) = \sum_{k \in \mathbb{Z}} a_k\, \phi(2x - k),$$

for coefficients $\ldots a_{-1}, a_0, a_1, \ldots$. In Subsection 3.3 we will see that the choice for b_k given by

$$b_k = (-1)^k a_{1-k} \tag{11}$$

yields the desired orthogonality $W_0 \perp V_0$. The task of choosing a_k so that the translates of ϕ are mutually orthogonal, in addition to possessing other properties, is also addressed in Subsection 3.3.

Example. For the Haar basis, the scaling function satisfies the recurrence equation

$$\phi(x) = \phi(2x) + \phi(2x - 1)$$

and the wavelet function is given by the formula

$$w(x) = \phi(2x) - \phi(2x - 1),$$

following Equation (11). The inner products $\int \phi(x-k)\, w(x)\, dx$, as well as $\int \phi(x-k)\, \phi(x)\, dx$ and $\int w(x-k)\, w(x)\, dx$, vanish for $k \neq 0$.

3.3. Daubechies wavelets

Constraints can be placed on the coefficients a_k in the expansion for the scaling function ϕ,

$$\phi(x) = \sum_{k \in \mathbb{Z}} a_k\, \phi(2x - k), \tag{12}$$

so that ϕ satisfies various properties. I. Daubechies [9] constructed a class $\{ {}_N\phi \mid N \in \mathbb{N} \}$ of scaling functions, such that each ${}_N\phi$ vanishes outside a finite interval, satisfies the orthonormality requirement

$$\int_{-\infty}^{\infty} {}_N\phi(x-k)\, {}_N\phi(x-l)\, dx = \delta_{kl}, \qquad k, l \in \mathbb{Z}, \tag{13}$$

and has regularity ${}_N\phi \in C^{\mu N}(\mathbb{R})$, where $\mu \approx .35$. An important additional property of her scaling function ${}_N\phi$ is that low-order polynomials can be expressed as a linear combination of its translates:

$$x^j = \sum_{k\in\mathbb{Z}} \alpha_{jk}^N \, {}_N\phi(x-k), \qquad j = 0, \ldots, N-1, \tag{14}$$

where $\alpha_{jk}^N = \int {}_N\phi(x-k)\, x^j \, dx$. This latter property leads to good approximation properties; functions with several continuous derivatives have rapidly convergent expansions in bases of Daubechies wavelets.

We now explore these properties of ${}_N\phi$ and the corresponding constraints on the dilation coefficients a_k. We also present a method for computing the values of ${}_N\phi$. (We drop the prefixed subscript for ϕ where convenient.)

3.3.1. Compact support

If only a finite number of the coefficients a_k for ϕ are nonzero, then ϕ vanishes outside a finite interval. In particular, if all coefficients other than $a_m, a_{m+1}, \ldots, a_n$ vanish, then ϕ vanishes outside the interval $[m, n]$. To see this, observe that the iteration of

$$\phi_{i+1}(x) = \sum_{k=m}^{n} a_k \, \phi_i(2x-k)$$

maps a function ϕ_i supported on $[a, b]$ to the function ϕ_{i+1} supported on [(a+m)/2,(b+n)/2].

3.3.2. Consistency

Integration of the dilation equation (Equation (12)) determines $\sum a_k$:

$$\begin{aligned}\int_{-\infty}^{\infty} \phi(x)\, dx &= \int_{-\infty}^{\infty} \sum_{k\in\mathbb{Z}} a_k \, \phi(2x-k)\, dx \\ &= \frac{1}{2} \sum_k a_k \int_{-\infty}^{\infty} \phi(2x-k)\, 2dx \\ &= \frac{1}{2} \sum_k a_k \int_{-\infty}^{\infty} \phi(x)\, dx.\end{aligned}$$

But $\int \phi(x)\, dx \neq 0$, since the sum of translates of ϕ is a nonzero constant (Equation (14) with $j = 0$). Dividing through by $\int \phi(x)\, dx$ yields

$$\sum_k a_k = 2. \tag{15}$$

The dilation equation is homogeneous, so it determines ϕ only up to a constant factor. It is convenient to choose the factor so that

$$\int_{-\infty}^{\infty} \phi(x)\, dx = 1, \tag{16}$$

which will be assumed in the following discussion.

3.3.3. Orthogonality

We examine the consequences of Equation (13) by combining it with the dilation equation:

$$\begin{aligned}
\delta_{kl} &= \int_{-\infty}^{\infty} \phi(x-k)\,\phi(x-l)\,dx \\
&= \int_{-\infty}^{\infty} \sum_{m\in\mathbb{Z}} a_m\,\phi(2(x-k)-m) \sum_{n\in\mathbb{Z}} a_n\,\phi(2(x-l)-n)\,dx \\
&= \frac{1}{2}\sum_m\sum_n a_m\,a_n \int_{-\infty}^{\infty} \phi(2(x-k)-m)\,\phi(2(x-l)-n)\,2dx \\
&= \frac{1}{2}\sum_m\sum_n a_m\,a_n\,\delta_{2k+m,2l+n} \\
&= \frac{1}{2}\sum_m a_m\,a_{2(k-l)+m},
\end{aligned}$$

hence,

$$\frac{1}{2}\sum_{m\in\mathbb{Z}} a_{2k+m}\,a_{2l+m} = \delta_{kl}, \qquad k,l\in\mathbb{Z}. \tag{17}$$

Equation (17) is the fundamental relation which ensures the orthogonality of the translates of the scaling function ϕ. To establish the orthogonality between the translates of the wavelet function

$$w(x) = \sum_{k\in\mathbb{Z}} b_k\,\phi(2x-k), \tag{18}$$

we use the coefficients b_k given by Equation (11) and apply Equation (17) to obtain

$$\begin{aligned}
&\int_{-\infty}^{\infty} w(x-k)\,w(x-l)\,dx \\
&\qquad = \int_{-\infty}^{\infty} \sum_m b_m\phi(2(x-k)-m) \sum_n b_n\,\phi(2(x-l)-n)\,dx \\
&\qquad = \frac{1}{2}\sum_m b_{2k+m}\,b_{2l+m} \\
&\qquad = \frac{1}{2}\sum_m (-1)^{2k+m} a_{1-2k-m}\,(-1)^{2l+m} a_{1-2l-m} \\
&\qquad = \frac{1}{2}\sum_m a_{1-2k-m}\,a_{1-2l-m} \\
&\qquad = \delta_{kl}.
\end{aligned}$$

Finally, the orthogonality between the ϕ and w translates is established (without Equation (17)!) by

$$\begin{aligned}\int_{-\infty}^{\infty} \phi(x-k)\, w(x-l)\, dx &= \frac{1}{2}\sum_m a_{2l+m}\, b_{2k+m} \\ &= \frac{1}{2}\sum_m a_{2l+m}\, (-1)^{2k+m} a_{1-2k-m} \\ &= 0.\end{aligned}$$

The last statement follows from the observation that in the summation, each product $a_i\, a_j$ occurs twice, with opposite signs.

To summarize, Equation (17), combined with the coefficient definition $b_k = (-1)^k a_{1-k}$, ensures the orthogonality relations $\int \phi(x-k)\, \phi(x-l)\, dx = \delta_{kl}$, $\int \phi(x-k)\, w(x-l)\, dx = 0$, and $\int w(x-k)\, w(x-l)\, dx = \delta_{kl}$.

3.3.4. Approximation

The projection of a function $f \in L^2(\mathbb{R})$ on the space V_n is defined by the formula

$$f_n(x) = \sum_{k\in\mathbb{Z}} \alpha_{nk}\, \phi_{nk}(x),$$

where the (orthonormal) basis functions for V_n are given by

$$\phi_{nk}(x) = 2^{-n/2}\phi(2^{-n}x - k)$$

and the coefficients are

$$\alpha_{nk} = \int_{-\infty}^{\infty} \phi_{nk}(x)\, f(x)\, dx.$$

The rate of convergence of the sequence $\ldots, f_1, f_0, f_{-1}, \ldots \to f$ is of interest, for it dictates the computational cost of representing f to a prescribed accuracy using some f_n. A theory of approximation by translates was developed by G. Strang and G. Fix [20]. Although their result was derived in the context of finite element analysis, it is directly applicable here. They proved that for an arbitrary $f \in C^N(\mathbb{R})$ the sequence $f_0, f_{-1}, f_{-2}, \ldots$ has order of convergence N if and only if polynomials of degree less than N are contained in V_0. This is the property given in Equation (14), for $V_0 = \text{span}\{_N\phi(x-k) \mid k \in \mathbb{Z}\}$. Thus, the Daubechies scaling function $_N\phi$ gives rise to wavelets whose expansions are Nth-order convergent.

The low-order polynomials are contained in V_0, so they are orthogonal to W_0, hence to w. In terms of the dilation equation coefficients, we may write for $j = 0, \ldots, N-1$

$$\begin{aligned} 0 &= \int_{-\infty}^{\infty} x^j \, w(x) \, dx \\ &= \int_{-\infty}^{\infty} \sum_k \left(\frac{2x - k + k}{2} \right)^j b_k \, \phi(2x-k) \, dx \\ &= 2^{-j-1} \sum_{r=0}^{j} \binom{j}{r} \sum_k k^{j-r} \, b_k \int_{-\infty}^{\infty} (2x-k)^r \, \phi(2x-k) \, 2dx \\ &= 2^{-j-1} \sum_{r=0}^{j} \binom{j}{r} \sum_k k^{j-r} \, b_k \int_{-\infty}^{\infty} x^r \, \phi(x) \, dx. \end{aligned} \tag{19}$$

The integrals $\int x^r \, \phi(x) \, dx$ are all nonzero (since x^r can be written as a linear combination of translates of ϕ for $r = 0, \ldots, N-1$). Consecutively substituting $j = 0, 1, \ldots, N-1$ into Equation (19) yields $\sum_k b_k \, k^j = 0$, $j = 0, \ldots, N-1$, from which we obtain

$$\sum_k (-1)^k a_k \, k^j = 0, \qquad j = 0, \ldots, N-1. \tag{20}$$

Daubechies [9] constructs the scaling function ${}_N\phi$ by the choice of nonzero coefficients $a_0, \ldots, a_{2N-1}$ to satisfy Equations (11) and (17), as well as a regularity condition on ${}_N\phi$. It is remarkable that the regularity requirement leads to good approximation properties, and is equivalent to (20).

Examples. The scaling function ${}_1\phi$ coincides with the scaling function $\chi_{[0,1)}$ of the Haar basis.

The function ${}_2\phi$ is given by the formula

$$\begin{aligned} {}_2\phi(x) = \frac{1}{4} \Big[& \left(1 + \sqrt{3}\right) {}_2\phi(2x) + \left(3 + \sqrt{3}\right) {}_2\phi(2x-1) \\ & + \left(3 - \sqrt{3}\right) {}_2\phi(2x-2) + \left(1 - \sqrt{3}\right) {}_2\phi(2x-3) \Big]. \end{aligned}$$

Figure 1 shows the graphs of ${}_2\phi$ and the corresponding wavelet function ${}_2w$. Note that, unlike ${}_1\phi$, they are continuous. They do not, however, possess continuous derivatives. What is not obvious from the figure is their second-order approximation property: the functions $f(x) = 1$ and $f(x) = x$ are given by linear combinations of the scaling functions

$$\left.\begin{aligned} 1 &= \sum_{k \in \mathbb{Z}} {}_2\phi(x-k), \\ x &= \sum_{k \in \mathbb{Z}} \left(k + \frac{3 - \sqrt{3}}{2} \right) {}_2\phi(x-k), \end{aligned}\right. \qquad x \in \mathbb{R}.$$

For increasing N, the scaling functions ${}_N\phi$ are increasingly regular. Figure 2 shows the graphs of ${}_5\phi$ and ${}_5w$, each of which has a continuous first derivative.

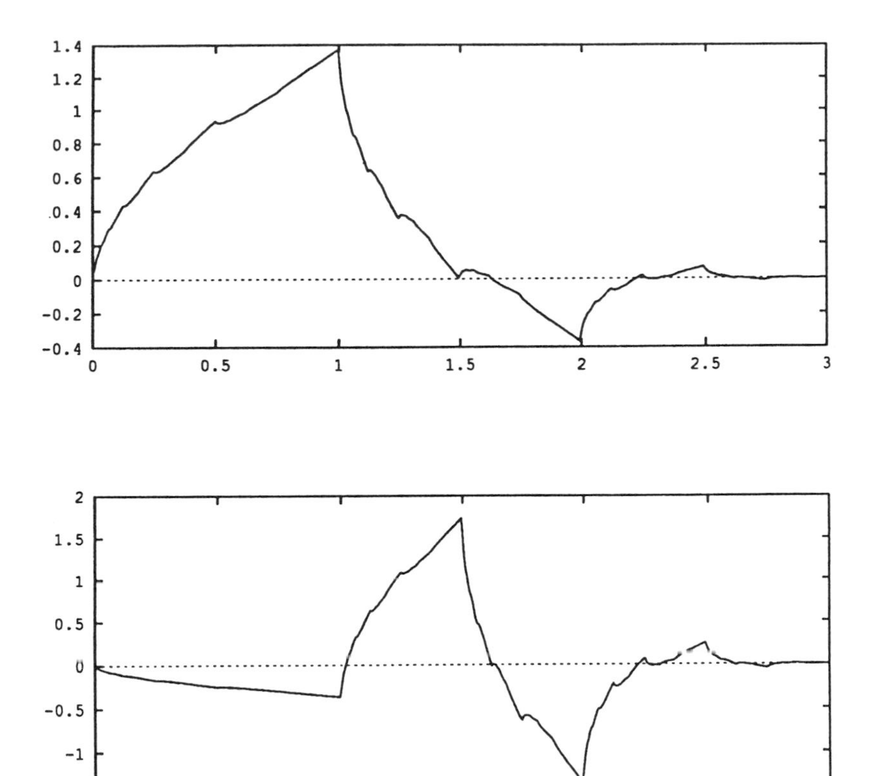

Figure 1. Daubechies scaling function ${}_2\phi$ (top) and corresponding wavelet function ${}_2w$. Here ${}_2w$ has been translated to the interval $[0, 3]$ by using coefficients $b_k = (-1)^k a_{2N-1-k}$.

3.3.5. Computation of ϕ

We have produced no explicit representation for ϕ; how it is computed? There are at least three approaches. The most obvious is to begin with some initial estimate ϕ_0 and calculate $\phi_1, \phi_2, \ldots$ by iteration of the dilation equation (12). Another method [9] calculates the Fourier transform $\hat{\phi}$ from an explicit formula involving the dilation coefficients. In the third, a very elegant approach, suggested by Strang [19], the values $\phi(1), \ldots, \phi(2N-2)$ at integer nodes are obtained by solving a linear system and are used to obtain the values at half-integer nodes, then quarter-integer nodes, and so forth, by the dilation

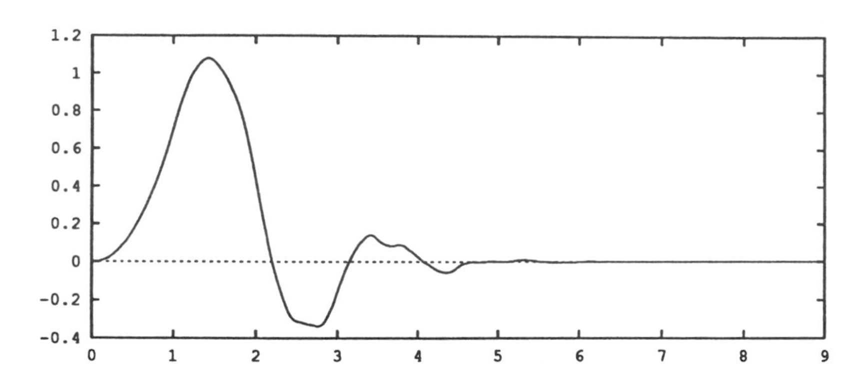

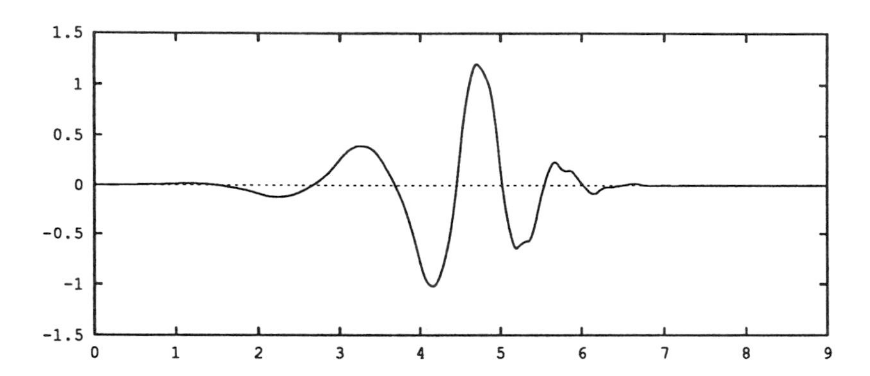

Figure 2. Scaling function ${}_5\phi$ (top) and corresponding wavelet function ${}_5w$, translated. In addition to their fifth-order approximation, they have greater smoothness than the second-order functions (Figure 1).

equation. The linear equations

$$\phi(j) = \sum_{k=0}^{2N-1} a_k \, \phi(2j-k), \qquad j = 1, \ldots, 2N-2$$

are deficient in rank by one (they determine ϕ only within a constant factor). Discarding one of them and adding the equation

$$1 = \sum_{j=1}^{2N-2} \phi(j),$$

we obtain a nonsingular system; it is solved to get $\phi(1), \ldots, \phi(2N-2)$.

This technique is fast, easy, and accurate; it was used to generate the graphs in Figures 1 and 2.

3.4. Multi-wavelets

We introduce now another class of bases for $L^2(\mathbb{R})$ which differ from wavelets in that instead of a single scaling function ϕ, there are several functions $\phi_0, \ldots, \phi_{N-1}$ whose translates span the space V_0. This difference enables high-order approximation with basis functions supported on non-overlapping intervals; the price is the lack of regularity of the functions.

In the framework of multiresolution analysis, this class is very simple. On the interval $[0,1)$, each scaling function ϕ_i is a dilated, translated, and normalized Legendre polynomial:

$$\phi_i(x) = \begin{cases} \sqrt{2i+1}\, P_i(2x-1), & \text{if } x \in [0,1); \\ 0, & \text{otherwise}; \end{cases} \qquad i = 0, \ldots, N-1,$$

where $P_0, P_1, \ldots$ are the Legendre polynomials. The rest of the basis construction (almost) follows, just by turning the "crank" of multiresolution analysis. The space V_0 consists of the span of the integer translates of the scaling functions $\phi_0, \ldots, \phi_{N-1}$. The spaces V_n for $n \in \mathbb{Z}$ are dilates of V_0, and the difference spaces W_n are defined, as before, by (8).

3.4.1. Construction of basis

The basis functions for W_n are not unique, of course, until we add additional constraints. Following [2], we construct orthonormal basis functions ${}_Nw_0, \ldots, {}_Nw_{N-1}$, which vanish outside the interval $[0,1)$, whose integer translates span W_0, and which are orthogonal to polynomials of maximum degree,

$$\int_{-\infty}^{\infty} {}_Nw_j(x)\, x^i \, dx, \qquad i = 0, \ldots, N-1+j.$$

We start with $2N$ dilates and translates of $\phi_0, \ldots, \phi_{N-1}$, which span the space of functions that are polynomials of degree less than N on the interval $[0, \frac{1}{2})$ and on $[\frac{1}{2}, 1)$, then orthogonalize N of them, first to the functions $\phi_0, \ldots, \phi_{N-1}$, then to the functions $\phi_N, \ldots, \phi_{2N-1}$, and finally among themselves. We define $f_0^1, f_1^1, \ldots, f_{N-1}^1$ by the formula

$$f_j^1(x) = \begin{cases} -\phi_j(2x), & \text{if } x \in [0, 1/2); \\ \phi_j(2x-1), & \text{if } x \in [1/2, 1); \\ 0, & \text{otherwise.} \end{cases}$$

Note that the $2N$ functions $\phi_0, \ldots, \phi_{N-1}, f_0^1, \ldots, f_{N-1}^1$ are linearly independent, hence span the space of functions that are polynomials of degree less than N on $[0, \frac{1}{2})$ and on $[\frac{1}{2}, 0)$.

1. By the Gram-Schmidt process we orthogonalize f_j^1 with respect to $\phi_0, \ldots, \phi_{N-1}$, to obtain f_j^2, for $j = 0, \ldots, N-1$. This orthogonality is preserved by the remaining orthogonalizations, which only produce linear combinations of the f_j^2.
2. The next sequence of steps yields $N-1$ functions orthogonal to ϕ_N, of which $N-2$ functions are orthogonal to ϕ_{N+1}, and so forth, down to 1 function which is orthogonal to ϕ_{2N-2}. First, if at least one of f_j^2 is not orthogonal to ϕ_N, we reorder the functions so that it appears first, $\langle f_0^2, \phi_N \rangle \neq 0$. We then define $f_j^3 = f_j^2 - a_j \cdot f_0^2$ where a_j is chosen so $\langle f_j^3, \phi_N \rangle = 0$ for $j = 1, \ldots, N-1$, achieving the desired orthogonality to ϕ_N. Similarly, we orthogonalize to $\phi_{N+1}, \ldots, \phi_{2N-2}$, each in turn, to obtain $f_0^2, f_1^3, f_2^4, \ldots, f_N^{N+1}$ such that $\langle f_j^{j+2}, \phi_i \rangle = 0$ for $i = 0, \ldots, N-1+j$.
3. Finally, we do Gram-Schmidt orthogonalization on $f_{N-1}^{N+1}, f_{N-2}^{N}, \ldots, f_0^2$, in that order, and normalize to obtain ${}_N w_{N-1}, {}_N w_{N-2}, \ldots, {}_N w_0$.

The resulting basis functions ${}_N w_0, \ldots, {}_N w_{N-1}$ are polynomials of degree $N-1$ on the intervals $[0, \frac{1}{2})$ and $[\frac{1}{2}, 1)$. The basis for $L^2(\mathbb{R})$ consists of the translates and dilates of ${}_N w_0, \ldots, {}_N w_{N-1}$. In addition, we can construct in this fashion a basis for a finite interval. For instance, $L^2[0,1]$ has an orthonormal basis consisting of $\phi_0, \ldots, \phi_{N-1}$ plus those translates and dilates of ${}_N w_0, \ldots, {}_N w_{N-1}$ that are supported in the interval $[0,1]$.

3.4.2. Approximation

The space V_0 consists of functions which are piecewise polynomial, of degree less than N. The expansion of an arbitrary function $f \in C^N(\mathbb{R})$ in a multi-wavelet basis of order N is Nth-order convergent. The argument for this case is simpler than the general argument of Strang and Fix [20], and we present it here. We define translates and dilates $\phi_j^{n,k}$ by the formula

$$\phi_j^{n,k}(x) = 2^{-n/2} \phi_j(2^{-n}x - k), \qquad n, k \in \mathbb{Z},$$

such that the set $\{ \phi_j^{n,k} \mid j = 0, \ldots, N-1;\ k \in \mathbb{Z} \}$ is a basis for V_n. The projection of f on V_n is defined (as before) by the formula

$$f_n(x) = \sum_{k \in \mathbb{Z}} \sum_{j=0}^{N-1} \alpha_j^{n,k} \, \phi_j^{n,k}(x), \tag{21}$$

where

$$\alpha_j^{n,k} = \int_{-\infty}^{\infty} \phi_j^{n,k}(x) \, f(x) \, dx.$$

The approximation error $\|f_n - f\|$ is bounded, according to the following lemma.

Lemma 1. *Suppose that the function $f : [0,1] \to \mathbb{R}$ is N times continuously differentiable, $f \in C^N[0,1]$. Then f_n* (given by Equation (21)) *approximates f with mean error bounded as follows:*

$$\|f_n - f\| \leq 2^{-nN} \frac{2}{4^N N!} \sup_{x\in[0,1]} |f^{(N)}(x)|. \tag{22}$$

Proof: We divide the interval $[0,1)$ into subintervals on which f_n is a polynomial; the restriction of f_n to one such subinterval $I_{n,k}$ is the polynomial of degree less than N that approximates f with minimum mean error. We then use the maximum error estimate for the polynomial which interpolates f at Chebyshev nodes of order N on $I_{n,k}$.

We define $I_{n,k} = [2^n k, 2^n(k+1))$ for $n = 0, -1, \ldots$; $k = 0, \ldots, 2^n - 1$, and obtain

$$\begin{aligned}
\|f_n - f\|^2 &= \int_0^1 [f_n(x) - f(x)]^2 \, dx \\
&= \sum_k \int_{I_{n,k}} [f_n(x) - f(x)]^2 \, dx \\
&\leq \sum_k \int_{I_{n,k}} \left[C^N_{n,k} f(x) - f(x)\right]^2 \, dx \\
&\leq \sum_k \int_{I_{n,k}} \left(\frac{2^{1-nN}}{4^N N!} \sup_{x\in I_{n,k}} \left|f^{(N)}(x)\right| \right)^2 dx \\
&\leq \left(\frac{2^{1-nN}}{4^N N!} \sup_{x\in[0,1]} \left|f^{(N)}(x)\right| \right)^2,
\end{aligned}$$

and by taking square roots we have bound (22). Here $C^N_{n,k} f$ denotes the polynomial of degree $N-1$ which agrees with f at the Chebyshev nodes of order N on $I_{n,k}$, and we have used the well-known maximum error bound for Chebyshev interpolation (see, *e.g.*, [1]). ■

The error of the approximation f_n of f therefore decays like 2^{-nN} and, since it requires $2^n N$ basis elements, we have convergence of order N. Despite the lack of regularity of the scaling functions ϕ_j (and the projections f_n), the convergence is similar to that obtained with the Daubechies wavelets.

Examples. The multi-wavelets case $N = 1$ coincides with the Haar basis (again).

For $N = 2$, the wavelet functions are given by

$$_2w_0(x) = \begin{cases} \sqrt{3}\,(1-4x), & \text{if } x \in [0,1/2); \\ \sqrt{3}\,(4x-3), & \text{if } x \in [1/2,1); \\ 0, & \text{otherwise;} \end{cases}$$

$$_2w_1(x) = \begin{cases} 6x-1, & \text{if } x \in [0,1/2); \\ 6x-5, & \text{if } x \in [1/2,1); \\ 0, & \text{otherwise.} \end{cases}$$

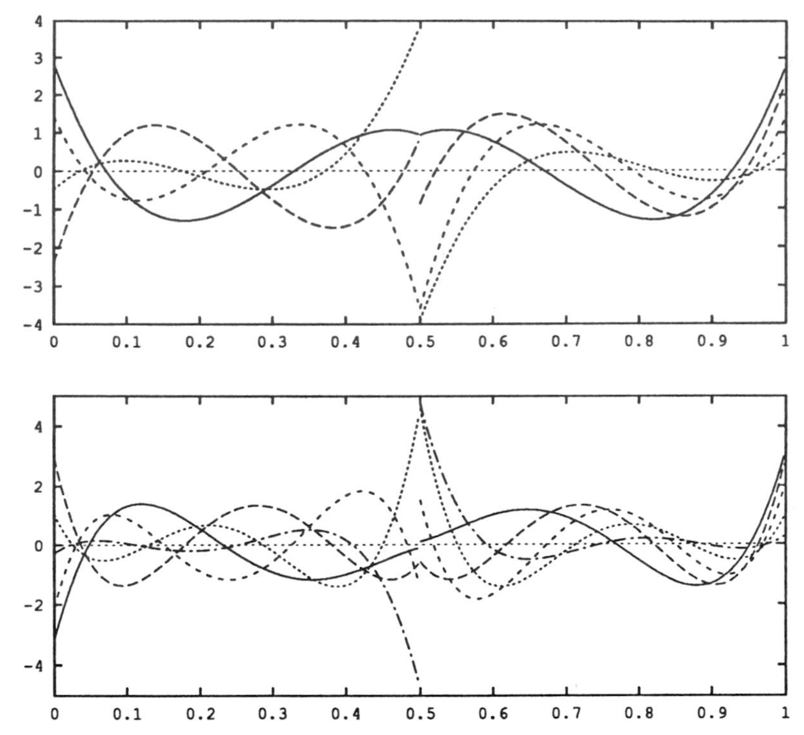

Figure 3. Wavelet functions for the piecewise polynomial multiwavelet bases for $N = 4$ (top) and $N = 5$. Note that each function is even or odd about the interval midpoint.

For larger N, the wavelet functions can also be derived explicitly, but the expressions naturally get rather long. Figure 3 shows the graphs of the wavelet functions for $N = 4$ and $N = 5$.

3.5. Discrete wavelet-like bases

The Daubechies wavelet bases of Subsection 3.3 and the multi-wavelet bases of Subsection 3.4 lie in the space of square-integrable functions $L^2(\mathbb{R})$. We consider now an analogue of the multi-wavelet bases which lies in the space of functions defined on a discrete set of points $\{x_1, \ldots, x_n\} \subset \mathbb{R}$. The structure of this analogue is essentially similar to that of the multi-wavelet bases, but the discrete construction is more convenient when the representation of a function (and its related operators) is based on its values at a finite set of points. Such representations arise in finite-difference computations and in Nyström discretizations for integral equations; the latter are discussed in Section 5.

The primary difference in the present construction from the multi-wavelet bases is the lack of complete scale invariance. In other words, the spaces V_n in the discrete construction are not dilates of a single space V_0, rather only nearly so.

3.5.1. Construction of bases

This construction follows that in [6]. The discrete set of points $\{x_1, \ldots, x_n\}$ is ordered so that $x_1 < \cdots < x_n$. We assume that the number n of points satisfies $n = 2^m N$, where N is the order of approximation required, and m is a positive integer. We define V_0 to be the N-dimensional vector space of polynomials of degree less than N on $\{x_1, \ldots, x_n\}$,

$$V_0 = \text{span}\left\{ \langle {x_1}^j, \ldots, {x_n}^j \rangle \;\middle|\; j = 0, \ldots, N-1 \right\}.$$

We define V_{-1} to be the $2N$-dimensional space of vectors which are polynomial of degree less than N on $\{x_1, \ldots, x_{n/2}\}$ and on $\{x_{n/2+1}, \ldots, x_n\}$. In general, V_{-j} is the $2^j N$-dimensional space consisting of vectors which are polynomial of degree less than N on $\{x_1, \ldots, x_{n/2^j}\}$, on $\{x_{n/2^j+1}, \ldots, x_{2n/2^j}\}$, and so forth, up to $\{x_{n-n/2^j+1}, \ldots, x_n\}$, for $j = 0, \ldots, m$. Thus V_{-m} is the entire n-dimensional vector space.

The difference space W_{-j}, for $j = 0, 1, \ldots, m-1$, is again defined by Equation (8). The procedure to construct a basis for V_{-m}, which exploits the decomposition

$$V_{-m} = W_{1-m} \oplus W_{2-m} \oplus \cdots \oplus W_0 \;\oplus\; V_0,$$

consists of an orthogonalization procedure to construct a basis for each of the W_{-j}. The result is an orthogonal matrix with rows which are the basis vectors, as shown in Figure 4. We construct the basis by the construction of a sequence of bases, for the decompositions

$$\begin{array}{l} W_{1-m} \oplus V_{1-m} \\ W_{1-m} \oplus W_{2-m} \oplus V_{2-m} \\ \quad\vdots \\ W_{1-m} \oplus W_{2-m} \oplus \cdots \oplus W_{-1} \oplus V_{-1} \\ W_{1-m} \oplus W_{2-m} \oplus \cdots \oplus W_{-1} \oplus W_0 \oplus V_0. \end{array}$$

The bases are given by the finite sequence of matrix products U_1, $U_2 U_1, \ldots$, $U_m \cdots U_1$. Before we can specify the matrices $U_1, \ldots, U_m$, we require some additional notation.

Suppose S is a matrix whose columns $s_1, \ldots, s_{2N}$ are linearly independent. We define $T = \text{Orth}(S)$ to be the matrix which results from the column-by-column Gram-Schmidt orthogonalization of S. Namely, denoting the columns of T by $t_1, \ldots, t_{2N}$, we have

$$\left.\begin{array}{r} \text{linear span}\{t_1, \ldots, t_i\} = \text{linear span}\{s_1, \ldots, s_i\} \\ {t_i}^T t_j = \delta_{ij} \end{array}\right. \qquad i, j = 1, \ldots, 2N.$$

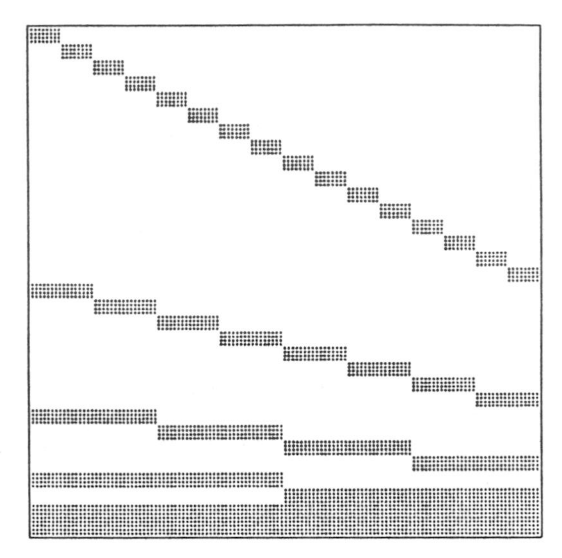

Figure 4. The matrix represents a wavelet-like basis for a discretization with $n = 128$ points, for $N = 4$. Each row denotes one basis vector, with the dots depicting nonzero elements. The first $n/2$ rows form a basis for W_{1-m}, the next $n/4$ form a basis for W_{2-m}, and so forth. All but the final N rows are orthogonal to polynomials of degree less than N, of which V_0 is composed.

For a $2N \times 2N$-matrix S we let S^U and S^L denote two $N \times 2N$-matrices, S^U consisting of the upper N rows and S^L the lower N rows of S,

$$S = \begin{pmatrix} S^U \\ S^L \end{pmatrix}.$$

Now we proceed to the definition of the basis matrices. Given the set of points $\{x_1, \ldots, x_n\} \in \mathbb{R}$ with $x_1 < \cdots < x_n$, where $n = 2^m N$, we define the $2N \times 2N$ moments matrices $M_{1,i}$ for $i = 1, \ldots, n/(2N)$ by the formula

$$M_{1,i} = \begin{pmatrix} 1 & x_{u_i+1} & \cdots & {x_{u_i+1}}^{2N-1} \\ 1 & x_{u_i+2} & \cdots & {x_{u_i+2}}^{2N-1} \\ \vdots & & & \vdots \\ 1 & x_{u_i+2N} & \cdots & {x_{u_i+2N}}^{2N-1} \end{pmatrix}, \tag{23}$$

where $u_i = (i-1)2N$. The first basis matrix U_1 is the $n \times n$-matrix given by the formula

$$U_1 = \begin{pmatrix} {U_{1,1}}^L & & & \\ & {U_{1,2}}^L & & \\ & & \ddots & \\ & & & {U_{1,n_1}}^L \\ {U_{1,1}}^U & & & \\ & {U_{1,2}}^U & & \\ & & \ddots & \\ & & & {U_{1,n_1}}^U \end{pmatrix},$$

where ${U_{1,i}}^T = \mathrm{Orth}(M_{1,i})$ and $n_1 = n/(2N)$. The second basis matrix is U_2U_1, with U_2 defined by the formula

$$U_2 = \begin{pmatrix} I_{n/2} & \\ & U_2' \end{pmatrix},$$

where I_j is the $j \times j$ identity matrix and the $n/2 \times n/2$-matrix U_2' is given by the formula

$$U_2' = \begin{pmatrix} {U_{2,1}}^L & & & \\ & {U_{2,2}}^L & & \\ & & \ddots & \\ & & & {U_{2,n_2}}^L \\ {U_{2,1}}^U & & & \\ & {U_{2,2}}^U & & \\ & & \ddots & \\ & & & {U_{2,n_2}}^U \end{pmatrix},$$

where $n_2 = n/(4N)$, ${U_{2,i}}^T = \mathrm{Orth}(M_{2,i})$, and the $2N \times 2N$ matrix $M_{2,i}$ is given by

$$M_{2,i} = \begin{pmatrix} {U_{1,2i-1}}^U M_{1,2i-1} \\ {U_{1,2i}}^U M_{1,2i} \end{pmatrix},$$

for $i = 1, \ldots, n/(4N)$. In general, the jth basis matrix, for $j = 2, \ldots, m$, is $U_j \cdots U_1$ with U_j defined by the formula

$$U_j = \begin{pmatrix} I_{n-n/2^{j-1}} & \\ & U_j' \end{pmatrix},$$

where U_j' is given by the formula

$$U_j' = \begin{pmatrix} {U_{j,1}}^L & & & \\ & {U_{j,2}}^L & & \\ & & \ddots & \\ & & & {U_{j,n_j}}^L \\ {U_{j,1}}^U & & & \\ & {U_{j,2}}^U & & \\ & & \ddots & \\ & & & {U_{j,n_j}}^U \end{pmatrix},$$

where $n_j = n/(2^j N)$, $U_{j,i}$ is given by

$$U_{j,i}{}^T = \mathrm{Orth}(M_{j,i}), \tag{24}$$

and $M_{j,i}$ is given by

$$M_{j,i} = \begin{pmatrix} U_{j-1,2i-1}{}^U M_{j-1,2i-1} \\ U_{j-1,2i}{}^U M_{j-1,2i} \end{pmatrix}, \tag{25}$$

for $i = 1, \ldots, n/(2^j N)$. The final basis matrix $U = U_m \cdots U_1$ represents the wavelet-like basis of parameter N on $x_1, \ldots, x_n$. Note that the matrices U and U_j are of dimension $n \times n$, U_j' is $n/2^{j-1} \times n/2^{j-1}$, $U_{j,i}$ and $M_{j,i}$ are $2N \times 2N$, and $U_{j,i}^L$ and $U_{j,i}^U$ are $N \times 2N$. In the computation of the basis, an adjustment must be made to these formulae to ensure numerical stability (see [6]).

§4. Linear algebraic operations

Very little indication has been presented, so far, of the value of wavelets for numerical linear algebra. Before embarking on the body of this section, we present an example showing the use of wavelets.

Example. For problems in electromagnetics, it is often necessary to determine the potential field due to a given distribution of charges. This is done by convolving the potential due to a point charge with the actual charge density. For a charge at the origin, this potential has the form $1/r$ in three dimensions and $\log r$ in two dimensions. We consider a simplified example, in which the function $f_a : \mathbb{R} \to \mathbb{R}$, given by $f_a(x) = \log|x - a|$, is to be represented. On any interval *separated* from a, the expansion of f_a in orthogonal polynomials converges rapidly. An interval I is separated from the point a if its distance to a is at least as great as its length. Expanding f_a on such an interval I, we obtain

$$f_a(x) = \sum_{j=0}^{\infty} \alpha_j \, p_j(x),$$

where $\alpha_j = \int_I p_j(x)\, f_a(x)\, dx$, and $p_0, p_1, \ldots$ are the orthonormal polynomials for the interval I. If we approximate f_a on I by keeping just the first n terms of the expansion, the truncation error decays exponentially in n (this is easy to show). For instance, keeping 8 terms yields six-digit accuracy; keeping 18 gives fifteen digits.

How does this connect with wavelets? In constructing the multi-wavelet bases, we built spaces V_n consisting of functions which are locally low-order polynomials. The multi-wavelet basis functions are orthogonal to these low-order polynomials, and are themselves locally supported. When f_a is expanded in a multi-wavelet basis, all basis functions lying on intervals separated from a have small coefficients, and can be neglected, up to a precision which depends on the order N. This property is illustrated in Figure 5. The representation of the function consists only of those basis functions supported near the singularity. It is this representational parsimony which leads to the usefulness of wavelets in a variety of applications.

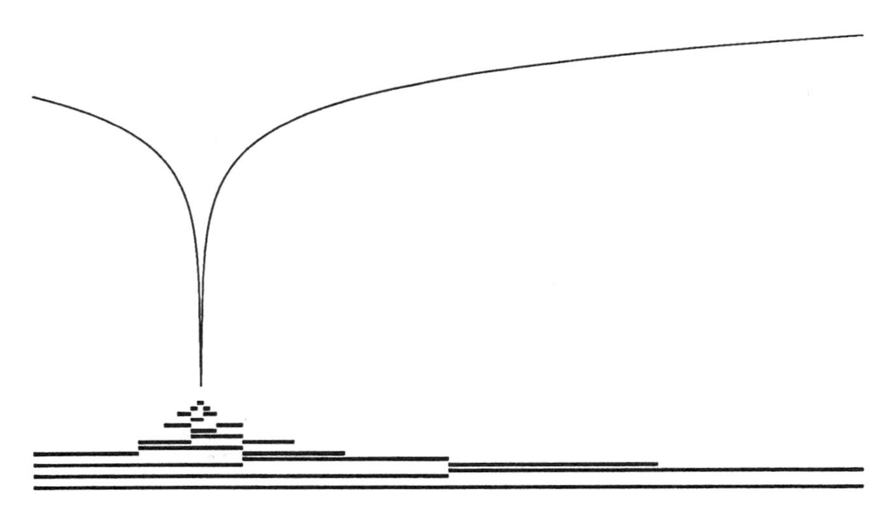

Figure 5. The function $f(x) = \log|x - .2|$, defined on the interval $[0, 1]$, is expanded in the multi-wavelet basis of order 8 to six-digit accuracy. The solid line segments indicate the intervals of support of the basis functions in the resulting representation.

4.1. Integral equations

A linear integral equation on the interval $[a, b]$ is an equation of the form

$$f(x) \cdot \gamma + \int_a^b K(x, y)\, f(y)\, dy = g(x), \tag{26}$$

where $f : [a, b] \to \mathbb{R}$ is the unknown, $K : [a, b] \times [a, b] \to \mathbb{R}$ is the kernel, and $g : [a, b] \to \mathbb{R}$ is the right hand side. The equation is an equation of the first kind if $\gamma = 0$, otherwise of the second kind. There is a rather complete theory for the existence and uniqueness of solutions to first- and second-kind integral equations, developed by Fredholm (see, *e.g.*, [12] or [14]). Integral equations form a powerful tool for the mathematical formulation of a wide range of physical problems; their relative neglect, compared with differential equations, is due in part to the historical lack of efficient solution techniques. Recently there has been substantial progress toward eliminating this deficiency.

Any equation to be solved numerically must be reduced to a finite dimensional problem, or discretized. There are two basic approaches to the discretization of integral equations. In one, often called the Galerkin method, expansions of f, K, and g are made in some basis, the expansions are truncated, and the resulting finite system of linear equations is solved numerically. In the second, developed by Nyström, the integral is approximated, at each of n points, by an n point quadrature, yielding again a system of equations that is solved numerically.

4.1.1. Galerkin method

Suppose that $\{b_1, b_2, \ldots\}$ is an orthonormal basis for $L^2[a,b]$. The expansions of $f, g \in L^2[a,b]$ in the basis are given by

$$f(x) = \sum_{j=1}^{\infty} f_j \, b_j(x), \qquad g(x) = \sum_{j=1}^{\infty} g_j \, b_j(x), \tag{27}$$

where the coefficients f_j and g_j are given by

$$f_j = \int_a^b b_j(x) \, f(x) \, dx, \qquad g_j = \int_a^b b_j(x) \, g(x) \, dx, \qquad j \in \mathbb{N}.$$

Similarly, the expansion for $K \in L^2([a,b] \times [a,b])$ is the integral in both coordinates

$$K(x,y) = \sum_{i=1}^{\infty} \sum_{j=1}^{\infty} K_{ij} \, b_i(x) \, b_j(y), \tag{28}$$

where the coefficient K_{ij} is the double integral

$$K_{ij} = \int_a^b \int_a^b b_i(x) \, b_j(y) \, K(x,y) \, dx \, dy, \qquad i,j \in \mathbb{N}.$$

Substitution of Equations (27) and (28) into integral Equation (26) yields an infinite system of equations in the coefficients f_j, g_j, and K_{ij}, namely,

$$f_i \cdot \gamma + \sum_{j=1}^{\infty} K_{ij} \, f_j = g_i, \qquad i \in \mathbb{N}. \tag{29}$$

The expansion for K may be truncated at a finite number of terms, producing the finite system of equations

$$f_i \cdot \gamma + \sum_{j=1}^{n} K_{ij} \, f_j = g_i, \qquad i = 1, \ldots, n. \tag{30}$$

For most applications, with classical bases (*e.g.*, Fourier or orthogonal polynomials) the matrix $\bar{K}^n = \{K_{ij}\}_{i,j=1,\ldots,n}$ is dense (nearly all of its elements are nonzero). There may be substantial cost in computing all elements K_{ij}. Even after the elements are computed, the cost of the application of the $n \times n$-matrix $\bar{K}^n$ to a vector is of order $O(n^2)$. The solution of (30), if obtained by a direct scheme such as Gaussian elimination, requires order $O(n^3)$ operations. If (30) is solved by an iterative method, which requires an order $O(n^2)$ matrix-vector product on each iteration, the number of iterations may be large, depending on the conditioning of the original integral equation. For large-scale problems, these costs are often prohibitive.

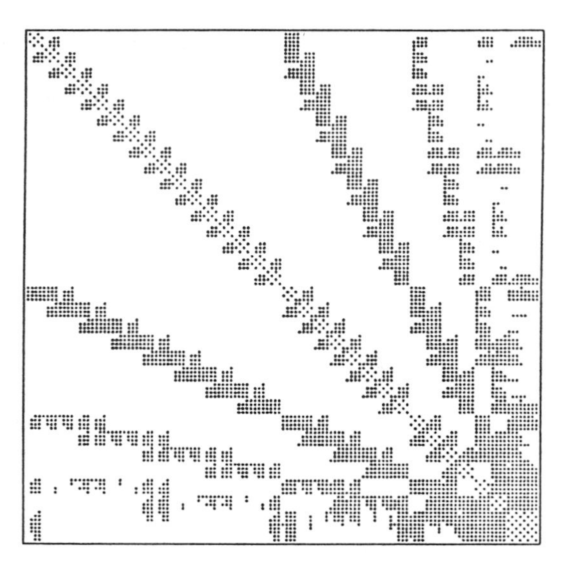

Figure 6. The integral operator with kernel $K(x,y) = \log|x-y|$ is discretized by the Galerkin method, with the multi-wavelet basis of order $N = 4$. The truncation is set at $n = 128$ basis functions. The dots represent elements above a threshold, determined so that the relative error ϵ is bounded at 10^{-3}.

The denseness of $\bar{K}^n$ depends on both the kernel K and the basis. For a wide variety of kernels arising in problems of potential theory, the matrix $\bar{K}^n$ is sparse, to high precision, if the basis is chosen to be a wavelet basis. This theme was developed by G. Beylkin, R. Coifman, and V. Rokhlin [8] for Daubechies wavelets. For these problems, the kernel $K(x,y) = \log|x-y|$ serves as a good model. The value of the kernel varies smoothly as a function of x and y away from the diagonal $x = y$, where the kernel is singular. In the example at the start of this section we saw what happens when $f_a(x) = \log|x-a|$ is expanded in the x-coordinate in a multi-wavelet basis. Now we propose to expand the kernel $K(x,y) = \log|x-y|$ in both coordinates in a multi-wavelet basis; one example of the matrices which result is shown in Figure 6.

4.1.2. Nyström method

An alternative to the Galerkin method for the discretization of integral equations, the Nyström method approximates the integral operator

$$(\mathcal{K}f)(x) = \int_a^b K(x,y)\, f(y)\, dy$$

by a quadrature for selected values of x. We define the operator R by the formula

$$(Rf)(x) = \sum_{j=1}^{n} \omega_j\, K(x, x_j)\, f(x_j),$$

which approximates $\mathcal{K}$ for appropriate choice of quadrature weights ω_j and nodes x_j. The values of x are chosen to coincide with the quadrature nodes $x_1, \ldots, x_n$, and the original integral Equation (26) is approximated by the system of equations

$$f(x_i) \cdot \gamma + \sum_{j=1}^{n} \omega_j \, K(x_i, x_j) \, f(x_j) = g(x_i), \qquad i = 1, \ldots, n. \tag{31}$$

(Compare to the Galerkin discretization (30).) For the trapezoidal rule, the quadrature weights ω_j are equal, except at the ends $j = 1$ and $j = n$. For kernels with singularities, however, the trapezoidal rule does not provide a good approximation of the integral. Quadratures have been developed in which the weights near the ends and near the singularities are altered to yield rapidly-convergent schemes [3]. In this case each weight depends on the argument x_i as well as the quadrature node x_j, so it becomes ω_{ij}, for $i, j = 1, \ldots, n$.

Even with these adjustments to the trapezoidal rule, most of the quadrature weights have constant value and the smoothness of the matrix $\tilde{K}^n = \{\omega_{ij} K(x_i, x_j)\}_{i,j=1,\ldots,n}$ depends primarily on the smoothness of the kernel K. For a kernel which is smooth except for diagonal singularities, the matrix $\tilde{K}^n$ can be transformed by a change of basis to a sparse matrix, to high precision. In particular, the wavelet-like basis matrix U defined in Subsection 3.5 can be used to obtain the similarity-transformed matrix $U \, \tilde{K}^n \, U^T$, which has the desired sparse structure. In fact, a picture of $U \, \tilde{K}^n \, U^T$ is nearly indistinguishable from Figure 6.

Remark. The Galerkin method and Nyström method are two techniques for the discretization of integral operators: which is preferred? Both are conceptually straightforward and an error analysis has been developed for each method (see, *e.g.*, [12]); the Nyström method offers, however, some computational benefits. Using the Nyström method with the trapezoidal rule, or high-order corrected trapezoidal rule, the kernel is evaluated just once for each element in the computed matrix. With the Galerkin method, on the other hand, a matrix element corresponds to an integral of the kernel with the basis elements in both coordinates. An appropriate quadrature must be applied to compute each of these elements, generally requiring many kernel evaluations. This complication usually makes the Galerkin method uncompetitive with the Nyström method.

Use of the Nyström method was the primary motivation behind the development of the discrete wavelet-like bases.

4.2. Sparsity in wavelet bases

The example at the beginning of the section suggests that the number of basis functions required to represent the function $f_a(x) = \log|x-a|$ to precision ϵ is of order $O(\log(1/\epsilon))$. This is indeed the case for this function, as well as other functions analytic except at separated, integrable singularities. Since the Nyström discretization $\tilde{K}^n$ consists of columns with elements described by functions like f_a, one might expect that its transformation $U\tilde{K}^n$ would be sparse (to high precision), containing only $O(n \log n)$ non-negligible elements. This is the case. Furthermore, the complete similarity transformation $U\tilde{K}^n U^T$, which exploits the smoothness in the rows of $\tilde{K}^n$, as well as the columns, is yet more sparse, containing only $O(n)$ non-negligible elements.

The story for the matrix $\bar{K}^n$ is similar, but here the matrix contains $O(n \log n)$ non-negligible elements. This sparsity is proved for several specific examples in [2].

4.3. Multiplication of integral operators

The product of two integral operators with smooth kernels itself has a smooth kernel, and it can be represented as a sparse matrix in wavelet coordinates.

We define integral operators $\mathcal{K}_1$ and $\mathcal{K}_2$ by the formula

$$(\mathcal{K}_i f)(x) = \int_a^b K_i(x,y)\, f(y)\, dy, \qquad i = 1,2.$$

The product operator $\mathcal{K}_3 = \mathcal{K}_1\mathcal{K}_2$ is given by the formula

$$\begin{aligned}(\mathcal{K}_1\mathcal{K}_2 f)(x) &= \int_a^b K_1(x,t) \int_a^b K_2(t,y)\, f(y)\, dy\, dt \\ &= \int_a^b \left(\int_a^b K_1(x,t)\, K_2(t,y)\, dt \right) f(y)\, dy \\ &= \int_a^b K_3(x,y)\, f(y)\, dy,\end{aligned}$$

where the kernel $\mathcal{K}_3$ of the product has the form

$$K_3(x,y) = \int_a^b K_1(x,t)\, K_2(t,y)\, dt.$$

If kernels K_1 and K_2 are analytic except along the diagonal $x = t$, where they have integrable singularities, then the same is true of the product kernel K_3. As a result, the product operator $\mathcal{K}_3$ also has a sparse representation in a wavelet basis.

4.4. Solution of integral equations

The representation of integral operators as sparse matrices, via transformation to wavelet coordinates, leads to new methods for the solution of integral equations. The integral equation (26), written in operator notation as

$$(\gamma + \mathcal{K})f = g,$$

has (formal) solution

$$f = (\gamma + \mathcal{K})^{-1} g.$$

The operator $(\gamma+\mathcal{K})^{-1}$ can be applied to g with the conjugate gradient method (conjugate residual if $A = \gamma + \mathcal{K}$ is nonsymmetric). This well-established method for sparse matrices is very fast if A is well conditioned. The number of iterations, which grows as the square root of the condition number (linearly in the condition number for conjugate residual), becomes rather large for poorly-conditioned problems.

Alternatively, one can directly invert A, obtaining a sparse inverse, or compute a sparse LU-factorization of A.

4.4.1. Schulz method of matrix inversion

Schulz's method [18] is an iterative, quadratically convergent algorithm for computing the inverse of a matrix. Its performance is characterized as follows.

Lemma 2. *Suppose that A is an invertible matrix, X_0 is the matrix given by $X_0 = A^T/\|A^T A\|$, and for $m = 0, 1, 2, \ldots$ the matrix X_{m+1} is defined by the recursion*

$$X_{m+1} = 2X_m - X_m A X_m.$$

Then X_{m+1} satisfies the formula

$$I - X_{m+1}A = (I - X_m A)^2. \tag{32}$$

Furthermore, $X_m \to A^{-1}$ as $m \to \infty$ and for any $\epsilon > 0$ we have

$$\|I - X_m A\| < \epsilon \quad \text{provided} \quad m \geq 2\log_2 \kappa(A) + \log_2 \log(1/\epsilon), \tag{33}$$

where $\kappa(A) = \|A\| \cdot \|A^{-1}\|$ is the condition number of A and the norm is given by $\|A\| =$ (largest eigenvalue of $A^T A)^{1/2}$.

Proof: Equation (32) is obtained directly from the definition of X_{m+1}. Bound (33) is equally straightforward. Noting that $A^T A$ is symmetric positive-definite and letting λ_0 denote the smallest and λ_1 the largest eigenvalue of $A^T A$ we have

$$\begin{aligned} \|I - X_0 A\| &= \left\| I - \frac{A^T A}{\|A^T A\|} \right\| \\ &= 1 - \lambda_0/\lambda_1 \\ &= 1 - \kappa(A)^{-2}. \end{aligned} \tag{34}$$

From Equation (32) we obtain $I - X_m A = (I - X_0 A)^{2^m}$, which in combination with Equation (34) and simple manipulation yields bound (33). ■

The Schulz method is a notably simple scheme for matrix inversion and its convergence is extremely rapid. It is rarely used, however, because it involves matrix-matrix multiplications on each iteration; for most problem formulations, this process requires order $O(n^3)$ operations for an $n \times n$-matrix. As

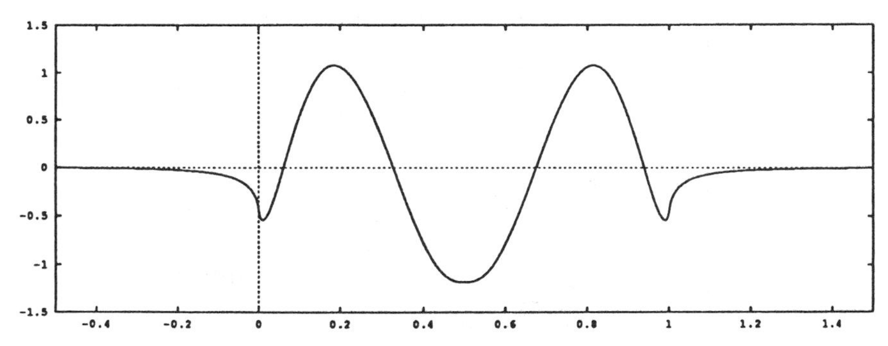

Figure 7. One of the multi-wavelet basis functions for $N = 4$ is convolved with $\log|x|$ and the image graphed. The function decays as the Nth power of the distance from its center, so effectively vanishes beyond the neighbors of the interval of support of the basis function.

we have seen above, on the other hand, an integral operator A represented in a discrete wavelet-like basis has only $O(n)$ elements (to finite precision). In addition, $A^T A$ and $(A^T A)^m$ are similarly sparse. This property enables us to employ the Schulz algorithm to compute A^{-1} in order $O(n)$ operations.

4.4.2. Sparse LU factorization

For dense matrices, computation of the inverse is almost never desirable. The decomposition into lower-triangular and upper-triangular (LU) factors requires roughly one third as many operations, and is equally useful. One might suppose that this advantage would also hold for sparse matrices with sparse inverses: perhaps it is possible to factor the sparse matrix $A = U\,\tilde{K}^n\,U^T$ into LU factors which are themselves sparse.

Unfortunately, direct factorization of A produces substantial fill-in of zero elements, and lower and upper triangular factors that are not sparse. This fill-in results from a "smearing" of the near-diagonal blocks. These blocks represent the interactions ${u_i}^T\,\tilde{K}^n\,u_j$ of basis elements u_i and u_j that are supported on adjacent intervals (see Figure 7).

Reordering the basis elements of U, we can construct a basis for which the elements are sorted into "levels" such that the basis elements on different intervals, but on one level, are separated from each other, and only interact with the elements of a single interval on each higher level. This ordering is illustrated in Figure 8. The reordered basis can then be used to transform the Nyström discretization $\tilde{K}^n$ of the integral operator into a sparse matrix lacking subdiagonal and superdiagonal blocks. In this form, shown in Figure 9, direct Gaussian elimination produces sparse lower- and upper-triangular factors.

The technique of basis reordering for sparse LU factorization of integral operators, as described, works very well for one-dimensional problems, even outperforming the Schulz method. For numerical examples, the reader is re-

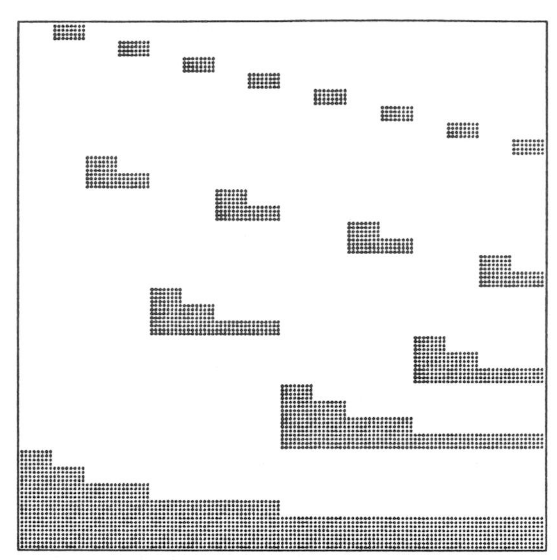

Figure 8. The matrix represents a reordering of the rows (basis vectors) of the matrix in Figure 4. In this order the basis is used to transform $\tilde{K}^n$ into a matrix supporting sparse LU factorization (shown below).

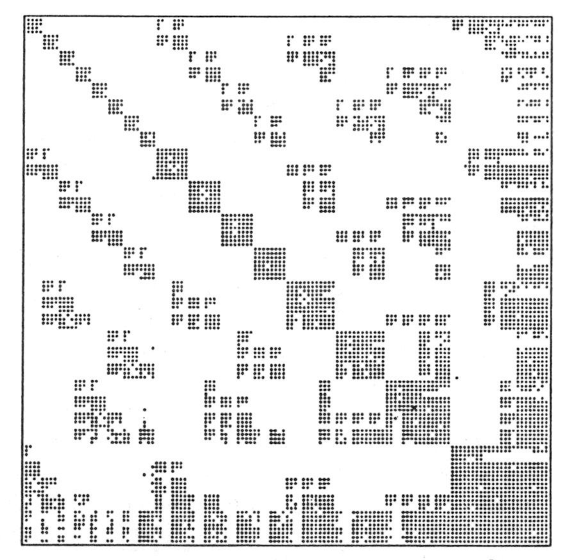

Figure 9. The non-negligible matrix elements ($\epsilon = 10^{-3}$) are shown for the integral operator with kernel $\log|x - y|$ expressed in the basis shown above. The matrix can be factored into lower- and upper-triangular matrices with no increase in the number of elements.

ferred to [4] and [6]. It does not appear, however, that the method of factorization cleanly generalizes to higher dimensional problems, where the Schulz method is expected to maintain its good performance.

4.5. Representation of differential operators

We have seen that integral operators can be expressed in wavelet bases as sparse matrices, to high precision. Certain differential operators, by contrast, are represented exactly by (infinite) sparse matrices. In [7] G. Beylkin determines the representations of various operators using Daubechies wavelets. Here we illustrate his technique for the derivative operator d/dx.

For a function represented as a wavelet expansion

$$f(x) = \sum_{n,k} \alpha_{nk}\, w_{nk}(x),$$

the derivative

$$\frac{df}{dx}(x) = \sum_{n,k} \alpha_{nk}\, \frac{d}{dx} w_{nk}(x)$$

is determined by the derivative of the basic wavelet function w. Through the definition (18) of w, its derivative is given in turn by the equation

$$\frac{dw}{dx}(x) = \sum_k b_k \frac{d}{dx}\phi(x-k).$$

Differentiation, therefore, is reduced to the determination of coefficients c_{nk} in the expansion for $d\phi/dx$,

$$\frac{d\phi}{dx}(x) = \sum_{n,k} c_{nk}\, w_{nk}(x),$$

where c_{nk} is given by the inner product

$$c_{nk} = \int_{-\infty}^{\infty} w_{nk}(x)\, \frac{d\phi}{dx}(x)\, dx, \qquad n,k \in \mathbb{Z}.$$

The application of the definitions of ϕ and w are used to reduce the coefficients c_{nk} to the coefficients

$$c_k = \int_{-\infty}^{\infty} \phi(x-k)\, \frac{d\phi}{dx}(x)\, dx, \qquad k \in \mathbb{Z}. \tag{35}$$

We apply the dilation Equation (12) to Equation (35) to obtain

$$\begin{aligned} c_k &= \int_{-\infty}^{\infty} \sum_l a_l \frac{d\phi}{dx}(2x-l)\, 2 \sum_m a_m\, \phi(2(x-k)-m)\, dx \\ &= \sum_l \sum_m a_l\, a_m \int_{-\infty}^{\infty} \frac{d\phi}{dx}(2x-l)\, \phi(2x-2k-m)\, 2dx \qquad (36) \\ &= \sum_l \sum_m a_l\, a_m\, c_{2k+m-l}, \qquad k \in \mathbb{Z}. \end{aligned}$$

The scaling function $\phi = {}_N\phi$ is supported on the finite interval $[0, 2N-1]$. As a result, we see from Equation (35) that $c_k = 0$ for $|k| \geq 2N-1$ (the integrand vanishes). Also, integration by parts yields $c_{-k} = -c_k$. Combining these two observations, (36) becomes a system of $2N-2$ equations in the $2N-2$ unknowns $c_1, \ldots, c_{2N-2}$. Due to its homogeneity, the rank of the system is deficient by one, and another equation is required to determine the scale of the c_k. The supplemental equation is obtained from the fact that the function $f(x) = x$ is a linear combination of translates of ϕ, for $N \geq 2$. In particular,

$$x = \sum_k \mu_k \phi(x-k), \tag{37}$$

where

$$\begin{aligned}
\mu_k &= \int_{-\infty}^{\infty} \phi(x-k)\, x\, dx \\
&= \int_{-\infty}^{\infty} \phi(x-k)\,(x-k)\, dx + \int_{-\infty}^{\infty} \phi(x-k)\, k\, dx \\
&= \int_{-\infty}^{\infty} \phi(x)\, x\, dx + k \\
&= \mu_0 + k.
\end{aligned}$$

Combining Equation (37) with

$$1 = \sum_k \phi(x-k)$$

and differentiating by x yields

$$1 = \sum_k k \frac{d}{dx} \phi(x-k),$$

which in combination with Equation (35) gives the desired supplemental equation

$$\sum_k k\, c_k = -1. \tag{38}$$

Equations (36) and (38) may be solved directly to obtain the coefficients $c_1, \ldots, c_{2N-2}$.

As mentioned above, similar techniques can be used to obtain higher derivatives and certain other differential and integral operators, including, for instance, the Hilbert transform. This development suggests that the evolution of a variety of differential equations in wavelet bases may become efficient, which would strengthen the arsenal of the numerical analyst attacking problems requiring highly adaptive schemes.

§5. Summary

We have illustrated the construction of wavelets and similar wavelet-like bases, their properties of orthogonality, approximation, compact support, and most distinguishingly, time-frequency localization through dilation invariance (and near-invariance). These bases lead to the sparse representation of integral operators and the rapid solution of integral equations. Differential operators are also represented as sparse matrices in wavelet bases, which permits the construction of adaptive algorithms for time-dependent partial differential equations.

References

1. Abramowitz, M. and I. Stegun, *Handbook of Mathematical Functions*, National Bureau of Standards, Washington, D.C., 1972.
2. Alpert, B., *Sparse Representation of Smooth Linear Operators*, Ph.D. thesis, Yale University Department of Computer Science Report #814, August, 1990.
3. Alpert, B., Rapidly-convergent quadratures for integral operators with singular kernels, Lawrence Berkeley Laboratory Report LBL-30091, December, 1990.
4. Alpert, B., Construction of simple multiscale bases for fast matrix operations, in *Wavelets and Their Applications*, G. Beylkin, R. Coifman, I. Daubechies, S. Mallat, Y. Meyer, L. Raphael, and M. Ruskai (eds.), Jones and Bartlett, Cambridge, MA, 1992, to appear.
5. Alpert, B. and V. Rokhlin, A fast algorithm for the evaluation of Legendre expansions, *SIAM J. Sci. Statist. Comput.* **12** (1991), 158–179.
6. Alpert, B., G. Beylkin, R. Coifman, and V. Rokhlin, Wavelet bases for the fast solution of second-kind integral equations, Yale University Department of Computer Science Report #837, December, 1990. *SIAM J. Sci. Statist. Comput.*to appear.
7. Beylkin, G., On the representation of operators in compactly supported wavelet bases, August, 1990, preprint.
8. Beylkin, G., R. Coifman, and V. Rokhlin, Fast wavelet transforms and numerical algorithms I, *Comm. Pure and Appl. Math.* **44** (1991), 141–183.
9. Daubechies, I., Orthonormal bases of compactly supported wavelets, *Comm. Pure and Appl. Math.* **41** (1988), 909–996.
10. Daubechies, I., The wavelet transform, time-frequency localization and signal analysis, *IEEE Trans. on Information Theory* **36** (1990), 961–1005.
11. Daubechies, I., A. Grossmann, and Y. Meyer, Painless non-orthogonal expansions, *J. Math. Phys.* **27** (1986), 1271–1283.
12. Delves, L. and J. Mohamed, *Computational Methods for Integral Equations*, Cambridge University Press, 1985.
13. Grossmann, A. and J. Morlet, Decomposition of Hardy functions into square integrable wavelets of constant shape, *SIAM J. Math. Anal.* **15** (1984), 723–736.

14. Hochstadt, H., *Integral Equations*, Wiley, 1989.
15. Mallat, S., Multiresolution approximation and wavelets, *Trans. Amer. Math. Soc.* **315** (1989), 69–88.
16. Meyer, Y., Principe d'incertitude, bases Hilbertiennes et algèbres d'opérateurs, *Séminaire Bourbaki* **662** (1985–1986).
17. Meyer, Y., Ondelettes, functions splines, et analyses graduées, *Rapport CEREMADE* **8703**, 1987.
18. Schulz, G., Iterative berechnung der reziproken matrix, *Zeitschrift für Angewandte Mathematik und Mechanik* **13** (1933), 57–59.
19. Strang, G., Wavelets and dilation equations: a brief introduction, *SIAM Review* **31** (1989), 614–627.
20. Strang, G. and G. Fix, A Fourier analysis of the finite element variational method, in *Constructive Aspects of Functional Analysis*, G. Geymonant (ed.), 1973, 793–840.

Supported in part by the Applied Mathematical Sciences subprogram of the Office of Energy Research, U.S. Department of Energy under Contract DE-AC03-76SF-00098.

Bradley K. Alpert
National Institute of Standards and Technology
Mail Code 881
325 Broadway
Boulder, CO 80303
alpert@bldr.nist.gov

Wavelets with Boundary Conditions on the Interval

Pascal Auscher

Abstract. We discuss the construction of wavelet bases with preassigned boundary value conditions on the unit interval. These boundary conditions arise naturally from boundary value differential equations, and the wavelets are Riesz bases of the Sobolev spaces and of the Hölder spaces where the solutions of these equations live. Our construction rests on the theory of Hermite interpolation and on the theory of multiresolution analyses.

§1. Introduction

In [8], S. Jaffard and Y. Meyer constructed a wavelet basis for an open set in $\mathbb{R}^n$. In the special case of the interval (0,1), these wavelets are all supported in [0,1]. Thus, they provide a wavelet analysis for the Sobolev spaces $H_0^s[0,1]$, $s > 0$, which is the closure in $H^s[0,1]$ of the C^∞ functions with support in (0,1).

Later, Y. Meyer showed in [13] that one can consider wavelets on [0,1] that do not show any conditions at the boundary of the interval. These wavelets, therefore, are suitable for the analysis of $H^s[0,1]$ itself.

If $s > 1/2$, $H^s[0,1]$ has proper subspaces that strictly contain $H_0^s[0,1]$. These subspaces are defined by boundary conditions. Let us give an example. For $s = 2$, $H_0^2[0,1]$ is defined as a subspace of $H^2[0,1]$ by the 4 conditions $f(0) = f'(0) = f(1) = f'(1) = 0$. We may define V by, *e.g.*, $f(0) = f(1) = 0$ (or by any other homogeneous system of rank 1, 2, or 3 of the 4 unknowns $f(0), f'(0), f(1)$, and $f'(1)$).

The problem we shall address in this chapter is a wavelet analysis of proper subspaces of $H^s[0,1]$ that strictly contain $H_0^s[0,1]$: is it possible to characterize these spaces through a wavelet basis?

It turns out that none of the wavelet bases in [8] and [13] characterizes the space V of our example. Indeed, each wavelet should vanish at 0 and at 1 but, at the same time, its first derivative should take arbitrary values at the same points.

Wavelets–A Tutorial in Theory and Applications
C. K. Chui (ed.), pp. 217–236.

ISBN 0-12-174590-2

Before going further and to motivate the above question, let us make a digression toward differential equations. Consider the following Dirichlet problem with data f in L^2:

$$-u'' + u = f \quad \text{on} \quad (0,1),$$
$$u(0) = u(1) = 0\,.$$

It is well known using the variational formulation that this problem has a unique solution u in $H_0^1[0,1]$. Furthermore, u belongs to $H^2[0,1] \cap H_0^1[0,1]$, which is precisely V. Had we changed the boundary conditions in the problem, we obtain (when there is a solution, which is not always the case) a different solution in a different subspace of $H^2[0,1]$ (see [4] for a thorough exposition on this topic).

If we are to expand the solution in a wavelet series, our wish is that partial sums (which provide approximations to the solution) belong to V and that the convergence be in the topology of V, that is for the $H^2[0,1]$-norm. Hence, we need a wavelet basis of V; *i.e.*, a wavelet basis on the interval, subject to the boundary conditions that define V and, of course, of regularity larger than 2.

There are already examples of such wavelets. For example, the periodic wavelets of [11] characterize the subspace of $H^2[0,1]$ defined by $f(0) = f(1)$ and $f'(0) = f'(1)$. Furthermore, in order to deal with boundary value problems on the interval, Ph. Tchamitchian and the author had to construct wavelets with various boundary conditions precisely [3].

We show in this chapter how to obtain wavelets with general boundary conditions. This extends the method developed in [3]. Our main tool to obtain these wavelets is taken from approximation theory: the Hermite interpolation. The basic properties of this interpolation is reviewed in the next section. Then, we turn in Section 3 to the construction of the wavelets. There is no real novelty in this construction since it uses algorithms mostly taken from [8] and applied to the appropriate setting. To avoid repetition, we suggest to the reader to get this reference. Section 4 presents the characterization of subspaces of Sobolev spaces and of Hölder spaces defined by boundary conditions.

§2. Cardinal Hermite B-spline functions

Let us start with a quick review on the Hermite interpolation, which can be found in, *e.g.*, [14]. We apologize to those specialists of interpolation for not following the standard notations in order to simplify the exposition. In what follows r will denote a nonnegative integer.

We begin with the following lemmas.

Lemma 1. *Let α, β, with $\alpha \neq \beta$, be two real numbers. Let a_k and b_k, $k = 0, 1, \ldots, r-1$, be two sequences of real or complex numbers. Then there exists a unique polynomial $P \in \Pi_{2r-1}$ such that $P^{(k)}(\alpha) = a_k$ and $P^{(k)}(\beta) = b_k$ for all $k = 0, 1, \ldots, r-1$.*

Here $P^{(k)}$ denotes the kth derivative of the function P and Π_n the space of (real or complex accordingly to the context) polynomials of degree not exceeding n.

Proof: It suffices to notice that P is the unique solution of the initial value problem

$$P^{(2r)}(x) = 0, \quad x \in \mathbb{R},$$
$$P^{(k)}(\alpha) = a_k \text{ and } P^{(k)}(\beta) = b_k, \quad k = 0, 1, \ldots, r-1. \qquad \blacksquare$$

Lemma 2. *There exists a unique r-tuple of function* $(L_0, \ldots, L_{r-1})$ *with the following requirements: for all* $i = 0, \ldots, r-1$,

(1) L_i *is of class* C^{r-1} ;

(2) $L_i \,|_{[\ell,\ell+1]} \in \Pi_{2r-1}$ *for all* $\ell \in \mathbb{Z}$;

(3) $L_i^{(j)}(\ell) = \begin{cases} 1 & \text{if } \ell = 0 \text{ and } i = j, \\ 0 & \text{otherwise for } \ell \in \mathbb{Z} \text{ and } j = 0, \ldots, r-1. \end{cases}$

The proof is an easy consequence of the previous lemma, which we apply on each interval $[l, l+1]$, $l \in \mathbb{Z}$, with the various sequences a_k and b_k implicitly given by conditions (3).

Let us also mention that *supp* L_i is contained in the interval $[-1, 1]$, L_i is real valued and Lipschitz of order r.

The functions L_i are called cardinal Hermite B-spline functions. Since r will be fixed most of the time we do not show the dependence in r in the notation. As an example, for $r = 1$, $L_0(x)$ is the triangle function $\Delta(x) := \sup(1 - |x|, 0)$. For $r = 2$, $L_0(x)$ is the even function give by $(1 + 2x)\Delta(x)^2$ for $x \geq 0$ and $L_1(x)$ is the odd function given by $x\Delta(x)^2$ for $x \geq 0$.

Corollary 3. *The functions defined on* $\mathbb{R}$, *of class* C^{r-1} *and such that their restrictions to each interval* $[k, k+1]$, $k \in \mathbb{Z}$, *belong to* Π_{2r-1}, *are of the form*

$$f(x) = \sum_{k=-\infty}^{+\infty} \sum_{i=0}^{r-1} f^{(i)}(k) L_i(x-k) . \tag{4}$$

When $r = 1$ Equation (4) is nothing but an interpolation by piecewise linear splines. For larger r the generalization is known as (a special case of) the Hermite interpolation. Notice that for each x the series in Equation (4) contains no more than $2r$ nonzero terms. Hence, this series converges at each point. Its sum is completely determined by the set of its values and the values of its derivatives up to order $r-1$ on the lattice $\mathbb{Z}$. We shall call $\mathcal{H}$ the set of functions given by Equation (4).

In order to make a connection with the multiresolution analysis framework (in short: MRA) we need the following result.

Lemma 4. *There exist two constants* A *and* B *depending only on* r, *with* $0 < A \leq B < +\infty$, *such that for any* $f \in \mathcal{H}$ *with compact support*

$$A \sum_{k=-\infty}^{+\infty} \sum_{i=0}^{r-1} |f^{(i)}(k)|^2 \leq \int_{-\infty}^{+\infty} |f(x)|^2 \, dx \leq B \sum_{k=-\infty}^{+\infty} \sum_{i=0}^{r-1} |f^{(i)}(k)|^2 . \tag{5}$$

Proof: Let us first work on the interval [0,1]. It follows from Lemmas 1 and 2 that the $2r$ polynomials $L_i(x)$ and $L_i(x-1)$ are linearly independent. Therefore, they form a basis of Π_{2r-1}. Using the facts that all norms defined on Π_{2r-1} are equivalent and that the restriction of $f(x)$ to $[0,1]$ is given by $\sum_{i=0}^{r-1} f^{(i)}(0)L_i(x) + f^{(i)}(1)L_i(x-1)$, we deduce that $\int_0^1 |f(x)|^2\,dx$ is equivalent to $\sum_{i=0}^{r-1} |f^{(i)}(0)|^2 + |f^{(i)}(1)|^2$ in the sense that their ratio is bounded above and below independently of f. By a translation argument we obtain the analogous comparison with the same constants when replacing by $[k, k+1]$, k integer, the interval $[0,1]$. We finally derive the inequalities (5) by adding up the inequalities obtained for each k. ■

This argument does not provide numerical values of the best constants A and B in inequalities (5). For this matter the reader may consult [7]. Also the support hypothesis can easily be removed from the statement of this lemma to allow more general $L^2(\mathbb{R})$ functions.

From these considerations we can define a multiresolution analysis for $L^2(\mathbb{R})$ with Hermite functions as in [4] or [7].

Define $V_0 := L^2(\mathbb{R}) \cap \mathcal{H}$. It follows from Lemma 4 that V_0 is a closed subspace of $L^2(\mathbb{R})$ that is generated by a Riesz basis, namely the collection of all the functions $L_i(x-k)$, $i = 0, \ldots, r-1$ and $k \in \mathbb{Z}$.

For $j \in \mathbb{Z}$ we let V_j be the space of functions $f(2^j x)$, $f \in V_0$. It is easily seen that this space is also the set of $L^2(\mathbb{R})$ functions, of class C^{r-1}, and such that their restriction to each dyadic interval $[k2^{-j}, (k+1)2^{-j}]$ belongs to Π_{2r-1}. From this observation follow the chain inclusions

$$V_j \subset V_{j+1}, \quad \forall\, j \in \mathbb{Z}.$$

A change of variable in Equations (4) and (5) gives us that any $f(x) \in V_j$ is the sum of a series converging pointwisely and in $L^2(\mathbb{R})$:

$$f(x) = \sum_{k=-\infty}^{+\infty} \sum_{i=0}^{r-1} a_j(i,k) L_i(2^j x - k) \ , \tag{6}$$

with

$$A \sum_{k=-\infty}^{+\infty} \sum_{i=0}^{r-1} 2^{-j} |a_j(i,k)|^2 \le \int_{-\infty}^{+\infty} |f(x)|^2\,dx \le B \sum_{k=-\infty}^{+\infty} \sum_{i=0}^{r-1} 2^{-j} |a_j(i,k)|^2 \ , \tag{7}$$

where $a_j(i,k) = 2^{-ji} f^{(i)}(k2^{-j})$. This means that the collection of functions $2^{j/2} L_i(2^j x - k)$, $i = 0, \ldots, r-1$, $k \in \mathbb{Z}$, forms a Riesz basis of V_j.

Eventually, the intersection of all these spaces reduces to $\{0\}$ and the union is dense in $L^2(\mathbb{R})$. The proofs of these two facts are as follows.

From the "geometric" characterization of each V_j we see that their intersection is made of polynomials on $\mathbb{R}^+$ and on $\mathbb{R}^-$. Since these functions also belong to $L^2(\mathbb{R})$, they must be 0. Thus, this intersection is $\{0\}$.

Let us now look at the density of the union. It is enough to show that there is in the union a sequence of functions converging to any compactly supported C^1 function. Let $f(x)$ be such a function and set $f_j(x) = \sum_{-\infty}^{+\infty} f(k2^{-j})L_0(2^j x - k)$. Obviously, $f_j(x) \in V_j$. We claim that $\sup |f(x) - f_j(x)| \leq C2^{-j}$ for some finite constant C. Suppose this holds, then by letting j increase to $+\infty$ we see that f_j converges uniformly to f. Moreover, both f and f_j are supported in a fixed compact, say K. Since uniform convergence on K implies convergence in $L^2(K)$ we obtain the desired conclusion.

To check the above estimate we first observe that, as a consequence of Equation (4) we have $1 = \cdots + L_0(x-1) + L_0(x) + L_0(x+1) + \cdots$. Indeed, the constants belong to $\mathcal{H}$. Therefore,

$$\begin{aligned} |f(x) - f_j(x)| &= \left|\sum (f(x) - f(k2^{-j}))L_0(2^j x - k)\right| \\ &\leq \sup |f'(x)| 2^{-j} \sum |2^j x - k||L_0(2^j x - k)| \\ &\leq C2^{-j}, \end{aligned}$$

using the fact that L_0 is compactly supported.

Hence, we have seen the properties that makes V_j, $j \in \mathbb{Z}$, an MRA in the sense of [10]. The only difference with the classical framework of Mallat is that V_0 contains r "father functions" (!) instead of 1. This causes no problem to compute the associated Hermite wavelets [6].

Our point here is not the same. We wish to restrict this framework to the interval [0,1] and to impose boundary conditions. As we shall see the geometric definition of our MRA will be very useful.

§3. Wavelets with boundary conditions on [0,1]

Let us start by explaining the boundary conditions on [0,1]. By this we mean a set of homogeneous linear equations involving, for a given, smooth enough function f, the values $f^{(i)}(0)$ and $f^{(i)}(1)$, $i = 0, 1, \ldots, r-1$. To make this precise, let s be an integer with $0 \leq s \leq 2r$ and M be a s-rows$\times 2r$-columns matrix with real or complex entries. We shall always suppose that the rank of M equals s. If $s = 0$, however, M will be assumed to be the null $2r \times 2r$ matrix.

Definition. We say that a C^{r-1} function f defined on [0,1] satisfies the boundary conditions associated with M (for short, we shall write $f \in BC(M)$) if the column vector $Vf = {}^t(f(0), f'(0), \ldots, f^{(r-1)}(0), f(1), f'(1), \ldots, f^{(r-1)}(1))$ solves the homogeneous equation

$$M(Vf) = 0. \tag{8}$$

Let us give various examples that motivate this definition.

The case where f satisfies no boundary conditions corresponds to $M = 0$ and $s = 0$. On the contrary, f and all its derivatives vanish at the boundary when M is of rank $2r$: we can take for M the identity matrix. But we may consider other choices for M. In fact, two matrices M and M' will lead to the same boundary conditions provided they share the same kernel in $\mathbb{R}^{2r}$ (or $\mathbb{C}^{2r}$).

The matrix associated with the Dirichlet problem in Section 1 can be chosen to be of the form

$$M = \begin{pmatrix} 1 & 0 & \cdots & 0 & 0 & 0 & \cdots & 0 \\ 0 & 0 & \cdots & 0 & 1 & 0 & \cdots & 0 \end{pmatrix}.$$

The number of columns may be adjusted to the *a priori* regularity of the solution $u(x)$ since its regularity increases with that of the data. We shall come back to this point later on.

There are 2 classes of boundary conditions:

(1) Class 1: uncorrelated BC at 0 and at 1.

This is when the BC at 0 are not linked to the BC at 1. The matrices of this class can be arranged to have the following structure

$$M = \begin{pmatrix} A & 0 \\ 0 & B \end{pmatrix} \tag{9}$$

where A is a $s_1 \times r$-matrix and B is a $s_2 \times r$-matrix and $s_1 + s_2 = s$ (either A or B can be 0).

(2) Class 2: correlated BC at 0 and at 1.

This class consists of all the other cases. For example when $r = 2$: $f(0) + f(1) = 0$ and $f'(0) - f(1) = 0$.

An important subclass here is the one of periodic boundary conditions.

Let us say at this point that all these BC do not necessarily come from a solvable boundary value problem. For the moment we are only interested in the construction of wavelets satisfying boundary conditions. We come to this now.

For an interval I and $j \geq 0$ define V_j^I to be the space of restrictions to I of the functions in V_j. In the sequel I will denote the interval [0,1] unless it is explicitly mentioned. Let M be a matrix as above and define $V_j^I(M) := V_j^I \cap BC(M)$.

Proposition 5.

(i) *$V_j^I(M)$ coincides with the space of $C^{r-1}(I)$ functions satisfying the boundary conditions associated with M and such that their restrictions to each dyadic interval $[k2^{-j}, (k+1)2^{-j}]$ contained in I belong to Π_{2r-1}.*

(ii) *$V_j^I(M)$, $j \geq 0$, is an increasing sequence of finite dimensional spaces whose union is dense in $L^2(I)$.*

Proof: (i) follows immediately from the geometric description of V_j. We deduce from (i) the chain inclusions $V_j^I(M) \subset V_{j+1}^I(M)$. Furthermore, $V_j^I(M)$ can be considered as a subspace of $L^2(I)$ and is equipped with the induced topology.

The density in $L^2(I)$ of the union of the $V_j^I(M)$'s can be obtained with the same argument as for the density in $L^2(\mathbb{R})$ of the union of the V_j's. ■

Let us make an important remark. None of the functionals $f \mapsto f^{(i)}(0)$ is continuous for the $L^2(I)$-topology. Hence, no matter which BC we impose on the V_j^I's, the union is automatically dense for this topology. Things are different for the topology of Sobolev spaces of positive order because some or all of the above linear functionals become bounded. The purpose of this chapter is precisely based on this fact.

We shall now describe a basis of $V_j^I(M)$. We shall assume that the matrix M is of Class 1 and has a form given by Equation (9). We postpone the other case to the end of this construction.

$V_0^I(M)$ is nothing but $\Pi_{2r-1} \cap BC(M)$ and its dimension is $2r-s$ since M has rank s. If $2r-s>0$ we select an orthonormal basis (for the inner product in $L^2(I)$) $\varphi_1(x), \ldots, \varphi_{2r-s}(x)$ in $V_0^I(M)$.

For $j \geq 1$, a function $f(x)$ in V_j^I can be written as the sum of three independent functions $f_1(x) + f_2(x) + f_3(x)$ where

$$f_1(x) = \sum_i a(i) L_i(2^j x),$$

$$f_2(x) = \sum_{i, 1 \leq k \leq 2^j - 1} a(i,k) L_i(2^j x - k),$$

and

$$f_3(x) = \sum_i b(i) L_i(2^j x - 2^j).$$

As usual, the index i ranges through $0, 1, \ldots, r-1$.

We have supp $f_2 \subset [0,1]$. Therefore, $f_2 \in V_j^I(M)$ for any boundary condition. Moreover, supp $f_1 \subset [-2^{-j}, 2^{-j}]$ and supp $f_3 \subset [1-2^{-j}, 1+2^{-j}]$. Since we have assumed uncorrelated BC at 0 and at 1, we obtain that $f \in V_j^I(M)$ if and only if $f_1 \in BC(A)$ and $f_2 \in BC(B)$ where A and B are related to M as in Equation (9).

Also we have written $BC(A)$ for the condition $A(V_0 f) = 0$ where $V_0 f = {}^t(f(0), \ldots, f^{(r-1)}(0))$ and the analog for $BC(B)$.

Let us study the condition : $f_1 \in BC(A)$. Given the form of $f_1(x)$ we can rescale the problem and assume that $j=0$. Thus, we look for functions of the form $\sum_{i=0}^{r-1} f^{(i)}(0) L_i(x)$ that solve the system $A(V_0 f) = 0$. By hypothesis, this system has rank s_1, so one finds $r - s_1$ independent solutions. If $g(x)$ is one of them, then $g^j(x) := \sum_i 2^{-ji} g^{(i)}(0) L_i(2^j x) \in V_j^I(M)$. Notice that, although $g^j(x)$ may not be given by $g(2^j x)$, it is obtained by the simple and explicit rule above.

There are, nonetheless, cases where the g^j's are given by the rescaling of a single function g. This happens when the BC at 0 do not mix derivatives of different order. When $r=3$ the equation $f'(0)=0$ is one example. This means that $f(0)$ and $f''(0)$ are arbitrary and we can choose $L_0(2^j x)$ and $L_2(2^j x)$ as two independent solutions.

We operate similarly for the BC at 1. Summing up, we have obtained the following result.

Proposition 6. *Suppose that the boundary conditions at 0 and at 1 are uncorrelated. There exist $r - s_1$ independent functions $g_1(x), \ldots, g_{r-s_1}(x)$, and $r - s_2$ independent functions $h_1(x), \ldots, h_{r-s_2}(x)$, all linear combination of $L_0(x), \ldots, L_{r-1}(x)$, such that for $j \geq 1$ the functions $g_1^j(x), \ldots, g_{r-s_1}^j(x)$, $h_1^j(x-1), \ldots, h_{r-s_2}^j(x-1)$ and $L_i(2^j x - k)$, $i = 0, 1, \ldots, r-1$, and $k = 1, \ldots, 2^j - 1$, form a basis of $V_j^I(M)$. For a function $g(x)$ defined by $g(x) = \sum_i g^{(i)}(0) L_i(x)$ we have set $g^j(x) = \sum_i 2^{-ji} g^{(i)}(0) L_i(2^j x)$.*

The next step is the construction of the wavelets which we obtain by following closely the method of [8].

Observe that the functions of $V_j^I(M)$ are completely determined by their values and the values of their derivatives up to order $r-1$ at the dyadic points $0, 2^{-j}, \ldots, 1$. Let U_j be the subspace of $V_{j+1}^I(M)$ composed of functions $u(x)$ that, together with their $r-1$ first derivatives, vanish at these points. Clearly $V_j^I(M) + U_j = V_{j+1}^I(M)$ where the sum is direct.

Furthermore, there is in U_j an independent family composed of $r2^j$ functions: $2^{j/2} L_i(2^{j+1}x - 2k - 1)$, $i = 0, \ldots, r-1$, and $k = 0, \ldots, 2^j - 1$ (we have renormalized these functions so that their L^2-norm is independent of the scaling parameter j). But $r2^j$ is precisely the dimension of U_j since, as a consequence of Proposition 6, $\dim V_j^I(M) = r(2^j - 1) + 2r - s_1 - s_2 = r2^j + r - s$. Hence, the above family is a basis of U_j.

For $j \geq 0$, let W_j be the orthogonal complement of $V_j^I(M)$ in $V_{j+1}^I(M)$. We have $\dim W_j = r2^j$ and we obtain a basis of W_j by performing the orthogonal projection of $V_{j+1}^I(M)$ onto W_j of the basis elements in U_j. This projection is explicit and the basis thus computed is made orthonormal in $L^2[0,1]$ following an explicit orthogonalization algorithm. Let us call $\psi_{i,J}(x)$ the elements of this orthonormal basis, where i ranges through the integers $0, 1, \ldots, r-1$ and J ranges through the dyadic intervals of length 2^{-j} contained in [0,1]. Notice that $\psi_{i,J}(x) \in BC(M)$ comes from $W_j \subset V_{j+1}^I(M)$.

These functions will be the wavelets with boundary conditions announced in Section 1.

In particular, these wavelets are of class C^{r-1} (in fact, Lipschitz of order r) and, for some $C \geq 0$ and $\gamma > 0$ independent of i and J, they satisfy the estimates

$$|\psi_{i,J}^{(m)}(x)| \leq C\ell(J)^{-1/2}\ell(J)^{-m}\exp(-\gamma\ell(J)^{-1}|x - x_J|) \,, \tag{10}$$

where $m = 0, 1, \ldots, r$, and $\ell(J)$ and x_J denote respectively the length and the midpoint of J. We have also estimates for the moments of these functions: for $0 \leq m \leq 2r - 1$,

$$\left|\int_0^1 (x - x_J)^m \, \psi_{i,J}(x)\, dx\right| \leq C\ell(J)^{1/2}\ell(J)^m \exp(-\gamma\ell(J)^{-1}d(J)) \,, \tag{11}$$

where $d(J) = \inf(x_J, 1 - x_J)$ is the distance of x_J to the boundary of [0,1].

A last remark is that these wavelets are real-valued whenever M has real entries.

Since we recover all of $L^2(I)$ from

$$L^2(I) = V_0^I(M) \oplus W_0 \oplus W_1 \oplus \cdots ,$$

we have proved the following result.

Theorem 7. *Let r and s be two positive integers with $s \leq 2r$, and M be an $s \times 2r$-matrix that induces uncorrelated boundary conditions at 0 and at 1. Then, the collection of functions $\varphi_1(x), \ldots, \varphi_{2r-s}(x)$, $\psi_{i,J}(x)$, $i = 0, 1 \ldots, r-1$, J dyadic interval in [0,1], is an orthonormal basis of $L^2[0,1]$. Furthermore, these functions are Lipschitz of order r, they satisfy the boundary conditions associated with M and they are subject to estimates (10) and (11).*

What has to be changed in this statement when the BC at 0 and at 1 are correlated? Almost nothing. The only difference can be found in estimates (10). Here they must read

$$|\psi_{i,J}^{(m)}(x)| \leq C\ell(J)^{-1/2}\ell(J)^{-m}\exp(-\gamma\ell(J)^{-1}\delta(x, x_J)) , \tag{12}$$

where $\delta(x, x_J) = \inf(x - x_J, x - (1 - x_J))$. This tells us that the wavelets have two peaks of localization that are located symmetrically about the midpoint of [0,1], reflecting the correlation between the BC at 0 and at 1.

This fact can already be observed in the analog of Proposition 6 for this class of BC. In the statement of Proposition 6, we had functions "affected" to 0: the $g^j(x)$'s with a small support about 0; and functions "affected" to 1: the $h^j(x-1)$'s with a small support about 1. This separation is no longer possible in the case of correlated BC. A basis of $V_j^I(M)$ already contains functions with a support in two parts, one about 0 and the other one about 1; *i.e.*, with two peaks of localization. During the projection-orthogonalization process, this twofold localization will propagate so that the wavelets satisfy (12).

Remarks. (1) As we mentioned earlier, periodic BC is a special case of the correlated BC class, but the method that is suggested here is rather long and clumsy for periodic BC. Indeed, estimates (12) are not periodic. A shorter way to calculate these "periodic" wavelets is to periodize the ordinary wavelets on $\mathbb{R}$. This is a well known technique that can be found in, *e.g.*, [11].

(2) The proof of Theorem 7 has been rather sketchy, but the projection-orthogonalization process explained above originates from [8] and by now, it can be found in many places (*e.g.*, [2],[3], and [12]). We invite the readers to review there. The only novelty here is to obtain estimates (12). Let us give a few explanations.

The projection-orthogonalization process involves the computation of the square root inverse of matrices having exponentially decaying entries away from the first diagonal. Estimates (10) then follow from the fact that the square root inverse of such matrices enjoys the same property. The precise statement is Lemma 2 in [8].

In the case of correlated BC, our matrices will have exponential fall-off away from the diagonals; *i.e.*, after a suitable change of indices, their entries satisfy estimates like

$$|m(k,\ell)| \leq C\exp(-\gamma\delta(k,\ell))$$

for some $\gamma > 0$, where $k = (k_1, \ldots, k_n)$ and $\ell = (\ell_1, \ldots, \ell_n)$ belong to some $\mathbb{Z}^n$ (or some symmetric subset of $\mathbb{Z}^n$) and

$$\delta(k,\ell) = \inf(|k_1 \pm \ell_1| + \cdots + |k_n \pm \ell_n|).$$

The infimum is taken over all the choices of plus or minus signs. Hence, δ is not a distance function but is still symmetric, and satisfies the triangular inequality and the exponential condition: for all $\gamma > 0$ there exists a $C = C(\gamma) < \infty$ such that

$$\sup_k \sum_\ell \exp(-\gamma\delta(k,\ell)) \leq C\ .$$

This is in fact all we need for the proof of Lemma 2 in [8] to apply without a change.

§4. Sobolev and Hölder spaces

In order to simplify the notations and the writing of this section, we shall assume from now on that the BC at 0 and at 1 do not mix derivatives; *i.e.*, the $BC(M)$ can be written as a set of $2r$ equations

$$\begin{aligned} a_0 f(0) + b_0 f(1) &= 0 \\ c_0 f(0) + d_0 f(1) &= 0 \\ &\vdots \\ a_{r-1} f^{(r-1)}(0) + b_{r-1} f^{(r-1)}(1) &= 0 \\ c_{r-1} f^{(r-1)}(0) + d_{r-1} f^{(r-1)}(1) &= 0\ . \end{aligned}$$

This is not too much a restriction since these conditions are the most often considered in practice. For each i, the 2×2 matrix with entries a_i, b_i, c_i, d_i is arbitrary.

The φ_k's and the wavelets $\psi_{i,J}$'s used in this section are those of Theorem 7 (or the wavelets of the comment following this theorem) with the boundary conditions associated with M and r is their Lipschitz regularity.

We shall discuss the characterization of various functional spaces with this wavelet basis.

Let us start with the case of Sobolev spaces with boundary conditions on [0,1].

For a positive integer s, define $H^s = H^s[0,1]$ as the space of restrictions to [0,1] of the functions in the inhomogeneous Sobolev space $H^s(\mathbb{R})$. We endow this space with the norm $\|f\|+\|f^{(s)}\|$ where $\|f\|^2 = \int_0^1 |f|^2$. Note that $H^0 = L^2$.

The Sobolev injection theorem tells us that $H^s \subset C^{s-1/2}$ if $s > 1/2$ (see the definition of the Hölder spaces below and see [4] for the easy proof of this fact if s is an integer). Hence, for any $x \in [0,1]$ the linear functionals $f \mapsto f^{(i)}(x)$ are bounded on H^s provided $0 \le i \le s-1$. If M is the matrix inducing the boundary conditions of the beginning of this section, we let $H^s(M)$, $s \le r$, be the closed subspace of H^s defined by the $2s$ first equations of $BC(M)$. For example, $H_0^s = H^s(M)$ whenever these equations have rank $2s$.

For noninteger values of s, we define $H^s(M)$ by complex interpolation: if $p < s < p+1$ and $\theta \in (0,1)$ are such that $s = (1-\theta)p + \theta(p+1)$, then $H^s(M) = [H^p(M), H^{p+1}(M)]_\theta$. There are equivalent ways to define $H^s(M)$. See the comments at the end of Theorem 9.

Since we shall try to apply the classical theory on the real line, it is worth recalling what the situation is for Sobolev spaces [12].

Theorem 8. *Let m be a nonnegative integer, and let $\{m_J\}$ a family of functions defined on $\mathbb{R}$, where J ranges through the dyadic intervals of length not exceeding 1 in $\mathbb{R}$.*

(i) *Assume that*

$$|m_J^{(i)}(x)| \le C\ell(J)^{-1/2}\ell(J)^{-i}\exp(-\gamma\ell(J)^{-1}|x - x_J|),$$

for $0 \le i \le m$. Then, for all $s \in (0,m)$, any series $\sum_J \alpha_J m_J(x)$ belongs to $H^s(\mathbb{R})$ provided $\sum_J |\alpha_J|^2 \ell(J)^{-2s} < \infty$.

(ii) *Assume that m_J satisfies the estimate just above (with no conditions on the derivatives) and that*

$$\int_{\mathbb{R}} x^i m_J(x)\,dx = 0$$

for $0 \le i \le m$. Then, $\sum_J |\langle f, m_J\rangle|^2 \ell(J)^{-2s} < \infty$ for all $f \in H^s(\mathbb{R})$, $s \in (0,m)$.

Remark. Here, $\langle\ ,\ \rangle$ stands for the complex inner product in $L^2(\mathbb{R})$. Parts (i) and (ii) apply when $s = 0$ provided we simultaneously assume to $\{m_J\}$ regularity, decay estimates and vanishing moments. However, both cases (i) and (ii) are not valid in general when $s = m$ without additional properties to the m_J's.

Theorem 9. *Let $0 \le s \le r$ and $f \in H^s(M)$. Then*

$$\int_0^1 f(x)\,\overline{\varphi}_k(x)\,dx = O(1)$$

$$\sum_{i,J} \left| \int_0^1 f(x)\,\overline{\psi}_{i,J}(x)\,dx \right|^2 \ell(J)^{-2s} < \infty\,. \tag{13b}$$

Reciprocally, if $0 \le s \le r$ and if $f \in L^2$ fulfills estimates (13), then $f \in H^s(M)$.

Proof: It goes as follows. First, by complex interpolation, it suffices to restrict our attention to integer values of s, which we assume from now on. We also suppose $s > 0$ since the $s = 0$ case is already taken care of by Theorem 7.

We split the argument into two parts depending on whether the $BC(M)$ are of Class 1 or of Class 2. Eventually, for each class we treat separately the case $s < r$ from the case $s = r$. That this latter case holds will heavily depend on the structure of the wavelets resulting from the use of Hermite interpolation and multiresolution analysis theory.

(1) $BC(M)$ of Class 1.

In this case the wavelets satisfy estimates (10) and (11).

(i) $0 < s < r$.

The easy part is the reciprocal. It is a mere application of part (i) in Theorem 8 and of the representation formula

$$f(x) = \sum \int_0^1 f \,\overline{\varphi}_k \, \varphi_k(x) + \sum \int_0^1 f \,\overline{\psi}_{i,J} \, \psi_{i,J}(x)$$

a priori valid for L^2 functions.

For the other part, we first observe that Equation (13a) is a consequence of the Cauchy-Schwarz inequality. To obtain estimates (13b), we need the following result.

Lemma 10. *For each i and J, $\psi_{i,J}(x)$ has a compactly supported extension $\widetilde{\psi}_{i,J}(x)$ to the real line with*

$$|\widetilde{\psi}_{i,J}^{(m)}(x)| \le C\ell(J)^{-1/2}\ell(J)^{-m}\exp(-\gamma\ell(J)^{-1}|x - x_J|) ,$$

where $m = 0, 1, \ldots, r$; and

$$\int_{\mathbb{R}} x^m \, \widetilde{\psi}_{i,J}(x) \, dx = 0 ,$$

where $m = 0, 1, \ldots, 2r - 1$.

Assuming that this lemma holds, we can proceed as follows. Let $f \in H^s(M)$. There is a polynomial $P \in V_0^I(M)$ such that $g = f - P$ satisfies $g^{(k)}(0) = g^{(k)}(1) = 0$ for all $0 \le k \le s - 1$. Hence, we may extend by 0 this function outside of [0,1] and this extension, which we keep calling g, belongs to $H^s(\mathbb{R})$ as one can easily check.

Using the orthogonality between $\psi_{i,J}$ and $V_0^I(M)$ and then, the definitions of the various extensions, we have

$$\begin{aligned} \int_0^1 f(x) \,\overline{\psi}_{i,J}(x) \, dx &= \int_0^1 g(x) \,\overline{\psi}_{i,J}(x) \, dx \\ &= \langle g, \widetilde{\psi}_{i,J} \rangle. \end{aligned} \tag{14}$$

Therefore, we can directly apply part (ii) of Theorem 8 (the index i occurs no problem since it ranges through a finite set). Hence, the integrals of the left hand-side of Equation (14) satisfy estimates (13*b*).

Now, we come to the proof of Lemma 10. It follows [8] very closely.

It follows from the construction that $\psi_{i,J}$ is naturally defined on $\mathbb{R}$ with *supp* $\psi_{i,J} \subset [-2^{-j}, 1+2^{-j}]$ when $\ell(J) = 2^{-j}$ and that estimates (10) are not only valid for x in [0,1] but for all x in *supp* $\psi_{i,J}$. It goes the same for estimates (11) : the domain of integration can be extended to the whole real line, since the contributions brought by the intervals $[-2^{-j}, 0]$ and $[1, 1+2^{-j}]$ are, by direct calculation, on the order of magnitude of the second term in estimates (11). Hence, we first extend the wavelets on their full support.

The second extension consists in adding to $\psi_{i,J}$ small contributions, supported in an interval outside of $[-2^{-j}, 1+2^{-j}]$, of length $\ell(J)$ and at a distance $Cd(J)$ from J. Recall that $d(J)$ is essentially the distance of J to the boundary of [0,1]. C denotes a numerical constant independent of J. These contributions can be chosen so as to make the $2r$ first moments vanish.

From this point, it suffices to literally copy the argument in [8, pp. 106-107]. This proves the lemma. ■

(ii) $s = r$.

We assume first $f \in H^r(M)$ and study the coefficients of f in the wavelet basis. We start as above by letting g be in $H^r(\mathbb{R})$ so that

$$\int_0^1 f(x)\,\overline{\psi}_{i,J}(x)\,dx = \langle g, \widetilde{\psi}_{i,J}\rangle = (-1)^r \ell(J)^{-r} \langle g^{(r)}, h_{i,J}\rangle,$$

where we have integrated by parts r times. The function $h_{i,J}$ denotes the primitive of order r of $\widetilde{\psi}_{i,J}$ that vanishes at $-\infty$ together with all its derivatives. If we set $m_{i,J} = \ell(J)^{-r} h_{i,J}$, then the family $\{m_{i,J}\}$ satisfies the hypothesis of Theorem 8, part (i). Furthermore,

$$\int m_{i,J}(x)\,dx = (-1)^r \ell(J)^r \int x^r \widetilde{\psi}_{i,J}(x)\,dx = 0$$

by Lemma 10.

Thus, we may apply the remark following Theorem 8 and $\{\langle g^{(r)}, m_{i,J}\rangle\} \in \ell^2$ since $g^{(r)} \in L^2$. Going back to f, we can see that this property is precisely the one we were looking for. This proves the first part of Theorem 9 (in this case) and we turn to the proof of the second part.

We have to show that $f(x) = \sum_{i,J} \alpha_{i,J} \psi_{i,J}(x)$ belongs to $H^r(M)$ provided $\sum_{i,J} |\alpha_{i,J}|^2 \ell(J)^{-2r} < \infty$. To simplify the exposition we forget about the index i and it suffices to prove the following.

Lemma 11.

$$\int_0^1 \Big|\sum_J \alpha_J H_J(x)\Big|^2 dx \le C \sum_J |\alpha_J|^2 \,, \tag{15}$$

where we have set $H_J(x) = \ell(J)^r \psi_J^{(r)}(x)$.

This inequality will strongly depend on the structure of the ψ_J's. By construction, $\psi_J \in V_{j+1}^I(M) \subset V_j^I$ when $\ell(J) = 2^{-j}$. Thus, by Equation (6) we obtain

$$\psi_J(x) = \sum_{k=0}^{2^{j+1}} \sum_{i=0}^{r-1} 2^{-(j+1)i} \psi_J^{(i)}(k2^{-j-1}) L_i(2^{j+1}x - k) \ . \tag{16}$$

Here too, we forget about the index i. Set $\sigma_Q(x) = 2^{(j+1)/2} L^{(r)}(2^{j+1}x - k)$ where $Q = [k2^{-j-1}, (k+1)2^{-j-1}] \subset [0,1]$. By taking the rth derivative in Equation (16) and summing over the range of J with $\ell(J) = 2^{-j}$, we see that

$$\sum_{J\,;\,\ell(J)=2^{-j}} \alpha_J H_J(x) = \sum_{Q\,;\,\ell(Q)=2^{-j-1}} \beta_Q \sigma_Q(x) \ ,$$

where the β_Q's are linear combination of the α_J's. The coefficients of these combinations are given by $2^{-(j+1)i} \psi_J^{(i)}(k2^{-j-1})$ and using decay estimates (10) it is not difficult to establish

$$\sum_{Q\,;\,\ell(Q)=2^{-j-1}} |\beta_Q|^2 \le C \sum_{J\,;\,\ell(J)=2^{-j}} |\alpha_J|^2 \ ,$$

where C is a constant independent of j (notice that in these sums Q and J are both contained in [0,1]). Therefore, inequality is a consequence of

$$\int_0^1 \Big| \sum_{Q\,;\,\ell(Q)\le 1/2} \beta_Q \sigma_Q(x) \Big|^2 dx \le C \sum_{Q\,;\,\ell(Q)\le 1/2} |\beta_Q|^2 \ . \tag{17}$$

To derive this estimate, we first enlarge the domain of integration to $\mathbb{R}$ (the σ_Q's are naturally defined on $\mathbb{R}$). By developing the square term inside the integral, we see that the inequality (17) is equivalent to the boundedness on ℓ^2 of the Gram matrix $G = (\langle \sigma_Q, \sigma_R \rangle)$ which, by an application of the Schur lemma, follows from

$$\sup_Q \sum_R |\langle \sigma_Q, \sigma_R \rangle| + \sup_R \sum_Q |\langle \sigma_Q, \sigma_R \rangle| < \infty. \tag{18}$$

Observe that by definition σ_Q is compactly supported in the interval $\widetilde{Q}$ which is the union of Q and of the dyadic interval Q^* with the same length as Q and $\inf Q = \sup Q^*$. Note that $\widetilde{Q}$ is no longer a dyadic interval.

This observation and the Cauchy-Schwarz inequality yield

$$|\langle \sigma_Q, \sigma_R \rangle| \le C \inf \left\{ \left(\frac{\ell(Q)}{\ell(R)} \right)^{1/2}, \left(\frac{\ell(R)}{\ell(Q)} \right)^{1/2} \right\} \ . \tag{19}$$

This, of course, is not sufficient to derive the inequality (18).

Furthermore, we claim that most of the coefficients are 0. More precisely, this happens when either $\widetilde{Q} \subset R$ or R^*, or symmetrically $\widetilde{R} \subset Q$ or Q^*, or $\widetilde{Q} \cap \widetilde{R} = \emptyset$.

It is then an easy exercise to obtain the inequality (18) from this previous claim and estimates (19). We leave it to the reader and turn to the proof of the claim.

The third condition : $\widetilde{Q} \cap \widetilde{R} = \emptyset$, tells us that σ_Q and σ_R have disjoint supports. Hence, $\langle \sigma_Q, \sigma_R \rangle = 0$ in this case.

Next, we consider the case $\widetilde{Q} \subset R$ or R^*. The remaining case is dealt with symmetrically. By construction, the restrictions of σ_R to R and to R^* are polynomials of degree not exceeding $r-1$. Therefore, σ_R restricted to the support of σ_Q is one of these two polynomials. On the other hand, as the rth derivative of a compactly supported function, σ_Q is orthogonal to the polynomials of degree not exceeding $r-1$. This implies immediately that $\langle \sigma_Q, \sigma_R \rangle = 0$ in this situation as well. ■

(2) $BC(M)$ of Class 2.

In this situation, the wavelets satisfy estimates (11) and (12).

(i) $s < r$.

The easy part in Theorem 9 is still the reciprocal. Estimates (12) tell us that $\psi_{i,J}$ can be written as the sum of two smooth functions verifying estimates (11): the one for the interval J, the other for the interval $1-J$. Thus, part (i) of Theorem 8 can still be applied.

We turn to the first part of Theorem 9. Let $f \in H^s(M)$. The coefficients $\int_0^1 f(x)\,\overline{\varphi}_k(x)\,dx$ and $\int_0^1 f(x)\,\overline{\psi}_{i,J}(x)\,dx$ for $\ell(J) = 1$ are $O(1)$ by an application of the Cauchy-Schwarz inequality. Thus, we are left with proving Equation (13*b*), the dyadic intervals J being this time of length smaller than 1/2.

There is a function $P \in V_1^I(M)$ such that $g = f - P$ satisfies $g^{(k)}(0) = g^{(k)}(1/2) = g^{(k)}(1) = 0$ for all $0 \le k \le s-1$. Let us denote by g_1 (resp. g_2) the function that coincides with g on $[0, \frac{1}{2}]$ (resp. $[\frac{1}{2}, 1]$) and that is 0 elsewhere. Obviously, g_1 (resp. g_2) belongs to $H^s(\mathbb{R})$. From the orthogonality between $\psi_{i,J}$ and $V_1^I(M)$ if $\ell(J) \le 1/2$, we obtain

$$\begin{aligned}\int_0^1 f(x)\,\overline{\psi}_{i,J}(x)\,dx &= \int_0^1 g(x)\,\overline{\psi}_{i,J}(x)\,dx \\ &= \int_0^{1/2} g_1(x)\,\overline{\psi}_{i,J}(x)\,dx + \int_{1/2}^1 g_2(x)\,\overline{\psi}_{i,J}(x)\,dx \\ &= \langle g_1, \widetilde{\psi}_{i,J}^1 \rangle + \langle g_2, \widetilde{\psi}_{i,J}^2 \rangle \,, \end{aligned} \tag{20}$$

where $\widetilde{\psi}_{i,J}^1$ (resp. $\widetilde{\psi}_{i,J}^2$) is an extension of $\psi_{i,J}$ outside of $[0, \frac{1}{2}]$ (resp. $[\frac{1}{2}, 1]$) that satisfies estimates (11) for the interval J (resp. $1-J$). Furthermore, these extensions have vanishing moments as in Lemma 10. They are obtained exactly as in Lemma 10 and we leave the details to the reader.

We are now in a good position to apply part (ii) of Theorem 8. This gives us the inequality (13*b*) as desired.

(ii) $s = r$

Using Equation (20) and the argument given for $BC(M)$ of Class 1, one proves that $f \in H^r(M)$ implies Equation (13).

A mild adaptation of the proof of Lemma 11 furnishes the converse implication.

This completes the proof of Theorem 9. ■

Remarks. (1) $H^s(M)$, defined by interpolation for noninteger s, has a natural equivalent definition : this space is also, with an equivalent topology, the space of restrictions to [0,1] of the functions in $H^s(\mathbb{R})$ that satisfy boundary conditions $BC(M)$ up to order $\lfloor s - 1/2 \rfloor$ (*i.e.*, the $2\lfloor s - 1/2 \rfloor + 2$ first conditions of $BC(M)$ are considered and we have denoted by $\lfloor t \rfloor$ the largest integer strictly smaller than t). In particular, this is a closed subspace of H^s. An exclusion case, however, is that this holds only when $s - 1/2$ is not an integer. This characterization is a consequence of the results in [9] but can also be obtained directly from Theorem 9.

Indeed, if $f \in H^s(M)$, then the series

$$\sum \int_0^1 f \overline{\varphi}_k \ \varphi_k(x) + \sum \int_0^1 f \overline{\psi}_{i,J} \ \psi_{i,J}(x) \ , x \in \mathbb{R},$$

converges in $H^s(\mathbb{R})$ to a compactly supported function, which we call $Pf(x)$. The operator P thus defined is linear and bounded from $H^s(M)$ to $H^s(\mathbb{R})$. Moreover, the restriction to [0,1] of $Pf(x)$ is exactly $f(x)$. Finally, the above series converges pointwisely as well as its derivatives up to the order $\lfloor s - 1/2 \rfloor$ and it is easy to check the boundary conditions.

This is where the condition $s - 1/2$ noninteger comes in. For $s - 1/2 = m \in \mathbb{Z}$, the situation is more subtle. One can show, indeed, that $H^s(M)$ is in some cases a dense and strict subspace of H^s (the interested reader may consult [9] for details), hence no simple equivalent formulation in terms of boundary conditions can be given to $H^s(M)$ in this case. To avoid this difficulty we have chosen to define $H^s(M)$ by complex interpolation, this choice, in some sense, being justified by Theorem 9.

(2) Let us return to our Dirichlet problem $-u'' + u = f$. Theorem 9 implies that the solution $u(x)$ of this problem can be written as

$$\sum \int_0^1 u \overline{\varphi}_k \ \varphi_k(x) + \sum \int_0^1 u \overline{\psi}_{i,J} \ \psi_{i,J}(x) \ ,$$

where $r = 2$ and the wavelets all satisfy: $g(0) = g(1) = 0$, and $g'(0), g'(1)$ arbitrary (we do not intend here to provide a way of computing the coefficients of this series). We have only convergence of this series in $H^2 \cap H_0^1$. If f already belongs to a Sobolev space H^s, $s > 0$, then $u \in H^{2+s} \cap H_0^1$. To obtain this information from the above wavelet series, it suffices to change $r = 2$ to $r \geq 2+s$ and to select the appropriate wavelets. This yields another wavelet series that converges to u, this time in $H^{2+s} \cap H_0^1$.

Let us turn to the study of Hölder spaces. First, we recall their definition.

For $s > 0$, $C^s = C^s[0,1]$ denotes the space of restrictions to [0,1] of the functions in $C^s(\mathbb{R})$. This latter space is defined by

$$|f(x) - f(x+h)| \leq C|h|^s$$

whenever $0 < s < 1$. For $s = 1$, $C^1(\mathbb{R})$ is the Zygmund class defined by

$$|f(x+h) + f(x-h) - 2f(x)| \leq C|h|.$$

If $s > 1$ and $m = \lfloor s \rfloor$, $C^s(\mathbb{R})$ is defined by $f^{(m)} \in C^{s-m}(\mathbb{R})$. Eventually, to obtain a norm for C^s, one adds to the $C^s(\mathbb{R})$-norm the L^∞[0,1]-norm.

The characterization of $C^s(\mathbb{R})$ is as follows [12]:

Theorem 12. *Let m be a non-negative integer, and let $\{m_J\}$ be a family of functions defined on $\mathbb{R}$, where J ranges through the dyadic intervals of length not exceeding 1 in $\mathbb{R}$.*

(i) *Assume that*

$$|m_J^{(i)}(x)| \leq C\ell(J)^{-1/2}\ell(J)^{-i}\exp(-\gamma\ell(J)^{-1}|x - x_J|)$$

for $0 \leq i \leq m$. Then, for all $s \in (0,m)$, a series $\sum_J \alpha_J m_J(x)$ belongs to $C^s(\mathbb{R})$ provided $\alpha_J = O(\ell(J)^{s+1/2})$. Moreover, this series converges to a bounded function uniformly on compact sets.

(ii) *Assume that m_J satisfies the estimate in (i) above (with no conditions on the derivatives) and that*

$$\int_{\mathbb{R}} x^i m_J(x)\, dx = 0$$

for $0 \leq i \leq m$. Then, $\langle f, m_J \rangle = O(\ell(J)^{s+1/2})$ for all $f \in C^s(\mathbb{R})$, $s \in (0,m)$.

Let us state the result on the interval. For $0 < s < r$ a noninteger and $m = \lfloor s \rfloor$, since a function of C^s is m times continuously differentiable, we denote by $C^s(M)$ the subspace of C^s composed of the functions that fulfill the $2m+2$ first equations of $BC(M)$. This is obviously a closed subspace of C^s.

If s is an integer, we define $C^s(M)$ by complex interpolation: if $s-1 < s_0 < s < s_1 < s+1$ and $s = s_0(1-\theta)+s_1\theta$, then we set $C^s(M) = [C^{s_0}(M), C^{s_1}(M)]_\theta$. We shall discuss this choice later.

Theorem 13. *Let $0 < s < r$. If $f \in C^s(M)$, then*

$$\int_0^1 f(x)\overline{\varphi}_k(x)\,dx = O(1); \tag{21a}$$

$$\int_0^1 f(x)\overline{\psi}_{i,J}(x)\,dx = O(\ell(J)^{1/2+s})\,. \tag{21b}$$

Reciprocally, assuming that $f \in L^2$, these conditions characterize $C^s(M)$.

The proof of this theorem is a simple adaptation of the argument of Theo rem 9. One advantage of this proof is in not using the geometric properties of the wavelets. Thus, it could be used if other constructions of boundary value wavelets were available. It is almost a pity, however, not to use Hermite interpolation when the situation is favorable. This is why we present a simple argument for the proof of the first part in Theorem 13 when s is a noninteger. This argument relies on the following form of Taylor's theorem .

Proposition 14. *Let s, with $0 < s < r$, be a noninteger. Let $f \in C^s(\mathbb{R})$ and $\alpha \in \mathbb{R}$. Suppose that $P(x)$ is a polynomial of degree larger than $\lfloor s \rfloor$ with $f^{(i)}(\alpha) = P^{(i)}(\alpha)$ for $0 \le i \le \lfloor s \rfloor$. Then $f(x) - P(x)$ is $O(|x-\alpha|^s)$ uniformly in x.*

Let $f(x) \in C^s(M)$, s being a noninteger. Inequalities (21a) are trivial since f is bounded.

Let J be a dyadic interval of length 2^{-j} in [0,1]. Define

$$P_j(x) = \sum_{i=0}^{\lfloor s \rfloor} \sum_{k=0}^{2^j} 2^{-ji} f^{(i)}(k2^{-j}) L_i(2^j x - k)$$

and $f_j(x) = f(x) - P_j(x)$. The key fact here is that $P_j \in V_j^I(M)$ hence,

$$\int_0^1 f(x)\overline{\psi}_{i,J}(x)\,dx = \int_0^1 f_j(x)\overline{\psi}_{i,J}(x)\,dx \tag{22}$$

because P_j is orthogonal to $\psi_{i,J}$. Furthermore, at each dyadic point $k2^{-j} \in [0,1]$, f and P_j satisfy $f^{(i)}(k2^{-j}) = P_j^{(i)}(k2^{-j})$ for $0 \le i \le \lfloor s \rfloor$ and the restriction of P_j to dyadic intervals of length 2^{-j} is a polynomial. An application of Proposition 14, therefore, gives

$$|f_j(x)| \le C 2^{-js} \sum_{k=0}^{2^j} |2^j x - k|^s \chi(2^j x - k)\,.$$

This estimate is uniform in x and χ denotes the characteristic function of the interval $[-1,1]$.

From this estimate, we see that the second integral in (22) is

$$O\left(2^{-js}\int_0^1 \sum_{k=0}^{2^j} |2^j x - k|^s \chi(2^j x - k)|\psi_{i,J}(x)|\, dx\right),$$

and the latter integral is $O(2^{-j/2})$. This yields estimate (21*b*). ■

Remark. As concerns the definition of $C^s(M)$, s an integer, Theorem 13 shows that it does not depend on the particular values of s_0 and s_1. We could also discuss the inclusion and density relation between these spaces (as the BC change) as we did for the Sobolev $H^{m+1/2}$. It is likely that a similar situation happens (we have not checked it).

§5. Concluding remarks

An issue we have not talked about is the numerical interest of these wavelets. This construction is probably too heavy to be efficiently used, *e.g.*, for solving numerically boundary value problems. First, we have lost the dilation and translation invariance structure, which implies fast pyramidal algorithms, of the real line case. Nevertheless, it is the case for any construction of wavelets on the interval. The second point is that the more smoothness we wish the more functions we need to compute: the complexity is necessarily linearly dependent of the smoothness parameter.

A way to overcome this last drawback is to consider a construction "à la Meyer" [11] that makes use of the compactly supported wavelets of Daubechies [4]. This would also minimize the first inconvenience since "most" of the wavelets are in this case the Daubechies wavelets on the real line and since, in some sense, the dilation invarience is preserved. The author is actually convinced that this works. The details will appear elsewhere [1].

Another issue is how to handle higher dimensional open sets. Apart from cubes for which all this construction extends by usual tensor product arguments (one can even take, in dimension 2, different BC for the horizontal and the vertical edge ...) it is not clear what to do for more general sets with BC. In our opinion, it is a very interesting question that, as far as we know, has not been considered.

References

1. Auscher, P., Ondelettes à support compact et conditions aus limites, preprint.
2. Auscher, P. and Ph. Tchamitchian, Bases d'ondelettes sur les courbes corde-arc, noyau de Cauchy et espaces de Hardy associés, *Rev. Mat. Iberoamericana* **5** (1989), 139–170.
3. Auscher, P. and Ph. Tchamitchian, Conjecture de Kato sur des ouverts de $\mathbb{R}$, preprint.

4. Brézis, H., *Analyse Fonctionnelle*, Masson, Paris, 1983
5. Daubechies, I., Orthonormal bases of compactly supported wavelets, *Comm. Pure and Appl. Math.* **41** (1988), 909–991.
6. Goodman, T. N. T., S. L. Lee, and W. S. Tang, Wavelets in wandering subspaces, preprint.
7. Hervé, L., Ph.D. Thesis, University of Rennes I, France, 1991.
8. Jaffard, J. and Y. Meyer, Bases d'ondelettes dans des ouverts de $\mathbb{R}^n$, *J. Math. Pures et Appl.* **68** (1989), 95–108.
9. Lions, J. L. and E. Magenes, *Problèmes aux Limites non Homogènes*, Volume I, Dunod, Paris, 1968.
10. Mallat, S., Multiresolution approximations and wavelet orthonormal bases of $L^2(\mathbf{R})$, *Trans. Amer. Math. Soc.* **315** (1989), 69–87.
11. Meyer, Y., Wavelet and operators, *Proc. Special Year in Modern Analysis*, W. Beckner, A. P. Calderón, R. Fefferman, and P. W. (eds.), Wadsworth, NY, 1981, 475–493.
12. Meyer, Y., *Ondelettes et Opérateurs*, Volumes I and II, Hermann, Paris, 1990.
13. Meyer, Y., Ondelettes sur l'intervalle, *Rev. Mat. Iberoamericana*, to appear.
14. Schumaker, L., *Spline Functions: Basic Theory*, Wiley-Interscience, New York, 1981.

Pascal Auscher
IRMAR
Université de Rennes I
Campus de Beaulieu
35042 Rennes Cedex
France
auscher@cicb.fr

Local Sine and Cosine Bases of Coifman and Meyer and the Construction of Smooth Wavelets

Pascal Auscher, Guido Weiss, and M. Victor Wickerhauser

Abstract. We give a detailed account of the local cosine and sine bases of Coifman and Meyer. We describe some of their applications; in particular, based on an approach by Coifman and Meyer, we show how these local bases can be used to obtain arbitrarily smooth wavelets. The understanding of this material requires only a minimal knowledge of the Fourier transform and classical analysis. It is our intention to make this presentation accessible to all who are interested in Wavelets and their applications.

§1. Introduction

It is often useful to focus on local properties of a signal. In precise mathematical language this means that if we are given a function f on $\mathbb{R}$ and want to consider its properties on a finite interval I, we can analyze the function multiplied by χ_I, the characteristic function of I. We can, for example, form the Fourier series of $\chi_I f$ with respect to a complete orthonormal system for $L^2(I)$. An example of such a system is

$$\left\{\sqrt{\frac{2}{|I|}}\chi_I(x)\sin\frac{2k+1}{2}\frac{\pi}{|I|}(x-\alpha)\right\}, \tag{1.1}$$

where α is the left end point of I, $k = 0, 1, 2...$(further discussion of this and other systems will be given in Section 3). If $-\infty < \cdots < \alpha_j < \alpha_{j+1} < \cdots < \infty$, with $\alpha_j \to \pm\infty$ as $j \to \pm\infty$, and $I_j = [\alpha_j, \alpha_{j+1}]$, we obtain an orthonormal system

$$\varphi_{j,k}(x) = \sqrt{\frac{2}{|I_j|}}\chi_{I_j}(x)\sin\frac{2k+1}{2}\frac{\pi}{|I_j|}(x-\alpha_j), \tag{1.2}$$

where j ranges through the integers $\mathbf{Z}$ and k through the non-negative integers $\mathbf{Z}^+$, that is a basis for $L^2(\mathbb{R})$. Expansions in terms of such bases are referred to

Wavelets–A Tutorial in Theory and Applications
C. K. Chui (ed.), pp. 237–256.

ISBN 0-12-174590-2

as "windowed" or "short time" Fourier transforms. Though such systems are appropriate for focusing on local properties (*i.e.*, , what happens on the interval I_j), the abrupt "cutoff" effected by the multiplication by the characteristic functions χ_{I_j} involve some undesirable artifacts. In [1], Coifman and Meyer introduced orthonormal bases of this general type that involve an arbitrarily smooth cut off. It is the purpose of this note to present the construction of such bases and, in addition, show some of their uses. In particular, we shall describe, in Section 4, how the smooth wavelets of Lemarié and Meyer [2] can be obtained from these bases. We hasten to add that this application was also pointed out to us by Coifman and Meyer.

Let us begin by trying to construct a projection P_I, given an interval I, that is similar to the one obtained by multiplication by χ_I but is "smoother". Clearly, P_I cannot have the form $f \to \rho_I f$, with $\rho = \rho_I$ a smooth function with support "close" to I, since the requirement that P_I be idempotent forces ρ to have values that are either 0 or 1. Perhaps a smooth version of χ_I can be "corrected" near the end points of I so that we have a projection. In order to reduce the problem to only one end point, let us try to carry out this idea on the infinite ray $I = [0, \infty)$ and let ρ be a smooth nonnegative function supported on $[-\epsilon, \infty)$ such that $\rho(x) = 1$ if $x \geq \epsilon$. Let us also assume that ρ shares with χ_I the property

$$\rho(x) + \rho(-x) = 1 \tag{1.3}$$

for all $x \in \mathbb{R}$. In order to perform the above "correction" of the operator $f \to \rho f$ so as to obtain an orthogonal projection we pose the question: can we find a function t such that

$$(P_I f)(x) = (Pf)(x) \equiv \rho(x)f(x) + t(x)f(-x)$$

is an orthogonal projection? One immediate calculation shows that P is self-adjoint if and only if $t(x) = \overline{t(-x)}$. If, for the sake of simplicity, we assume t is real-valued, this relation becomes, simply, that t is an even function. The idempotent property $P^2 = P$, because of (1.3), becomes

$$\{\rho^2(x) + t(x)t(-x)\}f(x) + t(x)f(-x) = (P^2 f)(x)$$

$$= (Pf)(x) = \rho(x)f(x) + t(x)f(-x)$$

for all $f \in L^2(\mathbb{R})$. Again using (1.3) this equality holds if and only if $t(x)t(-x) = \rho(x)(1 - \rho(x)) = \rho(x)\rho(-x)$. Since t is even this is equivalent to

$$t(x) = \pm\sqrt{\rho(x)\rho(-x)}.$$

This shows that, under these assumptions on ρ and t, P is a projection if and only if

$$(Pf)(x) = \rho(x)f(x) \pm \sqrt{\rho(x)\rho(-x)}f(-x). \tag{1.4}$$

It is not hard to make an explicit construction of such functions ρ that are as smooth as desired: we begin by choosing an even non-negative function ψ with $\text{supp}\,\psi \subset [-\epsilon, \epsilon]$ so normalized that

$$\int_{\mathbb{R}} \psi = \frac{\pi}{2}.$$

Then let

$$\theta(x) = \int_{-\infty}^{x} \psi(t)dt.$$

An immediate consequence of the fact that ψ is even is that

$$\theta(x) + \theta(-x) = \frac{\pi}{2}. \tag{1.5}$$

We now put $s_\epsilon(x) \equiv \sin\theta(x)$ and $c_\epsilon(x) \equiv \cos\theta(x)$. It follows from (1.5) that $c_\epsilon(x) = \cos[\frac{\pi}{2} - \theta(-x)] = s_\epsilon(-x)$. That is,

$$c_\epsilon(x) = s_\epsilon(-x). \tag{1.6}$$

Thus, the graph of c_ϵ is the mirror image, through the vertical axis $x = 0$, of the graph of s_ϵ. We also have the relation

$$s_\epsilon^2(x) + c_\epsilon^2(x) = 1. \tag{1.7}$$

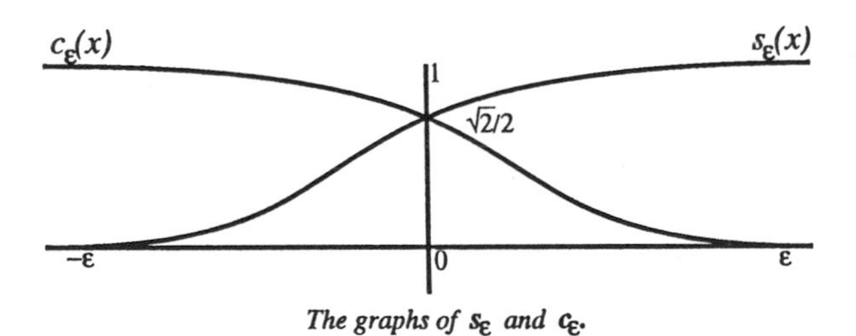

The graphs of s_ε and c_ε.

If we now let $\rho(x) = s_\epsilon^2(x)$ we see that ρ enjoys the above properties; in particular, (1.3) is an immediate consequence of (1.6) and (1.7). Equality (1.4) then becomes

$$(Pf)(x) \equiv (P_0f)(x) = s_\epsilon^2(x)f(x) \pm s_\epsilon(x)c_\epsilon(x)f(-x), \tag{1.8}$$

where we use the notation P_0 to indicate that P is associated with the ray $[0, \infty)$ (of course, P_0 also depends on $\epsilon > 0$ and, in a few occasions, we show this additional dependence by writing $P_{0,\epsilon}$ instead of P_0.)

Had we chosen the half ray $(-\infty, 0]$ instead of $[0, \infty)$, completely analogous reasoning would lead us to the orthogonal projection P^0 given by the formula

$$(P^0 f)(x) = c_{\epsilon'}^2(x)f(x) \pm c_{\epsilon'}(x)s_{\epsilon'}(x)f(-x), \tag{1.9}$$

where, again, we do not explicitly denote the dependence of P^0 on ϵ'. We should also observe that, in each case, the projections also depend on the choice of sign before the second summand. Thus, we have introduced the four projections

$$P_+^{0,\epsilon'}, P_-^{0,\epsilon'}, P_{0,\epsilon}^+, P_{0,\epsilon}^-. \tag{1.10}$$

§2. The construction of the projections P_I associated with the interval $I = [\alpha, \beta]$

We begin by translating P_0 and P^0 to any two points α, β in $\mathbb{R}$. Letting t_γ be the translation operator defined by $(t_\gamma f)(x) = f(x - \gamma)$, we introduce the *translates* (by α and γ) of P_0 and P^0 by letting

$$P_\alpha \equiv t_\alpha P_0 t_{-\alpha} \quad \text{and} \quad P^\beta \equiv t_\beta P^0 t_{-\beta}. \tag{2.1}$$

It is easy to check that

$$\begin{aligned} (P_\alpha f)(x) &= s_\epsilon^2(x-\alpha)f(x) \pm s_\epsilon(x-\alpha)c_\epsilon(x-\alpha)f(2\alpha - x), \\ (P^\beta f)(x) &= c_{\epsilon'}^2(x-\beta)f(x) \pm c_{\epsilon'}(x-\beta)s_{\epsilon'}(x-\beta)f(2\beta - x). \end{aligned} \tag{2.2}$$

Since $t_\gamma^* = t_{-\gamma} = t_\gamma^{-1}$, we see immediately that P_α and P^β are orthogonal projections for each α and β. Observe that x and $2\gamma - x$ are symmetric with respect to the line $x = \gamma$ (they lie on opposite sides of γ and at a distance $|x-\gamma|$ from γ). We say that a function g is *even with respect to* γ if $g(2\gamma - x) = g(x)$ for all x. It is an immediate consequence of (2.2) that $gP_\alpha = P_\alpha g$ (and $gP^\beta = P^\beta g$) when g is even with respect to $\alpha(\beta)$. Using this commutativity with $g = \chi_{[\alpha-\epsilon,\alpha+\epsilon]}$ and $g = \chi_{[\beta-\epsilon',\beta+\epsilon']}$, the properties of $s_\epsilon, c_\epsilon, s'_\epsilon, c'_\epsilon$, and (2.2) we see that

$$P_\alpha P^\beta f = \chi_{[\alpha-\epsilon,\alpha+\epsilon]} P_\alpha f + \chi_{[\alpha+\epsilon,\beta-\epsilon']} f + \chi_{[\beta-\epsilon',\beta+\epsilon']} P^\beta f = P^\beta P_\alpha f \tag{2.3}$$

as long as $\alpha + \epsilon \leq \beta - \epsilon'$. This allows us to define $P_I = P_{[\alpha,\beta]}$ by letting

$$P_{[\alpha,\beta]} \equiv P_\alpha P^\beta = P^\beta P_\alpha \tag{2.4}$$

whenever $-\infty < \alpha < \beta < \infty$. Because of this commuting property it is clear that P_I is an orthogonal projection. In view of (1.10) we remark that P_I depends on the choices of $+$ and $-$ in P_α, P^β, and on ϵ, ϵ' (as long as $\alpha+\epsilon \leq \beta-\epsilon'$). We shall discuss the importance of this dependency a little later. Before doing this we introduce the *bell over* I. This is the function b_I, that depends on $\alpha, \beta, \epsilon, \epsilon'$, but *not* on the choice of signs, defined by

$$b_I(x) \equiv s_\epsilon(x - \alpha)c_{\epsilon'}(x - \beta) \tag{2.5}$$

for all $x \epsilon \mathbb{R}$. We list the basic properties of this bell function; each is an easy consequence of (2.5) and the properties of the functions $s_\epsilon, c_\epsilon, s_{\epsilon'}, c_{\epsilon'}$ developed in Section 1 (in particular (1.6) and (1.7)):

The function b_I *satisfies*

(1) Supp $b_I = [\alpha - \epsilon, \beta + \epsilon']$.
On $[\alpha - \epsilon, \alpha + \epsilon]$:
(2) $b_I(x) = s_\epsilon(x - \alpha)$;
(3) $b_I(2\alpha - x) = s_\epsilon(\alpha - x) = c_\epsilon(x - \alpha)$;
(4) $\beta_I^2(x) + b_I^2(2\alpha - x) = 1$.
(5) Supp $b_I(x)b_I(2\alpha - x) = [\alpha - \epsilon, \alpha + \epsilon]$. (2.6)
(6) $b_I(x) = 1$ *when* $x \in [\alpha + \epsilon, \beta - \epsilon']$.
On $[\beta - \epsilon', \beta + \epsilon']$:
(7) $b_I(x) = c_{\epsilon'}(x - \beta)$;
(8) $b_I(2\beta - x) = c_{\epsilon'}(\beta - x) = s_{\epsilon'}(x - \beta)$;
(9) $b_I^2(x) + b_I^2(2\beta - x) = 1$.
(10) Supp $b_I(x)b_I(2\beta - x) = [\beta - \epsilon', \beta + \epsilon']$.
(11) *When* $x \in$ Supp $b_I = [\alpha - \epsilon, \beta + \epsilon']$

$$b_I^2(x) + b_I^2(2\beta - x) + b_I^2(2\alpha - x) = 1.$$

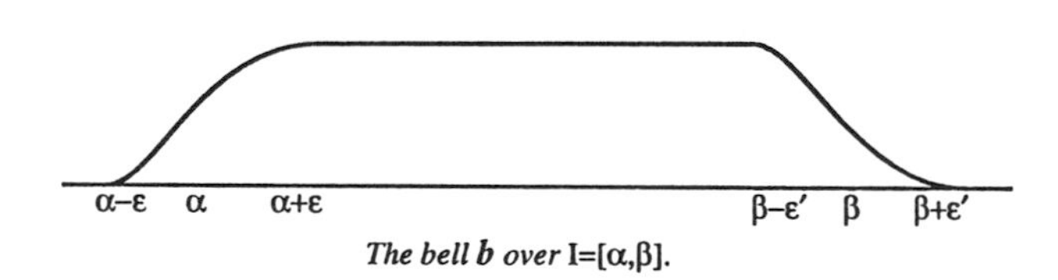

***The bell b over* I=[α,β].**

The last property, obviously an immediate consequence of (4), (6), and (9), is perhaps, the most useful. Observe that the most important feature of these properties is the focus on the behaviour of b_I on the three intervals $[\alpha - \epsilon, \alpha + \epsilon]$, $[\alpha + \epsilon, \beta - \epsilon']$, and $[\beta - \epsilon', \beta + \epsilon']$.

The projection $P_I - P_{[\alpha,\beta]}$ has a simple expression in terms of the bell function b_I:

$$(P_I f)(x) =$$
$$b_I^2(x)f(x) \pm b_I(x)b_I(2\alpha - x)f(2\alpha - x) \pm b_I(x)b_I(2\beta - x)f(2\beta - x). \quad (2.7)$$

This formula is an immediate consequence of (2.3) and (2.6). It exhibits very clearly the dependence of P_I on the choice of signs associated with the endpoints α and β of the interval I. When ϵ and ϵ' are fixed we are dealing with four projections that are dependent on the two *polarities* (that is, the choice of signs) at each endpoint. The polarities are particularly important when we want to study the properties of P_I and P_J when I and J are adjacent intervals. The dependence of these projections on the choice of ϵ and ϵ' is also important in these considerations. Let us examine this in detail.

Two adjacent intervals $I = [\alpha, \beta]$ and $J = [\beta, \gamma]$ are *compatible* and have *bells that are compatible* if $\alpha - \epsilon < \alpha < \alpha + \epsilon \le \beta - \epsilon' < \beta < \beta + \epsilon' \le \gamma - \epsilon'' < \gamma < \gamma + \epsilon''$ and

$$b_I(x) = s_\epsilon(x - \alpha)c_{\epsilon'}(x - \beta), \qquad b_J(x) = s_{\epsilon'}(x - \beta)c_{\epsilon''}(x - \gamma).$$

Clearly if we apply (3) of (2.6) to b_J (with J, β, ϵ' replacing I, α, ϵ) we have $b_J(2\beta - x) = c_{\epsilon'}(x - \beta)$ when $x \in [\beta - \epsilon', \beta + \epsilon']$; by (7) of (2.6), on the other hand, $b_I(x) = c_{\epsilon'}(x - \beta)$ when $x \in [\beta - \epsilon', \beta + \epsilon']$ (assuming I and J are compatible). Thus, when I and J are compatible,

$$b_I(x) = b_J(2\beta - x) \tag{2.8}$$

if $x \in [\beta - \epsilon', \beta + \epsilon']$. A similar use of (6) and (9) of (2.7) gives us

$$b_I^2(x) + b_J^2(x) = 1 \tag{2.9}$$

when $x \in [\alpha + \epsilon, \gamma - \epsilon'']$. This last relation extends to the equality

$$\sqrt{b_I^2(x) + b_J^2(x)} = s_\epsilon(x - \alpha)c_{\epsilon''}(x - \gamma)$$

for all x, which is equivalent to

$$b_I^2 + b_J^2 = b_{I\cup J}^2 \tag{2.10}$$

whenever I and J are compatible adjacent intervals. As mentioned before, the bell functions b_I are independent of the choices of sign associated with the endpoints of I. The polarity of two adjacent intervals I and J, however, plays an important role if we desire that the projections P_I and P_J satisfy an additive property analogous to (2.10). More precisely we have the following result:

Proposition 1. *Suppose $I = [\alpha, \beta]$ and $J = [\beta, \gamma]$ are adjacent compatible intervals and P_I, P_J have opposite polarity at β. Then $P_I + P_J$ is the orthogonal projection $P_{I\cup J}$:*

$$P_I + P_J = P_{I\cup J}. \tag{2.11}$$

Moreover, P_I and P_J are orthogonal to each other:

$$P_I P_J = P_J P_I = 0. \tag{2.12}$$

Proof: Equality (2.11) is an immediate consequence of (2.7), (2.8), and (2.10) (since the terms involving the end point β cancel each other). Equality (2.12) is a consequence of the general result in Hilbert space theory: if P and Q are orthogonal projections such that $P+Q$ is an orthogonal projection, then $PQ = 0$. Here is a simple proof of this fact. If $(P+Q)^2 = P+Q$, the idempotent properties of P and Q give us $PQ = -QP$. Thus, $PQ = P^2Q = -PQP = QP^2 = QP$. Since $PQ = -QP$ and $PQ = QP$ it follows that $PQ = QP = 0$. ■

Another consequence of formula (2.7) is a simple characterization of the image $\mathcal{H}_I = P_I L^2(\mathbb{R})$ of P_I. Let us say that a function f is *even (odd) with respect to* α *on* $[\alpha-\epsilon, \alpha+\epsilon]$ if and only if $f(2\alpha - x) = f(x)$ ($f(2\alpha - x) = -f(x)$) when $x \in [\alpha-\epsilon, \alpha+\epsilon]$. Observe that (2.7) can be written in the form

$$(P_I f)(x) = b_I(x) S(x), \tag{2.13}$$

where $S(x) = b_I(x)f(x) \pm b_I(2\alpha - x)f(2\alpha - x) \pm b_I(2\beta - x)f(2\beta - x)$. But this function is odd (even) with respect to α on $[\alpha-\epsilon, \alpha+\epsilon]$ if $-(+)$ is chosen before the second term in (2.7) and is odd (even) with respect to β on $[\beta-\epsilon', \beta+\epsilon']$ if $-(+)$ is chosen before the third term. If this odd/even property corresponds, at an end point of I, with the $-/+$ polarity of I at this end point, we say that f has the *same polarity as* I has at this point. Our characterization of $\mathcal{H}_I$, then, is given by:

Theorem 2. *$f \in \mathcal{H}_I = P_I L^2(\mathbb{R})$ if and only if $f = bS$, where S is a function in $L^2(\mathbb{R})$ having the same polarity as I at its end points.*

Proof: From (2.13) we see that each element of $\mathcal{H}_I$ has the form $f = bS$. Now suppose f has this form. Apply (2.7) to $f = bS$ and we obtain:

$$\begin{aligned}(P_I bS)(x) &= b^2(x)b(x)S(x) \pm b(x)b^2(2\alpha - x)S(2\alpha - x) \pm b(x)b^2(2\beta - x)S(2\beta - x)\\ &= \chi_{[\alpha-\epsilon,\alpha+\epsilon]}(x)b(x)S(x)\{b^2(x) + b^2(2\alpha - x)\} + \chi_{[\alpha+\epsilon,\beta-\epsilon']}(x)b(x)S(x) +\\ &\qquad \chi_{[\beta-\epsilon',\beta+\epsilon']}(x)b(x)S(x)\{b^2(x) + b^2(2\beta - x)\} = b(x)S(x)\end{aligned}$$

by (4), (5), (6), (9), and (10) of (2.6). ■

Finally, let us show how to use these projections to decompose $L^2(\mathbb{R})$ into a direct sum of mutually orthogonal subspaces that are images under such projections:

$$L^2(\mathbb{R}) = \bigoplus_{k \in \mathbf{Z}} \mathcal{H}_k. \tag{2.14}$$

We do this as follows: choose a sequence $\{\alpha_k\}_{k\in\mathbf{Z}}$ of reals and accompanying positive numbers $\{\epsilon_k\}$ such that

$$\alpha_k + \epsilon_k < \alpha_{k+1} - \epsilon_{k+1}$$

for all k. Thus, each pair of adjacent intervals $I_{k-1} = [\alpha_{k-1}, \alpha_k]$ and $I_k = [\alpha_k, \alpha_{k+1}]$ is a compatible pair. Let $b_k = b_{I_k}$ be the bell over I_k and $P_k \equiv P_{I_k}$. Let us also choose these projections so that they have opposite polarity at α_k. We then have $\mathbb{R} = \bigcup_{k\in\mathbf{Z}} I_k$ if $\lim_{k\to\pm\infty} \alpha_k = \pm\infty$. Since

$$\bigcup_{k=-N}^{N} I_k = [\alpha_{-N}, \alpha_{N+1}],$$

it follows from (2.11) that

$$\sum_{k=-N}^{N} P_k = P_{[\alpha_{-N},\alpha_{N+1}]} \tag{2.15}$$

Letting $\mathcal{H}_k \equiv P_k L^2(\mathbb{R})$, (2.12) assures us that $\mathcal{H}_k \perp \mathcal{H}_l$ if $k \neq l$. From (2.3) we see that

$$P_{[-\alpha_N,\alpha_{N+1}]} f = \chi_{[\alpha_{-N}+\epsilon_{-N},\alpha_{N+1}-\epsilon_{N+1}]} f + E_N f,$$

where $E_N f$ is a function supported in the two intervals

$$[\alpha_{-N} - \epsilon_{-N}, \alpha_{-N} + \epsilon_{-N}] \text{ and } [\alpha_{N+1} - \epsilon_{N+1}, \alpha_{N+1} + \epsilon_{N+1}].$$

On the first interval $|(E_N f)(x)|$ is dominated by $|f(x)| + |f(2\alpha_{-N} - x)|$ and, on the second, by $|f(x)| + |f(2\alpha_{N+1} - x)|$. Hence, $\|E_N f\|_2 \to 0$ as $N \to \infty$ and it follows that

$$\lim_{N\to\infty} \|f - P_{[-\alpha_N,\alpha_{N+1}]} f\|_2 = 0.$$

These considerations clearly give us the decomposition (2.14).

§3. The local cosine and sine bases for $L^2(\mathbb{R})$

Let us fix an interval I and consider the problem of constructing "natural" orthonormal bases for the spaces $\mathcal{H}_I = P_I L^2(\mathbb{R})$. There are four such spaces if we take into account the two possible polarities at each end point of I. We ask, first, the simpler question: if the projection is the one obtained by multiplication by χ_I, what are the "natural" bases of the image space $L^2(I)$ from the point of view of a harmonic analyst? Simplifying further, let us assume $I = [0, 1]$. Motivated by the polarity properties we have been discussing, we seek some orthonormal bases of $L^2(0, 1)$ that reflect these properties. Given $f \in L^2(0, 1)$ let us extend it to the interval [0,2] so that it gives us a function

$\tilde{f}$ that is even with respect to 1 on this larger interval. Analytically this means $\tilde{f}(x) = f(2-x)$ for $x \in [1,2]$. We then extend $\tilde{f}$ to an odd function on $[-2,2]$. This last function can then be developed into a Fourier series on $[-2,2]$ by means of the orthonormal basis

$$\{\frac{1}{2}, \frac{1}{\sqrt{2}} \sin \frac{k\pi x}{2}, \frac{1}{\sqrt{2}} \cos \frac{k\pi x}{2}\}, k = 1, 2, \ldots. \tag{3.1}$$

Since we are dealing with an odd function on $[-2,2]$, the even part $\{\frac{1}{2}, \frac{1}{\sqrt{2}} \cos \frac{k\pi x}{2}\}, k = 1, 2, \ldots$, plays no role in this expansion. For the same reason, among the remaining terms, only those that are even with respect to 1 on $[0,2]$ give us (possibly) non-zero coefficients in this Fourier series development. This shows that f can be expanded on $[0,1]$ in terms of the orthogonal family $\{\sin \frac{2k+1}{2} \pi x\}, k = 0, 1, 2, \ldots$.

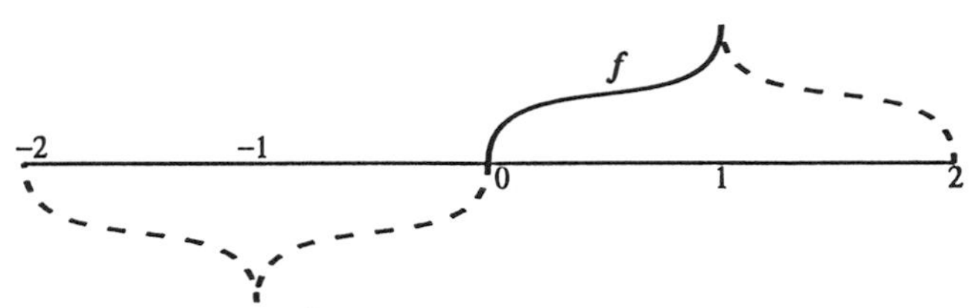

The extension of f on [0,1] *to* [-2,2] *so that it is even at* 1 *and odd at* 0.

Had we extended f to obtain the other three pairs of polarities at the points 0 and 1, we would have obtained three other subcollections of the family (3.1). These considerations give us

Proposition 3. *Each of the following four systems forms an orthonormal basis for* $L^2(0,1)$:

(i) $\{\sqrt{2} \sin \frac{2k+1}{2} \pi x\}, k = 0, 1, 2, \ldots$;
(ii) $\{\sqrt{2} \sin k\pi x\}, k = 1, 2, 3, \ldots$;
(iii) $\{\sqrt{2} \cos \frac{2k+1}{2} \pi x\}, k = 0, 1, 2, \ldots$;
(iv) $\{1, \sqrt{2} \cos k\pi x\}, k = 1, 2, 3, \ldots$.

The polarities of each of the functions in the first basis are $(-,+)$ *at* $(0,1)$, *in the second basis they are* $(-,-)$, *in the third they are* $(+,-)$, *and in the fourth they are* $(+,+)$.

Let us now return to the study of the space $\mathcal{H}_I = \mathcal{H}_{[0,1]} = P_I L^2(\mathbb{R})$. To fix our ideas let us assume P_I is chosen with negative polarity at 0 and positive polarity at 1. The bell over $I = [0,1]$ in this case is $b(x) = b_I(x) = s_\epsilon(x) c_{\epsilon'}(x-1)$ and $P = P_I$ is given (see (2.7)) by

$$(Pf)(x) = b(x)\{b(x)f(x) - b(-x)f(-x) + b(2-x)f(2-x)\} \equiv b(x)S(x). \tag{3.2}$$

If we restrict S to $[0,1]$, using the first basis in Proposition 3, we have

$$S(x) = \sqrt{2}\sum_{k=0}^{\infty} c_k \sin\frac{2k+1}{2}x, \tag{3.3}$$

where the equality may be interpreted in the norm of $L^2(I)$, and the coefficients c_k are given by the equality

$$c_k = \sqrt{2}\int_0^1 S(x)\sin\frac{2k+1}{2}\pi x dx. \tag{3.4}$$

But each of the functions $\sin\frac{2k+1}{2}\pi x$ satisfy the same polarity properties as S. It follows that equality (3.3) is valid on $[-\epsilon, 1+\epsilon']$ and the convergence can be taken in the norm of $L^2(-\epsilon, 1+\epsilon')$. Multiplying this new equality on both sides by $b(x)$ we see that any $f \in \mathcal{H}_I$ satisfies

$$f(x) = \sqrt{2}\sum_{k=0}^{\infty} c_k b(x)\sin\frac{2k+1}{2}x \tag{3.5}$$

in $L^2(-\epsilon, 1+\epsilon')$.

We claim the system $\{\sqrt{2}b(x)\sin\frac{2k+1}{2}\pi x\}, k = 0,1,2,\ldots$, is an orthonormal basis of $\mathcal{H}_I$ and $\{c_k\}$, given by (3.4), is the sequence of coefficients of $f \in \mathcal{H}_I$ with respect to this basis. It is clear from our discussion that in order to establish this claim all we need to show is that this system is orthonormal in $\mathcal{H}_I$. This is done by a simple calculation in which (2.6) and (2.7) play an important role: Let $f(x) = \sqrt{2}b(x)\sin\frac{2l+1}{2}\pi x$, then, by two changes of variables,

$$\begin{aligned}
&\int_{-\epsilon}^{\epsilon} f(x)\sqrt{2}b(x)\sin\frac{2k+1}{2}\pi x dx \\
&\qquad = \sqrt{2}\int_0^{\epsilon}\{f(x)b(x) - f(-x)b(-x)\}\sin\frac{2k+1}{2}\pi x dx \\
&\qquad = 2\int_0^{\epsilon}\{s_\epsilon^2(x) + c_\epsilon^2(x)\}(\sin\frac{2l+1}{2}\pi x)(\sin\frac{2k+1}{2}\pi x)dx \\
&\qquad = 2\int_0^{\epsilon}(\sin\frac{2l+1}{2}\pi x)(\sin\frac{2k+1}{2}\pi x)dx
\end{aligned}$$

and

$$\begin{aligned}
&\int_{1-\epsilon'}^{1+\epsilon'} f(x)\sqrt{2}b(x)\sin\frac{2k+1}{2}\pi x dx \\
&\qquad = \sqrt{2}\int_{1-\epsilon'}^{1}\{f(2-x)b(2-x) + f(x)b(x)\}\sin\frac{2k+1}{2}\pi x dx \\
&\qquad = 2\int_{1-\epsilon'}^{1}\{s_{\epsilon'}^2(x-1) + c_{\epsilon'}^2(x-1)\}(\sin\frac{2l+1}{2}\pi x)(\sin\frac{2k+1}{2}\pi x)dx \\
&\qquad = 2\int_{1-\epsilon'}^{1}(\sin\frac{2l+1}{2}\pi x)(\sin\frac{2k+1}{2}\pi x)dx.
\end{aligned}$$

Thus, since $b(x) = 1$ on $[\epsilon, 1-\epsilon']$,

$$\int_{-\epsilon}^{1+\epsilon'} f(x)\sqrt{2}b(x)\sin\frac{2k+1}{2}\pi x dx$$
$$= 2\int_0^1 (\sin\frac{2l+1}{2}\pi x)(\sin\frac{2k+1}{2}\pi x)dx = \delta_{kl}$$

the last equality being a consequence of (i) of Proposition 3.

Had we chosen the other polarities, $(-,-),(+,-),(+,+)$, in the definitions of $P = P_{[0,1]}$, we reach the same conclusions if, instead of the system (i) of Proposition 3, we use the systems (ii), (iii), and (iv) respectively. The case of the general interval $I = [\alpha, \beta]$ now follows from these results by simple translation and dilation arguments. More precisely, we obtain

Theorem 4. *If $P_I = P_{[\alpha,\beta]}$ is the projection associated with negative polarity at α and positive polarity at β, then the system*

$$\text{(i)} \left\{\sqrt{\frac{2}{|I|}}b_I(x)\sin\frac{2k+1}{2}\frac{\pi}{|I|}(x-\alpha)\right\}, k = 0,1,2,\ldots, \tag{3.6}$$

is an orthonormal basis for $\mathcal{H}_I = P_I L^2(\mathbb{R})$. If we choose the polarities $(-,-)$, $(+,-)$, and $(+,+)$ at (α,β), the same result is true if we use (respectively) the systems

$$\text{(ii)} \left\{\sqrt{\frac{2}{|I|}}b_I(x)\sin k\frac{\pi}{|I|}(x-\alpha)\right\}, k = 1,2,3,\ldots;$$

$$\text{(iii)} \left\{\sqrt{\frac{2}{|I|}}b_I(x)\cos\frac{2k+1}{2}\frac{\pi}{|I|}(x-\alpha)\right\}, k = 0,1,2,\ldots; \tag{3.6}$$

$$\text{(iv)} \left\{\frac{1}{\sqrt{|I|}}b_I(x), \sqrt{\frac{2}{|I|}}b_I(x)\cos k\frac{\pi}{|I|}(x-\alpha)\right\}, k = 1,2,3,\ldots,$$

instead of (i) *of (3.6).*

We can now combine the results obtained at the end of Section 2 with this theorem to obtain the *local sine and cosine bases of Coifman and Meyer*. Let a sequence $\{\alpha_j\}$ be selected as described at the end of Section 2. Thus, $\alpha_j < \alpha_{j+1}$, $\lim_{j\to\pm\infty}\alpha_j = \pm\infty$ and there exists an accompanying sequence $\{\epsilon_j\}$ such that

$$\alpha_j + \epsilon_j \le \alpha_{j+1} - \epsilon_{j+1}$$

for all $j \in \mathbf{Z}$. Let $P_j = P_{[}\alpha_j, \alpha_{j+1}]$ be constructed with, say, negative polarity at α_j and positive polarity at α_{j+1}. Let

$$b_{jk}(x) = b_{[\alpha_j,\alpha_{j+1}]}(x)\sqrt{\frac{2}{\alpha_{j+1}-\alpha_j}}\sin\frac{2k+1}{2}\frac{\pi}{\alpha_{j+1}-\alpha_j}(x-\alpha_j), \tag{3.7}$$

$j \in \mathbf{Z}, k = 0, 1, 2, \ldots$. Our discussion (in particular (2.14) and Theorem 4 allows us to conclude that the *functions in (3.7) form an orthonormal basis for* $L^2(\mathbb{R})$. We can change the polarity for each projection P_j and arrive at the same conclusion as long as the polarity for P_{j-1} at α_j is opposite to that of P_j at α_j; moreover, each such choice must be accompanied by exchanging the basis (3.7) with the basis in Theorem 4 that corresponds to these choices in polarity.

Here are some further observations that follow easily from the results we have obtained:

Theorem 5 . *Suppose $f \in \mathcal{H}_I$, where $I = [\alpha, \beta]$ is a finite interval in $\mathbb{R}$ and the polarities are, say, $-+$ at α and β, then the series*

$$\sqrt{\frac{2}{|I|}} \sum_{k=0}^{\infty} c_k b(x) \sin \frac{2k+1}{2} \frac{\pi}{|I|}(x-\alpha)$$

converges to $f(x)$ a.e. in $[\alpha - \epsilon, \beta + \epsilon']$. If $g \in L^2(\mathbb{R})$, then the development of g in terms of the basis (3.7) converges a.e. to $g(x)$.

Proof: If $f = b_I S_I = bS \in \mathcal{H}_I$ we have shown that,

$$f(x) = \sum_{k=0}^{\infty} c_k b(x) \sqrt{\frac{2}{|I|}} \sin \frac{2k+1}{2} \frac{\pi}{|I|}(x-\alpha) \tag{3.8}$$

with convergence in the $L^2(I)$ norm. But we observed that the coefficients

$$c_k = \sqrt{\frac{2}{|I|}} \int_{\alpha-\epsilon}^{\beta+\epsilon'} f(x) b(x) \sin \frac{2k+1}{2} \frac{\pi}{|I|}(x-\alpha) dx,$$

$k = 0, 1, 2, \ldots$, can also be calculated as the sine series coefficients

$$c_k = \sqrt{\frac{2}{|I|}} \int_{\alpha}^{\beta} S(x) \sin \frac{2k+1}{2} \frac{\pi}{|I|}(x-\alpha) dx. \tag{3.9}$$

This is a consequence of (3.4) (after an appropriate translation and dilation). Since $f \in L^2(\mathbb{R})$, $S = b^{-1} f$ is square integrable over $[\alpha, \beta]$. It then follows from Carleson's theorem that the series

$$\sqrt{\frac{2}{|I|}} \sum_{k=0}^{\infty} c_k \sin \frac{2k+1}{2} \frac{\pi}{|I|}(x-\alpha) = S(x) \tag{3.10}$$

converges to $S(x)$ a.e. in $[\alpha, \beta]$. Multiplying both sides by $b(x)$ we then have the validity of (3.8) for a.e. $x \in [\alpha, \beta]$. Since both sides are odd with respect to α on $[\alpha - \epsilon, \alpha + \epsilon]$ and even with respect to β on $[\beta - \epsilon', \beta + \epsilon']$, this a.e. convergence extends to $[\alpha - \epsilon, \beta + \epsilon']$. ■

Clearly there are versions of Theorem 5 for each choice of polarities associated with the interval $[\alpha, \beta]$. If the sequence $\{\alpha_j\}$ (and the accompanying sequence $\{\epsilon_j\}$) gives us a family of mutually orthogonal projections P_j whose ranges span $L^2(0,\infty)$ (by choosing $\alpha_j \to 0$ as $j \to -\infty$ and $\alpha_j \to \infty$ as $j \to \infty$), we obtain an orthonormal basis of those functions in $L^2(\mathbb{R})$ that are the Fourier transforms of the elements of the classical Hardy space H^2. In the next section we describe another application in which it is useful to think of these local bases as the Fourier transform images of interesting bases of $L^2(\mathbb{R})$.

We make one final comment before presenting the application that was just mentioned. The projections P_I are, indeed, "smoother" than the ones obtained by multiplication by χ_I. They are not, however, necessarily "smoothing." For example, let $f = \chi_{[0,1]}$ and $P_I = P_{[0,1]}$. If I has negative polarity at 0, $P_I f$ still has a jump at 0. Positive polarity at 0 on the other hand, "smooths out" f near 0.

§4. The construction of the Lemarié-Meyer smooth wavelets

Lemarié and Meyer [2] constructed a wavelet basis

$$\{w_{k,n}(x)\} = \{2^{-k/2} w(2^{-k}x - n)\}, \tag{4.1}$$

$k, n \in \mathbf{Z}$, where the "mother function" w belongs to $\mathcal{S}(\mathbb{R})$ (in fact, it is the restriction of an entire function on $\mathbb{C}$) and

$$\text{Supp}\ \widehat{w} \subset [-\frac{8\pi}{3}, -\frac{2\pi}{3}] \cup [\frac{2\pi}{3}, \frac{8\pi}{3}]. \tag{4.2}$$

This means, in particular, that $\{w_{k,n}\}$ is an orthonormal basis for $L^2(\mathbb{R})$. This basis furnishes us with an example of a smooth "Multiresolution Analysis" as described in [3] and [5]. We show that this construction can be carried out in an easy and natural way by using the local bases we have just presented.

Let $I = [\pi, 2\pi], \epsilon = \frac{\pi}{3}, \epsilon' = 2\epsilon = \frac{2\pi}{3}$. Consider the orthogonal projection P_I with polarity $(+,-)$. The bell function $b = b_I$ associated with P_I will not have an interval of constancy since $\pi + \epsilon = 2\pi - \epsilon'$. Its construction is explicitly given by (2.5) with $\alpha = \pi, \beta = 2\pi, \epsilon = \frac{\pi}{3} = \frac{\epsilon'}{2}$. The range of this projection is the subspace generated by the orthonormal basis

$$\psi(n;\xi) = \sqrt{\frac{2}{\pi}}\, b(\xi) \cos \frac{2n+1}{2}(\xi - \pi),$$

$n = 0, 1, 2, \ldots$ (see Theorem 4 and, in particular, (iii) of (3.6)). The dilations by $2^k, k \in \mathbf{Z}$, then give us the projections $P_k = P_{[2^{-k}\pi, 2^{-k+1}\pi]}$ with ranges spanned by the orthonormal basis

$$\psi_k(n;\xi) \equiv 2^{k/2}\psi(n; 2^k\xi), \tag{4.3}$$

$k \in \mathbf{Z}$. It follows from the material in Section 3. (see, in particular, the observation following Theorem 5, concerning the spanning of $L^2(0,\infty)$), that, with this choice of polarity, the collection defined by (4.3) forms an orthonormal basis of $L^2(0,\infty)$.

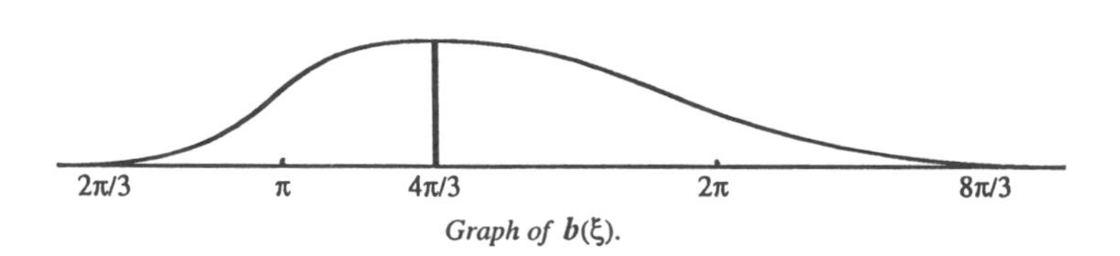

Graph of $b(\xi)$.

Let us also carry out the completely analogous construction based on the local sine basis (i) of (3.6):

$$\varphi(n;\xi) = \sqrt{\frac{2}{\pi}}\, b(\xi) \sin \frac{2n+1}{2}(\xi - \pi),$$

$n = 0, 1, 2, \ldots$. We then obtain the orthonormal basis of $L^2(0,\infty)$

$$\varphi_k(n;\xi) = 2^{k/2}\varphi(n; 2^k\xi), \tag{4.4}$$

$k \in \mathbf{Z}$.

In order to obtain an orthonormal basis for $L^2(\mathbb{R})$ we consider the even extensions of the functions (4.3) and the odd extensions of the functions (4.4) to all of $\mathbb{R}$. More precisely, let

$$\Psi_{k,n}(\xi) \equiv \frac{1}{\sqrt{2\pi}} 2^{k/2} b(2^k|\xi|) \cos \frac{2n+1}{2}(2^k|\xi| - \pi)$$

and

$$\Phi_{k,n}(\xi) \equiv \frac{1}{\sqrt{2\pi}} 2^{k/2} b(2^k|\xi|)(\mathrm{sgn}\xi) \sin \frac{2n+1}{2}(2^k|\xi| - \pi).$$

Theorem 6. *The collection of functions*

$$\alpha_{k,n} \equiv \Psi_{k,n} + i\Phi_{k,n}, \qquad \text{and} \qquad \beta_{k,n} = \Phi_{k,n} + i\Psi_{k,n},$$

$k = 0, \pm 1, \pm 2, \ldots, n = 0, 1, 2, \ldots$, *is an orthonormal basis for* $L^2(\mathbb{R})$.

Proof: Let $\langle f, g\rangle = \int_{\mathbb{R}} f\bar{g}$ be the standard inner product in $L^2(\mathbb{R})$. The orthonormality relations

$$\langle \alpha_{k,n}, \alpha_{k',n'}\rangle = \delta_{kk'}\delta_{nn'} = \langle \beta_{k,n}, \beta_{k,'n'}\rangle$$

follow from the orthonormality relations satisfied by the families (4.3), (4,4) and the fact that each product of the form $\Phi_{k,n}\Psi_{k',n'}$ is an odd function. The orthogonality relations

$$\langle \alpha_{k,n}, \beta_{k,'n'} \rangle = 0$$

for all $k \in \mathbf{Z}$ and $n = 0, 1, 2, \ldots$, follow by the simple calculation:

$$\begin{aligned}\langle \alpha_{k,n}, \beta_{k,'n'} \rangle &= \int_{\mathbb{R}} (\Psi_{k,n}\Phi_{k',n'} + \Phi_{k,n}\Psi_{k',n'}) + i(\Phi_{k,n}\Phi_{k',n'} - \Psi_{k,n}\Psi_{k',n'}) \\ &= 0 + \frac{i}{2}(\delta_{kk'}\delta_{nn'} - \delta_{kk'}\delta_{nn'}) = 0.\end{aligned}$$

It remains to be shown that the above system is complete. In order to do this it clearly suffices to show that for each real valued $f \in L^2(\mathbb{R})$

$$f = \sum_{k\in\mathbf{Z}} \sum_{n=0}^{\infty} \{\langle f, \alpha_{k,n}\rangle \alpha_{k,n} + \langle f, \beta_{k,n}\rangle \beta_{k,n}\}, \tag{4.5}$$

where the convergence is in the $L^2(\mathbb{R})$ norm. But, writing $f = f^{(e)} + f^{(o)}$, with $f^{(e)}$ the even part and $f^{(o)}$ the odd part of f, we see that

$$\begin{aligned}&\sum_{k\in\mathbf{Z}} \sum_{n=0}^{\infty} \langle f, \alpha_{k,n}\rangle \alpha_{k,n} \\ &\qquad = \sum_{k\in\mathbf{Z}} \sum_{n=0}^{\infty} \{\langle f^{(e)}, \Psi_{k,n}\rangle - i\langle f^{(o)}, \Phi_{k,n}\rangle\}(\Psi_{k,n} + i\Phi_{k,n})\end{aligned}$$

from which we obtain

$$\begin{aligned}&\sum_{k\in\mathbf{Z}} \sum_{n=0}^{\infty} \langle f, \alpha_{k,n}\rangle \alpha_{k,n} \\ &\qquad = \frac{1}{2} f + i \sum_{k\in\mathbf{Z}} \sum_{n=0}^{\infty} \{\langle f^{(e)}, \Psi_{k,n}\rangle \Phi_{k,n} - \langle f^{(o)}, \Phi_{k,n}\rangle \Psi_{k,n}\}.\end{aligned} \tag{4.6}$$

A completely analogous argument gives us

$$\begin{aligned}&\sum_{k\in\mathbf{Z}} \sum_{n=0}^{\infty} \langle f, \beta_{k,n}\rangle \beta_{k,n} \\ &\qquad = \frac{1}{2} f + i \sum_{k\in\mathbf{Z}} \sum_{n=0}^{\infty} \{\langle f^{(o)}, \Phi_{k,n}\rangle \Psi_{k,n} - \langle f^{(e)}, \Psi_{k,n}\rangle \Phi_{k,n}\}.\end{aligned} \tag{4.7}$$

Adding equalities (4.6) and (4.7) we obtain the desired result (4.5). ∎

We now show that by modifying this basis slightly (multiplying its members by scalars of absolute value 1) we obtain an orthonormal basis of $L^2(\mathbb{R})$

that is generated by the single function $\gamma(\xi) \equiv i\alpha_{0,0}(\xi)$. More precisely, we have

Theorem 7. *The functions*

$$\gamma_{k,n}(\xi) \equiv 2^{k/2} e^{-i2^k n\xi} \gamma(2^k \xi),$$

$k, n \in \mathbf{Z}$, *form an orthonormal basis of* $L^2(\mathbb{R})$, *where*

$$\gamma(\xi) = \frac{\operatorname{sgn} \xi}{\sqrt{2\pi}} e^{i\frac{\xi}{2}} b(|\xi|).$$

Proof: Let us put $\alpha_{k,-n}(\xi) \equiv \overline{\alpha_{k,n-1}(\xi)} = -i\beta_{k,n-1}$ for $k \in \mathbf{Z}$ and $n > 0$. It then follows from Theorem 6 that $\{\alpha_{k,n}\}, k, n \in \mathbf{Z}$, is an orthonormal basis for $L^2(\mathbb{R})$. Our theorem is established if we show

$$\alpha_{0,n}(\xi) = (-1)^n (-i) e^{in\xi} \gamma(\xi) \tag{4.8}$$

for all $n \in \mathbf{Z}$. To prove (4.8) assume, first, that $n \geq 0$, then

$$\begin{aligned}\alpha_{0,n}(\xi) &= \frac{1}{\sqrt{2\pi}} b(|\xi|)[\cos \frac{2n+1}{2}(|\xi| - \pi) + i(\operatorname{sgn} \xi) \sin \frac{2n+1}{2}(|\xi| - \pi)] \\ &= \frac{b(|\xi|)}{\sqrt{2\pi}} \begin{cases} e^{i(\xi-\pi)(2n+1)/2} & \text{if } \xi \geq 0 \\ e^{i(\xi+\pi)(2n+1)/2} & \text{if } \xi < 0 \end{cases} = (-i)(-1)^n e^{in\xi} \gamma(\xi).\end{aligned}$$

If $n < 0$, then

$$\begin{aligned}\alpha_{0,n}(\xi) = \overline{\alpha_{0,-n-1}(\xi)} &= i \frac{\operatorname{sgn}\xi}{\sqrt{2\pi}} (-1)^{-(n+1)} e^{i(n+1)\xi} e^{-i\frac{\xi}{2}} b(|\xi|) \\ &= (-i)(-1)^n e^{in\xi} \gamma(\xi).\end{aligned}$$

Thus, (4.8) holds for all integers n. ■

The promised wavelet basis is now immediately obtained from the functions in Theorem 7 by applying the inverse Fourier transform to them. Let us carry out the details. We adopt the following definition of the *Fourier transform* of a function $f \in L^1 \cap L^2$:

$$\hat{f}(\xi) = \int_{-\infty}^{\infty} f(x) e^{-i\xi x} dx.$$

The inverse Fourier transform formula is, in this case,

$$f(x) = \frac{1}{2\pi} \int_{-\infty}^{\infty} \hat{f}(\xi) e^{ix\xi} d\xi = \frac{1}{2\pi} \lim_{N \to \infty} \int_{-N}^{N} \hat{f}(\xi) e^{ix\xi} d(\xi),$$

the limit being an L^2-limit. A version of the Plancherel theorem, then tells us that if $f \in L^2(\mathbb{R})$, then

$$\|\sqrt{2\pi} f\|_2 = \|\hat{f}\|_2, \tag{4.9}$$

and, if g is another function in $L^2(\mathbb{R})$, we have

$$\langle f, g\rangle = \frac{1}{2\pi}\langle \hat{f}, \hat{g}\rangle. \tag{4.10}$$

Thus, if we define w by

$$w(x) = \frac{1}{\sqrt{2\pi}} \int_{\mathbb{R}} e^{i\xi x}\gamma(\xi)d\xi$$

we have $\widehat{w} = \sqrt{2\pi}\gamma$,

$$w_{k,n}(x) = 2^{-k/2}w(2^{-k}x - n) = \frac{1}{\sqrt{2\pi}} \int_{\mathbb{R}} e^{i\xi x}\gamma_{k,n}(\xi)d\xi,$$

for $k, n \in \mathbf{Z}$, and the collection $\{w_{k,n}\}$ is an orthonormal basis of $L^2(\mathbb{R})$. Since Supp $\widehat{w}$ = Supp $b(|\xi|) = [-\frac{8\pi}{3}, -\frac{2\pi}{3}] \cup [\frac{2\pi}{3}, \frac{8\pi}{3}]$ we see that w is, indeed, a mother function of the type described at the beginning of this section. That is, w generates a Lemarié-Meyer wavelet basis.

§5. Lemarié-Meyer wavelet bases with more general band limitations

In this section we shall examine to what extent we can generalize this construction if we consider other conditions on the support of $\widehat{w}$. We pose our problem in a way that is consistent with the basic properties of local sine/cosine bases. Instead of the interval $[\pi, 2\pi]$ let us consider the interval $[1, \lambda]$. Dilates by λ, then, will then give us a covering of the right half-line:

$$(0, \infty) = \bigcup_{k=-\infty}^{\infty} [\lambda^k, \lambda^{k+1}].$$

We construct a bell function b associated with the interval $[1, \lambda], \epsilon = \frac{\lambda-1}{\lambda+1}$ and $\epsilon' = \frac{\lambda(\lambda-1)}{(\lambda+1)} = \lambda\epsilon$. Observe that $1 + \epsilon = \lambda - \epsilon'$. The dilates $b(\lambda^{-k}\xi)$ are then bell functions corresponding to the intervals $[\lambda^k, \lambda^{k+1}], k \in \mathbf{Z}$. Observe that adjacent intervals are compatible.

In complete analogy with the construction that led into Theorem 6, we let $\alpha^{\lambda}_{k,n} = \Psi^{\lambda}_{k,n} + i\Phi^{\lambda}_{k,n}$ for $k \in \mathbf{Z}, n \geq 0$, and $\alpha^{\lambda}_{k,n} = \overline{\alpha^{\lambda}_{k,-n-1}}$ if $k \in \mathbf{Z}$ and $n < 0$, where

$$\Psi^{\lambda}_{k,n}(\xi) = \sqrt{\frac{1}{2(\lambda-1)}}\lambda^{k/2}b(\lambda^k|\xi|)\cos\pi(n+\frac{1}{2})(\frac{|\xi|\lambda^k - 1}{\lambda - 1}),$$

$k \in \mathbf{Z}, n \geq 0,$

$$\Phi_{k,n}^{\lambda}(\xi) = \sqrt{\frac{1}{2(\lambda-1)}}\lambda^{k/2}b(\lambda^k|\xi|)(\operatorname{sgn}\xi)\sin\pi(n+\frac{1}{2})(\frac{\lambda^k|\xi|-1}{\lambda-1}),$$

$k \in \mathbf{Z}, n \geq 0$. The argument establishing Theorem 6 can be easily extended to show

Theorem 8. *The collection of functions* $\{\alpha_{k,n}^{\lambda}\}, k, n \in \mathbf{Z}$ *is an orthonormal basis for* $L^2(\mathbb{R})$.

Let us now examine if the functions $\alpha_{k,n} = \alpha_{k,n}^{\lambda}$ are "generated", via dilations (by integral powers of λ) and multiplications by $e^{i\pi n\frac{\xi}{\lambda-1}}$ $(n \in \mathbf{Z})$ applied to $\alpha_{0,0}$ as was the case for the analogous basis in Theorem 7. We begin by trying to establish a formula similar to (4.8). If $n \geq 0$ we have

$$\begin{aligned}\sqrt{2(\lambda-1)}\alpha_{0,n}(\xi) &= b(|\xi|)\begin{cases} e^{i\pi(n+\frac{1}{2})\frac{(\xi-1)}{(\lambda-1)}}, \xi > 0 \\ e^{i\pi(n+\frac{1}{2})\frac{(\xi+1)}{(\lambda-1)}}, \xi < 0 \end{cases} \\ &= \sqrt{2(\lambda-1)}\alpha_{0,0}(\xi)e^{i\pi n\xi/(\lambda-1)}\rho^{n\operatorname{sgn}\xi},\end{aligned}$$

where $\rho = e^{-i\frac{\pi}{\lambda-1}}$. If $n < 0$ then

$$\begin{aligned}\alpha_{0,n}(\xi) &= \overline{\alpha_{0,-n-1}(\xi)} = \overline{\alpha_{0,0}(\xi)\rho^{-(n+1)\operatorname{sgn}\xi}e^{i\pi(n+1)\frac{\xi}{\lambda-1}}} \\ &= \frac{1}{\sqrt{2(\lambda-1)}}b(|\xi|)e^{-i\frac{\pi}{2}\frac{\xi}{\lambda-1}}e^{i\pi\frac{1}{2}\frac{\operatorname{sgn}\xi}{\lambda-1}}e^{-i\pi(n+1)\frac{\operatorname{sgn}\xi}{\lambda-1}}e^{i\pi(n+1)\frac{\xi}{\lambda-1}} \\ &= \frac{1}{\sqrt{2(\lambda-1)}}\{b(|\xi|)e^{i\frac{\pi}{2}\frac{\xi}{\lambda-1}}e^{-1\frac{\pi}{2}\frac{\operatorname{sgn}\xi}{\lambda-1}}\}e^{i\pi n\frac{\xi}{\lambda-1}}e^{-i\pi n\frac{\operatorname{sgn}\xi}{\lambda-1}} \\ &= \alpha_{0,0}(\xi)e^{i\pi n\frac{\xi}{\lambda-1}}\rho^{n\operatorname{sgn}\xi}.\end{aligned}$$

This shows

$$\alpha_{0,n}(\xi) = \rho^{n\operatorname{sgn}\xi}e^{in\pi\frac{\xi}{\lambda-1}}\alpha_{0,0}(\xi), \tag{5.1}$$

for all $n \in \mathbf{Z}$.

Thus, multiplication of $\alpha_{0,0}(\xi)$ by $e^{in\pi\frac{\xi}{\lambda-1}}$ gives us an orthonormal system, differing from $\alpha_{0,n}(\xi)$ by a constant multiple of absolute value 1, provided $\rho^{n\operatorname{sgn}\xi}$ is a constant of absolute value 1 for all ξ (and each $n \in \mathbf{Z}$). But this means that $\rho = e^{-i\frac{\pi}{\lambda-1}}$ must be real (either 1 or -1). Consequently, $\sin\frac{\pi}{\lambda-1} = 0$ and it follows that $\frac{\pi}{\lambda-1}$ must be an integral multiple of π. This and the condition $\lambda > 1$ force us to conclude that

$$\lambda = 1 + \frac{1}{m} \tag{5.2}$$

for m a positive integer. We thus obtain the following analog of Theorem 7:

Theorem 9. *The functions*

$$\gamma_{k,n}^{(\lambda)}(\xi) = \gamma_{k,n}(\xi) = \lambda^{\frac{k}{2}} e^{-in\pi \frac{\lambda^k \xi}{\lambda-1}} \gamma(\lambda^k \xi), \tag{5.3}$$

$k, n \in \mathbf{Z}$, *is an orthonormal basis of* $L^2(I\!R)$, *where*

$$\gamma(\xi) = \frac{e^{\frac{-i\pi \operatorname{sgn} \xi}{2(\lambda-1)}}}{\sqrt{2(\lambda-1)}} e^{i\pi \frac{\xi}{2(\lambda-1)}} b(|\xi|)$$

and $\lambda = 1 + \frac{1}{m}$, *m a positive integer.*

If we let $w(x) \equiv \frac{1}{\sqrt{2\pi}} \int_{-\infty}^{\infty} e^{ix\xi} \gamma(\xi) d\xi$, then w is a "mother function" that generates a wavelet basis (giving us a Multiresolution Analysis)

$$\{w(x\lambda^{-k} - n\frac{\pi}{\lambda - 1})\lambda^{\frac{-k}{2}}\},$$

for all $k, n \in \mathbf{Z}$, whenever $\lambda = 1 + \frac{1}{m}, m$ a positive integer.

§6. Concluding remarks

We repeat that the local bases we developed in Section 2 were introduced by Coifman and Meyer, and their use in obtaining the smooth wavelet bases were pointed out to us by these two authors. Some of the ideas involved in developing the properties of the local bases are also found in Malvar's paper [4]. The particular emphasis on the role played by the projections P_I, however, does not appear in Malvar's paper [4] and is not prominent in the exposition [1] of Coifman and Meyer. As mentioned in the Abstract, one of our goals is to make this material accessible to the largest possible audience. One can develop a parallel treatment connected with the discrete sine and cosine transforms. We shall present this and its connection with the corresponding discrete version of the smooth wavelet bases in a subsequent paper.

References

1. Coifman, R. R. and Y. Meyer, Remarques sur l'analyse de Fourier à fenêtre, série I, *C. R. Acad. Sci. Paris* **312** (1991), 259–261.
2. Lemarié, P. and Y. Meyer, Ondelettes et bases Hilbertiennes, *Rev. Mat. Iberoamericana* **2** (1986), 1–18.
3. Mallat, S., A theory for multiresolution signal decomposition: the wavelet reprsentation, *IEEE Pattern Anal. and Machine Intell.* **11** (1989), 674-693.
4. Malvar, H., Lapped transforms for efficient transform/subband coding, IEEE *Trans. Acoustics, Speech, and Signal Processing* **38** (1990), 969–978.
5. Meyer, Y., *Ondelettes et Opérateurs,* in two volumes, Hermann, Paris, 1990.

The research of the authors was supported in part by the following grants: Pascal Auscher's from the Southwestern Bell Telephone Company and AFOSR-90-0323; Guido Weiss' by AFSOR-90-0323; and M. Victor Wickerhauser's by ONR grant N00014-88-K0020.

Pascal Auscher
Department of Mathematics
University of Rennes, Campus of Beaulieu
35042 Rennes
Cedex, France
auscher@cicb.fr

Guido Weiss
Washington University
Campus Box 1146
One Brookings Drive
St. Louis, MO 63130-4899
c31801@wuvmd.bitnet

Mladen Victor Wickerhauser
Numerical Algorithms Research Group
Department of Mathematics
Yale University
New Haven, CT 06520
victor@math.yale.edu

Part IV

Multivariate Scaling Functions and Wavelets

Some Elementary Properties of Multiresolution Analyses of $L^2(R^n)$

W. R. Madych

Abstract. A multiresolution analysis is a family of subspaces which are generated by lattice translates of dilates of one function ϕ. Here we record several, hopefully useful, observations concerning such families. For example: the definitions are reviewed and examined, examples are given, functions ϕ whose translates and dilates generate such analyses are characterized, those multiresolution analyses which are invariant under all translations and those whose scaling functions are characteristic functions are completely described.

§1. Introduction

As is well known, orthogonal wavelet bases are usually generated via Meyer's notion of multiresolution analysis. The main elements of such analyses are the notions of dilation and translation relative to a lattice in R^n.

The point of this article is to communicate several elementary observations which may be useful to investigators and other individuals who work with wavelets and multiresolution analyses. Although we review the basic notions necessary for our development, we do not detail various well known and important facts which are not germane to these observations. In short, we assume that the reader is familiar with the fundamentals outlined in [43]. These observations are divided into three topics which we briefly describe as follows.

The first topic, which is treated in Section 2, concerns certain general but elementary properties of multiresolution analyses and scaling functions. For instance, we show that certain items on the usual axiom list are redundant, characterize those functions ϕ which can generate multiresolution analyses, and give examples which extend in a certain direction the multiresolution analyses composed of classical univariate piecewise polynomial splines. In the introductory subsection of Section 2 we recall the basic definitions of multiresolution analyses and give a more complete description of the questions we treat.

Wavelets–A Tutorial in Theory and Applications
C. K. Chui (ed.), pp. 259–294.

ISBN 0-12-174590-2

The second topic deals with those multiresolution analyses which are translation invariant. The classical example is the one generated by the cardinal sine function. In Section 3 we give a characterization of all such analyses.

The third topic concerns multiresolution analyses whose scaling functions are characteristic functions. Although the classical case is quite simple the general situation is not completely trivial and is related to the theory of self-similar tilings of R^n. In Section 4 we give a description of those scaling sequences which give rise to such scaling functions.

The second and third topic may be regarded as dual to each other in some sense. In one case the scaling functions are characteristic functions of tiles whereas in the other they are the Fourier transforms of tiles. Unfortunately, the nature of the tiles in each case is somewhat different.

Section 5 is devoted to miscellaneous remarks and citations.

We now briefly digress to list some of the conventions which are used here: The Fourier transform $\hat{f}$ of a function f is defined by

$$\hat{f}(\xi) = \int_{R^n} e^{-i\langle \xi, x\rangle} f(x)dx$$

whenever it makes sense and distributionally otherwise. Basic facts concerning Fourier transforms and distributions will be used without further elaboration in what follows. To avoid the pedantic repetition of "almost everywhere" and other modifying phrases which are inevitably necessary when dealing with functions defined almost everywhere, all equalities between functions and other related notions are interpreted in the distributional sense whenever possible. The term support is also used in the distributional sense; in particular the support of a function f in $L^2(R^n)$ is a well defined closed set. If W is a collection of tempered distributions then $\widehat{W}$ is the collection of Fourier transforms of elements of W, in other words $\widehat{W} = \{f : f = \hat{g}$ for some g in $W\}$. For a subset Ω of R^n, a linear transformation B on R^n, and an element y of R^n the sets $B\Omega$ and $\Omega + y$ are defined by $B\Omega = \{x : x = B\omega$ for some ω in $\Omega\}$ and $\Omega + y = \{x : x = \omega + y$ for some ω in $\Omega\}$; $L^2(\Omega)$ is the L^2 closure of the subspace of those functions in $L^2(R^n)$ whose support is contained in Ω. Given a measurable set Ω, χ_Ω denotes its characteristic or indicator function and $|\Omega|$ denotes its Lebesgue measure. The notation $\Omega_1 \simeq \Omega_2$ means that the sets Ω_1 and Ω_2 are equal up to a set of measure zero, in other words, $|\Omega_1 \setminus \Omega_2| = |\Omega_2 \setminus \Omega_1| = 0$. If $\Omega_1 \bigcap \Omega_2 \simeq \emptyset$ we say that Ω_1 and Ω_2 are essentially disjoint.

§2. Multiresolution analysis

2.1. Definitions and related background

Suppose Γ is a lattice in R^n, that is, Γ is the image of the integer lattice Z^n under some nonsingular linear transformation. We say that a linear transformation A on R^n is an *acceptable dilation for* Γ if it satisfies the following properties:

- A leaves Γ invariant. In other words, $A\Gamma \subset \Gamma$ where

$$A\Gamma = \{y \,:\, y = Ax \text{ and } x \in \Gamma\} \,.$$

- All the eigenvalues, λ_i, of A satisfy $|\lambda_i| > 1$.

These properties imply that $|\det A|$ is an integer q which is ≥ 2. For example, if $A = mI$ where m is an integer ≥ 2 and I is the identity then A is an acceptable dilation for any lattice Γ.

Such an A induces a unitary dilation operator $U_A \,:\, f \longrightarrow U_A f$ on $L^2(R^n)$, defined by

$$U_A f(x) = |\det A|^{-1/2} f(A^{-1}x) \,. \tag{1}$$

If V is a subspace of $L^2(R^n)$ we use the customary notation $U_A V$ to denote the image of V under U_A, that is,

$$U_A V = \{f \,:\, f = U_A g,\, g \in V\}.$$

The translation operator τ_y is defined by $\tau_y f(x) = f(x - y)$.

A *multiresolution analysis* $\mathcal{V}$ associated with (Γ, A) is a family $\{V_j\}_{j\in Z}$ of closed subspaces of $L^2(R^n)$ which enjoys the following properties:

A1. $V_j \subset V_{j+1}$ for all j in Z.

A2. $\bigcup_{j\in Z} V_j$ is dense in $L^2(R^n)$.

A3. $\bigcap_{j\in Z} V_j = \{0\}$.

A4. $f(x) \in V_j$ if and only if $f(Ax) \in V_{j+1}$. In other words

$$V_j = U_A^{-j} V_0, \quad j \in Z.$$

A5. V_0 is invariant under τ_γ. More specifically, if $f(x)$ is in V_0 then so is $f(x-\gamma)$ for all γ in Γ.

A6. There is a function $\phi \in V_0$, called the scaling function, such that $\{\tau_\gamma \phi, \gamma \in \Gamma\}$ is a complete orthonormal basis for V_0.

The case $A = 2I$ is often referred to as a *dyadic* multiresolution analysis. It is also the case to which most of the current literature is devoted and, except for certain technicalities, is representative of the general case.

We say that a dyadic multiscale analysis is *composed of generalized spline functions* in the sense of Meyer if all the elements of the subspace V_0 are continuous and the mapping which maps f into the sequence of values $\{f(\gamma)\}_{\gamma\in\Gamma}$ is an isomorphism from V_0 onto $l^2(\Gamma)$.

The following remarks and observations bring to the attention of the reader various basic facts concerning multiresolution analyses and also motivate the development in Subsections 2.2 and 2.3.

2.1.1

Every lattice Γ in R^n is the image of Z^n under an invertible linear transformation. Thus there is no loss of generality by restricting attention to the case $\Gamma = Z^n$. We do this in what follows to avoid unnecessary obfuscation.

The reader who is interested in the general statements of the results discussed below can systematically replace Z^n with Γ and $2\pi Z^n$ with the corresponding dual lattice Γ'.

Although more general lattices Γ are not more interesting than Z^n in our development, they may be of some practical interest. For instance, the corresponding scaling function may be more symmetric in one lattice than in another, see Figures 8 and 9.

Recall that if A is an acceptable dilation of Z^n then a coset of AZ^n is a set of the form

$$k + AZ^n = \{k + Aj : j \in Z^n\}$$

where k is any element of Z^n which is sometimes referred to as a representative of the coset. Any pair of cosets are either identical or disjoint so that the collection of all cosets, which is denoted by Z^n/AZ^n, consists of disjoint cosets whose union is Z^n. The number of disjoint cosets in Z^n/AZ^n is equal to $q = |\det A|$. A subset $\mathcal{K}$ of Z^n is said to be a full collection of representatives of Z^n/AZ^n if it contains exactly q elements and

$$\bigcup_{k \in \mathcal{K}} (k + AZ^n) = Z^n.$$

2.1.2

The orthonormality of the collection $\{\phi(x - j)\}_{j \in Z^n}$ is equivalent to

$$\sum_{k \in Z^n} |\hat{\phi}(\xi - 2\pi k)|^2 = 1. \tag{2}$$

Thus every scaling function ϕ must satisfy (2).

The existence of a scaling function is equivalent to the following:

A6$_1$. There is a function ϕ in V_0 such that $\{\phi(x - k)\}_{k \in Z^n}$ is a Riesz basis for V_0.

In many specific examples this property is much more apparent and easier to verify than A6.

We remind the reader that $\{\phi(x - k)\}_{k \in Z^n}$ is a Riesz basis for its closed linear span if and only if there are positive constants, c_1 and c_2, such that

$$c_1 \sum_{k \in Z^n} |a_k|^2 \leq \int_{R^n} |\sum_{k \in Z^n} a_k \phi(x - k)|^2 dx \leq c_2 \sum_{k \in Z^n} |a_k|^2.$$

2.1.3

In view of A1 and A4 the scaling function ϕ must satisfy the two scale difference equation

$$\phi(x) = \sum_{k \in Z^n} s_k \phi(Ax - k) \tag{3}$$

for some sequence $\{s_k\}_{k \in Z^n}$ in $l^2(Z^n)$; $\{s_k\}_{k \in Z^n}$ is called the scaling sequence. The Fourier transform of (3) is

$$\hat{\phi}(\xi) = S(B^{-1}\xi)\hat{\phi}(B^{-1}\xi) \tag{4}$$

where $B = A^*$ and $S(\xi)$ is the $2\pi Z^n$ periodic function

$$S(\xi) = \frac{1}{q} \sum_{k \in Z^n} s_k e^{-i\langle k, \xi \rangle}.$$

Here we remind the reader of the following facts:

- If $\hat{\phi}(\xi)$ and $S(\xi)$ are continuous at the origin then $S(0) = 1$. This follows from the fact that in this case $\hat{\phi}(0) = 1$, see Subsection 2.2. Another restriction on S, or equivalently on the sequence $\{s_k\}_{k \in Z^n}$, is

$$\sum_{\kappa \in \mathcal{K}} |S(\xi - 2\pi B^{-1}\kappa)|^2 = 1 \tag{5}$$

  where $\mathcal{K}$ is any full collection of representatives of Z^n/BZ^n. This follows from the fact that $\{\phi(x-k)\}_{k \in Z^n}$ is orthonormal via Formula (2).
- Note that in view of (3) ϕ may be considered as a fixed point of the operator $\phi \to \sum_{k \in Z^n} s_k \phi(Ax - k)$. Unfortunately the normalized solution of (3) is not unique since the function ψ defined by

$$\hat{\psi}(\xi) = h(\xi)\hat{\phi}(\xi)$$

  is also a normalized solution of (3) whenever h is any measurable function which enjoys $|h(\xi)| = 1$ for almost all ξ and $h(B\xi) = h(\xi)$. Such ψ are also scaling functions but not necessarily for the same multiresolution analysis. Further restrictions are needed on the scaling sequence $\{s_k\}_{k \in Z^n}$ and the scaling function ϕ to guarantee a unique solution of Equation (3).
- The scaling function ϕ is not unique since any function ψ whose Fourier transform satisfies

$$\hat{\psi}(\xi) = H(\xi)\hat{\phi}(\xi)$$

  is also a scaling function for the same multiresolution analysis whenever H is any measurable $2\pi Z^n$ periodic function which satisfies $|H(\xi)| = 1$ for almost all ξ. The scaling equation satisfied by ψ will, in general, be different from the one satisfied by ϕ.

2.1.4.

The simplest example of a dyadic multiresolution analysis $\mathcal{V} = \{V_j\}_{j\in Z}$ is the family of closed subspaces of $L^2(R)$ defined as follows:

$$V_j = \left\{ f \in L^2(R) : f = \text{ constant on } [k2^{-j}, (k+1)2^{-j}), \ k \in Z \right\}.$$

It is easy to verify that $\mathcal{V}$ is a dyadic multiresolution analysis with scaling function $\phi(x) = \chi_{[0,1)}(x)$. The two scale difference equation in this case is

$$\phi(x) = \phi(2x) + \phi(2x-1),$$

and S is given by

$$S(\xi) = (e^{-i\xi} + 1)/2.$$

Another scaling function for this multiresolution analysis is the function ϕ_0 whose Fourier transform is given by

$$\hat{\phi}_0(\xi) = \frac{2|\sin(\xi/2)|}{\xi}.$$

The corresponding scaling equation is

$$\phi_0(x) = 2\sum_{k\in Z} s_k \phi_0(2x-k)$$

where $\{s_k\}_{k\in Z}$ are the Fourier coefficients of

$$S_0(\xi) = |\cos(\xi/2)| = \sum_{k\in Z} s_k e^{-ik\xi}.$$

The relationship of ϕ_0 to ϕ is given by

$$\hat{\phi}_0(\xi) = H(\xi)\hat{\phi}(\xi)$$

where

$$H(\xi) = 2|\sin(\xi/2)|/i(e^{-i\xi} - 1) = \sum_{k\in Z} h_k e^{-ik\xi},$$

in other words

$$\phi_0(x) = \sum_{k\in Z} h_k \phi(x-k)$$

where $\{h_k\}_{k\in Z}$ are the Fourier coefficients of H.

If ψ is the function defined by

$$\psi(x) = c\left\{\log|x| - \log|x-1|\right\}$$

where c is a constant chosen so that $\int_{-\infty}^{\infty} |\psi(x)|^2 dx = 1$, then observe the following:

- ψ is a constant multiple of the Hilbert transform of ϕ.
- ψ satisfies the same two scale difference equation as ϕ, namely

$$\psi(x) = \psi(2x) + \psi(2x-1).$$

- ψ is a scaling function for a multiresolution analysis which is different from the one considered above.

2.1.5

It is clear that the item A5 in the list of axioms given above is an immediate consequence of the property A6 and thus its inclusion is unnecessary. What is not so apparent, but will be shown in what follows, is that the item A3 is a consequence of the other properties. Its inclusion on the list is also unnecessary and redundant.

2.1.6

Observe that item A4 on the list of axioms implies that $f(x)$ is in V_j if and only if $f(A^{-j}x)$ is in V_0. It follows that a multiresolution analysis is essentially completely determined by the closed subspace V_0. In other words, properties A1, A2, and A3 are consequences of A4 and certain features of the closed subspace V_0. Indeed, since V_0 is the closure of the linear span of the Z^n translates of the scaling function ϕ, the whole multiresolution analysis may be regarded as being generated by ϕ.

We have already noted that item A1 is equivalent to the two scale difference equation (3) which the scaling function ϕ must satisfy. In what follows we will examine other features of V_0 which give rise to and/or are implied by items A2, A3, and A6.

2.1.7.

Suppose that ϕ is a function in $L^2(R^n)$. Then it is well known and easy to verify that $\{\phi(x-j)\}_{j\in Z^n}$ is a Riesz basis for the closure of its linear span if and only if there are positive constants, c_1 and c_2, such that

$$c_1 \leq \sum_{j\in Z^n} |\hat{\phi}(\xi - 2\pi j)|^2 \leq c_2. \tag{6}$$

Orthonormality of $\{\phi(x-j)\}_{j\in Z^n}$ is equivalent to the case $c_1 = c_2 = 1$.

Thus the alternate item $A6_1$, in the definition of multiresolution analyses is equivalent to saying that V_0 is the closed linear span of the Z^n translates of one function ϕ which satisfies (6). The original item A6 means that the constants c_1 and c_2 satisfy $c_1 = c_2 = 1$.

In what follows we will show that the property A6 can be even further relaxed. Indeed it is equivalent to the following seemingly weaker property:

$A6_2$. V_0 is the closed linear span of the Z^n translates of one function ϕ which satisfies

$$\sum_{j\in Z^n} |\hat{\phi}(\xi - 2\pi j)|^2 > 0 \tag{7}$$

almost everywhere.

In passing we note that if ϕ decays sufficiently quickly at ∞ so that the Poisson summation formula can be applied then

$$\sum_{j\in Z^n} |\hat{\phi}(\xi - 2\pi j)|^2 = \sum_{j\in Z^n} \phi * \tilde{\phi}(j) e^{-i\langle j,\xi\rangle} \tag{8}$$

where $\tilde{\phi}(x) = \overline{\phi(-x)}$. In this case properties (6) and (7) can be expressed in terms of the right hand side of (8).

2.2. Some consequences of scaling

In this subsection we delve into the behavior, as j tends to $-\infty$ or ∞, of the sequences of subspaces $\{V_j\}_{j\in Z}$ which enjoy properties A4 and A6. In particular we are concerned with phenomena which imply and are implied by items A2 and A3. The main results are the conclusions of the proposition below.

Proposition 1. *Suppose $\{V_j\}_{j\in Z}$ is a sequence of closed subspaces of $L^2(R^n)$ which enjoys the following properties:*

- *$f(x)$ is in V_j if and only if $f(Ax)$ is in V_{j+1}.*
- *There is a function ϕ in V_0 such that $\{\phi(x-k)\}_{k\in Z^n}$ is a complete orthonormal system for V_0.*

  If $P_j f$ denotes the orthogonal projection of f into V_j then the following conclusions hold:

B1. *For all f in $L^2(R^n)$*

$$\lim_{j\to-\infty} \|P_j f\| = 0.$$

B2. *The following conditions are equivalent:*

- *For all f in $L^2(R^n)$*

$$\lim_{j\to\infty} \|f - P_j f\| = 0. \tag{9}$$

- *The function ϕ satisfies*

$$\lim_{k\to\infty} \frac{1}{|B^{-k}Q|} \int_{B^{-k}Q} |\phi(\xi)|^2 d\xi = 1 \tag{10}$$

for every cube Q of finite diameter in R^n.

We bring attention to the fact that the hypotheses of the above proposition do not include A1 and thus ϕ is not assumed to satisfy a scaling relation. On the other hand if A1 holds, that is, if $V_j \subset V_{j+1}$ for all j in Z it follows that A2 and (10) are equivalent. We emphasize this conclusion as follows:

Corollary 2. *If $\{V_j\}_{j\in Z}$ is a multiresolution analysis associated with (Z^n, A) then property A2 is equivalent to assuming that the scaling function ϕ satisfies property (10).*

Next observe that B1 implies A3, that is, $\cap V_j = \{0\}$. This validates the remark made in Subsection 2.1.5 concerning the redundance of A3. To be more precise we state the following:

Corollary 3. *If $\{V_j\}_{j\in Z}$ is a multiresolution analysis then property A3 is a consequence of the other properties enjoyed by $\{V_j\}_{j\in Z}$.*

2.2.1. Details

Proposition 1 is an easy consequence of the formula for the Fourier transform of $P_j f$:

$$\widehat{P_j f}(\xi) = \sum_{k \in Z^n} \left\{ \hat{f}(\xi - 2\pi B^j k)\overline{\hat{\phi}(B^{-j}\xi - 2\pi k)} \right\} \hat{\phi}(B^{-j}\xi) \tag{11}$$

where $B = A^*$ is the adjoint of A.

In what follows we will use the notation

$$\mathrm{per}_j(g(\xi)) = \sum_{k \in Z^n} g(\xi - 2\pi B^j k).$$

With this notation (11) may be re-expressed as

$$\widehat{P_j f}(\xi) = \mathrm{per}_j(\hat{f}(\xi)\overline{\hat{\phi}(B^{-j}\xi)})\hat{\phi}(B^{-j}\xi).$$

To see B1 use Plancherel's formula, Formula (11), and the fact that P_j is an orthogonal projection to write

$$\begin{aligned} \|P_j f\|^2 &= \langle P_j f, f\rangle \\ &= \frac{1}{(2\pi)^n} \int_{R^n} \mathrm{per}_j(\hat{f}(\xi)\overline{\hat{\phi}(B^{-j}\xi)})\hat{\phi}(B^{-j}\xi)\overline{\hat{f}(\xi)}d\xi. \end{aligned} \tag{12}$$

Observe that

$$\mathrm{per}_j(\hat{f}(\xi)\overline{\hat{\phi}(B^{-j}\xi)}) \le \{\mathrm{per}_j(|f(\xi)|^2)\}^{1/2}\{\mathrm{per}_j(|\hat{\phi}(B^{-j}\xi)|^2)\}^{1/2}$$

and since

$$\mathrm{per}_j(|\hat{\phi}(B^{-j}\xi)|^2) = 1,$$

by virtue of the fact that $\{\phi(x-k)\}_{k \in Z^n}$ is a complete orthonormal system for V_0, we may conclude that

$$\|P_j f\|^2 \le \frac{1}{(2\pi)^n} \int_{R^n} \{\mathrm{per}_j(q^j|\hat{f}(\xi)|^2)\}^{1/2} |q^{-j/2}\hat{\phi}(B^{-j}\xi)\hat{f}(\xi)| d\xi \tag{13}$$

where $q = |\det A|$.

Observe that $\mathrm{per}_j(q^j|\hat{f}(\xi)|^2)$ is essentially an approximating Riemann sum for $\int_{R^n} |\hat{f}(\xi - 2\pi\eta)|^2 d\eta$. Hence if $\hat{f}$ is continuous with compact support and if $j \le 0$,

$$\mathrm{per}_j(q^j|\hat{f}(\xi)|^2) \le C \tag{14}$$

where C is a constant which depends on $\hat{f}$ but not on j. Thus, in view of (13) and (14), for such f we may write

$$\|P_j f\|^2 \le C \int_{R^n} |q^{-j/2}\hat{\phi}(B^{-j}\xi)\hat{f}(\xi)| d\xi, \tag{15}$$

whenever $j \leq 0$, where C is a constant independent of j. Now, if $\hat{f}$ vanishes in a neighborhood of the origin say $\{\xi : |\xi| < \varepsilon\}$, we may write

$$\int_{R^n} |q^{-j/2}\hat{\phi}(B^{-j}\xi)\hat{f}(\xi)|d\xi \leq \|\hat{f}\|\{\int_{|\xi|\geq\varepsilon} q^{-j}|\hat{\phi}(B^{-j}\xi)|^2 d\xi\}^{1/2} \tag{16}$$

and note that, since $|\hat{\phi}(\xi)|^2$ is in $L^1(R^n)$, the integral involving $\hat{\phi}$ on the right hand side of (16) goes to zero as $j \to -\infty$. Thus from (15) and (16) we may conclude that B1 holds for all f such that $\hat{f}$ is continuous, compactly supported, and vanishes in a neighborhood of the origin. Since such f are dense in $L^2(R^n)$ and $\|P_j\| \leq 1$, we may make the stronger conclusion that B1 holds for all f in $L^2(R^n)$.

Next we prove the second assertion. Recall that

$$\|f - P_k f\|^2 = \langle (I - P_k)f, f\rangle = \langle f, f\rangle - \langle P_k f, f\rangle . \tag{17}$$

Now, if $\hat{f}$ has compact support and k is sufficiently large then by virtue of (12) we may write

$$\langle P_k f, f\rangle = \frac{1}{(2\pi)^n}\int_{R^n} |\hat{\phi}(B^{-k}\xi)|^2|\hat{f}(\xi)|^2 d\xi. \tag{18}$$

Since f's whose Fourier transforms are finite linear combinations of characteristic functions of essentially disjoint cubes are dense in $L^2(R^n)$ the desired result is an immediate consequence of (9), (17), and (18).

2.3. Pseudo scaling functions

In this subsection we investigate the closed linear span of $\{\phi(x-k)\}_{k\in Z^n}$ where ϕ is any function in $L^2(R^n)$. In particular we are interested in the question of when such a subspace can play the role of V_0 for some multiresolution analysis. We begin by introducing the following notation and definitions:

Given a function ϕ in $L^2(R^n)$ the subspace $V(\phi)$ is defined to be the closed linear span of $\{\phi(x-k)\}_{k\in Z^n}$.

A subspace V of $L^2(R^n)$ is said to generate the multiresolution analysis $\mathcal{V} = \{V_j\}_{j\in Z}$ associated with (Z^n, A) if and only if $V_j = U_A^{-j}V$ for all j in Z. In other words, $f(x)$ is in V_j if and only if $f(A^{-j}x)$ is in V.

A function ϕ in $L^2(R^n)$ is said to generate a multiresolution analysis $\mathcal{V}$ if and only if $V(\phi)$ generates $\mathcal{V}$.

The conclusions of the following lemma hold the key to the answer.

Lemma 4. *Suppose ϕ is any element of $L^2(R^n)$.*

C1. *The series $\sum_{j\in Z^n} |\hat{\phi}(\xi - 2\pi j)|^2$ converges for almost all ξ in R^n and the function Φ defined by*

$$\Phi(\xi) = \{\sum_{j\in Z^n} |\hat{\phi}(\xi - 2\pi j)|^2\}^{1/2}$$

is in $L^2(R^n/2\pi Z^n)$.

C2. *The function* ϕ_0 *is well defined by the formula for its Fourier transform*

$$\hat{\phi}_0(\xi) = \frac{\hat{\phi}(\xi)}{\Phi(\xi)},$$

is in $L^2(R^n)$, *and satisfies* $|\hat{\phi}_0(\xi)| \leq 1$.

C3. *If* $\Omega = \{\xi \in R^n : \Phi(\xi) > 0\}$ *and* χ_Ω *is the characteristic function of* Ω *then*

$$\chi_\Omega(\xi) = \sum_{j \in Z^n} |\hat{\phi}_0(\xi - 2\pi j)|^2.$$

C4. *Every element* f *in* $V(\phi_0)$ *can be expressed as*

$$\hat{f}(\xi) = F(\xi)\hat{\phi}_0(\xi)$$

where F *is in* $L^2(R^n/2\pi Z^n)$.

C5. $V(\phi) = V(\phi_0)$.

C6. *If* ψ *is any function in* $L^2(R^n)$ *such that*

$$\hat{\psi}(\xi) = F(\xi)\hat{\phi}(\xi)$$

for some measurable $2\pi Z^n$ *periodic function* F *which satisfies* $|F(\zeta)| > 0$ *almost everywhere, then* $V(\psi) = V(\phi)$.

C7. *The subspace* $V(\phi)$ *has an orthogonal basis of the form* $\{\psi(x - j)\}_{j \in Z^n}$ *for some* ψ *in* $V(\phi)$ *if and only if* $\Phi(\xi) > 0$ *almost everywhere.*

C8. *If* $\Phi(\xi) > 0$ *almost everywhere then* $\{\phi_0(x - j)\}_{j \in Z^n}$ *is an orthonormal basis for* $V(\phi)$. *Furthermore if* $\{\psi(x - j)\}$ *is another orthonormal basis for* $V(\phi)$ *then*

$$\hat{\psi}(\xi) = H(\xi)\hat{\phi}_0(\xi)$$

where $H(\xi)$ *is a measurable* $2\pi Z^n$ *periodic function which satisfies* $|H(\xi)| = 1$ *almost everywhere.*

Suppose ψ is a scaling function for a multiresolution analysis $\mathcal{V}$ associated with (Z^n, A). Consider the function ϕ defined by

$$\hat{\phi}(\xi) = F(\xi)\hat{\psi}(\xi)$$

where F is in $L^2(R^n/2\pi Z^n)$ and $|F(\xi)| > 0$ almost everywhere. In view of C6 of the Lemma, $V(\phi) = V(\psi)$ and we may conclude that ϕ also generates the multiscale analysis $\mathcal{V}$. Now, since

$$\hat{\psi}(\xi) = T(B^{-1}\xi)\hat{\psi}(B^{-1}\xi)$$

for some T in $L^2(R^n/2\pi Z^n)$, it follows that

$$\hat{\phi}(\xi) = S(B^{-1}\xi)\hat{\phi}(B^{-1}\xi)$$

where

$$S(B^{-1}\xi) = \{F(\xi)/F(B^{-1}\xi)\}T(B^{-1}\xi).$$

Observe that $S(\xi)$ can be a rather wild $2\pi Z^n$ periodic measurable function; it need not even be locally integrable.

Conversely, suppose ϕ is in $L^2(R^n)$ and satisfies

$$\hat{\phi}(\xi) = S(B^{-1}\xi)\hat{\phi}(B^{-1}\xi)$$

where $S(\xi)$ is a $2\pi Z^n$ periodic measurable function. Define $\Phi(\xi)$ as in C1 of the Lemma and assume $\Phi(\xi) > 0$. Look at the corresponding function ϕ_0, as defined in C2 of the Lemma , and write

$$\hat{\phi}_0(\xi) = S_0(B^{-1}\xi)\hat{\phi}_0(B^{-1}\xi)$$

where

$$S_0(B^{-1}\xi) = \{S(B^{-1}\xi)\Phi(B^{-1}\xi)\}/\Phi(\xi).$$

Note that $S_0(\xi)$ is $2\pi Z^n$ periodic and in view of the fact that $\sum_{k\in Z^n} |\hat{\phi}_0(\xi - 2\pi k)|^2 = 1$ it follows that

$$\sum_{k\in\mathcal{K}} |S_0(B^{-1}(\xi - 2\pi k))|^2 = 1$$

where $\mathcal{K}$ is any full collection of distinct coset representatives of Z^n/BZ^n. If, in addition, ϕ_0 satisfies (10) it is clear that ϕ_0 generates a multiresolution analysis associated with (Z^n, A). Finally, by virtue of C5 of the Lemma, we may conclude that the original function ϕ also generates the same multiresolution analysis.

We summarize these observations as follows:

Proposition 5. *Suppose ϕ is in $L^2(R^n)$. Then ϕ generates a multiresolution analysis associated to (Z^n, A) if and only if ϕ enjoys the following three properties:*

D1. *There is a $2\pi Z^n$ periodic measurable function $S(\xi)$ such that*

$$\hat{\phi}(\xi) = S(B^{-1}\xi)\hat{\phi}(B^{-1}\xi).$$

D2. *For almost all ξ*

$$\sum_{k\in Z^n} |\hat{\phi}(\xi - 2\pi k)|^2 > 0.$$

D3. *For all cubes Q of finite diameter in R^n*

$$\lim_{j\to\infty} \frac{1}{|B^{-j}Q|}\int_{B^{-j}Q} \{|\hat{\phi}(\xi)|^2 / \sum_{k\in Z^n} |\hat{\phi}(\xi - 2\pi k)|^2\} d\xi = 1.$$

This proposition provides an answer to the questions concerning $V(\phi)$ raised earlier in this section and validates the remark concerning A6$_2$ in Subsection 2.1.7. Furthermore, simple examples show that properties D1, D2, and D3 are mutually exclusive.

Condition D3 is not as inconvenient to verify as it appears to be. For example, if ϕ is in $L^2(R^n) \cap L^1(R^n)$ and satisfies $\hat{\phi}(0) \neq 0$ and $\hat{\phi}(2\pi k) = 0$ for all k in $Z^n \backslash \{0\}$ then ϕ enjoys property D3.

On the other hand, D3 does not necessarily follow if $\hat{\phi}$ is continuous in a neighborhood of $Z^n, \hat{\phi}(0) \neq 0$, and $\hat{\phi}(2\pi k) = 0$ for all k in $Z^n \backslash \{0\}$. The neighborhood must be "sufficiently fat" in a certain sense.

Finally we remark that the above results can be used to give an alternate and more compact characterization of multiresolution analyses. This can be stated more precisely as follows:

Corollary. *Suppose $\mathcal{V} = \{V_j\}_{j \in Z}$ is a sequence of closed subspaces of $L^2(R^n)$. Then $\mathcal{V}$ is a multiresolution analysis associated with (Z^n, A) if and only if $\mathcal{V}$ satisfies properties* A1, A4, *and $V_0 = V(\phi)$ where ϕ is a function in $L^2(R^n)$ which enjoys properties* D2 *and* D3.

2.3.1. Details

Here we outline the proof of the Lemma and substantiate the remarks made after the statement of the Proposition.

Since C1, C2, and C3 are quite clear, we begin with C4. To see this item recall that if f is in $V(\phi_0)$ then there is a sequence $\{P_k\}_{k=1}^{\infty}$ of $2\pi Z^n$ periodic trigonometric polynomials such that $P_k(\xi)\hat{\phi}_0(\xi)$ converges to $\hat{f}(\xi)$ in $L^2(R^n)$. If one writes

$$\int_{R^n} |P_j(\xi)\hat{\phi}_0(\xi) - P_k(\xi)\hat{\phi}_0(\xi)|^2 d\xi = \int_Q |P_j(\xi) - P_k(\xi)|^2 \chi_\Omega(\xi) d\xi$$

where Q is the cube $[-\pi, \pi]^n$ and $\Omega = \{\xi \in R^n : \Phi(\xi) > 0\}$, then it is clear that $P_k(\xi)$ converges in $L^2(\Omega/2\pi Z^n)$ to a function $F(\xi)$ as k tends to ∞. Extending F to be 0 on $R^n \backslash \Omega$ if necessary, it is quite evident that F is in $L^2(R^n/2\pi Z^n)$ and $F(\xi)\hat{\phi}_0(\xi) = \hat{f}(\xi)$ almost everywhere.

C5: Clearly $V(\phi) \subset V(\phi_0)$. To see that $V(\phi) = V(\phi_0)$ suppose f is an element in the orthogonal complement of $V(\phi)$ in $V(\phi_0)$. Then $\hat{f}(\xi) = F(\xi)\hat{\phi}_0(\xi)$ where F is in $L^2(R^n/2\pi Z^n)$ and

$$\int_{R^n} P(\xi)\phi(\xi)\overline{F(\xi)\phi_0(\xi)} d\xi = 0 \tag{19}$$

for all $2\pi Z^n$ periodic trigonometric polynomials P. Since

$$\int_{R^n} P(\xi)\phi(\xi)\overline{F(\xi)\phi_0(\xi) d\xi} = \int_Q P(\xi)\overline{F(\xi)}\Phi(\xi) d\xi$$

where $Q = [-\pi, \pi]^n$ and $\overline{F(\xi)}\Phi(\xi)$ is in $L^1(R^n/2\pi Z^n)$, it follows from (19) that $\overline{F(\xi)}\Phi(\xi) = 0$ almost everywhere which implies that $f = 0$.

C6: Observe that $\hat{\psi}_0(\xi) = \{F(\xi)/|F(\xi)|\}\hat{\phi}_0(\xi)$ and the hypothesis on F implies that $\hat{\phi}_0(\xi) = \{|F(\xi)|/F(\xi)\}\hat{\psi}_0(\xi)$. This last relation implies the desired result.

C7: Suppose ψ is a function in $V(\phi)$. Then $\hat{\psi}(\xi) = F(\xi)\hat{\phi}_0(\xi)$ for some F in $L^2(R^n/2\pi Z^n)$ and we may write

$$\begin{aligned} \int_{R^n} \psi(x-j)\overline{\psi(x-k)}dx &= (2\pi)^{-n} \int_{R^n} e^{-i\langle j-k,\xi\rangle}|F(\xi)|^2|\hat{\phi}_0(\xi)|^2 d\xi \\ &= (2\pi)^{-n} \int_Q e^{-i\langle j-k,\xi\rangle}|F(\xi)|^2 \chi_\Omega(\xi) d\xi \end{aligned} \tag{20}$$

where $Q = [-\pi, \pi]^n$ and $\Omega = \{\xi \in R^n : \Phi(\xi) > 0\}$. The string of equalities (20) makes it clear that $\{\psi(x-j)\}_{j \in Z^n}$ is orthonormal if and only if

$$|F(\xi)|^2 \chi_\Omega(\xi) = 1$$

almost everywhere. The last identity is possible if and only if $\Omega \simeq R^n$.

C8: is a transparent consequence of the argument used to prove C7, conclusion C4, and the fact that ψ must satisfy $\sum_{j \in Z^n} |\hat{\psi}(\xi - j)|^2 = 1$.

To see that items D1, D2, and D3 in Proposition 5 are mutually exclusive consider the following examples:

$$\hat{\phi}(\xi) = \frac{2\sin(\xi/2)}{\xi} \tag{21}$$

$$\hat{\phi}(\xi) = \chi_{[-\varepsilon,\varepsilon]}(\xi) \tag{22}$$

$$\hat{\phi}(\xi) = \chi_{[0,2\pi]}(\xi). \tag{23}$$

In example (22) ε can be taken to be any positive number less than π. Note that example (21) fails to satisfy D1 but enjoys properties D2 and D3, (22) fails to satisfy D2 but enjoys properties D1, and D3, and (23) fails to satisfy D3 but enjoys properties D1 and D2.

Now suppose ϕ is in $L^2(R^n) \cap L^1(R^n)$ and satisfies $\hat{\phi}(0) \neq 0$ and $\hat{\phi}(2\pi k) = 0$ for all k in $Z^n \backslash \{0\}$. Given any positive ε we may choose c sufficiently large so that

$$S_2(\xi) = \sum_{|j|>c} |\hat{\phi}(\xi - 2\pi j)|^2$$

satisfies $S_2(\xi) < \varepsilon$ for all ξ in the cube $Q_0 = [-\pi, \pi]^n$. Write

$$S_1(\xi) = \sum_{|j|\leq c} |\hat{\phi}(\xi - 2\pi j)|^2$$

and let Q be any cube with finite diameter. By choosing the integer m sufficiently large we can be assured that $S_1(\xi) > |\hat{\phi}(0)|^2/2$ and $|\hat{\phi}(\xi)|^2/S_1(\xi) > 1-\varepsilon$ for all ξ in $B^{-m}Q$. In this case

$$\int_{B^{-m}Q} \frac{S_2(\xi)}{S_1(\xi)} \frac{|\hat{\phi}(\xi)|^2}{S_1(\xi)+S_2(\xi)} d\xi < \frac{2\varepsilon}{|\hat{\phi}(0)|^2} |B^{-m}Q| \tag{24}$$

and

$$\int_{B^{-m}Q} \frac{|\hat{\phi}(\xi)|^2}{S_1(\xi)} d\xi > (1-\varepsilon)|B^{-m}Q|. \tag{25}$$

Since

$$\frac{|\hat{\phi}(\xi)|^2}{S_1(\xi)+S_2(\xi)} = \frac{|\hat{\phi}(\xi)|^2}{S_1(\xi)} - \frac{S_2(\xi)}{S_1(\xi)} \frac{|\hat{\phi}(\xi)|^2}{S_1(\xi)+S_2(\xi)}$$

inequalities (24) and (25) allow us to conclude that for sufficiently large m

$$\frac{1}{|B^{-m}Q|} \int_{B^{-m}Q} \frac{|\hat{\phi}(\xi)|^2}{S_1(\xi)+S_2(\xi)} d\xi > 1 - (1 + 2/|\phi(0)|^2)\varepsilon. \tag{26}$$

Since ε was arbitrary, (26) implies that ϕ satisfies D3.

Finally consider the function ϕ in $L^2(R)$ defined by

$$\hat{\phi}(\xi) = \chi_{[-\pi,\pi]}(\xi) + \sum_{m-1}^{\infty} \chi_E(2^{m-1}(\xi - 2\pi m))$$

where $E = \{\xi \in R : \pi/2 < |\xi| \leq \pi\}$. Clearly $\hat{\phi}$ is continuous in a neighborhood of $2\pi Z$, $\hat{\phi}(0) = 1$, and $\hat{\phi}(2\pi k) = 0$ for all k in Z. Note that

$$\Phi(\xi)^2 = \sum_{k \in Z} |\hat{\phi}(\xi - 2\pi k)|^2 = 2$$

almost everywhere. Thus $|\hat{\phi}(\xi)|^2/|\Phi(\xi)|^2 = 1/2$ in a neighborhood of the origin and hence it should be clear that ϕ does not enjoy property D3.

2.4. Examples

2.4.1. Similarities

There are many examples of matrices which are acceptable dilations for Z^n. An interesting class consists of matrices of the form $A = \rho A_0$ where A_0 is an orthogonal matrix and ρ is a real number such that $|\det A| = |\rho|^n$ is an integer ≥ 2. In the case $n = 2$ the matrices are all of the form

$$\begin{pmatrix} m & -p \\ p & m \end{pmatrix} \quad \text{or} \quad \begin{pmatrix} m & p \\ p & -m \end{pmatrix}$$

where m and p are integers such that $m^2+p^2 \geq 2$. Such matrices are sometimes referred to as similarities.

In the next subsection we consider functions ϕ which generate multiresolution analyses associated with (Z^n, A) for any acceptable similarity.

2.4.2. A space of splines

Let $P(\xi)$ be any $2\pi Z^n$ periodic trigonometric polynomial which enjoys the property that

$$P_m(\xi) = O(|\xi|^m) \text{ as } |\xi| \to 0. \tag{27}$$

An example of such P_m is given by $P_m(\xi) = \{\exp(-i\langle u, \xi \rangle) - 1\}^m$ where u is any point in $R^n \backslash \{0\}$. Consider the distribution $\phi_{\alpha,m}$ defined by

$$\hat{\phi}_{\alpha,m}(\xi) = P_m(\xi)/|\xi|^\alpha. \tag{28}$$

If $\alpha > n/2$ then $\hat{\phi}_{\alpha,m}$ is in L^2 in the complement of every neighborhood of the origin; its behavior at the origin depends on m. Note that

$$\hat{\phi}_{\alpha,m}(\xi) = \begin{cases} O(|\xi|^{m-\alpha}), & \text{as } |\xi| \to 0 \\ O(|\xi|^{-\alpha}), & \text{as } |\xi| \to \infty. \end{cases}$$

Hence if $m > \alpha - n/2$ then $\hat{\phi}_{\alpha,m}$ is in $L^2(R^n)$. Since

$$\Phi_\alpha(\xi) = \{\sum_{j \in Z^n} |\xi - 2\pi j|^{-2\alpha}\}^{1/2}$$

is positive almost everywhere it follows that

$$\Phi_{\alpha,m}(\xi) = |P_m(\xi)|\Phi_\alpha(\xi) = \{\sum_{j \in Z^n} |\hat{\phi}_{\alpha,m}(\xi - 2\pi j)|^2\}^{1/2}$$

can vanish only on $\{\xi : |P_m(\xi)| = 0\}$, which is a set of measure zero. Thus, if

$$\alpha > n/2 \text{ and } m > \alpha - n/2 \tag{29}$$

we see that the following holds:

- There is a $2\pi Z^n$ periodic measurable function $S_{\alpha,m}(\xi)$ such that

  $$\hat{\phi}_{\alpha,m}(\xi) = S_{\alpha,m}(B^{-1}\xi)\hat{\phi}_{\alpha,m}(B^{-1}\xi,$$

  whenever B is a similarity. Indeed, $S_{\alpha,m}(\xi) = \rho^{-\alpha}P_m(B\xi)/P_m(\xi)$, where $B = \rho B_0$ and B_0 is an orthogonal matrix.
- $\Phi_{\alpha,m}(\xi) > 0$, almost everywhere.
- $|\hat{\phi}_{\alpha,m}(\xi)/\Phi_{\alpha,m}(\xi)|^2$ is a continuous function which is equal to 1 at the origin.

By virtue of Proposition 5 we may conclude that if (29) holds then $\phi_{\alpha,m}$ generates a multiresolution analysis associated with (Z^n, A) whenever A is a similarity. Indeed, item C8 of Lemma 4 implies that $V(\phi_{\alpha,m}) = V(\phi_\alpha)$ where

$$\hat{\phi}_\alpha(\xi) = |\xi|^{-\alpha}/\Phi_\alpha(\xi). \tag{30}$$

Next we consider the smoothness properties of ϕ_α in order to determine whether it generates a multiresolution analysis consisting of splines in the sense of Meyer.

This is easy. If $\alpha \leq n$, the formula for $\hat{\phi}_{\alpha,m}$ implies that in the neighborhood of certain points j in Z^n we have

$$\phi_{\alpha,m}(x) = \begin{cases} O(\log|x-j|) & \text{if } \alpha = n \\ O(|x-j|^{\alpha-n}) & \text{if } \alpha < n. \end{cases}$$

Since ϕ_α is the limit of linear combinations of Z^n translates of $\phi_{\alpha,m}$ it is clear that it exhibits the same behavior. On the other hand, since $\hat{\phi}_\alpha = O(|\xi|^{-\alpha})$ as $|\xi| \to \infty$ it follows that if $\alpha > n$ and $\beta < \alpha - n$ then ϕ_α is in $C^\beta(R^n)$. Indeed, in this case we may write

$$\Lambda_\alpha(\xi) = \sum_{j \in Z^n} |\xi - 2\pi j)|^{-\alpha}$$

and observe that the function λ_α defined by

$$\hat{\lambda}_\alpha(\xi) = |\xi|^{-\alpha}/\Lambda_\alpha(\xi) \tag{31}$$

enjoys the following properties:

- $\{\lambda_\alpha(x-j)\}_{j \in Z^n}$ is a Riesz basis for $V(\phi_\alpha)$
- $$\lambda_\alpha(j) = \begin{cases} 1 & \text{if } j = 0 \\ 0 & \text{if } j \in Z^n \backslash \{0\}. \end{cases}$$
- Every element f in $V(\phi_\alpha)$ can be uniquely expressed as

$$f(x) = \sum_{j \in Z^n} f(j)\lambda_\alpha(x-j).$$

The properties of λ_α imply that the resulting multiresolution analysis generated by ϕ_α consists of splines in the sense of Meyer.

Finally we consider the behavior of ϕ_α at ∞. First observe that $\hat{\phi}$ is in $C^\infty(R^n \backslash 2\pi Z^n)$. In a neighborhood of $2\pi Z^n$ its behavior can be described as follows: if ξ is in $Q + 2\pi j$ where Q is the cube $[-\pi, \pi]^n$ and $2\pi j$ is any element in $2\pi Z^n$ then

$$|\hat{\phi}_\alpha(\xi) - \hat{\phi}_\alpha(2\pi j)| \leq c\frac{|\xi - 2\pi j|^\alpha}{1 + |\xi|^\alpha}$$

where c is a constant independent of j and ξ. If $\alpha > n$ this allows us to conclude that

$$|\phi_\alpha(x)| \leq c_\alpha(1 + |x|)^{-n-\alpha} \tag{32}$$

where c_α is a constant independent of x. In the case $\alpha = 2k$ where k is an integer greater than $n/2$ we can say more. Namely, $\hat{\phi}_\alpha$ has a holomorphic extension to a tube $\{\xi + i\eta : \xi \in R^n \ |\eta| < \varepsilon\}$ for some positive ε. Hence we may conclude that whenever k is an integer greater than $n/2$

$$|\phi_{2k}(x)| \leq ce^{-\varepsilon|x|} \tag{33}$$

where the positive constants c and ε depend only on k.

We summarize these observations as follows:

Proposition 6. *If A is any similarity which is an acceptable dilation for Z^n then the function ϕ_α, $\alpha > n/2$, defined by (30) is a scaling function for a multiresolution analysis $\mathcal{V}$ associated to (Z^n, A). Furthermore if $\alpha > n$ then ϕ_α and $\mathcal{V}$ enjoy the following properties:*

- *$\mathcal{V}$ is a multiresolution analysis composed of splines in the sense of Meyer.*
- *ϕ_α is in $C^\beta(R^n)$ for any positive β which satisfies $\beta < \alpha - n$.*
- *Every f in $V_0 = V(\phi_\alpha)$ can be uniquely expressed in terms of its values on Z^n via*

$$f(x) = \sum_{j \in Z^n} f(j)\lambda_\alpha(x - j)$$

where λ_α is defined by formula (31). Moreover $\{\lambda_\alpha(x - j)\}_{j \in Z^n}$ is a Riesz basis for V_0.
- *Both $\phi_\alpha(x)$ and $\lambda_\alpha(x)$ are $O(|x|^{-n-\alpha})$ as $|x| \to \infty$.*
- *If $\alpha = 2k$ where k is an integer then both $\phi_\alpha(x)$ and $\lambda_\alpha(x)$ are $O(e^{-\varepsilon|x|})$ as $|x| \to \infty$ where ε is a positive constant which depends only on k. Moreover, in this case every element f in V_0 satisfies a differential equation of the form*

$$\Delta^k f(x) = \sum_{j \in Z^n} a_j \delta(x - j)$$

where Δ^k is the k-th iterate of the Laplacian, $\{a_j\}_{j \in Z^n}$ is a sequence in $l^2(Z^n)$, and $\delta(x)$ is the unit Dirac distribution at the origin.

In the case $n = 1$ and $\alpha = 2k$, $k = 1, 2, \ldots$, these examples reduce to the well known classical univariate cardinal splines which are piecewise polynomials of degree $2k-1$ (see [43] and [46]). We bring attention to the fact that the case $n = 1$ and $\alpha = 2k - 1$, $k = 1, 2, \ldots$, does not result in piecewise polynomial functions. This is clarified by the last example considered in Subsection 2.1.4 where an explicit formula for $\phi_{1,1}$ is given.

Figures 1, 2 and 3 allow the reader to compare the behavior of λ_6, ϕ_6, and ϕ_7 in the case $n = 1$. In the case $n = 2$ the function λ_8 is illustrated in Figure 4.

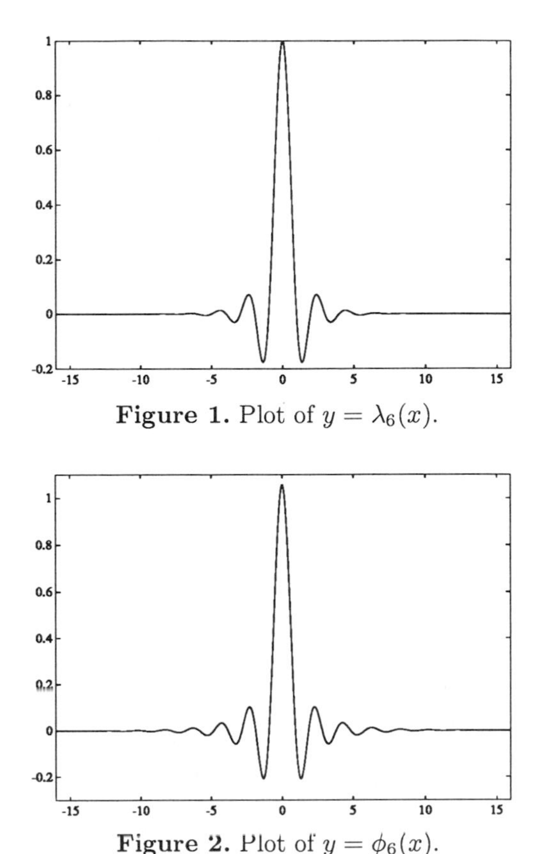

Figure 1. Plot of $y = \lambda_6(x)$.

Figure 2. Plot of $y = \phi_6(x)$.

§3. Translation invariant multiresolution analyses

3.1. Introduction

A multiresolution analysis $\mathcal{V} = \{V_j\}_{j\in Z}$ is said to be translation invariant if each subspace V_j is invariant under translation by arbitrary elements of R^n. Most examples fail to enjoy this property. The point of this section is to provide a characterization of those that do.

First, to be precise, we say that a multiresolution analysis is *translation invariant* if all the translates of $f(x)$, $\{f(x-y) \,:\, y \in R^n\}$, are in V_0 whenever $f(x)$ is in V_0.

The canonical example of a translation invariant dyadic multiresolution analysis of $L^2(R)$ is the case when V_0 is the collection of those functions in $L^2(R)$ whose Fourier transforms are supported in the interval $[-\pi, \pi]$. A natural choice of ϕ in this case is given by

$$\phi(x) = \frac{\sin \pi x}{\pi x} \, .$$

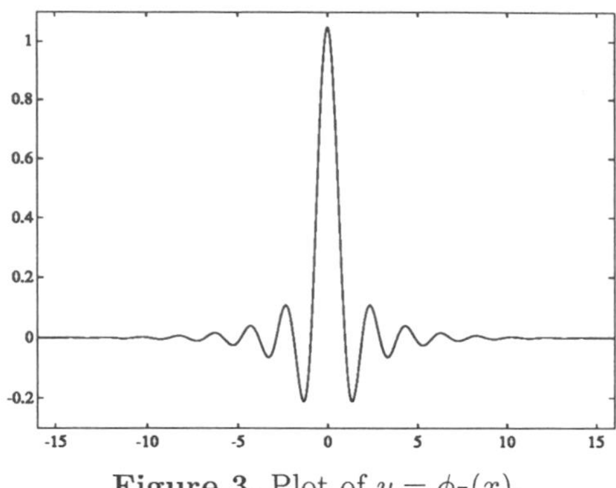

Figure 3. Plot of $y = \phi_7(x)$.

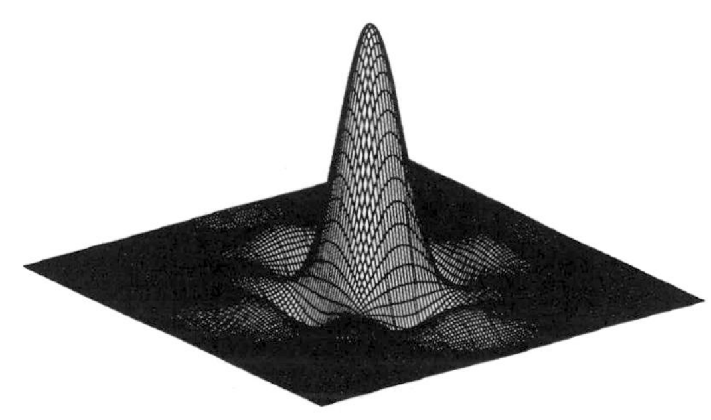

Figure 4. Plot of the surface $z = \lambda_8(x, y)$ where $-4 \leq x \leq 4$ and $-4 \leq y \leq 4$.

It turns out that all such multiresolution analyses have scaling functions whose Fourier transforms are characteristic functions of certain tiles.

Proposition 7. *Suppose $\mathcal{V}$ is a translation invariant multiresolution analysis of $L^2(R^n)$ associated with (Z^n, A) and $B = A^*$ is the adjoint of A. Then $\widehat{V}_0 = L^2(\Omega)$ where Ω is a closed subset of R^n which has the following properties:*

T1. $\Omega \subset B\Omega$.

T2. $\Omega \bigcap \{\Omega + 2\pi k\} \simeq \emptyset$ *for any element k in $Z^n \setminus \{0\}$.*

T3.

$$\bigcup_{k \in Z^n} \{\Omega + 2\pi k\} \simeq R^n .$$

T4. *For all cubes Q of finite diameter in R^n*

$$\lim_{j \to \infty} \frac{1}{|B^{-j}Q|} \int_{B^{-j}Q} \chi_\Omega(\xi) d\xi = 1 .$$

Conversely, if V_j, $j \in Z$ is defined by $\widehat{V}_j = L^2(B^j\Omega)$ where Ω is a closed subset of R^n which satisfies the properties above then the sequence of subspaces $\{V_j\}_{j\in Z}$ is a translation invariant multiresolution analysis of $L^2(R^n)$ associated with (Z^n, A).

3.1.1

Consider

$$\begin{aligned}\Omega_1 &= [-1,1]\bigcup[2\pi+1, 4\pi-1]\\ \Omega_2 &= [-5,5]\\ \Omega_3 &= [-1,1]\\ \Omega_4 &= [0,2\pi]\,.\end{aligned}$$

It is not difficult to verify that each Ω_j listed above is a closed set which fails to satisfy condition Tj but satisfies the remaining conditions in the Theorem in the univariate dyadic case. These examples show that conditions T1,..., T4 are not redundant.

3.1.2

Note that condition T1 implies that 0 is contained in Ω. In addition to this it is clear that if Ω contains a neighborhood of the origin then it satisfies condition T4. In view of this it seems reasonable to suspect that subsets Ω which satisfy the conditions of the Proposition must contain an open neighborhood of the origin. That this is not the case can be seen by considering the following example of Ω in the univariate dyadic case:

$$\left\{\bigcup_{k=1}^{\infty}[-(2-2^{-k})2^{-k}\pi, -2^{-k}\pi]\right\}\bigcup[0,\pi]\left\{\bigcup_{k=1}^{\infty}[(2-(2-2^{-k})2^{-k})\pi, (2-2^{-k})\pi]\right\}.$$

3.1.3.

Property T4 follows from and implies that $\bigcup V_j$ is dense in $L^2(R^n)$. It can be replaced by any one of the following equivalent properties:

- $\bigcup_{j\in Z} L^2(B^j\Omega)$ is dense in $L^2(R^n)$.
- $\bigcup_{j\in Z}(B^j\Omega) \simeq R^n$.

3.2. Details

For the sake of clarity in what follows we will restrict our attention to the univariate dyadic case. The statements and arguments in the general case are completely analogous to this basic case.

The next lemma is an essential step in the proof of the Proposition.

Lemma 8. *Suppose $\mathcal{V}$ is a translation invariant dyadic multiresolution analysis of $L^2(R)$ and ϕ is a scaling function for $\mathcal{V}$. Then*

$$|\hat{\phi}| = \chi_\Omega$$

where χ_Ω is the characteristic function of a closed set Ω which has properties T1-T4 *in the statement of Proposition 7.*

Let Ω be the support of ϕ. To prove the lemma we will first show that Ω satisfies property T2.

Recall that A6 implies that for all f in V_0 we may write

$$\hat{f}(\xi) = F(\xi)\hat{\phi}(\xi)$$

where F is 2π periodic and square integrable over the interval $[-\pi, \pi]$. In particular, since V_0 is translation invariant, $\phi(x-y)$ as a function of x is in V_0 so setting $\alpha = -y$ we may write

$$e^{i\alpha\xi}\hat{\phi}(\xi) = F(\xi)\hat{\phi}(\xi)$$

for some such F. Hence

$$e^{i\alpha(\xi-2\pi m)}\hat{\phi}(\xi - 2\pi m) = F(\xi - 2\pi m)\hat{\phi}(\xi - 2\pi m) = F(\xi)\hat{\phi}(\xi - 2\pi m)$$

which implies that

$$e^{i\alpha(\xi-2\pi m)} = F(\xi)$$

on $\Omega + 2\pi m$. For two different values of m the last equality implies that

$$e^{i\alpha(\xi-2\pi j)} = e^{i\alpha(\xi-2\pi k)}$$

on $\{\Omega + 2\pi j\} \bigcap \{\Omega + 2\pi k\}$. Re-expressing the last relation as

$$e^{i\alpha(\xi-2\pi k)}(e^{i2\pi\alpha(k-j)} - 1) = 0$$

it is clear that either α is an integer, j is equal to k, or $\{\Omega + 2\pi j\} \bigcap \{\Omega + 2\pi k\}$ is a set of measure zero. Since α may be any real number we conclude that $\{\Omega + 2\pi j\} \bigcap \{\Omega + 2\pi k\}$ has measure zero whenever $j \neq k$.

Now, since $\{\phi(x-k)\}_{k\in Z}$ is orthonormal, it must satisfy

$$\sum_{j\in Z} |\hat{\phi}(\xi - 2\pi j)|^2 = 1$$

on R and since $\{\Omega + 2\pi j\} \bigcap \{\Omega + 2\pi k\}$ has measure zero whenever $j \neq k$ we may conclude that

$$|\hat{\phi}(\xi)| = \chi_\Omega(\xi)$$

and

$$\bigcup_{k\in Z} \{\Omega + 2\pi k\} \simeq R\,.$$

To see T1 observe that the scaling relationship and the facts demonstrated above imply that

$$\chi_\Omega(\xi) = S(\xi/2)\chi_\Omega(\xi/2)$$

where S is 2π periodic and square integrable over $[-\pi, \pi]$. Since $\chi_{2\Omega}(\xi) = \chi_{\Omega}(\xi/2)$, the last equality involving S implies that χ_{Ω} vanishes whenever $\chi_{2\Omega}$ does so $\Omega \subset 2\Omega$.

Finally observe that T4 is an easy consequence of Corollary 2 to Proposition 1.

This completes the proof of Lemma 8.

Now, suppose ϕ and Ω are as in the Lemma and its proof. Then, since $\chi_{\Omega}(\xi) = H(\xi)\hat{\phi}(\xi)$ where $H(\xi)$ is the 2π periodic function

$$\sum_{k \in Z} \frac{|\hat{\phi}(\xi - 2\pi k)|}{\hat{\phi}(\xi - 2\pi k)},$$

it follows by virtue of item D8 of Lemma 4 that $\hat{V}_0 = L^2(\Omega)$. Thus the Lemma 8 together with the last observation imply the first assertion of Proposition 7.

The converse is an easy consequence of the definition of $\mathcal{V}$ and Corollary 2 to Proposition 1.

§4. Multiresolution analyses generated by tiles

4.1. Characteristic functions and scaling functions

Recall the simple example of a dyadic multiresolution analysis of $L^2(R)$ whose scaling function is $\chi_{[0,1]}(x)$ given in Subsection 2.1.4. In this section we concern ourselves with such simple analogues in the case of general acceptable dilations A. In particular, we describe those multiresolution analyses associated to (Z^n, A) whose scaling functions are characteristic functions of measurable sets Q.

Proposition 9. *Suppose $\phi = c\chi_Q$ is a scaling function for a multiresolution analysis associated with (Z^n, A). Here χ_Q is the characteristic function of a measurable set Q and $c = |Q|^{-1/2}$. Then Q satisfies the following properties:*

E1.

$$Q \cap (Q + k) \simeq \emptyset \quad \textit{for } k \in Z^n \backslash \{0\}.$$

E2.

$$\bigcup_{k \in Z^n} (Q + k) \simeq R^n \,.$$

E3. *There is a collection of q lattice points $k_1, \ldots, k_q$ which are representatives of distinct cosets in Z^n/AZ^n such that*

$$AQ \simeq \bigcup_{i=1}^{q} (k_i + Q) \,.$$

E4. *There is a compact set K such that $Q \simeq K$.*

Conversely, the characteristic function of a bounded measurable set Q

which satisfies properties E1, E2, *and* E3 *is the scaling function of a multiresolution analysis associated with* (Z^n, A).

Next we describe the nature of the sets Q which arise in Proposition 9.

Properties E1 and E2 mean that translates of Q by the integer lattice form a tiling of R^n. Sets Q which enjoy property E3 are sometimes said to be self-similar in the affine sense.

Note that properties E1 and E2 imply that $|Q| = 1$. Thus $c = 1$ and the function $\phi = \chi_Q$ satisfies the two scale difference equation

$$\phi(x) = \sum_{i=1}^{q} \phi(Ax - k_i) \tag{34}$$

where the k_i's are those lattice points whose existence is implied by property E3.

In view of the fact that the order of Z^n/AZ^n is q, the k_i's in E3 are a full set of coset representatives, namely,

$$\bigcup_{i=1}^{q} (k_i + AZ^n) = Z^n \,. \tag{35}$$

If $\mathcal{K} = \{k_1, \ldots, k_q\}$ is a full collection of representatives of distinct cosets of Z^n/AZ^n, we say that such a collection is a *full collection of digits* and refer to the elements of such a set $\mathcal{K}$ as *digits*.

Since the solution of (34) is a fixed point of the mapping

$$\psi(x) \to \sum_{k \in \mathcal{K}} \psi(Ax - k)$$

it is quite natural to apply fixed point iteration to solve for ϕ. Namely, start with an initial function ϕ_0 and define the sequence ϕ_1, ϕ_2, ϕ_3, ... via

$$\phi_{N+1}(x) = \sum_{k \in \mathcal{K}} \phi_N(Ax - k) \tag{36}$$

and hope that the sequence converges to ϕ. Since the desired solution is the characteristic function of a set Q whose Z^n translates tile, it is reasonable to begin the iteration with $\phi_0 = \chi_{Q_0}$ where Q_0 has the same properties.

Suppose Q_0 satisfies E1 and E2 and $\mathcal{K} = \{k_1, \ldots, k_q\}$ is a collection of representatives of distinct cosets of Z^n/AZ^n. If $\phi_0 = \chi_{Q_0}$ and ϕ_1 is related to ϕ_0 via (36) then $\phi_1 = \chi_{Q_1}$ where

$$Q_1 = \bigcup_{k \in \mathcal{K}} A^{-1}(Q_0 + k)$$

also satisfies E1 and E2; that Q_1 satisfies E2 follows from the fact that $\mathcal{K}$ is a full collection of distinct representatives of Z^n/AZ^n and that it satisfies E1 follows from $|Q_1| = 1$. By induction we may conclude that $\phi_{N+1}, N = 0, 1, 2, \ldots$, is the characteristic function of the set Q_{N+1} defined by

$$Q_{N+1} = \bigcup_{k \in \mathcal{K}} A^{-1}(Q_N + k). \tag{37}$$

Observe that (37) looks like an iterated function system in the sense of Barnsley [4]. Convergence of schemes like (37) is usually considered in terms of the following metric defined on the space of subsets of R^n:

$$\rho(P, Q) = \max\{r(P, Q), r(Q, P)\} \tag{38}$$

where

$$r(P, Q) = \sup_{x \in P} \inf_{y \in Q} |x - y| \,.$$

It is well known that when equipped with the metric ρ the class of compact subsets of R^n is a complete metric space.

Indeed, starting with a compact set Q_0 the sequence defined by (37) converges to the set defined in the following proposition.

Proposition 10. *Suppose $\mathcal{K} = \{k_1, \ldots, k_q\}$ is a full collection of digits and suppose the compact set Q is defined by*

$$Q = \{x \in R^n : x = \sum_{j=1}^{\infty} A^{-j} \epsilon_j, \ \epsilon_j \in \mathcal{K}\} \,. \tag{39}$$

Then the set Q has the following properties:

F1. *If Q_0 is any compact set then the sequence of sets $Q_1, Q_2, \ldots$, defined by (37) converges in the metric ρ to Q.*

F2. $AQ \simeq \bigcup_{i=1}^{q} (k_i + Q)$.

F3. $\bigcup_{k \in Z^n} (Q + k) \simeq R^n$.

F4. *$(Q + k_i) \cap (Q + k_j) \simeq \emptyset$ whenever both k_i and k_j are in $\mathcal{K}$ and $k_i \neq k_j$.*

F5. *$\phi = |Q|^{-1/2} \chi_Q$ is the unique solution, in the $L^1(R^n)$ sense, of*

$$\phi(x) = \sum_{k \in \mathcal{K}} \phi(Ax - k)$$

which has $L^2(R^n)$ norm 1.

F6. *If $\{Q_N\}$, $N = 0, 1, 2, \ldots$, is the sequence of sets generated via (37) with $Q_0 = [-1/2, 1/2]^n$, then the corresponding sequence of characteristic functions $\{\chi_{Q_N}\}$ converges weakly to $|Q|^{-1}\chi_Q$. In other words,*

$$\lim_{N \to \infty} \int_{R^n} \psi(x) \chi_{Q_N}(x) d(x) = \frac{1}{|Q|} \int_{R^n} \psi(x) \chi_Q(x) dx$$

for all functions ψ which are continuous and bounded on R^n.

F7. *$|Q|$ is equal to a positive integer.*

We remark that property F6 holds if $Q_0 = [-1/2, 1/2]^n$ is replaced by any compact set Q_0 which satisfies E1 and E2.

It should now be clear how to construct scaling functions ϕ for multiresolution analyses associated with (Z^n, A) which have characteristic functions as scaling functions.

- Start with a compact set Q_0 and a full collection $\mathcal{K} = \{k_1, \ldots, k_q\}$ of representatives of distinct cosets of Z^n/AZ^n.
- Find Q as the limit of the iteration (37).
- If $|Q| = 1$ the algorithm is successful and $\phi = \chi_Q$ is the scaling function of a multiresolution analysis associated with (Z^n, A). Otherwise the algorithm fails.

The reason one must check the condition $|Q| = 1$ is that the requirement that $\mathcal{K}$ be a full collection of representatives of distinct cosets of Z^n is a necessary but not sufficient condition on this set of indices. Indeed, examples show that Q need not satisfy this condition and the algorithm may fail.

The following proposition gives various equivalent conditions which guarantee that the above algorithm be successful.

Proposition 11. *Suppose $\mathcal{K} = \{k_1, \ldots, k_q\}$ is a collection of representatives of distinct cosets of Z^n/AZ^n and the compact set Q is defined by (39). Then the following statements are equivalent:*

G1. *χ_Q is a scaling function for a multiresolution analysis associated with (Z^n, A).*

G2. *$|Q| = 1$.*

G3. *$Q \bigcap (k + Q) \simeq \emptyset$ for all k in Z^n which are different from 0.*

G4. *If $\{Q_N\}$, $N = 0, 1, 2, \ldots$, is the sequence of sets generated via (37) with $Q_0 = [0, 1]^n$, then the corresponding sequence of characteristic functions $\{\chi_{Q_N}\}$ converges to χ_Q in measure.*

G5. *(Cohen's condition.) Let $S(\xi)$ be defined by*

$$S(\xi) = \frac{1}{q} \sum_{k \in \mathcal{K}} e^{-i\langle k, \xi \rangle} .$$

There exists a compact set K which contains a neighborhood of the origin and which satisfies

- *$\bigcup_{k \in Z^n} (2\pi k + K) = R^n$,*
- *$K \cap (2\pi k + K) \simeq \emptyset$, whenever $k \neq 0$, such that if $B = A^*$ then*

$$|S(B^{-j}\xi)| > 0$$

holds for all $\xi \in K$ and $j \geq 1$.

G6. *$Q \bigcap (k + Q) \simeq \emptyset$ whenever k is in $\mathcal{B}$ where*

$$\mathcal{B} = \{k \in Z^n : |k| < 2b\},$$

$|k|$ *denotes the Euclidean norm of* k, *and* $b = \sup\{|x| : x \in Q\}$.

G7. $|Q| < 2$.

We remark that the last item may be the most useful on the list. One only has to obtain a relatively rough estimate of the measure of Q.

To obtain other useful criteria for determining whether the set Q results in the desired tiling we proceed as follows:

If $\mathcal{K} = \{k_1, \ldots, k_q\}$ is a collection of digits as in the hypothesis of Proposition 11, then for any nonnegative integer N let

$$\mathcal{AK}_N = \sum_{j=0}^{N} A^j \mathcal{K}. \tag{40}$$

Thus $\mathcal{AK}_N$ is a finite subset of Z^n consisting of q^{N+1} sums k of the form

$$k = \sum_{j=0}^{N} A^j \varepsilon_j \tag{41}$$

where the ε_j's are in $\mathcal{K}$. Let

$$\mathcal{AK}_\infty = \bigcup_{N=0}^{\infty} \mathcal{AK}_N \tag{42}$$

and let

$$\mathcal{DAK}_\infty = \mathcal{AK}_\infty - \mathcal{AK}_\infty. \tag{43}$$

In other words every element in $\mathcal{AK}_\infty$ is a finite sum of the form (41) for some N and every element k in $\mathcal{DAK}_\infty$ is of the form

$$k = k_1 - k_2 \tag{44}$$

where k_1 and k_2 are in $\mathcal{AK}_\infty$. Finally let $\mathcal{B}$ be the set described in item G6 above. We are now ready to state the promised conditions.

Proposition 12. *If* $\mathcal{DAK}_\infty = Z^n$ *then* Q *satisfies* G3. *If* $\mathcal{B} \subset \mathcal{DAK}_\infty$ *then* Q *also satisfies* G3.

4.2. Details

Since complete details to most of the above observations can be found in [20] we provide outlines only to those arguments which cannot be found there.

4.2.1. Proof of F7 and G7

Let $Q_N, N = 0, 1, 2, \ldots$, be the sequence of sets in F6. Then the characteristic functions of Q_N satisfy $\sum_{k \in Z^n} \chi_{Q_N}(x - k) = 1$ which, by virtue of F6, implies that

$$\frac{1}{|Q|} \sum_{k \in Z^n} \chi_Q(x - k) = 1 \quad \text{a.e.} \tag{45}$$

Now, the sum in (45) is integer valued for almost all x so, since $1 \leq |Q| < \infty$ by virtue of F3, we may conclude that $|Q|$ is equal to a positive integer.

The equivalence of G7 and G2 is an immediate consequence of F7.

4.2.2. Proof of G6 and Proposition 12

If $B_r = \{x \in R^n : |x| \le r\}$ then a routine estimate shows that $Q \subset B_r$ whenever $r \ge b$. Since

$$\{k + B_r\} \bigcap B_r \simeq \emptyset$$

whenever $|k| \ge 2r$, we may conclude that in order to show that Q satisfies G3 it suffices to check that Q satisfies $Q \bigcap (k + Q) \simeq \emptyset$ for all k in $\mathcal{B}$. Hence the equivalence of G6 and G3 should be clear.

To see Proposition 12 recall that F4 says that

$$\{k_1 + Q\} \bigcap \{k_2 + Q\} \simeq \emptyset \tag{46}$$

whenever k_1 and k_2 are in $\mathcal{AK}_0$. Hence

$$A(k_1 + Q) \bigcap A(k_2 + Q) \simeq \emptyset \tag{47}$$

for such k_i. In view of F3 we may write

$$A(k_i + Q) = Ak_i + \bigcup_{k \in \mathcal{K}} \{k + Q\}.$$

Since the union in the right hand side of the above identity is taken over pairwise disjoint sets, we may conclude that this identity together with (47) implies (46) whenever k_1 and k_2 are in $\mathcal{AK}_1$. By induction it is clear that (46) is valid whenever k_1 and k_2 are in $\mathcal{AK}_N$ for any nonnegative integer N. In other words, since $k \in \mathcal{DAK}_\infty$ can be expressed as $k = k_1 - k_2$ with k_1 and k_2 in $\mathcal{AK}_N$ for some N we may conclude that the first statement of this proposition is valid.

The second statement is an immediate consequence of this and the equivalence of G6 and G3.

4.2.3. Examples

Given an acceptable dilation matrix A the particular choice of digits influences both the size and the shape of the set generated via fixed point iteration (37). For example, suppose $\mathcal{K}_1$ and $\mathcal{K}_2$ are two collections of digits related to each other via $\mathcal{K}_2 = m\mathcal{K}_1$ where m is an integer, in other words

$$\mathcal{K}_2 = \{k : k = m\nu,\ \nu \in \mathcal{K}_1\}.$$

It is not difficult to see that if Q_1 and Q_2 are the sets corresponding to $\mathcal{K}_1$ and $\mathcal{K}_2$ respectively via (37) then $Q_2 = mQ_1$.

The choice of digits can affect the shape of the corresponding set Q in some rather interesting ways. We illustrate this in R^2 by considering the similarity

$$A = \begin{pmatrix} 2 & -2 \\ 2 & 2 \end{pmatrix}$$

and the collections of digits $\mathcal{K}_1$, $\mathcal{K}_2$, and $\mathcal{K}_3$ whose elements are the columns of the following matrices:

$$\begin{pmatrix} 0 & 0 & 0 & 0 & 1 & 1 & -1 & -1 \\ 0 & 1 & 2 & 3 & 1 & 2 & 1 & 2 \end{pmatrix}$$

$$\begin{pmatrix} 0 & 0 & 0 & 0 & 1 & 1 & 1 & 1 \\ 0 & 1 & 2 & 3 & 0 & 1 & 2 & 3 \end{pmatrix}$$

$$\begin{pmatrix} 0 & 0 & 0 & 0 & 1 & 1 & -1 & -1 \\ 0 & 1 & 6 & 3 & 1 & 2 & 1 & 3 \end{pmatrix}.$$

The corresponding Q's are plotted in Figures 5, 6, and 7. We leave it as an exercise for the reader to verify that the Z^2 translates of these Q's result in a tiling of the plane.

§5. Miscellaneous Remarks

5.1. Multiresolution analyses

The basic definition of multiresolution analysis as used here was formulated by Meyer and his collaborators (see [23], [41], and [43], and the pertinent references cited there).

The conclusions of Proposition 1 are essentially folklore but the formulation found here seems to be new. Various hints and other perspectives may be found in the literature, for example see [9], [15], [39], [41], and [53].

Proposition 5, although quite elementary, appears to be new.

The class of tempered distributions which are essentially linear combinations of $\{\phi_\alpha(x-j)\}_{j\in Z^n}$ where ϕ_α is the scaling function defined by (30) and $\alpha = 2k$ with k an integer greater than $n/2$ were considered by the author as a natural multivariate extension of the theory of univariate cardinal splines [46]. The connection with multiresolution analyses was made in [37] where it is also shown that, by allowing k to tend to infinity, these so-called polyharmonic splines can be used in a summability method to recover multivariate band limited distributions from discrete samples. However, Lemarie [28] was apparently the first to observe the connection between these splines and wavelets (see also [29] and [30]).

Another source of examples is provided by box splines. These are functions ϕ whose Fourier transform is given by

$$\hat{\phi}(\xi) = \prod_{k=1}^{N} \{i(e^{-i\langle u_k,\xi\rangle} - 1)/\langle u_k, \xi\rangle\}$$

where $u_1, \ldots, u_N$ are elements of Z^n. Clearly such ϕ satisfy condition D1 of Proposition 5; additional restrictions on $u_1, \ldots, u_N$ are used to guarantee that ϕ be in $L^2(R^n)$ and satisfy conditions D2 and D3. These splines provide nice

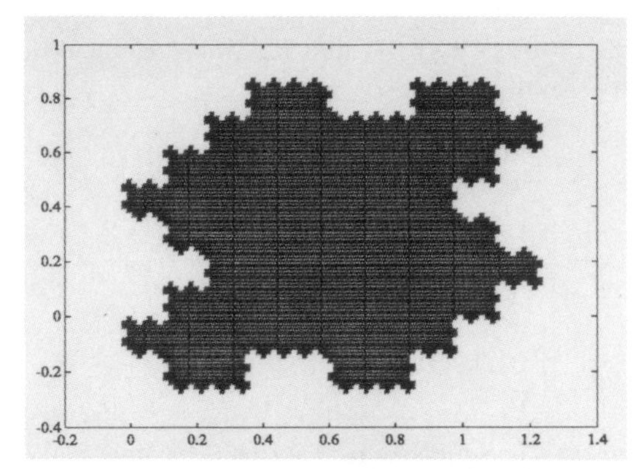

Figure 5. Tile corresponding to the digits $\mathcal{K}_1$.

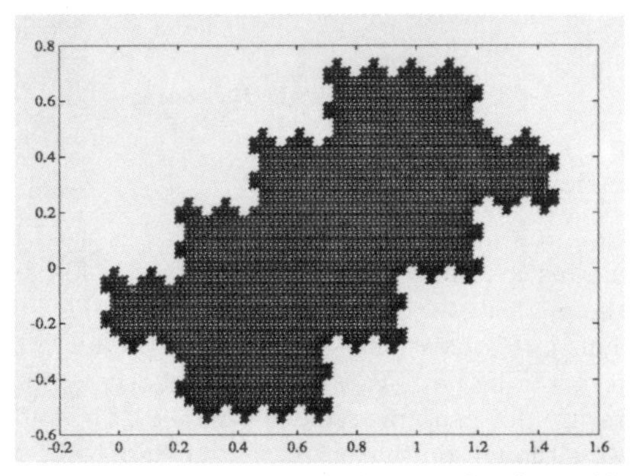

Figure 6. Tile corresponding to the digits $\mathcal{K}_2$.

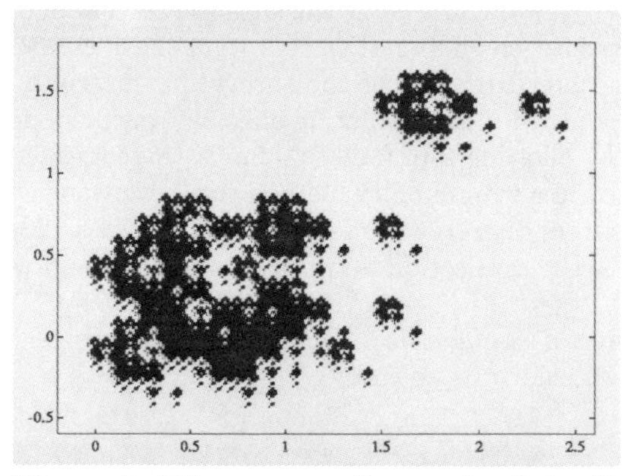

Figure 7. Tile corresponding to the digits $\mathcal{K}_3$.

examples of how a change of lattice can result in practical improvements. For instance consider the bivariate case where ϕ is determined by $u_1 = (1,0)$, $u_2 =$

$(0,1)$, and $u_3 = (-1,-1)$. This ϕ is the continuous piecewise linear function which satisfies $\phi(0) = 1$ and is supported in the polygon plotted in Figure 8. Its counterpart ϕ_Γ in the hexagonal lattice $\Gamma = \{(-k/2, l\sqrt{3}/2) : k, l \in Z\}$ is determined by the directions $u_1 = (1,0)$, $u_2 = (-1/2, \sqrt{3}/2)$, and $u_3 = (-1/2, -\sqrt{3}/2)$; it is the continuous piecewise linear function which satisfies $\phi_\Gamma(0) = 1$ and is supported in the hexagon plotted in Figure 9.

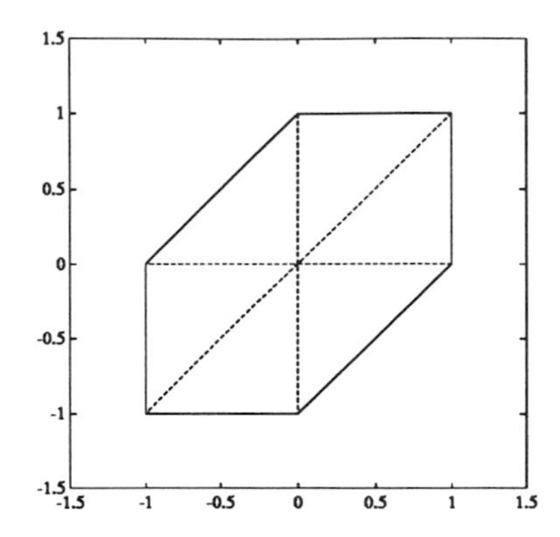

Figure 8. Support of the boxspline ϕ.

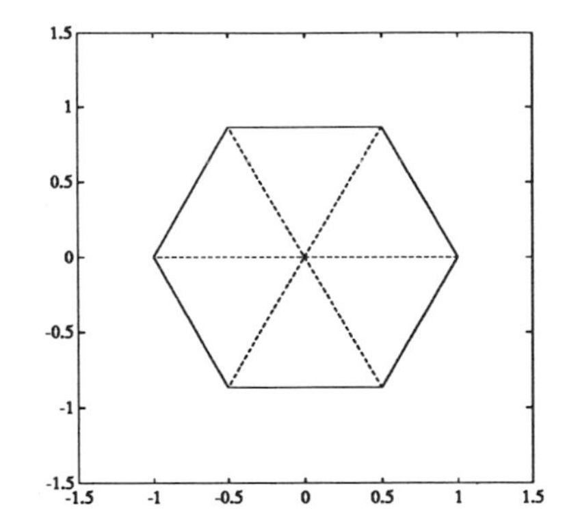

Figure 9. Support of the boxspline ϕ_Γ.

The amount of literature devoted to box splines is huge. For a recent survey see [44]; for some recent papers relating to multiresolution analyses see [7], [8], [25], [31] and [45]. Noting that the k fold convolution of a box spline with itself is also a box spline, we mention that, by allowing k to tend to infinity as in the polyharmonic case, these splines can also be used in a summability method to recover certain multivariate band limited distributions from discrete samples (see [6] and [44]).

An important class of scaling functions arises as solutions of the two scale difference equation

$$\phi(x) = \sum_{k \in Z^n} s_k \phi(Ax - k) \tag{48}$$

with appropriately chosen scaling sequences $\{s_k\}_{k \in Z^n}$. The univariate dyadic multiresolution analyses generated by the compactly supported scaling functions discovered by Daubechies [15] are a celebrated example. In principle this approach allows one to construct scaling functions with desired properties, for example see [10], [11], and [17]. On the other hand, although necessary and sufficient conditions to ensure that solutions of (48) result in scaling functions have been obtained by Cohen [9], in general this paradigm is not easy to apply, see [12], [20], [26], [27], and [47]. For recent work concerning solutions of univariate dyadic two scale difference equations see [13], [16], and [53]. There are however nontrivial examples of A's for which all the constructions from the univariate dyadic case carry over without any apparent difficulty; for instance if $n = 2$ consider

$$A = \begin{pmatrix} 0 & 1 \\ 2 & 0 \end{pmatrix}$$

or for general n

$$A = \begin{pmatrix} 0 & 1 & 0 & \dots & 0 \\ 0 & 0 & 1 & \dots & 0 \\ \cdot & \cdot & \cdot & \dots & \cdot \\ 0 & 0 & 0 & \dots & 1 \\ 2 & 0 & 0 & \dots & 0 \end{pmatrix}.$$

A preliminary version of Proposition 7 may be found in the report [38].

More details concerning the subject matter in Section 4 as well as additional examples may be found in [20] and [39]. Lawton and Reshnikoff [27] independently discovered such tiles as the simplest examples in a general theory of scaling functions ϕ which arise as solution of (48) with finite scaling sequences $\{s_k\}$. These types of tilings were also independently observed [3] outside the framework of multiresolution analyses and scaling functions; for their connection to generalized Gaussian integers see [19] and [50].

5.2. Wavelets

A *wavelet basis* associated to (Γ, A), is a complete orthonormal system for $L^2(R^n)$ whose members are Γ translates of A dilates of a finite collection $\psi_1, \dots, \psi_m$ of orthonormal functions. More specifically, the members of the basis are the functions

$$|\det A|^{k/2} \psi_j(A^k x - \gamma), \quad \gamma \in \Gamma, \quad k \in Z, \quad j = 1, \dots, m.$$

Wavelets are simply the elements of such a basis. We refer to the ψ_i's as basic wavelets. The classical Haar basis is the wavelet basis associated with the first example in Subsection 2.1.4.

Wavelets, of course, are the raison d'être for multiresolution analyses ([23], [41], and [43]). The minimum number of basic wavelets is related to A via the formula $m = |\det A| - 1$, ([20], [40], and [42]). The relationship of the scaling function ϕ to wavelets generated via the multiresolution analysis paradigm is given by

$$\psi_j(x) = \sum_{k \in Z^n} s_{jk}\phi(Ax - k), \quad j = 1, \ldots m, \tag{49}$$

where $\{s_{jk}\}_{k \in Z^n}$ are certain sequences in $l^2(Z^n)$ in the case $\Gamma = Z^n$. These sequences together with the scaling sequence, (48), form a set of *quadrature mirror filters* which are the basis of fast decomposition and reconstruction algorithms useful in signal and image processing ([10], [15], [41], and [52]) and numerical analysis ([5], [24], and [32]).

The relationship of the sequences $\{s_{jk}\}_{k \in Z^n}$, $j = 1, \ldots, m$, to the scaling sequence $\{s_k\}_{k \in Z^n}$ is best described via their Fourier transforms

$$S_j(\xi) = \frac{1}{q} \sum_{k \in Z^n} s_{jk} e^{i\langle k, \, \xi \rangle}, \quad j = 0, \ldots, m \tag{50}$$

where $q = |\det A|$, $m = q - 1$, and $s_{0k} = s_k$ for all $k \in Z^n$. If $q = 2$ then there are formulas for S_1. One reads

$$S_1(\xi) - e^{-i\langle \kappa, \xi \rangle} \overline{S_0(\xi - 2\pi B^{-1}\kappa)}$$

where κ is any element in Z^n such that $BZ^n \cup \{\kappa + BZ^n\} = Z^n$. If $q = 4$ and $S_0(\xi)$ is real valued then similar formulas hold for S_1, S_2, and S_3, see [43]. If $S_0(\xi) \geq 0$ then formulas which hold for any q can be found in [40]. Although S_j's are known to exist [42], to my knowledge, there are no known formulas in the general case.

Cardinal spline wavelets were first observed by Strömberg [48] via a different construction not related to scaling functions. Other more recent theoretical applications may be found in [1], [43], and [49] and many of the pertinent references cited there. Most of the articles cited in Subsection 5.1 contain constructions of wavelets and various applications.

Finally we mention that the term *wavelet* has been applied to various classes of not necessarily orthogonal expansions and other representations of functions. For examples of recent work and further references see [2], [14], [18], [21], [22], and [51].

References

1. Auscher, P. and Ph. Tchamitchian, Bases d'ondelettes sur les courbes corde-arc, noyau de Cauchy et espaces de Hardy associés, Rev. Mat. Iberoame ricana **5** (1989), 139–170.
2. Auslander, L. and I. Gertner, Wide-band ambiguity function and $ax + b$ group, in *Signal Processing Part I: Signal Processing Theory*, L. Auslander,

F. A. Grünbaum, J. W. Helton, T. Kailath, P. Khargonekar, S. Mitter (eds.), Springer-Verlag, New York, 1990, 1–12.

3. Bandt, C., Self-similar sets 5-Integer matrices and fractal tilings of R^n, *Proc. Amer. Math. Soc.* **112** (1991), 549–562.
4. Barnsley, M., *Fractals Everywhere*, Academic Press, 1988.
5. Beylkin, G., R. Coifman, and V. Rokhlin, Fast wavelet transforms and numerical algorithms I, *Comm. Pure and Appl. Math.* **44** (1991), 141–183.
6. de Boor, C., C. K. Hollig, and S. Riemenschneider, Convergence of cardinal series, *Proc. Amer. Math. Soc.* **98** (1986), 457–460.
7. Chui, C. K. and Wang, I., A general framework of compactly supported splines and wavelets, CAT Report #219, Texas A&M University, 1990.
8. Chui, C. K., J. Stöckler, and J. D. Ward, Compactly supported box-spline wavelets, CAT Report #230, Texas A& M University, 1991.
9. Cohen, A., Ondelettes, analyses multirésolutions et filtres miroirs en quadrature, *Ann. Inst. Henri Poincaré* **7** (1990), 439–459.
10. Cohen, A., I. Daubechies, and J. C. Feauveau, Biorthogonal bases of compactly supported wavelets, *Comm. Pure and Appl. Math.*, 1991, to appear.
11. Cohen, A. and I. Daubechies, Nonseparable bidimensional wavelet bases, AT&T Bell Laboratories, New Jersey, 1991, preprint.
12. Cohen, A. and I. Daubechies, Orthonormal Bases of compactly supported wavelets, III: better frequency resolution, preprint.
13. Colella, D. and C. Heil, Characterizations of scaling functions, I: continuous solutions, MITRE Corporations, McLean, Virginia, 1991.
14. Combes, J. M., A. Grossmann, and Ph. Tchamitchian, *Wavelets, Time-Frequency Methods and Phase Space, Lecture Notes on IPTI*, Springer-Verlag, Berlin, New York, 1989.
15. Daubechies, I., Orthonormal bases of compactly supported wavelets, *Comm. in Pure and Appl. Math.* **41** (1988), 909–996.
16. Daubechies, I. and J. C. Legarias, Two-scale difference operations I: existence and global regularity of solutions, *SIAM J. Math. Anal.* **22** (1991), 1388–1410.
17. Daubechies, I., Orthonormal bases of compactly supported wavelets, II: variations on a theme, AT&T Bell Laboratories, New Jersey, 1989, preprint.
18. Feichtinger, H. G. and K. Gröchenig, Nonorthogonal wavelet and Gabor expansions, and group representations, preprint.
19. Gilbert, W. J., Gaussian integers as bases for exotic number systems, in *The Mathematical Heritage of C. F. Gauss*, G. M. Rassias (ed.), World Scientific Publishing, Co., to appear.
20. Gröchenig, K. and W. R. Madych, Multiresolution analyses, Haar bases, and self-similar tilings of R^n, BRC technical report, *Trans. IEEE*, to appear.
21. Grossmann, A., G. Saracco, and Ph. Tchamitchian, Study of propagation of transient acoustic signals across a plane interface with the help of the wavelet transform, preprint.

22. Heil, C. E. and D. F. Walnut, Continuous and discrete wavelet transforms, *SIAM Review* **31** (4) (1989), 628–666.
23. Jaffard, S., P. G. Lemarie, S. Mallat, and Y. Meyer, Dyadic multiscale analysis of $L^2(R^n)$, manuscript.
24. S. Jaffard, Wavelet methods for fast resolution of elliptic problems, *SIAM J. Numer. Anal.*, to appear.
25. Jia, R. -Q. and C. A. Micchelli, Using the refinement equations for the construction of pre-wavelets II: powers of two, in *Curves and Surfaces*, P.-J. Laurent, A. Le Mehaute, L. Schumaker (eds.), Academic Press, San Diego, 1991.
26. Lawton, W. M., Tight frames of compactly supported affine wavelets, *J. Math. Phys.* **31** (8) (1990), 1898–1901.
27. Lawton, W. M. and H.L. Resnikoff, Multidimensional wavelet bases, Aware, Inc. Technical Report, *SIAM J. Math. Anal.*, submitted.
28. Lemarié, P. G., Theorie L^2 des surfaces-splines, 1987, preprint.
29. Lemarié, P. G., Bases d'ondelettes sur les groupes de Lie stratifiés, *Bull. Soc. Math. France,* to appear.
30. Lemarié, P. G., Some remarks on wavelet theory and interpolation, Univ. Paris-Sud, preprint.
31. Lorentz, R. A. H. and W. R. Madych, Wavelets and generalized box splines, GMD Technical Report, *Applicable Analysis*, to appear.
32. Lorentz, R. A. H. and W. R. Madych, Spline wavelets for ordinary differential equations, GMD Technical Report.
33. Madych, W. R. and S. A. Nelson, Polyharmonic cardinal splines, *Abstracts Amer. Math. Soc.* **7** (1986), 378.
34. Madych, W. R. and S. A. Nelson, Polyharmonic cardinal splines, *J. Approx. Theory* **60** (1990), 141–156.
35. Madych, W. R. and S. A. Nelson, Polyharmonic cardinal splines: a minimization property, *J. Approx. Theory* **63** (1990), 303–320.
36. Madych, W. R., Cardinal interpolation with polyharmonic splines, in *Multivariate Approximation IV*, C. Chui, W. Schepp, and K. Zeller (eds.), *Int. Ser. Num. Math.* **90**, Birkhäuser Verlag, Basel, 1989, 241–248.
37. Madych, W. R., Polyharmonic splines, multiscale analysis, and entire functions, in *Multivariate Approximation and Interpolation*, W. Haußmann and K. Jetter (eds.), Birkhäuser Verlag, Basel, 1990, 205–216.
38. Madych, W. R., Translation invariant multiscale analysis, in *Recent Advances in Fourier Analysis and its Applications*, J. S. Byrnes and J. L. Byrnes (eds.), *NATO ASI Series C* **315**, Kluwer, Dordrecht, 1990, 455–462.
39. Madych, W. R., Multiresolution analyses, tiles, and scaling functions, in *Proc. of NATO ASI*, Il Ciocco, 1991, to appear.
40. Madych, W. R., Multiresolution analysis, wavelets, and homogeneous functions, BRC, preprint.
41. Mallat, S., Multiresolution approximations and wavelet orthonormal bases of $L^2(R)$, *Trans. Amer. Math. Soc.* **315** (1989), 69–88.

42. Meyer, Y., Ondelettes, functions splines et analyses graduées, *Rapport CEREMADE* **8703**, 1987.
43. Meyer, Y., *Ondelettes et Opérateurs*, Hermann, Paris, 1990.
44. Riemenschneider, S. D., Multivariate cardinal interpolation, in *Approximation Theory VI*, C. K. Chui, L. L. Schumaker, and J. G. Ward (eds.), Academic Press, New York, 1989, 561–584.
45. Riemenschneider, S. D. and Z. W. Shen, Wavelets and pre-wavelets in low dimensions, Univ. Alberta, preprint.
46. Schoenberg, I. J., *Cardinal Spline Interpolation*, CBMS -NSF Series in Appl. Math. #12, SIAM Publ., , Philadelphia, 1973.
47. Strichartz, R. S., Wavelets and self-affine tilings, preprint.
48. Strömberg, J. -O., A modified Franklin system and higher order spline systems on $\mathbf{R}^n$ as unconditional bases for Hardy spaces, in *Proc. Conf. in Honor of Antoni Zygmund*, Vol. II, W. Beckner, A. P. Calderón, R. Fefferman, and P. W. Jones (eds.), Wadsworth, NY, 1981, 475–493.
49. Tchamitchian, Ph., Ondelettes et intégrale de Cauchy sur les Courbes Lipschitziennes, *Annals of Math.* **129** (1989), 641–649.
50. Thurston, W. P., *Groups, Tilings, and Finite State Automata*, Lecture Notes, Summer Meeting of the Amer. Math. Soc., Boulder, 1989.
51. Torrésani, B., Wavelet analysis of asymptotic signals, Centre de Physique Théorique, Marseille, preprint.
52. M. Vetterli, Multirate filter banks for subband coding, Technical Report CU/CTR/TR-165-90-02, Center for Telecommunication Research, Columbia University, New York, 1989, 1–57.
53. Volkmer, H., Distributional and square summable solutions of dilation equations, preprint.

This research is partially supported by a grant from the Air Force Office of Scientific Research, AFOSR-90-311.

W. R. Madych
Department of Mathematics
University of Connecticutt
Storrs, CT 06269-3009
madych@uconnvm

Multidimensional Two–Scale Dilation Equations

Marc A. Berger and Yang Wang

Abstract. This work involves the study of multidimensional scaling functions. These functions satisfy a two–scale dilation equation, just as the 1–D scaling functions which are used to construct 1–D wavelets. Two frameworks are presented for studying these dilation equations. Fourier methods are used to analyze existence, uniqueness and regularity of solutions. Expansions with infinite matrix products are used to derive an iterated function system (IFS) algorithm for generating the scaling function. Application to subdivision schemes for surface generation is presented, as a genuine multidimensional example where scaling functions arise and the IFS algorithm can be applied.

§1. Introduction

Recent work on subdivision schemes for surface generation (see [12]–[17], and [6]) and on wavelet surfaces [19] has led to the study of multidimensional *dilation equations*

$$f(\mathbf{x}) = \sum_{\boldsymbol{\eta}} c_{\boldsymbol{\eta}} f(2\mathbf{x} - \boldsymbol{\eta}). \tag{1}$$

Here $\boldsymbol{\eta} \in \mathbb{Z}^d$ is a multi–index, $\mathbf{x} \in \mathbb{R}^d$ and $\{c_{\boldsymbol{\eta}}\}$ are finitely supported *scaling coefficients.* In most applications the scaling coefficients satisfy the 2^d sum conditions

$$\sum_{\boldsymbol{\gamma}} c_{\boldsymbol{\eta}-2\boldsymbol{\gamma}} = 1, \quad \forall\, \boldsymbol{\eta}. \tag{2}$$

Note that these scaling coefficients are allowed to be negative. subdivision schemes proceed from a finitely supported set of control points $\mathbf{p}^0_{\boldsymbol{\eta}} \in \mathbb{R}^m$ defined on the integer lattice $\mathbb{Z}^d$. Points $\mathbf{p}^k_{\boldsymbol{\eta}}$ are defined on successively finer and finer grids $2^{-k}\mathbb{Z}^d$ through the recursion

$$\mathbf{p}^k_{\boldsymbol{\eta}/2} = \sum_{\boldsymbol{\gamma}} c_{\boldsymbol{\eta}-2\boldsymbol{\gamma}} \mathbf{p}^{k-1}_{\boldsymbol{\gamma}}.$$

Wavelets–A Tutorial in Theory and Applications
C. K. Chui (ed.), pp. 295–323.

ISBN 0-12-174590-2

Observe that there are *two* spaces involved here: the *index* (or *parameter*) *space* in $2^{-k}\mathbb{Z}^d$ which is sequentially being refined, over which the indices $\boldsymbol{\eta}$ of the points range; and the *physical space* in $\mathbb{R}^m$ where the actual points $\{\mathbf{p}^k_{\boldsymbol{\eta}} : \boldsymbol{\eta} \in 2^{-k}\mathbb{Z}^d\}$ reside. Both of these spaces are compact, but they are not related; *i.e.*, the points $\mathbf{p}^k_{\boldsymbol{\eta}}$ need not be positioned over the index space in $2^{-k}\mathbb{Z}^d$ where $\boldsymbol{\eta}$ ranges.

If the denser and denser defined sets $\{\mathbf{p}^k_{\boldsymbol{\eta}}\}$ converge to form a continuous surface, then the scheme is said to be a *convergent subdivision scheme.* Algorithmically, since it is natural to work with integer and not rational indices, the grid is rescaled each level, and the subdivision scheme is actually implemented on the rescaled (fixed) grid $\mathbb{Z}^d$ as

$$\mathbf{p}^k_{\boldsymbol{\eta}} = \sum_{\boldsymbol{\gamma}} c_{\boldsymbol{\eta}-2\boldsymbol{\gamma}} \mathbf{p}^{k-1}_{\boldsymbol{\gamma}}. \tag{3}$$

Then, the scheme is convergent, if there exists a continuous function $g : \mathbb{R}^d \to \mathbb{R}^m$ such that

$$\sup_{\boldsymbol{\eta}} \left| \mathbf{p}^k_{\boldsymbol{\eta}} - g\left(\frac{\boldsymbol{\eta}}{2^k}\right) \right| \to 0 \quad \text{as } k \to \infty.$$

When the sets $\{\mathbf{p}^k_{\boldsymbol{\eta}} : \boldsymbol{\eta} \in \mathbb{Z}^d\}$ are nested, in the sense that $\{\mathbf{p}^0_{\boldsymbol{\eta}}\} \subseteq \{\mathbf{p}^1_{\boldsymbol{\eta}}\} \subseteq \cdots$, then the scheme is *interpolatory.* The *normalized scaling function* f satisfying (1) and

$$\int_{\mathbb{R}^d} f = 1 \tag{4}$$

can be used to represent g as

$$g(\mathbf{x}) = \sum_{\boldsymbol{\eta}} f(x - \boldsymbol{\eta}) \mathbf{p}^0_{\boldsymbol{\eta}}. \tag{5}$$

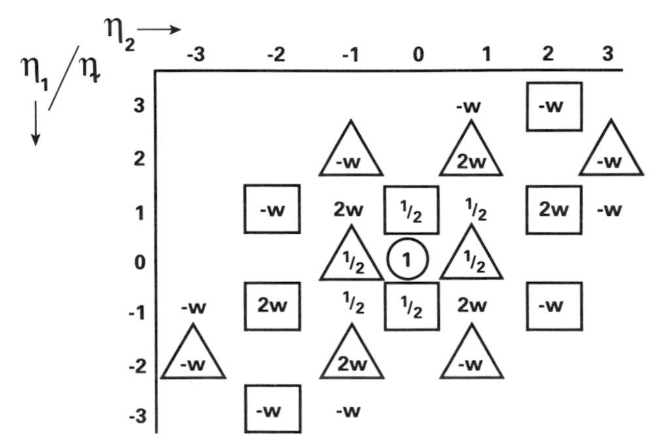

η_1 / η_2	-3	-2	-1	0	1	2	3
3					-w	-w	
2			-w		2w		-w
1		-w	2w	1/2	1/2	2w	-w
0			1/2	1	1/2		
-1	-w	2w	1/2	1/2	2w	-w	
-2	-w		2w		-w		
-3		-w	-w				

Table 1. Scaling coefficients $c_{\boldsymbol{\eta}}$ for Dyn et al.'s butterfly scheme [14].

Examples of subdivision schemes abound. There is de Rham's construction [24], Chaikin's scheme [7], the schemes of Lane and Riesenfeld [21], Dyn et al.'s 4– and 6–point interpolatory schemes (see [13] and[16]), Dyn et al.'s surface schemes (see [14] and [16]), etc.

Just to illustrate the setup described above, we present the scaling coefficients for Dyn et al.'s "butterfly scheme" in Table 1. Observe that the scaling coefficients have been partitioned into four groups, depending on the parity of $\boldsymbol{\eta}$ (even/even, even/odd, odd/even, odd/odd). Each set of coefficients sums to one, according to (2). The even/even set, marked by a circle, corresponds simply to the interpolatory condition

$$\mathbf{p}^k_{2i,2j} = \mathbf{p}^{k-1}_{i,j} \tag{6}$$

(or $\mathbf{p}^k_{i,j} = \mathbf{p}^{k-1}_{i,j}$ in terms of the non–rescaled grid $2^{-k}\mathbb{Z}^d$). The even/odd set, marked by triangles, corresponds to the refinement

$$\begin{aligned}\mathbf{p}^k_{2i,2j+1} = \frac{1}{2}\left(\mathbf{p}^{k-1}_{i,j} + \mathbf{p}^{k-1}_{i,j+1}\right) + 2w\left(\mathbf{p}^{k-1}_{i-1,j} + \mathbf{p}^{k-1}_{i+1,j+1}\right) \\ - w\left(\mathbf{p}^{k-1}_{i-1,j-1} + \mathbf{p}^{k-1}_{i-1,j+1} + \mathbf{p}^{k-1}_{i+1,j} + \mathbf{p}^{k-1}_{i+1,j+2}\right).\end{aligned} \tag{7}$$

In terms of the non–rescaled grid $2^{-k}\mathbb{Z}^d$, this refinement really defines the point $\mathbf{p}^k_{i,j+1/2}$, and is illustrated as follows:

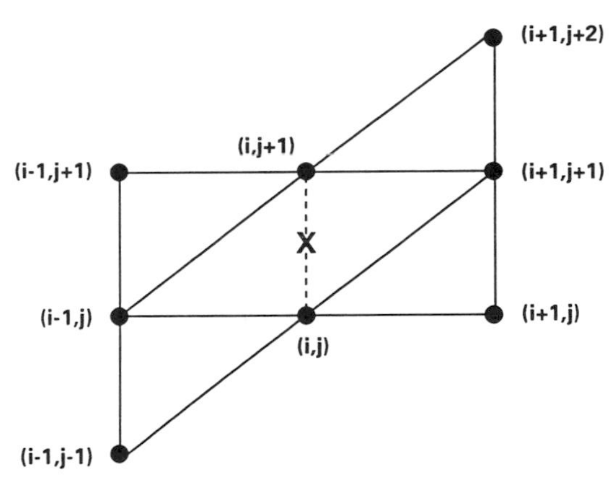

Figure 2. The new point at $(i, j + 1/2)$ is defined by (7) in terms of its neighbors. Here $i, j \in 2^{-k+1}\mathbb{Z}$.

Similarly the odd/even set, marked by squares, corresponds to the refinement

$$\begin{aligned}\mathbf{p}^k_{2i+1,j} = \frac{1}{2}\left(\mathbf{p}^{k-1}_{i,j} + \mathbf{p}^{k-1}_{i+1,j}\right) + 2w\left(\mathbf{p}^{k-1}_{i,j-1} + \mathbf{p}^{k-1}_{i+1,j+1}\right) \\ - w\left(\mathbf{p}^{k-1}_{i-1,j-1} + \mathbf{p}^{k-1}_{i+1,j-1} + \mathbf{p}^{k-1}_{i,j+1} + \mathbf{p}^{k-1}_{i+2,j+1}\right)\end{aligned} \tag{8}$$

and the odd/odd set, unmarked, corresponds to the refinement

$$\begin{aligned}\mathbf{p}^k_{2i+1,2j+1} = \frac{1}{2}\left(\mathbf{p}^{k-1}_{i,j} + \mathbf{p}^{k-1}_{i+1,j+1}\right) + 2w\left(\mathbf{p}^{k-1}_{i,j+1} + \mathbf{p}^{k-1}_{i+1,j}\right) \\ - w\left(\mathbf{p}^{k-1}_{i-1,j} + \mathbf{p}^{k-1}_{i,j-1} + \mathbf{p}^{k-1}_{i+1,j+2} + \mathbf{p}^{k-1}_{i+2,j+1}\right).\end{aligned} \tag{9}$$

In terms of the non–rescaled grid, these refinements really define the points $\mathbf{p}^k_{i+1/2,j}$ and $\mathbf{p}^k_{i+1/2,j+1/2}$, respectively, as illustrated below.

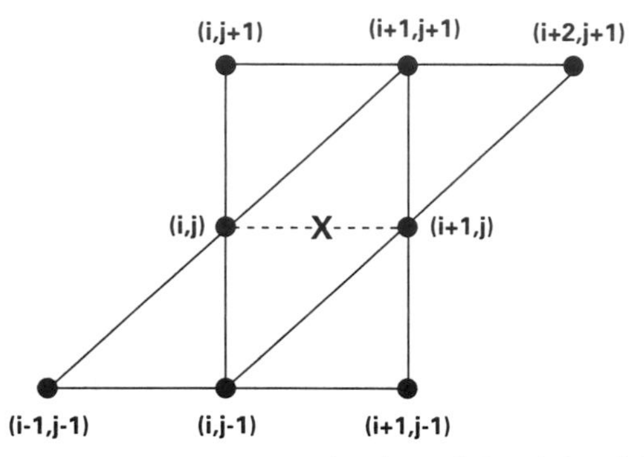

Figure 3. The new point at $(i + 1/2, j)$ is defined by (8) in terms of its neighbors. Here $i, j \in 2^{-k+1}\mathbb{Z}$.

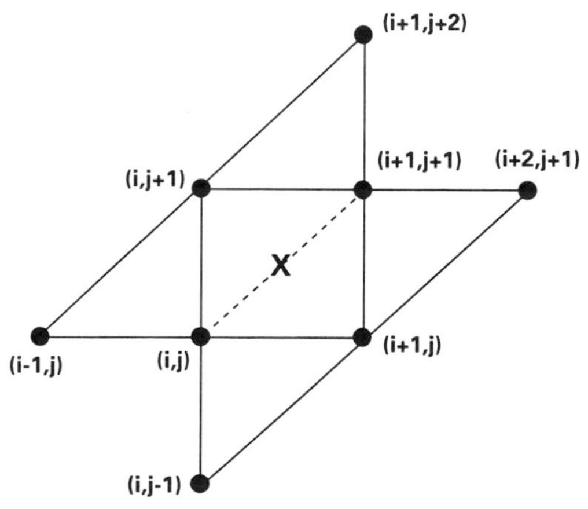

Figure 4. The new point at $(i + 1/2, j + 1/2)$ is defined by (9) in terms of its neighbors. Here $i, j \in 2^{-k+1}\mathbb{Z}$.

The one refinement equation (3), then, actually gives rise to the four equations (6)–(9). The shape of Figures 2–4 accounts for the name "butterfly scheme."

If we start with the control points

$$\{\mathbf{p}^0_{i,j} : -2 \le i, j \le 3\},$$

then the evolution of the index grids (for the non–rescaled refinement) is depicted in Figure 5.

Sub–Division Methods for Surface Generation

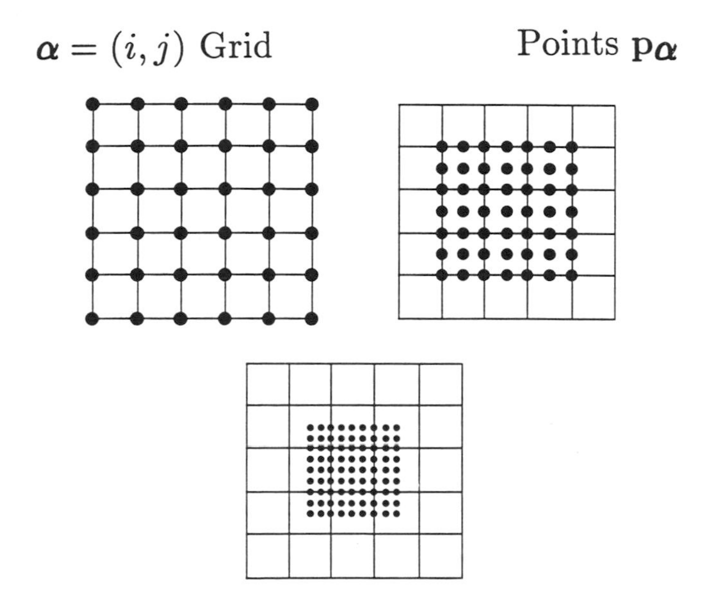

Coarse–to–Fine Interpolation:

$$\mathbf{p}_{\boldsymbol{\alpha}/2} = \sum_{\boldsymbol{\beta}} a_{\boldsymbol{\alpha}-2\boldsymbol{\beta}} \mathbf{p}_{\boldsymbol{\beta}}$$

Figure 5. Evolution of the index grid as the subdivision algorithm proceeds.

Since some of the original boundary gets lost, to render closed surfaces for example, one needs to have sufficient "wrap–around." That is, the physical points $\mathbf{p}$ near the boundary of the index grid have to coincide. In the limit as $k \to \infty$ one expects a surface to be generated, as a function $g : [0,1]^2 \to \mathbb{R}^3$. Observe that for this "butterfly scheme," the dilation equation (1) takes the form

$$f(x,y) = f(2x,2y) + \frac{1}{2}f^{(1,0)}(x,y) + 2w[f^{(1,-1)}(x,y) + f^{(1,2)}(x,y)] \\ - w[f^{(1,-2)}(x,y) + f^{(1,3)}(x,y) + f^{(2,3)}(x,y)] \tag{10}$$

where

$$f^{(i,j)}(x,y) = \sum_{(i',j')\in S_{i,j}} f(2x-i', 2y-j')$$

and

$$S_{i,j} = \{(i,j), (j,i), (-i,-j), (-j,-i)\}.$$

There is, of course, a lot of symmetry here. Some examples of surfaces constructed from the "butterfly scheme" are given in Figure 6 (at the end of this article).

In Section 2 below we study existence, uniqueness and regularity of solutions to (1). We show that if $\sum_{\eta} c_{\eta} = 2^{d+k}$, then one typically has $\binom{k+d-1}{k}$ linearly independent solutions to (1) — the same as the number of kth order partial derivatives a function of d variables has. (See Theorem 2(c).)

In Section 3 we describe an affine *iterated function system* (IFS) algorithm for constructing the normalized solution of (1) through an IFS attractor. This relates to the matrix product expansion approach developed in [4], [10], [11], [22], [23]. An affine IFS [1] consists of affine transformations $T_i : \mathbb{R}^M \to \mathbb{R}^M$, $i = 1, \ldots, N$; and it generates a discrete–time dynamical system $(\mathbf{X}_n)$ in $\mathbb{R}^M$ according to

$$\mathbf{X}_n = T_{\omega_n}\mathbf{X}_{n-1} ,$$

where (ω_n) is an appropriately chosen sequence of indices $\omega \in \{1, \ldots, N\}$. This sequence (ω_n) is said to *drive* the dynamics. In [2], [3], [5] we developed an IFS algorithm for the 1–D dilation equation, and applied it to construct compactly supported wavelets [8] and to generate curves through subdivision schemes. Here we present the multidimensional version of this work, geared to construct multidimensional wavelets and to generate surfaces through subdivision schemes.

In Section 4 we prove the results which are necessary to justify the IFS algorithm. They involve a representation for solutions to (1) using matrix product expansions. This section also contains a discussion of convergence of iterates, based on successively iterating the right–hand side of (1). The results in this section are analogues and extensions of their 1–D counterparts, which appeared in [5]. They are presented here in entirety, so as to make this work self–contained.

Haar–type multidimensional wavelets are considered by Gröchenig and Madych [19], and they too work with IFS and fractal sets. But their IFS is different from ours — it is used to describe the support of f, as in Lemma 1 below, rather than generate the function f itself. Thus, for Equation (1), the IFS producing *supp* f is

$$w_{\eta} : \mathbf{x} \mapsto \frac{\mathbf{x}+\eta}{2} ,$$

where $\boldsymbol{\eta} \in \mathbb{Z}^d$ is in the support for $(c_{\boldsymbol{\eta}})$. (That is, $\boldsymbol{\eta}$ is such that $c_{\boldsymbol{\eta}} \neq 0$.) In Gröchenig and Madych's case where all the scaling coefficients are positive (in fact they are all equal), one can conclude that *supp* f is precisely the attractor for the IFS $\{w_{\boldsymbol{\eta}}\}$ (see the remark following Lemma 1), but in general one can only infer inclusion.

Gröchenig and Madych have a nice extension of (1) which they study,

$$f(\mathbf{x}) = \sum_{\boldsymbol{\eta}} c_{\boldsymbol{\eta}} f(A\mathbf{x} - \boldsymbol{\eta}) \tag{11}$$

where $A \in Lin(\mathbb{R}^d)$ is an "acceptable dilation." We do not pursue this generalization here, but the interested reader will find that our techniques and results in Section 2 carry over to this setting as well.

§2. Existence, uniqueness, and regularity

Consider the d–dimensional *two–scale dilation equation*

$$f(\mathbf{x}) = \sum_{\boldsymbol{\eta}} c_{\boldsymbol{\eta}} f(\alpha\mathbf{x} - \boldsymbol{\beta}_{\boldsymbol{\eta}}), \tag{12}$$

where $\boldsymbol{\eta} \in \mathbb{Z}^d$ is a multi–index, $\mathbf{x} \in \mathbb{R}^d$, and the *scaling coefficients* $(c_{\boldsymbol{\eta}})$ are finitely supported. The $c_{\boldsymbol{\eta}}$'s can be complex numbers, but the $\boldsymbol{\beta}_{\boldsymbol{\eta}}$'s and α are assumed to be real, and moreover $\alpha > 1$. Correspondingly, let P be the multidimensional trigonometric polynomial

$$P(\boldsymbol{\xi}) = \frac{1}{\alpha^d} \sum_{\boldsymbol{\eta}} c_{\boldsymbol{\eta}} e^{i\langle \boldsymbol{\beta}_{\boldsymbol{\eta}}, \boldsymbol{\xi}\rangle}, \quad \boldsymbol{\xi} \in \mathbb{C}^d. \tag{13}$$

If $f \in L^1(\mathbb{R}^d)$ satisfies (12) then its Fourier transform $\widehat{f}$ satisfies

$$\widehat{f}(\boldsymbol{\xi}) = P(\boldsymbol{\xi}/\alpha)\widehat{f}(\boldsymbol{\xi}/\alpha) \tag{14}$$

and conversely if $\widehat{f}$ satisfies (14) then its inverse transform (which may be a distribution) satisfies (12). Set

$$\Delta = P(\mathbf{0}) = \frac{1}{\alpha^d} \sum_{\boldsymbol{\eta}} c_{\boldsymbol{\eta}}.$$

If $f \in L^1(\mathbb{R}^d)$ satisfies (12) then $\int f = 0$ whenever $\Delta \neq 1$.

Suppose $\Delta = 1$. We can estimate

$$\begin{aligned} \left|e^{i\langle \boldsymbol{\beta}, \boldsymbol{\xi}\rangle} - 1\right| = \left|i\langle \boldsymbol{\beta}, \boldsymbol{\xi}\rangle \int_0^1 e^{i\langle \boldsymbol{\beta}, \boldsymbol{\xi}\rangle t} dt\right| &\leq |\langle \boldsymbol{\beta}, \boldsymbol{\xi}\rangle| \int_0^1 |e^{i\langle \boldsymbol{\beta}, \boldsymbol{\xi}\rangle t}| dt \\ &\leq |\langle \boldsymbol{\beta}, \boldsymbol{\xi}\rangle| e^{B(-\operatorname{Im} \boldsymbol{\xi})} \qquad (\boldsymbol{\beta} \in \mathbb{R}^d, \boldsymbol{\xi} \in \mathbb{C}^d), \end{aligned}$$

where $B(\mathbf{x}) = \max(0, \langle \boldsymbol{\beta}, \mathbf{x} \rangle)$; and also

$$|e^{i\langle \boldsymbol{\beta}, \boldsymbol{\xi} \rangle} - 1| \le 1 + e^{B(-\mathrm{Im}\,\boldsymbol{\xi})} \le 2e^{B(-\mathrm{Im}\,\boldsymbol{\xi})}.$$

These lead to

$$|P(\boldsymbol{\xi}) - 1| \le A_1 |\boldsymbol{\xi}| e^{B(-\mathrm{Im}\,\boldsymbol{\xi})} \tag{15}$$

$$|P(\boldsymbol{\xi}) - 1| \le A_2 e^{B(-\mathrm{Im}\,\boldsymbol{\xi})} \tag{16}$$

where $A_1 = \sum_{\boldsymbol{\eta}} |c_{\boldsymbol{\eta}} \boldsymbol{\beta}_{\boldsymbol{\eta}}|$, $A_2 = 2\sum_{\boldsymbol{\eta}} |c_{\boldsymbol{\eta}}|$ and $B(\mathbf{y}) = \max_{\boldsymbol{\eta}}(0, \langle \boldsymbol{\beta}_{\boldsymbol{\eta}}, \mathbf{y} \rangle)$. The estimate (15) guarantees that $\prod_{m=1}^{\infty} P(\boldsymbol{\xi}/\alpha^m)$ converges uniformly on compact subsets of $\mathbb{C}^d$ to an entire function. If $f \in L^1(\mathbb{R}^d)$ satisfies (12) then it follows from (14) that

$$\widehat{f}(\boldsymbol{\xi}) = \widehat{f}(\mathbf{0}) \prod_{m=1}^{\infty} P(\boldsymbol{\xi}/\alpha^m). \tag{17}$$

Let C be a bound for $\prod_{m=1}^{\infty} P(\boldsymbol{\xi}/\alpha^m)$ on the unit disk $|\boldsymbol{\xi}| \le 1$, say from (15), $C = \exp\left(\dfrac{A_1 e^{\beta}}{\alpha - 1}\right)$, $\beta = \max |\boldsymbol{\beta}_{\boldsymbol{\eta}}|$. For $|\boldsymbol{\xi}| > 1$, let k be such that $\alpha^k < |\boldsymbol{\xi}| \le \alpha^{k+1}$. Then using (16) we can estimate

$$\begin{aligned}\left|\prod_{m=1}^{\infty} P(\boldsymbol{\xi}/\alpha^m)\right| &\le C \prod_{m=1}^{k+1} \left[1 + A_2 e^{\frac{B(-\mathrm{Im}\,\boldsymbol{\xi})}{\alpha^m}}\right] \\ &\le C(1 + A_2)^{k+1} e^{\frac{B(-\mathrm{Im}\,\boldsymbol{\xi})}{\alpha - 1}} \le C' |\boldsymbol{\xi}|^M e^{\frac{B(-\mathrm{Im}\,\boldsymbol{\xi})}{\alpha - 1}},\end{aligned}$$

where $M = \log_{\alpha}(1 + A_2)$. Thus, we have the global estimate

$$\left|\prod_{m=1}^{\infty} P(\boldsymbol{\xi}/\alpha^m)\right| \le C'(1 + |\boldsymbol{\xi}|)^M e^{\frac{B(-\mathrm{Im}\,\boldsymbol{\xi})}{\alpha - 1}}. \tag{18}$$

Let us now recall the *Payley-Wiener Theorem*, stated as follows:

Suppose that K is a convex compact subset of $\mathbb{R}^d$ with supporting function

$$H(\mathbf{y}) = \max_{\mathbf{x} \in K} \langle \mathbf{x}, \mathbf{y} \rangle.$$

Then every entire analytic function $u(\boldsymbol{\xi})$ in $\mathbb{C}^d$ satisfying an estimate

$$|u(\boldsymbol{\xi})| \le C(1 + |\boldsymbol{\xi}|)^M e^{H(-\mathrm{Im}\,\boldsymbol{\xi})}$$

is the Fourier transform of a distribution of order M supported in K.

Hence, it follows from (17) and (18) that if $f \in L^1(\mathbb{R}^d)$ satisfies (12), then

$$\mathit{supp}\ f \subseteq \frac{K}{\alpha - 1}, \quad \text{where } K = \mathit{conv{-}hull}(\boldsymbol{\beta}_{\boldsymbol{\eta}}).$$

Indeed, shifting f by $\dfrac{\boldsymbol{\beta}_{\boldsymbol{\eta}}}{\alpha-1}$ effectively shifts $\boldsymbol{\beta}_{\boldsymbol{\eta}}$ to zero, in which case $B(\mathbf{y}) = \max_{\boldsymbol{\eta}}\langle\boldsymbol{\beta}_{\boldsymbol{\eta}}, \mathbf{y}\rangle$ is the support function for K.

We can say more about the support of compactly supported solutions of (12). For $\boldsymbol{\eta} \in \mathbb{Z}^d$, let $w_{\boldsymbol{\eta}} : \mathbb{R}^d \to \mathbb{R}^d$ be the strictly contractive affine transformation

$$w_{\boldsymbol{\eta}} : \mathbf{x} \mapsto \frac{\mathbf{x} + \boldsymbol{\beta}_{\boldsymbol{\eta}}}{\alpha}.$$

The system $\{w_{\boldsymbol{\eta}}\}$ forms an *iterated function system* (IFS) in $\mathbb{R}^d$ [1]. Such a system of strictly contractive affine transformations always possesses an *attractor* $\mathcal{A} \subseteq \mathbb{R}^d$, which is the unique non–empty compact set satisfying

$$\mathcal{A} = \bigcup_{\boldsymbol{\eta}} w_{\boldsymbol{\eta}}(\mathcal{A}).$$

This set is the minimal non–empty closed set which is invariant under each $w_{\boldsymbol{\eta}}$. It can be constructed as $\mathcal{A} = \lim \mathcal{A}_k$ in the Hausdorff metric, where $\mathcal{A}_{k+1} = \bigcup_{\boldsymbol{\eta}} w_{\boldsymbol{\eta}}(\mathcal{A}_k)$ and $\mathcal{A}_0 \subseteq \mathbb{R}^d$ is any arbitrary non–empty compact set.

Lemma 1. *Let f be a measurable compactly supported solution of (12). Then supp $f \subseteq \mathcal{A}$, where $\mathcal{A}$ is the attractor for the IFS $\{w_{\boldsymbol{\eta}}\}$.*

Proof: Let $\Omega =$ *supp* f. Observe that *supp* $f(\alpha\mathbf{x} - \boldsymbol{\beta}_{\boldsymbol{\eta}}) = w_{\boldsymbol{\eta}}(\Omega)$. Thus, (12) implies that $\Omega \subseteq \bigcup_{\boldsymbol{\eta}} w_{\boldsymbol{\eta}}(\Omega)$, from which it follows that $\Omega \subseteq \mathcal{A}$. ■

Clearly, $\dfrac{1}{\alpha-1}$ *conv–hull*$(\boldsymbol{\beta}_{\boldsymbol{\eta}})$ is invariant under each $w_{\boldsymbol{\eta}}$, since $\dfrac{\boldsymbol{\beta}_{\boldsymbol{\eta}}}{\alpha-1}$ is the fixed point of $w_{\boldsymbol{\eta}}$. Thus, $\mathcal{A} \subseteq \dfrac{K}{\alpha-1}$ where $K =$ *conv–hull*$(\boldsymbol{\beta}_{\boldsymbol{\eta}})$, and Lemma 1 sharpens the support estimate we picked up from the Payley-Wiener Theorem. In fact, $\mathcal{A}$ is often a fractal Cantor–like set. For example if $\alpha = d = 2$ and if the $\boldsymbol{\beta}_{\boldsymbol{\eta}}$'s are $\binom{0}{0}, \binom{1}{0}, \binom{0}{1}$ then $\mathcal{A}$ is a totally disconnected Sierpinski triangle with vertices $\boldsymbol{\beta}_{\boldsymbol{\eta}}$ (see [1]); whereas if α is reduced to 3/2 then $\mathcal{A}$ is the solid triangle with vertices $\binom{0}{0}, \binom{2}{0}, \binom{0}{2}$. In either case the Payley-Wiener Theorem only gives a solid triangle as bounding *supp* f. Gröchenig and Madych [19] have some nice illustrations of *supp* f. When $d = 1$ we find that *supp* $f \subseteq \left[\frac{\beta_{\min}}{\alpha-1}, \frac{\beta_{\max}}{\alpha-1}\right]$. The IFS $\{w_{\boldsymbol{\eta}}\}$ is typically *overlapping* (see [1]). For example the IFS $\left\{\frac{x+n}{\alpha} : n = 0, \ldots, N\right\}$, which comes from the $1-D$ dilation equation

$$f(x) = \sum_{n=0}^{N} c_n f(\alpha x - n),$$

is overlapping whenever $N \geq \alpha$. When the coefficients $c_{\boldsymbol{\eta}}$ are all non–negative, $\boldsymbol{\eta} \in \mathbb{Z}^d$, and $\Delta = 1$, then the normalized solution to (12), $\int f = 1$, is the invariant pdf for the IFS with probabilities [1]

$$\mathbb{P}(w = w_{\boldsymbol{\eta}}) = \frac{c_{\boldsymbol{\eta}}}{\alpha^d}.$$

We next study existence and uniqueness of compactly supported solutions $f \in L^1(\mathbb{R}^d)$ to (12). It is no longer assumed that $\Delta = 1$, but in any event $\frac{1}{\Delta}P(\boldsymbol{\xi})$ satisfies the estimates (15), (16), and (18).

Theorem 2. (a) *If $|\Delta| \leq 1$ and $\Delta \neq 1$, then the only $L^1(\mathbb{R}^d)$ solution to (12) is $f \equiv 0$.*
(b) *If $\Delta = 1$ and (12) has a non-trivial $L^1(\mathbb{R}^d)$ solution f, then f is unique up to scale and $\widehat{f}$ is given by (17).*
(c) *If $|\Delta| > 1$, then a necessary condition for (12) to have a non-trivial compactly supported $L^1(\mathbb{R}^d)$ solution is $\Delta = \alpha^k$, for some $k \in \mathbb{Z}_+$. In this case*

$$\widehat{f}(\boldsymbol{\xi}) = h(\boldsymbol{\xi}) \prod_{m=1}^{\infty} \frac{P(\boldsymbol{\xi}/\alpha^m)}{\Delta} \tag{19}$$

where h is a homogeneous polynomial of degree k.

Proof: (a) If $f \in L^1(\mathbb{R}^d)$ satisfies (12), then it follows from (14) that

$$\widehat{f}(\boldsymbol{\xi}) = \lim_{\ell \to \infty} \left[\prod_{m=1}^{\ell} P(\boldsymbol{\xi}/\alpha^m) \right] \widehat{f}(\boldsymbol{\xi}/\alpha^\ell). \tag{20}$$

If $|\Delta| < 1$ then $\prod_{m=1}^{\ell} P(\boldsymbol{\xi}/\alpha^m) \to 0$, and so this limit is zero. If $|\Delta| = 1$ and $\Delta \neq 1$, we write

$$\widehat{f}(\boldsymbol{\xi}) = \lim_{\ell \to \infty} \Delta^\ell \left[\prod_{m=1}^{\ell} \frac{P(\boldsymbol{\xi}/\alpha^m)}{\Delta} \right] \widehat{f}(\boldsymbol{\xi}/\alpha^\ell).$$

Since $\Delta \neq 1$, we have $\widehat{f}(\mathbf{0}) = 0$, and since $\widehat{f}$ is continuous we obtain $\widehat{f}(\boldsymbol{\xi}/\alpha^\ell) \to 0$, and so this limit is zero also.
(b) Iterating (14) leads to the formula for $\widehat{f}$.
(c) Suppose $\Delta = \alpha^k \delta$, $1 < |\delta| \leq \alpha$, $\delta \neq \alpha$. We use induction on $n = k + d$ to show that f must be identically zero. If $k < 0$, then we can use (a) above, and so this covers the case $n \leq 0$.

If $d = 1$ then $\int f = 0$ since $\Delta \neq 1$. If $d > 1$, we then define

$$f_i(\widehat{\mathbf{x}}) = \int_{-\infty}^{\infty} f(\mathbf{x}) dx_i,$$

where $\widehat{\mathbf{x}} = (x_1, \ldots, x_{i-1}, x_{i+1}, \ldots, x_n)$. Observe that f_i is a compactly supported $L^1(\mathbb{R}^{d-1})$ solution to the $(d-1)$–dimensional dilation equation

$$f_i(\widehat{\mathbf{x}}) = \frac{1}{\alpha} \sum_{\boldsymbol{\eta}} c_{\boldsymbol{\eta}} f_i(\alpha \widehat{\mathbf{x}} - \widehat{\boldsymbol{\beta}}_{\boldsymbol{\eta}}),$$

where $\widehat{\boldsymbol{\beta}}$ is defined analogously to $\widehat{\mathbf{x}}$. For this equation the Δ would be equal to $\frac{1}{\alpha^{d-1}}\frac{1}{\alpha}\sum_{\boldsymbol{\eta}} c_{\boldsymbol{\eta}}$, which is the same as the original $\Delta = \alpha^k\delta$. But the dimension drops from d to $d-1$, so that $n = k+d-1$. Hence, the induction hypothesis applies, and we get that $f_i \equiv 0$ for each $i = 1, \ldots, d$.

Next define

$$F(\mathbf{x}) = \int_{-\infty}^{x_1} \cdots \int_{-\infty}^{x_d} f(\mathbf{y})d\mathbf{y}.$$

Since $f_i \equiv 0$, $\forall\ i$, it follows that F has compact support. Moreover, F is in $L^1(\mathbb{R}^d)$ since f is compactly supported, and it satisfies the dilation equation (12) with scaling coefficients $\frac{1}{\alpha^d}c_{\boldsymbol{\eta}}$. This brings the n down from $k+d$ to k, and so the induction hypothesis again applies to conclude that $F \equiv 0$. From this follows $f \equiv 0$.

Suppose next that $\Delta = \alpha^k$ for some $k \in \mathbb{Z}_+$. Define

$$h(\boldsymbol{\xi}) = \widehat{f}(\boldsymbol{\xi}) / \prod_{m=1}^{\infty} \frac{P(\boldsymbol{\xi}/\alpha^m)}{\Delta}.$$

Observe that h satisfies

$$h(\alpha\boldsymbol{\xi}) = \alpha^k h(\boldsymbol{\xi}) \tag{21}$$

at all points $\boldsymbol{\xi}$ where $\prod_{m=1}^{\infty}\frac{P(\boldsymbol{\xi}/\alpha^m)}{\Delta}$ is non–zero, and that $\prod_{m=1}^{\infty}\frac{P(\boldsymbol{\xi}/\alpha^m)}{\Delta}$ is bounded away from zero near $\boldsymbol{\xi} = \mathbf{0}$. Let $S \subseteq \mathbb{C}^d$ denote the zero set

$$S = \{\boldsymbol{\xi} \in \mathbb{C}^d : \prod_{m=1}^{\infty} \frac{P(\boldsymbol{\xi}/\alpha^m)}{\Delta} = 0\}.$$

If $(\boldsymbol{\xi}_n)$ is a sequence in $\mathbb{C}^d\backslash S$ with $\lim \boldsymbol{\xi}_n = \boldsymbol{\xi}^* \in S$ then $\lim h(\boldsymbol{\xi}_n)$ exists. Indeed $\alpha^{-\ell}\boldsymbol{\xi}_n \to \alpha^{-\ell}\boldsymbol{\xi}^*$, and for ℓ large enough these points will all be in a neighborhood of $\boldsymbol{\xi} = \mathbf{0}$, where h is continuous. So by (21) $h(\boldsymbol{\xi}_n) \to \alpha^{k\ell}h(\alpha^{-\ell}\boldsymbol{\xi}^*)$. By the Riemann Removable Singularity Theorem, we conclude that h is analytic. Then by matching coefficients in the power series expansions of both sides in (21), we find that h must be a homogeneous polynomial of degree k. ■

Remark. If we extend our considerations to non–compactly supported solutions $f \in L^1(\mathbb{R}^d)$ of (12), then the restriction $\Delta = \alpha^k$ in Theorem 2(c) is removed. The same argument as above can be used to show that if $f \in L^1(\mathbb{R}^d)$ satisfies (12), with $|\Delta| > 1$, then

$$\widehat{f}(\boldsymbol{\xi}) = |\boldsymbol{\xi}|^{\log_\alpha \Delta} G(\boldsymbol{\xi}) \prod_{m=1}^{\infty} \frac{P(\boldsymbol{\xi}/\alpha^m)}{\Delta}, \tag{22}$$

where G is continuous in $\mathbb{C}^d\backslash\mathbf{0}$ and satisfies $G(\alpha\boldsymbol{\xi}) = G(\boldsymbol{\xi})$. If $d = 1$, this means that

$$G(x) = g_{\operatorname{sgn} x}(\log_\alpha |x|), \quad x \in \mathbb{R},$$

where $g_{\pm}$ are continuous periodic functions of period one. This is consistent with [9, Thm. 2.1].

From Theorem 2(c) we conclude that *the solution space in* $L^1(\mathbb{R}^d)$ *for (12) is at most* $\binom{k+d-1}{k}$*-dimensional when* $\Delta = \alpha^k$, since this is the number of different monomials of degree k in d variables. One case where this dimension is achieved occurs when the dilation equation is a *tensor product*; *i.e.*, when the scaling coefficients factor as

$$c_{\boldsymbol{\eta}} = c^{(1)}_{\eta_1} \cdots c^{(d)}_{\eta_d}$$

and also

$$\boldsymbol{\beta}_{\boldsymbol{\eta}} = \left(\beta^{(1)}_{\eta_1}, \ldots, \beta^{(d)}_{\eta_d}\right).$$

In this case the scaling function itself also factors as

$$f(\mathbf{x}) = f^{(1)}(x_1) \cdots f^{(d)}(x_d),$$

where each $f^{(i)}$ is the univariate scaling function satisfying

$$f^{(i)}(x) = \sum_{\eta_j} c^{(i)}_{\eta_j} f^{(i)}(\alpha x - \beta^{(i)}_{\eta_j}).$$

By choosing the scaling coefficients so that $\Delta = 1$ and so that each $f^{(i)}$ is k times differentiable (see [8], [25] for how to ensure this), the kth order partial derivatives of f will each satisfy the dilation equation with coefficients $\alpha^k c_{\boldsymbol{\eta}}$. By choosing the $c^{(i)}$'s so that fewer than k derivatives of $f^{(i)}$ exist, one can contrive things so that the solution space for the dilation equation with coefficients $\alpha^k c_{\boldsymbol{\eta}}$ has fewer than $\binom{k+d-1}{k}$ independent solutions in $L^1(\mathbb{R}^d)$.

Suppose that f satisfies (12). If $f \in C^k$, then each of its kth order partial derivatives is a solution of the corresponding dilation equation with scaling coefficients $\alpha^k c_{\boldsymbol{\eta}}$. The Δ for this latter dilation equation is α^k times the Δ for the equation which f satisfies; and this is consistent with (19) since differentiation with respect to x_j corresponds to multiplication by $-i\xi_j$ in the spectral domain. The converse is more involved, however — at least in higher dimensions. *Suppose* $f \in L^1(\mathbb{R}^d)$ *is a compactly supported solution of (12) with* $\Delta \geq \alpha^k$, *and* $\widehat{f}(\boldsymbol{\xi}) = h(\boldsymbol{\xi})u(\boldsymbol{\xi})$ *where* h *is a homogeneous polynomial of degree* k *and* u *is entire analytic. Is* u *necessarily the Fourier transform of a compactly supported function* $g \in L^1(\mathbb{R}^d)$ *satisfying the corresponding dilation equation with scaling coefficients* $\frac{c_{\boldsymbol{\eta}}}{\alpha^k}$? If so, then, we would have, of course, $f = h(i\nabla)g$ in the sense of distribution.

If $h(\boldsymbol{\xi}) = \langle \boldsymbol{\lambda}, \boldsymbol{\xi} \rangle$ then such a g always exists; namely,

$$g(\mathbf{x}) = \int_{-\infty}^{0} f(\mathbf{x} + \boldsymbol{\lambda} t)\,dt.$$

Since f is $L^1(\mathbb{R}^d)$ and compactly supported, it follows that $g \in L^1(\mathbb{R}^d)$. Moreover, since $\widehat{f}(\boldsymbol{\xi}) = 0$ whenever $\langle \boldsymbol{\lambda}, \boldsymbol{\xi} \rangle = 0$, it follows that $\int_{-\infty}^{\infty} f(\mathbf{x} + \boldsymbol{\lambda} t)dt \equiv 0$, and so g is also compactly supported — since f is.

Suppose, next, that h is elliptic, so that $h(\mathbf{y}) = 0$, $\mathbf{y} \in \mathbb{R}^d \Rightarrow \mathbf{y} = \mathbf{0}$. Say $\deg h = 2m$. Then $|h(\boldsymbol{\xi})| \geq C|\boldsymbol{\xi}|^{2m}$ for some $C > 0$, and so

$$|u(\boldsymbol{\xi})| \leq C(1 + |\boldsymbol{\xi}|^2)^{-m}. \tag{23}$$

If $2m \geq d$ then $u \in L^2(\mathbb{R}^d)$ and its inverse transform $g \in L^2(\mathbb{R}^d)$. It follows from the Payley-Wiener Theorem that g is compactly supported, and so in fact $g \in L^1(\mathbb{R}^d)$.

If $d > 2m$ then it follows from [20, Thm. 7.1.20] that $h(\boldsymbol{\nabla})$ has a fundamental solution E which is homogeneous of degree $2m - d$ and C^∞ in $\mathbb{R}^d\{\mathbf{0}\}$. In particular,

$$|E(\mathbf{x})| \leq \frac{C}{|\mathbf{x}|^{d-2m}} \tag{24}$$

for some $C > 0$. Since g is compactly supported, and $\lim_{|\mathbf{x}| \to \infty} E * f(\mathbf{x}) = 0$, it follows from the ellipticity of $h(\boldsymbol{\nabla})$ that

$$g = E * f. \tag{25}$$

Moreover, since by (24) E is locally integrable in $\mathbb{R}^d$, and $f \in L^1(\mathbb{R}^d)$, it follows from (25) that g is (locally) integrable.

These arguments have led us to the following

Theorem 3. *Let $f \in L^1(\mathbb{R}^d)$ be compactly supported, with*

$$\widehat{f}(\boldsymbol{\xi}) = h(\boldsymbol{\xi})u(\boldsymbol{\xi}),$$

where h is a homogeneous polynomial and u is entire analytic. Then under either of the conditions below, it follows that $u = \widehat{g}$ where $g \in L^1(\mathbb{R}^d)$ and is compactly supported:

(i) *h is linear; i.e., $h(\boldsymbol{\xi}) = \langle \boldsymbol{\lambda}, \boldsymbol{\xi} \rangle$;*

(ii) *h is elliptic.*

Remark. Since every homogeneous polynomial in $d = 2$ variables factors into a product of linears and irreducible quadratics, the conclusion of Theorem 3 always holds. That is, for any homogeneous polynomial h in two variables, it is always true that $u = \widehat{g}$ where $g \in L^1(\mathbb{R}^2)$ and is compactly supported. In the elliptic case we can get more information about the integrability of g. Observe from (23) that $u \in L^p(\mathbb{R}^d)$ for any $p > \dfrac{d}{2m}$. In particular, we have that

(i) $g \in L^2(\mathbb{R}^d)$, if $d < 4m$;

(ii) $g \in L^\infty(\mathbb{R}^d)$, if $d < 2m$; and

(iii) $g \in L^q(\mathbb{R}^d)$, for any $q < \dfrac{d}{d - 2m}$, if $2m \leq d < 4m$.

Here we use Parseval's formula to derive (i), and the Hausdorff–Young Inequality to derive (ii) and (iii).

The significance of Theorem 3 as concerns the dilation equation (12) is as follows. Suppose $\Delta = \alpha^k$ and that $f \in L^1(\mathbb{R}^d)$ is a compactly supported solution of (12). Then according to Theorem 2(c), we have $\widehat{f}(\boldsymbol{\xi}) = h(\boldsymbol{\xi})u(\boldsymbol{\xi})$, where h is a homogeneous polynomial of degree k, and u is entire analytic. If h can be factored into linear and elliptic factors, then we are able to conclude that $f = h(\boldsymbol{\nabla})g$, where g satisfies the corresponding dilation equation with scaling coefficients $c_{\boldsymbol{\eta}}/\alpha^k$. This allows us to identify solutions of (12) with $\Delta = \alpha^k$ as coming from k^{th} order partial derivatives of solutions of the corresponding scaled equation with $\Delta = 1$.

§3. IFS algorithm

Algorithmically, subdivision schemes are easy to implement using a recursive 2^d–array tree traversal. For example, Figure 7 illustrates the case for the "butterfly scheme" which was described above. The root of the tree is a 6×6 index grid, where i, j range from -2 to 3, on which control points $\mathbf{p}_{\boldsymbol{\eta}}^0$ are defined. (See Figure 5.) In general, the root grid in $\mathbb{Z}^d$ can be constructed as follows: Let $\{M_1, \ldots, N_1\} \times \cdots \times \{M_d, \ldots, N_d\} \subseteq \mathbb{Z}^d$ be the smallest (lattice) box containing the support $\{\boldsymbol{\eta} \in \mathbb{Z}^d : c_{\boldsymbol{\eta}} \neq 0\}$. Then the root grid can be taken as $-S$, where

$$S = \{M_1, \ldots, N_1 - 1\} \times \cdots \times \{M_d, \ldots, N_d - 1\}. \tag{26}$$

The first subdivision level extends the grid to 7×7, where i, j now range from -2 to 4. (Note that in Figure 5 the grid is not rescaled, but in this figure it is.) This is broken up into four "overlapping" transformations $T_{0,0}, T_{1,0}, T_{0,1}, T_{1,1}$ as shown. Transformation $T_{0,0}$ maps the parent 6×6 points indexed by the root grid into the child 6×6 points in the lower left of the 7×7 sub–divided grid. Similarly $T_{1,0}$ maps the parent 6×6 points indexed by the root grid into the child 6×6 points in the lower right of the 7×7 sub–divided grid, etc. Each $T_{\boldsymbol{\omega}}$ is thus a linear transformation

$$T_{\boldsymbol{\omega}} \in Lin(\mathbb{R}^S); \qquad \boldsymbol{\omega} \in \{0,1\}^d$$

whose action is determined by

$$\mathbf{p}_{-\boldsymbol{\eta}}^{\text{child}} = \sum_{\boldsymbol{\gamma} \in S} (T_{\boldsymbol{\omega}})_{\boldsymbol{\eta},\boldsymbol{\gamma}} \mathbf{p}_{-\boldsymbol{\gamma}}^{\text{parent}}; \quad \boldsymbol{\eta} \in S. \tag{27}$$

Here, $S \subseteq \mathbb{Z}^d$ is the above finite set; and $-S = \{-2, \ldots, 3\}^2$ for the "butterfly scheme." Observe in (27) the convention that $-\boldsymbol{\eta}$ and $-\boldsymbol{\gamma}$ are the respective child and parent subscripts. Under this convention the coordinate description of $T_{\boldsymbol{\omega}}$ becomes

$$(T_{\boldsymbol{\omega}})_{\boldsymbol{\eta},\boldsymbol{\gamma}} = c_{2\boldsymbol{\gamma}-\boldsymbol{\eta}+\boldsymbol{\omega}}; \quad \boldsymbol{\eta}, \boldsymbol{\gamma} \in S \tag{28}$$

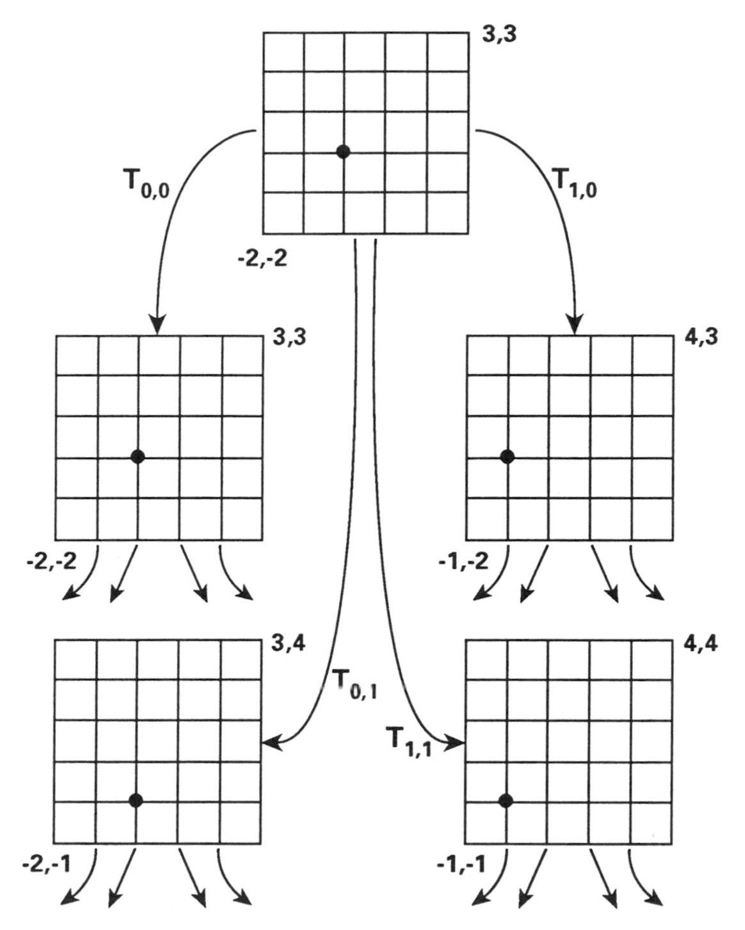

Figure 7. Tree for implementation of subdivision algorithm. At each level the 6×6 grid is replaced by four staggered grids.

where we have shifted each 6×6 grid here so that i, j consistently range from -2 to 3. This is important so that each child can in turn be treated like the root. The tree continues to evolve this way, with four edges sprouting out of each node. *At every stage of the subdivision scheme the relationship of the parent* 6×6 *points to the child* 6×6 *points along the* $(0,0)$ *edge is the same;* namely, $T_{0,0}$ — and similarly for the other three types of edges. Thus, the surface can be generated by recursively traversing this quad–tree, until a "deep enough" level ℓ is reached — at which stage the points can be plotted. The plotting depth ℓ is determined by the desired resolution of the graphics output.

Under condition (2) the transformations $T_{\boldsymbol{\omega}}$ have row sums all equal to one; *i.e.*, $\sum_{\boldsymbol{\gamma} \in S} (T_{\boldsymbol{\omega}})_{\boldsymbol{\eta},\boldsymbol{\gamma}} = 1$ for all $\boldsymbol{\eta} \in S$. When the subdivision algorithm converges, longer and longer paths of the tree produce limiting singletons. This manifests itself in that

$$\lim_{n \to \infty} T_{\boldsymbol{\omega}_n} \cdots T_{\boldsymbol{\omega}_1} = T_\infty \tag{29}$$

exists, for any sequence $(\boldsymbol{\omega}_n) \in \Omega^d$ where Ω is the sequence space (or "code space") $\Omega = \{0,1\}^\infty$; and, moreover, this limit is of rank one. Identify $\mathbf{x} = \sum \boldsymbol{\omega}_n 2^{-n} \in [0,1]^d$ with this sequence $(\boldsymbol{\omega}_n)$; *i.e.*, through the binary expansions

$$x_1 = .\omega_{1,1}\omega_{2,1}\cdots, \quad \ldots, \quad x_d = .\omega_{d,1}\omega_{d,2}\cdots$$

and denote correspondingly T_∞ by $T_\infty(\mathbf{x})$. Applying $T_\infty(\mathbf{x})$ to the original 6×6 set of control points gives 36 *identical* points — namely, that point on the surface corresponding to $\mathbf{x}$; *i.e.*, the point

$$\sum_{\boldsymbol{\eta} \in S} f(\mathbf{x} + \boldsymbol{\eta}) \mathbf{p}^0_{-\boldsymbol{\eta}},$$

where f is the solution to (1) with normalization $\int f = 1$.

It can be shown as in [5] that, on account of the shift relationship between the various $T_{\boldsymbol{\omega}}$'s, $T_\infty(\mathbf{x})$ is well–defined. That is, if some component of $\mathbf{x}$ is dyadic, then its two binary expansions (terminating and non–terminating) give rise to the same limiting product. Equivalently, for the first index component

$$\lim_{n \to \infty} T_{0,\boldsymbol{\omega}'_n} \cdots T_{0,\boldsymbol{\omega}'_2} T_{1,\boldsymbol{\omega}'_1} = \lim_{n \to \infty} T_{1,\boldsymbol{\omega}'_n} \cdots T_{1,\boldsymbol{\omega}'_2} T_{0,\boldsymbol{\omega}'_1} \tag{30}$$

for any sequence $(\boldsymbol{\omega}'_n) \in \Omega^{d-1}$; and it is similar for any other index component. From the fact that $T_\infty(\mathbf{x})$ exists and is well–defined, it follows from its definition that

$$T_\infty(\mathbf{x}) = T_\infty(\tau\mathbf{x}) T_{\boldsymbol{\omega}_1}, \tag{31}$$

where

$$\tau\mathbf{x} = 2\mathbf{x} (\text{mod } 1), \qquad \mathbf{x} \in [0,1]^d,$$

and $\boldsymbol{\omega}_1$ is the first vector of bits in the binary expansion of $\mathbf{x}$. From this it follows as in [5] that the rows of $T_\infty(\mathbf{x})$ comprise a solution to (1). That is, let $\mathbf{v} \in \mathbb{R}^S$, $\sum_{\boldsymbol{\eta} \in S} v_{\boldsymbol{\eta}} = 1$. Then $\mathbf{v}^t T_\infty(\mathbf{x})$ is the row vector

$$(f(\mathbf{x} + \boldsymbol{\eta}) : \boldsymbol{\eta} \in S),$$

where f is the solution to (1) with normalization $\int f = 1$.

According to the recipe in [2] and [5], this then gives rise to the following *IFS algorithm for surface generation.*

IFS Algorithm

```
  initialize X = (x_η) ∈ IR^S to be the fixed point of T_0, normalized
  so that Σ_{η∈S} x_η = 1
for n = 1, L
    plot Σ_{η∈S} x_η p^0_{-η}
    ω ← ω_n (from the bit string)
    X ← T_ω X
endfor
```

The choice of the bit string $\boldsymbol{\omega}_1, \ldots, \boldsymbol{\omega}_L$ must be done as in [5]. Given a desired resolution ℓ, the string must have the property that

($\mathcal{P}$) *as we slide a window of length ℓ across the string, every possible ℓ-bit vector pattern in $(\{0,1\}^d)^\ell$ should appear.*

Since there are $2^{d\ell}$ such patterns, it is clear that $L \geq 2^{d\ell} + \ell - 1$ (the last term is due to the window spillover). The bound $L = 2^{d\ell} + \ell - 1$ can in fact be attained by constructing a de–Bruijn sequence for 2^d symbols. With $\ell = d = 2$, for example, we can use the string

$$a\ a\ b\ c\ d\ b\ a\ d\ d\ c\ b\ b\ d\ a\ c\ c\ a$$

with $L = 2^4 + 1$ terms, where the symbols a, b, c, d stand for

$$a = \begin{pmatrix}0\\0\end{pmatrix} \quad b = \begin{pmatrix}1\\0\end{pmatrix} \quad c = \begin{pmatrix}0\\1\end{pmatrix} \quad d = \begin{pmatrix}1\\1\end{pmatrix}.$$

Observe that as we slide a window of length 2 across this string, we encounter the pairs

$$aa,\ ab,\ bc,\ cd,\ \ldots,\ ca$$

and indeed each of the 16 possible pairs occurs here.

In general if the initial set of control points spans more than a 6×6 index grid, then several trees can be traversed in parallel, each starting with a different 6×6 block of the initial grid as root — so long as the entire grid gets processed. These root 6×6 grids need to be offset in order to fit the full initial grid. This is illustrated in Figure 8 for $d = 1$. *Thus, there is never any need to work in a larger space than* $\mathbb{R}^{6\times 6}$; *i.e.,* $\mathbb{R}^S$ for the S given by (26). For $d > 2$ the tree in Figure 7 generalizes to a 2^d–array tree in the obvious way.

In the next section we develop the appropriate mathematical setting and prove that under a suitable contractivity condition on the $T_{\boldsymbol{\omega}}$'s the product expansion (29) indeed converges, and the consistency condition (30) holds. We show how the product expansion leads to a continuous solution of (1), and justify the IFS algorithm above.

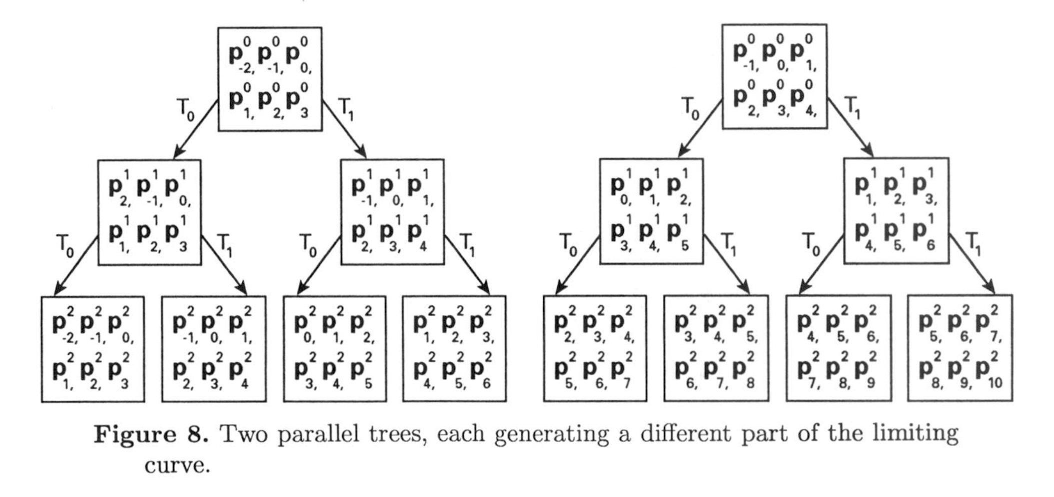

Figure 8. Two parallel trees, each generating a different part of the limiting curve.

§4. Product expansion

It follows from Theorem 2(c) above that when $\Delta = 2^k$, compactly supported L^1–solutions to (1) are k^{th} derivatives of solutions to the scaled version of (1), with coefficients $c_{\boldsymbol{\eta}}/2^k$. So we shall concentrate in what follows on the case $\Delta = 1$. In fact *we shall assume that the scaling coefficients satisfy the sum conditions* (2), which is the case in typical applications.

Let $C = [M_1, N_1] \times \cdots \times [M_d, N_d]$ be the smallest box containing the support $\{\boldsymbol{\eta} \in \mathbb{Z}^d : c_{\boldsymbol{\eta}} \neq 0\}$, and define $S \subseteq \mathbb{Z}^d$ as above, by

$$S = \{M_1, \ldots, N_1 - 1\} \times \cdots \times \{M_d, \ldots, N_d - 1\}.$$

As in [5] define the *multiscale operator* $\Psi : L^1(\mathbb{R}^d) \to L^1(\mathbb{R}^d)$ by

$$\Psi f(\mathbf{x}) = \sum_{\boldsymbol{\eta}} c_{\boldsymbol{\eta}} f(2\mathbf{x} - \boldsymbol{\eta}). \tag{32}$$

Observe that if $\sigma_{\mathbf{y}}$ denotes the *shift operator*

$$\sigma_{\mathbf{y}} f(\mathbf{x}) = f(\mathbf{x} - \mathbf{y}),$$

then

$$\Psi \sigma_{\mathbf{y}} = \sigma_{\mathbf{y}/2} \Psi. \tag{33}$$

For $f \subset L^1(\mathbb{R}^d)$ define $\Phi f \in L^1([0,1)^d)$ by

$$\Phi f(\mathbf{x}) = \sum_{\boldsymbol{\gamma} \in \mathbb{Z}^d} f(\mathbf{x} + \boldsymbol{\gamma}).$$

Observe that this *summation operator* $\Phi : L^1(\mathbb{R}^d) \to L^1([0,1)^d)$ is a linear contraction, $\|\Phi\| = 1$, and

$$\int_{[0,1)^d} \Phi f = \int_{\mathbb{R}^d} f. \tag{34}$$

Moreover, on account of (2), for any $f \in L^1(\mathbb{R}^d)$, we have

$$\Phi \Psi f = \Phi f \circ \tau, \tag{35}$$

where $\tau : [0,1)^d \to [0,1)^d$ is given by $\tau \mathbf{x} = 2\mathbf{x} (\text{mod } 1)$, as above.

For any function $f \in L^1(\mathbb{R}^d)$ with *supp* $(f) \subseteq C$, denote by $\boldsymbol{\Gamma} f \in L^1([0,1)^d; \mathbb{R}^S)$ the vector valued map

$$(\boldsymbol{\Gamma} f(\mathbf{x}))_{\boldsymbol{\gamma}} = f(\mathbf{x} + \boldsymbol{\gamma}), \quad \boldsymbol{\gamma} \in S. \tag{36}$$

Conversely, for any vector–valued map $\mathbf{v} \in L^1([0,1)^d; \mathbb{R}^S)$ denote by $F\mathbf{v} \in L^1(\mathbb{R}^d)$ the function

$$F\mathbf{v}(\mathbf{x}) = \begin{cases} v_{\boldsymbol{\gamma}}(\mathbf{x} - \boldsymbol{\gamma}), & \mathbf{x} \in \boldsymbol{\gamma} + [0,1)^d, \quad (\boldsymbol{\gamma} \in S) \\ 0, & \text{otherwise}, \end{cases} \tag{37}$$

with *supp* $F\mathbf{v} \subseteq C$. Observe that if $f \in C(\mathbb{R}^d)$ with *supp* $f \subseteq C$ then

$$\mathbf{\Gamma}\Psi f(\mathbf{x}) = T^t_{\boldsymbol{\omega}_1}\mathbf{\Gamma} f(\tau\mathbf{x}), \tag{38}$$

where $\boldsymbol{\omega}_1$ is the vector of the first bits in the dyadic expansion of $\mathbf{x} \in [0,1)^d$. To establish (38), we just use the formula (28) for $T_{\boldsymbol{\omega}}$, along with the relationship

$$\tau\mathbf{x} = 2\mathbf{x} - \boldsymbol{\omega}_1. \tag{39}$$

The *continuity assumption* on f enters here, using $f = 0$ on the boundary of C. More precisely, we need to use the fact that $f(\boldsymbol{\alpha} + \mathbf{x}) = 0$ whenever $\boldsymbol{\alpha} \notin S$, $\mathbf{x} \in [0,1)^d$, in arriving at (38).

The relationship (39) also resolves the issue about what to do in case some component(s) $x_i = 1/2$ in (38). Since $1/2$ has two dyadic expansions $(.1\overline{000}, .0\overline{111})$, it is not clear what to use as first bit. The answer is that *as long as* τ *is defined consistently with* $\boldsymbol{\omega}_1$, *so that* (39) *holds, it does not matter.* In fact τ is meant to left–shift the dyadic expansion of $\mathbf{x}$. Thus if $.1\overline{000}$ is used for $1/2$, then $\tau(1/2)$ must be 0, and if $.0\overline{111}$ is used then $\tau(1/2)$ must be 1.

The dilation equation (1) amounts to

$$\Psi f = f. \tag{40}$$

According to Theorem 2(b), there can be at most one solution $f \in L^1(\mathbb{R}^d)$ to (40), up to normalization. Moreover, since $\widehat{f}(\mathbf{0}) = \int f$, for f to be non–trivial in (17), it follows that $\int f \neq 0$. So we can always normalize f by requiring $\int f = 1$.

It also follows from (35) that if $f \in L^1(\mathbb{R}^d)$ satisfies (40), then Φf must be constant a.e. This is simply because Φf is τ–invariant and τ is ergodic. Furthermore, according to (34) this constant must be $\int f$; *i.e.*,

$$\Phi f \equiv \int f \quad \text{a.e.} \tag{41}$$

We study below compactly supported *continuous* solutions f to (40). It follows from Lemma 1 that necessarily *supp* $f \subseteq C$. It follows from (38) that $\Psi f = f$ if and only if

$$\mathbf{\Gamma} f(\mathbf{x}) = T^t_{\boldsymbol{\omega}_1}\mathbf{\Gamma} f(\tau\mathbf{x}), \quad \mathbf{x} \in [0,1)^d. \tag{42}$$

The continuity assumption on f enters through (38), where we need to know that $f = 0$ on the boundary of C, as mentioned above.

Recall from [11] the following setup. If Σ is a *bounded* set of linear transformations $\mathbb{R}^m \to \mathbb{R}^m$, and $\|\cdot\|$ is an *operator* norm, set

$$\|\Sigma\| = \sup\{\|T\| : T \in \Sigma\}, \quad \Sigma^n = \{T_1 \cdots T_n : T_i \in \Sigma, 1 \le i \le n\}.$$

Define

$$\widehat{\rho}(\Sigma) = \limsup_{n\to\infty} \|\Sigma^n\|^{1/n}.$$

It follows from [4, Lemma II(b)] and [11] that

$$\widehat{\rho}(\Sigma) = \inf_{\|\cdot\|} \|\Sigma\|,$$

where the inf is over *all* operator norms $\|\cdot\|$.

As mentioned above, the matrices $T_{\boldsymbol{\omega}}$ have row sums all equal to one. Thus, the constant vector $\mathbf{e} \in \mathbb{R}^S$ given by

$$e_{\boldsymbol{\gamma}} = 1, \qquad \forall \boldsymbol{\gamma} \in S,$$

satisfies $T_{\boldsymbol{\omega}}\mathbf{e} = \mathbf{e}$, $\forall \boldsymbol{\omega} \in \{0,1\}^d$. Let $\mathcal{E} \subseteq \mathbb{R}^S$ be the invariant subspace

$$\mathcal{E} = \{\mathbf{v} \in \mathbb{R}^S : \ T_{\boldsymbol{\omega}}\mathbf{v} = \mathbf{v}, \forall \boldsymbol{\omega} \in \{0,1\}^d\},$$

and set $r = \dim\ \mathcal{E}$. Then we can simultaneously reduce

$$T_{\boldsymbol{\omega}} \sim \begin{pmatrix} I & \mathbf{b}_{\boldsymbol{\omega}}^t \\ & \\ & \\ 0 & A_{\boldsymbol{\omega}} \end{pmatrix} \begin{matrix} \uparrow \\ r \\ \downarrow \\ \uparrow \\ |S| - r \\ \downarrow \end{matrix} \tag{43}$$

under a change–of–basis transformation. In what follows we make the *contractivity assumption*

$$\widehat{\rho}(A_{\boldsymbol{\omega}} : \boldsymbol{\omega} \in \{0,1\}^d) < 1. \tag{44}$$

Under this assumption it follows that for any sequence $\underline{\omega} = (\boldsymbol{\omega}_n) \in \Omega^d$ the limit

$$\lim_{n\to\infty} T_{\boldsymbol{\omega}_n} \cdots T_{\boldsymbol{\omega}_1} = T_\infty(\underline{\omega}) \tag{45}$$

exists, and is of rank r. Moreover, this limit is a continuous function of $\underline{\omega} \in \Omega^d$. (A "typical" metric on Ω^d is $d(\underline{\omega}', \underline{\omega}'') = \sum_n \|\boldsymbol{\omega}_n' - \boldsymbol{\omega}_n''\| 2^{-n}$ for sequences $\underline{\omega}' = (\boldsymbol{\omega}_n')$, $\underline{\omega}'' = (\boldsymbol{\omega}_n'') \in \Omega^d$. The only essential feature here is that "closeness" of two sequences means agreement of their first N components, where N is "large.")

We want to be able to identify the sequence $\underline{\omega} = (\boldsymbol{\omega}_n) \in \Omega^d$ with the point $\mathbf{x} \in [0,1]^d$ through the dyadic expansion

$$\mathbf{x} = \sum_n \boldsymbol{\omega}_n 2^{-n}, \tag{46}$$

in order to obtain a matrix–valued function $T_\infty(\mathbf{x})$. That is, we want to define

$$T_\infty(\mathbf{x}) = T_\infty(\underline{\omega}),$$

if $\mathbf{x}$ has the expansion (46). This is not necessarily well–defined, however, since dyadic components x_i have two different expansions (terminating and non–terminating). In order for $T_\infty(\mathbf{x})$ to be well–defined, we need to establish the consistency condition (30). This condition will be proved below, but in the meantime we adopt the convention that *the terminating expansion is used whenever some component x_i is dyadic.* This pins down a unique dyadic expansion for $\mathbf{x}$, and skirts the well–definedness issue for now.

Lemma 4. *Under the contractivity assumption (44), it follows that necessarily $r = 1$ in (43).*

Proof: Let $\mathbf{w} \in \mathbb{R}^S$ and define

$$\mathbf{v}(\mathbf{x}) = T_\infty^t(\mathbf{x})\mathbf{w}, \qquad \mathbf{x} \in [0,1)^d.$$

Then $\mathbf{v}(\mathbf{x})$ satisfies (42), and so the function $f = F\mathbf{v}$ defined above in (37) satisfies (40). Since $\mathbf{w}$ is arbitrary, the uniqueness of solutions $f \in L^1(\mathbb{R}^d)$ to (40) then implies that any two rows of $T_\infty(\mathbf{x})$ must be linearly dependent, for a.e. $\mathbf{x} \in [0,1)^d$. Thus, $T_\infty(\mathbf{x})$ must be rank one for a.e. $\mathbf{x}$. This in turn implies that $r = 1$. ■

Now we prove the consistency.

Theorem 5. *Let the contractivity assumption (44) hold. If $\sum_n \boldsymbol{\omega}_n' 2^{-n} = \sum_n \boldsymbol{\omega}_n'' 2^{-n}$, so that the sequences $\underline{\omega}' = (\boldsymbol{\omega}_n')$ and $\underline{\omega}'' = (\boldsymbol{\omega}_n'')$ correspond to the same $\mathbf{x} \in [0,1]^d$, then $T_\infty(\underline{\omega}') = T_\infty(\underline{\omega}'')$.*

Proof: Since the $T_{\boldsymbol{\omega}}$'s have row sums equal to one, and since the contractivity assumption (44) holds, it follows that each $T_{\boldsymbol{\omega}}^t$ has a unique normalized eigenvector $\mathbf{v} = \mathbf{v}(\boldsymbol{\omega})$, $\sum_{\boldsymbol{\gamma} \in S} v_{\boldsymbol{\gamma}} = 1$. Our first task is to find $\mathbf{v}$.

Let

$$S_+ = \{M_1 + 1, \ldots, N_1 - 1\} \times \cdots \times \{M_d + 1, \ldots, N_d - 1\}.$$

Observe that the matrix $T_+ \in Lin(\mathbb{R}^{S_+})$ given by

$$(T_+)_{\boldsymbol{\alpha},\boldsymbol{\beta}} = c_{2\boldsymbol{\beta}-\boldsymbol{\alpha}}; \qquad \boldsymbol{\alpha}, \boldsymbol{\beta} \in S_+,$$

also has row sums equal to one. Thus, there is an eigenvector $\mathbf{w}_+ \in \mathbb{R}^{S_+}$, $\sum_{\boldsymbol{\gamma} \in S_+} (\mathbf{w}_+)_{\boldsymbol{\gamma}} = 1$, satisfying $T_+^t \mathbf{w}_+ = \mathbf{w}_+$. Extend $\mathbf{w}_+ \in \mathbb{R}^{S_+}$ to $\mathbf{w} \in \mathbb{R}^{\mathbb{Z}^d}$ by defining

$$w_{\boldsymbol{\gamma}} = \begin{cases} (\mathbf{w}_+)_{\boldsymbol{\gamma}}, & \boldsymbol{\gamma} \in S_+ \\ 0, & \boldsymbol{\gamma} \notin S_+ \,. \end{cases}$$

Then it is easy to check that

$$v_{\boldsymbol{\gamma}} = w_{\boldsymbol{\gamma}+\boldsymbol{\omega}}, \qquad \boldsymbol{\gamma} \in S, \tag{47}$$

is the normalized eigenvector $\mathbf{v} = \mathbf{v}(\boldsymbol{\omega}) \in \mathbb{R}^S$ of $T^t_{\boldsymbol{\omega}}$.

The relationship (47) between the eigenvectors $\mathbf{v}(\boldsymbol{\omega})$ leads to the following notion of *shift* on $\mathbb{R}^S$. Given $\mathbf{v} \in \mathbb{R}^S$, extend it to $\widehat{\mathbf{v}} \in \mathbb{R}^{\mathbb{Z}^d}$ by defining $\widehat{v}_{\boldsymbol{\gamma}} = 0$, $\boldsymbol{\gamma} \notin S$. Then for any $\boldsymbol{\omega} \in \{0,1\}^d$ the *left $\boldsymbol{\omega}$–shift* $L_{\boldsymbol{\omega}} \in Lin(\mathbb{R}^S)$ is given by

$$(L_{\boldsymbol{\omega}}\mathbf{v})_{\boldsymbol{\gamma}} = \widehat{v}_{\boldsymbol{\gamma}+\boldsymbol{\omega}}, \quad \boldsymbol{\gamma} \in S.$$

Observe that $L^t_{\boldsymbol{\omega}}$ corresponds to the *right ω–shift*

$$(L^t_{\boldsymbol{\omega}}\mathbf{v})_{\boldsymbol{\gamma}} = \widehat{v}_{\boldsymbol{\gamma}-\boldsymbol{\omega}}, \quad \boldsymbol{\gamma} \in S.$$

On account of the way $\mathbf{v}$ was extended to be zero outside of S, it turns out that $L^t_{\boldsymbol{\omega}}$ is not quite the inverse of $L_{\boldsymbol{\omega}}$. In fact, $L^t_{\boldsymbol{\omega}} L_{\boldsymbol{\omega}}\mathbf{v} = \mathbf{v}$ if and only if $v_{\boldsymbol{\gamma}} = 0$ whenever $\boldsymbol{\gamma} \notin S \cap (\boldsymbol{\omega} + S)$ (*i.e.*, $\mathbf{v}$ is supported on $S \cap (\boldsymbol{\omega} + S)$).

It follows from (47) that

$$\mathbf{v}(\boldsymbol{\omega}_2) = L_{\boldsymbol{\omega}_2 - \boldsymbol{\omega}_1}\mathbf{v}(\boldsymbol{\omega}_1), \quad \forall \boldsymbol{\omega}_1, \boldsymbol{\omega}_2 \in \{0,1\}^d. \tag{48}$$

(Observe here that since $\mathbf{w}$ is supported on S_+, no non–zero elements are "lost" due to spill–over outside of S in (48).) It can also be checked that the identity

$$T^t_{\boldsymbol{\omega}_1} L_{\boldsymbol{\omega}_3} = L_{\boldsymbol{\omega}_4} T^t_{\boldsymbol{\omega}_2} L^t_{\boldsymbol{\omega}_3} L_{\boldsymbol{\omega}_3} \tag{49}$$

holds, whenever $\boldsymbol{\omega}_1 + \boldsymbol{\omega}_3 - \boldsymbol{\omega}_2 + 2\boldsymbol{\omega}_4$.

To prove the theorem it suffices, by symmetry, to establish (30); *i.e.*,

$$\lim_{n\to\infty} T_{0,\boldsymbol{\rho}_n} \cdots T_{0,\boldsymbol{\rho}_2} T_{1,\boldsymbol{\rho}_1} = \lim_{n\to\infty} T_{1,\boldsymbol{\rho}_n} \cdots T_{1,\boldsymbol{\rho}_2} T_{0,\boldsymbol{\rho}}$$

for any sequence $(\boldsymbol{\rho}_n) \in \Omega^{d-1}$. Suppose first that the sequence $(\boldsymbol{\rho}_n)$ is *eventually constant*, so that

$$\boldsymbol{\rho}_n = \boldsymbol{\rho} \quad \text{whenever } n > N.$$

Since $T^\infty_{0,\boldsymbol{\rho}}$ is the matrix

$$T^\infty_{0,\boldsymbol{\rho}} = \begin{pmatrix} \mathbf{v}^t(0,\boldsymbol{\rho}) \\ \cdots\cdots\cdots \\ \mathbf{v}^t(0,\boldsymbol{\rho}) \end{pmatrix}$$

with identical row vectors, and similarly for $T^\infty_{1,\boldsymbol{\rho}}$, it suffices to show that

$$\mathbf{v}^t(0,\boldsymbol{\rho}) T_{0,\boldsymbol{\rho}_N} \cdots T_{0,\boldsymbol{\rho}_2} T_{1,\boldsymbol{\rho}_1} = \mathbf{v}^t(1,\boldsymbol{\rho}) T_{1,\boldsymbol{\rho}_N} \cdots T_{1,\boldsymbol{\rho}_2} T_{0,\boldsymbol{\rho}_1}.$$

Let $L = L_{\boldsymbol{\omega}}$ be the left $\boldsymbol{\omega}$–shift for $\boldsymbol{\omega} = (1,0,\ldots,0)^t$. By (48), we have

$$\mathbf{v}^t(1,\boldsymbol{\rho}) = \mathbf{v}^t(0,\boldsymbol{\rho}) L^t,$$

and so by (49) , it follows that

$$\begin{aligned} \mathbf{v}^t(1,\boldsymbol{\rho}) T_{1,\boldsymbol{\rho}_N} \cdots T_{1,\boldsymbol{\rho}_2} T_{0,\boldsymbol{\rho}_1} &- \mathbf{v}^t(0,\boldsymbol{\rho}) L^t L T_{0,\boldsymbol{\rho}_N} L^t T_{1,\boldsymbol{\rho}_{N-1}} \cdots T_{1,\boldsymbol{\rho}_2} T_{0,\boldsymbol{\rho}_1} \\ &= \mathbf{v}^t(0,\boldsymbol{\rho}) L^t L T_{0,\boldsymbol{\rho}_N} L^t L T_{0,\boldsymbol{\rho}_{N-1}} L^t L \cdots T_{0,\boldsymbol{\rho}_2} L^t L T_{1,\boldsymbol{\rho}_1} L^t_{\mathbf{0}}. \end{aligned}$$

Now, as mentioned above, $L^tL\mathbf{v} = \mathbf{v}$ if and only if $\mathbf{v}$ belongs to the subspace

$$\{\mathbf{v} \in \mathbb{R}^S : v_{\boldsymbol{\gamma}} = 0, \forall \boldsymbol{\gamma} = (0, *)\};$$

and, moreover, this subspace is invariant under $T_{\boldsymbol{\omega}}$, for any $\boldsymbol{\omega} = (0, *)$. Thus,

$$\begin{aligned}\mathbf{v}^t(0, \boldsymbol{\rho})L^tLT_{0,\boldsymbol{\rho}_N}L^tLT_{0,\boldsymbol{\rho}_{N-1}}L^tL\cdots T_{0,\boldsymbol{\rho}_2}L^tLT_{1,\boldsymbol{\rho}_1}\\ = \mathbf{v}^t(0, \boldsymbol{\rho})T_{0,\boldsymbol{\rho}_N}\cdots T_{0,\boldsymbol{\rho}_2}T_{1,\boldsymbol{\rho}_1},\end{aligned}$$

which gives the desired result.

Since the limit $T_\infty(\boldsymbol{\omega})$ depends continuously on $\underline{\omega}$, we can obtain the above result in general by approximating $\underline{\omega}$ by sequences which are eventually constant. Indeed the only components of $\mathbf{x} = \sum_n \boldsymbol{\omega}'_n 2^{-n}$ which do not have eventually constant dyadic expansions are the non–dyadic components. These components have unique dyadic expansions, and so the corresponding components of $\boldsymbol{\omega}'_n$ and $\boldsymbol{\omega}''_n$ must agree for all n. Thus we can approximate these component sequences by eventually constant sequences. ■

From the consistency in Theorem 5 and the fact that T_∞ is continuous in $\boldsymbol{\omega} \in \Omega^d$, we get that T_∞ is also continuous in $\mathbf{x} \in [0, 1]^d$.

Lemma 6. *Under the contractivity assumption (44), there exists a non–trivial continuous compactly supported solution f to (1).*

Proof: Let $\mathbf{w} \in \mathbb{R}^S$ satisfy $\sum_{\boldsymbol{\gamma}\in S} w_{\boldsymbol{\gamma}} \neq 0$. Then, as in the proof of Lemma 4 above, $\mathbf{v}(\mathbf{x}) = T^t_\infty(\mathbf{x})\mathbf{w}$ satisfies (42), and so the function $f = F\mathbf{v}$ defined above in (37) satisfies (40). Since $T_\infty(\mathbf{x})$ has row sums all equal to one and is rank–one, it follows that all of its rows consist of the same row vector $\mathbf{r}^t(\mathbf{x})$. Thus, $\mathbf{v}(\mathbf{x}) = \left(\sum_{\boldsymbol{\gamma}\in S} w_{\boldsymbol{\gamma}}\right)\mathbf{r}(\mathbf{x})$, and so, $\mathbf{v}$ and $f = F\mathbf{v}$ are non–trivial. As mentioned above, under assumption (44) the function $\mathbf{v}(\mathbf{x})$ is continuous in $\mathbf{x} \in [0, 1]^d$, and it remains to show that $f = F\mathbf{v}$ is continuous in $\mathbb{R}^d$.

It can be seen from (37) that the "match–up" conditions at the boundaries of $\boldsymbol{\gamma}+[0, 1)^d$ are $v_{\boldsymbol{\gamma}}(\mathbf{x}) = v_{\boldsymbol{\gamma}+\boldsymbol{\omega}}(\mathbf{x}-\boldsymbol{\omega})$ *whenever* $\boldsymbol{\omega} \in \{0, 1\}^d$ *and* $\mathbf{x}, \mathbf{x}-\boldsymbol{\omega} \in [0, 1]^d$. Our convention here is that $v_{\boldsymbol{\gamma}} = 0$ if $\boldsymbol{\gamma} \notin S$. Equivalently, then, we need to show that

$$T^t_\infty(\mathbf{x}) = L_{\boldsymbol{\omega}}T^t_\infty(\mathbf{x} - \boldsymbol{\omega}), \tag{50}$$

whenever $\boldsymbol{\omega} \in \{0, 1\}^d$ and $\mathbf{x}, \mathbf{x} - \boldsymbol{\omega} \in [0, 1]^d$. By stepping from $\mathbf{x} - \boldsymbol{\omega}$ to $\mathbf{x}$ one coordinate at a time, we may assume, by symmetry, that $\boldsymbol{\omega} = (1, 0, \ldots, 0)^t$. Denote $L = L_{\boldsymbol{\omega}}$. Then (50) becomes

$$\lim_{n\to\infty} T_{0,\boldsymbol{\rho}_N}\cdots T_{0,\boldsymbol{\rho}_1}L^t = \lim_{n\to\infty} T_{1,\boldsymbol{\rho}_n}\cdots T_{1,\boldsymbol{\rho}_1}$$

for any sequence $(\boldsymbol{\rho}_n) \in \Omega^{d-1}$. This conclusion follows exactly as in the proof of Theorem 5 above. Indeed, suppose first that the sequence $(\boldsymbol{\rho}_n)$ is eventually constant, so that

$$\boldsymbol{\rho}_n = \boldsymbol{\rho} \text{ whenever } n > N.$$

Then it suffices to show that

$$\mathbf{v}^t(0,\boldsymbol{\rho})T_{0,\boldsymbol{\rho}_N}\cdots T_{0,\boldsymbol{\rho}_1}L^t = \mathbf{v}^t(1,\boldsymbol{\rho})T_{1,\boldsymbol{\rho}_N}\cdots T_{1,\boldsymbol{\rho}_1}.$$

By (48), we have

$$\mathbf{v}^t(1,\boldsymbol{\rho}) = \mathbf{v}^t(0,\boldsymbol{\rho})L^t,$$

and so by (49), we obtain

$$\mathbf{v}^t(1,\boldsymbol{\rho})T_{1,\boldsymbol{\rho}_N}\cdots T_{1,\boldsymbol{\rho}_1} = \mathbf{v}^t(0,\boldsymbol{\rho})L^tLT_{0,\boldsymbol{\rho}_N}L^tT_{1,\boldsymbol{\rho}_{N-1}}\cdots T_{1,\boldsymbol{\rho}_1},$$

etc. ■

Our final concern in this paper involves convergence of the iterates $\Psi^n f_0$.

Theorem 7. *Let the contractivity assumption (44) hold, and let f_0 be bounded measurable, with supp $f_0 \subseteq C$. Then $\Psi^n f_0$ converges uniformly to a continuous function f if and only if $\Phi f_0 \equiv$ const.*

Proof: To argue the necessity, we suppose $\Psi^n f_0$ converges uniformly to f. In particular $\Psi^n f_0 \to f$ in $L^1(\mathbb{R}^d)$ since these functions are all supported in C. Since $\Phi : L^1(\mathbb{R}^d) \to L^1([0,1)^d)$ is bounded, it follows from (35) and (41) that $\Phi f_0 \circ \tau^n - \int f \to 0$ in measure, and so since τ is measure-preserving $\Phi f_0 \equiv \int f$ a.e.

To argue the sufficiency, we suppose $\Phi f_0 \equiv$ const., and assume without loss of generality that $\Phi f_0 \equiv 1$. Observe that by (38), for any $\mathbf{x} \in [0,1)^d$

$$\boldsymbol{\Gamma}^t f(\mathbf{x}) - \boldsymbol{\Gamma}^t f_n(\mathbf{x}) = \boldsymbol{\Gamma}^t f_0(\tau^n \mathbf{x})[T_\infty(\mathbf{x}) - T_{\boldsymbol{\omega}_n}\cdots T_{\boldsymbol{\omega}_1}],$$

where $f_n = \Psi^n f_0$, f is the solution to (1) constructed in Lemma 6 (normalized so that $\int f = 1$) and $\boldsymbol{\omega}_1, \ldots, \boldsymbol{\omega}_n$ are the n first vectors of bits in the dyadic expansion of $\mathbf{x}$. We write

$$T_\infty(\mathbf{x}) - T_{\boldsymbol{\omega}_n}\cdots T_{\boldsymbol{\omega}_1} = [T_\infty(\mathbf{x}) - T_\infty(\mathbf{x}^n)] + (T_0^\infty - I)T_{\boldsymbol{\omega}_n}\cdots T_{\boldsymbol{\omega}_1},$$

where $\mathbf{x}^n$ is the (terminating) dyadic number

$$\mathbf{x}^n = .\boldsymbol{\omega}_1\cdots\boldsymbol{\omega}_n = \sum_{k=1}^n \boldsymbol{\omega}_k 2^{-k}.$$

Since $T_\infty(\mathbf{x})$ is uniformly continuous in $\mathbf{x}$, and since

$$(T_0^\infty - I)T_{\boldsymbol{\omega}_n}\cdots T_{\boldsymbol{\omega}_1} \sim \begin{pmatrix} 0 & \mathbf{b}_\infty^t(\mathbf{0})A_{\boldsymbol{\omega}_n}\cdots A_{\boldsymbol{\omega}_1} \\ 0 & -A_{\boldsymbol{\omega}_n}\cdots A_{\boldsymbol{\omega}_1} \end{pmatrix}$$

by (43), it follows from (44) that $T_\infty(\mathbf{x}) - T_{\boldsymbol{\omega}_n}\cdots T_{\boldsymbol{\omega}_1}$ converges uniformly to zero. Thus $\boldsymbol{\Gamma}^t f_n \to \boldsymbol{\Gamma}^t f$ uniformly, from which it follows that $f_n \to f$ uniformly. The continuity of f was established in Lemma 6. ■

If f_0 is supported in some shifted box $C + \mathbf{y}$ and satisfies $\Phi f_0 \equiv$ const., then the conclusion of Theorem 7 still holds; namely, $\Psi^n f_0 \to f$ uniformly. This follows from (33) and the uniform continuity of f.

The assumption (44) cannot be relaxed in Theorem 7, as the next result shows.

Lemma 8. *Let $f_0 = \chi_{[0,1)^d}$. If the sequence of iterates $(\Psi^n f_0)$ converges uniformly to a non–trivial continuous solution of (1), then (44) must hold.*

Proof: As mentioned just above, it follows from the assumption of this lemma that in fact $\Psi^n \sigma_{\boldsymbol{\beta}} f_0 \to f$ uniformly, for every $\boldsymbol{\beta} \in \mathbb{Z}^d$, where $\sigma_{\boldsymbol{\beta}} f_0$ is the shifted function. Let $(\boldsymbol{\omega}_n) \in \Omega^d$ be an arbitrary sequence, and correspondingly let $\mathbf{x} \in [0,1]^d$ have the dyadic expansion $\mathbf{x} = .\boldsymbol{\omega}_1 \boldsymbol{\omega}_2 \cdots = \sum_{n=1}^{\infty} \boldsymbol{\omega}_n 2^{-n}$. Observe that for any $\boldsymbol{\beta} \in \mathbb{Z}^d$ and for sufficiently large n, the vector $\boldsymbol{\Gamma} \sigma_{\boldsymbol{\beta}} f_0(\tau^n \mathbf{x}) \in \mathbb{R}^S$ is constant, independent of n. In fact if $\boldsymbol{\omega} \in \{0,1\}^d$ is such that $\tau^n \mathbf{x} - \boldsymbol{\omega} \in [0,1)^d$ (*i.e.*, all components strictly less than one), then $\mathbf{u} = \boldsymbol{\Gamma} \sigma_{\boldsymbol{\beta}} f_0(\tau^n \mathbf{x})$ is the vector

$$u_{\boldsymbol{\gamma}} = \begin{cases} 1, & \boldsymbol{\gamma} = \boldsymbol{\beta} - \boldsymbol{\omega} \\ 0, & \text{otherwise} . \end{cases}$$

Thus by (38), for sufficiently large n, we have

$$\mathbf{u}^t T_{\boldsymbol{\omega}_n} \cdots T_{\boldsymbol{\omega}_1} = \boldsymbol{\Gamma}^t \sigma_{\boldsymbol{\beta}} f_0(\tau^n \mathbf{x}) T_{\boldsymbol{\omega}_n} \cdots T_{\boldsymbol{\omega}_1} = \boldsymbol{\Gamma}^t \Psi^n \sigma_{\boldsymbol{\beta}} f_0(\mathbf{x}) \to \boldsymbol{\Gamma}^t f(\mathbf{x}). \tag{51}$$

By varying $\boldsymbol{\beta}$ over $S + \boldsymbol{\omega}$, we generate a basis of unit vectors $\mathbf{u} \in \mathbb{R}^S$ in (51), allowing us to conclude that $\lim_{n\to\infty} T_{\boldsymbol{\omega}_n} \cdots T_{\boldsymbol{\omega}_1}$ exists, and is of rank one. It follows then from (43) that $r = 1$ and $\lim_{n\to\infty} A_{\boldsymbol{\omega}_n} \cdots A_{\boldsymbol{\omega}_1} = 0$, for any sequence $(\boldsymbol{\omega}_n) \in \Omega^d$. Using a result of [4], [11], we conclude that (44) necessarily holds. ■

One can try to relax the assumption (44) and carry out an analysis of (1) and the sequence of iterates $\Psi^n f_0$. Rather than making the strict contractivity assumption, one can merely assume that the set $\{T_{\boldsymbol{\omega}} : \boldsymbol{\omega} \in \{0,1\}^d\}$ is LCP (left convergent products). This means that $\lim T_{\boldsymbol{\omega}_n} \cdots T_{\boldsymbol{\omega}_1}$ exists for every sequence $(\boldsymbol{\omega}_n) \in \Omega^d$. Under this weaker assumption it can still be shown that $r = 1$, as in Lemma 4, but in general one no longer has continuity of the solution f, nor uniform convergence of the iterates $\Psi^n f_0$. This analysis was carried out in the 1–D case in [5] (cf. Lemma II and Theorem V there), and can be extended to the d–dimensional case. The interested reader can go through the examples, arguments and constructions in [5], and see how they extend to the multi-dimensional setting.

Figure 6. Surfaces generated by the "butterfly" division scheme.

References

1. Barnsley, M. F., *Fractals Everywhere*, Academic Press, NY, 1988.
2. Berger, M. A., Random affine iterated function systems: smooth curve generation, *SIAM Review*, to appear.
3. Berger, M. A., Wavelets as attractors of random dynamical systems, in *Stochastic Analysis: Liber Amicorum for Moshe Zakai*, E. Mayer-Wolf, M. Zakai, E. Merzbach, and A. Shwartz (eds.), Academic Press, NY, 1991.
4. Berger, M. A. and Y. Wang, Bounded semi-groups of matrices, *Linear Algebra Appl.*, to appear.
5. Berger, M. A. and Y. Wang, Multi-scale dilation equations and iterated function systems, Report Math: 011491-001, Georgia Institute of Technology, preprint.
6. Cavaretta, A. S., W. Dahmen, and C. A. Micchelli, Stationary subdivision, *Mem. Amer. Math. Soc.*, to appear.
7. Chaikin, G. M., An algorithm for high speed curve generation, *Comp. Graphics and Image Proc.* **3** (1974), 346–349.
8. Daubechies, I., Orthonormal bases of compactly supported wavelets, *Comm. Pure Appl. Math.* **41** (1988), 909–996.
9. Daubechies, I. and J. C. Lagarias, Two-scale difference equations I: global regularity of solutions, *SIAM J. Appl. Math.*, to appear.
10. Daubechies, I. and J. C. Lagarias, Two-scale difference equations II: infinite matrix products, local regularity and fractals, *SIAM J. Appl. Math.*, to appear.
11. Daubechies, I. and J. C. Lagarias, Sets of matrices all infinite products of which converge, *Linear Algebra Appl.*, to appear.
12. Dyn, N., Subdivision schemes in CAGD, in *Advances in Numerical Analysis II: Wavelets, Subdivision Algorithms, and Radial Functions*, Oxford University Press, W. A. Light, (ed.), Oxford, 1991, 36–104.
13. Dyn, N., J. A. Gregory, and D. Levin, A 4-point interpolatory subdivision scheme for curve design, *Comput. Aided Geom. Design* **4** (1987), 257–268.
14. Dyn, N., J. A. Gregory, and D. Levin, A butterfly subdivision scheme for surface interpolation with tension control, *ACM Trans. on Graphics* **9** (1990), 160–169.
15. Dyn, N., J. A. Gregory, and D. Levin, Analysis of uniform binary subdivision schemes for curve design, *Const. Approx.*, to appear.
16. Dyn, N. and D. Levin, Interpolating subdivision schemes for the generation of curves and surfaces, in *International Series of Numerical Mathematics*, Vol. 94, Birkhäuser, Basel, 1990.
17. Dyn, N., D. Levin, and C. A. Micchelli, Using parameters to increase smoothness of curves and surfaces generated by subdivision, *Comput. Aided Geom. Design* **7** (1990), 129–140.
18. Golumb, S., *Shift Register Sequences*, Holden-Day, San Francisco, 1967.
19. Gröchenig, K. and W. R. Madych, Multiresolution analysis, Haar bases and self-similar tilings of $\mathbb{R}^n$, BRC technical report, *Trans. IEEE*, to appear.

20. Hörmander, L., *The Analysis of Linear Partial Differential Operators I*, Springer–Verlag, NY, 1983.
21. Lane, J. M. and R. F. Riesenfeld, A theoretical development for the computer generation of piecewise polynomial surfaces, *IEEE Trans. Pattern Anal. Machine Intell.* **2** (1980), 35–46.
22. Micchelli, C. A. and H. Prautzsch, Uniform refinement of curves, *Linear Algebra Appl.* **114/115** (1989), 841–870.
23. Micchelli, C. A. and H. Prautzsch, Refinement and subdivision for spaces of integer translates of a compactly supported function, in *Numerical Analysis*, D. A. Griffiths and G. A. Watson (eds.), 1987, 192–222.
24. de Rham, G., Sur une courbe plane, *J. Math. Pure Appl.* **39** (1956), 25–42.
25. Strang, G., Wavelets and dilation equations: a brief introduction, *SIAM Review* **31** (1989), 614–627.

This Research was supported by Air Force Office of Scientific Research Grant No. AFOSR–90–0288 and NSF Grant No. DMS–8915322.

Marc A. Berger
Micro-Electronics Building
Georgia Institute of Technology
791 Atlantic Drive
Atlanta, GA 30332-0269
berger@math.gatech.edu

Yang Wang
School of Mathematics
Georgia Institute of Technology
Atlanta, GA 30332-0269
wang@math.gatech.edu

Multivariate Wavelets

Joachim Stöckler

Abstract. The general concept of multivariate wavelet bases is described with emphasis on the construction procedure. A universal tool to identify special properties of a wavelet basis is provided by an associated symbol matrix. Conditions on the inheritance of decay properties, vanishing moments and generalized linear phase from the multiresolution analysis are presented. An explicit construction of compactly supported box-spline wavelets in arbitrary dimension is included.

§1. Introduction

Our interest is the description of multivariate wavelet bases and of their explicit construction. Many results in this direction as well as applications of orthonormal wavelets to functional spaces (e.g., L^p, Hölder- and Besov-spaces) are included in Meyer's book [27]. For univariate wavelets Chui [6] gives a comprehensive and unified description which highlights the importance of wavelets from the viewpoint of transform theory. We follow the more general definition of wavelets in [6]. First we introduce the following notations.

If I is a countable index set the set $\{f^i : i \in I\}$ of functions $f^i \in L^2(\mathbb{R}^s)$ is called *ℓ^2-stable*, if there exist constants $A_2 \geq A_1 > 0$ such that

$$A_1 \|(c_i)\|_{\ell^2} \leq \Big\| \sum_{i \in I} c_i f^i \Big\|_{L^2} \leq A_2 \|(c_i)\|_{\ell^2} \tag{1}$$

for any sequence $(c_i) \in \ell^2(I)$. The inner product of $f, g \in L^2(\mathbb{R}^s)$ is denoted by (f, g), and for $k \in \mathbb{Z}$ and $\alpha \in \mathbb{Z}^s$ we let

$$f_{k,\alpha} := 2^{ks/2} f(2^k \cdot - \alpha).$$

Definition 1.

(a) *A finite family $\{\psi^i : 1 \leq i \leq N\}$ is called a wavelet basis of $L^2(\mathbb{R}^s)$, if the collection $\{\psi^i_{k,\alpha} : 1 \leq i \leq N,\ k \in \mathbb{Z},\ \alpha \in \mathbb{Z}^s\}$ is a Riesz basis of $L^2(\mathbb{R}^s)$; i.e., it is ℓ^2-stable and its linear span is dense in $L^2(\mathbb{R}^s)$.*

Wavelets–A Tutorial in Theory and Applications
C. K. Chui (ed.), pp. 325–355.

ISBN 0-12-174590-2

(b) *The wavelet basis* $\{\psi^i\}$ *is called orthonormal if*

$$\left(\psi^i_{k,\alpha}\,,\,\psi^j_{l,\beta}\right) = \delta_{ij}\delta_{kl}\delta_{\alpha\beta} \tag{2}$$

for all $1 \leq i,j \leq N$, $k,l \in \mathbb{Z}$ *and* $\alpha,\beta \in \mathbb{Z}^s$. *It is called semi-orthogonal, if*

$$\left(\psi^i_{k,\alpha}\,,\,\psi^j_{l,\beta}\right) = 0 \quad \textit{whenever} \quad k \neq l.$$

A family $\{\widetilde{\psi}^i : 1 \leq i \leq N\} \subset L^2(\mathbb{R}^s)$ is called a *dual wavelet basis* if $\{\widetilde{\psi}^i_{k,\alpha}\}$ is also a Riesz basis of $L^2(\mathbb{R}^s)$ and if (2) holds with $\widetilde{\psi}^j_{l,\beta}$ instead of $\psi^j_{l,\beta}$. An orthonormal wavelet basis is therefore self-dual. In the literature the word "pre-wavelet" is often used for the elements of a semi-orthogonal wavelet basis. It is known from the univariate case that not every wavelet basis has a dual wavelet basis. If both a wavelet basis $\{\psi^i\}$ and its dual $\{\widetilde{\psi}^i\}$ exist, we can represent any $f \in L^2$ in the form

$$f = \sum_{i,k,\alpha} \left(f\,,\,\widetilde{\psi}^i_{k,\alpha}\right)\psi^i_{k,\alpha}$$

which we call a *complete wavelet decomposition*.

Special interest has been devoted to the construction of compactly supported smooth wavelets since they provide unconditional bases for a large variety of Banach spaces. Daubechies [13] developped orthonormal wavelet bases $\{{}_N\psi\}$ in $\mathbb{R}$ with these properties. Chui and Wang [9] introduced the notion of semi-orthogonal wavelet bases and obtained cardinal spline functions $\{\psi_m\}$ with compact support, which have many desirable properties such as symmetry and computational efficiency.

The main tool in merely all constructions of wavelet bases is the multiresolution analysis by [25] and [26], see also [17].

Definition 2. *A sequence* $(V_k)_{k\in\mathbb{Z}}$ *of closed subspaces of* $L^2(\mathbb{R}^s)$ *forms a (dyadic) multiresolution analysis if*

(i) $V_k \subset V_{k+1}$ *for all* $k \in \mathbb{Z}$,

(ii) $\bigcap_{k\in\mathbb{Z}} V_k = \{0\}$ *and* $\overline{\bigcup_{k\in\mathbb{Z}} V_k} = L^2(\mathbb{R}^s)$,

(iii) $f \in V_k \iff f(2\,.) \in V_{k+1}$ *for all* $k \in \mathbb{Z}$, *and*

(iv) *there is an element* $\phi \in V_0$ *such that* $\{\phi_{0,\alpha} : \alpha \in \mathbb{Z}^s\}$ *is a Riesz basis of* V_0.

The function ϕ is called the *generator* of the multiresolution analysis. It follows from (iii) that $\{\phi_{k,\alpha} : \alpha \in \mathbb{Z}^s\}$ is a Riesz basis of V_k. The word "dyadic" is associated with the dilation in (iii). A more general situation is considered in [11].

Let W_k, $k \in \mathbb{Z}$, be the orthogonal complement of V_k in V_{k+1}. Then (ii) gives the decomposition relation

$$L^2(\mathbb{R}^s) = \bigoplus_{k\in\mathbb{Z}} W_k$$

where '$\oplus$' denotes an orthogonal sum of vector spaces. Note that property (iii) is inherited by the sequence of spaces $(W_k)_{k\in\mathbb{Z}}$. The task of finding a semi-orthogonal wavelet basis is therefore reduced to the construction of a Riesz basis $\{\psi^i_{0,\alpha} : \alpha \in \mathbb{Z}^s, 1 \leq i \leq N\}$ of W_0. The following theorem was first proved by Gröchenig [15].

Theorem 3. *If $\phi \in L^2(\mathbb{R}^s)$ generates a multiresolution analysis and is regular, i.e., $|\phi(x)| \leq C_m(1 + \|x\|)^{-m}$ for all $m \in \mathbb{N}$ as $\|x\|$ tends to infinity, then there exists an orthonormal wavelet basis $\{\psi^i : 1 \leq i \leq 2^s - 1\}$ of functions ψ^i in W_0.*

A proof of this result under milder conditions on ϕ is included in Section 3.4. A dual wavelet basis is obtained by constructing a dual basis of W_0. With both wavelet basis and dual wavelet basis in W_0 we can decompose a function $f_N \in V_N$ on a fine scale as

$$f_N = f_{N-M} + \sum_{k=N-M}^{N-1} g_k \quad \text{with} \quad f_{N-M} \in V_{N-M},\ g_k \in W_k. \tag{3}$$

This is done recursively by taking the projections $f_k \in V_k$ and $g_k \in W_k$ of f_{k+1} for $N - M \leq k \leq N - 1$. The above decomposition is called *wavelet decomposition* of f_N. The practical importance of semi-orthogonal wavelets is based on the fact that (3) can be performed in a very efficient way by virtue of the "pyramidal scheme" (see [5], [13], and [25]).

This paper is organized as follows. Section 2 gives a detailed study of multiresolution analysis. In Section 3 a general procedure for the construction of semi-orthogonal wavelet bases and their duals is given, which is based on an "initial decomposition" of V_1. The derivation of dual wavelet bases and of the decomposition relation (3) is included. Fast decay properties or compact support of wavelet functions ψ^i are obtained by this procedure. Section 4 deals with further properties of the so constructed wavelet basis, which are generalized linear phase and vanishing moments. Section 5 is devoted to special cases where initial decompositions of V_1 can be given explicitly. In particular, the construction of a wavelet basis from box-splines in arbitrary dimension [30] is presented, since it provides a large class of wavelet bases with many desirable properties.

We finish this introduction with some notations. The Fourier transform of $f \in L^2(\mathbb{R}^s)$ is given by $\widehat{f}(y) = \int_{\mathbb{R}^s} f(x)e^{-ix\cdot y}\,dx$. For a sequence $c = (c_\alpha) \in \ell^2(\mathbb{Z}^s)$ we define the symbol

$$C(z) := \sum_{\alpha\in\mathbb{Z}^s} c_\alpha z^\alpha, \qquad z \in \mathbb{C}^s.$$

If $c \in \ell^1$ then $C(z)$ is continuous on the torus $\mathbb{T}^s = \{z \in \mathbb{C}^s : |z_k| = 1\}$. The semi-discrete convolution is defined as $c \tilde{*} f = \sum_{\alpha \in \mathbb{Z}^s} c_\alpha f(\,.\, - \alpha)$ whenever the series converges pointwise. For $c \in \ell^1$, $f \in L^2$ we obtain

$$(c \tilde{*} f)\widehat{\ }(y) = C(z^2)\widehat{f}(y), \tag{4}$$

where we assume that the relations $z_k = e^{-iy_k/2}$ hold. This correspondence between z and y is valid throughout this paper. We let $\mathcal{C} = (-\pi, \pi)^s$ and $E = \{0,1\}^s$. Note that

$$\|C(z^2)\|_{L^2(\mathcal{C})} = (2\pi)^{s/2}\|c\|_{\ell^2}, \tag{5}$$

if we use the standard Lebesgue measure on $\mathcal{C}$. For a symbol function $C(z)$ (which is always $4\pi-$periodic) we use the subscript notation

$$C_d(z) := C(e^{-i(y+2\pi d)/2}) = \sum_{\alpha \in \mathbb{Z}^s} (-1)^{d \cdot \alpha} c_\alpha z^\alpha \tag{6}$$

for any $d \in E$, where the expression e^{-iy} means $(e^{-iy_1}, \ldots, e^{-iy_s})$. The set E has a natural ordering $d \le d'$ if $\sum_{k=1}^s d_k 2^{k-1} \le \sum_{k=1}^s d'_k 2^{k-1}$.

§2. Multiresolution analysis

In this section we investigate some properties of a multiresolution analysis generated by $\phi \in V_0$. We also find weak conditions on the generator ϕ such that it satisfies all conditions in Definition 2.

2.1. The general case

Since $\{\phi_{1,\alpha} : \alpha \in \mathbb{Z}^s\}$ is a Riesz basis for the space V_1 we can find a sequence $(p_\alpha) \in \ell^2(\mathbb{Z}^s)$ such that

$$\phi(x) = \sum_{\alpha \in \mathbb{Z}^s} p_\alpha \phi(2x - \alpha). \tag{7}$$

From Theorem 4(a) below it becomes clear that the symbol calculus (4) can be applied to (7) giving

$$\widehat{\phi}(y) = 2^{-s} P(z)\widehat{\phi}(y/2). \tag{8}$$

We call P the *two-scale function* of ϕ. Another important periodic function associated with ϕ is

$$B(z^2) := \sum_{\alpha \in \mathbb{Z}^s} |\widehat{\phi}(y + 2\pi\alpha)|^2 \tag{9}$$

which is in $L^1(\mathcal{C})$. A different representation of B is obtained from the function

$$\Phi(x) := \int_{\mathbb{R}^s} \phi(y)\overline{\phi(y-x)}\, dy, \qquad x \in \mathbb{R}^s. \tag{10}$$

Since $\widehat{\Phi} = |\widehat{\phi}|^2$ holds, the function Φ is continuous and vanishes at infinity by the Riemann-Lebesgue theorem. Using the Fourier inversion formula and the symmetry $\Phi(x) = \Phi(-x)$ we obtain

$$\Phi(x) = (2\pi)^{-s} \int_{\mathbb{R}^s} |\widehat{\phi}(y)|^2 e^{-ix\cdot y}\, dy = (2\pi)^{-s} \int_C B(z^2) e^{-ix\cdot y}\, dy.$$

This gives

$$B(z^2) = \sum_{\alpha\in\mathbb{Z}^s} \Phi(\alpha) z^{2\alpha} \qquad \text{a.e. on } \mathbb{T}^s. \tag{11}$$

Well known properties of B and P are given in the following theorem.

Theorem 4. *Let $\phi \in L^2(\mathbb{R}^s)$ be the generator of a multiresolution analysis.*
(a) *$\{\phi_{0,\alpha} : \alpha \in \mathbb{Z}^s\}$ satisfies (1) with $A_2 \geq A_1 > 0$, if and only if*

$$A_1 \leq \operatorname*{ess\,inf}_{z\in\mathbb{T}^s} B(z^2) \leq \operatorname*{ess\,sup}_{z\in\mathbb{T}^s} B(z^2) \leq A_2. \tag{12}$$

In particular, the symbol calculus (4) applies to $c\tilde{*}\phi$ with any $c \in \ell^2$.
(b) *$\{\phi_{0,\alpha} : \alpha \in \mathbb{Z}^s\}$ is an orthonormal basis of V_0 if and only if $B(z^2) \equiv 1$ a.e. on $\mathbb{T}^s$.*
(c) *With the subscript notation from (6)*

$$B(z^2) = 2^{-2s} \sum_{d\in E} B_d(z)|P_d(z)|^2 \quad \textit{for all} \quad z \in \mathbb{T}^s. \tag{13}$$

Proof: (a) For any finite sequence c we obtain by using Plancherel's identity and the symbol calculus (4)

$$\|c\tilde{*}\phi\|_{L^2} = (2\pi)^{-s/2}\|C(z^2)\hat{\phi}(y)\|_{L^2} = (2\pi)^{-s/2}\|C(z^2)B(z^2)\|_{L^2(C)}. \tag{14}$$

Standard arguments show the necessity of (12). On the other hand, if (12) holds, then (4) can be applied to any $c \in \ell^2$, since the upper bound in (12) assures convergence of the right hand side in (4) with respect to the L^2–norm. So (14) is valid for any $c \in \ell^2$, and the bounds A_1, A_2 in (1) are established. Part (b) is a direct consequence of (10) and (11). (c) follows from the periodicity of P and (8). In order to see this we use the notation $e^{-iy} := (e^{-iy_1}, \ldots, e^{-iy_s})$ for $y \in \mathbb{R}^s$ and obtain for the right hand side in (13)

$$2^{-2s} \sum_{d\in E} \sum_{\alpha\in\mathbb{Z}^s} \left|P\left(e^{-i(y+2\pi d+4\pi\alpha)/2}\right)\right|^2 \left|\widehat{\phi}\left((y+2\pi d+4\pi\alpha)/2\right)\right|^2$$

$$= \sum_{d\in E} \sum_{\alpha\in\mathbb{Z}^s} \left|\widehat{\phi}\left(y+2\pi d+4\pi\alpha\right)\right|^2 = B(z^2). \quad \blacksquare$$

The above proof is given in analogy to the univariate case in [10].

By Theorem 4(a) both functions ϕ^o and $\widetilde{\phi}$ with

$$\widehat{\phi^o}(y) = \frac{\widehat{\phi}(y)}{\sqrt{B(z^2)}} \qquad \text{and} \qquad \widehat{\widetilde{\phi}}(y) = \frac{\widehat{\phi}(y)}{B(z^2)} \tag{15}$$

are elements of V_0, and the symbol B^o corresponding to ϕ^o in (9) satisfies $B^o(z^2) \equiv 1$ for almost all $z \in \mathbb{T}^s$. Therefore, $\{\phi^o_{0,\alpha} : \alpha \in \mathbb{Z}^s\}$ is an orthonormal basis of V_0. Furthermore we have

$$(\phi_{0,\alpha}\,,\,\widetilde{\phi}_{0,\beta}) = (\phi^o_{0,\alpha}\,,\,\phi^o_{0,\beta}) = \delta_{\alpha\beta}$$

for any $\alpha, \beta \in \mathbb{Z}^s$. So the family $\{\widetilde{\phi}_{0,\alpha} : \alpha \in \mathbb{Z}^s\}$ is a dual basis of V_0. It can be used to express the two-scale sequence (p_α) in (7) as

$$p_\alpha = 2^s \int_{\mathbf{R}^s} \phi(x)\overline{\widetilde{\phi}(2x-\alpha)}\,dx. \tag{16}$$

An immediate consequence of part (c) in the above theorem is the following quantitative estimate.

Corollary 5. *Let $\mathbf{P}(z) := (P_d(z),\, d \in E)$. Then the Euclidian norm of $\mathbf{P}$ satisfies*

$$\frac{2^s B(z^2)}{A_2} \le \|\mathbf{P}(z)\| \le \frac{2^s B(z^2)}{A_1}$$

with A_1, A_2 as in (12).

Next we look at property (ii) in the definition of a multiresolution analysis. Cohen [11] was the first who gave a proof that the intersection property in (ii) is automatically satisfied. In fact, as is shown below, the stability condition implies this property.

Theorem 6. *If $\phi \in L^2(\mathbb{R}^s)$ has ℓ^2–stable integer translates, then $\bigcap_{k\in\mathbb{Z}} V_k = \{0\}$.*

Proof: We show that the orthogonal projection $P_k f$ onto V_k converges to 0 as k tends to $-\infty$ for any fixed $f \in L^2$. Since the operators P_k are uniformly bounded we can restrict ourselves to a dense subspace of L^2, namely to functions with compact support. With the dual basis ϕ in (15) we obtain

$$P_k f = \sum_{\alpha\in\mathbb{Z}^s} c_{k,\alpha}\phi_{k,\alpha}, \quad \text{where} \quad c_{k,\alpha} = \left(f\,,\,\widetilde{\phi}_{k,\alpha}\right).$$

Since $\{\phi_{k,\alpha} : \alpha \in \mathbb{Z}^s\}$ satisfies the stability condition (1) with the same constants A_1, A_2 independent of k, this gives

$$\|P_k f\|^2 \le A_2 \|(c_{k,\alpha})\|^2_{\ell^2} = A_2 2^{ks} \sum_{\alpha\in\mathbb{Z}^s} \left| \int_{\mathbf{R}^s} f(x)\overline{\widetilde{\phi}(2^k x-\alpha)}\,dx \right|^2.$$

If $\operatorname{supp} f \subset [-R,R]^s$ for $R > 0$, then the Cauchy-Schwarz inequality implies

$$\|P_k f\|^2 \leq A_2 2^{ks} \|f\|_{L^2}^2 \sum_{\alpha \in \mathbb{Z}^s} \int_{[-R,R]^s} |\widetilde{\phi}(2^k x - \alpha)|^2 \, dx$$
$$= A_2 \|f\|_{L^2}^2 \int_{X_k} |\widetilde{\phi}(x - \alpha)|^2 \, dx,$$

where $X_k = \bigcup_\alpha (\alpha + 2^k[-R,R]^s)$ for small $k < 0$. By uniform boundedness the last integral converges to 0 as $k \to -\infty$. This proves the theorem. ■

There are many cases where even the stability condition is not needed for the proof of the intersection property. A result in this direction is given in [30].

Our next task is to find some simple criterion for the denseness condition in (ii) of Definition 2.

Theorem 7. *Let $\phi \in L^2(\mathbb{R}^s)$ be the generator of a multiresolution analysis. If $\phi \in L^1(\mathbb{R}^s)$ and if the symbols $P(z)$ and $B(z)$ have ℓ^1-coefficient sequences, then $\bigcup_{k \in \mathbb{Z}} V_k$ is dense in $L^2(\mathbb{R}^s)$.*

A result similar to the above theorem was obtained by Jia and Micchelli [20]. But our assumptions on ϕ are less restrictive and more natural. The proof of Theorem 7 is given in two steps. The two-scale relation (8) is used to find the following result.

Lemma 8. [20] *If ϕ satisfies the conditions of Theorem 7, then $\widehat{\phi}(2\pi\alpha) = 0$ for all $\alpha \in \mathbb{Z}^s \setminus \{0\}$ and $\widehat{\phi}(0) \neq 0$. Furthermore, $|P(1)| = 2^s$ and $P_d(1) = 0$ for all $d \in E \setminus \{0\}$.*

Proof: We again use the notation $e^{-iy} := (e^{-iy_1}, \ldots, e^{-iy_s})$. By iteration we obtain from (8)

$$\widehat{\phi}(y) = \prod_{j=1}^{k} \left(2^{-s} P(e^{-iy/2^j})\right) \widehat{\phi}(y/2^k). \tag{17}$$

By our assumption $P(z)$, $B(z)$ and $\widehat{\phi}$ are continuous. Let us first consider the case $|P(1)| < 2^s$. Then (8) for $y = 0$ implies $\widehat{\phi}(0) = 0$. If we choose q between $|P(1)|$ and 2^s, then $|2^{-s} P(e^{-iy/2^j})| \leq q$ holds for all sufficiently large $j \in \mathbb{N}$. Letting $k \to \infty$ in (17) gives $\widehat{\phi}(y) = 0$ for all $y \in \mathbb{R}^s$. So $\phi = 0$ in this case, which contradicts the stability condition.

The case $|P(1)| > 2^s$ is excluded by Theorem 4(c). So the only possibility is $|P(1)| = 2^s$, and the second assertion of the lemma follows from Theorem 4(c). Now we choose $y = 2^{k+1}\alpha\pi$ in (17), where $\alpha \in \mathbb{Z}^s \setminus \{0\}$, and obtain

$$\widehat{\phi}(2^{k+1}\alpha\pi) = (2^{-s} P(1))^k \widehat{\phi}(2\pi\alpha).$$

So the moduli of $\widehat{\phi}(2\pi\alpha)$ and $\widehat{\phi}(2^{k+1}\alpha\pi)$ are the same for all $k \geq 1$. Therefore, the Riemann-Lebesgue theorem implies $\widehat{\phi}(2\pi\alpha) = 0$. ■

Usually, the denseness of $\bigcup_k V_k$ in $L^2(\mathbb{R}^s)$ is proved based on Lemma 8 by an application of the Strang and Fix conditions [31] or by an explicit approximation scheme for the scaled spaces

$$V_h(\phi) := \{f : f(./h) \in V_0\}, \qquad h > 0.$$

Both approaches require more restrictive conditions on ϕ. The recent work [3] made the formulation of Theorem 7 in this general form possible.

Lemma 9. [3] *The following statements are equivalent for any* $\phi \in L^2(\mathbb{R}^s)$.

(a) $\lim_{h\to 0} \mathrm{dist}_{L^2}(f, V_h(\phi)) = 0$ *for all* $f \in L^2(\mathbb{R}^s)$,

(b) $\displaystyle \lim_{h\to 0} h^{-s} \int_{hC} \left(1 - \frac{|\widehat{\phi}(y)|^2}{\sum_{\alpha\in\mathbb{Z}^s} |\widehat{\phi}(y+2\pi\alpha)|^2}\right) dy = 0.$

For a complete proof of this result and the definition of the above quotient when the denominator vanishes see [3].

Proof of Theorem 7: We show that (b) in Lemma 9 is satisfied. A linear transformation and (10) show that (b) is equivalent to

$$\lim_{h\to 0} \int_C \left(1 - \frac{\widehat{\Phi}(hy)}{\sum_{\alpha\in\mathbb{Z}^s} \widehat{\Phi}(hy+2\pi\alpha)}\right) dy = 0. \tag{18}$$

We denote the integrand in (18) by $g(y;h)$. The positivity of $\widehat{\Phi}$ gives $0 \le g(y;h) \le 1$. The assumptions on ϕ guarantee that g is continuous on $\mathbb{R}^s \times \mathbb{R}$. Since $g(y;0) = 0$ by Lemma 8 we obtain pointwise convergence to 0 of the integrand in (18). Dominated convergence finally proves (18). ■

2.2. The case of regular ϕ

A multiresolution analysis is called r−regular [27, p. 22] for $r \in \mathbb{Z}$, $r \ge 0$, if it possesses a generator $\phi \in V_0$ such that

$$|D^\alpha \phi(x)| \le c_m (1 + \|x\|)^{-m} \tag{19}$$

for all multiintegers $0 \le |\alpha| \le r$ and all $m \in \mathbb{N}$, with constants c_m depending only on m. Here we use the notation $D^\alpha = \partial^{|\alpha|}/(\partial x_1^{\alpha_1} \ldots \partial x_s^{\alpha_s})$ and $|\alpha| = \alpha_1 + \ldots + \alpha_s$. In this case (10) implies that the coefficient sequence of $B(z^2)$ in (11) also decays faster than $(1 + \|\alpha\|)^{-m}$ for any $m \in \mathbb{N}$, so B is infinitely differentiable (see [27, Lemma 2.7]). Consequently, the Fourier coefficients of $1/B(z^2)$ and $1/\sqrt{B(z^2)}$ enjoy the same fast decay. This shows that both functions ϕ^o and $\widetilde{\phi}$ in (15) satisfy (19). If we use this in (16), we can finally conclude that the two-scale sequence $(p_\alpha)_{\alpha\in\mathbb{Z}^s}$ has the same fast decay. More generally, we obtain the following result from the decay properties of the dual basis function $\widetilde{\phi}$.

Theorem 10. *If $\phi \in L^2(\mathbb{R}^s)$ has ℓ^2−stable integer translates and satisfies (19), and if $f = c\tilde{*}\phi$ with $c \in \ell^2$ also satisfies (19), then*

$$|c_\alpha| = O\Big((1+\|\alpha\|)^{-m}\Big) \qquad \text{for any } m \in \mathbb{N}.$$

The two-scale relation (8) is used to derive the following result, which we include without a proof.

Theorem 11. [27, p. 56] *If ϕ is the generator of a multiresolution analysis and satisfies (19), then $D^\alpha\widehat{\phi}(2\pi\beta) = 0$ for all $0 \le |\alpha| \le r$ and all $\beta \in \mathbb{Z}^s \setminus \{0\}$.*

The proof in [27] is given for the function ϕ^o in (15), but immediately applies to ϕ by the above-mentioned differentiability of $B(z^2)$. Theorem 11 will be needed in Section 4 where we derive vanishing moment properties for wavelets.

If ϕ and its derivatives $D^\alpha\phi$ for $0 \le |\alpha| \le r$ decay exponentially as $\|x\| \to \infty$, then the multiresolution analysis is obviously r−regular. In this case the sequences (p_α) and $(\Phi(\alpha))$ also decay exponentially, see [20, Theorem 3.4].

2.3. The case of compactly supported ϕ

If $\phi \in L^2(\mathbb{R}^s)$ is compactly supported, then the semi-discrete convolution $c\tilde{*}\phi$ can be defined for any sequence $(c_\alpha)_{\alpha\in\mathbb{Z}^s}$ since the series is locally finite. The following notion is used in this case: The integer translates of ϕ are called *(algebraically) linearly independent*, if

$$c\tilde{*}\phi \equiv 0 \quad \text{a.e.} \qquad \Longleftrightarrow \qquad c \equiv 0. \tag{20}$$

Jia and Micchelli [20] prove that this property is stronger than the stability in (1). Relation (11) shows that the coefficient sequence of $B(z^2)$ is finite, so B is a trigonometric polynomial in this case. A similar reasoning as in 2.2 gives that ϕ^o, $\tilde{\phi}$, and (p_α) all decay exponentially. From an analogous result to Theorem 10 we can even conclude that (p_α) is a finite sequence.

Theorem 12. [1] *Suppose ϕ is a compactly supported tempered distribution whose integer translates are linearly independent. If $f = c\tilde{*}\phi$ is compactly supported then c is a finite sequence.*

Several examples of multiresolution analyses with r−regular ϕ or compactly supported ϕ are given in Section 5. The finiteness of the sequence (p_α) is not only of great importance for practical applications of the multiresolution analysis, but it also allows the construction of a finitely supported 'biorthogonal basis' as in [12].

§3. Semi-orthogonal wavelet bases and their duals

From now on we assume that $\phi \in L^2(\mathbb{R}^s)$ is the generator of a multiresolution analysis and that the coefficient sequences of $P(z)$ and $B(z)$ in (8), (11) are in ℓ^1. W_0 will always denote the orthogonal complement of V_0 in V_1. We describe the construction of semi-orthogonal and orthonormal wavelet bases and their duals from the given multiresolution analysis. Different aspects of this procedure were described by several authors [8], [20], and [27].

3.1. Decomposition of V_1

Direct sum decompositions of V_1 can be described in terms of matrices whose entries are symbol functions on $\mathbb{T}^s$. Given a family of functions $\psi^i \in V_1$, $0 \le i \le N$,

$$\psi^i(x) = \sum_{\alpha\in\mathbb{Z}^s} q^i_\alpha \phi(2x-\alpha)$$

with $(q^i_\alpha) \in \ell^1(\mathbb{Z}^s)$, we define the $N+1$ by 2^s matrix

$$\mathcal{Q}(z) := \left(Q^i_d(z)\right)_{0\le i\le N,\, d\in E}. \tag{21}$$

Note that $\widehat{\psi}^i(y) = 2^{-s}Q^i(z)\widehat{\phi}(y/2)$ as in (8). We also use the notation

$$W^i := \operatorname{clos}_{L^2} \operatorname{span}\{\psi^i_{0,\alpha} :\ \alpha\in\mathbb{Z}^s\}$$

and the "inner product" notation

$$[C^1, C^2](z^2) := \sum_{d\in E} B_d(z) C^1_d(z)\overline{C^2_d(z)} \tag{22}$$

for two symbol functions C^1, C^2. The function in (22) is 2π–periodic with respect to $y \in \mathbb{R}^s$. This fact is pointed to by the argument z^2. An important representation of the L^2–inner product of functions in V_1 is obtained by Plancherel's formula and by periodizing, namely

$$(c^i \tilde{*} \psi^i\,,\, c^j \tilde{*}\psi^j) = 2^{-2s}(2\pi)^{-s}\int_C C^i(z^2)\overline{C^j(z^2)}[Q^i, Q^j](z^2)\,dy. \tag{23}$$

Theorem 13. [8] *Let $q^i \in \ell^1(\mathbb{Z}^s)$, $0\le i\le 2^s-1$, be given, and let $\mathcal{Q}(z)$ be the matrix (21). Then the following statements hold.*

(a) *$\{\psi^i_{0,\alpha} : \alpha\in\mathbb{Z}^s,\, 0\le i\le 2^s-1\}$ is a Riesz basis of V_1 if and only if $\mathcal{Q}(z)$ is invertible for all $z\in\mathbb{T}^s$.*

(b) *For any $0\le i<j\le 2^s-1$ the spaces W^i and W^j are orthogonal if and only if*

$$[Q^i, Q^j](z^2) = 0 \qquad \textit{for all } z\in\mathbb{T}^s. \tag{24}$$

Proof: (a) First we note that all entries of $\mathcal{Q}(z)$ are continuous on $\mathbb{T}^s$ by the assumption $q^i\in\ell^1$. Since $\{\phi_{1,\alpha} : \alpha\in\mathbb{Z}^s\}$ is a Riesz basis of V_1 we can apply the symbol calculus (4) to a function $f^i = c^i\tilde{*}\psi^i$ with $c^i\in\ell^2$. We obtain from (23)

$$\Big\|\sum_{i=0}^{2^s-1} c^i\tilde{*}\psi^i\Big\|^2_{L^2} = 2^{-2s}(2\pi)^{-s}\int_C \sum_{i,j=0}^{2^s-1} C^i(z^2)\overline{C^j(z^2)}[Q^i, Q^j](z^2)dy. \tag{25}$$

The above integrand is a quadratic form described by the hermitian matrix

$$\mathcal{A}(z^2) := \left([Q^i, Q^j](z^2)\right)_{0 \le i,j \le 2^s - 1}$$

with 2π–periodic entries. Since $B_d(z) > 0$ for all $d \in E$ the above matrix is positive definite if and only if $\mathcal{Q}(z)$ is invertible. The sufficiency part in (a) is now proved as follows. By continuity we can find constants $A_2 \ge A_1 > 0$ such that

$$A_1 v^T \overline{v} \le v^T \mathcal{A}(z^2) \overline{v} \le A_2 v^T \overline{v} \quad \text{for all } v \in \mathbb{C}^{2^s} \text{ and all } z \in \mathbb{T}^s.$$

Since $\sum_i \|c^i\|_{\ell^2}^2 = (2\pi)^{-s} \sum_i \|C^i\|_{L^2(C)}^2$ by (5) it follows from (25) that

$$2^{-2s} A_1 \sum_{i=0}^{2^s-1} \|c^i\|_{\ell^2}^2 \le \Big\| \sum_{i=0}^{2^s-1} c^i \tilde{*} \psi^i \Big\|_{L^2}^2 \le 2^{-2s} A_2 \sum_{i=0}^{2^s-1} \|c^i\|_{\ell^2}^2.$$

But this is condition (1) for the set $\{\psi_{0,\alpha}^i : \alpha \in \mathbb{Z}^s, 0 \le i \le 2^s - 1\}$. The denseness of its linear span in V_1 is a consequence of Theorem 14 below.

Let us now prove the necessity part of (a). Note that $\mathcal{A}(z^2)$ is a normal matrix with continuous entries. If $\mathcal{Q}(z)$ is not invertible at $z_0 \in \mathbb{T}^s$, then for any $\epsilon > 0$ we can find an open neighbourhood $U \in \mathbb{T}^s$ of z_0 and a continuous selection of eigenvectors $v(z^2)$ of $\mathcal{A}(z^2)$ on U with Euclidian norm 1 such that

$$0 \le v^T(z^2) \mathcal{A}(z^2) \overline{v(z^2)} \le \epsilon \qquad \text{for all } z \in U.$$

If we let $C^i(z^2)$ be the i–th component of $v(z^2)$ in U and 0 outside U, then we see from (5) that

$$\sum_{i=0}^{2^s-1} \|c^i\|_{\ell^2}^2 = \lambda(U),$$

where λ denotes normalized Lebesgue measure on $\mathbb{T}^s$. On the other hand the integral in (25) becomes

$$2^{-2s}(2\pi)^{-s} \int_U v^T(z^2) \mathcal{A}(z^2) \overline{v(z^2)} dy \le 2^{-2s} \epsilon \lambda(U).$$

Therefore, a lower bound in (1) does not exist.

(b) is an immediate consequence of (23) and the continuity of the symbols Q^i, Q^j, and B. ■

The matrix $\mathcal{Q}(z)$ can be viewed as a basis transformation from the Riesz basis $\{\phi_{1,\alpha} : \alpha \in \mathbb{Z}^s\}$ to $\{\psi_{0,\alpha}^i : \alpha \in \mathbb{Z}^s, 0 \le i \le 2^s - 1\}$. This becomes clear from the following decomposition result which also proves the denseness assertion in part (a) of the above theorem.

Theorem 14. [8] *If the matrix $\mathcal{Q}(z)$ in (21) is invertible for all $z \in \mathbb{T}^s$, then*

$$\left(\mathcal{Q}^T(z)\right)^{-1} = \left(R_d^i(z)\right)_{0\le i\le 2^s-1,\ d\in E} \tag{26}$$

with symbol functions $R^i(z)$ whose coefficient sequences r^i lie in ℓ^1. Furthermore,

$$\phi(2x-d) = 2^s \sum_{i=0}^{2^s-1} \sum_{\alpha\in\mathbb{Z}^s} r^i_{2\alpha-d}\psi^i(x-\alpha) \tag{27}$$

for all $d \in E$.

Proof: We first prove that $\left(\mathcal{Q}^T(z)\right)^{-1}$ has the structure as in (26). If we let $S_{i,d}$ be the minor of $\mathcal{Q}(z)$ obtained by cancelling row i and column d, then the relation $S_{i,d}(z) = (-1)^{d_1}[S_{i,0}(z)]_d$ is obtained from (21). This shows that the matrix $\mathcal{S}(z) := \left((-1)^{i+d_1}S_{i,d}(z)\right)$ has the form (21) and that

$$\Delta(z^2) = \det \mathcal{Q}(z) = \sum_{d\in E}(-1)^{d_1}Q^0_d(z)S_{0,d}(z)$$

is 2π-periodic. Since all entries of $\mathcal{Q}(z)$ have ℓ^1- coefficients this is also true for $S_{i,d}(z)$, $\Delta(z^2)$ and $\Delta^{-1}(z^2)$ by Wiener's lemma. The first assertion therefore follows from $\left(\mathcal{Q}^T(z)\right)^{-1} = \Delta^{-1}(z^2)\mathcal{S}(z)$.

In order to prove (27) we observe that by absolute convergence

$$\sum_{\alpha\in\mathbb{Z}^s} r^i_{2\alpha-d}z^{2\alpha} = 2^{-s}z^d \sum_{d'\in E}(-1)^{d\cdot d'}R^i_{d'}(z), \quad d \in E. \tag{28}$$

Taking the Fourier transform on the right hand side in (27) we obtain with (4) and (28)

$$\sum_{i=0}^{2^s-1}\left(z^d\sum_{d'\in E}(-1)^{d\cdot d'}R^i_{d'}(z)\right)\widehat{\psi^i}(y)$$
$$= 2^{-s}z^d\widehat{\phi}(\frac{y}{2})\sum_{d'\in E}(-1)^{d\cdot d'}\left(\sum_{i=0}^{2^s-1}R^i_{d'}(z)Q^i(z)\right).$$

The last expression in parentheses is equal to $\delta_{0d'}$ by (26), so (27) is established since we have $(\phi(2.-d))\widehat{\ }(y) = 2^{-s}z^d\widehat{\phi}(y/2)$. ■

Because of (27) the sequences $r^i \in \ell^1$ are called *decomposition sequences*, whereas q^i are called *reconstruction sequences*. This terminology is chosen in connection with the pyramidal scheme described in [5], [13], and [25]. If we assume for a moment that $\psi^0 = \phi$ and that $\{\psi^i_{0,\alpha} : \alpha \in \mathbb{Z}^s,\ 1 \le i \le 2^s-1\}$ is a Riesz basis for W_0 then the sequences r^i give an efficient way to find the wavelet decomposition (3). For $f = f_N = \sum_\alpha a_\alpha^N \phi_{N,\alpha} \in V_N$ we obtain from (27)

$$f_N = 2^{3s/2}\Big[\sum_{\alpha\in\mathbb{Z}^s} \left(r^0 * a^N\right)_{2\alpha} \phi_{N-1,\alpha} + \sum_{i=1}^{2^s-1} \sum_{\alpha\in\mathbb{Z}^s} \left(r^i * a^N\right)_{2\alpha} \psi^i_{N-1,\alpha}\Big] \tag{29}$$

where '$*$' denotes the convolution of sequences. Note that the first part in (29) is of the same form as f_N, except it is in V_{N-1}. Further steps of the decomposition (29) lead therefore to the wavelet decomposition (3). (29) also explains that fast decay (or finiteness) of the sequences r^i is a desirable property for computational efficiency. In [8] a truncation method is devised for infinite decomposition sequences and a bound on the truncation error is given.

Relation (28) in the above proof is valid for any ℓ^1–sequence r^i. It can be used in order to describe decompositions of V_1 with a different type of matrix which we call $\mathcal{M}(z)$. Let

$$\mathcal{U}(z) := 2^{-s/2}\left((-1)^{d\cdot d'} z^d\right)_{d,d'\in E}, \qquad z \in \mathbb{T}^s, \tag{30}$$

which obviously has the form (21). Furthermore, $\mathcal{U}(z)$ is unitary since it is a multiple of a $(1,-1)$–Hadamard matrix. In view of Theorem 13 this matrix describes the trivial decomposition of V_1 into the subspaces W^d generated by integer translates of $\phi_{1,d}$, $d \in E$.

Now we associate with any matrix $\mathcal{Q}(z)$ of type (21) the matrix

$$\mathcal{M}(z^2) := \left(M^{i,d}(z^2)\right)_{0\leq i\leq 2^s \quad 1, d\subset E} := \mathcal{Q}(z)\mathcal{U}^*(z), \tag{31}$$

where $\mathcal{U}^*$ is the adjoint of $\mathcal{U}$. For later reference we mention

$$M^{0,d}(z^2) = 2^{-s/2}\overline{z}^d \sum_{d'\in E} (-1)^{d\cdot d'} Q^0_{d'}(z). \tag{32}$$

The argument z^2 points to the fact that all entries of $\mathcal{M}$ are 2π–periodic functions of y. On the other hand there are no more interrelations between entries in the same row. A decomposition of V_1 can therefore be described by 2^{2s} symbols $M^{i,j}$, each of which is a 2π–periodic function on $\mathbb{R}^s$. Theorems 13 and 14 can also be formulated in terms of $\mathcal{M}(z^2)$. This is done in [20].

3.2. Dual basis for V_1

If a decomposition of V_1 is given by a matrix $\mathcal{Q}(z)$ in (21) then we can easily find a dual basis to the Riesz basis $\{\psi^i_{0,\alpha}\}$ of V_1. We let

$$\widetilde{\mathcal{Q}}(z) := \left(\widetilde{Q}^i_d(z)\right) := 2^{2s}\left(\mathcal{Q}^*(z)\right)^{-1} \operatorname{diag}\left(1/B_d(z),\ d \in E\right). \tag{33}$$

Then $\widetilde{Q}_d^i(z) = 2^{2s}\overline{R_d^i(z)}/B_d(z)$ with R^i in (26). The corresponding functions $\widetilde{\psi}^i$, $0 \le i \le 2^{s-1}$, with

$$\widehat{\widetilde{\psi}^i}(y) = 2^{-s}\widetilde{Q}^i(z)\widehat{\phi}(\frac{y}{2}) \tag{34}$$

define a dual basis for V_1.

Theorem 15. *If $\mathcal{Q}(z)$ in (21) defines a decomposition of V_1, then the functions $\widetilde{\psi}^i$ in (34) satisfy*

$$\left(\psi_{0,\alpha}^i, \widetilde{\psi}_{0,\beta}^j\right) = \delta_{ij}\delta_{\alpha\beta} \tag{35}$$

for all $0 \le i,j \le 2^s - 1$ and $\alpha, \beta \in \mathbb{Z}^s$.

Proof: The symbols $\widetilde{Q}^j(z)$ are chosen in (33) such that $[Q^i, \widetilde{Q}^j](z^2) = 2^{2s}\delta_{ij}$. Relation (23) gives

$$\left(\psi_{0,\alpha}^i, \widetilde{\psi}_{0,\beta}^j\right) = (2\pi)^{-s}2^{-2s}\int_C z^{2\beta-2\alpha}[Q^i, \widetilde{Q}^j](z^2)\,dy$$

which proves (35). ■

A more general setting of *biorthogonal* bases is investigated in [7] and [12]. This idea is based on the search for functions $\widetilde{\psi}^i$ with property (35) which need not be elements of V_1, but which still satisfy a two-scale relation (7).

3.3. Semi-orthogonal and orthonormal wavelets

We now describe the construction procedure for obtaining wavelet bases from a multiresolution analysis. The starting point for this process is an "initial decomposition" given by an invertible matrix $\mathcal{Q}(z)$ in (21) where the first element ψ^0 generates a Riesz basis of V_0. Therefore we always assume that $\psi^0 = \phi$ and $Q^0 = P$. The functions $\{\psi_{0,\alpha}^i : \alpha \in \mathbb{Z}^s, 1 \le i \le 2^s - 1\}$ provide a Riesz basis for the space W_0 if and only if the orthogonality relations in Theorem 13(b) hold; *i.e.*,

$$[Q^i, P](z^2) = 0 \qquad \text{for all } 1 \le i \le 2^s - 1 \text{ and all } z \in \mathbb{T}^s. \tag{36}$$

So (36) is a sufficient condition for the set $\{\psi^i : 1 \le i \le 2^s - 1\}$ to form a semi-orthogonal wavelet basis of $L^2(\mathbb{R}^s)$. If in addition to (36) the relations

$$[Q^i, Q^j](z^2) = 2^{2s}\delta_{ij} \qquad \text{for all } 1 \le i,j \le 2^s - 1 \text{ and all } z \in \mathbb{T}^s \tag{37}$$

are satisfied then by (23) this basis is an orthonormal wavelet basis. Moreover, if (37) also holds for $i = j = 0$ we conclude from Theorem 4(c) that $B(z^2) \equiv 1$ on $\mathbb{T}^s$, so $2^{-s}\mathcal{Q}(z)$ is a unitary matrix. The description of orthonormal wavelet bases in [15] and [27, Corollary 4.2] is given in terms of these unitary matrices $\mathcal{Q}(z)$ and their associated unitary matrices $\mathcal{M}(z^2)$ in (31).

For the purpose of constructing an orthonormal wavelet basis from a given decomposition of V_1 the usual Gram-Schmidt procedure applied to the rows of $\mathcal{Q}(z)$ is feasible. As described in [8] and [20] a refined version of this procedure is useful for the construction of semi-orthogonal wavelet bases with functions of compact support.

Theorem 16. *Assume that $\mathcal{Y}(z)$ is invertible for all $z \in \mathbb{T}^s$, its entries have coefficient sequences in ℓ^1, and $Y^0 = P$.*
(a) *Let $Q^0 := P$ and*

$$Q^i := [P, P]Y^i - [Y^i, P]P \qquad \text{for } 1 \le i \le 2^s - 1. \tag{38}$$

Then the symbols $Q^i(z)$ have ℓ^1−coefficients and define a semi-orthogonal wavelet basis $\{\psi^i : 1 \le i \le 2^s - 1\}$.
(b) *Let $Q^0 := P$, $T^0 := [P, P]$, and for $1 \le i \le 2^s - 1$ let*

$$\widetilde{Y}^i := T^{i-1}Y^i - \sum_{j=0}^{i-1} \frac{T^{i-1}}{T^j}[Y^i, \widetilde{Y}^j]\widetilde{Y}^j, \qquad T^i := [\widetilde{Y}^i, \widetilde{Y}^i]. \tag{39}$$

Then the symbols $Q^i := \widetilde{Y}^i/\sqrt{T^i}$ have ℓ^1−coefficients and define an orthonormal wavelet basis $\{\psi^i : 1 \le i \le 2^s - 1\}$.
Furthermore, if all functions in the initial decomposition decrease rapidly (as in (10) with $r = 0$) or exponentially, then the same is true for the wavelet basis in (a) or (b). If ϕ has compact support and linearly independent integer translates as in (20), and if all functions in the initial decomposition have compact support, then the symbols $Q^i(z)$ in (a) and $\widetilde{Y}^i(z)$ in (b) define semi-orthogonal wavelets with compact support.

Proof: Since in (a) and (b) an analogue to the usual Gram-Schmidt procedure is used the orthogonality relations (36) and (37) are immediate for Q^i in (a) resp. (b). The non-vanishing of the symbols T^i is caused by the invertibility of $\mathcal{Y}(z)$, so Wiener's lemma gives the ℓ^1−condition on the sequences. The decay (resp. finiteness) of the initial coefficient sequences y^i is implied by Theorems 10 and 12, resp. They are carried over to the symbol functions Q^i in (a) and $\widetilde{Y}^i$, T^i in (b) since the constructions (38) and (39) are 'division-free'. For (b) this is seen inductively by the relation

$$\frac{T^i}{T^{i-1}} = T^{i-1}[Y^i, Y^i] - \sum_{j=0}^{i-1} \frac{T^{i-1}}{T^j} \left|[Y^i, \widetilde{Y}^j]\right|^2, \quad i \ge 1. \tag{40}$$

Therefore, $T^i(z^2)$ is infinitely differentiable. As pointed out in Section 2.2 the decay properties of the coefficient sequences also hold for $1/\sqrt{T^i(z^2)}$ and for $Q^i(z)$ in part (b). ■

With the same arguments as in the above proof it follows that the symbols R^i_d in (26) and R^i_d/B_d in (33) have rapidly (resp. exponentially) decaying sequences. So the decomposition sequences as well as the dual basis functions inherit the decay properties from the initial decomposition. The same principle is not true for compactly supported functions. While the semi-orthogonal wavelets in (a) of the above theorem are still compactly supported, the matrix inversion in (26), (33) requires the division by a trigonometric polynomial. This leads to exponentially decaying decomposition sequences r^i and exponentially decaying dual functions $\widetilde{\psi}^i$.

The pairwise orthogonality of the spaces $\widetilde{W}^i$ generated by the functions $\widehat{\psi}^i(y) = 2^{-s}\widetilde{Y}^i(z)\widehat{\phi}(y/2)$ with $\widetilde{Y}^i$ in Theorem 16(b) has some advantage over the semi-orthogonal case in (a). If we denote the corresponding matrix by $\widetilde{\mathcal{Y}}(z)$ then the inverse can be found from the relations

$$[\widetilde{Y}^i, \widetilde{Y}^j](z^2) = \delta_{ij} T^i(z^2).$$

We obtain

$$\widetilde{\mathcal{Y}}^{-1}(z) = \left(B_d(z)\widetilde{Y}_d^i(z)/T^i(z^2)\right)^*_{0\le i\le 2^s-1, d\in E},$$

which leads to simple expressions for the decomposition sequences and the dual basis.

The power of semi-orthogonal wavelet bases lies in the fact that they give rise to a dual wavelet basis while they are more flexible than orthonormal wavelet bases. Theorem 15 above applied to a matrix $\mathcal{Q}(z)$ which satisfies the orthogonality relation (36) gives the dual basis elements

$$\widetilde{\psi}^0 = \widetilde{\phi} \in V_0 \quad \text{as in (14) and}$$
$$\widetilde{\psi}^i \in W_0,$$

hence the functions $\{\widetilde{\psi}^i : 1 \le i \le 2^s - 1\}$ generate a Riesz basis of W_0 as well. Therefore they provide a dual wavelet basis. A further orthogonalization as in Theorem 16(b) leads to a further decoupling of the dual basis functions until $\widetilde{\psi}^i \in W^i$ is obtained.

3.4. Existence of semi-orthogonal wavelet bases

We conclude this section with two theoretical results on the existence of initial decompositions of V_1 described by an invertible matrix $\mathcal{Q}(z)$ of type (21) with $Q^0 = P$. Prescribing the first row of this matrix makes an explicit construction very difficult in general.

The first result was already mentioned in Theorem 3. We repeat it here with more general assumptions on ϕ.

Theorem 17. *If $\phi \in L^2(\mathbb{R}^s)$ generates a multiresolution analysis, P and B have coefficient sequences in ℓ^1, and P is lipschitz continuous, then there exists a semi-orthogonal wavelet basis $\{\psi^i : 1 \le i \le 2^s - 1\}$ of functions $\psi^i \in W_0$.*

Proof: [27, p. 92] For $s = 1$ the matrix

$$\mathcal{Q} = \begin{pmatrix} P_0 & P_1 \\ -\overline{P_1}B_1 & \overline{P_0}B_0 \end{pmatrix}$$

has determinant $\det \mathcal{Q} = B_0|P_0|^2 + B_1|P_1|^2 > 0$. For $s \geq 2$ we let $\mathbf{M}(z^2) := (M^{0,d}(z^2),\ d \in E)$ be as in (32) and try to extend this vector to an invertible matrix $\mathcal{M}(z^2)$. Since the Euclidian norm of $\mathbf{M}(z^2)$ is bounded from below (Corollary 5) we can assume that it is a point on the unit sphere $S \subset \mathbb{C}^s$ (or $\mathbb{R}^s$) for all $z \in \mathbb{T}^s$. But $s < 2^s - 1$, so as a well known topological fact the set of all these points has measure 0 in S. Let us denote the set of all such points by $K \subset S$. Without loss of generality we can assume that $(-1, 0, \ldots, 0) \notin K$. Since K is compact we find $0 < r < 1$ such that $M^{0,0}(z^2) \geq -r$ for all $z \in \mathbb{T}^s$. Finally we obtain for the matrix

$$\mathcal{M}(z^2) := \begin{pmatrix} M^{0,0} & M^{0,1} & \ldots & M^{0,2^s-1} \\ -\overline{M^{0,1}} & \epsilon & & \mathbf{0} \\ \vdots & & \ddots & \\ -\overline{M^{0,2^s-1}} & \mathbf{0} & & \epsilon \end{pmatrix} \tag{41}$$

by elementary calculations

$$\det \mathcal{M}(z^2) = \epsilon^{2^s-2}\big(\epsilon M^{0,0}(z^2) + \sum_{d \neq 0} |M^{0,d}(z^2)|^2\big). \tag{42}$$

The expression (42) is positive for all $0 < \epsilon < r^{-1} - r$ and all $z \in \mathbb{T}^s$. ∎

Although the proof looks very theoretical a detailed analysis of the two-scale function for multivariate box-splines is given in Section 5.2 to find a matrix of the form (41). Note that the initial decomposition (41) provides compactly supported functions whenever P and ϕ are finitely supported. In special cases even more desirable properties of semi-orthogonal wavelets which are studied in Section 4 can be obtained from (41). For later reference we mention that the corresponding $\mathcal{Q}$–matrix to $\mathcal{M}(z^2)$ in (41) is given by $Q^0 = P$ and, with (32),

$$Q^d(z) = 2^{-s/2}\big(-\overline{M^{0,d}(z^2)} + \epsilon z^d\big) = 2^{-s} z^d \left(2^{s/2}\epsilon - \sum_{d' \in E} (-1)^{d \cdot d'} \overline{P_{d'}(z)}\right). \tag{43}$$

The second result is due to Jia and Micchelli [20, Theorem 8.1]. We formulate it slightly differently from [20].

Theorem 18. *Assume that $\phi \in L^2(\mathbb{R}^s)$ generates a multiresolution analysis, is compactly supported and has linearly independent integer translates (see (20)). Then there exists a matrix $\mathcal{Q}(z)$ with $Q^0 = P$ whose determinant is a non-zero monomial.*

The theorem in the above form shows that the decomposition sequences in (27) and the dual basis (34) obtained from $\mathcal{Q}(z)$ are also finite (resp. compactly supported). This is used in the construction of biorthogonal wavelets, see [7] and [12]. While the orthogonalization procedure in Theorem 16(a) leads to a compactly supported semi-orthogonal wavelet basis, it sacrifices the divisibility by the determinant, which means that the dual wavelet basis and the decomposition sequences of the wavelet basis will in general be exponentially decaying.

The proof of the above theorem is based on the Quillen-Suslin theorem for Laurent polynomials. Details can be found in [20].

§4. Generalized linear phase and vanishing moments

In many applications of wavelets, especially in signal processing and image processing, it is important that the generator ϕ of the multiresolution analysis has *generalized linear phase*, see [5]. This means that

$$\widehat{\phi}(y) = e^{-i(a\cdot y+b)} A(y) \tag{44}$$

with real valued A and $a \in \mathbb{R}^s$, $b \in \mathbb{R}$. This is by the Fourier inversion formula equivalent to

$$\phi(x) = e^{-2ib}\overline{\phi(2a-x)} \qquad \text{a.e. in } \mathbb{R}^s. \tag{45}$$

Since ϕ also satisfies a two-scale relation (7), which we assume to hold with $(p_\alpha) \in \ell^1$, the parameter vector a must lie in $\frac{1}{2}\mathbb{Z}^s$ and (8) gives

$$P(z) = z^{2a}\overline{P(z)} \qquad \text{for all } z \in \mathbb{T}^s. \tag{46}$$

More generally, a function $\psi(x) = \sum_\alpha q_\alpha \phi(2x-\alpha)$ with $q \in \ell^1$ has generalized linear phase with parameters $\widetilde{a}$, $\widetilde{b}$ in (44) if and only if its symbol $Q(z)$ satisfies

$$Q(z) = cz^{4\widetilde{a}-2a}\overline{Q(z)}, \tag{47}$$

where $c = e^{2i(b-\widetilde{b})}$ and $\widetilde{a} \in \frac{1}{4}\mathbb{Z}^s$. This is seen by using the symbol calculus (4) and the stability condition for ϕ.

So the generalized linear phase of the functions in an initial decomposition of V_1 can be described in terms of the symbols. The next theorem gives a condition on the parameter vectors a^i such that the orthogonalization in Theorem 16 does not destroy this property.

Theorem 19. *Assume that the symbols $Y^i(z)$ of an initial decomposition of V_1 satisfy*

$$Y^i(z) = c^i z^{a^i}\overline{Y^i(z)} \quad \text{with } c^i \in \mathbb{T} \text{ and } a^i \in \mathbb{Z}^s. \tag{48}$$

If the vectors a^i satisfy the "parity condition"

$$a^i - a^j \in 2\mathbb{Z}^s \qquad \text{for all } 0 \le i,j \le 2^s-1, \tag{49}$$

then (48) holds for Q^i and $\widetilde{Y}^i$ in Theorem 16.

Proof: It is sufficient to prove the assertion for a typical step of the modified Gram-Schmidt method in Theorem 16(b). Note that T^i and B are real valued by their definition. Assuming (48) for $\widetilde{Y}^j$, $0 \le j < i$, we obtain directly from the definition (22)

$$[Y^i, \widetilde{Y}^j](z^2) = c^i \overline{c^j} z^{a^i - a^j} \sum_{d \in E} (-1)^{d \cdot (a^i - a^j)} B_d(z) \overline{Y_d^i(z)} \widetilde{Y}_d^j(z).$$

The parity condition (49) gives

$$[Y^i, \widetilde{Y}^j](z^2) = c^i \overline{c^j} z^{a^i - a^j} \overline{[Y^i, \widetilde{Y}^j](z^2)}.$$

Multiplying this by $\widetilde{Y}^j$ confirms (48) for $\widetilde{Y}^i$. ■

Next we consider the property of vanishing moments. Assume that $\mathcal{Q}(z)$ with $Q^0 = P$ defines a semi-orthogonal wavelet basis; *i.e.*, orthogonality (36) is true. The first result deals with a general multiresolution analysis of $L^2(\mathbb{R}^s)$.

Proposition 20. *Let $\phi \in L^2 \cap L^1$ generate a multiresolution analysis with the cofficient sequences of P and B in ℓ^1. Then $\widehat{f}(0) = 0$ for any function $f = \sum_{\alpha \in \mathbb{Z}^s} c_\alpha \phi(2. - \alpha) \in W_0$ with $c \in \ell^1$.*

Proof: By our assumptions $P(z)$ and $B(z)$ are continuous on $\mathbb{T}^s$. For f as above $[C, P](z^2) = 0$ holds for all $z \in \mathbb{T}^s$, since $f \in W_0$. Especially, for $z = 1$ we obtain $0 = [C, P](1) = 2^s C(1) B(1)$ by Lemma 8, and this gives $\widehat{f}(0) = 2^{-s} C(1) \widehat{\phi}(0) = 0$. ■

If ϕ is r−regular in the sense of (19), $r \ge 0$, then also higher order moments of the functions ψ^i in a semi-orthogonal wavelet basis vanish.

Theorem 21. *Let $\mathcal{Q}(z)$ be a matrix of type (21) with $Q^0 = P$ and Q^i, $1 \le i \le 2^s - 1$, such that the functions $\{\psi^i : 1 \le i \le 2^s - 1\}$ generate a Riesz basis of W^0. If ϕ and all ψ^i satisfy (19) for $r \ge 0$, then*

$$\int_{\mathbb{R}^s} x^\alpha \psi^i(x)\, dx = 0 \qquad \text{for all } 0 \le |\alpha| \le r. \tag{50}$$

Proof: From the discussion in Section 2.2 we conclude that P, B and all symbols Q^i, $1 \le i \le 2^s - 1$, are infinitely differentiable, see Theorem 10. Under the regularity assumption on ψ^i equation (50) is equivalent to

$$D^\alpha \widehat{\psi}^i(0) = 0 \qquad \text{for all } 0 \le |\alpha| \le r. \tag{51}$$

First we prove (51) under the additional assumption that $2^{-s}\mathcal{Q}(z)$ is unitary, *i.e.*, (37) holds and $\{\phi_{0,\alpha}\}$ is an orthonormal basis of V_0. This gives

$$B(z^2) \equiv 1 \qquad \text{and} \qquad \sum_{i=0}^{2^s - 1} |Q^i(z)|^2 \equiv 2^{2s}$$

on $\mathbb{T}^s$. Together with Theorem 11 the first identity gives $D^\alpha\widehat{\phi}(0) = 0$ for $0 < |\alpha| \le r$ and $\widehat{\phi}(0) = 1$. The second identity implies

$$\sum_{i=0}^{2^s-1} |\widehat{\psi}^i(y)|^2 = |\widehat{\phi}(\frac{y}{2})|^2 = 1 + O(\|y\|^{2r+2}) \quad \text{as } \|y\| \to 0.$$

So we obtain $|\widehat{\psi}(y)|^2 = O(\|y\|^{2r+2})$ near 0 for all $1 \le i \le 2^s - 1$, which gives (51).

The general case now follows from Theorem 10. If $\mathcal{Q}(z)$ is the given matrix we denote by $\widetilde{\mathcal{Q}}(z)$ the completely orthogonalized matrix after applying the procedure of Theorem 16(b). We pointed out in Section 3.3 that all functions $\widetilde{\psi}^i$ given by the new matrix satisfy (19) as well. From the above proof we see that (51) holds for all $\widetilde{\psi}^i$. Since the original functions ψ^i are elements of $W_0 \perp V_0$ we have

$$\left(\widehat{\widetilde{\psi}^i}\right)_{1\le i\le 2^s-1} = \mathcal{A}(z^2)\left(\widehat{\psi^i}\right)_{1\le i\le 2^s-1}, \tag{52}$$

where $\mathcal{A}(z^2)$ is a lower triangular $(2^s-1)\times(2^s-1)$ matrix with non-vanishing diagonal elements and infinitely differentiable entries. Inversion of the linear system (52) represents $\widehat{\psi}^j$ as a linear combination $\widehat{\psi}^j(y) = \sum_i C^{i,j}(z^2)\widehat{\widetilde{\psi}^i}(y)$ with infinitely differentiable functions $C^{i,j}$. This proves that (51) also holds for $\widehat{\psi}^j$, $1 \le j \le 2^s - 1$. ■

The first part of the above proof is taken from [27, p. 93]. We finish this section with a remark on the decomposition (41) of V_1 which appeared in the proof of the general existence result. The matrix $\mathcal{Q}(z) = \mathcal{M}(z^2)\mathcal{U}(z)$ describes the initial decomposition with $Q^0 = P$ and Q^d as in (43). So all functions in this initial decomposition have the decay and smoothness properties of ϕ, if ϕ is $r-$regular (or decays exponentially or has compact support, resp.). Moreover, if ϕ has generalized linear phase with parameter $a = 0$, then (43) and (46) give

$$Q^d(z) = z^{2d}\overline{Q^d(z)} \qquad \text{for all } d \in E \setminus \{0\}. \tag{53}$$

Therefore, the functions ψ^d in this decomposition have generalized linear phase with parameter $\widetilde{a} = d/2$, see (47). Even more is true: the parity condition (49) is satisfied. This shows that (41) provides a powerful tool for constructing semi-orthogonal or orthonormal wavelet bases, as soon as a quantified version of the topological argument in 3.4 is available.

§5. Initial decompositions

This section deals with four different classes of multiresolution analysis where initial decompositions with many desirable properties can be found explicitly.

5.1. Tensor product wavelets

Let $\phi \in L^2(\mathbb{R})$ generate a multiresolution analysis. As mentioned in 3.4 the matrix

$$\mathcal{Q} = \begin{pmatrix} P_0 & P_1 \\ -\overline{P_1}B_1 & \overline{P_0}B_0 \end{pmatrix}$$

provides a semi-orthogonal wavelet basis $\{\psi\}$ of $L^2(\mathbb{R})$. This matrix describes the construction of orthonormal wavelets in [13] and of spline wavelets in [9] and [10]. Higher dimensional wavelet bases can be obtained by choosing

$$\Psi^d(x_1, \ldots, x_s) := \psi_{d_1}(x_1)\psi_{d_2}(x_2)\cdots\psi_{d_s}(x_s), \qquad d \in E \setminus \{0\},$$

with $\psi_0 = \phi$ and $\psi_1 = \psi$, see [27, Chapter 3.3]. The corresponding dual wavelets are built in the same way from the univariate dual wavelet. If $\{\psi\}$ is an orthonormal wavelet basis in $\mathbb{R}$ then $\{\Psi^d : d \in E \setminus \{0\}\}$ is also orthonormal. The inheritance of decay properties, vanishing moments and generalized linear phase is immediate. In [18] tensor product wavelets on finite domains are studied.

5.2. Box-spline wavelets in arbitrary dimension

An important class of functions which satisfy a two-scale relation (7) are multivariate box-splines. They were introduced by de Boor and DeVore [2]. More recent surveys can be found in [4], [16], and [19]. Given an integer matrix $X = (\xi_1, \ldots, \xi_n) \in \mathbb{Z}^{s\times n}$ with rank s the box-spline N_X is defined by

$$\widehat{N}_X(y) = \prod_{\xi \in X} \frac{1 - e^{-i\xi\cdot y}}{i\xi \cdot y}, \qquad y \in \mathbb{R}^s. \tag{54}$$

Here we use the set notation $\xi \in X$ in order to denote that the product is taken over all columns of X. We also use $Y \subset X$ and $X \setminus Y$ for a set of columns and the cancellation of columns of X. Note that (54) defines N_X as a regular distribution up to a set of measure 0. We always assume that the function is chosen globally continuous whenever the right hand side in (54) is in $L^1(\mathbb{R}^s)$. It is well known that N_X is a piecewise polynomial of degree $n-s$, has compact support with center

$$c_X := \frac{1}{2}\sum_{\xi \in X} \xi, \tag{55}$$

and is $(\rho - 1)$–times continuously differentiable, where

$$\rho := \max\{r : \operatorname{rank}(X \setminus Y) = s \text{ for any set } Y \text{ of } r \text{ columns of } X\}. \tag{56}$$

Furthermore, N_X satisfies the Strang and Fix conditions [31]

$$D^\alpha \widehat{N}_X(2\pi\beta) = \delta_{\alpha 0}\delta_{\beta 0} \quad \text{for all } 0 \le |\alpha| \le \rho \text{ and } \beta \in \mathbb{Z}^s. \tag{57}$$

We also use the "centered" box-spline $N_X^* = N_X(\,.\,+c_X)$ with real valued Fourier transform

$$\widehat{N}_X^*(y) = \prod_{\xi\in X} \frac{\sin(\xi\cdot y/2)}{(\xi\cdot y/2)}. \tag{58}$$

It is readily seen from (54) that N_X satisfies a two-scale relation (8) with

$$P(z) = 2^{-n+s} \prod_{\xi\in X} (1+z^\xi). \tag{59}$$

For achieving our goal of constructing a wavelet basis we restrict ourselves to the cases where the direction matrix X has the following properties:

(X1) every $s\times s$ minor of X is either 0, 1, or -1,

(X2) the center c_X lies in $\mathbb{Z}^s$.

It is again well known that (X1) is equivalent to algebraic linear independence of the integer translates of N_X as defined in (20). Therefore, N_X is the generator of a multiresolution analysis in $\mathbb{R}^s$.

For our construction we assume more generally that $\phi\in L^2$ is the generator of a multiresolution analysis, and that its two-scale function has the form

$$P(z) = 2^{-n+s} \prod_{\xi\in X} (1+z^\xi)W(z), \qquad z\in\mathbb{T}^s, \tag{60}$$

where $W(z)$ has ℓ^1-coefficients and is strictly positive on $\mathbb{T}^s$. This includes both cases $\phi = N_X$ and $\phi = N_X^o$, the orthogonalized function in (15). By condition (X2) the function $\phi^* := \phi(\,.\,+c_X)$ also satisfies a two-scale relation (8), and its two-scale function

$$P^*(z) = z^{-c_X}P(z) = 2^s W(z) \prod_{\xi\in X} \cos\frac{\xi\cdot y}{4} \tag{61}$$

is real-valued. The corresponding symbol function B is not affected by this shift. Under these general conditions the following result holds.

Theorem 22. [30] *Let ϕ^* and P^* be defined as above, and B be the corresponding symbol function (9). With $\gamma := \left(\max_{z\in\mathbb{T}^s} B(z^2)\right)^{-1}$, the functions ψ^d given by*

$$\widehat{\psi}^d(y) := 2^{-2s} z^d \left(2^{s/2}\gamma B(z^2) - \sum_{d'\in E} (-1)^{d\cdot d'} P_{d'}^*(z)\right) \widehat{\phi}^*(\tfrac{y}{2}) \tag{62}$$

for $d\in E\setminus\{0\}$ generate, together with $\psi^0 := \phi^*$, a Riesz basis of V_1. In particular, if ϕ^* is $r-$regular (or decays exponentially or has compact support) then all ψ^d have the same property.

Before we give the proof of this result we mention some consequences of the special form of ψ^d in (62). The $\mathcal{M}$–matrix of the above decomposition (see (31)) can be expressed in terms of the symbols

$$\mathbf{M}(z^2) := (M^d(z^2),\ d \in E) := (P_d^*(z),\ d \in E) \cdot \mathcal{U}^*(z). \tag{63}$$

Since the symbol function in (62) has exactly the same form as (43) we find that

$$\mathcal{M}(z^2) := \begin{pmatrix} M^0 & M^1 & \dots & M^{2^s-1} \\ -\overline{M^1} & \gamma B(z^2) & & \mathbf{0} \\ \vdots & & \ddots & \\ -\overline{M^{2^s-1}} & \mathbf{0} & & \gamma B(z^2) \end{pmatrix}.$$

(Here we used $i(d) = \sum_k d_k 2^{k-1}$ as superscript.) Therefore, its determinant is given by (42), namely

$$\det \mathcal{M}(z^2) = [\gamma B(z^2)]^{2^s-2} \left(\sum_{d \neq 0} |M^d(z^2)|^2 + \gamma B(z^2) M^0(z^2) \right). \tag{64}$$

Since ϕ^* has generalized linear phase with parameters $a = b = 0$ the discussion at the end of Section 4 and Theorem 21 have the following immediate consequence.

Corollary 23. *The functions ψ^d in (62) have generalized linear phase with parameter vector $a^d = d/2$ and, in particular, the parity condition (49) is true. The resulting semi-orthogonal (or orthonormal) wavelets η^d, constructed from ψ^d as in Theorem 16, have generalized linear phase. Furthermore, if $W(z)$ is infinitely differentiable, then η^d has vanishing moments*

$$\int_{\mathbf{R}^s} x^\alpha \eta^d(x)\, dx = 0 \qquad \text{for all } 0 \le |\alpha| \le \rho$$

with ρ in (56).

We now turn to the proof of Theorem 22. First we give a brief outline of the proof. The central part is the study of the sign structure of the functions $P_d^*(z)$. For notational ease we now use $y \in \mathbb{R}^s$ instead of z. Then the definition (6) of the subscript notation and (61) give

$$P_d^*(y) = P^*(y + 2\pi d) = 2^s W(y + 2\pi d) \\ \times \prod_{\substack{\xi \in X \\ \xi \cdot d \text{ even}}} (-1)^{\xi \cdot d/2} \cos \frac{\xi \cdot y}{4} \cdot \prod_{\substack{\xi \in X \\ \xi \cdot d \text{ odd}}} (-1)^{(\xi \cdot d + 1)/2} \sin \frac{\xi \cdot y}{4} \tag{65}$$

for all $d \in E$.

Lemma 24. *For any $y \in \mathbb{R}^s$ there exists $d \in E$ with $P_d^*(y) \ge 0$.*

The proof of this lemma is given by using the converse assumption

$$P_d^*(y^*) < 0 \qquad \text{for all } d \in E \text{ and one } y^* \in \mathbb{R}^s, \tag{66}$$

in order to show that the direction matrix X, up to a transformation by a unimodular matrix, contains a submatrix X' with the following properties: $X' \in \mathbb{Z}^{s\times n'}$ with $n' \geq 2$ and

- (X1') X' is totally unimodular; *i.e.*, all $k \times k$–minors are either 0, 1, or -1 for all $1 \leq k \leq \min\{s, n'\}$,
- (X2') each row sum of X' is even,
- (X3') for any $1 \leq k < l \leq s$ the number of matching pairs $\binom{\pm 1}{\pm 1}$ in rows k and l (regardless of the signs) is even,
- (X4') the columns of X' are distinct, and any two non-zero columns are linearly independent.

Condition (X1') is stronger than the earlier condition (X1), while (X2') and (X2) are equivalent. The proof of the first lemma is then completed by an inductive proof of the following combinatorial result.

Lemma 25. [30] *A totally unimodular matrix $X' \in \mathbb{Z}^{s\times n}$, $n \geq 2$, with properties (X1') to (X4') does not exist for any $s \geq 1$.*

Let us now assume the validity of Lemma 24 and proceed in the proof of Theorem 22.

Proof of Theorem 22: We have to show that the expression in (64) never vanishes. We do this by proving its positivity. The first factor in (64) can be omitted by the stability assumption, see Theorem 4. For the second factor we perform a detailed analysis based on Lemma 24. Let $n = 2^s$ and $\mathbf{P}(z) = (P_d^*(z),\ d \in E)$. A direct application of this lemma gives for the Euclidian norm of the vector $\mathbf{P}$

$$\left\|\mathbf{P}(z) - \frac{1}{\sqrt{n}}\|\mathbf{P}(z)\|(-1,\ldots,-1)\right\|^2 \geq \frac{1}{n}\|\mathbf{P}(z)\|^2. \tag{67}$$

Since $\mathcal{U}^*(z)$ is a unitary matrix and maps $(1/\sqrt{n})(-1,\ldots,-1)$ to $(-1,0,\ldots,0)$, inequality (67) is equivalent to

$$\left\|\mathbf{M}(z^2) - \|\mathbf{M}(z^2)\|(-1,0,\ldots,0)\right\|^2 \geq \frac{1}{n}\|\mathbf{M}(z^2)\|^2.$$

Since M^0 is real valued, see (32), this can be equivalently expressed as

$$M^0(z^2) \geq -\left(1 - \frac{1}{2n}\right)\|\mathbf{M}(z^2)\|. \tag{68}$$

In the case $M^0(z^2) \geq 0$ the determinant (64) is positive. So we only consider the case $M^0(z^2) < 0$ in the rest of the proof. In that case inequality (68) implies

$$\sum_{d\neq 0}|M^d(z^2)|^2 = \|\mathbf{M}(z^2)\|^2 - |M^0(z^2)|^2 \geq \frac{1}{n}\left(1 - \frac{1}{4n}\right)\|\mathbf{M}(z^2)\|^2.$$

So for any $0 < \epsilon(z) \le \|\mathbf{M}(z^2)\|/n = \|\mathbf{P}(z)\|/n$ we obtain

$$\epsilon M^0 + \sum_{d \neq 0} |M^d|^2 \ge \frac{1}{4n^2} \|\mathbf{P}\|^2 > 0. \tag{69}$$

The inequality in Corollary 5 allows us to choose $\epsilon(z) := \gamma B(z^2)$ in (69). This completes the proof of Theorem 22. ■

We still need to verify the assertions of Lemma 24 and Lemma 25.

Proof of Lemma 24: We assume that $y^* \in \mathbb{R}^s$ can be found such that (66) holds. Since P^* in (65) is 4π–periodic by assumption (X2) this is equivalent to

$$P^*(y^* + 2\pi\alpha) < 0 \qquad \text{for all } \alpha \in \mathbb{Z}^s. \tag{70}$$

In three steps we show that (70) implies the existence of a submatrix X' of X with properties (X1') to (X4').

Step 1. Since X has rank s and satisfies condition (X1) we can choose a unimodular submatrix B of X. Then the matrix $X_1 := B^{-1}X$ is an integer matrix and contains all unit vectors e_ν, $1 \le \nu \le s$, and still satisfies (X1) and (X2). Moreover, X_1 is totally unimodular. The function $\theta(x) := \phi^*(Bx)$ also defines a multiresolution analysis, and its two-scale function is

$$P^\theta(y) = P^*\left((B^T)^{-1}y\right) = 2^s W\left((B^T)^{-1}y\right) \prod_{\xi \in X} \cos \frac{(B^{-1}\xi) \cdot y}{4}, \quad y \in \mathbb{R}^s.$$

But B^T leaves the lattice $2\pi\mathbb{Z}^s$ invariant, so that (70) is equivalent to

$$P^\theta(B^T y^* + 2\pi\alpha) < 0 \qquad \text{for all } \alpha \in \mathbb{Z}^s.$$

Note also that P^* in (61) does not change when we replace any column ξ of X by $-\xi$. Summarizing we may conclude that X itself can be assumed to be totally unimodular and does not contain any two non-zero columns of the form $(\xi, -\xi)$.

Step 2. We can cancel all factors of P_d^* in (65) which have constant positive sign for all $d \in E$. Omitting the factors $2^s W(y + 2\pi d)$, all even powers of the same cos- or sin-factor in (65), and all factors that correspond to a zero column of X we obtain a reduced matrix X' with pairwise linearly independent columns (X4'), such that (70) holds for X'. While this may affect the rank of X', conditions (X1') and (X2') are inherited from X. Furthermore, X' must have at least two non-zero columns as a consequence of (X1'), (X2') and (70).

Step 3. We now prove that the submatrix X' obtained in Step 2 also satisfies (X3'). In order to achieve this we compare the reduced inequalities (70) where X is replaced by X' with each other. Inequalities (70) for d, $d + e_\nu \in E$, where d has ν–th component 0, yield

$$\prod_{\substack{\xi \cdot d \text{ even} \\ \xi \cdot e_\nu \text{ odd}}} (-1)^{\frac{\xi \cdot e_\nu + 1}{2}} \tan \frac{\xi \cdot y^*}{4} \cdot \prod_{\substack{\xi \cdot d \text{ odd} \\ \xi \cdot e_\nu \text{ odd}}} (-1)^{\frac{\xi \cdot e_\nu - 1}{2}} \cot \frac{\xi \cdot y^*}{4} > 0.$$

The same inequality for $d = 0$ is

$$\prod_{\xi \cdot e_\nu \ odd} (-1)^{\frac{\xi \cdot e_\nu + 1}{2}} \tan \frac{\xi \cdot y^*}{4} > 0.$$

The last two inequalities for $d = e_\mu$, $\mu \neq \nu$, combined finally give

$$\prod_{\substack{\xi \cdot e_\nu \ odd \\ \xi \cdot e_\mu \ odd}} (-1) > 0.$$

It can be readily seen that this last inequality is equivalent to (X3'). This completes the proof of Lemma 24. ■

Proof of Lemma 25: We use an inductive argument in order to show that an integer matrix X' with at least two columns cannot have all properties (X1') to (X4'). For $s = 1$ there is no such matrix. Now let $s \geq 1$ and $X' \in \mathbb{Z}^{(s+1)\times n}$, $n \geq 2$, satisfy all conditions (X1') to (X4'). Then X' has at least one non-zero column by (X4'), and all its entries are 0, +1, or −1 by (X1'). Rearranging rows and columns gives a matrix of the form

$$X' = \left(\begin{array}{cccc|ccc} \pm 1 & \pm 1 & \dots & \pm 1 & 0 & \dots & 0 \\ \hline & & & & & & \\ & & Y & & & Z & \\ & & & & & & \end{array} \right)$$

with submatrices $Y \in \mathbb{Z}^{s \times r}$, $r \geq 2$, and $Z \in \mathbb{Z}^{s \times (n-r)}$, where Z is possibly empty. The fact $r \geq 2$ is a consequence of (X2'). We now demonstrate that Y inherits all properties from X' which leads to a contradiction to the case $s = 1$. Condition (X1') is inherited directly, while (X2') follows from condition (X3') which was assumed to be true for X'.

Next we establish (X4') for Y. Since columns 1 to r of X' are pairwise linearly independent by (X4'), Y cannot have two zero-columns. Assuming linear dependence of two non-zero columns of Y, together with linear independence of these two columns of X' shows the existence of a 2×2 submatrix of X' of the form

$$\begin{pmatrix} 1 & 1 \\ 1 & -1 \end{pmatrix} \quad \text{or} \quad \begin{pmatrix} 1 & -1 \\ 1 & 1 \end{pmatrix}$$

whose determinant is 2 or −2. This contradicts (X1'), so (X4') holds for Y.

(X3') is trivially satisfied if $s = 1$. So let us assume that $s \geq 2$ and that two rows $1 \leq k < l \leq s$ of Y exist where (X3') is violated. We will demonstrate that this contradicts (X1') by finding either a 2×2 or a 3×3 minor of X' which is 2 or −2. With k, l as above we consider the $3 \times n$ submatrix of X' which consists of rows 1, $k + 1$, and $l + 1$. Since (X3') is violated for Y, but holds for X', there are two columns

$$c_1 = (\pm 1, \pm 1, \pm 1)^T \quad \text{and} \quad c_2 = (0, \pm 1, \pm 1)^T.$$

Since we find an odd number of matching pairs between rows k and l of Y, but the row sums of Y are even, there must be two further columns of the form

$$c_3 = (\pm 1, \pm 1, 0)^T \qquad \text{and} \qquad c_4 = (\pm 1, 0, \pm 1)^T.$$

But for any sign pattern the 3×4 matrix (c_1, c_2, c_3, c_4) has either a 2×2 or a 3×3 minor, which is 2 or -2. This esablishes (X3') for Y and completes the proof of Lemma 25. ■

5.3. The case $\widehat{\phi} \geq 0$

If the Fourier transform of $\phi \in L^2(\mathbb{R}^s)$ is nonnegative then the symbol functions $Q^0 = P$, $Q^d = 2^{-s/2} z^d$ for $d \in E \backslash \{0\}$ define an initial decomposition of V_1. The corresponding matrix of symbols is always invertible. With $\mathcal{U}(z)$ in (30) we have

$$\mathcal{M}(z^2) = \mathcal{Q}(z)\mathcal{U}^*(z) = \begin{pmatrix} & M^{0,d}(z^2) & \\ \hline 0 & 1 & \mathbf{0} \\ \vdots & \ddots & \\ 0 & \mathbf{0} & 1 \end{pmatrix}, \tag{71}$$

where $M^{0,d}(z^2)$ is given in (32). In particular,

$$M^{0,0}(z^2) = 2^{-s/2} \sum_{d \in E} P_d(z) > 0,$$

since $P_d(z) = \widehat{\phi}(y + 2\pi d)/\widehat{\phi}(\frac{y}{2} + \pi d)$ is nonnegative and the Euclidian length of the vector $\mathbf{P}(z) = (P_d(z),\, d \in E)$ is always positive, see Corollary 5.

A first special case is given by box-splines with a direction matrix X which additionally to the conditions (X1) and (X2) in 5.2 satisfies

(X3) each column of X is repeated an even number of times.

If we let $\phi = N_X^*$, then the positivity $\widehat{\phi}(y) \geq 0$ can be immediately seen from (58). Analogous to the description in 5.2, we can construct compactly supported semi-orthogonal wavelets which have generalized linear phase and vanishing moments of order ρ, where ρ is defined in (56). The construction of the wavelet basis in this case was first performed in [22].

A second important example is provided by *polyharmonic cardinal splines* which are generalizations of the thin-plate spline. Let the function ϕ be defined by

$$\widehat{\phi}(y) = \frac{\|y\|^{-2m}}{\sum_{\alpha \in \mathbb{Z}^s} \|y + 2\pi\alpha\|^{-4m}} \qquad \text{for} \quad m > \frac{s}{4}. \tag{72}$$

Madych [23] shows that $\{\phi_{0,\alpha} : \alpha \in \mathbb{Z}^s\}$ is an orthonormal basis of V_0 and that ϕ and all its (distributional) derivatives of order $< 2m - s/2$ lie in $L^2(\mathbb{R}^s)$ and decay exponentially at infinity. So ϕ is r-regular in the sense of (19),

$$A_8 = \begin{pmatrix} a & b & c & d & e & f & g & h \\ b & -a & d & -c & f & -e & h & -g \\ c & -d & -a & b & g & -h & -e & f \\ d & c & -b & -a & -h & -g & f & e \\ e & -f & -g & h & -a & b & c & -d \\ f & e & h & g & -b & -a & -d & -c \\ g & -h & e & -f & -c & d & -a & b \\ h & g & -f & -e & d & c & -b & -a \end{pmatrix}, \qquad z^{\eta(d)} = \begin{pmatrix} 1 \\ z_1 \\ z_1 z_2 \\ z_2 z_3 \\ z_1 z_2 z_3 \\ z_3 \\ z_1 z_3 \\ z_2 \end{pmatrix}$$

Table 1. Orthogonal design matrix.

where $0 \le r < 2m - s/2$. Theorem 7 assures that ϕ generates a multiresolution analysis. Its two-scale function is given by

$$P(z) = 2^{-2m+s} \frac{\sum_{\alpha \in \mathbb{Z}^s} \|y + 2\pi\alpha\|^{-4m}}{\sum_{\alpha \in \mathbb{Z}^s} \|y/2 + 2\pi\alpha\|^{-4m}}.$$

It is also shown in [23] that the coefficient sequence of P decays exponentially. This is also true for the functions ψ^i after the orthogonalization process. These functions have vanishing moments of order $\le r$ in virtue of the general Theorem 21. Since ϕ has (generalized) linear phase with parameters $a = b = 0$ in (44) and the parity condition (49) is satisfied ψ^i has generalized linear phase. This wavelet basis was also studied by Lemarié [21]. A similar function as in (72) arises in the minimization of the functional

$$\int_{\mathbb{R}^s} |\Delta^m f(x)|^2 \, dx \tag{73}$$

over all functions in a Sobolev space with prescribed values on the lattice $\mathbb{Z}^s$. Here Δ is the Laplace operator. The function ϕ_I with

$$\widehat{\phi_I}(y) = \frac{\|y\|^{-2m}}{\sum_{\alpha \in \mathbb{Z}^s} \|y + 2\pi\alpha\|^{-2m}} \qquad \text{for} \quad m > \frac{s}{2}$$

interpolates the fundamental sequence $(\delta_{0\alpha})_{\alpha \in \mathbb{Z}^s}$. Its minimal properties with respect to (73) are studied in [24], see also [21] and [27, p. 86ff]. Note that the orthonormalized function (15) to ϕ_I is the function ϕ in (72).

5.4. Wavelets in low dimensions

Several authors have investigated the construction of semi-orthogonal and orthonormal wavelets of $L^2(\mathbb{R}^s)$ for $s \le 3$, *e.g.*, [8], [20], [28], and [29]. The restriction on the dimension is inherent to the construction method. Let us describe this method in terms of *orthogonal designs* [14]. Assume that the function $\phi \in L^2(\mathbb{R}^s)$ has generalized linear phase, *i.e.*,

$$P(z) = z^{2a} \overline{P(z)} \qquad \text{for some } a \in \tfrac{1}{2}\mathbb{Z}^s,$$

see (46). We define $S^d(z) := i^{d\cdot 2a}P_d(z)$. Then from the above relation we obtain

$$S^d(z) = (-i)^{d\cdot 2a} z^{2a}\overline{P_d(z)} = z^{2a}\overline{S^d(z)} \tag{74}$$

for all $d \in E$. Hereby we achieve commutativity $S^d\overline{S^{d'}} = S^{d'}\overline{S^d}$. The matrix A_8 in Table 1 and its 2×2 and 4×4 submatrices A_2 and A_4 in the upper left corner define an orthogonal design; *i.e.*,

$$AA^T = A^T A = kI \tag{75}$$

where k is the sum of squares of each row and I is the identity matrix. They can be found in [14, p. 74]. Matrices of this form do not exist for dimensions different from $n = 1, 2, 4, 8$ (which corresponds to $s = 1, 2, 3$ by $n = 2^s$).

If we use $S^d(z)$ for the variables in the given matrix A, then (74), (75) give

$$A(z)A^*(z) = \sum_{d\in E} |S^d(z)|^2 \cdot I,$$

so $A(z)$ is invertible for any $z \in \mathbb{T}^s$ by Corollary 5. In order to obtain a matrix $Q(z)$ of type (21) we first note that the matrix of monomials $\left(z_{d'}^{\eta(d)}\right)_{d,d'\in E}$ with $\eta(d)$ as in Table 1 has the same sign pattern as A. In order to eliminate the complex factors in $A(z)$ we let

$$\widetilde{A}(z) := \left((-i)^{(d+d')\cdot 2a} A^{d,d'}(z)\right),$$

which again is invertible. The general element of this matrix is

$$\widetilde{A}^{d,d'} = i^{2a\cdot(d+d'-(d\oplus d'))}(-1)^{\eta(d)\cdot d'} P_{d\oplus d'}$$

where the symbol '$\oplus$' denotes addition modulo $2\mathbb{Z}^s$. The newly introduced factors $i^{2a\cdot(d+d'-(d\oplus d'))} \in \{-1, 1\}$ can be written in the form $\prod_{k=1}^{s}(-1)^{2a_k d_k d'_k}$. So we find $Q(z)$ by choosing

$$Q^d(z) = \prod_{k=1}^{s} z^{2a_k d_k} z^{\eta(d)} P_d(z)$$

and obtain an invertible matrix.

The decay properties of ψ^i as $\|x\| \to \infty$ as well as the smoothness and generalized linear phase are, as before, inherited from the function ϕ. They also hold for the wavelet basis constructed in Theorem 16, since the parity condition (49) is satisfied. This method has been successfully applied to box-splines [8], [28], and [29] and piecewise linear hexagonal splines [27, p. 85]. In the bivariate case generators ψ^i of pairwise orthogonal spaces $W^i \subset W_0$ with fairly small support were obtained by a refined initial decomposition in [8].

References

1. Ben-Artzi, A. and A. Ron, On the integer translates of a compactly supported function: dual bases and linear projectors, *SIAM J. Math. Anal.* **21** (1990), 1550-1562.
2. de Boor, C. and R. DeVore, Approximation by smooth multivariate splines, *Trans. Amer. Math. Soc.* **276** (1983), 775-788.
3. de Boor, C., R. DeVore, and A. Ron, Approximation from shift-invariant subspaces of $L_2(\mathbb{R}^d)$, preprint.
4. Chui, C. K., *Multivariate Splines*, CBMS-NSF Regional Conference Series in Applied Math., # 54, SIAM, Philadelphia, 1988.
5. Chui, C. K., An overview of wavelets, in *Approximation Theory and Functional Analysis*, C. K. Chui (ed.), Academic Press, Boston, 1991, 47-72.
6. Chui, C. K., *An Introduction to Wavelets*, Academic Press, Boston, 1992.
7. Chui, C. K. and C. Li, A general framework of multivariate compactly supported wavelets and dual wavelets, in manuscript.
8. Chui, C. K., J. Stöckler, and J. D. Ward, Compactly supported box-spline wavelets, CAT Report #230, Texas A&M University, 1990; supplement, 1991.
9. Chui, C. K. and J. Z. Wang, A cardinal spline approach to wavelets, *Proc. Amer. Math. Soc.*, 1991, to appear.
10. Chui, C. K. and J. Z. Wang, A general framework of compactly supported splines and wavelets, CAT Report #219, Texas A&M University, 1990.
11. Cohen A., Ondelettes, analyses multirésolutions et traitement numérique du signal, Doctoral Thesis, Univ. Paris-Dauphine, 1990.
12. Cohen, A., I. Daubechies, and J. C. Feauveau, Biorthogonal bases of compactly supported wavelets, *Comm. Pure and Appl. Math.*, 1991, to appear.
13. Daubechies, I., Orthonormal bases of compactly supported wavelets, *Comm. Pure and Appl. Math.* **41** (1988), 909-996.
14. Geramita, A. V. and J. Seberry, *Orthogonal Designs, Quadratic Forms and Hadamard Matrices*, Marcel Dekker, New York, 1979.
15. Gröchenig, K., Analyse multiéchelles et bases d'ondelettes, *CRAS Paris, Série 1*, **305** (1987), 13-15.
16. Höllig, K., Box splines, in *Approximation Theory V*, C. K. Chui, L. L. Schumaker, and J. D. Ward (eds.), Academic Press, Boston, 1986, 71-95.
17. Jaffard, S., P. G. Lemarié, S. Mallat, and Y. Meyer, Dyadic multiscale analysis of $L^2(\mathbb{R}^n)$, in manuscript.
18. Jaffard, J. and Y. Meyer, Bases d'ondelettes dans des ouverts de $\mathbb{R}^n$, *J. Math. Pures et Appl.* **68** (1989), 95-108.
19. Jetter, K., A short survey on cardinal interpolation by box-splines, in *Topics in Multivariate Approximation*, C. K. Chui, L. L. Schumaker, and F. I. Utreras (eds.), Academic Press, Boston, 1987, 125-139.
20. Jia, R.-Q. and C. A. Micchelli, Using the refinement equations for the construction of pre-wavelets II: powers of two, in *Curves and Surfaces*,

P.-J. Laurent, A. Le Méhauté, L. L. Schumaker (eds.), Academic Press, San Diego, 1991, 209-246.
21. Lemarié, P. G., Bases d'ondelettes sur les groupes de Lie stratifiés, *Bull. Soc. Math. France*, to appear.
22. Lorentz, R. A. H. and W. R. Madych, Wavelets and generalized box splines, GMD Technical Report, *Applicable Analysis*, to appear.
23. Madych, W. R., Polyharmonic splines, multiscale analysis, and entire functions, in *Multivariate Approximation and Interpolation*, W. Haußmann and K. Jetter (eds.), Birkhäuser Verlag, Basel, 1990, 205-216.
24. Madych, W. R. and S. A. Nelson, Polyharmonic cardinal splines: a minimization property, *J. Approx. Theory* **63** (1990), 303-320.
25. Mallat, S., Multiresolution approximations and wavelet orthonormal bases of $L^2(\mathbb{R})$, *Trans. Amer. Math. Soc.* **315** (1989), 69-87.
26. Meyer, Y., Ondelettes et fonctions splines, *Séminaire EDP*, Ecole Polytechnique, Paris, December 1986.
27. Meyer, Y., *Ondelettes et Operateurs*, vol. 1, Hermann, Paris, 1990.
28. Riemenschneider, S. D. and Z. W. Shen, Box splines, cardinal series, and wavelets, in *Approximation Theory and Functional Analysis*, C. K. Chui (ed.), Academic Press, Boston, 1991, 133-150.
29. Riemenschneider, S. D. and Z. W. Shen, Wavelets and pre-wavelets in low dimensions, Univ. Alberta, preprint.
30. Stöckler, J., The construction of box-spline wavelets in arbitrary dimension, preprint.
31. Strang, G. and G. Fix, A Fourier analysis of the finite element variational method, in *Constructive Aspects of Functional Analysis*, G. Geymonat (ed.), C.I.M.E., 1973, 793-840.

This research was supported by NATO Grant CRG 900158.

Joachim Stöckler
Fachbereich Mathematik
Universität Duisburg
Lotharstr. 65
4100 Duisburg 1
Germany
hn277st@math.uni-duisburg.de

Part V

Short-time Fourier and Window-Radon Transforms

Gabor Wavelets and the Heisenberg Group: Gabor Expansions and Short Time Fourier Transform from the Group Theoretical Point of View

Hans G. Feichtinger and Karlheinz Gröchenig

Abstract. We study series expansions of signals with respect to Gabor wavelets and the equivalent problem of (irregular) sampling of the short time Fourier tranform. Using Heisenberg group techniques rather than traditional Fourier analysis allows to design stable iterative algorithms for signal analysis and synthesis. These algorithms converge for a variety of norms and are compatible with the time-frequency localization of signals.

§1. Introduction and Notations

Among scientist and mathematicians it is a well known that the Fourier transform (FT) is a perfect tool to analyze periodic functions, to represent them as a superposition of pure frequencies and to consider the behavior of the FT as indication about the contribution of certain frequencies to the total signal. This makes sense for stationary signals, but even for such nice signals as a piece of classical music it would not make sense to consider the global FT. Instead, the function has to be considered locally, and some local Fourier analysis has to be carried out (much in the same way as a musician would describe a piece of music as a sequence of pure tones at certain times). Thus, the idea of time-frequency representations of signals came up. Among the pioneers of this approach were Wigner and Gabor. Wigner proposed a function, now named Wigner distribution, over the time frequency plane (TF-plane for short) which shows many properties that one would like to have for a joint time-frequency energy distribution (which actually does not exist). For many purposes the Short Time Fourier Transform (STFT) of a signal is easier to handle, because it depends in a linear way on the analyzed signal. It has become a standard tool in signal analysis (see [1],[3],[40] and in particular [46] for survey notes). As we shall see it requires the choice of a *window function* used to localize the analyzed signal in a decent way. The concept of the TF-plane was also

Wavelets–A Tutorial in Theory and Applications
C. K. Chui (ed.), pp. 359–397.

ISBN 0-12-174590-2

motivating for D. Gabor's [34] suggestion to use a *coherent family* of Gauss functions, *i.e.*, a family of functions generated from the ordinary Gauss function by time-frequency shifts along the integer lattice in Z^2 (the standard lattice in $I\!R^2$) as building blocks for a series representation (now called *Gabor series*) for arbitrary signals. Although his suggestion dates back to 1946 it has found serious interest by mathematicians only in the last decade. Meanwhile it is known that it is *not* possible to expand arbitrary L^2-functions exactly in the way suggested by Gabor if one wants to have square summable coefficients, or at least convergence of the series in L^2 (see [43] or [35]).

On the other hand it turned out that much more is true. Following the current trend in the mathematical community, we call a collection of functions of the form $M_y T_x g$, obtained from a single function g by TF-shifts, a collection of *Gabor wavelets*. The established terminology in the engineering is literature is *Gabor type functions* and in quantum physics these functions are referred to as (generalized) *coherent states* (see [49] and [45]). The great interest in Gabor wavelet expansions may be documented by a selection of recent articles, mostly in the applied literature, due to Bastiaans [4], Cenker [8], Daubechies [11], [12], [14], Daugman [16], Dufaux and Kunt [17], Gertner and Zeevi [36], Porat and Zeevi [50],[51], Super an Bovik [58], or Walnut [62], [63].

As we will point out in detail in this note Gabor wavelets are a special case of the general approach described by the authors in [9], [22] and [25]–[28].

In this article we treat a general theory of Gabor wavelets. This theory emerges as a special case from the group theoretical approach ([25]-[27],[38]). It is our intention to give it a formulation concrete enough so that this very general theory can be useful to applied mathematicians. Besides this we shall discuss in some detail the connections between various concepts related to the TF-plane and those using the Heisenberg group. In addition to the known results we also state several new results (such as Theorems 15, 18 and 25 and most of section 7) which have not yet appeared in print.

Our exposition will focus on the following *main problems*:

A) **Atomic representation problem** (Gabor wavelet expansions)

We are looking for sufficient conditions on the basic building block g and the discrete family $(z_i)_{i\in I} = (x_i, y_i)_{i\in I}$ in $I\!R^{2d}$ such that any f in the Hilbert space $L^2(\mathbb{R}^d)$ allows a norm convergent series representation of the form (also called *Gabor type expansion* or *Gabor series expansion*):

$$f(t) = \sum_{i\in I} c_i e^{2\pi i y_i (t-x_i)} g(t-x_i) \ , \tag{1.1}$$

Of course we would like to have stability in the following sense:

$$(\sum_{i\in I} |c_i|^2)^{1/2} \leq C \cdot \|f\|_2 \quad \forall\, f \in L^2(\mathbb{R}^d). \tag{1.2}$$

One cannot expect uniqueness of the representation because coherent families are in general linear dependent. Therefore a *constructive* method of choosing the coefficients $(c_i)_{i\in I}$, depending linearly on f, is asked for. Finally good locality and stability of the expansion are desirable (*i.e.*, small local changes should effect only few coefficients; small errors in the coefficients should only result in a minimal distortion of the corresponding series,...).

B) **Irregular sampling of the Short Time Fourier Transform**

There are formulas (see Bastiaans [4],[5], Heil/Walnut [39], Daubechies et al. [13]) which allow to recover the signal or equivalently the full STFT from its sampling values over a sufficienty fine *regular* grid in the TF-plane. For this case the Zak-transform is a very useful tool (see [44], [39]). If the signal f belongs to the Hilbert space of L^2-functions (signals of finite energy) this question is equivalent to the question, whether a *coherent* family is a so-called *frame*.

Again we are interested in efficient iterative reconstruction algorithms, which should be stable as well. Thus, replacing the window by 'similar' one, subsequent reconstruction of the signal from samples of the STFT should still give a good approximation to the signal in discussion. As we will show, the algorithms derived by our group theoretical approach satisfy all these requirements.

The plan of this paper is as follows. First we recall some notations, mainly motivated by applications in signal analysis in Chap. 2. In particular, concepts that are related to time-frequency representations of signal such as the STFT, are explained. We define two families of function spaces (or spaces of tempered distributions) which are formed according to the behavior of the STFT of their elements, and which are called *generalized modulation spaces* and formulate two main questions. In Section 3 we describe our solution of the two main problems. In Chap. 4 we explain the role of the reduced Heisenberg group and how the concepts of signal analysis are directly related to the representation coefficients for the Schrödinger representation. This point of view also allows us to identify modulation spaces with *coorbit spaces*. Section 5 is devoted to coorbit spaces related to L^1-spaces, which are important as spaces of test functions. The kernel-theorem for the Segal algebra $S_0(\mathbb{R}^d)$ is the main result of this section. In Section 6 we show how the main results can be obtained from the general theory, and some further applications of the group theoretical point of view are given. Recent results concerning Wilson bases are mentioned among others. The iterative methods used to solve the two problems is compared to the standard Hilbert space approach using frames, and the *method of adaptive weights* persented. The final chapter contains results on the stability of both the atomic representation method and the reconstruction methods for the STFT. Invariance properties of modulation spaces (e.g., under chirp operators) are proved, but it is also shown how seemingly different spaces (such as Bargmann-Fock spaces) can be identified in a very natural way with modulation spaces.

§2. Signal Analysis, Function Spaces Defined by Means of the STFT

Let the $\mathcal{H}$ be the Hilbert space $L^2(\mathbb{R}^d)$ of square integrable functions ("signals of finite energy") over the d–dimensional Euclidian space $\mathbb{R}^d$. Following the tradition of physicists and consistent with [25]–[28] we describe the inner product by the formula $\langle g, f\rangle := \int_{\mathbb{R}^d} \overline{g(x)} f(x)dx$, which is linear with respect to the *second* argument. This will be convenient later. For the FT $\mathcal{F}$ we choose the following normalization:

$$\mathcal{F}f(y) = \hat{f}(y) := \int_{\mathbb{R}^d} f(x)e^{-2\pi i x\cdot y}dx \ , \tag{2.1}$$

where $x \cdot y$ denotes the inner product on $\mathbb{R}^d$ given as $x \cdot y := \sum_{l=1}^{d} x_l y_l$. By *Plancherel's theorem* $\mathcal{F}$ is a unitary mapping on $L^2(\mathbb{R}^d)$, in particular, one has the *fundamental relation* $\langle f, g\rangle = \langle \hat{f}, \hat{g}\rangle$. The inversion formula is given by $\mathcal{F}^{-1}g(x) = \int_{\mathbb{R}^d} \hat{g}(y)e^{2\pi i x\cdot y}dy$.

Time-frequency shifts are denoted by T_x for the *translation operators* and M_y for the *modulation operator* with $x, y \in \mathbb{R}^d$, i.e.,

$$T_x f(z) := f(z - x), \qquad M_y f(z) := e^{2\pi i y\cdot z} f(z) \tag{2.2}$$

and we have

$$\widehat{T_x f} = M_x \hat{f} \ , \qquad \widehat{M_y f} = T_y \hat{f} \ , \tag{2.3}$$

and the important *commutation relation*

$$M_y T_x = e^{2\pi i x\cdot y} T_x M_y \ . \tag{2.4}$$

By $\mathcal{S}(\mathbb{R}^d)$ we denote the *Schwartz space* of rapidly decreasing functions on $\mathbb{R}^d$. It is invariant under the FT. The dual space $\mathcal{S}'(\mathbb{R}^d)$ consists of *tempered distributions* and the FT can be extended to $\mathcal{S}'(\mathbb{R}^d)$ by $\hat{\sigma}(f) := \sigma(\hat{f})$.

$(L^1(\mathbb{R}^d), \|\cdot\|_1)$ denotes the Banach space of all Lebesgue integrable functions with the natural norm $\|f\|_1 := \int_{\mathbb{R}^d} |f(x)|dx$. It contains $\mathcal{K}(\mathbb{R}^d)$, the space of all continuous, complex-valued functions k with compact support $(supp(k))$ as a dense subspace. $(L^1(\mathbb{R}^d), \|\cdot\|_1)$ is a Banach algebra with respect to convolution, which is defined as the integral:

$$f * g(x) := \int_{\mathbb{R}^d} g(x-y)f(y)dy \ = \left(\int_{\mathbb{R}^d} T_y g\, f(y)\, dy\right)(x). \tag{2.5}$$

For $1 \le p < \infty$ set $L^p_m(\mathbb{R}^d) := \{f | \|f\|_{p,m} := (\int_{\mathbb{R}^d} |f(y)|^p m(y)^p\, dy)^{1/p} < \infty\}$. For the trivial weight $m(y) \equiv 1$ we just write $(L^p, \|\cdot\|_p)$. If m is a moderate weight (see below) then L^p_m is translation invariant. Using the fact that $\|T_x f\|_p = \|f\|_p$ for all $x \in \mathbb{R}^d$ and $\|T_x f - f\|_p \to 0$ for $|x| \to 0$ and all $f \in L^p(\mathbb{R}^d)$ is also follows (by vector-valued integration)

$$L^1 * L^p \subseteq L^p \ , \text{ and } \|g * f\|_p \le \|g\|_1 \|f\|_p \ \forall\, g \in L^1(\mathbb{R}^d), f \in L^p(\mathbb{R}^d) \ . \tag{2.6}$$

The involution in L^1, given by $g^*(x) = \overline{g(-x)}$, satisfies $(g * f)^* = f^* * g^*$.

The *Short Time Fourier Transform* (or *Sliding Window Fourier Transform*), for short STFT, is a well-known tool for the time-frequency representation of signals, especially of non-stationary signals (see Allen-Rabiner [1], Nawab and Quatieri [46] , Papoulis [47] , Flandrin [32] , Hlawatsch [40] for the use of STFT in signal analysis). It is also found under the names of *continuous Gabor transform* in [39] and *holographic transform* (see Schempp [54],[55]). By means of a *window function* g , usually a plateau-like real-valued functions with compact support centered near the origin a piece of the signal f is "cut out" and its frequencies are then analyzed with the Fourier transform.

The STFT of f with respect to the window g is given by

$$S_g f(x,y) := \int_{\mathbb{R}^d} e^{-2\pi i y \cdot z}\, \bar{g}(z-x) f(z) dz = \langle M_y T_x g, f\rangle \text{ for } (x,y) \in \mathbb{R}^{2d}. \quad (2.7)$$

The STFT is symmetric with respect to f and g in the following sense:

$$S_g f(x,y) = e^{-\pi i x \cdot y} \cdot \overline{S_f g(-x,-y)}\ . \quad (2.8)$$

This is important as it shows that decay properties of $S_g f$ are joint properties of f *and* g. In particular, if g is *not* smooth, one cannot expect good decay of $S_g f$ in the frequency direction, even if f is very smooth (or even constant). On the other hand much smoothness of g requires a large support which destroys the locality properties of $S_g f$. This dilemma makes "window design" so interesting. However, any window g which is sufficiently well localized in time and frequency will allow to identify a "piece of music", *i.e.*, a superposition of a couple of pure tones with smooth envelopes by its time-frequency behavior.

Due to its good time-frequency concentration (it gives equality in the Heisenberg uncertainty relation) and its invariance under the FT the Gauss kernel g_0, $g_0(x) := e^{-\pi x^2}$ is a very good choice. The formula $S_g f(x,y) = e^{-2\pi i x \cdot y} S_{\hat{g}} \hat{f}(y,-x)$ implies

$$|S_{g_0} f(x,y)| = |S_{g_0} \hat{f}(y,-x)| \quad \text{for } f \in L^2(\mathbb{R}^d)\ , \quad (2.9)$$

showing that the behavior of $S_{g_0}\hat{f}$ is exactly the same as that of $S_{g_0} f$, rotated by 90° in the TF-plane.

Lemma 1. *(Inversion Formula for the STFT)*
a. *Given two non-orthogonal elements $g, g_1 \in L^2(\mathbb{R}^d)$ any signal f can be recovered (for some $C_1 \neq 0$) by*

$$f(z) = C_1 \int_{\mathbb{R}^d} T_x(\bar{g} g_1)(z) f(x) dx = C_1 \int_{\mathbb{R}^d}\int_{\mathbb{R}^d} S_g f(x,y) M_y T_x g_1(z) dy dx\ , \quad (2.10)$$

or in vector-valued form

$$f = C_1 \int_{\mathbb{R}^d}\int_{\mathbb{R}^d} \langle M_y T_x g, f\rangle M_y T_x g_1 dy dx. \quad (2.11)$$

b. *The mapping $f \mapsto S_g f$ defines an isometry from $L^2(\mathbb{R}^d)$ onto $L^2(\mathbb{R}^{2d})$, if $g \in L^2(R^d)$ satisfies $\|g\|_2 = 1$, i.e.,*

$$\|S_g(f)\|_2^2 = C\|f\|_2^2 \;\; \forall \; f \in L^2(\mathbb{R}^d). \tag{2.12}$$

Sketch of Proof: The Fourier inversion formula allows to recover $(T_x\bar{g})f$ from $S_g f(x,.)$. If $\sum_{n\in\mathbb{Z}^d} T_n g \equiv 1$ one recovers f as $\sum_{n\in\mathbb{Z}^d} T_n g \cdot f$. For general $g \in L^1 \cap L^2$ summation is replaced by an integral. Since

$$\int_{\mathbb{R}^d} \bar{g}(z-x)dx = \int_{\mathbb{R}^d} g^*(x-z)dx = \int_{\mathbb{R}^d} g(u)du = \hat{g}(0) \; , \tag{2.13}$$

one derives

$$\hat{g}(0)\cdot f(z) = \int_{\mathbb{R}^d} (T_x\bar{g}(z))f(z)dx = \int_{\mathbb{R}^d}\int_{\mathbb{R}^d} e^{2\pi i y\cdot z} S_g f(x,y)dxdy \; . \tag{2.14}$$

Since the *same* argument holds if we replace $\bar{g}$ above by some other function $\bar{g}g_1$, with $\int_{\mathbb{R}^d} \bar{g}(z)g_1(z)dz =: (C_1)^{/1} \neq 0$, the result is proved.

b) Take $g = g_1$ and $\|g\|_2 = 1$ and take the inner product with f on both sides of (2.10).

The lemma indicates how the signal f can be recovered from complete knowledge of the STFT, as a (continuous) superposition of copies of g_1, shifted in the time-frequency sense. If we want to write f as such a superposition, with basic building block g, we just have to interchange the roles of g_1 and g. From this point of view the two main questions are: Whether reconstruction is possible if the STFT is only given over a discrete set of points in the TF-plane, and whether the continuous collection $\{M_y T_x g\}$ of "atoms" can be replaced by a discrete (but sufficiently rich) subcollection of Gabor wavelets.

Following Folland [33] or Heil/Walnut [39] (see more detailed references there, see also Auslander/Tolimieri [2]) we also introduce the *radar ambiguity function* $A(f,g)$

$$A(f,g)(x,y) := \int_{\mathbb{R}^d} e^{2\pi i y\cdot z} f(z+x/2)\bar{g}(z-x/2)dz, \tag{2.15}$$

It has the same absolute value as the STFT. For the description of good window functions we also need the *(cross) Wigner distribution* $W(f,g)$

$$W(f,g)(x,y) := \int_{\mathbb{R}^d} e^{2\pi i y\cdot z} f(x+z/2)\bar{g}(x-z/2)dz. \tag{2.16}$$

Despite their deceiptive similarity these two functions can be shown to be mapped into each other via the Fourier transform. For most applications in signal analysis the quadratic expressions $A_f := A(f,f)$ and $W_f := W(f,f)$ are used (see Hlawatsch [40] and references there).

Lemma 1.b) suggests to think of $\|S_g f\|^2$ as an indicator for the "energy distribution" of the signal f on the TF-plane. The behavior of $S_g f$ also reflects local smoothness of f in some domain $U \in \mathbb{R}^d$, by good decay of $S_g f(x,y)$, for $|y| \to \infty$, for all $x \in U$ (at least if g is smooth enough). If g is well concentrated near the origin then $S_g f$ will reflect decay (or growth) of f itself. It is therefore a natural idea to use the STFT as a tool to measure decay and smoothness properties of a signal as well as its time-frequency concentration. We do this by considering the following class of *generalized modulation spaces.* But first we need an appropriate class of weight functions.

Definition 2. *A strictly positive and continuous function m on the TF-plane will be called a−moderate if for some $a \geq 0$ there exists $C > 0$ such that*

$$(M1) \qquad m(z+v) \leq C(1+|v|)^a m(z) \ \forall\, z, v \in \mathbb{R}^d\ , \tag{2.17}$$

For any such weight we define the *generalized modulation space* $M^m_{p,q}(\mathbb{R}^d)$:

Definition 3. *Let $g \in \mathcal{S}$ and m be an a-moderate function, then we define the generalized modulation space $M^m_{p,q}(\mathbb{R}^d)$ as*

$$M^m_{p,q}(\mathbb{R}^d) = \{f \in \mathcal{S}'(\mathbb{R}^d)\ ,\ \|f|M^m_{p,q}(\mathbb{R}^d)\| :=$$
$$= [\int_{\mathbb{R}^d} (\int_{\mathbb{R}^d} |S_g f(x,y)|^p m^p(x,y) dx)^{q/p} dy]^{1/q} < \infty\}. \tag{2.18}$$

For the special case

$$m(x,y) = (1+|x|)^r (1+|y|)^s \ \forall\, x, y \in \mathbb{R}^d, \tag{2.19}$$

with $r, s \in \mathbb{R}$ we shall use the symbol $M^{r,s}_{p,q}$. The "classical" modulation spaces $M^s_{p,q}$ (see [22] and Triebel [61]) are identical with the spaces $M^{0,s}_{p,q}$ in the new terminology. On the other hand the spaces $M^{r,s}_{p,q}$ may be considered as a weighted version of $M^s_{p,q}$.

By reversing the order of integration we define another family of spaces, to be called $W^m_{p,q}$. We have chosen the letter W because this family contains the Wiener-amalgam spaces $W(\mathcal{F}L^p, L^q)$ as special cases, just as the classical modulation spaces are contained in the M-family. See [7], [20] and [21] for results about these spaces and their role in the description of the generalized FT (GFT).

Definition 4.

$$W^m_{p,q}(\mathbb{R}^d) = \{f \in \mathcal{S}'(\mathbb{R}^d)\ ,\ \|f|W^m_{p,q}(\mathbb{R}^d)\| :=$$
$$= [\int_{\mathbb{R}^d} (\int_{\mathbb{R}^d} |S_g f(x,y)|^q m^q(x,y) dy)^{p/q} dx]^{1/p} < \infty\}. \tag{2.20}$$

Obvious modifications take place if $p = \infty$ or $q = \infty$.

The definition of both families of spaces is inspired by the theory of Besov-Triebel-Lizorkin spaces $B^s_{p,q}$ and $F^s_{p,q}$ (see Stein [57], Peetre [48], and Triebel [60]). While the theory of modulation spaces is very similar to the theory of BTL-spaces, there are some fundamental differences. The $B-$ and $F-$spaces are (with few exceptions) *not* isomorphic as Banach spaces, on the other hand, the Fourier transform is an isomorphism between $M^m_{p,q}(\mathbb{R}^d)$ and $W^{\tilde{m}}_{p,q}$, where $\tilde{m}(x,y) = m(-y,x)$, as follows immediately from the definitions.

Lemma 5. *For $1 \le p, q \le \infty$ the $M^s_{p,q}(\mathbb{R}^d)$-spaces are Banach spaces of distribution on $\mathbb{R}^d$ with respect to their natural norms; they are invariant under translation and modulation operators and these operations are continuous on $M^s_{p,q}(\mathbb{R}^d)$ for $1 \le p, q < \infty$ in the following sense (writing $_B$ for any of them)*

$$\|T_x f - f\|_B \to 0 \text{ for } |x| \to 0, \forall\, f \in B; \tag{2.21}$$

as well as

$$\|M_y f - f\|_B \to 0 \text{ for } |y| \to 0, \forall\, f \in B; \tag{2.22}$$

The submultiplicativity of the weight w_a, given by $w_a(v) := (1 + |v|)^a$, for $a \ge 0$, i.e., the inequality $w_a(x + y) \le w_a(x) w_a(y)$ for $x, y \in \mathbb{R}^{2d}$ implies

$$\|T_x f\|_B \le w_1(x)\|f\|_B \text{ and } \|M_y f\|_B \le w_2(y)\|f\|_B \tag{2.23}$$

for all $x, y \in \mathbb{R}^d$, where $w_1(x) = w(x, 0)$ and $w_2(y) = w(0, y)$. It follows by vector-valued integration that $L^1_{w_1} * B \subseteq B$, and

$$\|g * f\|_B \le \|g\|_{1,w_1}\|f\|_B \quad \forall\, g \in L^1_{w_1}, f \in B. \tag{2.24}$$

and by a similar argument $\mathcal{F}(L^1_{w_2}) \cdot B \subseteq B$, and

$$\|h \cdot f\|_B \le \|\hat{h}\|_{1,w_2}\|f\|_B \text{ for } h \in \mathcal{F}L^1_{w_2} , f \in B . \tag{2.25}$$

It is also true for the same reason that for any Dirac sequence $(e_n)_{n=1}^\infty$ in $L^1_{w_1+w_2}(\mathbb{R}^d)$ one has $\|e_n * f - f\|_B \to 0$. More precisely, for any sequence of normalized functions in $L^1(\mathbb{R}^d)$ with supports shrinking to zero this is true. One also has $\|\hat{e_n} \cdot f - f\|_B \to 0$ by a similar argument. Combining both facts and using the inclusion $(\mathcal{S} * \mathcal{S}') \cdot \mathcal{S} \subseteq \mathcal{S}$ we derive also that $\mathcal{S}$ (or even $\mathcal{D} := \mathcal{S} \cap \mathcal{K}$) is dense in any of these spaces (for $1 \le p, q < \infty$).

The above definitions raise a couple of questions. First of all, are these spaces well defined, *i.e.*, actually independent of the choice of the window function g ? If so, how can a sufficiently large class of windows be described, such that any non-trivial choice of a window from that class gives the same spaces (with equivalent norms). This questions will be answered in detail in Lemma 26.

The following lemma gives a list of familiar spaces that are contained in the family of modulation spaces.

Lemma 6. *L^2 is contained in the family of modulation spaces as $M^0_{2,2}$. More generally the L^2 Bessel potential spaces $H^s = \{f \in \mathcal{S}' : \int |\hat{f}(t)|^2(1 + |t|)^{2s} dt < \infty\}$ (or $\mathcal{L}^2_s$ in Stein's notation, see Stein [57]) coincide with the modulation spaces $M^s_{2,2}$. The Segal algebra S_0 (see section 6) arises as $M^0_{1,1}$ (see [19]). Its algebra of multipliers (operators commuting with translations) can be identified with $M^0_{1,\infty}$, and the dual space $S'_0(\mathbb{R}^d)$ is just $M^0_{\infty,\infty}$ (with appropriate interpretations of the symbols for p or $q = \infty$).*

For details we refer to [20]–[22], [26], [27] . Atomic decompositions of these spaces are studied in [22], and [25] – [28] .

§3. Gabor Wavelets and Irregular Sampling of the STFT

In this section we describe expansions of signals in the modulation spaces by means of Gabor wavelets. These results are special cases of the general theory in our papers [9] and [25]–[29], obtained by choosing as group the reduced Heisenberg group with the Schrödinger representation . The relevance of the Heisenberg group for a wide range of problems in mathematical analysis and also for applications in signal analysis is meanwhile well documented (see Auslander and Tolimieri [2], Folland [33], Howe [41], Schempp [54],[55], and Taylor [59]). However, since the translation from the abstract results into the concrete form may not be obvious for "users", and since Gabor wavelet expansions can be stated independently and directly, we postpone the discussion of the background to section 4 and state the results in the language of signal analysis. We start with the characterization of modulation spaces by means of Gabor wavelet expansions.

In order to describe the 'density' of a discrete family $(z_i)_{i\in I}$ in $\mathbb{R}^d$ we use the following concept: Given some set W with nonvoid interior in $\mathbb{R}^d$ we call $(z_i)_{i\in I}$ W–dense, if $\bigcup_{i\in I} z_i + W = \mathbb{R}^d$.

Theorem 7. *(Gabor wavelet expansions for modulation spaces). Let us consider the family of all modulation spaces $M^m_{p,q}(\mathbb{R}^d)$, with m satisfying (M1) for the same $a \geq 0$ and $C > 0$. Then for any $g \in L^2(\mathbb{R}^d)$ with Wigner distribution W_g (or equivalently with ambiguity function A_g) in $L^1_{w_a}(\mathbb{R}^d)$ there exists a neighborhood W of $(0,0)$ in $\mathbb{R}^{2d}$ such that for any W-dense family $(z_i)_{i\in I} = (x_i, y_i)_{i\in I}$ in $\mathbb{R}^{2d}$ it is possible to write any $f \in M^m_{p,q}(\mathbb{R}^d)$ as*

$$f = \sum_{i\in I} c_i T_{x_i} M_{y_i} g \ , \tag{3.1}$$

the family $(c_i)_{i\in I}$ belonging to a sequence space connected with $M^m_{p,q}(\mathbb{R}^d)$. The mapping $f \mapsto (c_i)_{i\in I}$ is linear. If $(z_i)_{i\in I}$ is a subset of some lattice $\alpha\mathbb{Z}^d \times \beta\mathbb{Z}^d \subseteq \mathbb{R}^{2d}$ the coefficient mapping is bounded from $M^m_{p,q}(\mathbb{R}^d)$ into the sequence space

$$(\ell^q(\ell^p))_m := \{(c_{l,n})_{l,n\in Z^{2d}} : [\sum_{l\in\mathbb{Z}^d} (\sum_{n\in\mathbb{Z}^d} |c_{l,n}|^p m^p_{l,n})^{q/p}]^{1/q} < \infty\} \tag{3.2}$$

where $m_{l,n} := m(\alpha l, \beta n)$, for $k, n \in \mathbb{Z}^d$. Conversely, for any set of coefficients $(c_{l,n})_{l,n\in Z^{2d}} \in (\ell^q(\ell^p))_m$ the series

$$f = \sum_{n,l} c_{n,l} T_{\alpha l} M_{\beta n} g \tag{3.3}$$

is unconditionally convergent in $M^m_{p,q}(\mathbb{R}^d)$, since for every $\varepsilon > 0$ there exists a finite subset $F_0 \subseteq I$ such that for any finite set $F \supseteq F_0$ one has

$$\|f - \sum_{i\in F} c_{n,l} T_{\alpha l} M_{\beta n} g | M^m_{p,q}(\mathbb{R}^d)\| \leq \varepsilon \ . \tag{3.4}$$

In particular

$$f = \sum_n \sum_l c_{n,l} T_{\alpha l} M_{\beta n} g = \sum_l \sum_n c_{n,l} T_{\alpha l} M_{\beta n} g \,. \tag{3.5}$$

A version of the above result which gives more quantitative information about the set W reads as follows (details are to be given in a forth-coming note by Gröchenig). For the case of regular sampling along lattices in the TF-plane results of this kind are given by Walnut [62].

Theorem 7.b. *Suppose that $supp(g) \subseteq [-a, a]$ for some $g \in L^2(R)$. Let $(x_n)_{n\in Z}$ be any increasing sequence such that for some $A, B > 0$*

$$0 < A_1 \le \sum_{n\in Z} |g(t - x_n)|^2 \le B_1 < \infty \;\text{ a.e.} \tag{3.6}$$

Assume that $(y_{n,k})_{k\in Z}$ are increasing sequences with $sup_{k,n}(y_{k+1,n} - y_{k,n}) < 1/2a$. Then every $f \in L^2(R)$ has a stable expansions

$$f = \sum_{n,k} c_{k,n}(y_{k+1,n} - y_{k,n}) M_{y_{k,n}} T_{x_n} g \tag{3.7}$$

with

$$(c_{k,n})_{k,n\in Z} \in \ell^2 \;\text{ and }\; \|(c_{k,n})\|_2 \le C \cdot \|f\|_2. \tag{3.8}$$

Actually, the collection $\{(y_{k+1,n} - y_{k,n})^{1/2} M_{y_{k,n}} T_{x_n} g\}_{n,k\in Z^2}$ is a frame with frame constants $A = A_1 \times (1 - 2\delta a)$ and $B = B_1 \times (1 + 2\delta a)$.

Remark 8. Note that condition (3.6) implies $sup_{k\in Z}|x_{k+1} - x_k| < 2a$.

Remark 9. Of course it is possible the characterize the W-spaces in the same way, just reversing the order of summation compared to the M-spaces.

The results in [38] can be interpreted as statements, that sufficiently rich *coherent families*) are suitable as building blocks for atomic representations for large classes of function spaces if and only if they are *Banach frames* for the same class.

As answer to the second main question we present the following result.

Theorem 10. *Consider the same family of spaces as in Theorem 7, and let g be as above. Then there exists a neighborhood $\tilde{W}$ of zero in $\mathbb{R}^{2d}$ such that for any $\tilde{W}$-dense family $(z_i)_{i\in I}$ in $\mathbb{R}^{2d}$ there is a sequence of adaptive weights $(w_i)_{i\in I}$ such that the sequence $(f^{(n)})_{n=1}^{\infty}$, given recursively by*

$$f^{(0)} := Af := \sum_{i\in I} w_i S_g f(x_i, y_i) M_{y_i} T_{x_i} g \tag{3.9}$$

and

$$f^{(n+1)} := f^{(n)} + A(f - f^{(n)}) \;\text{ for } n \ge 1 \tag{3.10}$$

is convergent in $M^m_{p,q}(\mathbb{R}^d)$ *to* f *at a geometric rate, i.e., there is* $\gamma, 0 < \gamma < 1$ *and* $C > 0$ *, so that*

$$\|f - f^{(n)}\|_{M^m_{p,q}(\mathbb{R}^d)} < C \cdot \gamma^n \cdot \|f\|_{M^m_{p,q}(\mathbb{R}^d)}. \tag{3.11}$$

Remark 11. The input in this algorithm are the values $S_g f(z_i)$. The theorem is thus an irregular sampling theorem for the STFT and provides an easy reconstruction method for *all* modulation spaces. The proposed method will be called the *ADAPTIVE WEIGHTS METHOD* for the irregular sampling problem of the STFT. The actual rate of convergence depends mainly on the density of the sampling family in the TF-plane.

Corollary 12. *If in addition to the assumptions of Theorem 7 the sampling set belongs to a (sufficiently fine) lattice in* $\mathbb{R}^{2d}$*, then a function in* $L^2(\mathbb{R}^d)$ *belongs to* $M^m_{p,q}(\mathbb{R}^d)$ *if the sampling family* $(S_g f(x_i, y_i))_{i \in I}$ *belongs to the corresponding sequence space* $(\ell^q(\ell^p))_m$.

There are several immediate questions related to these results. If the coefficients are not uniquely determined, to what extent does the choice of the specific method, and in particular the choice of auxiliary parameters influence the result. Let us assume we want to replace the building block (atom) by another one which is close in some sense. Will the corresponding family of coefficients be similar to the original one? What is a good way to measure this similarity of atoms? For the STFT we have similar questions. What can be said about the reconstruction, if the window that has been used for the STFT is only know approximately (*i.e.*, close to some known window)? Can we still hope for approximate reconstruction? What, if there is jitter error, *i.e.*, if we do not have exact informations about the actual sampling positions in the time-frequency plane for the STFT? In the best case we can hope that the reconstruction error is small if the maximal jitter deviation from a family of sampling points is sufficiently small. Such results and many related questions can be answered affirmatively, based on our group theoretic point of view.

From a practial point of view one will have to look for a combined optimization of α and β simultanuously, given g, if one has the freedom to chose α and β independently. Also, both the speed of convergence as well as the degree of stability of the scheme will depend very much on the choice of g, α and β.

The following corollary is of great practical use, especially if the same sampling geometry and the same window occur for many different signals.

Corollary 13. *There are functions* $e_i \in M^{1,1}_{w_a}$ *, so that*

$$f = \sum_i S_g f(x_i, y_i) e_i, \tag{3.12}$$

with convergence in $M^m_{p,q}(\mathbb{R}^d)$ *, for any* $f \in M^m_{p,q}(\mathbb{R}^d))$.

The practical relevance of this Corollary is based on the fact that one may calculate the collection $(e_i)_{i \in I}$ offline, perhaps on a faster computer, and recombine them with the sampling values according to (3.12).

§4. The Group Theoretical Background, Correspondence Principle

Gabor wavelets are obtained from a single function – sometimes called mother wavelet – by means of time and frequency shifts T_x and M_y. These operators do not commute, but the family

$$\{\tau T_x M_y \mid \tau \in \mathbb{C}, |\tau| = 1;\ x, y \in \mathbb{R}^d\} \tag{4.1}$$

forms a natural group of unitary operators on $L^2(\mathbb{R}^d)$, which also acts continuously on a variety of other function spaces. Identifying these operators with their parameters, which are in $\mathbb{H}^d := \mathbb{R}^d \times \mathbb{R}^d \times \mathbb{T}$ we recognize the following *group law* for the *reduced Heisenberg group* $\mathbb{H}^d$:

$$h_1 \cdot h_2 := (x_1, y_1, \tau_1) \cdot (x_2, y_2, \tau_2) = (x_1 + x_2, y_1 + y_2, \tau_1 \cdot \tau_2 c^{2\pi i x_2 y_1})\ . \tag{4.2}$$

The inverse can be calculated as $h^{-1} = (x, y, \tau)^{-1} = (-x, -y, \bar{\tau} e^{2\pi i x y})$.
NOTE: Since we use the factor 2π in the definition of the modulation operators (because of our definition of F) this law differs slightly from the description in our earlier notes [25]–[27]. We refer to Folland [33], Taylor [59], and Heil/Walnut [40] and Reiter [53] for technical details concerning following facts : The Lebesgue measure $dh := dx dy d\tau$ is the Haar measure for $\mathbb{H}^d$. It is both left and right invariant, *i.e.*, $\mathbb{H}^d$ is *unimodular*. This means that for $F \in \mathcal{K}(\mathbb{H}^d)$

$$\int_{\mathbb{H}^d} F(h_0^{-1} h) dh = \int_{\mathbb{H}^d} F(h) dh = \int_{\mathbb{H}^d} F(h h_0) dh \quad \forall\ h_0 \in \mathbb{H}^d. \tag{4.3}$$

Unimodularity also implies that the *left* and *right translation operators* , given by $L_h F(v) := F(h^{-1} v)$ and $R_h F(v) := F(vh)$ respectively, act isometrically on $L^p(\mathbb{H}^d)$, for $1 \le p \le \infty$. For $F, G \in \mathcal{K}(\mathbb{H}^d)$ *convolution* is given (as point-wise or vector-valued integral)

$$F * G(h_0) := \int_{\mathbb{H}^d} G(h^{-1} h_0) F(h) dh \text{ or } \left(\int_{\mathbb{H}^d} F(h) L_h G dh \right)(h_0). \tag{4.4}$$

Convolution on $\mathbb{H}^d$ is associative, but *not* commutative. In the following we summarize basic facts concerning convolution. (1) Translation and convolution:

$$L_h(F * G) = L_h F * G, \qquad R_h(F * G) = F * R_h G. \tag{4.5}$$

(2) $(L^1(\mathbb{H}^d), \|\cdot\|_1)$ is a Banach algebra with respect to convolution.
(3) Young's inequality: since $\mathbb{H}^d$ is unimodular, the convolution relations

$$L^p(\mathbb{H}^d) * L^1(\mathbb{H}^d) \subseteq L^p(\mathbb{H}^d), \text{ and } L^2(\mathbb{H}^d) * L^2(\mathbb{H}^d) \subseteq C^0(\mathbb{H}^d), \tag{4.6}$$

hold, where $C^0(\mathbb{H}^d)$ is the space of continuous functions vanishing at infinity, endowed with the sup-norm $\|F\|_\infty := sup_{z \in \mathbb{H}^d} |F(z)|$.

(4) If F^* denotes the involution $F^*(h) := \bar{F}(h^{-1})$, then $(F*H)^* = H^* * F^*$. Convolution by G^* (from the right) is the adjoint operator of the convolution operator $F \mapsto F * G$, *i.e.*,

$$\langle H, F * G\rangle = \langle H * G^*, F\rangle \quad \text{for } F, H \in L^2(\mathbb{H}^d), G \in L^1(\mathbb{H}^d). \tag{4.7}$$

(5)

$$H * G(h) = \langle L_h G^*, H\rangle \ \ \forall\, h \in \mathbb{H}^d, \quad H * G^*(0) = \langle G, H\rangle. \tag{4.8}$$

(6) Convolution inequalities for weighted L^p-spaces on $\mathbb{H}^d$ are derived as on $\mathbb{R}^d$. If w is a *submultiplicative* (Beurling) weight w, *i.e.*,

$$w(hh') \leq w(h)w(h') \ \ \forall\, h, h' \in \mathbb{H}^d, \tag{4.9}$$

then $L^1_w(\mathbb{H}^d)$ is a Banach convolution algebra. An important example of weight functions on $\mathbb{H}^d$ is $w(h) := w_a(h) := (1 + |(x,y)|)^a$, with $a \geq 0$.

(7) A strictly positive and continuous function on $\mathbb{H}^d$ is called a-moderate if $m(hh') \leq m(h)w_a(h') \quad \forall\, h, h' \in \mathbb{H}^d$. Then the pointwise estimate

$$|F * G(h)|\ m(h) \leq (|F|m * |G|w)(h) \ \ \forall\, h \in \mathbb{H}^d \tag{4.10}$$

holds and Young's inequality implies that $L^p_m(\mathbb{H}^d) := \{F | Fm \in L^p(\mathbb{H}^d)\}$ satisfies $L^p_m * L^1_w \subseteq L^p_m$ for $1 \leq p \leq \infty$.

In order to apply the general theory of [26] to Gabor wavelet expansions, we proceed along the following steps:

(A) Check that the assumptions of the general theory are satisfied in the particular case of the Heisenberg group and the Schrödinger representation.

(B) Relate the abstract concepts of [25] and [26] to the language of signal analysis, as developed in Section 2.

(C) Translate the main result on series expansions in [26] into the concrete situation of Gabor wavelet expansions.

To verify the assumptions made in [26], we have to check that the Schrödinger representation is *irreducible*, *one-to-one*, and *integrable* with respect to all weights $w_a, a > 0$ on $\mathbb{H}^d$, *i.e.*, has *matrix coefficients* in $L^1_{w_a}(\mathbb{H}^d)$.

Recall that the matrix (or representation) coefficients or the *generalized wavelet transform* (with respect to the Schrödinger representation) are functions on $\mathbb{H}^d$ given by

$$V_g f(h) := \langle \pi(h)g, f\rangle \ \ \text{for } h \in \mathbb{H}^d. \tag{4.11}$$

We speak of g as *analyzing window* (since g may be any bump function we avoid the word "wavelet" at this point), and f is called the *analyzed signal*. We will call $f \mapsto V_g f$ the *Gabor(-Heisenberg) transform* with respect to g in this note in order to indicate the intimate connection to the Heisenberg group.

The Gabor transform is a linear mapping from the Hilbert space $L^2(\mathbb{R}^d)$ into a space of bounded and continuous functions on $\mathbb{H}^d$, since for $f, g \in L^2(\mathbb{R}^d)$

$$|V_g f(h)| \leq \|g\|_2 \|f\|_2 \ \ \forall\, h \in \mathbb{H}^d, \text{ by Cauchy-Schwartz} \tag{4.12}$$

$$|V_g f(h) - V_g f(h \cdot h')| \leq |\langle (\pi(h') - 1)g, \pi(h^{-1})f\rangle| \to 0 \text{ for } |h'| \to 0. \tag{4.13}$$

It will be crucial that V_g also satisfies the *intertwining property*

$$V_g(\pi(h)f) = L_h(V_g(f)) \ \text{ for } h \in \mathbb{H}^d. \tag{4.14}$$

Proof:

$$V_g(\pi(h)f)(\tilde{h}) = \langle \pi(\tilde{h})g, \pi(h)f\rangle =$$

$$= \langle \pi(h^{-1}\tilde{h})g, f\rangle = V_g f(h^{-1}\tilde{h}) = L_h(V_g(f))(\tilde{h}). \ ♣$$

Changing the role of f and g is the same as applying the involution to $V_g f$

$$(V_g f)^* = V_f(g). \tag{4.15}$$

Proof: $V_g f^*(h) = \overline{V_g f(h^{-1})} = \overline{\langle \pi(h^{-1})g, f\rangle} = \langle \pi(h)f, g\rangle = V_f g(h)$. ■

The existence of *admissible vectors* g, *i.e.*, of non-zero functions $g \in L^2(\mathbb{R}^d)$ for which $V_g g \in L^1(\mathbb{H}^d)$, is by definition the same as *integrability of the representation* π. Note that by the boundedness of $V_g g$ square integrability is a consequence of integrability. Since we have

$$|V_g f(x, y, \tau)| = |V_g f(x, y, 1)| \ \ \forall\, \tau \in \mathbb{T} \tag{4.16}$$

we may use the symbol $V_g f(x, y)$ for $V_g f(x, y, 1)$ and check integrability of $V_g f$ over R^{2d} only. To this end let us derive two pointwise estimates. One has

$$|V_g f(x, y)| = |\langle T_x M_y g, f\rangle| = |M_y g^* * f(x)| \leq (|g^*| * |f|)(x); \tag{4.17}$$

for $f, g \in L^2(\mathbb{R}^d)$, but also by Plancherel's theorem

$$|V_g f(x, y)| = |\langle M_{-x} T_y \hat{g}, \hat{f}\rangle| = |M_{-x}\hat{g}^* * \hat{f}(y)| \leq (|\hat{g}|^* * |\hat{f}|)(y). \tag{4.18}$$

Combining these two estimates one obtains

$$|V_g g(x, y)| \leq \min[(|g^*| * |g|)(x), |(\hat{g}^*| * |\hat{g}|)(y)]. \tag{4.19}$$

Since $g, g^*, \hat{g}, \hat{g}^*$ and therefore both convolution products are faster decaying than the inverse of any polynomial for any $g \in \mathcal{S}(\mathbb{R}^d)$, it follows that $V_g g \in L^1(\mathbb{H}^d)$ (or $L^1(R^{2d})$). Actually, we have shown that $g \in \mathcal{H}^1_{w_a} = \{g \mid V_g g \in L^1_{w_a}\}$ for any $a \geq 0$.

Among the basic consequences of the (square) integrability of the Schrödinger representation are the *orthogonality relations*, which are also known as Moyal's formulas in this case.

$$\langle V_{g_1} f_1, V_{g_2} f_2 \rangle_{L^2(\mathbb{H}^d)} = \langle f_1, f_2 \rangle_{L^2(\mathbb{R}^d)} \cdot \langle g_2, g_1 \rangle_{L^2(\mathbb{R}^d)}. \tag{4.20}$$

Taking $f_1 = f_2 = f$ and $g_1 = g_2 = g$, then we obtain

$$\|V_g f\|_2 = \|f\|_2 \|g\|_2 \quad \forall\ f, g \in L^2(\mathbb{R}^d). \tag{4.21}$$

Both formulas follow immediately from the inversion formula of the STFT (3.10), since for the representation coefficients the integration over the torus is trivial.

The orthogonality relations have several important consequences. In order to show that π is irreducible we have to verify that for any $g \neq 0$ in $L^2(\mathbb{R}^d)$ the finite linear combinations of elements of the form $T_x M_y g$ are dense in $L^2(\mathbb{R}^d)$.

Proof: Assume $f \perp \pi(h)g$ for all $h \in \mathbb{H}^d$. Then we have $V_g f = 0$, hence $\|V_g f\|_2 = \|f\|_2 \|g\|_2 = 0$, which in turn implies that $f = 0$. ■

Next we verify the *convolution formula* for the representation coefficients:

$$V_{g_1} f_1 * V_{g_2} f_2 = \langle g_1, f_2 \rangle \cdot V_{g_2} f_1 \quad \text{for } f_1, f_2, g_1, g_2 \in L^2(\mathbb{R}^d) \tag{4.22}$$

Proof:

$$V_{g_1} f_1 * V_{g_2} f_2(h) = \langle L_h (V_{g_2} f_2)^*, V_{g_1} f_1 \rangle =$$
$$= \langle L_h (V_{f_2} g_2), V_{g_1} f_1 \rangle = \langle V_{f_2}(\pi(h) g_2), V_{g_1} f_1 \rangle =$$
$$= \langle g_1, f_2 \rangle \langle \pi(h) g_2, f_1 \rangle = \langle g_1, f_2 \rangle V_{g_2} f_1(h). \quad \clubsuit$$

As a special case of (4.22) we observe the *reproducing convolution equation*, (assuming only that $\|g\|_2 = 1$):

$$V_g f * V_g g = V_g f \quad \forall\ f \in L^2(\mathbb{R}^d). \tag{4.23}$$

Writing $G := V_g g$ it follows that $G * G = G$ and $G = G^*$, *i.e.*, $F \mapsto F * G$ is an (orthogonal) projection operator on $L^2(\mathbb{H}^d)$. It also can be shown that $L^2(\mathbb{H}^d) * G$ is exactly $V_g(L^2(\mathbb{R}^d))$, the range of V_g.

The spaces $\mathcal{H}^1_w$ play an important role as spaces of test functions: For any functional f on one of these Banach spaces it is possible to define the Gabor transform $V_g f$ for any analyzing signal $g \in \mathcal{H}^1_w$. For reasons of convenience and consistency with our papers we work with the so-called anti-dual $(\mathcal{H}^1_w)^{-\prime}$ of all additive and continuous mappings σ from $\mathcal{H}^1_w$ into the complex numbers satisfying $\sigma(\lambda g) = \bar{\lambda}\sigma(g)$. This allows us to use the symbols $\langle \cdot, \cdot \rangle$, both for the new duality and the Hilbert space duality. For the definition of *coorbit spaces* we need also a solid Banach space $(Y, \|\cdot\|_Y)$ of locally integrable functions on $\mathbb{H}^d$ with

$$(Y1)\ |F(h)| \leq |G(h)| \text{ a.e. and } G \in Y \ \rightarrow\ F \in Y \quad \text{and} \quad \|F\|_Y \leq \|G\|_Y. \quad (4.24)$$

$(Y2)$ Y is left *and* right translation invariant and for some $a \geq 0$

$$\|L_h F\|_Y \leq w_a(h)\|F\|_Y, \text{and} \quad \|R_h F\|_Y \leq w_a(h)\|F\|_Y. \quad (4.25)$$

$(Y2')$ Furthermore we assume that for the same $a \geq 0$ we have

$$Y * L^1_a \subseteq Y \ \text{ with } \ \|F * G\|_Y \leq \|F\|_Y \|G\|_{1,a} \ \ \forall\ F \in Y, g \in L^1_a, \quad (4.26)$$

where convolution is in the sense of the Heisenberg group.

With these assumption made we are ready to define the *coorbit spaces with respect to the Schrödinger representation* :

Definition 14.

$$Co(Y) = \{f | f \in (\mathcal{H}^1_w)^{-\prime}\ , V_g f \in Y\}, \ \textit{with}\ \|f\|_Y := \|V_g f\|_Y. \quad (4.27)$$

Given the definition of generalized modulation spaces and W-spaces we have to establish the link between these spaces and coorbit spaces next. First a general remark. The above definition raises the question why we replace the handy and concrete spaces using only the time frequency plane and the STFT by objects related to the Heisenberg group and the matrix coefficients of the Schrödinger representation . This seems like an unnecessary abstraction and complication. The answer lies in the reproducing formula (4.22). It is a simple and ordinary convolution equation and can be treated by methods of non-commutative harmonic analysis. Staying in the TF-plane would lead to a "twisted" convolution (see [33]) and to a rather "twisted" analysis of the problem. Adding the trivial and almost redundant parameter of the torus component provides a fundamental group structure and thus makes life much easier. In fact, it makes the powerful machinery of general atomic representations (see [26]), which also covers many other "wavelet-theories", applicable to M- and W-spaces. The link consists of two main observations:

(1) The representation coefficents $V_g f$, which a priori "live" on the group $\mathbb{H}^d$, are uniquely determined by their values on $\mathbb{R}^d \times \mathbb{R}^d \times \{1\}$, since

$$V_g f(x, y, \tau) = \tau V_g f(x, y, 1) \quad \forall\ f, g \in L^2(\mathbb{R}^d). \quad (4.28)$$

The connection to the STFT (with window g) is given trough

$$V_g f(x, y, \tau) = \tau e^{2\pi i x \cdot y} S_g f(x, y), \ \forall\ f, g \in L^2(\mathbb{R}^d) \quad (4.29)$$

(2) There are several natural ways to relate a Banach space $(V, \|\cdot\|_V)$ on $\mathbb{R}^{2d}$ satisfying $(Y1)-(Y2')$ (and ordinary addition as group multiplication)

to a Banach space $(Y, \|\cdot\|_Y)$ on the reduced Heisenberg group $I\!H^d$ satisfying $(Y1) - (Y2')$ in such a way that for continuous functions F on $I\!H^d$ satisfying $F(x,y,\tau) = \tau F(x,y,1)$ membership of $F_{red} : (x,y) \mapsto F(x,y,1)$ in $(V, \|\cdot\|_V)$ is exactly the same as membership of F in $(Y, \|\cdot\|_Y)$ (with equivalence of norms).

Despite the possibility of giving a "general recipe" it seems to be better to treat this questions individually in order to preserve "natural" identifications. Thus, it is clear that $V = L^p(\mathbb{R}^{2d})$ matches with the space $Y = L^p(I\!H^d)$, for $1 \leq p < \infty$. There are also no serious problems with moderate weight functions m, because any moderate weight can only vary to a certain amount over the *compact* torus, *i.e.*, there are positive constants C_1 and C_2 such that

$$C_1 \cdot m(x,y,1) \leq m(x,y,\tau) \leq C_2 \cdot m(x,y,1) \quad \forall\, \tau \in I\!\!T. \tag{4.30}$$

Also moderateness of m with respect to the weight w_a on $I\!H^d$ is equivalent to moderateness of its restriction to $\mathbb{R}^{2d}$ with respect to the weight w_a (on $\mathbb{R}^{2d}$).

For more general spaces, such as the weighted, mixed norm spaces used for the definition of M-spaces, a good general recipe to generate $(Y, \|\cdot\|_Y)$ from $(V, \|\cdot\|_V)$ is to use the sublinear mapping

$$P_T F := \int_{I\!\!T} |F(x,y,\tau)| d\tau \tag{4.31}$$

and to define

$$Y := Y_V := \{F | F \text{ locally integrable on } I\!H^d, \; P_T F \in V\} \tag{4.32}$$

with the natural norm $\|F\|_Y := \|P_T F\|_V$. Note that whenever it is possible to give an estimate $\|T_{(x,y)}F\|_V \leq C \cdot w(x,y) \|F\|_V \;\forall\, F \in V$ it also follows that

$$\|L_{(x,y,\tau)}F\|_Y \leq C \cdot w(x,y) \|F\|_Y \;\forall\, F \in Y. \tag{4.33}$$

This is obvious for weighted L^p-spaces and follows directly for the spaces as defined in (4.32) , using the identity

$$|P_T(L_{(x,y,\tau)}F)| = |T_{(x,y)} P_T F|. \tag{4.34}$$

Furthermore, $\mathcal{K}(I\!H^d)$ is dense in Y if $\mathcal{K}(\mathbb{R}^{2d})$ is dense in V. In this case also continuity of $h \mapsto L_h F$ (or $R_h F$), from $I\!H^d$ to Y, and $\|L_h F\|_Y \leq w_a(h) \|F\|_Y$ follows, hence by vector-valued integration $Y * L^1_w \subseteq Y$ (now with Heisenberg convolution), *i.e.*, conditions $Y1) - Y2')$ are satisfied.

There remains one last (formal) difference in the definition of generalized modulation spaces and the corresponding coorbit spaces to be discussed. Whereas $M^m_{p,q}(I\!R^d)$ is defined as a subset of $\mathcal{S}'(\mathbb{R}^d)$, the abstract definition selects the elements from the antidual space $(\mathcal{H}^1_w)^{-\prime}$ ($w = w_a$ in our case). The reasons choosing the definition in this way have been the following ones:

i) For the abstract approach there is no such "natural" reservoir of "distributions" such as $\mathcal{S}'(\mathbb{R}^d)$ in the present case.

ii) The generality of coorbit spaces using the spaces $(\mathcal{H}^1_w)^{-\prime}$ allows to go beyond tempered distributions, and opens the way to the definition of Banach spaces of ultra-distributions.

Fortunately this ambiguity has been settled already by Theorem 4.2.iii) in [26] , because it is easy to check that $\mathcal{S}'(\mathbb{R}^d)$ is continuously embedded into $(\mathcal{H}^1_w)^{-\prime}$ for sufficiently strong weights, such as exponential weights on $\mathbb{H}^d$.

Altogether we have now verified that part (A) is OK. Let us come to to (B). The key to this the intertwining property $V_g(\pi(h)f) = L_h(V_g f)$ (4.14) of the Gabor transform. It allows to translate the two main problems into questions about functions on $\mathbb{H}^d$ which satisfy the reproducing convolution equation (4.23). We call this fact the *correspondence* or *transference principle.* As we shall see both problems involve *coherent* systems of functions (see Peremelov [49] for this notation), *i.e.*, families of the form $(\pi(h_i)g)_{i\in I}$, where $(h_i)_{i\in I}$ is some discrete family in $\mathbb{H}^d$ (or the TF-plane $\mathbb{R}^d \times \mathbb{R}^d \times \{1\} \subseteq \mathbb{H}^d$). They correspond to families $(L_{h_i} V_g g)_{i\in I}$ under the Gabor transform. Thus, we may reformulate the main questions in the group theoretical language:
i) *Atomic decomposition problem*: We are looking for sufficient conditions on g and $(h_i)_{i\in I}$ such that any $F = V_g f$, with $f \in L^2(\mathbb{R}^d)$ allows a series representation of the form (writing $G := V_g g$): $F = \sum_{i\in I} c_i L_{h_i} G$.
ii) The *irregular sampling problem* for the STFT translates by means of (4.29) into the question of reconstructing F from the family of coefficients

$$(F(h_i))_{i\in I} = (\langle \pi(h_i)g, f\rangle)_{i\in I} = (\tau e^{2\pi x_i y_i} S_g f(x_i, y_i))_{i\in I}. \tag{4.35}$$

The inversion formula (2.14) then allows to recover f from F. Of course we shall look for constructive ways of obtaining f more directly from the given data.

§5. $S_0(\mathbb{R}^d)$ and Harmonic Analysis

In this section we investigate the sapce $M^0_{1,1}(\mathbb{R}^d)$, traditionally denoted as $S_0(\mathbb{R}^d)$, in more detail. This space is of great interest in window design and can serve as a substitute for the Schwartz space $S_0(\mathbb{R}^d)$. Since the Banach space $S_0(\mathbb{R}^d)$ has a much simpler structure than the Frechet space $\mathcal{S}(\mathbb{R}^d)$, an alternative approach to abstract harmonic analysis can be based on $S_0(\mathbb{R}^d)$ [23]. The space $S_0(\mathbb{R}^d)$ has a great number equivalent characterizations. We present here mainly those which can be well formulated using notions from signal analysis.

Theorem 15. *For $f \in L^2(\mathbb{R}^d)$ the following conditions are equivalent:*

1. *For some $g \in \mathcal{S}(\mathbb{R}^d)$, $g \neq 0$: $S_g f \in L^1(\mathbb{R}^{2d})$, or $V_g f \in L^1(\mathbb{H}^d)$,* i.e., *$g \in Co(L^1)$.*
2. *For some (or any) non-zero $g \in S_0(\mathbb{R}^d)$: $S_g f \in L^1(\mathbb{R}^{2d})$;*
3. *$S_f f \in L^1(\mathbb{R}^{2d})$ (or equivalently $V_f f \in L^1(\mathbb{H}^d)$);*
4. *The Wigner distribution W_f is integrable (over the TF-plane $\mathbb{R}^{2d}$);*
5. *The ambiguity function A_f of f is integrable (over the TF-plane $\mathbb{R}^{2d}$);*

6. *For some/any non-zero $g \in S_0(\mathbb{R}^d)$, f has a Gabor series representation*

$$f = \sum_{i \in I} a_i T_{x_i} M_{y_i} g, \text{ with } \sum_{i \in I} |a_i| < \infty. \tag{5.1}$$

7. $f = \sum_{j \in J} f_j$, with $f_j \in L^1(\mathbb{R}^d)$ and $\mathrm{spec}(f_j) \subseteq y_j + Q$ for some fixed set Q (bounded with nonvoid interior), and $\sum_{j \in J} \|f_j\|_1 < \infty$.

Proof: Since $|V_g f(x, y, \tau)| = |S_g f(x, y)|$ for $(x, y) \in \mathbb{R}^{2d}$ the first two characterizations are more or less direct reformulations of the membership of $f \in \mathcal{H} = L^2(\mathbb{R}^d)$ in $Co(L^1(\mathbb{H}^d))$. Since $L^1(\mathbb{H}^d)$ is isometrically left and right transation invariant the weight function w required in the general theory can be taken as the trivial one ($w(h) \equiv 1$ for all $h \in \mathbb{H}^d$), and therefore the set of admissible vectors $\mathcal{A}_w$ (described by 3.) coincides with $S_0 = Co(Y)$.

That $S_f f \in L^1(\mathbb{R}^{2d})$ if and only if $W_f \in L^1(\mathbb{R}^{2d})$ follows from the identity (recall that $\check{g}(z) = g(-z)$)

$$W(f, g)(x, y) = 2^d S_{\check{g}} f(-2x, -2y) e^{-\pi i (2x)(2y)}, \tag{5.2}$$

which implies (choosing $f = g$) that W_f is obtained by dilation from

$$S_{\check{f}} f(-x, -y) e^{-\pi i x y}.$$

Therefore integrability of W_f is equivalent to integrability of $S_{\check{f}}$ or $V_{\check{f}}$.

Characterization 6. is just a reformulation of the atomic characterization of coorbit spaces. Note however, that in the ℓ^1 case under discussion it is possible to admit arbitrary point sets $(x_i, y_i)_{i \in I}$, without assuming that they are separated. On the other hand it is sufficient (according to the general theory) to use sufficiently dense discrete families.

In order to check 7. let f be given as described, and choose $g \in \mathcal{S}(\mathbb{R}^d) \subseteq Co(L^1)$ such that $\hat{g}(y) \equiv 1$ on Q. Using $L^1(\mathbb{R}^d) * S_0(\mathbb{R}^d) \subseteq S_0(\mathbb{R}^d)$ and

$$\|f_j\|_{S_0} = \|M_{y_j}(M_{-y_j} f_j * g)\|_{S_0} \leq \|M_{-y_j} f_j\|_1 \|g\|_{S_0} = \|f_j\|_1 \|g\|_{S_0}, \tag{5.3}$$

it follows that the functions as described in 7. can be represented as absolutely convergent series in $S_0(\mathbb{R}^d)$, and therefore belong to $S_0(\mathbb{R}^d)$, since

$$\|f\|_{S_0} \leq \sum_j \|f_j\|_{S_0} \leq \|g\|_{S_0} \sum_j \|f_j\|_1 < \infty. \tag{5.4}$$

For the converse let $f \in S_0(\mathbb{R}^d)$ be given, and choose some $g \in \mathcal{S}(\mathbb{R}^d)$ with compact spectrum satisfying $\sum_{n \in \mathbb{Z}^d} T_n \hat{g} \equiv 1$. It follows for $f \in S_0(\mathbb{R}^d)$

$$f = \sum_{n \in \mathbb{Z}^d} f * M_n g, \text{ with } \sum_{n \in \mathbb{Z}^d} \|f * M_n g\|_1 \leq C_2 \cdot \|f\|_{S_0}. \tag{5.5}$$

In fact, if $g_1 \in \mathcal{S}(\mathbb{R}^d)$ satisfies $\hat{g_1} \equiv 1$ on $supp(\hat{g}) + B_1(0)$ (the ball of radius one around zero), then $T_{s+n}\hat{g_1} \cdot T_n\hat{g} = T_n\hat{g}$ for $|s| \leq 1$, and therefore

$$\|M_n g * f\|_1 \leq \int_{n+Q} \|M_s g_1 * M_n g * f\|_1 ds \leq \|M_n g\|_1 \cdot \int_{n+Q} \|M_s g_1 * f\|_1 ds \quad (5.6)$$

and consequently

$$\sum_{n \in \mathbb{Z}^d} \|f * M_n g\|_1 \leq \|g\|_1 \sum_{n \in \mathbb{Z}^d} \int_{n+Q} \|M_s g_1 * f\|_1 ds = \|g\|_1 \|S_{g_1} f\|_1 = C_2 \|f\|_{S_0}, \quad (5.7)$$

as claimed. ■

Remark 16. Due to its special properties the Segal algebra $S_0(\mathbb{R}^d)$ plays a special role among modulation spaces. It shares many properties with the Schwartz space $\mathcal{S}(\mathbb{R}^d)$ Of rapidly decreasing functions, in particular the invariance under the Fourier transform. It is a Banach space, isometrically invariant under time-frequency shifts, and actually the smallest Banach space with this property. Applying the usual duality argument it is possible to define a *generalized Fourier transform* on $S_0'(\mathbb{R}^d)$. Although $S_0'(\mathbb{R}^d)$ is a proper subspace of $\mathcal{S}'(\mathbb{R}^d)$ it is large enough to contain all kinds of functions and measures (or distributions) that are of interest in signal anlalysis, such as Dirac functions, Dirac pulse trains (also called Shah-distributions), or functions in L^p-spaces on $\mathbb{R}^d$, for any $p \geq 1$. An elementary introduction to harmonic analysis (*i.e.*, an approach which is *not* based on Lebesgue integration theory) based on the pair $S_0(\mathbb{R}^d)$ and $S_0'(\mathbb{R}^d)$ has been proposed in [23]. $S_0(\mathbb{R}^d)$ (or more generally $S_0(\mathcal{G})$, for arbitrary locally compact Abelian groups $\mathcal{G}$) has been discoverd independently by J.P. Bertrandias [7] and the first author [19], who also was the first to point out the minimality properties of $S_0(\mathcal{G})$ and its consequences. The existence of a very weak form of atomic decompositions for $S_0(\mathcal{G})$ together with a hint concerning Gabor's classical expansions for L^2-signals (see Gabor [34]) motivated the search for more general Gabor-type expansions of distributions, which resulted in a characterization of modulation spaces in terms of Gabor expansions (see [22]). The observed similarity to the construction of Y.Meyers orthogonal (affine) wavelets sparked the development of the general group theoretical approach to atomic decompositions.

An easy consequence of the atomic characterization of $S_0(\mathbb{R}^d)$ is the "tensor-product stability". In order to describe it we need the space

$$S_0(\mathbb{R}^d)\hat{\otimes}S_0(\mathbb{R}^d) := \{f \in L^1(\mathbb{R}^{2d}) \mid f(x,y) = \sum_{n=1}^{\infty} f_n(x) g_n(y),$$

$$\text{with } \sum_{n=1}^{\infty} \|f_n\|_{S_0} \|g_n\|_{S_0} < \infty\} . \quad (5.8)$$

The infimum over these series expressions defines the natural norm on this space, which will be denoted by $\|f\|_{S_0 \hat{\otimes} S_0}$.

Corollary 17. *The spaces $S_0(\mathbb{R}^d)\hat{\otimes}S_0(\mathbb{R}^d)$ and $S_0(\mathbb{R}^{2d})$ coincide, and their natural norms are equivalent.*

The most direct argument (we leave it to the reader to check details, or to compare with [19]) uses atomic building blocks $g = g_1 \otimes g_2$, with $g_1, g_2 \in S_0(\mathbb{R}^d)$ for the characterization of $S_0(\mathbb{R}^{2d})$, where we use the notation $g(x,y) = g_1 \otimes g_2(x,y) = g_1(x)g_2(y)$.

As the main consequence of this property we shall verify the so-called *kernel theorem* for $S_0(\mathbb{R}^d)$. The possibility of proving a kernel theorem (announced already in [20]) is remarkable, because usually the "nuclearity" of a topological vector space is the basis for such a result.

Theorem 18. *(Kernel Theorem) For every bounded linear operator T from $S_0(\mathbb{R}^d)$ into $S_0'(\mathbb{R}^d)$ there exists exactly one (distributional kernel) $\sigma \in S_0(\mathbb{R}^{2d})$ such that $Tf(g) = \sigma(f \otimes g)$, for all $f, g \in S_0(\mathbb{R}^d)$.*

Proof:

i) Given σ it is clear that for fixed f the mapping $\sigma_f : g \mapsto \sigma(f \otimes g)$ defines a bounded linear functional, *i.e.*, some element in $S_0'(\mathbb{R}^d)$. Moreover, the mapping $T_\sigma : f \mapsto \sigma_f$ is linear and satisfies $|\sigma(f \otimes g)| \leq \|\sigma\|_{S_0'}\|f\|_{S_0}\|g\|_{S_0}$. This allows to estimate the operator norm of T_σ by $|||T||| \leq \|\sigma\|_{S_0'}$.

ii) In order obtain σ from the operator T it seems to be sufficient to use the tensor product property of $S_0(\mathbb{R}^d)$ and to define σ on $f = \sum_{n=1}^{\infty} f_n \otimes g_n$ through $\sigma(f) := \sum_{n=1}^{\infty}[T(f_n)](g_n)$. However, it is hard to verify directly that this definition makes sense, *i.e.*, that it is independent from the representation of f. It is therefore helpful to remember that a 'kernel theorem' is a distributional analogue of a matrix representation. If T is an integral operator of the form

$$T_h(f)(x) = \int_{\mathbb{R}^d} h(x,y)f(y)dy, \tag{5.9}$$

for some continuous function h, it is possible to recover $h(x,y)$ by calculating $T(\delta_x)(y) = \delta_y(T(\delta_x))$ (in the same way as the columns of a matrix are the images of the unit vectors of the induced linear mapping). Of course this does not work in the general setting, because Dirac measures are not in the domain of T and also the range of T consists in general of distributions (*i.e.*, elements of $S_0'(\mathbb{R}^d)$), so it does not make sense to evaluate them pointwise. However, it is possible to use impulse-like instead of Dirac measures elements from $S_0(\mathbb{R}^d)$, and to go to the limit afterwards (in $S_0'(\mathbb{R}^d)$). This motivates the following procedure:

Fix some Dirac sequence $(e_k)_{k=1}^{\infty}$ in $S_0(\mathbb{R}^d)$. For simplicity we may assume that $e_k(x) = 2^{dk}e_0(2^{-k}x)$ for $k \geq 1$, with $e_0 \in \mathcal{S}(\mathbb{R}^d)$ and compact support. We may also assume that $e_0(x) \geq 0$ for all $x \in \mathbb{R}^d$ and that $\|e_0\|_1 = 1$, and therefore $\|e_k\|_1 = 1$ for $k \geq 1$. Using this sequence we are able to define a sequence (of bounded, continuous functions) h_k on $\mathbb{R}^{2d}$ by

$$h_k(u,v) := [T(T_u e_k)](T_v(e_k)) \text{ for } k \geq 1. \tag{5.10}$$

Since the translation operators T_x, $x \in \mathbb{R}^d$ are isometric on $S_0(\mathbb{R}^d)$ and since $x \mapsto T_x(f)$ is a continuous mapping from $\mathbb{R}^d$ into $S_0(\mathbb{R}^d)$, both, boundedness and continuity of each single h_k, follows easily from the definition. In general this family of functions on $\mathbb{R}^{2d}$ will be unbounded with respect to the sup-norm. However, we show next that it is bounded with respect to the norm of $S_0'(\mathbb{R}^d)$. To this end we denote the functional generated by h_k by σ_k. Then

$$\sigma_k(f \otimes g) = [T(e_k * f)](e_k * g). \tag{5.11}$$

Since for any fixed k the functions $u \mapsto T_u\check{e_k}f(u)$ and $v \mapsto T_v\check{e_k}g(v)$ are bounded, continuous and integrable (even in the sense of a vector-valued Riemannian integral), and because both integration and the operator T are continuous mappings, this is verified as follows, for $f, g \in S_0(\mathbb{R}^d)$:

$$\sigma_k(f \otimes g) = \int_{\mathbb{R}^d}\int_{\mathbb{R}^d} h_k(u,v)(f \otimes g)(u,v) =$$

$$= T\left(\int_{\mathbb{R}^d} T_u\check{e_k}f(u)du\right)\left(\int_{\mathbb{R}^d} T_v\check{e_k}g(v)dv\right) = [T(e_k * f)](e_k * g) \; .$$

Using $\|e_k * f - f\|_{S_0} \to 0$ for $k \to \infty$ it follows therefrom that

$$\sigma_k(f \otimes g) \to Tf(g) \text{ for } f, g \in S_0(\mathbb{R}^d). \tag{5.12}$$

Using the tensor product representation we may write any $h \in S_0(\mathbb{R}^{2d})$ as $h = \sum_{n=1}^\infty f_n \otimes g_n$, with the appropriate norm estimate. It follows therefrom that

$$\begin{aligned} |\sigma_k(h)| &\leq \textstyle\sum_{n=1}^\infty \|T(e_k * f_n)\|_{S_0'(\mathbb{R}^d)} \, \|e_k * g\|_{S_0(\mathbb{R}^d)} \\ &\leq |||T||| \cdot \textstyle\sum_{n=1}^\infty \|f_n\|_{S_0} \|g_n\|_{S_0}, \end{aligned} \tag{5.13}$$

for any admissbile representation of h, and thus for some $C > 0$:

$$|\sigma_k(h)| \leq C \cdot |||T||| \cdot \|h\|_{S_0(\mathbb{R}^{2d})}. \tag{5.14}$$

Furthermore it follows that $\sigma_k(h)$ is a Cauchy-sequence of complex numbers for any $h \in S_0(\mathbb{R}^{2d})$ (this is obvious for functions h which are finite linear combinations of tensor products and follows therefrom for general elements of $S_0(\mathbb{R}^d)$ by approximation). We denote the pointwise (or w^* in the functional analytic description) limit by σ, i.e.

$$\sigma(h) := \lim_{k \to \infty} \sigma_k(h) \;\text{ for } h \in S_0(\mathbb{R}^{2d}). \tag{5.15}$$

It now follows from (5.12) that

$$\sigma(h) = \textstyle\sum_{n=1}^\infty [Tf_n](g_n) \text{ for } \; h \in S_0(\mathbb{R}^{2d}). \tag{5.16}$$

Thus, this expression is independent of the representation of h and

$$\|\sigma(h)\|_{S_0'(\mathbb{R}^{2d})} \leq C \cdot |||T|||, \tag{5.17}$$

i.e., σ meets all the requirements. ■

Remark 19. Looking once more at the sequence $(\sigma_k)_{k=1}^{\infty}$ we can relate it to σ more directly by verifying that

$$\sigma_k(f \otimes g) = [T(e_k * f)](e_k * g) = \sigma((e_k * f) \otimes (e_k * g)) = \\ = \sigma((e_k \otimes e_k) * (f \otimes g)) = [(e_k \otimes e_k) * \sigma](f \otimes g) \text{ for } f, g \in S_0(\mathbb{R}^d),$$

or in other words

$$\sigma_k = (e_k \otimes e_k) * \sigma, \tag{5.18}$$

where the convolution of σ is taken in the sense of $\mathbb{R}^{2d}$.

Remark 20. The above proof can be easily modified to give a proof for the case of general lca. groups.

Remark 21. Using Wilson bases (see below) it is possible to show that the space $S_0(\mathbb{R}^d)$ is isomorphic as a Banach space to ℓ^1.

§6. Proofs and further Results, Frames, Adaptive Weights

Proof of Theorem 7.

First observe that a weighted variant of Theorem 15 implies that $W_g \in L^1_{w_a}(\mathbb{R}^{2d})$ (or the same for A_g) if and only if $g \in \mathcal{H}^1_{w_a} = Co(L^1_{w_a})$. As has been pointed out at the end of section 4 the modulation space $M^m_{p,q}(\mathbb{R}^d)$ concides with a coorbit space $Co(Y)$, where Y is a suitable weighted mixed norm space on $\mathbb{H}^d$, with moderate weight m, hence satisfying $(Y1) - (Y2')$. Therefore the general theory applies (Theorem 6.1 in [26]), *i.e.*, for a sufficiently small neigborhood U of the neutral element in $\mathbb{H}^d$ any coherent family $\pi(h_i)g$ can be used as a family of atoms for this coorbit space, i.e., allows a series representation $f = \sum_{i\in I} \lambda_i \pi(g)$ with coefficients (λ_i) in the appropriate sequence space.

In the present situation we only have to observe that without loss of generality we may assume that U is of the form $W \times D$, where W and D are neighborhoods of the $(0,0) \in \mathbb{R}^{2d}$ and $1 \in \mathbb{T}$ respectively. If we fix a D−dense family $(\tau_l)_{l=1}^r$ in $\mathbb{T}$, e.g. a collection of roots of unity of sufficiently high order it is clear that $(x_i, y_i, \tau_l)_{i,l}$ is dense in $\mathbb{H}^d$ if (and only if) the given family $(x_i, y_i)_{i\in I}$ is W-dense on $\mathbb{R}^{2d}$.

Thus, by the general theory f can be represented as a series

$$f = \sum_{i,l} \lambda_{i,l} \pi(z_i, \tau_l) g, \tag{6.1}$$

with complex coefficients $(\lambda_{i,l})$ in a sequence space associated with the weighted mixed norm space Y generating the coorbit space.

If the additional partial lattice structure (or if more general conditions of a similar kind are fulfilled) this sequence space has the simple natural description used in the theorem. We have to leave this verification to the interested reader.

Observe, however, that we do *not* assume that the sampling set is a lattice by itself.

As a last argument in order to obtain the required series representation we add up over l, *i.e.*, we set $c_i := \sum_{l=1}^{r} \bar{\tau}_l \lambda_{i,l}$, in order to obatin

$$f = \sum_i \sum_l \lambda_{i,l} \bar{\tau}_l T_{x_i} M_{y_i} g = \sum_i c_i T_{x_i} M_{y_i} g, \tag{6.2}$$

with $(c_i)_i$ being in the same weighted mixed norm sequence space. ■

Proof of Theorem 10.
Once more we observe that knowing the STFT on some $\tilde{W}$-dense set is the same as knowing it on $\tilde{W} \times \mathbb{T}$, hence on a $\tilde{U} := \tilde{W} \times \mathbb{T}$-dense subset of $\mathbb{H}^d$. By the general theory (see [38], Theorem S and Theorem U for details) the operator $A := D_\Psi^+$ (as well as other operators of a similar form) will do the reconstruction job. It is given by

$$AF := \sum_{i \in I} F(h_i) w_i L_{h_i} G, \tag{6.3}$$

where the sequence of weights $(w_i)_{i \in I}$ has to be chosen in an appropriate way. Since we choose them according to the local density of points, *i.e.*, adaptively with respect to the distribution of sampling points, we call the resulting *iterative* method the *adaptive weights method.* The key estimate, which also decides about the size of $\tilde{W}$, is then of the following form (we have to omit the technical details here):

$$\|D_\Psi^+ F * G - F * G\|_Y < \gamma \|F\|_Y \tag{6.4}$$

for some $\gamma < 1$ and all $F \in Y$, if $\tilde{W}$ is sufficiently small.

However, we would like to mention here that the formula

$$S_g f(x,y) M_y T_x g = V_g f(z) \pi(z) g, \quad \text{for } z = (x,y) \in \mathbb{R}^{2d}. \tag{6.5}$$

can be used to reformulate the reconstruction algorithm in a way avoiding the use of the torus compenent entirely in the description of the method. ■

At this point few remarks concerning the required sampling density, *i.e.*, the size of the sets W and $\tilde{W}$ above are in order.

For general g and general modulation spaces it is still an open problem to find good estimates. For the reader's convenience we list some results in this direction:

- In Daubechies [11] (fairly complicated) estimates on $\alpha, \beta > 0$ are derived for a given $g \in L^2$, so that the family $\{e^{2\pi i k \alpha x} g(x - n\beta)\}_{k,n \in \mathbb{Z}}$ constitutes a frame for $L^2(\mathbb{R}^d)$.
- For the case of regular sampling sequences $S_g(k\alpha, n\beta)$ of the STFT it is known that $0 < \alpha\beta \leq 1$ is a necessary condition for stable reconstruction in $L^2(\mathbb{R}^d)$.

- If $g(x) = e^{-\pi x^2}$, then $\{e^{2\pi ik\alpha x}g(x - n\beta)\}_{k,n\in\mathbb{Z}}$ is a frame for $L^2(\mathbb{R}^d)$ if and only if $0 < \alpha\beta < 1$, (see Seip/Wallsten [56]).
- Let $Z_g(\tau, \omega) = \sum_{k\in\mathbb{Z}} f(\tau + k)e^{2\pi ik\omega}$ be the *Zak transform* (or Weil-Brezin transform) of $g \in S_0(R)$. If for some $g \in \mathcal{S}$ there exists $N \geq 2, A, B > 0$, so that

$$0 < A \leq \sum_{j=0}^{N-1} |Z_g(\tau, \omega - j/N)|^2 \leq B < \infty, \tag{6.6}$$

then all modulation spaces $M^m_{p,q}$ have Gabor wavelet expansions w.r.t. $\{e^{2\pi ik\alpha x}g(x - n\beta)\}_{k,n\in\mathbb{Z}}$, whenever $\alpha\beta = 1/N$, see Walnut [62] .
- Daubechies/Jaffard/Journe [15] prove the existence of functions $g \in \mathcal{S}$ so that both g and $\hat{g}$ are of exponential decay, *i.e.*, satisfy

$$|g(x)| \leq Ce^{-a|x|}, |\hat{g}(\xi)| \leq C'e^{-b|\xi|},$$

so that the following simple linear combinations of Gabor wavelets $g_{ln}, l \in \mathbb{N}, n \in \mathbb{Z}$ constitute an orthonormal basis for $L^2(\mathbb{R}^d)$: $g_{0n}(x) = g(x - n)$ and

$$g_{ln}(x) = \sqrt{2}g(x - n/2)\cos(2\pi lx) \quad \text{if} \quad l \neq 0, l + n \in 2\mathbb{Z}, \tag{6.7}$$

$$g_{ln}(x) = \sqrt{2}g(x - n/2)\sin(2\pi lx) \quad \text{if} \quad l \neq 0, l + n \in 2\mathbb{Z} + 1. \tag{6.8}$$

In [31] it is show that $\{g_{l,n}\}_{l\geq 0,n\in\mathbb{Z}}$, also constitutes an unconditional basis for all modulation spaces $M^m_{p,q}$ (and all spaces $W^m_{p,q}$). With a little more work it can be shown ([37]) that the collection $\{g_{l,n}\}_{l\geq 0.n\in\mathbb{Z}}$, where $g(x) = e^{-\pi x^2}$, is also an unconditional basis for the modulation spaces. In this case, however, the biorthogonal system is hard to compute whereas in the former case the coefficient of $g_{l,n}$ is just $\langle f, g_{l,n}\rangle$.

- If the Gabor wavelet g has compact support or if g is band-limited then it is easy to give sufficient conditions on the sampling density for $S_g f$ for certain sampling sets:
- Assume that $supp(\hat{g}) \subseteq [-\omega, \omega]$. Let (y_n) be an increasing sequence so that for two constants $A, B > 0$

$$0 < A \leq \sum_n |\hat{g}(\xi - 2\pi y_n)|^2 \leq B < \infty. \tag{6.9}$$

For each n let $(x_{k,n})_{k\in\mathbb{Z}}$ be an increasing sequence so that $sup_k(x_{k+1,n} - x_{k,n}) < \pi/\omega$ is true for all n. Then

$$\{e^{2\pi iy_n x}g(x - x_{k,n})|\ k, n \in \mathbb{Z}\} \tag{6.10}$$

is a frame for $L^2(\mathbb{R}^d)$.

We think that these known results give sufficient evidence that the required sampling density in Theorem 10 is close to the best possible value and that the condition "sufficiently dense" does not pose any problem in the application of these results. Having established the main results as special cases of theorems which hold true for a much more general situation (described in group theoretical terms) we give next a short explanation of *frames*. We shall discuss the advantages of the *adaptive weights method* over the frame approach which works for Hilbert spaces only, and present the functional analytic foundation (omitting technical details) of the new approach as "Main Lemma".

We start with the observation that the reproducing equation (4.8) and (4.22) establish the connection between our two problems by

$$F(h_i) = F * G(h_i) = \langle L_{h_i} G, F\rangle \quad \text{since} \quad G = G^*. \tag{6.11}$$

Based on this fact one can solve both questions simultanuously using the theory of *frames*. The notion of frames goes back to the theory of non-harmonic Fourier series (see Duffin/Schaeffer [18], and Young [64]), but has become popular recently in connection with wavelet theory (e.g. [13],[14], ...). The family $(L_{h_i}G)_{i\in I}$ is a *frame* for the Hilbert space $\mathcal{H} := \{F \in L^2(\mathbb{H}^d) |\ F * G = F\}$, *i.e.*, that there are positive constants $A, B > 0$ such that

$$A\|f\|^2 \leq \sum_{i\in I} |\langle L_{h_i} G, F\rangle|^2 \leq B\|f\|^2 \quad \forall\ F \in \mathcal{H}, \tag{6.12}$$

The optimal values are called *lower* and *upper frame constant.* On H the so-called *frame-operator* S, given by

$$S(F) := \sum_{i\in I} \langle L_{h_i} G, F\rangle \cdot L_{h_i} G = \sum_{i\in I} F(h_i) L_{h_i} G \tag{6.13}$$

is invertible because it satisfies the operator inequality $A \cdot Id_{\mathcal{H}} \leq S \leq B \cdot Id_{\mathcal{H}}$, and thus $(\lambda S)^{-1} = \sum_{n=0}^{\infty} (Id - \lambda S)^n$ is well defined through Neumann's series for a certain range of values λ, the optimal value being $\lambda = 2/(A+B)$ (see Daubechies [14]). The constant λ may be considered as a *relaxation parameter* for the recursive description of the partial sums arising by this method, which starts with $F^{(0)} := \lambda SF$ and is given as

$$F^{(n+1)} = (Id - \lambda S)F^{(n)} + \lambda SF = F^{(n)} + \lambda S(F - F^{(n)}) \quad \text{for } n \geq 1. \tag{6.14}$$

The rate of convergence can be estimated using the operator norm:

$$\|F - F^{(n)}\| \leq C |\!|\!| Id - \lambda S |\!|\!|^n \quad \text{for } n \geq 1. \tag{6.15}$$

Furthermore SF and therefore $F^{(n)}$ for any $n \geq 1$ can be obtained from the knowledge of the family $(F(h_i))_{i\in I}$ alone. Finally the formula

$$F = S(S^{-1}F) = \sum_{i\in I} \langle L_{h_i} G, S^{-1} F\rangle L_{h_i} G, \tag{6.16}$$

shows that F can be represented as a series with atoms $L_{h_i}G$.

Disadvantages of the Frame Approach

a) Only for special cases the frame bounds A, B have been estimated [11], but in general they are difficult to determine or even good estimates may not be available. As a consequence the relaxation parameter λ and the speed of convergence have to be determined by expensive trial and error experiments.

b) For very irregular sampling geometries, where the local density of the sampling points varies significantly within the TF-plane, the frame bounds are necessarily far apart. This entails extremly slow convergence. To improve the situation one might even throw away some information, but clearly this makes the reconstruction less stable and more sensitive to noise.

c) The frame approach works only for Hilbert spaces. In order to deal with more sensitive norms additional arguments, as given in [38] have to be given. Again they are based on group-theoretic concepts.

The Adaptive Weights Methods

The problems mentioned above suggests to look for alternatives with improved speed of convergence and stability. This also means that one should look for proofs which allow to check convergence with respect to other norms, not just the global L^2-norm. This family of norms should include at least the weighted L^p-spaces (on the group level), *i.e.*, convergence should be verified for the norm of coorbit spaces such as M- and W-spaces.

Convergence of several iterative methods which work for a variety of norms can be derived from the following **Main Lemma**.

Lemma 22. *Assume that $F \mapsto F * G$ is a bounded linear operator on a Banach space $(Y, \|\cdot\|_Y)$ of functions on a locally compact group $\mathcal{G}$. If A is a bounded, linear operator on $(Y, \|\cdot\|_Y)$ such that there exists some $\gamma < 1$ with*

$$\|F * G - AF\|_Y \leq \gamma \|F\|_Y \quad \forall\, F \in Y. \tag{6.17}$$

Then any F satisfying $F = F * G$ can be recovered from AF by the recursion

$$F^{(0)} = AF \quad , \quad F^{(n+1)} = F^{(n)} * G + AF - AF^{(n)} \quad \forall\, n \geq 1. \tag{6.18}$$

Moreover, we have

$$\|F - F^{(n)}\|_Y \leq \gamma^n \|F\|_Y \quad \forall\, n \geq 1. \tag{6.19}$$

Alternatively, one can show that there is a bounded linear operator D on Y such that $F = D(AF)$ if $F = F*G$. If furthermore $G = G*G$ and $AF = AF*G$ for all $F \in Y$ then it also true that $F = A(DF)$.

Proof: Write $RF := F*G - AF$, or $F*G = AF + RF$. By induction (replacing RF by $R(F * G) = R(AF + RF)$ over and over again we derive therefrom

$$F = F * G = \left(\sum_{k=0}^{n} R^k \right) AF + R^{n+1} F \quad \text{for } n \geq 1. \tag{6.20}$$

The sequence $F^{(n)} := (\sum_{k=0}^{n} R^k)AF$ can be described recursively through

$$F^{(0)} = AF \ , \quad F^{(n+1)} = RF^{(n)} + AF = F^{(n)} * G + A(F - F^{(n)}). \tag{6.21}$$

Since by assumption

$$\|R^{n+1}F\|_Y \leq \gamma \|R^n F\|_Y \leq \gamma^{n+1} \|F\|_Y \quad \text{for } n \geq 1. \tag{6.22}$$

the operator $D := \sum_{k=0}^{\infty} R^k$ is well defined and satisfies $D(AF) = F$ if $F = F * G \in B$.

Assuming now the additional condition one may apply the splitting operation $F = AF + RF$ also to $R^k F$ for any $k \geq 1$ and obtains by induction

$$F = F * G = A\left(\sum_{k=0}^{n} R^k F\right) + R^{n+1}. \tag{6.23}$$

Going with n to infinity gives the result. ■

Remark 23. The proof also shows that $DF * G = DF$ under the additional assumption.

According to our experience the most simple and efficient approximation is obtained by (6.3). A simple recipe concerning the choice of the family of weights $(w_i)_{i \in I}$ is the following one: Think of a partition of the Heisenberg-group of the form $(P_i)_{i \in I}$, the sets P_i being as small as possible, with $h_i \in P_i$. One might take for example as P_i the set of nearest neighbors of h_i. Then a good choice is to take as weights w_i the Haar measure (*i.e.*, practically the 3D-volume) of P_i. In general it seems to be only important that the weights are small for areas of high density of $(h_i)_{i \in I}$ and larger elsewhere.

We do not have enough space here to demonstrate how one can obtain the required estimates in order to apply the Main Lemma. Let us just indicate that a very natural argument can be based on the fact that the approximation operator A is a Riemannian sum to the vector-valued integral describing the convolution product $F * G$ (see (4.3)). Thus, convergence of those Riemannian sums with respect to a variety of norms stands in the background of our method. It is therefore *not limited to the Hilbert space case.* The group theoretical description also allows to make statements about uniformity with respect to the possible families $(h_i)_{i \in I} \in \mathbb{H}^d$. For example, one obtains *uniform* estimates on the speed of convergence for all families $(h_i)_{i \in I}$ which have the same *density*, *i.e.*, for which the family $(h_i \cdot Q)_{i \in I}$ defines a covering of $\mathbb{H}^d$ the same (small) set $Q \subseteq \mathbb{H}^d$.

Next we indicate how the Lemma can be used to answer the main questions. Observe first that the sequence defined iteratively in the Main Lemma is convergent to F at a geometric rate. Since AF can be built if only the sampling values $(F(h_i))_{i \in I}$ are known the irregular sampling problem has been solved based on the first part of the lemma.

In order to verify the result concerning Gabor wavelet expansions we note that the choice of A and the equation $G = G * G$ imply that the additional assumptions of the Lemma are fulfilled as well. Since any function of the form AF, for some $F \in Y$ has the required series representation the argument is complete.

We have to mention here that the convolution relations for the so-called Wiener amalgam spaces are a useful tool to verify that the sequence of coefficients in the series representation, which is just $(w_i F(h_i))_{i \in I}$, belongs to the sequence space corresponding the Y in a natural way. This can be shown using once more the reproducing property $F * G = F$ and the fact that G is a smooth function on $\mathbb{R}^{2d}$ for 'nice' analyzing Gabor wavelets, such as $g \in S_0(\mathbb{R}^d)$. However, we also have to omit a detailed discussion of the verification that the natural sequence spaces which correspond to a mixed norm space on $\mathbb{H}^d$ are the obvious ones.

We have to skip the discussion technical details of the crucial estimate: $\|F * G - AF\| \leq \gamma \|F\|$, for some $\gamma < 1$ and for all $F \in Y$, and refer to [26] , etc. for details. each step produces a series in the term $L_{h_i} G$ which finally leads to the required series representation. It is also possiple and sometimes convenient from a practical point of view that the recursion can also be described directly in terms of the cofficients:

$$f = \sum_{i \in I} \lambda_i \pi(z_i) g \tag{6.24}$$

we only have to recall that we start with the family Λ^1, with $\Lambda_i^1 = F(z_i) w_i$, for $i \in I$, and that the recursion delivers (for the coefficients):

$$\Lambda^{n+1} = \Lambda^n + ((F(z_i) - F^n(z_i)) \cdot w_i)_{i \in I}. \tag{6.25}$$

The geometric decay of the error terms in the recursion guarantees that $\Lambda = lim_{n \to \infty} \Lambda^n$ is an appropriate set of coefficients for (6.24) to hold true.

We have now shown that the general group theoretic approach can be used to give a positive answer to both of our main questions, not only for the Hilbert space case, but for a large family of norms. Now we want to indicate, how the convolution relations can be used to quickly answer a variety of related questions with almost no extra effort. First we will verify the independence of the definition of the $M^m_{p,q}(\mathbb{R}^d)$ spaces from the analyzing function g.

Lemma 24. *The definition of* $M^m_{p,q}(\mathbb{R}^d)$ *is independent of the choice of the non-zero analyzing function* g, *i.e., two non-zero functions* g *and* $\tilde{g}$ *with* W_g *and* $W_{\tilde{g}} \in L^1_{w_a}(\mathbb{R}^{2d})$ *define the same spaces and equivalent norms on* $M^m_{p,q}(\mathbb{R}^d)$.

Proof: We have seen that membership of $S_g f$ in the defining weighted, mixed norm space over the TF-plane is equivalent to membership of $V_g f$ in some Banach function space Y on $\mathbb{H}^d$, which satisfies $(Y1)$ and $Y * L^1_{w_a} \subseteq Y$.

Assuming that $V_g f \in Y$ we want to verify that $V_{\tilde{g}} f \in Y$. It follows directly from the convolution formula (4.22) that

$$V_g f * V_{\tilde{g}} \tilde{g} = \langle g, \tilde{g} \rangle \cdot V_{\tilde{g}} f \subseteq Y * L^1_{w_a} \subseteq Y \,. \tag{6.26}$$

If $\langle g, \tilde{g} \rangle \neq 0$ this implies $V_{\tilde{g}} f \in Y$. Otherwise we may replace $\tilde{g}$ by $\pi(h_0)\tilde{g}$. Since $\langle \pi(h_0)\tilde{g}, g \rangle \neq 0$ for at least some $h_0 \in \mathbb{H}^d$ we obtain in any case that $V_{\pi(h_0)\tilde{g}} f \in Y$ for some $h_0 \in \mathbb{H}^d$. But it follows therefrom that

$$V_{\pi(h_0)\tilde{g}} f(h) = \langle \pi(h)\pi(h_0) g, \tilde{g} \rangle = \langle \pi(hh_0) g, \tilde{g} \rangle = V_{\tilde{g}} f(hh_0), \tag{6.27}$$

i.e., some *right* translate (in the sense of $\mathbb{H}^d$) of $V_{\tilde{g}} f$ belongs to Y, and therefore $V_{\tilde{g}} f$ itself, since Y is left *and* right translation invariant. ■

The redundancy (linear dependence) of a coherent system of Gabor wavelets implies that the coefficients for a Gabor series representation of a signal is in general not uniquely determined. However, we will show that sometimes uniqueness can be achieved. Recall that we have shown in Theorem 7.3. [27] that it is possible to derive interpolation results for generalized wavelet transforms, if the sampling points are well separated. In order to motivate the name of the following theorem let us look at the following 1D-situation: Think of the signal f as an acoustic signal, produced by a piano player, who is only able to produce tones at a well- defined (and known) scale of frequencies, and assume further for simplicity of the model that everything can be thought to happen on a discrete time-scale. Then it should be possible to reconstruct the scores from the acoustic signal, *i.e.*, to obtain the *unique* Gabor expansion of the signal from the signal itself. Of course we have lost (with this restricted set of atoms) the ability to represent arbitrary signals, but nobody would try to produce an arbitrary sound by means of an ordinary piano.

Theorem 25. *(Piano Reconstruction Theorem) Consider the family of $M^m_{p,q}(\mathbb{R}^d)$-spaces described in Theorem 7 (for fixed $C > 0$ and $a \geq 0$), and fix also $g \in \mathcal{H}^1_{w_a}$. Then there exists a compact set S , such that for any fixed family $(z_i)_{i \in I}$ in the TF-plane which is S-separated, i.e., which satisfies $(z_i + S) \cap (z_j + S) = \emptyset$ for $i \neq j$ one has: If $f \in M^m_{p,q}(\mathbb{R}^d)$ has a representation as*

$$f = \sum_{i \in I} c_i T_{x_i} M_{y_i} g \,, \text{ subject to } \sup_{i \in I} |c_i| w_a^{-1}(z_i) < \infty \tag{6.28}$$

the sequence of coefficient is uniquely determined (assuming that the side-condition is satisfied).

Proof: The proof of Theorem 7.3 of [27] indicates how the correct coefficients for the atomic decomposition of f can be obtained constructively. Starting with the sampling values of the STFT one just has to iterate in a way similar to the general atomic representation algorithm. The extra information that only *certain* building blocks will be required in the atomic representation can be used in the algorithm by simply setting the other coefficients to zero at each step. ■

§7. Stability and invariance properties

In this section we indicate that the algorithms derived from our approach can be shown to be stable with respect to minor perturbations. Later on in this section we shall discuss the invariance properties of coorbit spaces, in particular, of M- and W-spaces. We start showing that the STFT changes only a little bit if the window is not modified too much (in the sense of $S_0(\mathbb{R}^d)$!). For simplicity we formulate the result for unweighted spaces $M^0_{p,p}(\mathbb{R}^d)$.

Lemma 26. *Let $g \in S_0(\mathbb{R}^d)$ be given. Then for any $\varepsilon > 0$ there exists some $\delta > 0$ such that for all $\tilde{g} \in S_0(\mathbb{R}^d)$ with $\|g - \tilde{g}\|_{S_0} \leq \delta$ one has*

$$\|S_g f - S_{\tilde{g}} f\|_{L^p(\mathbb{R}^{2d})} < \varepsilon \|S_g f\|_{L^p(\mathbb{R}^{2d})}.$$

for any distribution f with STFT $S_g f \in L^p(\mathbb{R}^{2d})$.

Proof: Since we are free to choose the analyzing wavelet we may suppose that $g_0 \in \mathcal{S}(\mathbb{R}^d)$ is some non-zero Schwartz-function with $\|g_0\|_2 = 1$. The involution $F \mapsto F^*$ being isometric on $L^1(\mathbb{H}^d)$ and since $(V_g f)^* = V_f(g)$ (see (4.15)) our assumption is equivalent to

$$\|V_g g_0 - V_{\tilde{g}} g_0\|_1 \leq \delta. \tag{7.1}$$

It follows therefrom by means of (4.22)

$$\begin{aligned} \|V_g f - V_{\tilde{g}} f\|_p &= \|V_{g_0} f * V_g g_0 - V_{g_0} f * V_{\tilde{g}} g_0\|_p \\ &\leq \|V_{g_0} f\|_p \|V_g g_0 - V_{\tilde{g}} g_0\|_1 \leq \delta \|V_{g_0} f\|_p. \end{aligned} \tag{7.2}$$

Since the $p-$norm of $S_g f$ over the TF-plane is a norm equivalent to the coorbit norm, *i.e.*, the norm of $V_{g_0} f$ in $L^p(\mathbb{H}^d)$ the result is proved. ■

Modifying the window g only a little bit in the S_0-sense has another consequence, which is more important. The L^p-norm of the corresponding STFT changes only a little bit, it is also true that for such function (*i.e.*, for $f \in M^0_{p,p}(\mathbb{R}^d)$ the sampling sequence over any relatively separated set (e.g. on which is carried by some fine lattice as described in Theorem 10) will only undergo a small change with respect to the ℓ^p-norm. Therefore imprecise knowledge concerning the window, which has been used for the calculation of the STFT, will not completely destroy the capability of our algorithms to reconstruct the signal from irregular sampling values of the STFT at least approximately. The same is true concerning the jitter error which arises, if sampling values $S_g f(\tilde{z}_i)_{i\in I}$ with $\tilde{z}_i = f(\tilde{x}_i, \tilde{y}_i)_{i\in I}$ are used instead of the sampling values at the points $(z_i)_{i\in I}$. This question comes up if only unprecise information about the sampling positions is available.

For simplicity we state the stability results (also indicating that such errors can be handled simultanuously) for both types of errors in one theorem (see [26], Theorem 6.5 for a proof, and [63] for a special case).

Theorem 27. *(Stability Theorem) Consider the family of $M^m_{p,q}(\mathbb{R}^d)$-spaces as in Theorem 7, and let $\varepsilon > 0$ be given. Then there exists some $\delta > 0$ such that $\|g - \tilde{g}\|_{\mathcal{H}^1_{w_a}} \le \delta$ and $|z_i - \tilde{z}_i| \le \delta$ implies that for any f in $M^m_{p,q}(\mathbb{R}^d)$ the distribution $\tilde{f}$ reconstructed from the sampling sequence $S_{\tilde{g}}(\tilde{z}_i)$ satisfies*

$$\|f - \tilde{f}|M^m_{p,q}(\mathbb{R}^d)\| \le \varepsilon \|f|M^m_{p,q}(\mathbb{R}^d)\| \ . \tag{7.3}$$

Remark 28. For the purpose of better numerical treatment it is often desirable to replace a window g with unbounded support by some compactly supported window $\tilde{g}$. That ordinary truncation is not a good idea will be clear to everyone familiar with Fourier analysis, as the gaps arising at the truncation points will introduce bad decay of the STFT $S_g f$, independently of the smoothness of f. However, if $\Psi = (\psi_i)_{i \in I}$ is a bounded partition of unity in the Fourier algebra $\mathcal{F}L^1$, then the finite partial sums $\tilde{g} := \sum_{i \in F} g\psi_i$ will approximate $g \in S_0(\mathbb{R}^d)$ in the sense of $S_0(\mathbb{R}^d)$. In particular, this holds for a partition of unity built up by triangular functions over $\mathbb{R}$. For higher smoothness partitions of unity arising from B-splines of higher order are useful.

We have already shown that the coorbit spaces are invariant with respect to time/frequency shifts. For the subsequent discussion of further invariance properties of the modulation spaces it is more convenient to use the Heisenberg group with the symmetric multiplication

$$(x_1, y_1, \tau_1) \cdot (x_2, y_2, \tau_2) = (x_1 + x_2, y_1 + y_2\,,\, \tau_1 \tau_2 e^{i\pi(x_1 y_2 - x_2 \cdot y_1)}) \tag{7.4}$$

and unitary representation $\pi'(x, y, \tau)f(t) = \tau e^{\pi i x y} e^{2\pi i y t} f(t - x), f \in L^2(\mathbb{R})$.

It is then easily seen that $(x, y, \tau) \to (x, y, \tau e^{i\pi x \cdot y})$ defines an isomorphism between the old and the new multiplication. Since $\langle f, \pi'(x, y, \tau) g\rangle$ and $V_g(f)(x, y, \tau)$ differ only by a phase factor, neither the definition of the modulation spaces nor any of the results do change.

The interesting part of the automorphism group of *automorphism* of $\mathbb{H}^d$ can now be described as follows (see [33] for details): Let A be a $2n \times 2n$-matrix, which leaves the bilinear form $(x_1, y_1), (x_2, y_2) \to x_1 \cdot y_2 - x_2 \cdot y_1$, $x_i, y_i \in \mathbb{R}^{2d}$ invariant, *i.e.*, a *symplectic* matrix. Then $h_A(x, y, \tau) = (A(x, y), \tau)$ for $(x, y, \tau) \in \mathbb{H}^d$ is an automorphism of $\mathbb{H}^d$, and h_A leaves the center $0 \times 0 \times \mathbb{T}$ of $\mathbb{H}^d$ invariant. Since the restrictions of the representations $\pi'(x, y, \tau)$ and $\pi' \circ h_A(x, y, \tau)$ to the center are indentical, π' and $\pi' \circ h_A$ are equivalent by the Stone-von Neumann uniqueness theorem. Thus, there exist unitary operators $\sigma(A)$, so that

$$\pi' \circ h_A(x, y, \tau) = \sigma(A)\pi'(x, y, \tau)\sigma(A)^{-1} \forall\ (x, y, \tau) \in \mathbb{H}^d\ . \tag{7.5}$$

The operators $\sigma(A)$ are called *metaplectic* operators and are determined only up to a phase factor. It can be shown that $A \to \sigma(A)$ can be defined in such a way that it yields a unitary representation of a two-fold covering group of the symplectic group (see Schempp [54] and Reiter [53] for facts concerning the metaplectic group).

For a large class of symplectic matrices the intertwining operators $\sigma(A)$ can be written out explicitly (see Folland [33] for an excellent exposition). If $A = \begin{pmatrix} C & 0 \\ 0 & (C^*)^{-1} \end{pmatrix}$ for some $C \in GL(n, \mathbb{R})$, then

$$\sigma(A)f(x) = |\det A|^{\pm 1/2} f(A^{\pm 1}x). \tag{7.6}$$

For $A = \begin{pmatrix} 0 & I \\ I & 0 \end{pmatrix}$, one obtains $\sigma(A)f = \hat{f}$, the Fourier transform.
If $A = \begin{pmatrix} I & 0 \\ C & I \end{pmatrix}$ for some symmetric $n \times n$-matrix C, then

$$\sigma(A)f(x) = \pm e^{i\pi x \cdot C_x} f(x), \tag{7.7}$$

i.e., multiplication with a chirp function.

Originally these operators are defined on $L^2(\mathbb{R}^d)$. It is an important consequence of a general principle ([26], Thm.4.8]) that the metaplectic operators can be extended to modulation spaces and leave an important class of them invariant. On an intuitive level the invariance statements assert that the time-frequency concentration of a signal is not affected by the action of the metaplectic representation. In the following theorem we collect some mapping properties of the $\sigma(A)$.

Theorem 29.

(i) *If the a-moderate function $m(x, y)$ satisfies $m(A(x, y)) \leq Cm(x, y)$ for some $C > 0$ and $\forall\ (x, y) \in \mathbb{R}^{2d}$, then $\sigma(A)$ extends to an isomorphism of M^m_{pp} (onto $M^m_{p,p}$).*

(ii) *If in particular $w \equiv 1$ then M^0_{pp}, $1 \leq p \leq \infty$, is invariant under all $\sigma(A)$.*

(iii) *Let $\tilde{m}(x, y) = m(-y, x)$, then the Fourier transform establishes an isomorphism between the spaces $M^m_{p,q}$ and $W^{\tilde{w}}_{q,p}$.*

(iv) *If $m(x, y) = m_0(x)$, then $M^m_{p,q}$ is invariant under multiplication by chirps $e^{2wix \cdot C_x}$.*

(v) *In dimension $n = 1$, the operators $\sigma(A)$, where A is an orthogonal matrix, are called Mehler transforms (They are diagonalized by the Hermite functions h_n, i.e., if $U_\vartheta = \begin{pmatrix} \cos\vartheta & \sin\vartheta \\ -\sin\vartheta & \cos\vartheta \end{pmatrix}$, then $\sigma(U_\vartheta)h_n = e^{i\vartheta_n}h_n$, and form a one parameter group which contains the Fourier transform). If $w(x, y)$ is rotation invariant, then $M^m_{p,p}$ is invariant under $\sigma(U_\vartheta)$, $0 \leq \vartheta < 2\pi$.*

Remark 30. Ad (i): A typical family of rotations invariant weights with polynomial growth of arbitrary order is given by $w_a(x, y) = (1 + |x| + |y|)^a$. For $a \geq 0$ these weights satisfy $w(A(x, y)) \leq C_A w(x, y)$ for all $A \in GL(2n, \mathbb{R})$.

Remark 31. Besides the abstract proof contained in [26], Thm.4.6. there are several other, often involved proofs of the invariance of S_0 or $M^m_{1,1}$ under chirps.

In order to give the flavor of the proof of the general intertwining principle, we give a short and direct proof for the chirp invariance of S_0 and $M^m_{1,1}$.

Let $M_g(x) = e^{i\pi x\cdot C_x} g(x)$ be the "chirp operator" for a symmetric $n \times n$ matrix C. Then

$$\begin{aligned} S_{M_g}(M_g)(x,y) &= \int e^{-2\pi i y t} e^{-i\pi(t-x)\cdot C(t-x)} \overline{g}(t-x) e^{-i\pi t\cdot Ct} g(t)\, dt \\ &= e^{i\pi x\cdot Cx} \int e^{-2\pi i(y+C_x)t} \overline{g}(t-x) g(t)\, dt = \\ &= e^{i\pi x\cdot Cx} S_g(g)(x, y - Cx) \end{aligned}$$

Now using the characterization 3. of Theorem 15 for $S_0(\mathbb{R}^d)$ or $M^w_{1,1}(\mathbb{R}^d)$ with $w(x,y) = w_0(x)$ and the above computation we obtain

$$\begin{aligned} e^{-i\pi x\cdot C_x} g(x) \in S_0(\text{ resp. } M^w_{1,1}) &\Leftrightarrow S_{M_g}(M_g)(x,y) L^1(\mathbb{R}^{2d})(\text{ resp. } L^1_w(\mathbb{R}^{2d}) \\ &\Leftrightarrow S_g(g) \in L^1(\mathbb{R}^{2d}) \ (\text{ resp.} \in L^1_w(\mathbb{R}^{2d})) \\ &\Leftrightarrow g \in S_0 \ (\text{ resp. } \in M^w_{1,1}) \ . \end{aligned}$$

Another instance of the intertwining theorem (Thm.4.8 in [26]) occurs for the *Bargmann–Fock spaces*. These are spaces of entire functions in $\mathbb{C}^n$ defined by

$$\begin{aligned} \mathcal{F}^m_{p,q}(\mathbb{C}^n) = \{F \ \text{ entire: } & \int \left(\int |F(x+iy)|^p m(x,y)^p e^{-p\pi|z|^2/2}\, dx \right)^{q/p} dy \\ &=: \|F\|^q_{\mathcal{F}^m_{p,q}} < \infty\} \end{aligned} \tag{7.8}$$

where $1 \leq p,\ q \leq \infty$, m is an a-moderate function and $z = x + iy \in \mathbb{C}^n$.

$\mathcal{F}^2$ is a Hilbert space on which the Heisenberg group $\mathbb{H}^d$ acts by means of the Fock representation β. Writing $w = r + is \in \mathbb{C}^n$, then β is

$$[\beta(r,s,\tau)F](z) = \tau e^{i\pi\overline{w}\cdot z} e^{-\pi|w|^2/2} F(z-w) \tag{7.9}$$

for $F \in \mathcal{F}^2$. Since $\beta(0,0,\tau)$ is just multiplication by τ, by the Stone-von Neumann uniqueness theorem β and the Schrödinger representation π must be equivalent, *i.e.*, there is a unitary operator

$$B : L^2(\mathbb{R}) \to \mathcal{F}^2, \quad \text{so that } \ B \circ \pi(r,s,\tau) = \beta(r,s,\tau) \circ B \ . \tag{7.10}$$

By the intertwining property (4.11) B extends to the modulation spaces and this extension is an isomorphism between $M^m_{p,q}$ and $\mathcal{F}^m_{p,q}$. Consequently all results for modulation spaces have a corresponding version in the Bargmann–Fock spaces.

Before we state the series expansions and sampling theorems for the $\mathcal{F}^m_{p,q}$, let us look at the so-called Bargmann transform B and its significance. $Bf(w) = 2^{n/4} \int f(x) e^{2\pi x w - \pi x^2 - (\pi/2) w^2} dx = S_g(f)(r,s) e^{\pi |w|^2/2}$, where $g(x) = e^{-\pi x^2}$, is essentially the STFT with the Gaussian as the window. The study of Bargmann-Fock spaces with tools of complex analysis can be and have been used to obtain results about Gabor expansions in the strict sense (see Janson/Peetre/Rochberg [42]). Theorem 7 translates into

Theorem 32. *Let m be a-moderate and $G \in \mathcal{F}^{1,1}_{w_a}$. Then there exists a neighborhood $W \subseteq \mathbb{C}^n$ of 0, such that for any W-dense family $(z_i) \subseteq \mathbb{C}^n$ every $F \in \mathcal{F}^m_{p,q}$, $1 \leq p, q < \infty$ can be written as*

$$F(z) = \sum_i c_i \beta(z_i) G \sum_i c_i e^{-\pi |z_i|^2/2} e^{i\pi \bar{z}_i \cdot z} G(z - z_i) \ . \tag{7.11}$$

The coefficients depend linearly and stably on F as in Theorem 7.

In particular, if $G(z) \equiv 1 \in \mathcal{F}^{1,1}_{w_a}$ $\forall\, a \geq 0$ is chosen, one obtains expansions in $\mathcal{F}^m_{p,q}$ in terms of the exponential functions $e^{i\pi \bar{z}_i \cdot z}$

$$F(z) = \sum c_i e^{-\pi |z_i| 2/2} e^{i\pi \bar{z}_i \cdot z} \ . \tag{7.12}$$

Using a duality argument (see [38] for details), one can derive that $F \in \mathcal{F}^m_{p,q}$ B is uniquely determined by the coefficients $< \beta(z_i) G, F >$. Since for $G \equiv 1$ $< \beta(z_i) 2, F >= e^{-\pi |z_i| 2/2} F(z_i)$, this yields a general sampling theorem for the Bargmann–Fock spaces (see [14] for the case of regular lattices).

Theorem 33. *Given $1 \leq p, q \leq \infty$ and m, there exists a neighborhood $W \subseteq \mathbb{C}^n$ of 0, such that $F \in \mathcal{F}^m_{p,q}$ is uniquely determined by its sampled values $\{F(z_i), i \in I\}$ from any W-dense set $\{z_i\}$ in $\mathbb{C}^n$.*

We leave it to the reader to rephrase the reconstruction algorithms for Theorem 10 for this case.

FINAL REMARKS. There are close similarites between the reconstruction problem and the irregular sampling problem for band-limited functions as discussed in [29] and [30]. In particular, the adpative weights for the 2D-band-limited reconstruction problem turn out to be the same as those for the STFT-problem. The analogy can be used to carry out an error analysis for the STFT-reconstruction algorithms in much the same way as in [30]. Theorem 27 above is just one of such statements. This analogy has been addressed in more detail in [24].

The above theory can be also formulated without serious changes for general locally compact Abelian groups, replacing the Euclidean space $\mathbb{R}^d$. This is *not* just of theoretical interest, but very useful for applications, because it readily provides a sound theoretical background for discrete implementations of the algorithms used to solve the two main questions, just by replacing $\mathbb{R}^d$ by some finite (e.g., cyclic) group $\mathcal{G}$.

The first named author wants to thank the Centre for Theoretical Physics in Marseille, in particular A.Grossman and B.Torresani, for their hospitality. The first draft of this paper was prepared during a one-month visit at Marseille in the summer of 1991. The second author acknowledges partial funding by the grant AFOSR-90-0311.

References

1. Allen, J. B. and L. R. Rabiner, A unified theory of short-time spectrum analysis and synthesis, *Proc. IEEE* **65** (11) (1977), 1558–1564.
2. Auslander, L. and R. Tolimieri, Radar ambiguity functions and group theory, *SIAM J. Math. Anal.* **16** (1985), 577–601.
3. Bastiaans,M. J., On the sliding-window representation in digital signal proc essing, *IEEE Trans. ASSP* **33** (1985), 868 – 873.
4. Bastiaans, M. J., Gabor's expansion of a signal into Gaussian elementary signals, *Proc. IEEE* **68** (1980), 538–539.
5. Bastiaans, M. J., A sampling theorem for the complex spectrogram, and Gabor's expansion of a signal in Gaussian elementary signals, *Opt. Eng.* **20** (1981), 594 – 598.
6. Bastiaans, M. J., Local-frequency description of optical signals and systems, EUT Report 88-E-191, Eindhoven University Technology, 1988.
7. Bertrandias, J. P., C. Datry, and C. Dupuis, Unions et intersections d'espaces L^p invariantes par translation ou convolution, *Ann. Inst. Fourier* **28** (2) (1978), 53–84.
8. Cenker, C., Master thesis, Kohärente reihendarstellungen von distributionen, Vienna, 1989.
9. Cenker, C., H. G. Feichtinger, and K. Gröchenig, Non-orthogonal expansions of signals and some of their applications, in *Image Acquisition and Real-Time Visualization, 14*, OEAGM Treffen, ÖCG, Bd. **56**, G. Bernroider and A. Pinz (eds.), May 1990, 129–138.
10. Daubechies, I., *Wavelets*, CBMS/NSF Regional Conference Series in Applied Mathematics 61, *SIAM*, 1992, to appear.
11. Daubechies, I., The wavelet transform, time-frequency localization and signal analysis, *IEEE Trans. Inform. Theory* **36** (1990), 961–1005.
12. Daubechies, I., Time-frequency localization operators: a geometric phase space approach, *IEEE Trans. Inform. Theory* **34** (1988), 605–612.
13. Daubechies, I., A. Grossmann, and Y. Meyer, Painless nonorthogonal expansions, *J. Math. Phys.* **27** (1986), 1271–1283.
14. Daubechies, I., and A. Grossmann, Frames in the Bargmann space of entire functions, *Comm. Pure and Appl. Math.* **41** (1988), 151–164.

15. Daubechies, I., S. Jaffard, and J. Journe, A simple Wilson orthonormal basis with exponential decay, *SIAM J. Math. Anal.* **22** (2) (1991), 554–572.
16. Daugman, J. G. , Complete discrete 2-D Gabor transforms by neural networks for image analysis and compression, *IEEE Trans. ASSP* **36** (7) (1988), 1169–1179.
17. Dufaux, F. and M. Kunt, Massively parallel implementation for real-time Gabor decompositions, *SPIE Conf. in Visual Comm. and Image Processing*, Boston, November 1991.
18. Duffin, R. and A. Schaeffer, A class of nonharmonic Fourier series, *Trans. Amer. Math. Soc.* **72** (1952), 341–366.
19. Feichtinger, H. G., On a new Segal algebra, *Monatsh. Math.* **92** (1981), 269–289.
20. Feichtinger, H. G., Un espace de Banach de distributions tempérées sur les groupes localement compacts abélians, *C. R. Acad. Sc. Paris Ser. A* **290** (1980), 791–794.
21. Feichtinger, H. G., Modulation spaces on locally compact abelian groups, Technical report, Wien, 1984.
22. Feichtinger, H. G., Atomic characterizations of modulation spaces through Gabor-type representations, *Proc. Conf. in Constructive Function Theory*, Edmonton, July 1986, *Rocky Mount. J. Math.* **19** (1989), 113–126.
23. Feichtinger, H. G., An elementary approach to the generalized Fourier transform, in *Topics in Mathematical Analysis*, Volume in the honour of Cauchy, 200th anniv., Th. Rassias (ed.), World Sci. Publ., 1988, 246–272.
24. Feichtinger, H. G., Coherent frames and irregular sampling, *Proc. Conf. in Recent Advances in Fourier Analysis and Its Applications*, NATO conference, PISA, July 1989, J. S. Byrnes and J. L. Byrnes (eds.), Kluwer Acad. Publ., 1990, *NATO ASI Series C* **315**, 427–440.
25. Feichtinger, H. G. and K. Gröchenig, A unified approach to atomic decompositions via integrable group representations. *Proc. Conf. in Function Spaces and Applications*, Lund, 1986, Lecture Notes in Math. 1302 (1988), 52–73.
26. Feichtinger, H. G. and K. Gröchenig, Banach spaces related to integrable group representations and their atomic decompositions, I, *J. Funct. Anal.* **86** (1989), 307–340.
27. Feichtinger, H. G. and K. Gröchenig, Banach spaces related to integrable group representations and their atomic decompositions II, *Monatsh. f. Math.* **108** (1989), 129–148.
28. Feichtinger, H. G. and K. Gröchenig, , Non-Orthogonal wavelet and Gabor expansions, and group representations, Extended version of the lecture presented at the Wavelet conference, Lowell, June 1990, in *Wavelets and their applications*, G. Beylkin, R. Coifman, I. Daubechies, L. Raphael, B. Ruskai (eds.), Jones and Bartlett, Cambridge, MA, 1992, to appear.
29. Feichtinger, H. G. and K. Gröchenig, Iterative Reconstruction of multivariate band-limited functions from irregular sampling values, *SIAM J. Math. Anal.*, to appear.

30. Feichtinger, H. G. and K. Gröchenig, Error analysis in regular and irregular sampling theory, *Applicable analysis*, to appear.
31. Feichtinger, H. G., K. Gröchenig, and D. Walnut, Wilson bases and modulation spaces. *Math. Nachrichten*, 1991, to appear.
32. Flandrin, P., Quelques méthodes temps–fréquence et temps–échelle en traitment du signal, in *Les Ondelettes en 1989*, LN 1438, P. G. Lemarié , (ed.), Springer- Verlag, 1990, 81–92.
33. Folland, G. B., *Harmonic Analysis in Phase Space, Annals of Math. Studies*, Princeton University Press, Princeton, NJ, 1989.
34. Gabor, D., Theory of communication. *J. Inst. Elect. Eng.* **93** (1946), 429–457.
35. Genossar, T. and M. Porat, Can one evaluate Gabor Expansion using Gabor's Iterative Algorithm?, Technion, Israel, 1990, preprint.
36. Gertner, I. and Y. Y. Zeevi, Image representation with position-frequency localization, *Proc. ICASSP*, Toronto, 1991, 2353 – 2356.
37. Gröchenig, K. and D. Walnut, A Riesz basis for Bargmann-Fock space related to sampling and interpolation, *Ark. f. Mat.*, to appear.
38. Gröchenig, K., Describing functions: atomic decompositions versus frames, *Monatsh. Math.* **112** (1991), 1–42.
39. Heil, C. and D. Walnut, Continuous and discrete wavelet transforms. *SIAM Rev.* **31** (4) (1989), 628–666.
40. Hlawatsch, F., *Methoden der Zeit/Frequenz-Analyse*, report, TU Wien, 1990.
41. Howe, R., On the role of the Heisenberg group in harmonic analysis, *Bull. Amer. Math. Soc. (N.S.)* **3** (1980), 821–843.
42. Janson, S., J. Peetre, and R. Rochberg, Hankel forms on the Fock space, *Rev. Mat. Iberoamer* **3** (1987), 61–138.
43. Janssen, A. J. E. M., Gabor representation of generalized functions, *J. Math. Anal. Appl.* **83** (1981), 377–394.
44. Janssen, A. J. E. M., Bargmann transform, Zak transform, and coherent states, *J. Math. Phys.* **23** (1982), 720–731.
45. Klauder, J. and K. Skagerstam (eds.), *Coherent States – Applications in Physics and Mathematical Physics*, Singapore, World Scientific Publ., 1985.
46. Nawab, S. H. and T. Quatieri, Short-time fourier transform, in *Advanced Topics in Signal Processing*, J. S. Lim and A. V. Oppenheim (eds.), Prentice Hall Signal Proc. Series, 1988.
47. Papoulis, A., *Signal Analysis*, McGraw-Hill, NY, 1977.
48. Peetre, J., *New thought on Besov spaces.* Duke University Press, Durham, 1976.
49. Perelomov, A., *Generalized Coherent States and Their Applications, Texts and Monographs in Physics*, Springer, Berlin-Heidelberg, 1986.
50. Porat, M., and Y. Y. Zeevi, The generalized Gabor scheme of image representation using elementary functions matched to human vision, in *Theor. Foundations of Computer Graphics*, Springer Verlag, NY, 1197–1241.

51. Porat, M. and Y. Y. Zeevi, The generalized Gabor scheme in biological and machine vision, *IEEE Trans. PAMI* **10** (4) (1988), 452–468.
52. Reiter, H., *Classical Harmonic Analysis and Locally Compact Groups.* Oxford University Press, 1968.
53. Reiter, H., *Metaplectic Groups and Segal Algebras*, Lect. Notes in Math. 1382, Springer-Verlag, NY, 1989.
54. Schempp, W., Harmonic analysis on the Heisenberg nilpotent Lie group with applications to signal theory, *Pitman Res. Notes in Math.* **147**, Harlow, Essex, Longman Scientific and Technical, 1986.
55. Schempp, W., Holographic image processing, *Coherent Optical Computing and Neural Computer Architecture for Pattern Recognition*, preprint.
56. Seip, K. and R. Wallsteèn , Sampling and interpolation in the Bargmann-Fock space, Report, Institut Mittag Leffler #4, 1990/91.
57. Stein, E. M., *Singular Integrals and Differentiability Properties of Functions*, Princeton University Press, Princeton, NJ, 1975.
58. Super, B. J. and A.C. Bovik, Shape from texture using Gabor wavelets, *SPIE Conf. in Visual Comm. and Image Processing*, Boston, November 1991.
59. Taylor, M. E., *Noncommutative Harmonic Analysis, Math. Surveys and Monographs 22*, Amer. Math. Soc., Providence, RI, 1986.
60. Triebel, H.,*Theory of Function Spaces*, Birkhäuser, Basel, 1983.
61. Triebel, H., Modulation spaces on the Euclidean n-space, *Zeitschr. f. Anal. und Anwendungen* **2** (1983), 443–457.
62. Walnut, D. F., Lattice size estimates for Gabor decompositions, preprint.
63. Walnut, D. F., Continuity properties of the Gabor frame operator, *JMAA*, 1991, to appear.
64. Young, R., *An Introduction to Nonharmonic Fourier Series*, Academic Press, NY, 1980.

Hans G. Feichtinger
Institute fur Mathematik
Universitat Wien
Strudlhorgf.4, A-1090, Vienna, Austria
a8131dan@awiuni11

Karlheinz Gröchenig
Department of Mathematics
University of Connecticut
Storrs, CT 06269-3009

Windowed Radon Transforms, Analytic Signals, and the Wave Equation

Gerald A. Kaiser and Raymond F. Streater

Abstract. The act of measuring a physical signal or field suggests a generalization of the wavelet transform that turns out to be a windowed version of the Radon transform. A reconstruction formula is derived which inverts this transform. A special choice of window yields the "Analytic–Signal transform" (AST), which gives a partially analytic extension of functions from $\mathbb{R}^n$ to $\mathbb{C}^n$. For n =1, this reduces to Gabor's classical definition of "analytic signals." The AST is applied to the wave equation, giving an expansion of solutions in terms of wavelets specifically adapted to that equation and parametrized by real space and imaginary time coordinates (the Euclidean region).

§1. Introduction

The ideas presented here originated in relativistic quantum theory ([13], [14], and [15]), where a method was developed for extending arbitrary functions from $\mathbb{R}^n$ to $\mathbb{C}^n$ in a semi–analytic way. This gave rise to the "Analytic–Signal transform" (AST) [16]. Later it was realized that the AST has a natural generalization to what we have called a Windowed X–Ray transform [17], and the latter is a special case of a Windowed Radon transform, to be introduced below. For $n = 1$, these transforms reduce to the (continuous) Wavelet transform. In the general case, they retain many of the properties of the Wavelet transform.

In Section 2 we motivate and define the d–dimensional Windowed Radon transform in $\mathbb{R}^n$ for $1 \leq d \leq n$ and derive reconstruction formulas which can be used to invert it. In Section 3 we define the AST in $\mathbb{R}^n$ and give some of its applications. In Section 4 we develop a new application of the AST by generalizing a construction in [16] to the wave equation in $\mathbb{R}^2$. This results in a representation of solutions of the wave equation as combinations of "dedicated" wavelets that are especially customized to that equation. In particular, these

Wavelets–A Tutorial in Theory and Applications
C. K. Chui (ed.), pp. 399–441.

ISBN 0-12-174590-2

wavelets are themselves solutions and represent *coherent wave packets,* being well-localized in both space (at any particular time) and frequency, within the limitations of the uncertainty principle. The parameters labeling these wavelets (*i.e.*, the variables on which the AST depends) have a direct geometrical significance: They give the initial position, direction of motion and average frequency, or *color* of the wavelets. The representation of a solution in terms of these wavelets therefore gives a *geometrical–optics* (ray) picture of the solution. It is suggested that this could be of considerable practical value in signal analysis, since many naturally occurring signals (*e.g.*, sound waves, electromagnetic waves) satisfy the wave equation away from sources and the geometrical–optics picture gives a readily accessible display of their informational contents.

§2. Windowed Radon transforms

2.1. The windowed x–ray transform

Suppose we wish to measure a physical field distributed in $\mathbb{R}^n$. This field could be a "signal," such as an electromagnetic field or the pressure distribution due to a sound wave. For simplicity, we assume to begin with that it is real–valued, such as pressure. (Our considerations easily extend to complex–valued, vector–valued, or tensor–valued signals, such as electromagnetic fields; we shall indicate later how this is done.) The given field is therefore a function $f : \mathbb{R}^n \to \mathbb{R}$. We may think of $\mathbb{R}^n$ as physical space (so that $n = 3$), in which case the field is time–independent, or as space–time (so that $n = 4$), in which case the field may be time–dependent. In the former case, $\mathbb{R}^n$ is endowed with a Euclidean metric, while in the latter case the appropriate metric is Lorentzian, as mandated by Relativity theory.

Actual measurements are never instantaneous, nor do they take place at a single point in space. A measurement is performed by reading an instrument, and the instrument necessarily occupies some region in space and must interact with the field for some time–interval before giving a meaningful reading. Let us assume, to begin with, that the spatial extension of the instrument is negligible, so that it can be regarded as being concentrated at a single point at any time. We allow our instrument to be in an arbitrary state of uniform motion, so that its position is given by $\mathbf{x}(t) = \mathbf{x} + \mathbf{v}t$, where $t \in \mathbb{R}$ is a "time" parameter and $\mathbf{x}, \mathbf{v} \in \mathbb{R}^n$. Note that t need not be the physical time. For example, if $\mathbb{R}^n$ is space–time, then each "point" $\mathbf{x}$ represents an *event,* i.e. a particular location in space at a particular time. In that case, the line $\mathbf{x}(t)$ is called a *world–line* and represents the entire history of the point–instrument. The "velocity" vector $\mathbf{v}$ then has one too many components and may be regarded as a set of *homogeneous coordinates* for the physical velocity. Note that in this case $\mathbf{v}$ cannot vanish, since this would correspond to an instrument not subject to the flow of time. Even if $\mathbb{R}^n$ is space, the case $\mathbf{v} = \mathbf{0}$ is not interesting since then the instrument can only measure the field at a single point. We therefore assume that $\mathbf{v} \neq \mathbf{0}$, hence $\mathbf{v} \in \mathbb{R}^n_* \equiv \mathbb{R}^n \backslash \{\mathbf{0}\}$.

Let us assume that the reading registered by the instrument at time s gives a weight $h(t-s)$ to the value of the field passed by the instrument at time t. (For motivational purposes we note that causality would demand that $h(t-s) = 0$ for $t > s$; moreover, $h(t-s)$ should be concentrated in some interval $s - \tau \leq t \leq s$, where τ is a "response time" or memory characteristic of the instrument. However, the results below do not depend on these assumptions.) Our model for the observed value of the field at the "point" $\mathbf{x}$, as measured by the instrument traveling with uniform velocity $\mathbf{v}$, is then

$$f_h(\mathbf{x}, \mathbf{v}) \equiv \int_{-\infty}^{\infty} dt\, h(t) f(\mathbf{x} + \mathbf{v}t). \tag{1}$$

To accomodate complex–valued signals, we allow the weight function h to be complex–valued. $h(t)^*$ will denote the complex conjugate of $h(t)$. In order to minimize analytical subtleties, we assume that h is smooth and bounded, and that f is smooth with rapid decay (say, a Schwartz test function).

Definition 1. *The Windowed X–Ray Transform of* $f : \mathbb{R}^n \to \mathbb{C}$ *is the function* $f : \mathbb{R}^n \times \mathbb{R}^n_* \to \mathbb{C}$ *given by*

$$f_h(\mathbf{x}, \mathbf{v}) = \int_{-\infty}^{\infty} dt\, h(t)^* f(\mathbf{x} + \mathbf{v}t). \tag{2}$$

Remarks.

1. In the special case $h(t) \equiv 1$ and $|\mathbf{v}| = 1$, f_h is known as the (ordinary) *X–Ray transform* of f (Helgason [11]), due to its applications in tomography. We may then regard f_h as being defined on the set of all lines in $\mathbb{R}^n$, independent of their parametrization. In the general case, we think of the function $h(t)$ as a *window,* which explains our terminology.
2. Some work along related lines was recently done by Holschneider [12]. He considers a two–dimensional wavelet transform which is covariant under translations, rotations and and dilations of $\mathbb{R}^2$. When the window function is supported on a line, say $h(t_1, t_2) = \delta(t_2)$, this becomes an X–Ray transform in $\mathbb{R}^2$. His inversion method is less direct than ours in that it involves a limiting process.
3. Note that f_h has the following *dilation property* for $a \neq 0$:

$$f_h(\mathbf{x}, a\mathbf{v}) = \int_{-\infty}^{\infty} dt\, |a|^{-1}\, h(t/a)^*\, f(\mathbf{x} + t\mathbf{v}) = f_{h_a}(\mathbf{x}, \mathbf{v}) \tag{3}$$

where $h_a(t) \equiv |a|^{-1}\, h(t/a)$. This may be used to study the behavior of f_h as $\mathbf{v} \to \mathbf{0}$. For the "forbidden" value $\mathbf{v} = \mathbf{0}$, the transform becomes $f_h(\mathbf{x}, \mathbf{0}) = \hat{h}(0)^* f(\mathbf{x})$, where $\hat{h}$ is the Fourier transform of h. (We shall see that $\hat{h}(0) = 0$ for "admissible" h.)

4. For $n = 1$ and $v \neq 0$, a change of variables gives

$$\begin{aligned} f_h(x,v) &= |v|^{-1} \int_{-\infty}^{\infty} dt'\, h\left(\frac{t'-x}{v}\right)^* f(t') \\ &= |v|^{-1/2}\, Wf(x,v), \end{aligned} \tag{4}$$

where Wf is the usual wavelet transform of f ([5], [7], and [23]), with v playing the role of a *dilation factor* and the window function $h(t)$ playing the role of a *basic wavelet*.

5. All our considerations extend to vector–valued signals. The cleanest approach is to let the window function h^* assume values in the *dual* vector space, so that $h(t)^* f(\mathbf{x}+\mathbf{v}t)$ and $f_h(\mathbf{x},\mathbf{v})$ are scalars. More than one window needs to be used (or rotated versions of a single window), in order to 'probe' the different components of f. The same applies to tensor–valued signals such as electromagnetic fields, since they may be regarded as being valued in a higher–dimensional vector space.

It will be useful to write f_h in another form by substituting the Fourier representation of f into f_h. Formally, this gives

$$\begin{aligned} f_h(\mathbf{x},\mathbf{v}) &= \int_{-\infty}^{\infty} dt \int_{\mathbf{R}^n} d\mathbf{p}\, e^{2\pi i \mathbf{p}\cdot(\mathbf{x}+t\mathbf{v})}\, h(t)^* \hat{f}(\mathbf{p}) \\ &= \int_{\mathbf{R}^n} d\mathbf{p}\, e^{2\pi i \mathbf{p}\cdot\mathbf{x}}\, \hat{h}(\mathbf{p}\cdot\mathbf{v})^* \hat{f}(\mathbf{p}) \\ &\equiv \langle \hat{h}_{\mathbf{x},\mathbf{v}}, \hat{f} \rangle_{L^2(d\mathbf{p})} = \langle h_{\mathbf{x},\mathbf{v}}, f \rangle_{L^2(d\mathbf{x})}, \end{aligned} \tag{5}$$

where $\hat{h}_{\mathbf{x},\mathbf{v}}$ is defined by

$$\hat{h}_{\mathbf{x},\mathbf{v}}(\mathbf{p}) = e^{-2\pi i \mathbf{p}\cdot\mathbf{x}}\, \hat{h}(\mathbf{p}\cdot\mathbf{v}), \tag{6}$$

so that

$$h_{\mathbf{x},\mathbf{v}}(\mathbf{x}') = \int_{\mathbb{R}^n} d\mathbf{p}\, e^{2\pi i \mathbf{p}\cdot(\mathbf{x}'-\mathbf{x})}\, \hat{h}(\mathbf{p}\cdot\mathbf{v}). \tag{7}$$

(We have adopted the convention used in the physics literature, where complex inner products are linear in the *second* factor and antilinear in the first factor.)

The functions $h_{\mathbf{x},\mathbf{v}}$ are n–dimensional "wavelets" and will be used in the next subsection to reconstruct the signal f. Note that $\hat{h}_{\mathbf{x},\mathbf{v}}$ (hence also $h_{\mathbf{x},\mathbf{v}}$) is not square–integrable for $n > 1$, since its modulus is constant along directions orthogonal to $\mathbf{v}$. But Equation (5) still makes sense provided $\hat{f}$ is sufficiently well–behaved. (This is one of the reasons we have assumed that f is a test function.)

A common method for the construction of n–dimensional wavelets consists of taking tensor products of one–dimensional wavelets. However, this means that not all directions in $\mathbb{R}^n$ are treated equally, and consequently the set of wavelets does not transform "naturally" (in a sense to be explained below) under the *affine group* G of $\mathbb{R}^n$, which consists of all transformations of the form

$$\mathbf{x} \mapsto g(A, \mathbf{b})\mathbf{x} \equiv A\mathbf{x} + \mathbf{b} \tag{8}$$

with A a non-singular $n \times n$ matrix and $\mathbf{b} \in \mathbb{R}^n$. Each such $g(A, \mathbf{b})$ defines a unitary operator on $L^2(\mathbb{R}^n)$, given by

$$(U(A, \mathbf{b})f)(\mathbf{x}) \equiv |A|^{-\frac{1}{2}} f\left(A^{-1}(\mathbf{x} - \mathbf{b})\right), \tag{9}$$

where $|A|$ denotes the absolute value of the determinant of A. The map $g(A, \mathbf{b}) \mapsto U(A, \mathbf{b})$ forms a *representation* of G on $L^2(\mathbb{R}^n)$, meaning that it preserves the group structure of G under compositions. To see how $h_{\mathbf{x},\mathbf{v}}$ transforms under U, note that the unitarity of U implies

$$\begin{aligned} \langle U(A, \mathbf{b})\, h_{\mathbf{x},\mathbf{v}}, f \rangle &= \langle h_{\mathbf{x},\mathbf{v}}, U(A, \mathbf{b})^{-1} f \rangle \\ &= \int_{-\infty}^{\infty} dt\; h(t)^* \, |A|^{\frac{1}{2}} f\left(A(\mathbf{x} + t\mathbf{v}) + \mathbf{b}\right) \\ &= |A|^{\frac{1}{2}} \langle h_{A\mathbf{x}+\mathbf{b}, A\mathbf{v}}, f \rangle. \end{aligned} \tag{10}$$

Hence

$$U(A, \mathbf{b})\, h_{\mathbf{x},\mathbf{v}} = |A|^{\frac{1}{2}} h_{A\mathbf{x}+\mathbf{b}, A\mathbf{v}}, \tag{11}$$

which states that affine transformations take wavelets to wavelets. Thus, for example, translations, rotations and dilations merely translate, rotate and dilate the labels $\{\mathbf{x}, \mathbf{v}\}$, while the factor $|A|^{\frac{1}{2}}$ preserves unitarity. By contrast, tensor products of one–dimensional wavelets are not transformed into one another by rotations.

2.2. A reconstruction formula

A reconstruction consists of a recovery of f from f_h or its restriction to some subset. In the one–dimensional case, for example, f can be reconstructed using *all* of $\mathbb{R} \times \mathbb{R}_*$ or (for certain choices of h) just a discrete subset ([2], [6], [18], [21], and [22].) For general n, the choice of reconstructions becomes even richer since various new possibilities arise. For example, h may have symmetries which imply that f_h is determined by its values on some lower–dimensional subsets of $\mathbb{R}^n \times \mathbb{R}^n_*$, making integration over the whole space unnecessary and, moreover, undesirable since it may lead to a divergent integral. Furthermore, f may satisfy some partial differential equation which implies that it is determined by its values on subsets of $\mathbb{R}^n$. For example, if $\mathbb{R}^n$ is space–time and f represents a pressure wave or an electromagnetic potential, it satisfies the wave equation away from sources, hence is determined by initial data on a Cauchy surface in $\mathbb{R}^n$, and it becomes both unnecessary and undesirable to use all of $\mathbb{R}^n \times \mathbb{R}^n_*$ in the reconstruction (see Subsection 3.2 and Section 4).

The reconstruction to be developed in this subsection is "generic" in that it does not assume any particular forms for $h(t)$ or $f(\mathbf{x})$. It uses all of $\mathbb{R}^n \times \mathbb{R}^n_*$, so it breaks down for certain choices of h or f. Again we emphasize that this is far from the only way to proceed; other types of reconstruction will be discussed below and elsewhere. The present reconstruction formula is interesting in part because it generalizes the one for the ordinary continuous wavelet transform $(n = 1)$.

To reconstruct f, we look for a *resolution of unity* in terms of the vectors $h_{\mathbf{x},\mathbf{v}}$. This means we need a measure $d\mu(\mathbf{x}, \mathbf{v})$ on $\mathbb{R}^n \times \mathbb{R}^n_*$ such that

$$\int_{\mathbb{R}^n \times \mathbb{R}^n_*} d\mu(\mathbf{x}, \mathbf{v})\, |f_h(\mathbf{x}, \mathbf{v})|^2 = \int_{\mathbb{R}^n} d\mathbf{x}\, |f(\mathbf{x})|^2 \equiv \|f\|^2_{L^2}. \tag{12}$$

(Such an identity is sometimes called a "Plancherel formula.") For then the map $T: f \mapsto f_h$ is an isometry from $L^2(d\mathbf{x})$ onto its range $\mathcal{H} \subset L^2(d\mu)$, and polarization gives

$$\langle g, T^*Tf \rangle_{L^2(d\mathbf{x})} \equiv \langle Tg, Tf \rangle_{\mathcal{H}} = \langle g, f \rangle_{L^2(d\mathbf{x})}. \tag{13}$$

This shows that $f = T^*Tf = T^*f_h$ in $L^2(d\mathbf{x})$, which is the desired reconstruction formula. (See [16] for background on resolutions of unity, generalized frames and related subjects.)

To obtain a resolution of unity, note that

$$f_h(\mathbf{x}, \mathbf{v}) = \left(\hat{h}(\mathbf{p} \cdot \mathbf{v})^* \hat{f}(\mathbf{p})\right)\check{}\,(\mathbf{x}), \tag{14}$$

where $\check{}$ denotes the inverse Fourier transform, so by Plancherel's theorem,

$$\int_{\mathbb{R}^n} d\mathbf{x}\, |f_h(\mathbf{x}, \mathbf{v})|^2 = \int_{\mathbb{R}^n} d\mathbf{p}\, |\hat{h}(\mathbf{p} \cdot \mathbf{v})|^2\, |\hat{f}(\mathbf{p})|^2. \tag{15}$$

We therefore need a measure $d\rho(\mathbf{v})$ on $\mathbb{R}^n_*$ such that

$$H(\mathbf{p}) \equiv \int_{\mathbb{R}^n_*} d\rho(\mathbf{v})\, |\hat{h}(\mathbf{p} \cdot \mathbf{v})|^2 \equiv 1 \quad \text{for almost all } \mathbf{p}, \tag{16}$$

since then $d\mu(\mathbf{x}, \mathbf{v}) = d\mathbf{x}\, d\rho(\mathbf{v})$ has the desired property. The solution is simple: Every $\mathbf{p} \neq \mathbf{0}$ can be transformed to $\mathbf{q} \equiv (1, 0, \cdots, 0)$ by a *dilation and rotation* of $\mathbb{R}^n$. That is, the orbit of $\mathbf{q}$ (in Fourier space) under dilations and rotations is $\mathbb{R}^n_*$. Thus we choose $d\rho$ to be invariant under rotations and dilations, which gives

$$d\rho(\mathbf{v}) = N|\mathbf{v}|^{-n} d\mathbf{v}, \tag{17}$$

where N is a normalization constant, $|\mathbf{v}|$ is the Euclidean norm of $\mathbf{v}$ and $d\mathbf{v}$ is Lebesgue measure in $\mathbb{R}^n$. Then for $\mathbf{p} \neq \mathbf{0}$,

$$
\begin{aligned}
H(\mathbf{p}) = H(\mathbf{q}) &= N \int |\mathbf{v}|^{-n} d\mathbf{v}\, |\hat{h}(v_1)|^2 \\
&= N \int_{-\infty}^{\infty} dv_1\, |\hat{h}(v_1)|^2 \int_{\mathbb{R}^{n-1}} \frac{dv_2 \cdots dv_n}{(v_1^2 + \cdots v_n^2)^{n/2}}.
\end{aligned}
\tag{18}
$$

Now a straightforward computation gives

$$
\int_{\mathbb{R}^{n-1}} \frac{dv_2 \cdots dv_n}{(v_1^2 + \cdots v_n^2)^{n/2}} = \frac{\pi^{n/2}}{|v_1|\, \Gamma(n/2)}. \tag{19}
$$

This shows that the measure $d\mu(\mathbf{x}, \mathbf{v}) \equiv d\mathbf{x}\, d\rho(\mathbf{v})$ gives a resolution of unity if and only if

$$
c_h \equiv \int_{-\infty}^{\infty} \frac{d\xi}{|\xi|}\, |\hat{h}(\xi)|^2 < \infty, \tag{20}
$$

which is precisely the *admissibility condition* for the usual (one–dimensional) wavelet transform [5]. (As mentioned above, admissibility implies that $\hat{h}(0) = 0$.) If h is admissible, the normalization constant is given by

$$
N = \frac{\Gamma(n/2)}{\pi^{n/2}\, c_h} \tag{21}
$$

and the reconstruction formula is

$$
f(\mathbf{x}') = (T^* f_h)(\mathbf{x}') = N \int_{\mathbb{R}^n \times \mathbb{R}^n_*} |\mathbf{v}|^{-n} d\mathbf{x}\, d\mathbf{v}\, h_{\mathbf{x},\mathbf{v}}(\mathbf{x}')\ f_h(\mathbf{x}, \mathbf{v}). \tag{22}
$$

The *sense* in which this formula holds depends on the behavior of f. The class of possible f's, in turn, depends on the choice of h. Note that in spite of the factor $|\mathbf{v}|^n$ in the denominator, there is no problem at $\mathbf{v} = \mathbf{0}$ since $f_h(\mathbf{x}, \mathbf{0}) = \hat{h}(0)^* f(\mathbf{x}) = 0$ by the admissibility condition, and a similar analysis can be made for small $|\mathbf{v}|$ by using the dilation property (Equation (3)).

2.3. The d–dimensional windowed Radon transform

Next, we allow the instrument to extend in $k \geq 0$ spatial dimensions. For example, $k = 1$ for a wire antenna whereas $k = 2$ for a dish antenna. If $\mathbb{R}^n$ is space, then $k \leq n$; if $\mathbb{R}^n$ is space–time, then $k \leq n - 1$. When moving through space with a uniform velocity, the instrument sweeps out a d–dimensional surface in $\mathbb{R}^n$, where $d = k + 1$ if the motion is transverse to its spatial extension and $d = k$ if it is not. If $k < n$, then the set of non–transversal motions is "non–generic" (has measure zero) and can thus be ignored; we therefore set $d = k + 1$ in that case. If $k = n$, then necessarily $d = n$. In either case, we represent the moving instrument by a window function $h : \mathbb{R}^d \to \mathbb{C}$.

The parameter $t \in \mathbb{R}$ has thus been replaced by $\mathbf{t} \in \mathbb{R}^d$. The velocity vector $\mathbf{v}$, which may be regarded as a linear map $t \mapsto \mathbf{v}t$ from $\mathbb{R}$ to $\mathbb{R}^n$, is now replaced by a linear map $A : \mathbb{R}^d \to \mathbb{R}^n$, which we call a *motion* of the instrument in $\mathbb{R}^n$. Denote the set of all such maps by $L(\mathbb{R}^d, \mathbb{R}^n)$. Later, when seeking reconstruction, we shall need to restrict ourselves to subsets of $L(\mathbb{R}^d, \mathbb{R}^n)$ ('rigid' motions); but this need not concern us presently.

Definition 2. *The d-dimensional Windowed Radon Transform of* $f : \mathbb{R}^n \to \mathbb{C}$ *is the function* $f_h : \mathbb{R}^n \times L(\mathbb{R}^d, \mathbb{R}^n) \to \mathbb{C}$ *given by*

$$f_h(\mathbf{x}, A) = \int_{\mathbf{R}^d} d\mathbf{t}\, h(\mathbf{t})^* f(\mathbf{x} + A\mathbf{t}). \tag{23}$$

Upon substituting the Fourier representation of f, a computation similar to the above yields the expression

$$\begin{aligned} f_h(\mathbf{x}, A) &= \int_{\mathbf{R}^d} d\mathbf{t} \int_{\mathbf{R}^n} d\mathbf{p}\, e^{2\pi i \mathbf{p}\cdot(\mathbf{x}+A\mathbf{t})}\, h(\mathbf{t})^* \hat{f}(\mathbf{p}) \\ &= \int_{\mathbf{R}^n} d\mathbf{p}\, e^{2\pi i \mathbf{p}\cdot\mathbf{x}}\, \hat{h}(A'\mathbf{p})^* \hat{f}(\mathbf{p}) \\ &\equiv \langle \hat{h}_{\mathbf{x},A}, \hat{f} \rangle_{L^2(d\mathbf{p})} = \langle h_{\mathbf{x},A}, f \rangle_{L^2(d\mathbf{x})}, \end{aligned} \tag{24}$$

where $A' : \mathbb{R}^n \to \mathbb{R}^d$ is the map dual to A. (For given bases in $\mathbb{R}^d$ and $\mathbb{R}^n$, A is represented by an $n \times d$ matrix; then A' is the transposed $d \times n$ matrix.) In the above equation we have set

$$\hat{h}_{\mathbf{x},A}(\mathbf{p}) = e^{-2\pi i \mathbf{p}\cdot\mathbf{x}}\, \hat{h}(A'\mathbf{p}), \tag{25}$$

which gives the generalized wavelets

$$h_{\mathbf{x},A}(\mathbf{x}') = \int_{\mathbf{R}^n} d\mathbf{p}\, e^{2\pi i \mathbf{p}\cdot(\mathbf{x}'-\mathbf{x})}\, \hat{h}(A'\mathbf{p}). \tag{26}$$

Let us now attempt to reconstruct f from f_h by generalizing the procedure in Subection 2.2. Equation (15) now becomes

$$\int_{\mathbf{R}^n} d\mathbf{x}\, |f_h(\mathbf{x}, A)|^2 = \int_{\mathbf{R}^n} d\mathbf{p}\, |\hat{h}(A'\mathbf{p})|^2\, |\hat{f}(\mathbf{p})|^2. \tag{27}$$

Again we need a measure $d\rho(A)$ which is invariant under dilations and rotations of $\mathbb{R}^n$. Now the *largest* set of maps A which can be considered consists of all those with rank d. (Otherwise the instrument sweeps out a surface of dimension lower than d.) Let us call this set $L_d(\mathbb{R}^d, \mathbb{R}^n)$. Then a measure on $L_d(\mathbb{R}^d, \mathbb{R}^n)$ which is invariant with respect to rotations and dilations of $\mathbb{R}^n$ has the form

$$d\tilde{\rho}(A) = |\det(A'A)|^{-n/2}\, dA, \tag{28}$$

where dA is the Haar measure on $L(\mathbb{R}^d, \mathbb{R}^n) \approx \mathbb{R}^{nd}$ as an additive group. However, no reconstruction is possible using the measure $d\mathbf{x}\, d\tilde{\rho}(A)$ on $\mathbb{R}^n \times L_d(\mathbb{R}^d, \mathbb{R}^n)$, because *no admissible window exists* in general when $d > 1$. (This can be easily verified when $d = n = 2$.) Thus $L_d(\mathbb{R}^d, \mathbb{R}^n)$ is too large, and we return to our imaginary measuring process for inspiration. On physical grounds, we are interested in *rigid motions* of the instrument. A map corresponding to such a motion must have the form $A = vRJ$, where $J : \mathbb{R}^d \to \mathbb{R}^n$ is the canonical inclusion map, R is a rotation of $\mathbb{R}^n$ (RJ gives the orientation of the instrument as well as its direction of motion), and $v > 0$ is the speed. (If $\mathbb{R}^n$ is space–time, then "rotations" involving the time axis are actually Lorentz transformations! For the present, assume that $\mathbb{R}^n$ is space, so R is a true rotation.) We therefore parametrize the set of permissible A's by $(v, R) \in \mathbb{R}^+ \times SO(n) \equiv G$, where $SO(n)$ is the group of unimodular orthogonal $n \times n$ matrices, which represent rotations in $\mathbb{R}^n$. This parametrization is redundant because two rotations of $\mathbb{R}^n$ which have the same effect on the subspace $\mathbb{R}^d$ give the same motion. A non–redundant parametrization of rigid motions is given by $\mathbb{R}^+ \times (SO(n)/SO(n-d))$. However, we use the redundant one here for simplicity. (We shall need the Haar measure on $SO(n)$.) Note that for $d = 1$, J is represented by the vector $\mathbf{q} = (1, 0, \cdots, 0)$ and the set of all maps $A = vRJ$ as above coincides with the set $\mathbb{R}^n_*$ of non–zero velocities considered in Subsections 2.1 and 2.2. A measure on G which is invariant under rotations and dilations of $\mathbb{R}^n$ (*i.e.*, under G itself) has the form

$$d\rho(A) = Nv^{-1}dv\, dR, \tag{29}$$

where N is a normalization constant and dR is the Haar measure on $SO(n)$. Thus for all $\mathbf{p} \neq \mathbf{0}$,

$$H(\mathbf{p}) \equiv \int_G d\rho(A)\, |\hat{h}(A'\mathbf{p})|^2 = H(\mathbf{q}). \tag{30}$$

Now

$$A'\mathbf{q} = vJ'R'\mathbf{q} = vJ'\mathbf{R}_1', \tag{31}$$

where $\mathbf{R}_1$ is the first row of R, which is a unit vector, and J' is the projection of $\mathbb{R}^n$ onto $\mathbb{R}^d$. The admissibility condition therefore reads

$$N^{-1} \equiv \int_0^\infty v^{-1}dv \int_{SO(n)} dR\, |\hat{h}(vJ'\mathbf{R}_1')|^2 < \infty. \tag{32}$$

For $d = 1$, this reduces to Equation (20). If h is admissible, we obtain the reconstruction formula

$$f(\mathbf{x}') = \int_{\mathbb{R}^n} d\mathbf{x} \int_G d\rho(A)\, h_{\mathbf{x},A}(\mathbf{x}')\ f_h(\mathbf{x}, A). \tag{33}$$

§3. Analytic–signal transforms

3.1. Analytic signals in one dimension

Suppose we are given a (possibly complex–valued) one–dimensional signal $f : \mathbb{R} \to \mathbb{C}$. For simplicity, assume that f is smooth with rapid decay. Consider the positive– and negative– frequency parts of f, defined by

$$\begin{aligned} f^+(x) &\equiv \int_0^\infty dp\, e^{2\pi ipx}\, \hat{f}(p) \\ f^-(x) &\equiv \int_{-\infty}^0 dp\, e^{2\pi ipx}\, \hat{f}(p). \end{aligned} \tag{34}$$

Then f^+ and f^- extend analytically to the upper–half and lower–half complex planes, respectively; *i.e.*,

$$\begin{aligned} f^+(x+iy) &= \int_0^\infty dp\, e^{2\pi ip(x+iy)}\, \hat{f}(p), \quad y>0 \\ f^-(x+iy) &= \int_{-\infty}^0 dp\, e^{2\pi ip(x+iy)}\, \hat{f}(p), \quad y<0, \end{aligned} \tag{35}$$

since the factor $e^{-2\pi py}$ decays rapidly for $p \to \pm\infty$ in the respective integrals. $f^+(z)$ and $f^-(z)$ are just the (inverse) *Fourier–Laplace transforms* of the restrictions of $\hat{f}$ to the positive and negative frequencies. If f is complex–valued, then f^+ and f^- are independent and the original signal can be recovered from them as

$$f(x) = \lim_{y\downarrow 0} \left[f^+(x+iy) + f^-(x-iy) \right]. \tag{36}$$

If f is real–valued, then

$$\hat{f}(p) = \hat{f}(-p)^*. \tag{37}$$

In that case, f^+ and f^- are related by reflection,

$$f^+(x+iy) = f^-(x-iy)^*, \quad y>0, \tag{38}$$

and

$$f(x) = 2\lim_{y\downarrow 0} \Re f^+(x+iy) = 2\lim_{y\downarrow 0} \Re f^-(x-iy). \tag{39}$$

When f is real, the function $f^+(z)$ is known as the *analytic signal* associated with $f(x)$. Such functions were first introduced and applied extensively by Gabor [8]. A complex–valued signal would have *two* independent associated analytic signals f^+ and f^-. What significance do $f^\pm$ have? For one thing, they are *regularizations* of f. Equation (36) states that f is jointly a "boundary–value" of the pair f^+ and f^-. As such, f may actually be quite singular while remaining the boundary–value of analytic functions. Also, $2f^{\pm}$ provide a kind of "envelope" description of f (see Born and Wolf [4], Klauder and Sudarshan [19]). For example, if $f(x) = \cos x$, then $2f^{\pm}(x) = e^{\pm ix}$.

In order to extend the concept of analytic signals to more than one dimension, let us first of all unify the definitions of f^+ and f^- by setting

$$\tilde{f}(x+iy) \equiv \int_{-\infty}^{\infty} dp\, \theta(py)\, e^{2\pi ip(x+iy)}\, \hat{f}(p) \tag{40}$$

for *arbitrary* $x + iy \in \mathbb{C}$, where θ is the unit step function, defined by

$$\theta(u) = \begin{cases} 0, & u < 0 \\ \frac{1}{2}, & u = 0 \\ 1, & u > 0. \end{cases} \tag{41}$$

Then we have

$$\tilde{f}(x+iy) = \begin{cases} f^+(x+iy), & y > 0 \\ \frac{1}{2} f(x), & y = 0 \\ f^-(x+iy), & y < 0. \end{cases} \tag{42}$$

Although this unification of f^+ and f^- may at first appear to be somewhat artificial, it turns out to be quite natural, as will now be seen. Note first of all that for any real u, we have

$$\theta(u)\, e^{-2\pi u} = \frac{1}{2\pi i} \int_{-\infty}^{\infty} \frac{d\tau}{\tau - i}\, e^{2\pi i\tau u}, \tag{43}$$

since the contour on the right–hand side may be closed in the upper–half plane when $u > 0$ and in the lower–half plane when $u < 0$. For $u = 0$, the equation states that

$$\begin{aligned} \theta(0) &= \frac{1}{2\pi i} \int_{-\infty}^{\infty} \frac{(\tau + i)\, d\tau}{\tau^2 + 1} \\ &= \frac{1}{2\pi} \int_{-\infty}^{\infty} \frac{d\tau}{\tau^2 + 1} = \frac{1}{2}, \end{aligned} \tag{44}$$

in agreement with our definition, if we interpret the integral as the limit when $L \to \infty$ of the integral from $-L$ to L. Therefore

$$\theta(py)\, e^{2\pi ip(x+iy)} = \frac{1}{2\pi i} \int_{-\infty}^{\infty} \frac{d\tau}{\tau - i}\, e^{2\pi ip(x+\tau y)}. \tag{45}$$

If this is substituted into our expression for $\tilde{f}(z)$ and the order of integrations on τ and p is exchanged, we obtain

$$\tilde{f}(x+iy) = \frac{1}{2\pi i}\int_{-\infty}^{\infty} \frac{d\tau}{\tau - i}\, f(x+\tau y) \tag{46}$$

for arbitrary $x + iy \in \mathbb{C}$. We have referred to the right–hand side as the *Analytic–Signal transform* of $f(x)$ ([16] and [17].) It bears a close relation to the *Hilbert transform,* which is defined by

$$Hf(x) = \frac{1}{\pi}\mathrm{PV}\int_{-\infty}^{\infty} \frac{du}{u}\, f(x-u), \tag{47}$$

where PV denotes the principal value of the integral. Consider the complex combination

$$\begin{aligned} f(x) - iHf(x) &= \frac{1}{\pi i}\int_{-\infty}^{\infty} du \left[\pi i\delta(u) + \mathrm{PV}\frac{1}{u}\right] f(x-u) \\ &= \frac{1}{\pi i}\lim_{\epsilon\downarrow 0}\int_{-\infty}^{\infty} \frac{du}{u - i\epsilon}\, f(x-u) \\ &= \frac{1}{\pi i}\lim_{\epsilon\downarrow 0}\int_{-\infty}^{\infty} \frac{d\tau}{\tau - i}\, f(x-\tau\epsilon) \\ &= 2\lim_{\epsilon\downarrow 0}\tilde{f}(x - i\epsilon). \end{aligned} \tag{48}$$

Similarly,

$$f(x) + iHf(x) = 2\lim_{\epsilon\downarrow 0}\tilde{f}(x+i\epsilon). \tag{49}$$

Hence

$$Hf(x) = i\lim_{\epsilon\downarrow 0}[\tilde{f}(x-i\epsilon) - \tilde{f}(x+i\epsilon)], \tag{50}$$

which for real–valued f reduces to

$$Hf(x) = 2\lim_{\epsilon\downarrow 0}\Im\tilde{f}(x+i\epsilon) = -2\lim_{\epsilon\downarrow 0}\Im\tilde{f}(x-i\epsilon). \tag{51}$$

3.2. Generalization to n dimensions

We are now ready to generalize the idea of analytic signals to an arbitrary number of dimensions. Again we assume initially that $f(\mathbf{x})$ belongs to the space of Schwartz test functions $\mathcal{S}(\mathbb{R}^n)$, although this assumption proves to be unnecessary.

Definition 3. *The Analytic–Signal Transform (AST) of $f \in \mathcal{S}(\mathbb{R}^n)$ is the function $\tilde{f} : \mathbb{C}^n \to \mathbb{C}$ defined by*

$$\tilde{f}(\mathbf{x}+i\mathbf{y}) = \frac{1}{2\pi i}\int_{-\infty}^{\infty} \frac{d\tau}{\tau - i}\, f(\mathbf{x}+\tau\mathbf{y}). \tag{52}$$

The same argument as above shows that for $\mathbf{z} = \mathbf{x} + i\mathbf{y} \in \mathbb{C}^n$,

$$\begin{aligned}\tilde{f}(\mathbf{z}) &= \int_{\mathbb{R}^n} d^n\mathbf{p}\, \theta(\mathbf{p}\cdot\mathbf{y})\, e^{2\pi i \mathbf{p}\cdot\mathbf{z}}\, \hat{f}(\mathbf{p}) \\ &= \int_{M_{\mathbf{y}}} d\mathbf{p}\, e^{2\pi i \mathbf{p}\cdot\mathbf{z}}\, \hat{f}(\mathbf{p}),\end{aligned} \tag{53}$$

where $M_{\mathbf{y}}$ is the half–space

$$M_{\mathbf{y}} \equiv \{\mathbf{p} \in \mathbb{R}^n : \; \mathbf{p}\cdot\mathbf{y} \geq 0\}, \qquad \mathbf{y} \neq \mathbf{0}. \tag{54}$$

We shall refer to the right–hand side of Equation (53) as the (inverse) *Fourier–Laplace transform* of $\hat{f}$ in $M_{\mathbf{y}}$. The integral converges absolutely whenever $\hat{f} \in L^1(\mathbb{R}^n)$, since $|e^{2\pi i \mathbf{p}\cdot\mathbf{z}}| \leq 1$ on $M_{\mathbf{y}}$, defining $\tilde{f}$ as a *function* on $\mathbb{C}^n$, although not an analytic one in general (see below). This shows that $\tilde{f}(\mathbf{z})$ can actually be defined for some distributions f, not only for test functions.

Note: In spite of the appearance of expressions such as $\mathbf{p}\cdot\mathbf{y}$, we have not assumed any particular metric structure in $\mathbb{R}^n$. The Fourier transform naturally takes functions on $\mathbb{R}^n$ (considered as an abelian group) to functions on the *dual* space $\mathbb{R}_n \equiv (\mathbb{R}^n)^*$ of linear functionals, and $\mathbf{p}\cdot\mathbf{y}$ merely denotes the value $\mathbf{p}(\mathbf{y})$. (See Rudin [25].) This remark becomes especially important when considering time–dependent signals, so that $\mathbb{R}^n$ is space–time, for then the natural structure on $\mathbb{R}^n$ is a Lorentzian metric rather than a Euclidean metric (see [16], Subsection 1.1.)

For $n = 1$, $\tilde{f}(z)$ was analytic in the upper– and lower–half planes. In more than one dimension, $\tilde{f}(\mathbf{z})$ need not be analytic, even though, for brevity, we still write it as a function of $\mathbf{z}$ rather than $\mathbf{z}$ and its complex conjugate $\mathbf{z}^*$. However, $\tilde{f}(\mathbf{z})$ does in general possess a *partial* analyticity which reduces to the above when $n = 1$. Consider the partial derivative of $\tilde{f}(\mathbf{z})$ with respect to $z_k^* \equiv x_k - iy_k$, defined by

$$2\bar{\partial}_k \tilde{f} \equiv 2\frac{\partial \tilde{f}}{\partial z_k^*} \equiv \frac{\partial \tilde{f}}{\partial x_k} + i\frac{\partial \tilde{f}}{\partial y_k}. \tag{55}$$

Then $\tilde{f}$ is analytic at $\mathbf{z}$ if and only if $\bar{\partial}_k \tilde{f} = 0$ for all k. But using our definition of $\tilde{f}(\mathbf{z})$, we find that

$$2\bar{\partial}_k \tilde{f}(\mathbf{z}) = \frac{1}{2\pi} \int_{-\infty}^{\infty} d\tau\, \frac{\partial f}{\partial x_k}(\mathbf{x} + \tau\mathbf{y}). \tag{56}$$

It follows that the complex $\bar{\partial}$–derivative in the direction of $\mathbf{y}$ vanishes, *i.e.*,

$$\begin{aligned}4\pi \mathbf{y}\cdot\bar{\partial}\tilde{f}(\mathbf{z}) \equiv 4\pi \sum_k y_k\, \bar{\partial}_k \tilde{f}(\mathbf{z}) &= \int_{-\infty}^{\infty} d\tau\, \sum_k y_k \frac{\partial f}{\partial x_k}(\mathbf{x} + \tau\mathbf{y}) \\ &= \int_{-\infty}^{\infty} d\tau\, \frac{\partial}{\partial \tau} f(\mathbf{x} + \tau\mathbf{y}) = 0,\end{aligned} \tag{57}$$

if f decays for large $|\mathbf{x}|$ (for example, if f is a test function, as we have assumed). Equivalently, using

$$\begin{aligned} 2\bar{\partial}_k\left[\theta(\mathbf{p}\cdot\mathbf{y})\,e^{2\pi i\mathbf{p}\cdot\mathbf{z}}\right] &= 2\bar{\partial}_k\left[\theta(\mathbf{p}\cdot\mathbf{y})\right]\,e^{2\pi i\mathbf{p}\cdot\mathbf{z}} \\ &= i\frac{\partial\theta(\mathbf{p}\cdot\mathbf{y})}{\partial y_k}\,e^{2\pi i\mathbf{p}\cdot\mathbf{z}} \\ &= ip_k\,\delta(\mathbf{p}\cdot\mathbf{y})\,e^{2\pi i\mathbf{p}\cdot\mathbf{z}} \\ &= ip_k\,\delta(\mathbf{p}\cdot\mathbf{y})\,e^{2\pi i\mathbf{p}\cdot\mathbf{x}}, \end{aligned} \tag{58}$$

we have for $\mathbf{y}\neq\mathbf{0}$

$$2\mathbf{y}\cdot\bar{\partial}\tilde{f}(\mathbf{z}) = i\int_{\mathbf{R}^n} d^n\mathbf{p}\,(\mathbf{p}\cdot\mathbf{y})\,\delta(\mathbf{p}\cdot\mathbf{y})\,e^{2\pi i\mathbf{p}\cdot\mathbf{x}}\,\hat{f}(\mathbf{p}) = 0. \tag{59}$$

Thus $\tilde{f}(\mathbf{z})$ *is analytic in the direction* $\mathbf{y}$. In the one–dimensional case, this reduces to

$$\frac{\partial\tilde{f}(z)}{\partial z^*} = 0 \quad \forall\, y\neq 0, \tag{60}$$

which states that $\tilde{f}(z)$ is analytic in the upper– and lower–half planes. In one dimension, there are only *two* imaginary directions, whereas in n dimensions, every $\mathbf{y}\neq\mathbf{0}$ defines an imaginary direction.

The multivariate AST is related to the *Hilbert transform in the direction* $\mathbf{y}$ (see [26], p. 49), defined as

$$H_{\mathbf{y}}f(\mathbf{x}) = \frac{1}{\pi}\mathrm{PV}\int_{-\infty}^{\infty}\frac{du}{u}\,f(\mathbf{x}-u\mathbf{y}), \quad \mathbf{x},\mathbf{y}\in\mathbb{R}^n,\ \mathbf{y}\neq\mathbf{0}. \tag{61}$$

(Usually, it is assumed that $\mathbf{y}$ is a unit vector; we do not make this assumption.) Namely, an argument similar to the above shows that

$$f(\mathbf{x}) \pm iH_{\mathbf{y}}f(\mathbf{x}) = 2\lim_{\epsilon\downarrow 0}\tilde{f}(\mathbf{x}\pm i\epsilon\mathbf{y}), \tag{62}$$

hence

$$H_{\mathbf{y}}f(\mathbf{x}) = i\lim_{\epsilon\downarrow 0}\,[\tilde{f}(\mathbf{x}-i\epsilon\mathbf{y}) - \tilde{f}(\mathbf{x}+i\epsilon\mathbf{y})]. \tag{63}$$

For $n=1$ and $y>0$, this reduces to the previous relation with the ordinary Hilbert transform.

As in the one–dimensional case, $f(\mathbf{x})$ is the boundary–value of $\tilde{f}(\mathbf{z})$ in the sense that

$$f(\mathbf{x}) = \lim_{\epsilon\to 0}\,[\tilde{f}(\mathbf{x}+i\epsilon\mathbf{y}) + \tilde{f}(\mathbf{x}-i\epsilon\mathbf{y})]. \tag{64}$$

For real–valued f, these equations become

$$
\begin{aligned}
f(\mathbf{x}) &= 2 \lim_{\epsilon \to 0} \Re \tilde{f}(\mathbf{x} + i\epsilon\mathbf{y}) \\
H_{\mathbf{y}} f(x) &= 2 \lim_{\epsilon \downarrow 0} \Im \tilde{f}(\mathbf{x} + i\epsilon\mathbf{y}).
\end{aligned} \tag{65}
$$

Two unresolved yet fundamental questions are:

(a) For what classes of 'functions' (possibly distributions) can the AST be defined, apart from $\mathcal{S}(\mathbb{R}^n)$; *i.e.*, what is the *domain* of the AST?
(b) Given a vector space $\mathcal{H}$ of 'functions' on $\mathbb{R}^n$ for which the AST is defined, what is the *range* of the AST on $\mathcal{H}$? That is, given a function F on $\mathbb{C}^n$, how can we tell whether F is the transform of some $f \in \mathcal{H}$?

A necessary, though probably not sufficient, condition for $F = \tilde{f}$ is that F satisfy the directional Cauchy–Riemann equation $\mathbf{y} \cdot \bar{\partial} F(\mathbf{x}, \mathbf{y}) = 0$. Complete answers to the above questions can be given in some specific cases: When f is a solution of the Klein–Gordon equation, then $\tilde{f}$ must satisfy a certain consistency condition (the reproducing property of the associated wavelets). This condition, when satisfied by F, also guarantees that $F = \tilde{f}$ for some f (see [16], Chapters 1 and 4). A similar result will be obtained in Section 4 for solutions of the wave equation in two space–time dimensions. The comments below apply to the general case and are, consequently, informal.

The most direct way to find if F is the AST of some f is to *construct* f from F and then check that $\tilde{f} = F$. The first part has already been done formally, since f has been shown to be the boundary–value of $\tilde{f}$. Here we suggest an alternative method which can be used to construct $\hat{f}(\mathbf{p})$ instead of $f(\mathbf{x})$. Assume that the Fourier transform is defined on $\mathcal{H}$. If $F(\mathbf{x}, \mathbf{y}) = \tilde{f}(\mathbf{x} + i\mathbf{y})$ for some $f(\mathbf{x}) \in \mathcal{H}$, then the $2n$–dimensional Fourier transform of F is seen (formally) to be

$$
\begin{aligned}
\hat{F}(\mathbf{p}, \mathbf{q}) &\equiv \int_{\mathbb{R}^{2n}} d\mathbf{x}\, d\mathbf{y}\, e^{-2\pi i(\mathbf{p}\cdot\mathbf{x} + \mathbf{q}\cdot\mathbf{y})}\, F(\mathbf{x}, \mathbf{y}) \\
&= \hat{f}(\mathbf{p}) \int_{\mathbb{R}^n} d\mathbf{y}\, \theta(\mathbf{p} \cdot \mathbf{y})\, e^{-2\pi i(\mathbf{q} - i\mathbf{p})\cdot\mathbf{y}} \\
&\equiv \hat{f}(\mathbf{p})\, \tilde{\delta}(\mathbf{q} - i\mathbf{p}),
\end{aligned} \tag{66}
$$

where $\tilde{\delta}$ is the AST, *in Fourier space,* of the Dirac measure $\delta(\mathbf{q})$. (This suggests that the AST, like the Fourier transform, exhibits some symmetry between space and Fourier space.) $\tilde{\delta}$ can be shown to be invariant under *real* rotations $(\mathbf{q} - i\mathbf{p} \to R(\mathbf{q} - i\mathbf{p})$, with $R \in SO(n))$ and to transform under dilations as

$$
\tilde{\delta}(\lambda(\mathbf{q} - i\mathbf{p})) = \lambda^{-n}\, \tilde{\delta}(\mathbf{q} - i\mathbf{p}), \qquad \lambda \neq 0. \tag{67}
$$

Given $\mathbf{p} \neq \mathbf{0}$, let R be a rotation such that $\mathbf{p} = |\mathbf{p}|\, R\mathbf{e}$, where $\mathbf{e} = (1, 0, \cdots, 0)$, and let $\mathbf{k} \equiv |\mathbf{p}|^{-1} R^{-1}\mathbf{q}$, so that $\mathbf{q} - i\mathbf{p} = |\mathbf{p}|\, R\,(\mathbf{k} - i\mathbf{e})$. Then

$$\begin{aligned}\tilde{\delta}(\mathbf{q}-i\mathbf{p}) &= |\mathbf{p}|^{-n}\,\tilde{\delta}(\mathbf{k}-i\mathbf{e})\\ &= |\mathbf{p}|^{-n}\int_{\mathbf{R}^n} d\mathbf{y}\,\theta(y_1)\,e^{-2\pi i(\mathbf{k}-i\mathbf{e})\cdot\mathbf{y}}\\ &= |\mathbf{p}|^{-n}\int_0^\infty dy_1\,e^{-2\pi(1+ik_1)y_1}\int_{\mathbf{R}^{n-1}} dy_2\cdots dy_n\,e^{-2\pi i(k_2y_2+\cdots+k_ny_n)}\\ &= \frac{\delta(k_2,\cdots,k_n)}{2\pi|\mathbf{p}|^n(1+ik_1)}.\end{aligned} \tag{68}$$

Let $P:\mathbb{R}^n\to\mathbb{R}$ and $Q:\mathbb{R}^n\to\mathbb{R}^{n-1}$ denote the projections $P\mathbf{k}=k_1$ and $Q\mathbf{k}=(k_2,\cdots,k_n)$. Then the numerator in the last expression is

$$\delta(Q\mathbf{k})=\delta(Q|\mathbf{p}|^{-1}R^{-1}\mathbf{q})|=|\mathbf{p}|^{n-1}\,\delta(QR^{-1}\mathbf{q}), \tag{69}$$

hence

$$\tilde{\delta}(\mathbf{q}-i\mathbf{p})=\frac{\delta(QR^{-1}\mathbf{q})}{2\pi(|\mathbf{p}|+iPR^{-1}\mathbf{q})},\qquad \mathbf{p}\neq\mathbf{0}. \tag{70}$$

Together with Equation (66), this gives an explicit formal condition for F to be the AST of f and, if so, to determine $\hat{f}(\mathbf{p})$. When $n=1$, $\tilde{\delta}$ takes the simple form

$$\tilde{\delta}(q-ip)=\frac{\operatorname{sgn}(p)}{2\pi(p+iq)},\qquad p\neq 0. \tag{71}$$

3.3. Some applications of the AST

The AST is an example of a Windowed X–Ray transform, introduced in Subsection 2.1, with the window function

$$h(\tau)^*=\frac{1}{2\pi i(\tau-i)}. \tag{72}$$

(This window function is not "admissible" in the sense of Subsection 2.2, hence the reconstruction formula developed there fails. However, that reconstruction was based on some assumptions which may not be appropriate in general; see the comments below Equation (88).) We now give two examples of the usefulness of the AST. An extensive use of this transform will also be made in Section 4.

Example 4. Hardy spaces

Suppose that $\hat{f}(\mathbf{p})$ vanishes outside of some closed convex cone V_+. The cone V_+' *dual* to V_+ is defined as

$$V_+'=\{\mathbf{y}\in\mathbb{R}^n:\ \mathbf{p}\cdot\mathbf{y}>0\ \forall\,\mathbf{p}\in V_+\}, \tag{73}$$

and it is clearly an open convex cone. Note that for $\mathbf{y} \in V'_+$, $\theta(\mathbf{p}\cdot\mathbf{y}) \equiv 1$ on the support of $\hat{f}$ (except at $\mathbf{p} = \mathbf{0}$, which has measure 0), hence if $f \in L^2(\mathbb{R}^n)$ and $\mathbf{y} \in V'_+$, then

$$\tilde{f}(\mathbf{x}+i\mathbf{y}) = \int_{V_+} d\mathbf{p}\, e^{2\pi i \mathbf{p}\cdot(\mathbf{x}+i\mathbf{y})}\, \hat{f}(\mathbf{p}) \tag{74}$$

and it follows (Stein and Weiss [27]) that $\tilde{f}(\mathbf{z})$ is analytic in the *tube domain*

$$\mathcal{T}^+ \equiv \{\mathbf{x}+i\mathbf{y} \in \mathbb{C}^n : \ \mathbf{y} \in V'_+\}. \tag{75}$$

The set $H^2 \equiv \{\tilde{f} : \ f \in L^2(V_+)\}$ is known as a *Hardy space.* Note also that $\tilde{f}$ vanishes in the tube

$$\mathcal{T}^- \equiv \{\mathbf{x}+i\mathbf{y} \in \mathbb{C}^n : \ -\mathbf{y} \in V'_+\}, \tag{76}$$

since there $\mathbf{p}\cdot\mathbf{y} < 0$ for all $\mathbf{0} \neq \mathbf{p} \in V_+$. Equation (74) gives

$$f(\mathbf{x}) = \lim_{\epsilon \downarrow 0} \tilde{f}(\mathbf{x}+i\epsilon\mathbf{y}), \qquad \mathbf{y} \in V'_+\,, \tag{77}$$

which states that f is a boundary–value of $\tilde{f}$. Since $\tilde{f}(\mathbf{z})$ is analytic, it may be regarded as a *regularization* of $f(\mathbf{x})$ (the latter, being merely square–integrable, is a distribution). Equation (77) can be viewed as a "reconstruction" of f from $\tilde{f}$, albeit a somewhat trivial one.

Example 5. The Klein–Gordon Equation

An important application of the AST is to signals that satisfy some partial differential equations. (In fact, it was in this contcxt that thc transform originated.) Suppose that f is a solution of the *Klein–Gordon equation* in $\mathbb{R}^n$,

$$\Box f + m^2c^4 f = 0, \tag{78}$$

where

$$\Box \equiv \frac{\partial^2}{\partial x_1^2} - c^2\frac{\partial^2}{\partial x_2^2} - \cdots - c^2\frac{\partial^2}{\partial x_n^2} \tag{79}$$

is the D'Alembertian or wave operator for waves with propagation speed c. Here $\mathbb{R}^n$ is interpreted as *space–time,* with x_1 the time coordinate and $(x_2, \cdots, x_n)$ the space coordinates, and $m > 0$ is a mass parameter. This equation describes free relativistic particles in quantum mechanics. The limit $m \to 0$ gives the wave equation, which will be discussed below. Define the *solid light cone* in Fourier space by

$$V = \{\mathbf{p} \in \mathbb{R}^n : \ \mathbf{p}^2 \equiv p_1^2 - c^2p_2^2 - \cdots c^2p_n^2 \geq 0\}. \tag{80}$$

(Note that we are now using a Lorentz metric!) V is the union of the *forward* and *backward* light cones V_+ and V_-, where $p_1 \geq 0$ and $p_1 \leq 0$, respectively.

Note that $V_\pm$ are convex but V is not. The fact that f satisfies the Klein–Gordon equation means that its Fourier transform $\hat{f}(\mathbf{p})$ is supported on the double mass hyperboloid

$$\Omega_m = \{\mathbf{p} \in \mathbb{R}^n : \ \mathbf{p}^2 = m^2 c^4\} = \Omega_m^+ \cup \Omega_m^-, \tag{81}$$

where $\Omega_m^\pm \subset V_\pm$. Thus $\hat{f} = \hat{f}^+ + \hat{f}^-$, where $\hat{f}^\pm$ are distributions supported on $\Omega_m^\pm$. Since $\Omega_m^\pm \subset V_\pm$, an argument similar to that used for Hardy spaces shows that the corresponding solutions $f^\pm(\mathbf{x})$ have AST's $\tilde{f}^\pm(\mathbf{z})$ which are analytic in $\mathcal{T}^\pm$ and vanish in $\mathcal{T}^\mp$, where

$$\mathcal{T}^\pm = \{\mathbf{x} + i\mathbf{y} \in \mathbb{C}^n : \ \mathbf{y} \in V'_\pm\} \tag{82}$$

and $V'_\pm$ are the cones dual to $V_\pm$, which can be seen to be

$$V'_\pm = \{\mathbf{y} \in \mathbb{R}^n : \ \mathbf{y}^2 \equiv c^2 y_1^2 - y_2^2 - \cdots - y_n^2 > 0, \quad \pm y_1 > 0\}. \tag{83}$$

Note that while V is a cone in *Fourier space* (*i.e.*, p_1 is a frequency and $p_2, \cdots p_n$ are wave numbers per unit length), $V' \equiv V'_+ \cup V'_-$ is a cone in space–time. Technically, these two spaces are dual and should not be identified with one another.

The AST of f, given by $\tilde{f}(\mathbf{z}) = \tilde{f}^+(\mathbf{z}) + \tilde{f}^-(\mathbf{z})$, is therefore analytic in the *double tube* $\mathcal{T} \equiv \mathcal{T}^+ \cup \mathcal{T}^-$, with $\mathcal{T}^+$ and $\mathcal{T}^-$ containing only the positive– and negative–frequency parts of f, respectively. This "polarization" of frequencies is important because it makes it possible to reconstruct the solution f from $\tilde{f}$ without approaching the singular boundary $\mathbb{R}^n$ ($\mathbf{y} \to \mathbf{0}$). Equation (74) shows that $\tilde{f}^\pm$ have the form

$$\tilde{f}^\pm(\mathbf{z}) = \int_{\Omega_m^\pm} d\tilde{p}\ \hat{e}_{\mathbf{z}}^\pm(\mathbf{p})^*\, a^\pm(\mathbf{p}), \qquad \mathbf{z} \in \mathcal{T}^\pm, \tag{84}$$

where

$$d\tilde{p} = \frac{dp_2\, dp_3 \cdots dp_n}{2|p_1|} \equiv \frac{dp_2\, dp_3 \cdots dp_n}{2c\sqrt{m^2c^2 + p_2^2 + \cdots + p_n^2}} \tag{85}$$

is the induced measure on Ω_m and

$$\hat{e}_{\mathbf{z}}^\pm(\mathbf{p})^* \equiv e^{2\pi i \mathbf{p}\cdot\mathbf{z}}, \qquad \mathbf{z} \in \mathcal{T}^\pm, \ \mathbf{p} \in \Omega_m^\pm. \tag{86}$$

The corresponding expression $e_{\mathbf{z}}^\pm$ in the space–time domain, defined by

$$e_{\mathbf{z}}^\pm(\mathbf{x}')^* = \int_{\Omega_m^\pm} d\tilde{p}\ e^{2\pi i \mathbf{p}\cdot(\mathbf{z}-\mathbf{x}')}, \tag{87}$$

is a solution of the Klein–Gordon equation which can be shown to be a *coherent wave-packet* whose parameters $\mathbf{z} = \mathbf{x} + i\mathbf{y}$ have a direct geometric interpretation: $\mathbf{x}$ is a point in space–time about which $e_{\mathbf{z}}^\pm$ is "focused" (*i.e.*, $e_{\mathbf{z}}^\pm(\mathbf{x}')$ converges toward the point $(x_2, \cdots, x_n)$ in space for times $x'_1 < x_1$ and diverges away from it for times $x_1' > x_1$), and $\mathbf{y}$ is a set of homogeneous coordinates for the *average velocity* at which $e_{\mathbf{z}}^{\pm}$ is traveling. Furthermore, the invariant $\lambda > 0$ defined by $\lambda^2 \equiv \mathbf{y}^2$ can be interpreted as a *scale parameter* which, roughly speaking, measures the spread (resolution) of the wave packet at the instant of its maximal focus ($x_1' = x_1$) in its rest–frame (the coordinate system in which $y_2 = y_3 = \cdots = y_n = 0$). For small λ, $e_{\mathbf{z}}^{\pm}(\mathbf{x}')$ is localized near $x_k' = x_k$ ($k = 2, 3, \cdots, n$) at time $x_1' = x_1$, whereas for large λ, it is spread out in space. (In Fourier space, on the other hand, it is λ^{-1} which measures the spread.) The positive– and negative–frequency packets $e_{\mathbf{z}}^{+}$ and $e_{\mathbf{z}}^{-}$ are interpreted in quantum theory as particles and antiparticles, respectively. (This agrees with the usual observation that antiparticles "go backward in time." See [16], Chapter 5.)

Since the window function $h(\tau)$ used in the AST has Fourier transform $\hat{h}(\xi) = \theta(\xi)\, e^{-2\pi\xi}$, it is not admissible in the general sense developed in Subsection 2.2; *i.e.*, we have

$$\int_{-\infty}^{\infty} \frac{d\xi}{|\xi|}\, |\hat{h}(\xi)|^2 = \infty. \tag{88}$$

However, the rules of the game have changed. Equation (88) was associated with a reconstruction formula which represents f as an integral of generalized wavelets parametrized by *all* of $\mathbb{R}^n \times \mathbb{R}^n_*$, *i.e.*, all $\mathbf{x} + i\mathbf{y}$ with $\mathbf{y} \neq \mathbf{0}$. This was acceptable when considering general functions $f(\mathbf{x})$ in $L^2(\mathbb{R}^n)$, since then we could define a representation of the affine group on such functions. But now we are dealing with a Hilbert space $\mathcal{H}$ of solutions of the Klein–Gordon equation,

$$f(\mathbf{x}) = \int_{\Omega_m} d\tilde{p}\, e^{2\pi i \mathbf{p}\cdot\mathbf{x}}\, a(\mathbf{p}), \tag{89}$$

with the *Sobolev* norm

$$\|f\|^2 \equiv \int_{\Omega_m} d\tilde{p}\, |a(\mathbf{p})|^2 \tag{90}$$

rather the L^2 norm used in Subsection 2.2. General affine transformations no longer map solutions to solutions, *i.e.*, they no longer define operators on $\mathcal{H}$. The mass m spoils the invariance of the equation under dilations. Only the subgroup $\mathcal{P}$ of translations together with Lorentz transformations (*i.e.*, linear maps $\mathbf{y} \mapsto A\mathbf{y}$ which preserve the Lorentz norm $\mathbf{y}^2$) maps solutions to solutions. $\mathcal{P}$ is called the inhomogeneous Lorentz or *Poincaré* group.

Recall that the measure used in the reconstruction formula of Subsection 2.2 was chosen to be invariant under dilations and rotations. Since dilations no longer define operators on $\mathcal{H}$, this measure is no longer appropriate. Rather, we now expect to reconstruct f by integrating in $\mathbf{y}$ over the double hyperboloid

$$\Omega_\lambda = \{\mathbf{y} \in V' : \mathbf{y}^2 = \lambda^2\} = \Omega_\lambda^+ \cup \Omega_\lambda^- \tag{91}$$

for an arbitrary fixed $\lambda > 0$. Furthermore, we do not expect to integrate over all $\mathbf{x} \in \mathbb{R}^n$, since a solution is determined by its data on any *Cauchy surface* $S \subset \mathbb{R}^n$. For simplicity, take S to be the hyperplane $x_1 = t$ for fixed $t \in \mathbb{R}$, though any Cauchy surface (spacelike $(n-1)$–dimensional submanifold of $\mathbb{R}^n$) will do ([16], Subsection 4.5). Thus consider the $(2n-2)$–dimensional submanifold

$$\sigma = \{\mathbf{x} + i\mathbf{y} \in \mathcal{T} : \; x_1 = t, \; \mathbf{y} \in \Omega_\lambda\} = \sigma_+ \cup \sigma_-, \tag{92}$$

where $\mathbf{y} \in \Omega_\lambda^\pm$ in $\sigma_\pm$. σ parametrizes all possible locations and velocities of a classical particle at the fixed time t, *i.e.*, it is a *phase space.* A reconstruction formula has been obtained in the form

$$f(\mathbf{x}') = \int_\sigma d\mu(\mathbf{z})\, e_{\mathbf{z}}(\mathbf{x}')\, \tilde{f}(\mathbf{z}), \tag{93}$$

where $e_{\mathbf{z}} \equiv e_{\mathbf{z}}^\pm$ on $\sigma_\pm$, $\sigma_\pm$ is parametrized by $(x_2, \cdots x_n, y_2, \cdots y_n)$, and

$$d\mu(\mathbf{z}) = A(\lambda, m)^{-1} dx_2 \cdots dx_n dy_2 \cdots dy_n. \tag{94}$$

$A(\lambda, m)$ is a certain constant related to the admissibility of the wavelets $e_{\mathbf{z}}$ with respect to the measure $dx_2 \cdots dy_n$. Note that this differs from the usual construction of a solution from its initial data, which uses the values of both f and $\partial f / \partial x_1$ on S. The intuitive explanation is that the dependence of $\tilde{f}(\mathbf{x}+i\mathbf{y})$ on $\mathbf{y} \in \Omega_\lambda$, for fixed $\mathbf{x} \in S$, gives the equivalent "velocity" information. The independence of the reconstruction from the choice of Cauchy surface is due to a conservation law satisfied by solutions (see [16]).

The above reconstruction formula bears a close resemblance to the standard representation of a function in terms of wavelets, for the following reason: In the hyperbolic geometry of spacetime, a moving object undergoes a *Lorentz contraction,* i.e. it shrinks in its direction of motion. Since $\mathbf{y}$ represents a velocity, Equation (93) expresses f as a linear combination of "wavelets" centered about all possible points in space (at the given time t) and in various states of compression. However, the analogy is incomplete since the $e_{\mathbf{z}}$'s can only contract and not dilate. (That is, they have a minimum width in their rest frames, determined by the choice of λ.) Their contraction is due to Lorentz transformations rather than ordinary dilations of the form $\mathbf{x} \mapsto a\mathbf{x}$, $a \neq 0$. As noted earlier, the Klein–Gordon equation is not invariant under such dilations, due to the presence of $m > 0$. On the other hand, the wave equation $(m \to 0)$ *is* invariant under dilations, hence the analogy with wavelets can be expected to be closer. This is the subject of our next section.

§4. Wavelets and the wave equation

4.1. Introduction

As explained in Section 1, a representation of solutions of the wave equation similar to that given for the Klein–Gordon equation in (93) should be of some interest in the analysis of naturally occurring signals, since it would automatically display much of their informational contents. Unfortunately, the reconstruction formula in (93) fails when $m \to 0$ because $A(\lambda, m)$ diverges in that limit. (The wavelet representation is no longer square–integrable in that limit.) This is probably related to the fact, well–known in quantum mechanics, that the Klein–Gordon equation with $m > 0$ has a very different group–theoretical structure from the wave equation. The symmetry group of the Klein–Gordon equation is the Poincaré group $\mathcal{P}$, while the symmetry group of the wave equation is the *conformal group* $\mathcal{C}$, which contains the Poincaré group as well as dilations and uniform accelerations. The fundamental difference between massive particles (such as electrons) and massless particles (such as photons) is that the former can be at rest while the latter necessarily travel at the speed of light. It may well be that once the conformal group is taken into account, an appropriate reconstruction formula can be found. In this section we confirm this hypothesis in two–dimensional space–time.

4.2. Symmetries of the wave and Dirac equations in $\mathbb{R}^2$

In this subsection, we examine some group–theoretical aspects of solutions of the wave equation in $\mathbb{R}^2$. To simplify the notation, we choose the units of length and time such that the propagation speed $c = 1$. The wave equation then reads

$$\begin{aligned} 0 = -\Box f(x,t) &\equiv \left(\partial_x^2 - \partial_t^2\right) f(x,t) \\ &= (\partial_x + \partial_t)(\partial_x - \partial_t) f(x,t), \end{aligned} \tag{95}$$

and we consider solutions which are possibly complex–valued. In terms of the *light–cone coordinates*

$$u = x + t, \qquad v = x - t, \tag{96}$$

the equation becomes $\partial_u \partial_v f = 0$, hence the general solution has the form first given by D'Alembert,

$$f(x,t) = f_+(u) + f_-(v). \tag{97}$$

$f_+(t + x)$ is a *left–moving* wave since it is constant on the characteristics $x = x_0 - t$, and similarly $f_-(t - x)$ is a *right–moving wave*. Later we shall find an appropriate family of Hilbert spaces $\mathcal{H}_s$ ($s \geq 0$) to which $f_\pm$ will be required to belong, so we now write $f_\pm \in \mathcal{H}_s$ without being specific. Note that we can let $f_\pm \to f_\pm \pm c$, where c is a constant, without affecting f. This ambiguity will not be a problem since $\mathcal{H}_s$ contains no nonzero constants. Hence the solutions are in one–to–one correspondence with the elements of the orthogonal sum $\mathcal{D}_s = \mathcal{H}_s \oplus \mathcal{H}_s$, whose elements can be written in the vector form

$$\psi = \begin{pmatrix} f_+ \\ f_- \end{pmatrix}. \tag{98}$$

The wave equation can be restated as $\partial_v f_+ = \partial_u f_- = 0$, or

$$\begin{pmatrix} 0 & \partial_t + \partial_x \\ \partial_t - \partial_x & 0 \end{pmatrix} \psi \equiv (\gamma_t \partial_t + \gamma_x \partial_x)\, \psi \equiv \partial\!\!\!/ \psi = 0, \tag{99}$$

where

$$\gamma_t = \begin{pmatrix} 0 & 1 \\ 1 & 0 \end{pmatrix}, \qquad \gamma_x = \begin{pmatrix} 0 & 1 \\ -1 & 0 \end{pmatrix}. \tag{100}$$

Equation (99) is known in quantum mechanics as the (two–dimensional, massless) *Dirac equation;* ψ is a *spinor*, and γ_t, γ_x are *Dirac matrices.* The Dirac equation may be viewed as a particular system of first–order equations equivalent to the wave equation which is, moreover, especially suited to the symmetries of the latter. The Dirac operator $\partial\!\!\!/$ is a "square root" of the wave operator in the sense that $\partial\!\!\!/^2 = -\Box I$, where I is the 2×2 identity matrix. (This is related to the Clifford algebra associated with the Lorentz metric.) In more than one space dimension, there is no such simple relation between scalar–valued solutions f and spinor–valued solutions ψ.

We denote by $\mathcal{D}_{s+}$ and $\mathcal{D}_{s-}$ the subspaces of $\mathcal{D}_s$ with vanishing second and first components, respectively. Thus $\mathcal{D}_{s\pm} \approx \mathcal{H}_s$.

A *symmetry* of the wave equation is a transformation which maps solutions to solutions. As symmetries can be composed and inverted, they form a group. We shall be particularly interested in "geometric" symmetries, which are induced from mappings on the underlying space–time that respect the wave equation. Some obvious ones are:

Translations: For each $(x_0, t_0) \in \mathbb{R}^2$, the map $(x, t) \to (x + x_0, t + t_0)$ induces a symmetry transformation $f(x, t) \to f(x - x_0, t - t_0)$. This maps the right– and left–moving waves by

$$T(u_0, v_0) : \; f_+(u) \to f_+(u - u_0), \qquad f_-(v) \to f_-(v - v_0), \tag{101}$$

where $u_0 = x_0 + t_0$ and $v_0 = x_0 - t_0$. Hence translations can be made to act independently on the left and right parts of solutions.

Lorentz transformations: For any real θ, the map

$$x \to x \cosh\theta + t \sinh\theta, \qquad t \to x \sinh\theta + t \cosh\theta \tag{102}$$

preserves the Lorentz metric $x^2 - t^2 = uv$. (It is a "rotation" by the imaginary angle $-i\theta$ and represents the space–time coordinates as measured by an observer moving with velocity $-\tanh\theta$.) This map has a particularly simple form in terms of the light–cone coordinates:

$$u \to e^{\theta} u \equiv \lambda u, \qquad v \to e^{-\theta} v \equiv \lambda^{-1} v. \tag{103}$$

Since this leaves the wave operator $\Box = -4\partial_u \partial_v$ invariant, it induces a symmetry on solutions. The simplest such transformation acts on $f_\pm$ by $f_+(u) \to f_+(\lambda^{-1}u)$, $f_-(v) \to f_-(\lambda v)$. We shall need a more general induced transformation, given by

$$L(\lambda): \; f_+(u) \to S_+(\lambda)\, f_+(\lambda^{-1}u), \qquad f_-(v) \to S_-(\lambda)\, f_-(\lambda v), \tag{104}$$

where $S_\pm : \mathbb{R}^+ \to \mathbb{C}_* \equiv \mathbb{C}\backslash\{0\}$ (called "multipliers") must satisfy

$$S_\pm(\lambda^{-1}) = S_\pm(\lambda)^{-1}, \qquad S_\pm(\lambda\lambda') = S_\pm(\lambda)\, S_\pm(\lambda') \tag{105}$$

in order that Lorentz transformations form a group. (This means that $S_\pm$ are group homomorphisms.) Continuity in $\lambda > 0$ then implies that $S_+(\lambda) = \lambda^{-j}$ and $S_-(\lambda) = \lambda^{-j'}$ for some $j, j' \in \mathbb{C}$.

The three–dimensional symmetry group of all maps $T(u_0, v_0)\, L(\lambda)$ is the *restricted Poincaré group* $\mathcal{P}_0$ in $1+1$ space–time dimensions. Note that $\mathcal{P}_0$ leaves invariant the subspaces $\mathcal{D}_{s\pm}$ of right– and left–moving waves. Since we shall be interested in *irreducible* representations of the symmetry group which characterize the complete wave equation, it is desirable to include a symmetry which mixes these two subspaces.

Space reflection: The map $(x, t) \to (-x, t)$ is a discrete symmetry of the wave equation, corresponding to $(u, v) \to (-v, -u)$. We take the induced mapping on solutions to be

$$P: \; f_+(u) \to f_-(-u), \qquad f_-(v) \to f_+(-v). \tag{106}$$

Thus, P interchanges right and left waves, as desired. The idea that right and left be on equal footing is expressed as

$$L(\lambda)\, P = P\, L(\lambda^{-1}), \tag{107}$$

which implies that

$$S_-(\lambda^{-1}) = S_+(\lambda), \tag{108}$$

hence $j' = -j$. We call j the *Lorentz weight* of the solution. (In more than one space dimension, it is related to *spin.)*

The group $\mathcal{P}^\uparrow$ obtained from $\mathcal{P}_0$ by adjoining the space reflection is called the *orthochronous* Poincaré group in the physics literature, since it still leaves the direction of time invariant. The full Poincaré group $\mathcal{P}$ is obtained by further adjoining *time reversal.* However, the latter must be antilinear for reasons which need not concern us here (see Streater and Wightman [28]). The fact that $\mathcal{P}^{\uparrow}$ leaves the direction of time invariant implies that the subspaces of positive–and negative–frequency solutions are invariant under it. To mix them, we introduce the following.

Total reflection: The map $(x,t) \to (-x,-t)$ is another discrete symmetry of the wave equation, corresponding to $(u,v) \to (-u,-v)$. We take the induced mapping on solutions to be

$$R: \ f_+(u) \to f_+(-u), \qquad f_-(v) \to f_-(-v). \tag{109}$$

Note that unlike translations, Lorentz transformations do not act independently on the left– and right–moving waves. This will be remedied by including the next symmetry.

Dilations: Since the wave equation is homogeneous in x and t, it is invariant under the map $(x,t) \to (\alpha x, \alpha t)$, for any $\alpha \neq 0$. Equivalently, $(u,v) \to (\alpha u, \alpha v)$. It suffices to confine our attention to $\alpha > 0$, since dilations with $\alpha < 0$ can be obtained by combining $D(\alpha)$ with R. As with Lorentz transformations, we shall allow a multiplier $M : \mathbb{R}^+ \to \mathbb{C}_*$ in the induced mapping. Thus $f(x,t) \to M(\alpha)\, f(\alpha^{-1}x, \alpha^{-1}t)$, or

$$D(\alpha): \ f_+(u) \to M(\alpha)\, f_+(\alpha^{-1}u), \qquad f_-(v) \to M(\alpha)\, f_-(\alpha^{-1}v). \tag{110}$$

In order that dilations form a group, we must have $M(\alpha^{-1}) = M(\alpha)^{-1}$ and $M(\alpha\alpha') = M(\alpha)\, M(\alpha')$. Again, continuity implies that $M(\alpha) = \alpha^{-\kappa}$ for some $\kappa \in \mathbb{C}$, called the *conformal weight* of the solution.

The following simple argument should convince the reader of the need to include non–trivial multipliers in Equation (110), *i.e.*, to consider conformal weights other than zero: Suppose that $f(x,t)$ is a solution with conformal weight κ and let $g(x,t) = (a\partial_x + b\partial_t)\, f(x,t)$, where a and b are constants. Then g is also a solution of the wave equation, and it is easily seen to have conformal weight $\kappa + 1$. In this way we can *shift* the conformal weight of a solution by any positive integer n by applying a homogeneous partial differential operator of order n with constant coefficients. (In three space dimensions, the electromagnetic potentials satisfy the wave equation in free space; the electromagnetic fields are obtained from them by applying first–order partial differential operators, hence are solutions but with higher weight.)

Note that dilations commute with Lorentz transformations. Composing $L(\lambda)$ and $D(\alpha)$, we obtain

$$f_+(u) \to \alpha^{-\kappa}\, \lambda^{-j}\, f_+(\lambda^{-1}\alpha^{-1}u), \qquad f_-(v) \to \alpha^{-\kappa}\, \lambda^{j}\, f_-(\lambda\alpha^{-1}v). \tag{111}$$

In order for the combination of dilations and Lorentz transformation to act independently on f_+ and f_-, we must therefore require that $j = \kappa$. We shall refer to $\kappa \equiv j \equiv -j'$ simply as the *weight* of f. Setting $\beta = \alpha\lambda$ and $\gamma = \alpha/\lambda$, we then obtain

$$f_+(u) \to \beta^{-\kappa} f_+(\beta^{-1}u), \qquad f_-(v) \to \gamma^{-\kappa} f_-(\gamma^{-1}v). \tag{112}$$

Consequently, the semi–direct product $\mathbb{R}^+ \times \mathcal{P}_0$ acts on f by

$$f_+(u) \to \beta^{-\kappa} f_+(\beta^{-1}(u-u_0)), \qquad f_-(v) \to \gamma^{-\kappa} f_-(\gamma^{-1}(v-v_0)). \tag{113}$$

This means that *the group $\mathcal{G}_0 \equiv \mathbb{R}^+ \times \mathcal{P}_0$ generated by the continuous transformations T, L and D can be represented as a direct product $\mathcal{A} \times \mathcal{A}$ of two copies of the affine group $\mathcal{A}$ acting independently on the right–moving and left–moving waves.* We denote by $\mathcal{G}_1$ the group obtained from $\mathcal{G}_0$ by adjoining R, and by $\mathcal{G}_2$ the group obtained by further adjoining P. $\mathcal{G}_1$ contains negative as well as positive dilations, but the signs of the dilations in the two components are equal. (They can be made independent by adjoining yet another discrete symmetry, namely $(x,t) \to (t,x)$, but we resist the temptation.) Note that the subspaces of right– and left–moving waves are invariant under $\mathcal{G}_1$ but not under $\mathcal{G}_2$.

Since the affine group is closely related to wavelets, the decomposition $\mathcal{G}_0 \approx \mathcal{A} \times \mathcal{A}$ suggests that we apply separate wavelet analyses to f_+ and f_-. Actually, we shall see that more can be done, due to the fact that the wave equation has another, quite unexpected, symmetry.

4.3. Hilbert structures on solutions

So far we have dealt exclusively with solutions in the space–time domain, where symmetries have a direct and intuitive meaning. We now wish to introduce inner products on solutions which make the above symmetry transformations unitary. For this purpose, it is necessary to venture into the Fourier domain and, later, also into the domain of complex space–time, invoking the Analytic–Signal transform introduced earlier.

Let $f(x,t)$ be a solution transforming under $\mathcal{G}_2$ with weight κ. Representing $f_\pm$ formally by Fourier integrals, we obtain

$$f(x,t) = \int_{-\infty}^{\infty} dp\, e^{2\pi i p u}\, \hat{f}_+(p) + \int_{-\infty}^{\infty} dp\, e^{2\pi i p v}\, \hat{f}_-(p). \tag{114}$$

The symmetry operations defined earlier can now be represented in Fourier space. Translations act by

$$T(u_0, v_0): \ \hat{f}_+(p) \to e^{-2\pi i p u_0}\, \hat{f}_+(p), \qquad \hat{f}_-(p) \to e^{-2\pi i p v_0}\, \hat{f}_-(p), \tag{115}$$

Lorentz transformations act by

$$L(\lambda): \ \hat{f}_+(p) \to \lambda^{1-\kappa}\, \hat{f}_+(\lambda p), \qquad \hat{f}_-(p) \to \lambda^{\kappa-1}\, \hat{f}_-(\lambda^{-1} p), \tag{116}$$

space reflection acts by

$$P: \ \hat{f}_+(p) \to \hat{f}_-(-p), \qquad \hat{f}_-(p) \to \hat{f}_+(-p), \tag{117}$$

total reflection acts by

$$R: \ \hat{f}_\pm(p) \to \hat{f}_\pm(-p) \tag{118}$$

and dilations act by

$$D(\alpha): \ \hat{f}_\pm(p) \to \alpha^{1-\kappa} \hat{f}_\pm(\alpha p). \tag{119}$$

Let $s \equiv 2\Re(\kappa) - 1 \geq 0$ and define $\mathcal{H}_s$ to be the the Hilbert space of all 'functions' $g(x)$ such that

$$\|g\|_s^2 \equiv \int_{-\infty}^{\infty} dp \ |p|^{-s} \, |\hat{g}(p)|^2 < \infty. \tag{120}$$

In the terminology of Battle [3], this is the "massless Sobolev space of degree $-s/2$." The norm of a solution $f(x,t)$ (which may be identified with the corresponding spinor $\psi \in \mathcal{D}_s$) is then given by

$$|||f|||_s^2 \equiv |||\psi|||_s^2 = \|f_+\|_s^2 + \|f_-\|_s^2. \tag{121}$$

From the above actions it easily follows that the transformations T, L, R, P and D act unitarily on $\mathcal{D}_s$, thus giving a unitary representation of $\mathcal{G}_2$. Denote by $\mathcal{H}_s^+$ and $\mathcal{H}_s^-$ the subspaces of $\mathcal{H}_s$ with support on $[0, \infty)$ and $(-\infty, 0]$, respectively. (Note that since $s \geq 0$, the value of $\hat{f}_\pm$ at $p = 0$ is unimportant; when $s = 0$, the origin has zero measure, and when $s > 0$, $f_\pm \in \mathcal{H}_s$ implies $\hat{f}_\pm(0) = 0$.) The continuous symmetries T, L and D all preserve the sign of p, hence the group $\mathcal{G}_0$ generated by them leaves invariant the subspaces of solutions

$$\begin{aligned}
f_+^+(x,t) &= \int_0^\infty dp\, e^{2\pi i p u}\, \hat{f}_+(p) \\
f_+^-(x,t) &= \int_{-\infty}^0 dp\, e^{2\pi i p u}\, \hat{f}_+(p) \\
f_-^-(x,t) &= \int_0^\infty dp\, e^{2\pi i p v}\, \hat{f}_-(p) \\
f_-^+(x,t) &= \int_{-\infty}^0 dp\, e^{2\pi i p v}\, \hat{f}_-(p),
\end{aligned} \tag{122}$$

where the superscipts denote positive– and negative–frequency components, as in Section 3. The complete solution is $f = f_+^+ + f_+^- + f_-^- + f_-^+$. Note that $\hat{f}_-(p)$ with $p > 0$ represents a *negative–frequency* component since $v = x - t$. Thus $f_+^\pm \in \mathcal{H}_s^\pm$ but $f_-^\pm \in \mathcal{H}_s^\mp$. The decomposition $\mathcal{H}_s = \mathcal{H}_s^+ \oplus \mathcal{H}_s^-$ therefore gives a corresponding decomposition

$$\mathcal{D}_s = \mathcal{D}_{s+}^{+} \oplus \mathcal{D}_{s+}^{-} \oplus \mathcal{D}_{s-}^{+} \oplus \mathcal{D}_{s-}^{-}\,, \tag{123}$$

where $\mathcal{D}_{s+}^{\pm} = \mathcal{H}_s^{\pm} \oplus \{0\}$ and $\mathcal{D}_{s+}^{\pm} = \{0\} \oplus \mathcal{H}_s^{\mp}$. Let $\mathcal{D}_s^{\pm} = \mathcal{D}_{s+}^{\pm} \oplus \mathcal{D}_{s-}^{\pm}$. The restriction of the representation of $\mathcal{G}_2$ to $\mathcal{G}_0$ leaves invariant all four subspaces $\mathcal{D}_{s\sigma}{}^{\tau}$ $(\sigma, \tau = \pm)$, hence it is reducible. When R is included, only the subspaces $\mathcal{D}_{s+}$ and $\mathcal{D}_{s-}$ remain invariant. When P is further included, no invariant subspaces remain and we have an irreducible unitary representation of $\mathcal{G}_2$ on $\mathcal{D}_s$. This shows that the discrete symmetries P and R serve to 'weave' the four representations on $\mathcal{D}_{s\pm}^{\pm}$, $\mathcal{D}_{s\pm}^{\mp}$ (which are associated with the affine group rather than the wave equation) into a single representation characterising the wave equation as a whole. This was the purpose for which they were introduced.

However, distinct values of κ are not significantly different at this point. For if f has weight κ, let $\kappa' \in \mathbb{C}$ with $s' \equiv 2\Re(\kappa') - 1$ and

$$\hat{g}_{\pm}(p) \equiv |p|^{\kappa' - \kappa}\, \hat{f}_{\pm}(p). \tag{124}$$

Then the corresponding solution $g(x,t)$ has weight κ' and, furthermore, the map $f \to g$ is unitary from $\mathcal{D}_s$ onto $\mathcal{D}_{s'}$. This shows that the UIR's with any two values of κ are unitarily equivalent. If no more could be said, we might as well restrict ourselves to a single value of κ, since solutions of arbitrary weight can be obtained by the above method. (Recall that we could shift the weight *up* by a positive integer by applying a differential operator; the above generalizes this process to arbitrary shifts.) Actually, it turns out that for certain special values of κ, the group of symmetries can be significantly enlarged, and distinct values of κ then give unitarily inequivalent representations of the larger group. Unlike $\mathcal{A}$, however, the larger group no longer acts simply on the Fourier space; instead, its natural domain of action is the *complexified space–time* associated with the AST. For this reason it becomes necessary to re–express the norm of $\mathcal{D}_s$ in terms of the AST $\tilde{f}$ of f, as will be done next.

4.4. Norms in terms of analytic signals

We wish to give the norms defined above expressions which are local in space, that is, involve only values of $f(x,t)$ at an arbitrary fixed time t, say $t = 0$. Since a knowledge of both $f(x,0)$ and $\partial_t f(x,0)$ is required to determine f, we cannot expect such expressions to characterize the complete solution f. On the other hand, the positive– and negative–frequency parts $f^{\pm}$ of f satisfy the first–order *pseudo–differential* equation

$$-i\partial_t f^{\pm} = \pm\sqrt{-\partial_x^2}\, f^{\pm}\,, \tag{125}$$

hence are determined by their initial values $f^{\pm}(x,0)$. (Thus, instead of using $f(x,0)$ and $\partial_t f(x,0)$ as initial data, we use the symmetric pair $f^{\pm}(x,0)$.) We therefore consider separately the norms on $\mathcal{H}_s^{+}$ and $\mathcal{H}_s^{-}$, given respectively by

$$
\begin{aligned}
|||f^+|||_s^2 &= \int_{-\infty}^{\infty} dp\, |p|^{-s} \left[\theta(p)\, |\hat{f}_+(p)|^2 + \theta(-p)\, |\hat{f}_-(p)|^2\right] \\
&= \int_0^{\infty} dp\, p^{-s} \left[|\hat{f}_+(p)|^2 + |\hat{f}_-(-p)|^2\right], \\
|||f^-|||_s^2 &= \int_{-\infty}^{\infty} dp\, |p|^{-s} \left[\theta(-p)\, |\hat{f}_+(p)|^2 + \theta(p)\, |\hat{f}_-(p)|^2\right] \\
&= \int_0^{\infty} dp\, p^{-s} \left[|\hat{f}_+(-p)|^2 + |\hat{f}_-(p)|^2\right].
\end{aligned}
\tag{126}
$$

The separation of positive and negative frequencies is characteristic of the AST; hence let us consider the AST of the complete solution f:

$$
\begin{aligned}
\tilde{f}(x + ix', t + it') &\equiv \frac{1}{2\pi i} \int_{-\infty}^{\infty} \frac{d\tau}{\tau - i}\, f(x + \tau x', t + \tau t') \\
&= \frac{1}{2\pi i} \int_{-\infty}^{\infty} \frac{d\tau}{\tau - i}\, [f_+(u + \tau u') + f_-(v + \tau v')] \\
&= \tilde{f}_+(u + iu') + \tilde{f}_-(v + iv'),
\end{aligned}
\tag{127}
$$

where the imaginary parts of the space–time and light–cone coordinates are related by $u' = x' + t'$, $v' = x' - t'$ and $\tilde{f}_\pm$ are the AST's of $f_\pm$. Recall from Section 3 that the AST's $\tilde{f}^\pm$ of $f^\pm(x,t)$ are obtained by restricting $\tilde{f}$ to the forward and backward tubes $\mathcal{T}^\pm$, where $\pm t' > |x'|$. This suggests that the AST is well–suited for the analysis of the above norms, since it naturally breaks f up into its four components.

The symmetry operations on real space–time extend to the complexified space–time by $\mathbb{C}$–linearity, and these extensions induce transformations on $\tilde{f}$ in the obvious way. For example,

$$
T(u_0, v_0):\ \tilde{f}_+(u + iu') \to \tilde{f}_+(u - u_0 + iu'), \quad \tilde{f}_-(v + iv') \to \tilde{f}_+(v - v_0 + iv'). \tag{128}
$$

(Note that the parameters of the symmetries are still real; *e.g.*, we do not consider complex translations.)

Since only the combinations $(x + ix') \pm (t + it')$ enter into $\tilde{f}$, it suffices to set $x' = 0$ and $t = 0$. This is called the *Euclidean region* in quantum field theory, since the Lorentzian metric $x^2 - t^2$ becomes $x^2 + t'^2$, which is Euclidean. (See Glimm and Jaffe [10].) Note that

$$
\tilde{f}(x, it') = \tilde{f}_+(x + it') + \tilde{f}_-(x - it') \equiv \tilde{f}_+(z) + \tilde{f}_-(z^*) \tag{129}
$$

is now presented as a sum of analytic and anti–analytic functions of $z \equiv x + it'$, as befits a solution of the "analytically continued" wave equation

$$\left[\frac{\partial^2}{\partial x^2}+\frac{\partial^2}{\partial t'^2}\right]\tilde{f}(x,it')=0, \tag{130}$$

which states that $\tilde{f}(x,it')$ is harmonic. In terms of the Fourier transforms,

$$\tilde{f}(x,it')=\int_{-\infty}^{\infty}dp\,\left[\theta(pt')\,e^{2\pi ipz}\,\hat{f}_+(p)+\theta(-pt')\,e^{2\pi ipz^*}\,\hat{f}_-(p)\right]. \tag{131}$$

Hence the initial values of $f^\pm$ are given by

$$\begin{aligned}F^\pm(x)\equiv f^\pm(x,0)&=\lim_{\pm t'\to 0+}\tilde{f}(x,it')\equiv\tilde{f}(x,\pm i0)\\&=\int_{-\infty}^{\infty}dp\,e^{2\pi ipx}\left[\theta(\pm p)\,\hat{f}_+(p)+\theta(\mp p)\,\hat{f}_-(p)\right],\end{aligned} \tag{132}$$

from which

$$\hat{F}^\pm(p)=\theta(\pm p)\,\hat{f}_+(p)+\theta(\mp p)\,\hat{f}_-(p). \tag{133}$$

Returning to the norms, consider first the simplest case $s=0$ (*i.e.*, $\kappa=\frac{1}{2}+i\mu$, with μ real):

$$\begin{aligned}|||f^\pm|||_0^2&=\int_{-\infty}^{\infty}dp\,\left[\theta(\pm p)\,|\hat{f}_+(p)|^2+\theta(\mp p)\,|\hat{f}_-(p)|^2\right]\\&=\int_{-\infty}^{\infty}dp\,|\hat{F}^\pm(p)|^2\\&=\int_{-\infty}^{\infty}dx\,|F^\pm(x)|^2\end{aligned} \tag{134}$$

by Plancherel's theorem. We have therefore proved

Theorem 6. *The norms in the positive– and negative–frequency subspaces $\mathcal{D}_0^\pm$ of $\mathcal{D}_0$ can be be written as*

$$|||f^\pm|||_0^2=\int_{-\infty}^{\infty}dx\,|\tilde{f}(x,\pm i0)|^2, \tag{135}$$

and the complete norm in $\mathcal{D}_0$ is therefore

$$|||f|||_0^2=\int_{-\infty}^{\infty}dx\,\left[|\tilde{f}(x,i0)|^2+|\tilde{f}(x,-i0)|^2\right]. \tag{136}$$

This is not quite local in space since it involves the boundary values $\tilde{f}(x,\pm i0)$. Rather, it is 'local' in the *complex* space generated by the AST. We refer to this property as *pseudo–locality;* it stands in the same relation to locality as the pseudo–differential equation (125) stands to the wave equation. Since

$f(x,0) = \tilde{f}(x,i0) + \tilde{f}(x,-i0)$ and $\tilde{f}(x,it')$ may be regarded as a regularized version of $f(x,0)$, the term 'pseudo–locality' seems appropriate.

Next, fix $\kappa \in \mathbb{C}$ with $s > 0$. Can we obtain pseudo–local expressions for the norms with $s > 0$? We have

$$\begin{aligned} |||f^{\pm}|||_s^2 &= \int_{-\infty}^{\infty} dp\, |p|^{-s} \left[\theta(\pm p)\, |\hat{f}_+(p)|^2 + \theta(\mp p)\, |\hat{f}_-(p)|^2\right] \\ &= \int_0^{\infty} dp\; p^{-s} \left[|\hat{f}_+(\pm p)|^2 + |\hat{f}_-(\mp p)|^2\right]. \end{aligned} \tag{137}$$

Theorem 7. *For arbitrary $s > 0$, the norms $|||f^{\pm}|||_s$ in $\mathcal{D}_s^{\pm}$ have the following pseudo–local expressions in terms of the restrictions of $\tilde{f}$ to the Euclidean region:*

$$|||f^{\pm}|||_s^2 = N_s \int_{-\infty}^{\infty} dx \int_0^{\infty} dt'\; t'^{s-1}\, |\tilde{f}(x, \pm it')|^2, \tag{138}$$

where

$$N_s = \frac{(4\pi)^s}{\Gamma(s)}. \tag{139}$$

Hence the norm in $\mathcal{D}_s$ is given by

$$|||f|||_s^2 = N_s \int_{-\infty}^{\infty} dx \int_{-\infty}^{\infty} dt'\; |t'|^{s-1}\, |\tilde{f}(x, it')|^2. \tag{140}$$

Proof: We prove the theorem for $|||f^{+}|||_s$. The proof for $|||f^{-}|||_s$ is similar. By Equation (131), $\tilde{f}(x,it')$ can be written as an inverse Fourier transform

$$\tilde{f}(x,it') = \left[\theta(pt')\, e^{-2\pi pt'}\, \hat{f}_+(p) + \theta(-pt')\, e^{2\pi pt'}\, \hat{f}_-(p)\right]\check{}\,(x), \tag{141}$$

hence by Plancherel's theorem,

$$\begin{aligned} \int_{-\infty}^{\infty} dx\, |\tilde{f}(x,it')|^2 &= \int_{-\infty}^{\infty} dp\, \left[\theta(pt')\, e^{-4\pi pt'} |\hat{f}_+(p)|^2 + \theta(-pt')\, e^{4\pi pt'} |\hat{f}_-(p)|^2\right] \\ &= \int_{-\infty}^{\infty} dp\; \theta(pt')\, e^{-4\pi pt'} \left[|\hat{f}_+(p)|^2 + |\hat{f}_-(-p)|^2\right]. \end{aligned} \tag{142}$$

When $t' > 0$, then $p > 0$ in the last integral. Integrating over t' with the factor t'^{s-1}, exchanging the order of integration on the right–hand side and using

$$\int_0^{\infty} dt'\; t'^{s-1}\, e^{-4\pi pt'} = (4\pi p)^{-s}\, \Gamma(s) = N_s^{-1}\, p^{-s} \tag{143}$$

for $s > 0$, we obtain

$$N_s \int_{-\infty}^{\infty} dx \int_0^{\infty} dt'\, t'^{s-1}\, |\tilde{f}(x, it')|^2 = \int_0^{\infty} dp\, p^{-s} \left[|\hat{f}_+(p)|^2 + |\hat{f}_-(-p)|^2 \right]$$
$$= |||f^+|||_s^2 \tag{144}$$

as claimed. ■

4.5. The wavelets e_z and their mother

Here we show that Theorem 7 establishes a resolution of the identity I_s in $\mathcal{D}_s$ $(s > 0)$ in terms of wavelets $e_z \in \mathcal{D}_s$ parametrized by $z = x + it'$, $t' \neq 0$, in the manner described in Subsections 2.2 and 3.3. We shall define e_z as an element of $\mathcal{D}_s$ whose inner product with $f \in \mathcal{D}_s$ is the *value* of the AST $\tilde{f}$ at (x, it'), *i.e.*,

$$\tilde{f}(x, it') = \langle\!\langle\, e_z, f \rangle\!\rangle_s \equiv \langle\, e_{z+}, f_+ \rangle_s + \langle\, e_{z-}, f_- \rangle_s\,, \tag{145}$$

where $\langle\, \cdot, \cdot\, \rangle_s$ and $\langle\!\langle \cdot, \cdot \rangle\!\rangle_s$ denote the inner products in $\mathcal{H}_s$ and $\mathcal{D}_s$, respectively. Since e_z is to belong to $\mathcal{D}_s$, it must itself be a solution. Equation (131) shows that its components in Fourier space are

$$\begin{aligned} \hat{e}_{z+}(p) &= \theta(pt')\, |p|^s\, e^{-2\pi i p z^*} \\ \hat{e}_{z-}(p) &= \theta(-pt')\, |p|^s\, e^{-2\pi i p z}. \end{aligned} \tag{146}$$

(Recall that according to our convention, $\langle\, g, f\, \rangle_s$ is anti–linear in g.) Note that $e_{z-} = e_{z^*+}$ as elements of $\mathcal{H}_s$, and that e_{z-} and e_{z+} are orthogonal not only as elements of the direct sum $\mathcal{D}_s = \mathcal{D}_{s+} \oplus \mathcal{D}_{s-}$ (since $e_{z\pm} \in \mathcal{D}_{s\pm}$) but also as elements of $\mathcal{H}_s$. General solutions clearly do not share this property. The positive– and negative–frequency components $e_z^{\pm}$ of e_z are obtained by choosing $t' > 0$ and $t' < 0$, respectively:

$$e_z = \begin{cases} e_z^+ & \text{if } t' > 0 \\ e_z^- & \text{if } t' < 0. \end{cases} \tag{147}$$

Hence

$$\begin{aligned} \hat{e}_{z+}^{+}(p) &= \theta(p)\, p^s\, e^{-2\pi i p z^*}, \quad t' > 0 \\ \hat{e}_{z-}^{+}(p) &= \theta(-p)\, |p|^s\, e^{-2\pi i p z}, \quad t' > 0 \\ \hat{e}_{z+}^{-}(p) &= \theta(-p)\, |p|^s\, e^{-2\pi i p z^*}, \quad t' < 0 \\ \hat{e}_{z-}^{-}(p) &= \theta(p)\, p^s\, e^{-2\pi i p z}, \quad t' < 0. \end{aligned} \tag{148}$$

For $t' > 0$, we have

$$\begin{aligned} |||e_z^+|||_s^2 &= \|e_{z+}^+\|_s^2 + \|e_{z-}^+\|_s^2 \\ &= \int_{-\infty}^{\infty} dp\,|p|^s \left[\theta(p)e^{-4\pi pt'} + \theta(-p)e^{4\pi pt'}\right] \\ &= 2\int_0^{\infty} dp\; p^s\, e^{-4\pi pt'} = \frac{2\Gamma(s+1)}{(4\pi t')^{s+1}}. \end{aligned} \tag{149}$$

Since $e_{z-} = e_{z^*+}$ in $\mathcal{H}_s$, we have for $t' < 0$

$$|||e_z^-|||_s^2 = \frac{2\Gamma(s+1)}{(-4\pi t')^{s+1}}, \tag{150}$$

hence

$$|||e_z|||_s^2 = |||e_z^+|||_s^2 + |||e_z^-|||_s^2 = \frac{2\Gamma(s+1)}{(4\pi|t'|)^{s+1}} \quad \forall\, t' \neq 0, \tag{151}$$

since one of the vectors $e_z^\pm$ vanishes. This proves that $e_z \in \mathcal{D}_s$ as claimed.

Theorem 7 can be given a suggestive form in terms of the orthogonal projection $P(z)$ onto the one–dimensional subspace of e_z in $\mathcal{D}_s$. For $t' \neq 0$,

$$|\tilde{f}(x, it')|^2 = |\langle\!\langle\, e_z, f \rangle\!\rangle_s|^2 = |||e_z|||_s^2 \,\langle\!\langle\, f, P(z) f \rangle\!\rangle_s = \frac{2\Gamma(s+1)}{(4\pi|t'|)^{s+1}} \langle\!\langle\, f, P(z) f \rangle\!\rangle_s\,. \tag{152}$$

Substituting this into Equation (140) and simplifying, we obtain

$$\frac{s}{2\pi}\int_{-\infty}^{\infty} dx \int_{-\infty}^{\infty} \frac{dt'}{t'^2} \langle\!\langle\, f, P(z) f \rangle\!\rangle_s = \langle\!\langle\, f, f \rangle\!\rangle_s\,. \tag{153}$$

Upon polarization, this proves the *weak* operator identity

$$\frac{s}{2\pi}\int_{-\infty}^{\infty} dx \int_{-\infty}^{\infty} \frac{dt'}{t'^2} P(z) = I_s\,, \tag{154}$$

giving a resolution of the identity in $\mathcal{D}_s$ in terms of the projections $P(z)$. Note that $dx\, dt'/t'^2$ is just the left–invariant measure on the affine group. We shall see below that x and t' can indeed be used to parametrize the affine group, with $t' < 0$ corresponding to *negative* dilations. The fact that Equation (154) involves the left–invariant measure on the affine group shows that weight function $|t'|^{s-1}$ in the original measure only served as a normalization factor for e_z, so as to make e_{z+} and e_{z-} analytic in z^* and z, respectively, which in turn makes $\langle\!\langle\, e_{z+}, f \rangle\!\rangle_s \equiv \tilde{f}_+(z)$ and $\langle\!\langle\, e_{z-}, f \rangle\!\rangle_s \equiv \tilde{f}_-(z^*)$ analytic in z and z^*, respectively.

The wavelet e_z can be regarded as a *regularization* of the point x in space, with $|t'|$ as a measure of its diffusion. Hence e_z transforms naturally under those symmetry operations which leave the Euclidean region invariant. These consist of space translations $T(x_0)$, dilations $D(\alpha)$ and the reflections P and R. Clearly the Euclidean region is not invariant under Lorentz transformations and time translations. (It is not difficult to extend the family of wavelets to one parametrized by general points $(x+ix', t+it')$ in complex space–time, with $x' \neq \pm t'$. This extended family is then invariant under $\mathcal{G}_2$, and the above inner products are obtained by simply integrating over the submanifold defined by $x' = t = 0$. Then the invariance of the inner product under the full group $\mathcal{G}_2$, known to hold in the Fourier domain, can be made manifest in the complex space-time domain by introducing conserved currents and using Stokes' theorem. See [16], Subsection 4.5 for a related treatment of the Klein–Gordon equation.)

To see how e_z transforms under a symmetry operation, we need only apply Equation (145). Thus for translations,

$$\begin{aligned}\langle\!\langle\, T(x_0)\, e_z, f \,\rangle\!\rangle_s &= \langle\!\langle\, e_z, T(-x_0) f \,\rangle\!\rangle_s = \left(T(-x_0)\,\tilde{f}\right)(x, it') \\ &= \tilde{f}(x+x_0, it') = \langle\!\langle\, e_{z+x_0}, f \,\rangle\!\rangle_s \,,\end{aligned} \tag{155}$$

hence

$$T(x_0)\, e_z = e_{z+x_0}. \tag{156}$$

Similarly,

$$\begin{aligned}\langle\!\langle\, D(\alpha)\, e_z, f \,\rangle\!\rangle_s &= \langle\!\langle\, e_z, D(\alpha^{-1})\, f \,\rangle\!\rangle_s \\ &= \alpha^{\kappa} \tilde{f}(\alpha x, i\alpha t') = \alpha^{\kappa} \langle\!\langle\, e_{\alpha z}, f \,\rangle\!\rangle_s \,,\end{aligned} \tag{157}$$

hence

$$D(\alpha)\, e_z = \alpha^{\kappa^*} e_{\alpha z}, \qquad \alpha > 0. \tag{158}$$

Since $P^* = P^{-1} = P$ and $(P\tilde{f})(x, it') = \tilde{f}(-x, it')$, we have

$$Pe_z = e_{-z^*}. \tag{159}$$

Finally, $R^* = R^{-1} = R$ and $(R\tilde{f})(x, it') = \tilde{f}(-x, -it')$ implies

$$Re_z = e_{-z}. \tag{160}$$

Thus $T(x_0)$ and $D(\alpha)$ generate a single copy of $\mathcal{A}$ (as opposed to $\mathcal{G}_0 \approx \mathcal{A} \times \mathcal{A}$), and P and R extend this to include space reflections and negative dilations. We may therefore begin with a single *basic* wavelet, say $\phi = e_{i+} \in \mathcal{D}_{s+}$ (left–moving wavelet with $z = i$). Its components in Fourier space are then

$$\hat{\phi}_+(p) = \theta(p)\, p^s\, e^{-2\pi p}, \qquad \hat{\phi}_-(p) \equiv 0. \tag{161}$$

For $z = x + it'$ with $t' > 0$ we obtain all the left–moving wavelets by applying T, D and R to ϕ:

$$
\begin{aligned}
e_{z+}^{+} &= (t')^{-\kappa^*}\, T(x)\, D(t')\, \phi \\
e_{z^*+}^{-} &= (t')^{-\kappa^*}\, T(x)\, R\, D(t')\, \phi,
\end{aligned}
\tag{162}
$$

and the right–moving wavelets are obtained in the same way from $P\phi = e_{i-}$. Note that the choice of ϕ depends only on s and not on the imaginary part of κ. We may regard Equations (162) as providing a *construction* of the UIR with weight κ by giving all of its wavelets.

Modified versions of the left component ϕ_+ of ϕ have already appeared in the literature, in connection with the representation theory of the affine group. Namely, if we absorb the Sobolev weight function $|p|^{-s}$ into the elements g of $\mathcal{H}_s$ by defining $\hat{G}(p) = |p|^{-s/2}\,\hat{g}(p)$ (so that $\hat{G} \in L^2(\mathbb{R}, dp)$), then $\hat{\phi}$ becomes

$$
\hat{\Phi}_+(p) = \theta(p)\, p^{s/2}\, e^{-2\pi p}, \qquad \hat{\Phi}_-(p) \equiv 0. \tag{163}
$$

The function $\hat{\Phi}_+(p)$ with $s = 1$ first appeared in the classical papers of Aslaksen and Klauder [1], where it was used as a 'fiducial vector' to obtain a 'continuous representation' of the affine group. Since those papers contain the first instance of what is now called continuous wavelet analysis, we suggest that ϕ with $s = 1$ deserves to be called the *Mother of all Wavelets.* For $s = 1, 2, 3, \cdots$, $\hat{\Phi}_+(p)$ also appears in the work of Paul [24] in connection with representations of the affine group and their extension to t (see below). It must be noted, however, that in our case the wavelets are *prescribed* by the problem at hand (solving the wave equation and extending the solutions to complex space–time, via the AST) rather than chosen arbitrarily as a convenient family of functions to be used in expansions. This is further discussed in Section 5.

Having established a resolution of unity in terms of the wavelets e_z, let us now investigate the associated *reconstruction* (see Subsection 2.2). The formula corresponding to Equation (22) is

$$
f(x,t) = N_s \int_{\mathbf{R}^2} dx_1\, dt_1'\, |t_1'|^{s-1}\, e_{x_1+it_1'}(x,t)\, \tilde{f}(x_1, it_1'). \tag{164}
$$

This gives an expansion of an arbitrary solution $f \in \mathcal{D}_s$ in terms of the wavelets e_{z_1}, this time expressed in the *real* space–time domain. Let us therefore compute the left–moving wavelet $e_{z_1+}(x,t)$. For simplicity, choose $x_1 = 0$ and $t' > 0$; the other cases ($x_1 \neq 0$, $t_1' < 0$ and right–moving wavelets) can be obtained easily by using $T(x_1)$, R and P. By Equation (146),

$$
e_{it_1'+}(x,t) = \int_0^\infty dp\; p^s\, e^{2\pi i p(u+it_1')} = \frac{\Gamma(s+1)}{(2\pi(t_1' - iu))^{s+1}}. \tag{165}
$$

Hence

$$
\left|e_{it_1'+}(x,t)\right|^2 = \frac{\Gamma(s+1)^2}{(2\pi)^{2s+2}\,({t_1'}^2 + u^2)^{s+1}}. \tag{166}
$$

At time t, this solution is localized in space around $x = -t$, and its width is proportional to t'_1. As $t'_1 \to 0$, the wavelet becomes an infinitely sharp spike at $x = -t$. This is not surprising, since t'_1 acts as a scale parameter for the affine group. Similarly, $e_{x_1+it'_1}(x,t)$ is centered near $x = x_1 - t$.

A direct and important interpretation of t' is that $1/t'$ is proportional to the average frequency $\nu_{s\pm}$, or *color,* in the frequency spectrum of $e_{z\pm}$. This can be seen most easily by noting that the function $p^s \exp(-2\pi p t')$ has a maximum at $p = s/(2\pi t')$. A more precise argument is based on the quantum mechanical notion of *expectation values,* where $|\hat{e}_{z\pm}(p)|^2$ is viewed as an (unnormalized) probability distribution for p with respect to the measure $|p|^{-s}\,dp$. Remembering that the frequency in $\mathcal{D}_{s\pm}$ is $\pm p$, this gives

$$\nu_{s\pm} \equiv \frac{\int_{-\infty}^{\infty} |p|^{-s}\,dp\,(\pm p)\,|\hat{e_{z}}_{\pm}(\mathbf{p})|^2}{\int_{-\infty}^{\infty} |p|^{-s}\,dp\,|\hat{e_{z}}_{\pm}(\mathbf{p})|^2} = \pm\left(-\frac{1}{4\pi}\frac{\partial}{\partial t'}\right)\log\|e_{z\pm}\|_s^2 = \frac{s+1}{4\pi t'}. \tag{167}$$

(Note that $\nu_{s+} = \nu_{s-} \equiv \nu_s$ and, as expected, ν_s has the same sign as t'.) Similarly, one computes the standard deviation in the frequency to be

$$\Delta\nu_s = \sqrt{\left(-\frac{1}{4\pi}\frac{\partial}{\partial t'}\right)^2 \log\|e_{z\pm}\|_s^2} = \frac{\sqrt{s+1}}{4\pi|t'|}. \tag{168}$$

Note that Equation (154) assumes an especially simple form in terms of the variables (x, ν_s):

$$\frac{2s}{s+1}\int_{-\infty}^{\infty} dx \int_{-\infty}^{\infty} d\nu_s\, P\left(x + i\frac{s+1}{4\pi\nu_s}\right) = I_s\,, \tag{169}$$

which gives *a resolution of the identity in $\mathcal{D}_s$ in terms of the orthogonal projections onto the wavelet subspaces parametrized by initial location and color, with all locations and colors given equal a priori probability,* since the measure is Lebesgue! (The appearance of Lebesgue measure is not an accident. The variables (x, ν_s) are *phase-space coordinates*, also famous in symplectic geometry as *Darboux canonical coordinates*, and $dx\,d\nu_s$ is the corresponding *Liouville measure*. For a similar analysis of the wavelets associated with the Klein–Gordon equation, see ref. [16], Subsection 4.4.)

We are now in a position to answer a question which was posed in a more general context in Subsection 3.2: Given a function $F(x,t')$, how can we tell whether $F = \tilde{f}(x, it')$ for some $f \in \mathcal{D}_s$? Suppose this were the case for some positive integer s. From the polarized version of Equation (140), *i.e.*,

$$\langle\!\langle g, f\rangle\!\rangle_s = N_s \int_{\infty}^{\infty} dx \int_{-\infty}^{\infty} dt'\,|t'|^{s-1}\,\tilde{g}(x,it')^*\,\tilde{f}(x,it')\,, \tag{170}$$

we obtain by choosing $g = e_{z_1}$ with $z_1 = x_1 + it'_1, z = x + it'$:

$$\tilde{f}(x_1, it'_1) = N_s \int_{-\infty}^{\infty} dx \int_{-\infty}^{\infty} dt' \, |t'|^{s-1} \langle\!\langle e_{z_1}, e_z \rangle\!\rangle_s \, \tilde{f}(x, it') \,. \tag{171}$$

The function

$$K(z_1, z) \equiv \langle\!\langle e_{z_1}, e_z \rangle\!\rangle_s \tag{172}$$

is a *reproducing kernel* for the function space $\mathcal{F}_s \equiv \{\tilde{f} : f \in \mathcal{D}_s\}$. A computation similar to that for $|||e_z|||_s$ gives

$$K(z_1, z) = \frac{2\theta(t'_1 t')\, \Gamma(s+1)}{(2\pi |t'_1 + t'|)^{s+1}} \Re \left[\left(1 - i \frac{x_1 - x}{t'_1 + t'} \right)^{-s-1} \right]. \tag{173}$$

The integral operator defined by K acts as the orthogonal projection from $L^2(|t'|^{s-1}\, dx\, dt')$ to $\mathcal{F}_s$. The given function F is therefore the AST of some $f \in \mathcal{D}_s$ if and only if it satisfies the *consistency condition*

$$F(x_1, t'_1) = N_s \int_{-\infty}^{\infty} dx \int_{-\infty}^{\infty} dt' \, |t'|^{s-1} K(z_1, z)\, F(x, t') \tag{174}$$

(see [16], Chapter 1 for an exposition of this general idea).

4.6. Extension to $SL(2, \mathbb{R})$

We are at last ready to extend the symmetry group to a larger one that will 'break' the unitary equivalence of the UIR's of $\mathcal{A}$ with different weights. We do so by showing that for $2\kappa = 1, 2, 3, \cdots$, the unitary action of $\mathcal{A}$ on each of the subspaces $\mathcal{D}_{s\pm}$ extends to the group of real 2×2 matrices of unit determinant, and that this extension preserves the positive– and negative–frequency subspaces $\mathcal{D}^+_{s\pm}$ and $\mathcal{D}^-_{s\pm}$ and is therefore irreducible when restricted to these subspaces. The extension can be implemented most easily on the AST's of solutions in $\mathcal{D}_s$. We concentrate on $\mathcal{D}_{s+} \approx \mathcal{H}_s$ for simplicity and denote its elements by g rather than f_+ to avoid a proliferation of indices.

The action of the affine group on $\mathcal{H}_s$ is induced by the map $u \to \alpha u + \beta$ on $\mathbb{R}$, which is but a very special case of an *order–preserving diffeomophism* $h : \mathbb{R} \to \mathbb{R}$; *i.e.*, an invertible map h such that both h and h^{-1} are C^∞ and $dh/du > 0$. Under composition, the set of all such maps forms a group $\mathrm{Diff}_+(\mathbb{R})$, sometimes called the *pseudo–conformal group*, of which $\mathcal{A}$ is a subgroup. Note that whereas it takes only two parameters to specify an affine map, an infinite number of parameters are needed to specify a general diffeomorphism. Nevertheless, for $\kappa = \frac{1}{2}$, the action of $\mathcal{A}$ on $\mathcal{H}_0 = L^2(\mathbb{R})$ can be extended to $\mathrm{Diff}_+(\mathbb{R})$ by defining

$$A(h) : \; g(u) \to \sqrt{\frac{dh^{-1}}{du}}\, g\left(h^{-1}(u)\right), \tag{175}$$

which is unitary and gives a representation since $A(h_1)\,A(h_2) = A(h_1 \circ h_2)$. However, this representation is problematic since general elements of $\text{Diff}_+(\mathbb{R})$ do not preserve the positive– and negative–frequency subspaces of $\mathcal{H}_0$. To see this, recall that g is the boundary–value of its AST $\tilde{g}(z)$ and $\tilde{g} = 0$ in $\mathbb{C}^\mp$ if $g \in \mathcal{H}_s^\pm$. Hence elements of $\mathcal{H}_s^\pm$ are boundary–values of functions analytic in $\mathbb{C}^\pm$. When composed with a diffeomorphism h^{-1} which is not the boundary–value of a function analytic in $\mathbb{C}^\pm$, the result cannot be the boundary–value of a function analytic in $\mathbb{C}^\pm$. ($\text{Diff}_+(\mathbb{R})$ plays important roles in string theory and quantum field theory; there, the fact that it mixes positive and negative frequencies leads to a difficulty known as the "conformal anomaly.") Since the decomposition into positive– and negative–frequency components is fundamental to our approach, the representation theory of $\text{Diff}_+(\mathbb{R})$ will not be pursued further. However, the above discussion suggests that we confine ourselves to diffeomorphisms of $\mathbb{R}$ which extend 'naturally' to $\mathbb{C}$ and whose restrictions to $\mathbb{C}^\pm$ are *holomorphisms* (analytic diffeomorphisms) of $\mathbb{C}^\pm$. That is, we consider symmetry operations on $\mathcal{H}_s$ which are induced from geometric maps of the complex space $\mathbb{C}^+ \cup \mathbb{C}^-$ associated with the AST rather than geometric maps of the real space $\mathbb{R}$. Since only the analytic functions $\tilde{g}(z)$ or their boundary–values $\tilde{g}(x \pm i0)$ enter into the pseudo–local inner products of $\mathcal{H}_s$, we might even get away with maps which have singularities on $\mathbb{R}$. Thus we look for the group of holomorphisms of $\mathbb{C}^\pm$. Now it is well–known that the group of holomorphisms of the Riemann sphere $\widehat{\mathbb{C}} \equiv \mathbb{C} \cup \{\infty\}$ is $SL(2,\mathbb{C})$, the set of all complex 2×2 matrices with unit determinant, acting by fractional–linear (Möbius) transformations. The subgroup preserving $\mathbb{C}^\pm$, hence also the one–point compactification $\widehat{\mathbb{R}} \equiv \mathbb{R} \cup \{\infty\}$, then consists of all such *real* matrices, and is denoted t. We shall write $G \equiv t$ throughout this subsection. An element of G, being a diffeomorphism of $\widehat{\mathbb{R}}$ rather than $\mathbb{R}$, will have a singulariy in $\mathbb{R}$ (the inverse image of ∞) unless it has ∞ as a fixed point.

We begin by constructing two inequivalent UIR's of G which extend the UIR's of $\mathcal{A}$ on $\mathcal{H}_0^\pm$. The matrix

$$\sigma = \begin{pmatrix} a & b \\ c & d \end{pmatrix}, \qquad ad - bc = 1, \tag{176}$$

acts on $\mathbb{C}^+ \cup \mathbb{C}^- = \hat{\mathbb{C}} \setminus \hat{\mathbb{R}}$ by the fractional–linear transformation

$$\sigma :\ z \to \sigma(z) \equiv \frac{az+b}{cz+d} \equiv w. \tag{177}$$

(This action will be derived below.) From the easily derived identity

$$w - w^* = \frac{z - z^*}{|cz+d|^2} \tag{178}$$

it follows that σ leaves $\mathbb{C}^\pm$ invariant. Note that the boundary map $\sigma(u)$ on $\mathbb{R}$ has a singularity at $u = -d/c$, in agreement with the above discussion.

Since $d\sigma(u)/du = (cu+d)^{-2}$, the induced map on boundary–values in $\mathcal{H}_0$ is, according to Equation (175),

$$A(\sigma^{-1}):\ g(u) \to |cu+d|^{-1} g\left(\frac{au+b}{cu+d}\right). \tag{179}$$

The singularity at $u = -d/c$ poses no problem, since C^∞ functions of compact support $g \in \mathcal{D}(\mathbb{R})$ vanish at infinity and $\mathcal{D}(\mathbb{R})$ is dense in $\mathcal{H}_0$. The set of σ's whose boundary maps are non–singular is, in fact, just the affine group $\mathcal{A}$, since $c = 0$ implies

$$\sigma(u) = a^2 u + ab \equiv \alpha u + \beta, \tag{180}$$

giving the relation between the parameters of $\mathcal{A}$ and those of G. To represent $A(\sigma^{-1})$ on $\tilde{g}(z)$, we first replace the multiplier $|cu+d|^{-1}$ with $(cu+d)^{-1}$, which also leads to a representation and, unlike the former, can be extended analytically. This does not affect the norm as $z \to u \in \mathbb{R}$. Thus define

$$B(\sigma^{-1}):\ \tilde{g}(z) \to (cz+d)^{-1} \tilde{g}\left(\frac{az+b}{cz+d}\right). \tag{181}$$

For $z \in \mathbb{C}^\pm$, we obtain the actions on $\mathcal{H}_0^\pm$, which are unitary by the same argument as used for $\mathrm{Diff}_+(\mathbb{R})$. This gives inequivalent UIR's of t on $\mathcal{H}_0^\pm$. Comparison with Bargmann's classification shows that these coincide with the two representations of the "mock discrete series" with $s = 0$ (Lang [20], pp. 120 and 123).

We can arrive at the action of G on $\mathbb{C}^\pm$ (Equation (177)) by the method of cosets. Consider the subroups T and H of G given by

$$\begin{aligned} T &= \left\{ t(u) \equiv \begin{pmatrix} 1 & u \\ 0 & 1 \end{pmatrix} :\ u \in \mathbb{R} \right\} \\ H &= \left\{ h(c,d) \equiv \begin{pmatrix} d^{-1} & 0 \\ c & d \end{pmatrix} :\ d \neq 0 \right\}. \end{aligned} \tag{182}$$

Then T is isomorphic to $\mathbb{R}$, and H is isomorphic to $\mathcal{A}$. Moreover, any element of G with $d \neq 0$ can be written uniquely in the form

$$\sigma \equiv \begin{pmatrix} a & b \\ c & d \end{pmatrix} = \begin{pmatrix} 1 & b/d \\ 0 & 1 \end{pmatrix} \begin{pmatrix} d^{-1} & 0 \\ c & d \end{pmatrix} = t(b/d)\, h(c,d). \tag{183}$$

Hence we may coordinatize G (except for a set of Haar measure zero) by the product of sets TH. (This does not mean that G is a direct product since T and H, as subgroups of G, do not commute.) The space $U \equiv G/H$ of right cosets is then parametrized by $u \in \widehat{\mathbb{R}}$, as follows: If $d \neq 0$, then

$$\sigma H \equiv \{\sigma k :\ k \in H\} = \{t(b/d)\, h(c,d)\, k :\ k \in H\} = t(b/d)\, H \equiv H_{b/d}. \tag{184}$$

The set of matrices with $d = 0$ forms a single coset, which will be denoted by H_∞. Thus $U \approx \widehat{\mathbb{R}}$. Now G acts on U by left multiplication, and this translates into an action of G on $\widehat{\mathbb{R}}$ as follows: If $u \neq \infty$, then

$$\begin{pmatrix} a & b \\ c & d \end{pmatrix} H_u = \begin{pmatrix} a & b \\ c & d \end{pmatrix} \begin{pmatrix} 1 & u \\ 0 & 1 \end{pmatrix} H = \begin{pmatrix} a & au+b \\ c & cu+d \end{pmatrix} H = H_{\sigma(u)}, \tag{185}$$

with $\sigma(u)$ as given by Equation (177). The action on H_∞ is similarly found to be

$$\begin{pmatrix} a & b \\ c & d \end{pmatrix} H_\infty = H_{a/c}. \tag{186}$$

(If $c = 0$, then H_∞ is invariant under σ.)

Thus we see that the action of G on $\mathbb{R}$ (more precisely, on $\widehat{\mathbb{R}}$) can be derived from the action of G on itself. Similarly, the representation of G on $L^2(\mathbb{R})$ obtained above can be viewed as a group–theoretical construction: The Lebesgue measure du on $\mathbb{R}$ can be extended to $U = \widehat{\mathbb{R}}$ by letting $\{\infty\}$ have measure zero; this extension $d\hat{u}$ is quasi–invariant under the action of G, and functions in $L^2(\mathbb{R}) = L^2(U, d\hat{u})$ may be interpreted as functions on G which are constant on each coset H_u. The above representation can then be interpreted as being "induced" (Lang [20], chapter 3) from the identity representation of H. Since H is not compact, non–zero functions on G which are constant on the cosets H_u are not square–integrable on G. This proves that the above representation is not square–integrable on G. That means that we cannot use it to obtain a resolution of the identity in $L^2(G)$.

We now construct some other representations of G, with weights κ other than $\frac{1}{2}$, which *do* turn out to be square–integrable on G. Let κ be real, so that $s = 2\kappa - 1$. Define the action of G on $\mathcal{H}_s$ by

$$B(\sigma^{-1}): \ \tilde{g}(z) \to (cz+d)^{-s-1}\, \tilde{g}\left(\frac{az+b}{cz+d}\right). \tag{187}$$

We must have $s \in \mathbb{Z}$ in order to preserve analyticity. For $\kappa = \frac{1}{2}$, this coincides with the earlier action. Note that for $a = 1/\sqrt{\alpha}$, $b = -\beta/\sqrt{\alpha}$, $c = 0$ and $d = \sqrt{\alpha}$ we obtain

$$B(\sigma^{-1}): \ \tilde{g}(z) \to \alpha^{-\kappa}\, \tilde{g}(\alpha^{-1}(z-\beta))\,, \tag{188}$$

which shows that the action restricts to that of the affine group on $\mathcal{H}_s$.

To show that $B(\sigma^{-1})$ acts unitarily, we need to prove that it preserves the pseudo–local norm introduced in Theorem 7. There we found that for G replaced by its subgoup $\mathcal{A}$, we needed $s > 0$. Since s must now, in addition, be an integer, we therefore have $2\kappa - 1 = s = 1, 2, 3, \cdots$. We must show that

$$\int_{\mathbb{C}} d^2z\, |\Im(z)|^{s-1}\, |cz+d|^{-2s-2} \left| \tilde{g}\left(\frac{az+b}{cz+d}\right)\right|^2 = \int_{\mathbb{C}} d^2z\, |\Im(z)|^{s-1}\, |\tilde{g}(z)|^2, \tag{189}$$

where d^2z denotes Lebesgue measure in $\mathbb{C}$. Let $w = \sigma(z) = (az+b)/(cz+d)$. Then $dw/dz = (cz+d)^{-2}$, hence

$$d^2w = |cz+d|^{-4}\, d^2z, \quad \text{and} \quad \Im(w) = |cz+d|^{-2}\Im(z), \tag{190}$$

where the second equality follows from Equation (178). Thus

$$\int_{\mathbb{C}} d^2z\, |\Im(z)|^{s-1}\, |cz+d|^{-2s-2}\, |\tilde{g}(w)|^2 = \int_{\mathbb{C}} d^2w\, |\Im(w)|^{s-1}\, |\tilde{g}(w)|^2 = \|g\|_s^2, \tag{191}$$

proving the result. The unitary representation on $\mathcal{H}_s$ decomposes into a direct sum of UIR's on $\mathcal{H}_s^{\pm}$. All these representations are inequivalent, and they are known collectively as the *discrete series* (Gelfand et al. [9], Lang [20]).

The action of t on the wavelets $e_{z+} \in \mathcal{D}_{s+} \approx \mathcal{H}_s$ is easily computed. By the unitarity of $B(\sigma)$, we have

$$\begin{aligned}\langle B(\sigma)\, e_{z+}, g\rangle_s &= \langle e_{z+}, B(\sigma^{-1})\, g\rangle_s = \left(B(\sigma^{-1})\, \tilde{g}\right)(z) \\ &= (cz+d)^{-s-1}\, \tilde{g}(\sigma(z)) = (cz+d)^{-s-1}\, \langle e_{\sigma(z)+}, g\rangle_s,\end{aligned} \tag{192}$$

hence

$$B(\sigma)\, e_{z+} = (cz^* + d)^{-s-1}\, e_{\sigma(z)+}\, . \tag{193}$$

The representations of the discrete series are square–integrable over G. The subgroup of G which leaves $z = \pm i$ invariant is $SO(2)$, hence $\mathbb{C}^{\pm} \approx G/SO(2)$. Since $SO(2)$ is compact (unlike its counterpart H for $s = 0$), the norms in $\mathcal{H}_s^{\pm}$ can be rewritten as integrals over all of G rather than just the homogeneous spaces $\mathbb{C}^{\pm}$.

Returning to the wave equation, we obtain mutually inequivalent UIR's of what we shall call the *restricted conformal group* $\mathcal{C}_0 \equiv G\times G$ on $\mathcal{D}_{s+}^{+}, \mathcal{D}_{s+}^{-}, \mathcal{D}_{s-}^{-}$ and $\mathcal{D}_{s-}^{+}$ $(s = 0, 1, 2, \cdots)$. When the total reflection R is included, we obtain inequivalent UIR's of the resulting group $\mathcal{C}_1$ on $\mathcal{D}_{s+}$ and $\mathcal{D}_{s-}$. When the space reflection P is further included, we obtain a single set of mutually inequivalent UIR's of the resulting group $\mathcal{C}_2$ on $\mathcal{D}_s$. As they did for $\mathcal{G}_0$, the reflections unify the four subspaces $\mathcal{D}_{s\pm}^{\pm}$ into a single one representing the wave equation as a whole.

§5. Concluding remarks: dedicated wavelets

The wavelet analysis developed in [16] for the Klein–Gordon equation has the interesting feature that the wavelets are "dedicated" to the equation rather than being merely a convenient set of functions to be used in expansions. (This is somewhat reminiscent of the situation in the spectral theorem, where expansions are customized to a given operator.) The reward for such dedication is that symmetry operations (such as translations, rotations, Lorentz transformations, and even time evolution) take wavelets to wavelets. This has the practical consequence of making the description economical and precise. For example, wavelets obtained by taking tensor products of one–dimensional wavelets cannot be rotated; consequently, a function consisting of but a few wavelets in one coordinate system is represented (inefficiently) by a combination of many wavelets in a rotated coordinate system. Similar considerations apply to the dedicated wavelets associated with the wave equation in $\mathbb{R}^2$ developed in the last section. But since there is now only one space dimension, the results are somewhat less dramatic: There are only two directions, left and right, giving rise to the labeling e_{z+} and e_{z-}. We believe that the results of Section 4 generalize to $d > 1$ space dimensions, where the set of directions is parametrized by S^{d-1}. (The case $d = 1$ is degenerate since $S^0 = \{\pm 1\}$ is disconnected.) Work on this is in progress.

References

1. Aslaksen, E. W. and J. R. Klauder, Unitary representations of the affine group, *J. Math. Phys.* **9** (1968), 206–211; Continuous representation theory using the affine group, *J. Math. Phys.* **10** (1969), 2267–2275.
2. Battle, G., A block spin construction of ondelettes. Part I: Lemarié functions, *Comm. Math. Phys.* **110** (1987), 601–615.
3. Battle, G., Wavelets: A renormalization point of view, in *Wavelets and Their Applications*, G. Beylkin, R. R. Coifman, I. Daubechies, S. Mallat, Y. Meyer, L. A. Raphael, and M. B. Ruskai (eds.), Jones and Bartlett, Cambridge, MA, 1992, to appear.
4. Born, M. and E. Wolf, *Principles of Optics*, Fifth Edition, Pergamon Press, Oxford, 1975.
5. Daubechies, I., A. Grossmann, and Y. Meyer, Painless nonorthogonal expansions, *J. Math. Phys.* **27** (1986), 1271–1283.
6. Daubechies, I., Orthonormal bases of compactly supported wavelets, *Comm. Pure and Appl. Math.* **41** (1988), 909–996.
7. Daubechies, I., *Wavelets*, Lecture notes of CBMS-NSF Regional Conference, University of Lowell, SIAM, 1992, to appear.
8. Gabor, D., Theory of communications, *J. IEE (London)* **93** (1946), 429–457.
9. Gelfand, I. M., M. I. Graev, and N. Ya. Vilenkin, *Generalized Functions*, vol. 5, Academic Press, New York, 1966.

10. Glimm, J. and A. Jaffe, *Quantum Physics: A Functional Integral Point of View,* Second Edition, Springer, New York, 1987.
11. Helgason, S., *Groups and Geometric Analysis,* Academic Press, New York, 1984.
12. Holschneider, M., Inverse Radon transforms through inverse wavelet transforms, CNRS–Luminy, 1990, preprint.
13. Kaiser, G., Phase–Space approach to relativistic quantum mechanics, Ph. D. Thesis, Mathematics Department, University of Toronto, 1977.
14. Kaiser, G, Phase–space approach to relativistic quantum mechanics. Part I: Coherent–state representation of the Poincaré group, *J. Math. Phys.* **18** (1977), 952–959; part II: Geometrical aspects, *J. Math. Phys.* **19** (1978), 502–507; part III: Quantization, relativity, localization and gauge freedom, *J. Math. Phys.* **22** (1981), 705–714.
15. Kaiser, G., Quantized fields in complex spacetime, *Annals of Physics* **173** (1987), 338–354.
16. Kaiser, G., *Quantum Physics, Relativity, and Complex Spacetime: Towards a New Synthesis,* North–Holland, Amsterdam, 1990.
17. Kaiser, G., Generalized wavelet transforms. Part I: The windowed X–Ray transform, Technical Reports Series #18, University of Lowell, 1990; part II: The multivariate analytic–signal transform, Technical Reports Series #19, University of Lowell, 1990.
18. Kaiser, G., An algebraic theory of wavelets. Part I: Operational calculus and complex structure, *SIAM J. Math. Anal.* **23** (1) (1992), to appear.
19. Klauder, J. R. and E. C. G. Sudarshan, *Fundamentals of Quantum Optics,* Benjamin, New York, 1968.
20. Lang, S., $SL_2(\mathbb{R})$, Springer, New York, 1985.
21. Lemarié, P., Ondelettes à localisation exponentielle, *J. Math. Pures et Appl.* **67** (1988), 227–236.
22. Meyer, Y. *Séminaire Bourbaki* **38** (1985–86), 662.
23. Meyer, Y., *Ondelettes et Opérateurs,* in two volumes, Hermann, Paris, 1990.
24. Paul, T., Functions analytic on the half–plane as quantum mechanical states, *J. Math. Phys.* **25** (1984), 3252–3263.
25. Rudin, W., *Fourier Analysis on Groups,* Interscience, New York, 1960.
26. Stein, E., *Singular Integrals and Differentiability Properties of Functions,* Princeton University Press, Princeton, 1970.
27. Stein, E. and G. Weiss, *Introduction to Fourier Analysis on Euclidean Spaces,* Princeton University Press, Princeton, 1971.
28. Streater, R. F. and A. S. Wightman, *PCT, Spin & Statistics, And All That,* Benjamin, New York, 1964.

Gerald A. Kaiser
Department of Mathematics
University of Massachusetts
Lowell, MA 01854, USA
kaiserg@woods.ulowell.edu

Raymond F. Streater
Department of Mathematics
King's College, London WC2 2LS
England
UDAH110@oak.cc.kcl.ac.uk

Part VI

Theory of Sampling and Interpolation

Irregular Sampling and Frames

John J. Benedetto

Abstract. A theory of irregular sampling is developed for the class of real sampling sequences for which there are L^2-convergent sampling formulas. The sampling sequences are effectively characterized, and the formulas are accompanied by methods of computing coefficients. These sampling formulas depend on the theory of coherent state (Gabor) frames and an analysis of the inverse frame operator. The results include regular sampling theory and the irregular sampling theory of Paley-Wiener, Levinson, Beutler, and Yao-Thomas. The chapter also presents a new aliasing technique, perspective on stability and uniqueness, and references to recent contributions by others.

This chapter consists of eleven sections. The titles of the sections are listed as follows.

1. Introduction
2. The classical sampling theorem
3. The Paley-Wiener theorem
4. Frames and exact frames
5. Regular sampling and frames
6. Irregular sampling and exact frames
7. The Duffin-Schaeffer theorem and frame conditions
8. Irregular sampling and frames
9. Stability and uniqueness
10. Irregular sampling - approaches and topics
11. Notation

§1. Introduction

The subject of sampling, whether as method, point of view, or theory, weaves its fundamental ideas through a panorama of engineering, mathematical, and scientific disciplines. Results have been discovered in one or another discipline independently of similar results in other disciplines. The spectacular expositions and research-tutorials of Bützer et al. [21] and Higgins [31] not

Wavelets–A Tutorial in Theory and Applications
C. K. Chui (ed.), pp. 445–507.

ISBN 0-12-174590-2

only establish the pervasiveness of sampling, but leave the reader with a sense that the time has arrived for an efficacious synthesis of the subject. As an example from the past, Schwartz's treatise [57] on distribution theory was a compendium of diverse past accomplishments, a unification of technologies, an original formulation of ideas and techniques both new and old, and a research manual leading to new mathematics and applications (see [47] for a similar phenomenon in wavelet theory at the present time). The stage is set for a comparable development in sampling theory.

Our more focused and realistic goal in this chapter is to use the theory of frames to formulate applicable sampling formulas in an elementary and unified way for irregularly spaced sampling sequences. The treatment is general with regard to the spacing of the sampling sequences, and the sampling formulas are mathematical theorems when mild hypotheses are made. Our basic technique involves an analysis of the inverse frame operator, and the formulas include a good deal of the existing theory as well as new material. We have also developed an algorithm associated with the sampling formulas, and there are undoubtedly other algorithms for various specific applications. The theory of frames is due to Duffin and Schaeffer [24], and there is an extraordinary presentation by Young [64] on the subject. Our basic formulas are stated in Section 8, and these results, as well as other observations throughout the chapter, are part of a collaborative venture with Heller, *e.g.*, [9], [10]. Item [9] is referenced in various sections, and item [10] deals with multidimensional sampling and applications of our algorithm.

We do not repeat material covered in [21] and [31], except to round out a discussion now and then. In particular, we reference, rather than prove, certain well-known and occasionally difficult theorems.

The titles for Sections 2-4 are self-explanatory. In Section 5 we utilize coherent state frames to obtain the results of Section 2 in another way. This leads to a new point of view on aliasing, and sets the stage for the case of irregular sampling in Sections 6 and 8. Section 6 presents the sampling theory of Paley- Wiener, Levinson, Beutler, and Yao-Thomas, but it does so in terms of the theory of exact frames and the inverse frame operator. Section 7 provides results so that the formulas in Section 8 can be implemented. Section 9 is perhaps the most idiosyncratic section of the chapter, and it deals with stability and uniqueness. In Section 9, we have also chosen to point out the deep work for Beurling [12], [13], Beurling and Malliavin [14], [15], and Landau [41], and to hint at exciting questions which remain to be answered.

Section 10 is divided into two parts. The first part is brief but important. It lists references to new ideas in irregular sampling. It is, of course, dangerous to make such a list, since I am undoubtedly unaware of other excellent work besides that listed. However, irregular sampling, in the context of coherent states and wavelets, is a subject whose time has come; and these new contributions fit into the theme of the chapter. The second part of Section 10 shows how aliasing problems provide a transition from the coherent state setting of this chapter to the threshold of the wavelet and wavelet packet setting of some of our other work.

Besides the usual notation in analysis as found in the books by Hörmander [32], Katznelson [38], Schwartz [57], and Stein and Weiss [59], we use the conventions and notation described in Section 11.

§2. The classical sampling theorem

$L^1(\mathbb{R})$ is the space of complex-valued integrable functions f defined on the real-line $\mathbb{R}$. The L^1*-norm* of $f \in L^1(\mathbb{R})$ is

$$\|f\|_1 = \int |f(t)|\, dt < \infty,$$

where "$\int$" designates integration over $\mathbb{R}$. The *Fourier transform* $\hat{f}$ of $f \in L^1(\mathbb{R})$ is defined as

$$\hat{f}(\gamma) = \int f(t) e^{-2\pi i t\gamma}\, dt$$

for each $\gamma \in \hat{\mathbb{R}}(= \mathbb{R})$. $A(\hat{\mathbb{R}})$ is the space of such Fourier transforms. The *A-norm* of $F = \hat{f} \in A(\hat{\mathbb{R}})$ is

$$\|F\|_A \equiv \|f\|_1.$$

$L^2(\mathbb{R})$ is the space of complex-valued square integrable functions, *i.e.*, the space of *finite-energy signals*. The L^2*-norm* of $f \in L^2(\mathbb{R})$ is

$$\|f\|_2 = \left(\int |f(t)|^2\, dt\right)^{\frac{1}{2}} < \infty;$$

and the Fourier transform of $f \in L^2(\mathbb{R})$ is a well-defined element of $L^2(\hat{\mathbb{R}})$. Katznelson's book [38] is a standard and excellent reference for the Fourier analysis we shall use.

The *classical sampling theorem* is as follows.

Theorem 1. *Let $f \in L^1(\mathbb{R}) \cap A(\mathbb{R})$, or let $f \in L^2(\mathbb{R})$. Assume there are constants $T, \Omega > 0$ such that*

$$\text{supp}\, \hat{f} \subseteq [-\Omega, \Omega], \tag{1}$$

i.e., $\hat{f} = 0$ off of the interval $[-\Omega, \Omega]$, and

$$0 < 2T\Omega \leq 1. \tag{2}$$

Then

$$f(t) = 2T\Omega \sum f(nT) \frac{\sin[2\pi\Omega(t - nT)]}{2\pi\Omega(t - nT)}, \tag{3}$$

where the convergence is pointwise on $\mathbb{R}$ *for* $f \in L^1(\mathbb{R}) \cap A(\mathbb{R})$*, or the convergence is uniform on* $\mathbb{R}$ *and in* L^2*-norm for* $f \in L^2(\mathbb{R})$.

Proof: Our proofs shall be honest but brief, to highlight their simplicity.

a. Let $f \in L^1(\mathbb{R}) \cap A(\mathbb{R})$. For each $t \in \mathbb{R}$,

$$\begin{aligned} f(t) &= \int \hat{f}(\gamma) e^{2\pi i t \gamma}\, d\gamma \\ &= \int_{-\Omega}^{\Omega} G(\gamma) e^{2\pi i t \gamma}\, d\gamma \\ &= \sum c_n \int_{-\Omega}^{\Omega} e^{2\pi i (t-nT)\gamma}\, d\gamma \\ &= \sum c_n \frac{\sin[2\pi\Omega(t-nT)]}{\pi(t-nT)}, \end{aligned} \tag{4}$$

where G is defined as

$$G(\gamma) = \begin{cases} \hat{f}(\gamma), & \text{if} \quad |\gamma| \le \Omega, \\ 0, & \text{if} \quad \Omega < |\gamma| \le \frac{1}{2T}, \end{cases}$$

extended $\frac{1}{T}$-periodically on $\hat{\mathbb{R}}$. The *Fourier series* of G is

$$G(\gamma) \equiv \sum c_n\, e^{-\pi i n \gamma (2T)},$$

where

$$c_n = T \int_{-\frac{1}{2T}}^{\frac{1}{2T}} G(\gamma) e^{\pi i n \gamma (2T)}\, d\gamma.$$

Thus, $c_n = Tf(nT)$.

The calculation (4) is justified as follows. The first equation is a consequence of a classical inversion theorem for $f \in L^1(\mathbb{R}) \cap A(\mathbb{R})$, the second equation follows by the definition of G, the third equation results from the fact that Fourier series of integrable functions can be integrated term by term, and the fourth equation is clear by a simple calculation.

b. Let $f \in L^2(\mathbb{R})$. By the definition of the L^2-Fourier transform,

$$\left\| f(t) - \int_{-\Omega}^{\Omega} G(\gamma) e^{2\pi i t \gamma}\, d\gamma \right\|_2 = 0. \tag{5}$$

If $S_N G$ is the N-th partial sum of the Fourier series of G then

$$\begin{aligned} &\left\| \int_{-\Omega}^{\Omega} G(\gamma) e^{2\pi i t \gamma}\, d\gamma - \int_{-\Omega}^{\Omega} (S_N G)(\gamma) e^{2\pi i t \gamma}\, d\gamma \right\|_2 \\ &\qquad = \left\| \int \mathbf{1}_{(\Omega)}(\gamma) \big(G(\gamma) - (S_N G)(\gamma)\big) e^{2\pi i t \gamma}\, d\gamma \right\|_2 \\ &\qquad = \left\| \mathbf{1}_{(\Omega)} (G - S_N G) \right\|_2 = \left\| G - S_N G \right\|_{L^2[-\Omega,\Omega]}, \end{aligned} \tag{6}$$

where the second equation is a consequence of the Plancherel theorem. Also, we have

$$\lim_{N\to\infty} \|G - S_N G\|_{L^2[-\Omega,\Omega]} = 0,$$

since

$$\lim_{N\to\infty} \|G - S_N G\|_{L^2[-\frac{1}{2T},\frac{1}{2T}]} = 0$$

because of properties of Fourier series of square integrable functions. Combining this information with (5), (6), Hölder's inequality, and the definition of c_n, we obtain (3) with convergence in the L^2- norm (see the proof of Proposition 9).

Assuming the result for L^2-convergence, we prove the uniform convergence on $\mathbb{R}$ by means of Hölder's inequality and the Plancherel theorem. ∎

Discussion 2.

a. A common and essential feature of both proofs of Theorem 1 is the interplay between Fourier series and Fourier transforms.

b. The space $L^1(\mathbb{R}) \cap A(\mathbb{R})$, with norm $\|f\| = \|f\|_1 + \|\hat{f}\|_1$, is a reflexive Banach space. Similarly, the space $L^1(\mathbb{R}) \cap A(\mathbb{R}) \cap L^2(\mathbb{R})$, with norm $\|f\| = \|f\|_1 + \|\hat{f}\|_1 + \|f\|_2$, is a reflexive Banach space. Along with the Hilbert space $L^2(\mathbb{R})$, both of these spaces are useful in sampling theory.

c. There is an intriguing and labyrinthine history associated with Equation (3), *e.g.*, [31].

Calculation 3. *Classical sampling and the Poisson summation formula.*

There is another attractive proof of (3) which we shall now outline. The proof uses the Poisson summation formula,

$$\left(\sum \delta_{nT}\right)^{\wedge} = \frac{1}{T}\sum \delta_{n/T}, \tag{7}$$

where δ_{nT} is the Dirac δ-measure sorted by the point $nT \in \mathbb{R}$. We are not awarding this (correct) proof "Theorem" status since we do not wish to stress smoothness requirements (in the function version of (7)) or distributions in this section, *e.g.*, [32], [38, pages 130-131, number 15], and [57].

Implementing (7), we see that if f and s are well-behaved functions, then

$$f = \left[\left(\sum \delta_{nT}\right) f\right] * s \tag{8}$$

if and only if

$$\hat{f} = \left(\frac{1}{T}\sum \tau_{n/T}\hat{f}\right)\hat{s}, \tag{9}$$

where $(\tau_{n/T}\hat{f})(\gamma) = \hat{f}(\gamma - \frac{n}{T})$. Assuming (1) and (2), we obtain the validity of (9) in the case $\hat{s} = T$ on $[-\Omega, \Omega]$ and $supp\,\hat{s} \subseteq [-\frac{1}{2T}, \frac{1}{2T}]$. For such an s and using the fact,

$$\delta_{nT} f = f(nT)\,\delta,$$

the equivalence of (8) and (9) allows us to conclude that

$$f = \sum f(nT)\, \tau_{nT}\, s. \tag{10}$$

Equation (10) reduces to (3) for the case,

$$\hat{s} = T\,\mathbf{1}_{(\Omega)},$$

or, equivalently,

$$s(t) = T\,\frac{\sin 2\pi\Omega t}{\pi t}. \tag{11}$$

Example 4. It is natural to investigate an appropriate converse of Theorem 1. *Suppose we assume Equation* (3),*with convergence in the* L^2*-norm, for all* $f \in L^2(\mathbb{R})$ *satisfying* (1). *Can we conclude the validity of* (2), *i.e., is*

$$2T\Omega \leq 1?$$

We shall answer this question in the *positive*. Assume $T, \Omega > 0$ satisfy $2T\Omega > 1$ and define

$$f(t) = \frac{\sin\left(\frac{\pi t}{T}\right)}{\frac{\pi t}{T}} \in L^2(\mathbb{R}).$$

Clearly,

$$\hat{f} = T\,\mathbf{1}_{(\frac{1}{2T})},$$

and, in particular, (1) is satisfied. Since

$$f(nT) = \begin{cases} 0, & \text{if} \quad n \neq 0, \\ 1, & \text{if} \quad n = 0, \end{cases}$$

then the right side of the sampling formula (3) is

$$g(t) = 2T\Omega \frac{\sin 2\pi\Omega t}{2\pi\Omega t}.$$

The functions f and g are not equal since both are continuous on $\mathbb{R}$, and $f(0) = 1$ and $g(0) = 2T\Omega > 1$. (For another proof that $f \neq g$, note that $\hat{g} = T\,\mathbf{1}_{(\Omega)}$.)

Definition 5.

a. In the sampling formula (3), the *sampling period* is T, the *sampling sequence* is $\{nT : n \in \mathbb{Z}\}$, and the sequence of *sampled values* is $\{f(nT) : n \in \mathbb{Z}\}$.

b. Let 2Ω be a given frequency bandwidth (of a linear time-invariant system having an even frequency response). For example, consider the ideal lowpass filter with cutoff frequency Ω. Because of Theorem 1 and Example 4, the minimum rate at which each element $f \in L^2(\mathbb{R})$, for which *supp* $\hat{f} \subseteq$

$[-\Omega, \Omega]$, must be sampled for exact reconstruction is 2Ω samples per unit time. This sampling rate, 2Ω, is the *Nyquist rate.*

If the unit time is seconds (for convenience) and we define

$$T \equiv \frac{1}{2\Omega},$$

then the Nyquist rate or *sampling frequency* is $1/T$ samples per second, *i.e.*, 1 sample per T seconds, from which we obtain the maximal sampling period of T.

The effect of undersampling continuous (in fact, analytic) signals $f \in L^2(\mathbb{R})$ satisfying (1), *vis a vis* the goal of exact reconstruction by means of (3), is called *aliasing.*

Definition/Discussion 6. *Aliasing.*

a. Suppose $2T\Omega > 1$. For simplicity, let us further assume that $1 \geq T\Omega$. Consider the Fourier series of G, defined in the proof of Theorem 1. Since G is $\frac{1}{T}$-periodic, it is of the form $\sum F(\gamma - \frac{n}{T})$, where, from the analysis of Theorem 1, F must vanish off of $[-\frac{1}{2T}, \frac{1}{2T}]$. By our hypotheses on T and Ω, $[\Omega - \frac{1}{T}, -\Omega + \frac{1}{T}]$ is the only subset of $[-\frac{1}{2T}, \frac{1}{2T}]$ where the Fourier series of G can be expected to converge to G. That this is the case follows since both $\tau_{1/T}F$ and $\tau_{-1/T}F$ are nonzero on subsets of $[-\frac{1}{2T}, \frac{1}{2T}]$. The ensuing phenomenon, caused by this nonconvergence of the Fourier series on all of $[-\frac{1}{2T}, \frac{1}{2T}]$, is *aliasing.* The term "aliasing", due to Tukey, catches the flavor of high and low frequencies from the Fourier series "assuming the alias of each other" (see the more quantitative discussion of aliasing in Sections 5 and 10 as well as [50]).

b. Old motion pictures of fast moving events produce "jumpy" video, and this is a classical example of aliasing. In fact, each frame of film is a sampled value, but the sampling rate is not sufficiently high to produce exact reconstruction of the event.

In Definition 5b, we referred to the notion of a "linear time-invariant system". This is an elementary engineering concept, *e.g.*, [48], [51], which also has a long history in mathematics, *e.g.*, [1, pages 216-217]. To be a little more precise, let X and Y be Banach sub-algebras of the convolution algebra $M_b(\mathbb{R})$ of bounded Radon-measures, let $\delta \in X$, assume $X \subseteq Y$, and let $L : X \to Y$ be a continuous linear map. L is a *linear time-invariant system* if the impulse response $L\delta = h$ has the property that

$$\forall\, f \in X, \quad Lf = h * f.$$

The Fourier transform $\hat{h}$ is the *frequency response* of the system, and L or h is called a *filter* if the inclusion *supp* $\hat{h} \subseteq \hat{\mathbb{R}}$ is proper.

We omit the proof of the following result. The proof is similar to that of Theorem 1, and it becomes valid when natural hypotheses are added to the statement of the "theorem". The "theorem", itself, provides the classical sampling formula for linear time-invariant systems.

Theorem 7. *Let $L : X \to Y$ be a linear time-invariant system with impulse response $h \in Y$ and frequency response $\hat{h}$. Assume there are constants $T, \Omega > 0$ such that*

$$0 < 2T\Omega \leq 1,$$

and define the function

$$s(t) = T \int_{-\Omega}^{\Omega} \frac{1}{\hat{h}(\gamma)} e^{2\pi i t \gamma} \, d\gamma. \tag{12}$$

Let $f \in X$ satisfy the support condition,

$$supp \hat{f} \subseteq [-\Omega, \Omega].$$

Then

$$f = \sum (Lf)(nT)\tau_{nT}s. \tag{13}$$

The function s, in Equations (10), (11), (12), and (13), is a *sampling function.* In the case of (11) we obtain the classical sampling formula (3). Our goal in the sequel is to obtain sampling formulas of the form,

$$f = \sum f(t_n)s_n \tag{14}$$

and

$$f = \sum c_n(f)\tau_{t_n}s, \tag{15}$$

where $\{t_n\}$ is an irregular sampling sequence, *i.e.*, $\{t_n\}$ is not uniformly spaced. Later, we shall comment on the other parameters in (14) and (15). Also, there are versions of (14) and (15) corresponding to (13).

§3. The Paley-Wiener theorem

In light of the support hypothesis in Theorem 1 and Theorem 7, we make the following definition.

Definition/Notation 8.

a. Let $\Omega > 0$, and define the space,

$$PW_\Omega = \{f \in L^2(\mathbb{R}) : supp \hat{f} \subseteq [-\Omega, \Omega]\}.$$

PW_Ω is the *Paley-Wiener space*, and it is a Hilbert space, considered as a closed subspace of $L^2(\mathbb{R})$ taken with the L^2- norm. The support condition, $supp \hat{f} \subseteq [-\Omega, \Omega]$, is described by saying that f is *Ω-bandlimited.*

b. Let

$$d(t) = \frac{\sin t}{\pi t},$$

and define the L^1-dilation

$$d_\lambda(t) = \lambda d(t\lambda).$$

(" d " is for Dirichlet.) Equation (3) can be written as

$$f(t) = T \sum f(nT)(\tau_{nT}\, d_{2\pi\Omega})(t).$$

PW_Ω is a *reproducing kernel Hilbert space* in the sense that

$$PW_\Omega = \{f \in L^2(\mathbb{R}) : f * d_{2\pi\Omega} = f\},$$

(see [30], [61], [64] for the usual development of reproducing kernel Hilbert space).

c. If $f \in PW_\Omega$ then $\hat{f} \in L^2[-\Omega, \Omega] \subseteq L^1[-\Omega, \Omega]$, where the inclusion is a consequence of Hölder's inequality. Further, if we define the function,

$$g(z) \equiv \int_{-\Omega}^{\Omega} \hat{f}(\gamma) e^{2\pi i z \gamma}\, d\gamma, \quad z = t + iy \in \mathbb{C},$$

then g is a continuous function on $\mathbb{C}$ and $g = f$ a.e. on $\mathbb{R}$.

In the other direction, if $F \in L^2[-\Omega, \Omega]$ and we define the continuous function

$$f(t) \equiv \int_{-\Omega}^{\Omega} F(\gamma) e^{2\pi i t \gamma} d\gamma, \quad t \in \mathbb{R},$$

then $f \in PW_\Omega$ and $F = \hat{f}$ a.e. on $\hat{\mathbb{R}}$. These observations are elementary facts from real and harmonic analysis, *e.g.*, [2], [38].

d. A function $f : \mathbb{C} \to \mathbb{C}$ which is analytic at every point of $\mathbb{C}$ is *entire*; and an entire function is of *exponential type* A if for each $B > A$

$$\exists\, C = C(B) > 0 \quad \text{such that} \quad \forall\, z \in \mathbb{C}, \qquad |f(z)| \leq Ce^{B|z|}. \tag{16}$$

Proposition 9. *If $f \in PW_\Omega$ then f is an entire function of exponential type $2\pi\Omega$, and*

$$\forall\, t \in \mathbb{R}, \qquad f(t) = \int_{-\Omega}^{\Omega} \hat{f}(\gamma) e^{2\pi i t \gamma}\, d\gamma,$$

i.e., pointwise equality as well as L^2-norm equality.

Proof: If $f \in PW_\Omega$ then

$$g(z) \equiv \int_{-\Omega}^{\Omega} \hat{f}(\gamma) e^{2\pi i z \gamma}\, d\gamma, \qquad z = t + iy \in \mathbb{C},$$

is a well-defined continuous function on $\mathbb{C}$. One proof that g is entire follows from Fubini's and Morera's theorems. Finally,

$$\begin{aligned} |g(z)| &\leq \int_{-\Omega}^{\Omega} |\hat{f}(\gamma)| e^{-2\pi y \gamma}\, d\gamma \\ &\leq e^{2\pi|y|\Omega} \int_{-\Omega}^{\Omega} |\hat{f}(\gamma)|\, d\gamma \leq e^{2\pi|z|\Omega} \int_{-\Omega}^{\Omega} |\hat{f}(\gamma)|\, d\gamma \\ &\equiv Ce^{2\pi|z|\Omega} < \infty. \end{aligned}$$

Now, the Fourier inversion theorem for $L^2(\mathbb{R})$ asserts that

$$f(t) = \int \hat{f}(\gamma) e^{2\pi i t \gamma} \, d\gamma$$

in $L^2(\mathbb{R})$. Consequently, $\|f - g\|_2 = 0$ so that $f = g$ a.e., and, hence, f can be identified with g on $\mathbb{R}$. ■

In this case, $\inf\{B > 2\pi\Omega : (16) \text{ is valid}\} = 2\pi\Omega$ also satisfies (16) with the constant $C = \|\hat{f}\|_1$ (see [19, page 104, Equations (6.8.4)-(6.8.6)]).

The converse of Proposition 9 is the Paley-Wiener theorem which plays a role in much of what follows.

Theorem 10. (Paley-Wiener, [49, Theorem X]). *Let $f \in L^2(\mathbb{R})$. Then $f \in PW_\Omega$ if and only if f is an entire function of exponential type $2\pi\Omega$.*

Discussion 11.

a. There are a number of similar formulations of Theorem 10; and, besides the proof in [49], proofs of Theorem 10 can be found in [19, pages 103-108], [56, pages 370-372]. The sufficient condition in Theorem 10, in order that $f \in L^2(\mathbb{R})$ be an element of PW_Ω, can be replaced by the condition,

$$\exists\, C > 0 \quad \text{such that} \quad \forall\, z \in \mathbb{C}, \quad |f(z)| \leq C e^{2\pi |z| \Omega}.$$

b. Because of our interest in d-dimensional sampling, we note that d-dimensional versions of the Paley-Wiener theorem were proved early-on, *e.g.*, [52]. Further, there are distributional versions in $\mathbb{R}^d$, *e.g.*, [32, volume I, pages 181-182 (see pages 21-22 of Hörmander's first edition)], [57]. A proof of the Paley-Wiener theorem for $L^2(\mathbb{R}^d)$ is found in [59, pages 112-114].

Example 12. There is an analogue of Theorem 10 for the usual Laplace transform, which was also proved by Paley and Wiener [49, Theorem V]. To state this result, we first define the classical *Hardy space* H^2 to be the set of functions f, analytic in the right half plane, with the property that

$$\sup_{t>0} \left(\int |f(t+iy)|^2 \, dy \right)^{\frac{1}{2}} < \infty,$$

(see Lemma 42a). Further, if $F : \hat{\mathbb{R}} \to \mathbb{C}$ is supported by $[0, \infty)$, its Laplace transform is

$$L(F)(z) = \int_0^\infty F(\gamma) e^{-z\gamma} \, d\gamma.$$

The Laplace transform analogue of Theorem 10 is the representation theorem: *if $f \in H^2$ then there is $F \in L^2(\hat{\mathbb{R}})$, supported by $[0, \infty)$, for which*

$$f(z) = L(F)(z), \qquad z = t + iy \quad \text{and} \quad t > 0.$$

The converse is true, and is not difficult to prove.

The L^p-version of this result is due to Doetsch (1936), using results of Hille and Tamarkin (1935). Weighted versions were initiated by Rooney in the 1960s; and a general theory for weighted L^p and H^p spaces was formulated by the author, with Heinig and Johnson, in the 1980s, *e.g.*, [6] for references.

§4. Frames and exact frames

Definition 13.

a. A sequence $\{g_n\} \subseteq H$, a separable Hilbert space, is a *frame* if there exist $A, B > 0$ such that

$$\forall\, f \in H, \quad A\|f\|^2 \leq \sum |\langle f, g_n\rangle|^2 \leq B\|f\|^2,$$

where $\langle , \rangle$ is the inner product on H and the norm of $f \in H$ is $\|f\| = \langle f, f\rangle^{\frac{1}{2}}$. For example, if $H = L^2(\mathbb{R})$ and $f, g \in H$ then $\langle f, g\rangle = \int f(t)\overline{g}(t)\, dt$. A and B are the *frame bounds*, and a frame $\{g_n\}$ is *tight* if $A = B$.

A frame $\{g_n\}$ is *exact* if it is no longer a frame when any one of its elements is removed.

Clearly, if $\{g_n\}$ is an orthonormal basis of H then it is a tight exact frame with $A = B = 1$.

b. The *frame operator* of the frame $\{g_n\}$ is the function $S : H \to H$ defined as $Sf = \sum\langle f, g_n\rangle\, g_n$. In Theorem 14a, the first expansion is the *frame expansion*, and the second is the *dual frame expansion*.

The theory of frames is due to Duffin and Schaeffer [24] in 1952. Expositions include [64] and [28], the former presented in the context of non- harmonic Fourier series and the latter in the setting of wavelet theory.

Theorem 14. *Let $\{g_n\} \subseteq H$ be a frame with frame bounds A and B.*

a. *S is a topological isomorphism with inverse $S^{-1} : H \to H$. $\{S^{-1}g_n\}$ is a frame with frame bounds B^{-1} and A^{-1}, and*

$$\forall\, f \in H, \quad f = \sum\langle f, S^{-1}g_n\rangle\, g_n = \sum\langle f, g_n\rangle\, S^{-1}\, g_n \quad \text{in } H.$$

b. *If $\{g_n\}$ is tight, $\|g_n\| = 1$ for all n, and $A = B = 1$, then $\{g_n\}$ is an orthonormal basis of H.*

c. *If $\{g_n\}$ is exact, then $\{g_n\}$ and $\{S^{-1}g_n\}$ are biorthonormal, i.e.,*

$$\forall\, m, n \quad \langle g_m, S^{-1}g_n\rangle = \delta_{mn},$$

and $\{S^{-1}g_n\}$ is the unique sequence in H which is biorthonormal to $\{g_n\}$.

d. *If $\{g_n\}$ is exact, then the sequence resulting from the removal of any one element is not complete in H, i.e., the linear span of this resulting sequence is not dense in H.*

Discussion 15. *Vitali's theorem.*

We comment on Theorem 14a because it is surprisingly useful and because of a stronger result by Vitali (1921).

To prove part b we first use tightness and $A = 1$ to write,

$$\|g_m\|^2 = \|g_m\|^4 + \sum_{n \neq m} |\langle g_m, g_n\rangle|^2;$$

and obtain that $\{g_n\}$ is orthonormal since each $\|g_n\| = 1$. To conclude the proof we then invoke the well-known result: if $\{g_n\} \subseteq H$ is orthonormal then it is an orthonormal basis of H if and only if

$$\forall\, f \in H, \quad \|f\|^2 = \sum |\langle f, g_n\rangle|^2.$$

In 1921, Vitali proved that an orthonormal sequence $\{g_n\} \subseteq L^2[a,b]$ is complete, and so $\{g_n\}$ is an orthonormal basis, if and only if

$$\forall\, t \in [a,b], \quad \sum \Big| \int_a^t g_n(u)\, du \Big|^2 = t - a. \tag{17}$$

For the case $H = L^2[a,b]$, Vitali's result is stronger than part b since (17) is tightness with $A = 1$ for functions $f = \mathbf{1}_{[a,t]}$.

Other remarkable and important contributions by Vitali are highlighted in [2].

Definition 16. Let H be a separable Hilbert space. A sequence $\{g_n\} \subseteq H$ is a *Schauder basis* or *basis* of H if each $f \in H$ has a unique decomposition $f = \sum c_n(f) g_n$. A basis g_n is an *unconditional basis* if

$$\exists\, C \quad \text{such that} \quad \forall\, F \subseteq \mathbb{Z}, \quad \text{where} \quad \text{card}\, F < \infty, \quad \text{and}$$

$$\forall\, b_n, c_n \in \mathbb{C}, \quad \text{where} \quad n \in F \quad \text{and} \quad |b_n| \le |c_n|,$$

$$\Big\| \sum_{n \in F} b_n g_n \Big\| \le C \Big\| \sum_{n\in F} c_n g_n \Big\|.$$

An unconditional basis $\{g_n\}$ is *bounded* if

$$\exists\, A, B > 0 \quad \text{such that} \quad \forall\, n,\ A \le \|g_n\| \le B.$$

Separable Hilbert spaces have orthonormal bases, and orthonormal bases are bounded unconditional bases.

Köthe (1936) proved the implication, *b* implies *c*, of the following theorem. The implication, *c* implies *b*, is straightforward; and the equivalence of a and *c* is found in [64, pages 188-189].

Theorem 17. *Let H be a separable Hilbert space and let $\{g_n\} \subseteq H$ be a given sequence. The following are equivalent:*

a. *$\{g_n\}$ is an exact frame of H;*

b. *$\{g_n\}$ is a bounded unconditional basis of H;*

c. *$\{g_n\}$ is a Riesz basis, i.e., there is an orthonormal basis $\{u_n\}$ and a topological isomorphism $L : H \to H$ such that $Lg_n = u_n$ for each n.*

Definition/Discussion 18. *Weyl-Heisenberg (Gabor) and Fourier frames.*

a. Let $g \in L^2(\mathbb{R})$ and suppose we are given sequences $\{a_n\}$, $\{b_n\} \subseteq \mathbb{R}$. Recall that translation is defined by $(\tau_a g)(t) = g(t-a)$ and, notationally, we write $e_b(t) = e^{2\pi itb}$. If $\{e_{b_m}\tau_{a_n}g\}$ is a frame for $L^2(\mathbb{R})$ it is called a *Gabor* or *Weyl-Heisenberg* frame. *Fourier frames* $\{e_{b_m}\}$ were defined in [24] for $L^2[-T,T]$. Precisely, if $\{e_{b_m}\}$ is a frame for $L^2[-T,T]$ it is called a *Fourier frame* for $L^2[-T,T]$.

b. Gabor's seminal paper [26] (1946) deals with "regularly latticed" systems $\{e_{mb}\tau_{na}g\}$, where g is the Gaussian; and it turns out that the Heisenberg group is fundamental in analyzing the structure of modulations and translations such as $\{e_{mb}\tau_{na}g\}$ for $g \in L^2(\mathbb{R})$. This explains our terminology in part a. Gabor won the Nobel prize for his conception and analysis of holography (1947), which is a method for photographically recording a three-dimensional image; and, as demonstrated by Schempp, the Heisenberg group also plays a role in this setting. The special kind of light required to demonstrate the capability of holography is a single frequency form called *coherent* light; and it became readily available after the laser was developed in 1960. Further, the *coherent states* of quantum mechanics are the elements of $\{e_{mb}\tau_{na}g\}$ in the case g is the Gaussian and $ab = 1$. For the Gaussian g and $ab = 1$, $\{e_{mb}\tau_{na}g\}$ is not a frame (see [58] for $ab < 1$).

Duffin and Schaeffer's Fourier frames were also part of a larger picture, dealing with problems in non-harmonic Fourier series and complex analysis and the work of Paley-Wiener, Levinson, Plancherel-Pólya, and Boas (see Section 7).

c. $\{e_{b_m}\tau_{a_n}g\}$ is a frame for $L^2(\mathbb{R})$ if and only if $\{\tau_{a_n}(e_{b_m}g)\}$ is a frame for $L^2(\mathbb{R})$.

Our Weyl-Heisenberg frames will often be defined for $L^2(\hat{\mathbb{R}})$. As such, we note that

$$\forall\, g \in L^2(\mathbb{R}), \quad (e_{a_n}\tau_{b_m}\hat{g})^{\vee} = e^{2\pi i a_n b_m}\, e_{b_m}\tau_{-a_n}g,$$

where "$\vee$" designates the inverse Fourier transform.

Theorem 19. *Let $g \in L^2(\mathbb{R})$ and suppose $a, b > 0$. Define*

$$G(t) = \sum |g(t - na)|^2.$$

Assume that there exist $A, B > 0$ such that

$$0 < A \le G(t) \le B < \infty \quad \text{a.e. on} \quad \mathbb{R}, \tag{18}$$

and that supp $g \subseteq I$ where I is an interval of length $1/b$. Then $\{e_{mb}\tau_{na}g\}$ is a frame for $L^2(\mathbb{R})$, with frame bounds $b^{-1}A$ and $b^{-1}B$, and

$$\forall\, f \in L^2(\mathbb{R}), \quad S^{-1}f = \frac{bf}{G}. \tag{19}$$

Theorem 20. *Let $g \in L^2(\mathbb{R})$ and suppose $a, b > 0$. Assume $\{e_{na}\tau_{mb}\hat{g}\}$ is a frame for $L^2(\hat{\mathbb{R}})$. Then*

$$S^{-1}(e_{na}\tau_{mb}\,\hat{g}) = e_{na}\tau_{mb}\,S^{-1}\,\hat{g}. \tag{20}$$

Example 21.

a. Let $g \in L^2(\mathbb{R})$ and suppose $a, b > 0$ satisfy $ab = 1$. If $\{e_{mb}\tau_{na}g\}$ is a frame then it is an exact frame. This remarkable fact (for $ab = 1$) can be proved using properties of the Zak transform which we now define.

b. The *Zak transform* of $f \in L^2(\mathbb{R})$ is defined as

$$Zf(x, \omega) = a^{1/2} \sum f(xa + ka)\ e^{2\pi ik\omega}$$

for $(x, \omega) \in \mathbb{R} \times \hat{\mathbb{R}}$ and $a > 0$. It turns out that the Zak transform is a unitary map of $L^2(\mathbb{R})$ onto $L^2(Q), Q = [0,1) \times [0,1)$.

c. If $\{e_{mb}\tau_{na}\, g\}$ is a frame for $ab = 1$, it is a bounded unconditional basis (part a and Theorem 17). In particular, the frame decomposition,

$$\forall\, f \in L^2(\mathbb{R}), \qquad f = \sum c_{m,n}\, e_{mb}\, \tau_{na}\, g,$$

from Theorem 14a is unique; and it is easy to verify that

$$\begin{aligned} \forall\, m, n, \quad c_{m,n} &= \langle f, S^{-1} e_{mb}\tau_{na} g\rangle \\ &= \int_0^1 \int_0^1 \frac{Zf(x,\omega)}{Zg(x,\omega)} e^{-2\pi imx} e^{-2\pi in\omega}\, dx d\omega. \end{aligned}$$

Results in the same spirit as the one stated in Example 21a have been formulated in the coherent state literature for many years; and, in this context, they seem to go back to the analysis of von Neumann found in [60, pages 405-407]. The following is a representative list of such results and the proofs are not difficult, *e.g.*, [5], [28].

Theorem 22. *Let $g \in L^2(\mathbb{R})$ and suppose $a, b > 0$ satisfy $ab = 1$.*

a. *$\{e_{mb}\tau_{na}\, g\}$ is complete in $L^2(\mathbb{R})$ if and only if* $|Zg| > 0$ a.e.

b. *$\{e_{mb}\tau_{na}\, g\}$ is an orthonormal basis of $L^2(\mathbb{R})$ if and only if* $|Zg| = 1$ a.e.

c. *$\{e_{mb}\tau_{na}\, g\}$ is a frame for $L^2(\mathbb{R})$ with frame bounds A and B if and only if $A \le |Zg|^2 \le B$* a.e. *In this case, $\{e_{mb}\tau_{na}\, g\}$ is an exact frame (Example 21a).*

We shall now state the Balian-Low Theorem, which can be proved using the Zak transform.

Theorem 23. *Let $g \in L^2(\mathbb{R})$ and suppose $a, b > 0$ satisfy $ab = 1$. If $\{e_{mb}\tau_{na}g\}$ is a frame then either $tg(t) \notin L^2(\mathbb{R})$ or $\gamma\hat{g}(\gamma) \notin L^2(\hat{\mathbb{R}})$.*

We first learned of the Balian-Low "phenomenon" in an early preprint of [23]. It turns out that this result has been proved at various levels of precision; and [8] contains an analysis of these proofs, as well as a complete, new proof based on some established ideas.

Discussion 24. *Exact frames: oversampling and undersampling.*

a. We have stated Theorems 22 and 23 since a comparison of the exact frame case ($ab = 1$ for regular lattices) and the more general frame case is fundamental in our approach to irregular sampling. For example, exact frames $\{e_{mb}\tau_{na}\, g\}$ (in particular, orthonormal bases) don't have the flexibility to take into account the oversampling occurring naturally in biological processes such as cochlear processing; our contention is that general frames do have such flexibility. The oversampling reflected in Theorem 26 generally involves sampling functions s with more rapid decay in the Fourier domain than the classical sampling function in Theorem 1. The goal is better computational efficiency for low pass filters; the price to be paid is a sampling rate greater than the Nyquist rate. Of course, we can sample greater than the Nyquist rate for the classical sampling function, but the slow decay in the Fourier domain is still a liability.

b. Using the notation from Section 2, recall that undersampling occurs if $2T\Omega > 1$, where $\{nT\}$ is the sampling sequence and 2Ω is the given frequency bandwidth. If $a = T$ and $b = 2\Omega$ then $ab > 1$ (see the proof of Theorem 25). It is an expected fact, with a surprisingly abstract proof, due to Rieffel [55], that *if $g \in L^2(\mathbb{R})$ and $a, b > 0$ satisfy $ab > 1$ then $\{e_{mb}\tau_{na}\, g\}$ is incomplete in $L^2(\mathbb{R})$* (see [23]).

The quantity $\Delta = \frac{1}{ab}$ is the natural density of the so-called von-Neumann lattices $\{(na, mb)\}$, *e.g.*, Definition 36 and Example 37c.

The most significant contribution in classical analysis related to Rieffel's theorem is due to Landau [43]. Using a theorem of Daubechies [23, Theorem 3.1], he proved that *if g and $\hat{g}$ have sufficient decay and $\{e_{mb}\tau_{na}\, g\}$ is a Gabor frame for $L^2(\mathbb{R})$, then $ab \leq 1$.* This is, of course, weaker than Rieffel's theorem, but the method, although intricate, is constructive. The following are sufficient decay conditions to implement Daubechies' theorem: there are constants $C > 0$ and $\alpha > \frac{1}{2}$ such that

$$\forall\ (t, \gamma) \in \mathbb{R} \times \hat{\mathbb{R}}, \quad |g(t)| \leq \frac{C}{(1+t^2)^\alpha} \quad \text{and} \quad |\hat{g}(\gamma)| \leq \frac{C}{(1+\gamma^2)^\alpha}.$$

Of course, if $\alpha > 3/4$ and $ab = 1$ then, by Theorem 23, there are no such g for which $\{e_{mb}\tau_{na}\, g\}$ is a Gabor frame.

§5. Regular sampling and frames

The theme of this section is to deal with classical sampling results by frame methods in the case that the inverse frame operator S^{-1} is a multiplier. The following is Theorem 1 proved by such methods [9, Theorem 3.1].

Theorem 25. *Let $T, \Omega > 0$ be constants for which*

$$0 < 2T\Omega \leq 1.$$

Then

$$\forall\ f \in PW_\Omega, \quad f = T \sum f(nT)\tau_{nT} d_{2\pi\Omega} \quad \text{in} \quad L^2(\mathbb{R}). \tag{21}$$

(The Dirichlet kernel d_λ was defined in Definition/Notation 8).

Proof: Let $g = (2\Omega)^{-1/2} d_{2\pi\Omega}$, and set $a = T$ and $b = 2\Omega$. Theorem 19 is applicable and so $\{e_{na}\tau_{mb}\hat{g}\}$ is a frame. Consequently, by Theorem 14a and Theorem 20,

$$\forall\, f \in L^2(\mathbb{R}), \quad \hat{f} = \sum \langle \hat{f}, e_{na}\tau_{mb} S^{-1}\hat{g}\rangle e_{na}\tau_{mb}\hat{g} \quad \text{in} \quad L^2(\hat{\mathbb{R}}). \tag{22}$$

Since $supp\, \hat{g}$ is compact, we have

$$\forall\, h \in L^2(\hat{\mathbb{R}}), \quad S^{-1}\hat{h} = 2T\Omega\hat{h}$$

by Theorem 20; and, hence, (22) becomes

$$\forall\, f \in L^2(\mathbb{R}), \quad f = 2T\Omega \sum \langle \hat{f}, e_{na}\tau_{mb}\hat{g}\rangle \tau_{-na} e_{mb} g \quad \text{in} \quad L^2(\mathbb{R}). \tag{23}$$

If $f \in PW_\Omega$ then

$$\langle \hat{f}, e_{na}\tau_{mb}\hat{g}\rangle = \begin{cases} (2\Omega)^{-1/2} f(-nT), & \text{if} \quad m = 0, \\ 0, & \text{if} \quad m \neq 0. \end{cases} \tag{24}$$

The sampling formula (21) follows from (23) and (24). ■

Using the same method from the theory of frames we can prove the following result [9, Theorem 3.3], which we "proved" by Poisson summation in Calculation 3 and which we had in mind in our comment about oversampling in Discussion 24a. The condition, $\hat{g} > 0$ on $(-\frac{1}{2T}, -\Omega] \cup [\Omega, \frac{1}{2T})$, can be relaxed, e.g., [29].

Theorem 26. *Let $T, \Omega > 0$ be constants for which*

$$0 < 2T\Omega \leq 1,$$

and let $g \in \mathcal{S}(\mathbb{R})$ have the following properties:

$$\hat{g} = 1 \quad on \quad [-\Omega, \Omega],$$

$$supp\, \hat{g} \subseteq [-\frac{1}{2T}, \frac{1}{2T}],$$

$$\hat{g} > 0 \quad on \quad (-\frac{1}{2T}, -\Omega] \cup [\Omega, \frac{1}{2T}).$$

Set

$$G(\gamma) = \sum |\hat{g}(\gamma - mb)|^2 \quad and \quad s(t) = T\Big(\frac{\hat{g}}{G}\Big)^\vee(t),$$

where $\Omega + \frac{1}{2T} \leq b < \frac{1}{T}$. Then $0 < A \leq G(\gamma) \leq B < \infty$, $s \in \mathcal{S}(\mathbb{R})$, $supp\, \hat{s} = supp\, \hat{g}$, $\hat{s} = T\frac{1}{G}$ on $[-\Omega, \Omega]$, and

$$\forall\, f \in PW_\Omega,\ f = \sum f(nT)\tau_{nT} s \quad in \quad L^2(\mathbb{R}).$$

($\mathcal{S}(\mathbb{R})$ is the Schwartz space, e.g., [57], and the result is true for g and s having significantly less smoothness.)

Example 27.

a. In Theorem 25, $\{e_{na}\tau_{mb}\hat{g}\}$ is a tight frame with frame bounds $A = B = 1$ in the case $2T\Omega = 1$, where $a = T$ and $b = 2\Omega$. From Theorem 14b, $\{e_{na}\tau_{mb}\hat{g}\}$ is an orthonormal basis if and only if $2T\Omega = 1$.

b. To construct functions $g \in \mathcal{S}(\mathbb{R})$ satisfying the hypotheses of Theorem 26, we proceed as follows for $2T\Omega < 1$.

We begin in the standard "distributional way" by defining

$$\psi_\epsilon(\gamma) = \frac{\phi(\epsilon - |\gamma|)}{\int \phi(\epsilon - |\gamma|)\, d\gamma},$$

where $\phi \in C^\infty(\hat{\mathbb{R}})$ vanishes on $(-\infty, 0]$ and equals $e^{-1/\gamma}$ on $[0, \infty)$. Thus, $\psi_\epsilon \in C_c^\infty(\hat{\mathbb{R}})$ is an even function satisfying the conditions, *supp* $\psi_\epsilon = [-\epsilon, \epsilon]$ and $\int \psi_\epsilon(\gamma)\, d\gamma = 1$. Next set

$$\psi_{U,V} = \frac{1}{|V|}\mathbf{1}_V * \mathbf{1}_{U-V}, \quad U, V \subseteq \hat{\mathbb{R}},$$

so that $\psi_{U,V}$ is 1 on U and vanishes off of $U + V - V$. The function g will be defined in terms of $\hat{g}$ as $\hat{g} = \psi_{U,V} * \psi_\epsilon$, where we shall now specify ϵ, U, and V given $2T\Omega < 1$. Let $U = [-u, u]$, where $u \in (\Omega, \frac{1}{2T})$ is arbitrary, and let $\epsilon = u - \Omega$. Choose $V = [-v, v]$ by setting $v = \frac{w-u}{2}$, where $w = \frac{1}{2T} + \epsilon$. These choices are necessitated by a simple geometrical argument, and the resulting function $\hat{g}$ satisfies the desired properties.

Definition/Discussion 28. *Aliasing.*

a. The proof of Theorem 25 required the support condition, $f \in PW_\Omega$, to verify both parts of (24). *When f is not Ω-bandlimited, so that aliasing occurs, phase information contributed by $m \neq 0$ in (23) is required in the frame decomposition of a signal* (see Definition/Discussion 6). To quantify this observation, we define the *aliasing pseudomeasure* $\alpha_{t,\Omega}$ on $\mathbb{R}$ (for the low pass filter $d_{2\pi\Omega}$) to be the distributional Fourier transform, $\alpha_{t,\Omega} = A^\vee_{t,\Omega}$, where each t is fixed and

$$A_{t,\Omega} \equiv \sum (e^{2\pi i(2m\Omega)t} - 1)\tau_{2m\Omega}\mathbf{1}_{(\Omega)} \in L^\infty(\hat{\mathbb{R}}).$$

The study of pseudomeasures is intimately related to spectral synthesis [1] (see Section 9). In equation (25) of part b we shall see the relation between the aliasing pseudomeasure and the manner in which aliasing is manifested in our proof of Theorem 25.

b. Let $f \in L^2(\mathbb{R})$ and assume $2T\Omega = 1$. Then, formally,

$$\begin{aligned} f(t) = {} & T\sum f(nT)\tau_{nT}d_{2\pi\Omega}(t) \\ & + T\sum (f * \alpha_{t,\Omega})(nT)\tau_{nT}d_{2\pi\Omega}(t). \end{aligned} \tag{25}$$

To verify (25), we write (23) as a sum, $\sum_{m=0,n} + \sum_{m\neq 0,n}$, and obtain

$$f(t) = 2T\Omega \sum_n \left(\frac{1}{(2\Omega)^{1/2}} \int \hat{f}(\gamma)\mathbf{1}_\Omega(\gamma) e^{-2\pi i n T\gamma} d\gamma \right) g(t+nT)$$
$$+ 2T\Omega \sum_{m\neq 0} \sum_n \left(\frac{1}{(2\Omega)^{1/2}} \int \hat{f}(\gamma)\mathbf{1}_\Omega(\gamma - 2m\Omega) e^{-2\pi i n T\gamma} d\gamma \right) \times$$
$$e^{2\pi i(2m\Omega)(t+nT)} g(t+nT), \qquad (26)$$

where $g = (2\Omega)^{-1/2} d_{2\pi\Omega}$. Equation (26) can be written as

$$2T\Omega \sum_n \frac{1}{(2\Omega)^{1/2}} \left(\int \hat{f}(\gamma) e^{-2\pi i n T\gamma} d\gamma - \int_{[-\Omega,\Omega]^\sim} \hat{f}(\gamma) e^{-2\pi i n T\gamma} d\gamma \right) g(t+nT)$$
$$+ 2T\Omega \sum_{m\neq 0} \sum_n \left(\frac{1}{(2\Omega)^{1/2}} \int \hat{f}(\gamma)\ \mathbf{1}_\Omega(\gamma - 2m\Omega) e^{-2\pi i n T\gamma} d\gamma \right) \times$$
$$e^{2\pi i(2m\Omega)(t+nT)} g(t+nT),$$

and so

$$f(t) = 2T\Omega \sum f(nT) \frac{1}{(2\Omega)}\ d_{2\pi\Omega}(t-nT)$$
$$+ 2T\Omega \Big\{ \sum_{m\neq 0} \sum_n \left(\frac{1}{(2\Omega)^{1/2}} \int \hat{f}(\gamma)\mathbf{1}_\Omega(\gamma - 2m\Omega) e^{-2\pi i n T\gamma} d\gamma \right) \times$$
$$e^{2\pi i(2m\Omega)(t+nT)} g(t+nT)$$
$$- \sum_n \left(\frac{1}{(2\Omega)^{1/2}} \int_{[-\Omega,\Omega]^\sim} \hat{f}(\gamma) e^{-2\pi i n T\gamma} d\gamma \right) g(t+nT) \Big\}. \qquad (27)$$

The bracketed term in (27) can be written as

$$\{\cdots\} = \frac{1}{(2\Omega)^{1/2}} \sum_n g(t+nT) \times$$
$$\Big(\sum_{m\neq 0} e^{2\pi i(2m\Omega)(t+nT)} \int \hat{f}(\gamma)\mathbf{1}_\Omega(\gamma - 2m\Omega) e^{-2\pi i n T\gamma} d\gamma$$
$$- \sum_{m\neq 0} \int \hat{f}(\gamma)\mathbf{1}_\Omega(\gamma - 2m\Omega) e^{-2\pi i n T\gamma} d\gamma \Big)$$
$$= \sum_n \frac{1}{(2\Omega)} \tau_{-nT} d_{2\pi\Omega}(t) \times$$
$$\Big(\sum_{m\neq 0} \Big(\int \hat{f}(\gamma)\mathbf{1}_\Omega(\gamma - 2m\Omega) e^{-2\pi i n T\gamma} d\gamma \Big) \times$$
$$\Big(e^{2\pi i(2m\Omega)(t+nT)} - 1 \Big) \Big).$$

At this point we use our hypothesis that $2T\Omega = 1$ so that the bracketed term becomes

$$\begin{aligned}\left\{\cdots\right\} &= T\sum_n \tau_{-nT}d_{2\pi\Omega}(t)\sum_{m\neq 0}(e^{2\pi i(2m\Omega)t}-1)\times \\ &\quad \int \hat{f}(\gamma)\mathbf{1}_\Omega(\gamma-2m\Omega)e^{-2\pi i nT\gamma}d\gamma \\ &= T\sum_n\left(\int \hat{f}(\gamma)e^{2\pi i nT\gamma}\left(\sum_{m\neq 0}(e^{2\pi i(2m\Omega)t}-1)\mathbf{1}_\Omega(\gamma-2m\Omega)\right)d\gamma\right)\tau_{nT}d_{2\pi\Omega}(t) \\ &= T\sum(f*\alpha_{t,\Omega})(nT)\tau_{nT}d_{2\pi\Omega}(t).\end{aligned}$$

Combining this calculation with (27) yields (25).

In general, we must assume $f \in L^2(\mathbb{R}) \cap A(\mathbb{R})$ to expect the convergence of (25) in $L^2(\mathbb{R})$.

c. The *aliasing error* of $f \in L^2(\mathbb{R})$ for the low pass filter $d_{2\pi\Omega}$ with sampling at the Nyquist rate 2Ω is the second term on the right side of (25), viz.,

$$T\sum(f*\alpha_{t,\Omega})(nT)\tau_{nT}d_{2\pi\Omega}(t) = \epsilon_f(t).$$

Formally, standard calculations give

$$\|\epsilon_f\|_\infty \leq 2\int_{|\gamma|\geq\Omega}|\hat{f}(\gamma)|d\gamma.$$

§6. Irregular sampling and exact frames

The theory of non-harmonic Fourier series was developed by Paley and Wiener [49, Chapters 6 and 7] and Levinson [44, Chapter 4]. Related work preceding [49] is due to G. D. Birkhoff (1917), J. L. Walsh (1921), and Wiener (1927). The Paley-Wiener and Levinson theory has been reformulated and analyzed in terms of *irregular sampling* by Beutler [16], [17], who obtained completeness results, and Yao and Thomas [63], who obtained an irregular sampling theorem. The Yao and Thomas expansion was discovered independently by Higgins (1976) using reproducing kernels, *e.g.*, [30].

In this section we shall state and prove the Yao-Thomas irregular sampling expansion by frame methods. The coefficients in the expansion are the values of the given signal at the given irregularly spaced sampling points, and, because of our frame methods, the setting is necessarily in terms of exact frames (see Section 8). Whereas we implemented S^{-1} as a multiplier in Section 5, in this section we shall invoke the formula for S^{-1} contained in the following result, which is a consequence of Theorem 14.

Proposition 29. *Let H be a separable Hilbert space and let $\{g_n\} \subseteq H$ be an exact frame with inverse frame operator S^{-1}. Then*

$$\forall\, f \in H, \quad S^{-1}f = \sum \langle f, S^{-1}g_n\rangle S^{-1}g_n, \tag{28}$$

and so S^{-1} is the frame operator of the dual frame $\{S^{-1}g_n\}$.

In the case of irregular lattices, we have the following result [9, Theorem 4.1], which is the analogue of Theorem 19 for $\hat{\mathbb{R}}$. Although this result (Theorem 30) and Corollary 31 are concerned with frames and not exact frames, they are required in our proof of Theorem 32.

Theorem 30. *Let $g \in PW_\Omega$ for a given $\Omega > 0$. Assume that $\{a_n\}$, $\{b_m\}$ are real sequences for which*

$$\{e_{a_n}\} \quad \text{is a frame for} \quad L^2[-\Omega, \Omega],$$

and that there exist $A, B > 0$ such that

$$0 < A \le G(\gamma) \le B < \infty \ \text{ a.e. on } \hat{\mathbb{R}},$$

where

$$G(\gamma) = \sum |\hat{g}(\gamma - b_m)|^2.$$

Then $\{e_{a_n}\tau_{b_m}\hat{g}\}$ is a frame for $L^2(\hat{\mathbb{R}})$ with frame operator S; and $\{e_{a_n}\tau_{b_m}\hat{g}\}$ is a tight frame for $L^2(\hat{\mathbb{R}})$ if and only if $\{e_{a_n}\}$ is a tight frame for $L^2[-\Omega, \Omega]$ and G is constant a.e. on $\hat{\mathbb{R}}$.

Corollary 31. *Let us assume the hypotheses and notation of Theorem 30, and set $I_m = [-\Omega, \Omega] + b_m$. Then, for each fixed m, $\{\tau_{b_m}e_{a_n}\}$ is a frame for $L^2(I_m)$ with frame operator S_m, and*

$$\forall\, h \in L^2(\mathbb{R}),\ S\hat{h} = \sum (\tau_{b_m}\hat{g}) S_m(\hat{h}\tau_{b_m}\overline{\hat{g}}) \quad \text{in} \quad L^2(\hat{\mathbb{R}}).$$

Proof: We compute

$$\begin{aligned} S\hat{h} &= \sum_m \sum_n \langle \hat{h}, e_{a_n}\tau_{b_m}\hat{g}\rangle e_{a_n}\tau_{b_m}\hat{g} \\ &= \sum_m (\tau_{b_m}\hat{g}) \Big(\sum_n \langle \hat{h}, e_{a_n}\tau_{b_m}\hat{g}\rangle e_{a_n}\Big) \mathbf{1}_{I_m} \\ &= \sum_m (\tau_{b_m}\hat{g}) \Big(\sum_n \langle \hat{h}\tau_{b_m}\overline{\hat{g}}, e_{a_n}\rangle_{I_m} e_{a_n} \mathbf{1}_{I_m}\Big) \\ &= \sum_m (\tau_{b_m}\hat{g}) S_m(\hat{h}\tau_{b_m}\overline{\hat{g}}). \quad \blacksquare \end{aligned}$$

We can now prove the irregular sampling theorem for exact frames. Classical properties of the sampling functions $\{s_n\}$ in Theorem 32 are recorded in the remainder of the section.

Theorem 32. *Let $\{e_{a_n}\}$ be an exact frame for $L^2[-\Omega,\Omega]$ for a given $\Omega > 0$ and real sequence $\{a_n\}$. Define the sampling function s_n in terms of its involution $\tilde{s}_n(t) = \overline{s}_n(-t)$, where*

$$\forall\, t \in \mathbb{R},\ \ \tilde{s}_n(t) = \int_{-\Omega}^{\Omega} \overline{h}_n(\gamma) e^{2\pi i t\gamma}\, d\gamma,$$

and where $\{h_n\} \subseteq L^2[-\Omega,\Omega]$ is the unique sequence for which $\{e_{a_n}\}$ and $\{h_n\}$ are biorthonormal. (In particular, $\tilde{s}_n \in PW_\Omega$.) If $t_n = -a_n$ then

$$\forall\, f \in PW_\Omega,\ \ f = \sum f(t_n) s_n \ \text{ in } \ L^2(\mathbb{R}).$$

Proof: [9, Proof of Theorem 5.2]. Let $g = (2\Omega)^{-1/2} d_{2\pi\Omega}$, and set $b_m = 2m\Omega$. Since $\{e_{a_n}\}$ is a frame we can apply Theorem 30, and, hence, $\{e_{a_n}\tau_{b_m}\hat{g}\}$ is a frame for $L^2(\hat{\mathbb{R}})$ with frame operator S. In particular,

$$\forall\, h \in L^2(\mathbb{R}),\ \ \hat{h} = \sum \langle \hat{h}, e_{a_n}\tau_{b_m}\hat{g}\rangle S^{-1}(e_{a_n}\tau_{b_m}\hat{g}) \ \text{ in } \ L^2(\hat{\mathbb{R}}). \tag{29}$$

Similarly to (24), we obtain

$$\langle \hat{f}, e_{a_n}\tau_{b_m}\hat{g}\rangle = \begin{cases} (2\Omega)^{-1/2} f(-a_n), & \text{if} \quad m = 0, \\ 0, & \text{if} \quad m \neq 0 \end{cases} \tag{30}$$

for $f \in PW_\Omega$.

By means of Corollary 31 we can then verify that

$$S^{-1} = 2\Omega S_0^{-1} \quad \text{on} \quad L^2[-\Omega,\Omega].$$

Thus, since $g \in PW_\Omega$, we compute

$$S^{-1}(e_{a_n}\tau_{b_0}\hat{g}) = (2\Omega)^{1/2}\ S_0^{-1}(e_{a_n}\mathbf{1}_{(\Omega)}),$$

so that, by the exactness hypothesis and Proposition 29, the right side is

$$(2\Omega)^{1/2} \sum_m \langle e_{a_n}, h_m\rangle_{[-\Omega,\Omega]}\ h_m = (2\Omega)^{1/2}\ h_n.$$

Combining these two equalities with (29) and (30) gives the reconstruction,

$$\forall\, f \in PW_\Omega,\ \hat{f} = \sum_n (2\Omega)^{-1/2} f(-a_n)(2\Omega)^{1/2} h_n \ \text{ in } \ L^2(\mathbb{R}),$$

and the result follows. ■

Definition/Discussion 33.

a. For a given $\Omega > 0$, a real sequence $\{a_n\}$ satisfies the *Kadec-Levinson condition* if

$$\sup_n |a_n - \frac{n}{2\Omega}| < \frac{1}{4}(\frac{1}{2\Omega}). \tag{31}$$

b. Levinson [44, Theorem 18] (1940) proved that if (31) is assumed then $\{e_{a_n}\}$ is complete in $L^2[-\Omega, \Omega]$, and there is a unique sequence $\{h_n\}$ for which $\{e_{a_n}\}$ and $\{h_n\}$ are biorthonormal (see Theorem 14c). Kadec (1964) provided the direct calculation proving that $\{e_{a_n}\}$ is an exact frame for $L^2[-\Omega, \Omega]$ (see Theorem 60 and [64, pages 42-44]). Levinson [44, Theorem 19, in particular, page 67] proved that the bound "$\frac{1}{4}$" in (31) is "best possible", *i.e.*, there exists $\{a_n\} \subseteq \mathbb{R}$ for which equality is obtained in (31) and $\{e_{a_n}\}$ is complete in $L^2[-\Omega, \Omega]$, but $\{e_{a_n}\}$ is not an exact frame for $L^2[-\Omega, \Omega]$.

c. There are exact frames of exponentials that do not satisfy the Kadec-Levinson condition, [24, page 362], [29]. On the other hand, it is not known whether there are bases of exponentials which are not exact frames for $L^2[-\Omega, \Omega]$, *e.g.*, [64, page 197].

The explicit formulas in the following result were proved in [49, pages 89-90 and pages 114-116] and [44, pages 48 ff.]. The calculations by Paley and Wiener were refined by Young (1979), *e.g.*, [64, pages 148-150].

Theorem 34. *Let the real sequence $\{a_n\}$ satisfy the Kadec-Levinson condition for a given $\Omega > 0$. Then $\{e_{a_n}\}$ is an exact frame for $L^2[-\Omega, \Omega]$, there is a unique sequence $\{h_n\} \subseteq L^2[-\Omega, \Omega]$ for which $\{e_{a_n}\}$ and $\{h_n\}$ are biorthonormal, and $\tilde{s}_n$, defined in Theorem 32, is of the form*

$$\tilde{s}_n(t) = \frac{\tilde{s}(t)}{\tilde{s}'(a_n)(t - a_n)} \tag{32}$$

where

$$\tilde{s}(t) = (t - a_0) \prod_{n=1}^{\infty} (1 - \frac{t}{a_n})(1 - \frac{t}{a_{-n}}).$$

Definition/Example 35.

a. Let $\{L_n\}$ be a sequence of functions defined on $\mathbb{R}$ and let $\{t_n\}$ be a real sequence. $\{L_n, t_n\}$ is a *Lagrange interpolating system of functions L_n with respect to the sampling sequence $\{t_n\}$* if

$$\forall\, m, n \in \mathbb{Z}, \quad L_m(t_n) = \delta_{mn}.$$

b. Historically, *Lagrange's interpolation* is the method of defining a specific polynomial L of degree $\deg L \leq N$ such that L interpolates from $N + 1$ given points $(t_j, f(t_j)) \in \mathbb{R} \times \mathbb{C}, j = 0, \ldots, N$. L is of the form

$$L(t) = \sum_{m=0}^{N} f(t_m) L_m(t),$$

where

$$L_m(t) = \frac{p(t)}{p'(t_m)(t - t_m)}$$

and

$$p(t) = \prod_{j=0}^{N} (t - t_j),$$

(see the form of equation (32)). Clearly,

$$\forall\, m, n = 0, \ldots, N, \quad L_m(t_n) = \delta_{mn},$$

since

$$p'(t_m) = \prod_{j \neq m} (t_m - t_j)$$

and

$$\frac{p(t)}{(t - t_m)} = \prod_{j \neq m} (t - t_j).$$

One goal of such interpolation is to provide polynomial approximation for a function f whose values are known at $\{t_j : j = 0, \ldots, N\}$. Such approximation is useful for a variety of reasons, e.g., to estimate $\int f(t)\, dt$ even when we have complete knowledge of f. Unfortunately, the Lagrange polynomials L_m tend to oscillate wildly for large values of N.

To deal with this oscillation, the notion of spline and natural spline were defined on the interval $[t_0, t_N]$ for a given set of points $\{t_j : j = 0, \ldots, N\}$. A function $S : [t_0, t_N] \to \mathbb{C}$ is a *spline* of degree M if $S \in C^{M-1}[t_0, t_N]$ and S, restricted to each interval $[t_{j-1}, t_j]$, is a polynomial of degree $\deg S \leq M$. Because of issues dealing with degrees of freedom and symmetry with respect to endpoints, we deal with $M = 2K - 1$ and suppose

$$\forall\, j = K,\ K+1, \ldots, 2K-2, \qquad S^{(j)}(t_0) = S^{(j)}(t_N) = 0.$$

If a spline satisfies this additional condition, it is a *natural spline*. A fundamental fact about natural splines is that for a given data set $\{(t_j, f(t_j)) : j = 0, \ldots, N\}$, there is a unique natural spline S which interpolates the data in the case $N - 1 \geq k$.

An important type of inequality, which we shall state for $M = 3$ and which addresses the oscillation problem of Lagrange polynomials, is the following. *Let S be the natural spline interpolant for a given data set $\{(t_j, f(t_j)) : j = 0, \ldots, N\}$. If $f : [t_0, t_N] \to \mathbb{C}$ is another interpolant of the data, f' is absolutely continuous, and $f^{(2)} \in L^2[t_0, t_N]$, then*

$$\int_{t_0}^{t_N} |S^{(2)}(t)|^2\, dt \leq \int_{t_0}^{t_N} |f^{(2)}(t)|^2\, dt.$$

The curvature of a function f at t is $f^{(2)}(t)/(1+(f'(t))^2)^{3/2}$. Consequently, if f' is small then $f^{(2)}$ approximates the curvature; and so the natural spline interpolant can be viewed as the interpolant of minimum curvature in this case.

c. Let $\{e_{a_n}\}$ be an exact frame for $L^2[-\Omega,\Omega]$, and set $t_n = -a$. We know there is a unique sequence $\{h_n\}$ for which $\{e_{a_n}\}$ and $\{h_n\}$ are biorthonormal. If we define $\{s_n\}$ as we did in Theorem 32, then $\{s_n, t_n\}$ is a Lagrange interpolating system since

$$\begin{aligned} s_m(t_n) &= \overline{\hat{s}}(-t_n) \\ &= \int_{-\Omega}^{\Omega} h_n(\gamma) e^{-2\pi i(-t_n)\gamma} d\gamma \\ &= \langle h_n(\gamma), e^{2\pi i a_n \gamma} \rangle. \end{aligned}$$

d. We shall not discuss Lagrange interpolation any further except to make the following observations:

i. Lagrange interpolating systems not only provide a backdrop for classical considerations with splines (part *b*), but also play a role in expanding on the obvious relations between multiresolution analysis in wavelet theory and the theory of splines, *e.g.*, [22], [46], [47, pages 24-25]. This is all the more interesting since we have used Lagrange interpolating systems in the context of exact Fourier frames associated with coherent states and the Heisenberg group, whereas wavelet theory is characterized by translations and dilations and the $ax + b$ group.

ii. Interpolation problems are a fundamental part of the theory of entire functions, *e.g.*, [19] (see [63]).

iii. Interpolation problems have a dual relationship with the ideas we shall sketch in Section 9, *e.g.*, [41].

iv. If, instead of PW_Ω, we consider the space of pseudo- measures supported by $[-\Omega,\Omega]$, then Beurling (1959-1960) has characterized the conditions in order that a discrete set $\{(t_j, f(t_j))\} \subseteq \mathbb{R} \times \mathbb{C}$ can be used to interpolate a Fourier transform f on $\mathbb{R}$ of such a pseudo-measure, *e.g.*, [12]; the conditions are in terms of uniformly discrete sequences (Section 7) and uniform Beurling density (Section 9).

At the end of Section 2 we set our goal of obtaining the sampling formulas (14) and (15). In Theorem 32, accompanied by the well-known formulas of Theorem 34, we have (14), viz.,

$$f = \Sigma f(t_n) s_n.$$

Because of the setting of exact frames and biorthonormal sequences, (14) has a dual frame formulation,

$$f = \Sigma c_n(f) \tau_{t_n} d_{2\pi\Omega},$$

e.g., [29]. We shall now proceed to obtain (15) in Sections 7 and 8.

§7. The Duffin-Schaeffer theorem and frame conditions

Definition 36.

a. A sequence $\{a_n\} \subseteq \mathbb{R}$ is *uniformly discrete* if

$$\exists\ d > 0, \text{ such that } \forall\ m \neq n,\ |a_m - a_n| \geq d. \tag{33}$$

A uniformly discrete sequence $\{a_n\} \subseteq \mathbb{R}$ is *uniformly dense* with *uniform density* $\Delta > 0$ if

$$\exists\ \Delta > 0 \text{ and } \exists\ L > 0, \text{ such that } \forall n \in \mathbb{Z},\ |a_n - \frac{n}{\Delta}| \leq L. \tag{34}$$

The description "uniform" is used since (34) has the form

$$\sup_{n\in\mathbb{Z}} |a_n - \frac{n}{\Delta}| < \infty.$$

b. Classically, an increasing sequence $\{a_n\} \subseteq \mathbb{R}$ has *natural density* $\Delta \geq 0$ if

$$\lim_{r\to\infty} \frac{n_0(r)}{r} = \Delta,$$

where

$$n_0(r) = \text{card}\{a_n \in [-\frac{r}{2}, \frac{r}{2}]\}$$

("card" is cardinality). If a sequence fails to have a natural density it always has (finite or infinite) upper and lower natural density defined in terms of $\overline{\lim}$ and $\underline{\lim}$, resp. *e.g.*, [19, Section 1.5]. In Section 9 we shall deal with

$$n_I(r) = \text{card}\{a_n \in I, |I| = r\},$$

where I is an interval, and with *upper* and *lower uniform Beurling* densities.

Example 37.

a. Let $a_n = nT$, where $T > 0$ is fixed and $n \in \mathbb{Z}$. Then $nT \notin [-N, N]$ if and only if $nT > N$ or $nT < -N$. Thus, $\text{card}\{nT \in [-N, N]\}$ is essentially $2N/T$ and so $\{nT\}$ *has natural density* $1/T$. Clearly, *this example is uniformly dense with uniform density* $1/T$; in fact, $d = T$ in (33), and any $L \geq 0$ can be used in (34) for the case $\Delta = 1/T$, i.e., $\sup |a_n - \frac{n}{\Delta}| = 0$ for $\Delta = 1/T$. If $\{nT\}$ is a sampling sequence then the natural density $\Delta = 1/T$ is the number of samples per second, *e.g.*, Section 2, Definition 5b.

In this case, if $2T\Omega = 1$ then $\{e_{nT}\}$ is an orthonormal basis of $L^2[-\Omega, \Omega]$ (see Theorem 38).

b. *Uniform density is more restrictive than natural density.* To see this, let $\{a_n\} \subseteq \mathbb{R}$ be an increasing sequence satisfying (34), and assume $\lim_{n\to\infty} a_n = \infty$. This latter assumption is weaker than (33). Suppose $\{a_n\}$ satisfies the convenient, but non-essential, symmetric distribution, $a_{-n} = -a_n < 0$ for each $n \geq 1$. We shall prove that

$$\lim_{r\to\infty} \frac{n_0(r)}{r} = \Delta,$$

with $\Delta > 0$ as in (34).

Because of (34) we have

$$\left| \frac{a_n}{n} - \frac{1}{\Delta} \right| \leq \frac{1}{n},$$

and, hence,

$$\lim_{n\to\infty} \frac{n}{a_n} = \Delta.$$

By definition,

$$\forall\, n \geq 1, \quad n_0(2a_n) = 2n + 1,$$

and, so,

$$\lim_{n\to\infty} \frac{n_0(2a_n)}{2a_n} = \lim_{n\to\infty} \frac{2n+1}{2a_n} = \Delta$$

by the previous step and the assumption that $\lim_{n\to\infty} a_n = \infty$. Of course, we really want to consider $n_0(r)/r$ instead of $n_0(2a_n)/(2a_n)$. For a given a_n, $n \geq 1$, let a_p be the first element of the sequence for which $a_p > a_n$, i.e., $a_n = a_{n+1} = \ldots = a_{p-1} < a_p$. Then, for any $r \in (a_n, a_p)$,

$$n_0(2r) = 2(p-1) + 1,$$

and so

$$\frac{n_0(2r)}{2r} < \frac{2p-1}{2a_n} = \frac{2(p-1)+1}{2a_{p-1}}$$

since $1/r < 1/a_n$. As indicated above,

$$\lim_{p\to\infty} \frac{2(p-1)+1}{2a_{p-1}} = \Delta$$

and so

$$\overline{\lim_{r\to\infty}} \frac{n_0(r)}{r} \leq \Delta.$$

Similarly,

$$\underline{\lim_{r\to\infty}} \frac{n_0(r)}{r} \geq \Delta,$$

and we have shown that the natural density of $\{a_n\}$ is Δ.

c. If we consider the von Neumann lattice $\{(na, mb) : (n,m) \in \mathbb{Z} \times \mathbb{Z}\}$ for fixed $a, b > 0$, then the expected definition of its natural density is

$$\lim_{r\to\infty} \frac{\text{card}\{(na, mb) \in [-r,r] \times [-r,r]\}}{4r^2} \equiv \Delta.$$

In this case the cardinality in the numerator is essentially $(2r/a)(2r/b)$, and so

$$\Delta = \frac{1}{ab}.$$

Thus, we can phrase the Balian-Low phenomenon (Theorem 23) in terms of the natural density of the von-Neumann lattice. Further, if the natural density Δ of the von Neumann lattice is less than 1, then $\{e_{mb}\tau_{na}g\}$ is not even complete in $L^2(\mathbb{R})$, no matter which $g \in L^2(\mathbb{R})$ is chosen, *e.g.*, Remark 24b.

d. Suppose $\{a_n\} \subseteq \mathbb{R}$ is a sequence satisfying the Kadec-Levinson condition,

$$\sup_{n\in\mathbb{Z}} |a_n - \frac{n}{2\Omega}| < \frac{1}{4}\left(\frac{1}{2\Omega}\right),$$

defined in Section 6, Definition/Remark 33. We observe that $\{a_n\}$ is uniformly dense with uniform density 2Ω.

To see this we note that (34) is immediate for $\Delta = 2\Omega$ and $L = \frac{1}{4}\left(\frac{1}{2\Omega}\right)$. To verify (33), we compute

$$\begin{aligned} |a_m - a_n| &= |a_m - \frac{m}{2\Omega} + \frac{m}{2\Omega} - \frac{n}{2\Omega} + \frac{n}{2\Omega} - a_n| \\ &\geq |\frac{m}{2\Omega} - \frac{n}{2\Omega}| - \left(|a_m - \frac{m}{2\Omega}| + |\frac{n}{2\Omega} - a_n|\right) \\ &\geq \frac{1}{2\Omega} - \frac{1}{2}\left(\frac{1}{2\Omega}\right) = \frac{1}{2}\left(\frac{1}{2\Omega}\right) \equiv d \end{aligned}$$

for $m \neq n$.

Theorem 38. (Duffin and Schaeffer, [24, Theorem I]). *Let $\{a_n\} \subseteq \mathbb{R}$ be a uniformly dense sequence with uniform density Δ. If $0 < 2\Omega < \Delta$ then $\{e_{a_n}\}$ is a Fourier frame for $L^2[-\Omega, \Omega]$.*

After 40 years, Theorem 38 is still difficult to prove, and, among other notions and estimates, its proof involves fundamental properties of entire functions of exponential type.

We shall prove the following component of Theorem 38 since it is useful in our strengthening of this theorem, since it is associated with related work by Plancherel and Pólya (1937-1938) and Boas (1940), and since it is not too difficult to verify.

Lemma 39. *Let $\{e_{a_n}\}$ be a Fourier frame for $L^2[-\Omega, \Omega]$ for a given sequence $\{a_n\} \subseteq \mathbb{R}$ and a given $\Omega > 0$, and let $\{b_n\} \subseteq \mathbb{R}$ satisfy the condition,*

$$\sup_{n\in\mathbb{Z}} |a_n - b_n| \equiv M < \infty. \tag{35}$$

Then

$$\begin{aligned} &\exists\, C = C(\{a_n\}, \Omega, M) \quad \textit{such that} \\ \forall\, f \in PW_\Omega, \quad &\sum |f(-b_n)|^2 \leq C \sum |f(-a_n)|^2. \end{aligned} \tag{36}$$

Proof:

a. We first note that

$$\forall\, f \in PW_\Omega \ \text{ and } \ \forall\, k \geq 0, \qquad \|f^{(k)}\|_2 \leq (2\pi\Omega)^k \left(\int_{-\Omega}^{\Omega} |\hat{f}(\gamma)|^2 d\gamma \right)^{1/2}. \tag{37}$$

This is a consequence of Plancherel's theorem and the Ω- bandlimitedness of f, *i.e.*,

$$\|f^{(k)}\|_2^2 = \|(2\pi i\gamma)^k \hat{f}(\gamma)\|_2^2 \leq (2\pi\Omega)^{2k} \int_{-\Omega}^{\Omega} |\hat{f}(\gamma)|^2 d\gamma.$$

b. By Taylor's theorem and the fact that $f \in PW_\Omega$ is entire, we have

$$f(-b_n) - f(-a_n) = \sum_{k=1}^{\infty} \frac{f^{(k)}(-a_n)}{k!}(-b_n + a_n)^k$$

for each n. We use Hölder's inequality to obtain the estimate,

$$|f(-b_n) - f(-a_n)|^2 \leq \left(\sum_{k=1}^{\infty} \frac{|f^{(k)}(-a_n)|^2}{p^{2k}k!} \right) \left(\sum_{k=1}^{\infty} \frac{(|a_n - b_n|p)^{2k}}{k!} \right),$$

for any positive number p; and hence

$$|f(-b_n) - f(-a_n)|^2 \leq (e^{(Mp)^2} - 1) \sum_{k=1}^{\infty} \frac{|f^{(k)}(-a_n)|^2}{p^{2k}k!}. \tag{38}$$

c. Because of (37) and the Fourier pairing $f^{(k)}(t) \leftrightarrow (2\pi i\gamma)^k \hat{f}(\gamma)$, we have $f^{(k)} \in PW_\Omega$ when $f \in PW_\Omega$. Thus, using (37) again and the (upper) frame hypothesis, we compute

$$\forall\, k \geq 1, \quad \sum_n |f^{(k)}(-a_n)|^2 \leq B\|f^{(k)}\|_2^2 \leq B(2\pi\Omega)^{2k}\|f\|_2^2, \tag{39}$$

where B is the upper frame bound of $\{e_{a_n}\}$. Consequently, combining (38), (39) and the lower frame estimate with lower frame bound A, we have

$$\begin{aligned}
\forall\, f \in PW_\Omega, \quad & \sum |f(-b_n) - f(-a_n)|^2 \\
& \leq (e^{(Mp)^2} - 1) \sum_{k=1}^{\infty} \left(\frac{2\pi\Omega}{p} \right)^{2k} \frac{B}{k!} \|f\|_2^2 \\
& \leq \frac{B}{A} \left(e^{(Mp)^2} - 1 \right) \left(\sum_{k=1}^{\infty} \left(\frac{2\pi\Omega}{p} \right)^{2k} \frac{1}{k!} \right) \sum |f(-a_n)|^2 \\
& \equiv K^2 \sum |f(-a_n)|^2,
\end{aligned} \tag{40}$$

which is finite for any fixed $p > 2\pi\Omega$.

d. Because of Minkowski's inequality, (40) allows us to write

$$\left(\sum |f(-b_n)|^2\right)^{1/2} \le (1+K)\left(\sum |f(-a_n)|^2\right)^{1/2},$$

and (36) follows. ■

Theorem 40. *Let $\{a_n\} = A_1 \cup A_2 \subseteq \mathbb{R}$ be a disjoint union, where A_1 is a uniformly dense sequence $\{a_{1,n}\}$ with uniform density Δ and where $A_2 = \{a_{2,n}\}$ satisfies the condition,*

$$\sup_{n\in\mathbb{Z}} |a_{1,n} - a_{2,n}| \equiv M < \infty.$$

If $0 < 2\Omega < \Delta$ then $\{e_{a_n}\}$ is a Fourier frame for $L^2[-\Omega,\Omega]$.

Proof: Since A_1 is uniformly dense with uniform density Δ and $0 < 2\Omega < \Delta$, we can apply Theorem 38 to assert that $\{e_a : a \in A_1\}$ is a Fourier frame for $L^2[-\Omega,\Omega]$ with frame bounds A and B_1. Thus,

$$\begin{aligned}
\forall\, f \in PW_\Omega, \quad A\|f\|_2^2 &\le \sum_{a\in A_1} |f(-a)|^2 \\
&\le \sum |f(-a_n)|^2 \equiv \sum_{a\in A_1} |f(-a)|^2 + \sum_{a\in A_2} |f(-a)|^2 \\
&\le (1+C)\sum_{a\in A_1} |f(-a)|^2 \le (1+C)B_1\|f\|_2^2, \qquad (41)
\end{aligned}$$

where the inequalities follow since $f(-a) = \langle \hat{f}, e_a\rangle$ for $a \in A_1$, by the inclusion $A_1 \subseteq A_1 \cup A_2$, because of Lemma 39, and by the upper frame condition, respectively.

Clearly,

$$\sum |\langle \hat{f}, e_{a_n}\rangle|^2 = \sum_{a\in A_1} |\langle \hat{f}, e_a\rangle|^2 + \sum_{a\in A_2} |\langle \hat{f}, e_a\rangle|^2;$$

and, hence, from (41) we see that $\{e_{a_n}\}$ is a Fourier frame for $L^2[-\Omega,\Omega]$ with frame bounds A and $B = (1+C)B_1$. ■

Example 41. In light of Theorem 40 and Theorem 44, and the usefulness of such results in applying Theorem 46 and some of the other material in Section 8, we shall now point out the generality of sampling sequences $\{t_n\}$ determined by the hypotheses of these theorems, letting $a_n = -t_n$.

Suppose $\{a_{1,n}\}$ has uniform density $\Delta = 1$, upper bound $L \in \mathbb{N}$, and lower uniformly discrete bound d. Let $\{a_n\} = \{a_{1,n}\} \cup \{a_{2,n}\}$, where $\{a_{2,n}\}$

satisfies the hypothesis of Theorem 40. There are two phenomena we wish to mention about such sequences.

First, *if d is small enough then $\{a_{1,n}\}$ can have the property that there is a subsequence of $\mathbb{Z}$ each of whose elements can have as many as $2L+1$ elements of $\{a_{1,n}\}$ close to it.* For example, let $L = 3$ and $d = 1/10$, and, for simplicity of exposition let $\{b_n\} \equiv \{a_{1,n}\}$ be symmetric. Set $b_1 = 1$, $b_2 = 5-3d$, $b_3 = 5-2d$, $b_4 = 5-d$, $b_5 = 5$, $b_6 = 5+d$, $b_7 = 5+2d$, $b_8 = 5+3d$. Hence, $2L+1 = 7$ elements of $\{b_n\}$ are within a distance of $3d = 3/10$ of $n = 5$. In this case, of course, it is necessary that $b_9 \geq 6$.

Second, *for properly chosen $\{a_{2,n}\}$, $\{a_n\}$ can have the property that* $\lim_{n\to\infty}(a_{n+1} - a_n) = 0$.

The types of sequences used in the hypotheses of Theorems 40 and 44 are more closely related than they appear, but we shall not pursue that topic here (see Formula 45). The main point is that the sampling sequences that can be used in our theory of Section 8 are much more general that those allowed by the Kadec-Levinson condition and the classical results of Section 6. Of course, one does have to show a little respect for Theorem 32 *vis a vis* the frame decompositions of Section 8, since there are exact frames which do not satisfy the Kadec-Levinson condition.

We can prove an important refinement of Theorem 40, by invoking the following result from the Plancherel-Pólya theory [52] (see [18], [19, pages 97–103]).

Lemma 42.

a. *For all $f \in PW_\Omega$,*

$$\int |f(t+iy)|^2 dt \leq e^{2\pi\Omega|y|}\|f\|_2^2.$$

b. Let $\{a_n\} \subseteq \mathbb{R}$ be a uniformly discrete sequence with lower bound d (as in (33)). Then

$$\forall\, f \in PW_\Omega, \quad \sum |f(-a_n)|^2 \leq B\|f\|_2^2, \tag{42}$$

where B can be taken as

$$B \equiv \frac{4}{\pi^2 \Omega d^2}\left(e^{\pi\Omega d} - 1\right).$$

Proof:

a. We shall omit the classical proof of part a which involves a Phragmén-Lindelöf argument, *e.g.*, [52], [64, pages 94–96]. Instead, in Remark 43 we shall hint at our verification of part a which is appropriate for our multidimensional work.

b. i. Let f be an element of PW_Ω. Then

$$\forall\, z_0 \in \mathbb{C} \text{ and } \forall\, r > 0, \qquad |f(z_0)|^2 \leq \frac{1}{2\pi}\int_{-\pi}^{\pi} |f(z_0 + re^{i\theta})|^2 d\theta, \tag{43}$$

i.e., the continuous function f has the property that $|f|^2$ is subharmonic. (Of course, we really know that f is an entire function.) The verification of (43) is standard. First, $\log|f|$ is upper semicontinuous, and one uses Jensen's formula to prove that $u = \log|f|$ satisfies an inequality analogous to (43). Then, since $\phi(t) = e^{2t}$ is an increasing convex function on $\mathbb{R}$, we obtain a similar inequality, *viz.*, (43) for $\phi \circ u$.

b. ii. We multiply both sides of (43) by r and integrate between 0 and $\frac{d}{2}$ to obtain

$$\begin{aligned}\forall\, z_0 \in \mathbb{C}, \quad |f(z_0)|^2 &\leq \frac{4}{\pi d^2} \int_0^{\frac{d}{2}} \int_{-\pi}^{\pi} |f(z_0 + re^{i\theta})|^2 d\theta dr \\ &= \frac{4}{\pi d^2} \iint_{|z-z_0| \leq \frac{d}{2}} |f(z)|^2 dt dy, \quad z = t + iy.\end{aligned}$$

Therefore, letting $t_n = -a_n$, we can make the estimate,

$$\begin{aligned}\sum |f(t_n)|^2 &\leq \frac{4}{\pi d^2} \sum \iint_{|z| \leq \frac{d}{2}} |f(z + t_n)|^2 dt dy \\ &\leq \frac{4}{\pi d^2} \int_{-\frac{d}{2}}^{\frac{d}{2}} \sum \int_{-\frac{d}{2}}^{\frac{d}{2}} |f(t + t_n + iy)|^2 dt dy \\ &= \frac{4}{\pi d^2} \int_{-\frac{d}{2}}^{\frac{d}{2}} \sum \int_{t_n - \frac{d}{2}}^{t_n + \frac{d}{2}} |f(t + iy)| dt dy \\ &\leq \frac{4}{\pi d^2} \int_{-\frac{d}{2}}^{\frac{d}{2}} \int |f(t + iy)| dt dy, \end{aligned} \tag{44}$$

where the last inequality follows by the definition of d.

b. iii. We apply part a to estimate the right side of (44), and obtain

$$\begin{aligned}\sum |f(-a_n)|^2 &\leq \frac{4}{\pi d^2} \int_{-\frac{d}{2}}^{\frac{d}{2}} e^{2\pi\Omega|y|} dy \; \|f\|_2^2 \\ &= \frac{4}{\pi^2 \Omega d^2} \left(e^{\pi\Omega d} - 1\right) \|f\|_2^2. \quad \blacksquare\end{aligned}$$

Discussion 43. *The Plancherel-Pólya theory.*

a. The following formal calculation for $f \in PW_\Omega$ can be made rigorous, not only to provide an alternative to the Phragmén- Lindelöf proof of Lemma 42a but to allow a means of dealing with analogous problems in d-dimensions [10]:

$$\left(\int |f(t+iy)|^2 dt\right)^{1/2} = \left(\int \left| \int_{-\Omega}^{\Omega} \hat{f}(\gamma) e^{2\pi i t \gamma} e^{-2\pi y \gamma} d\gamma \right|^2 dt\right)^{1/2}$$

$$= \left(\int_{-\Omega}^{\Omega} \int_{-\Omega}^{\Omega} \hat{f}(\gamma) \overline{\hat{f}}(\lambda) e^{-2\pi y(\lambda+\gamma)} \left(\int e^{2\pi i t(\gamma-\lambda)} dt \right) d\lambda d\gamma \right)^{1/2}$$

$$= \left(\int_{-\Omega}^{\Omega} \overline{\hat{f}}(\lambda) e^{-2\pi y \lambda} \int_{-\Omega}^{\Omega} \delta(\gamma - \lambda) \hat{f}(\gamma) e^{-2\pi y \gamma} d\gamma d\lambda \right)^{1/2}$$

$$= \left(\int_{-\Omega}^{\Omega} (|\hat{f}(\lambda)| e^{-2\pi y \lambda})^2 d\lambda \right)^{1/2} \leq e^{2\pi |y| \Omega} \|f\|_2,$$

where the manipulations with δ and $\int e^{2\pi i y(\gamma-\lambda)} dt$ can be dealt with properly by approximate identity arguments.

b. Boas [18, Theorems 1,2,3] has provided a proof of Lemma 42 different from the argument of Plancherel-Póyla (see [19, pages 100-101] for another exposition of the Plancherel-Pólya proof). An attractive feature of Boas' approach in 1940 is that it leads us to rewrite (42) as a *weighted Fourier transform norm inequality*, viz.,

$$\|f\|_{2,\nu}^2 \leq B \|\hat{f}\|_{2,\mu}^2. \tag{45}$$

In the case of (42), ν is the discrete measure

$$\nu = \sum \delta_{-a_n},$$

μ is the absolutely continuous measure $\mu = \mathbf{1}_{(\Omega)}$ (usually more comfortably written as "$d\mu(\gamma) = \mathbf{1}_{(\Omega)}(\gamma) d\gamma$"), and PW_Ω is the weighted L^2-space, $L^2_\mu(\hat{\mathbb{R}})$. Weights such as $\mu = \mathbf{1}_{(\Omega)}$ play a role in the Bell Labs uncertainty principle theory from the early 1960s, *e.g.*, [42], and measure weights play a significant role in the general uncertainty principle theory, *e.g.*, [6] as well as [19, Chapter 10].

In the same spirit as Theorem 40, Theorem 38 and the ideas of Plancherel-Pólya play a major role in the following recent observation by Jaffard [33].

Theorem 44. *Let $\{a_n\} \subseteq \mathbb{R}$ be the union of a finite number of uniformly discrete sequences and a uniformly dense sequence having uniform density Δ. If $0 < 2\Omega < \Delta$ then $\{e_{a_n}\}$ is a Fourier frame for $L^2[-\Omega, \Omega]$.*

Proof: Let $\{a_n\} = A_1 \cup \ldots \cup A_n$, where A_1 has uniform density Δ and each of the A_j is uniformly discrete. Take any $\Omega > 0$ for which $0 < 2\Omega < \Delta$. Then, by Theorem 38, $\{e_a : a \in A_1\}$ is a Fourier frame for $L^2[-\Omega, \Omega]$. Thus,

$$\forall\, f \in PW_\Omega, \qquad A\|f\|_2^2 \leq \sum_{a \in A_1} |\langle \hat{f}, e_a \rangle|^2$$

$$\leq \sum |\langle \hat{f}, e_{a_n} \rangle|^2 = \sum |f(-a_n)|^2. \tag{46}$$

For the upper frame bound, we invoke Lemma 42 and obtain

$$\forall\, j = 1, \ldots, m \quad \text{and} \quad \forall\, f \in PW_\Omega,$$

$$\sum_{a \in A_j} |\langle \hat{f}, e_a \rangle|^2 \leq B_j \|f\|_2^2.$$

Consequently,

$$\sum |f(-a_n)|^2 = \sum_{j=1}^{n} \sum_{a \in A_j} |\langle \hat{f}, e_a \rangle|^2 \leq B \|f\|_2^2, \tag{47}$$

where $B = B_1 + \ldots + B_n$. Combining (46) and (47), we obtain our result. ■

Formula 45. *The frame radius.*

We have just used the Duffin-Schaeffer theorem (Theorem 38) to prove Theorem 44. Theorem 44 provides a density/discreteness condition on a sequence $\{a_n\} \subseteq \mathbb{R}$ to ensure that $\{e_{a_n}\}$ is a Fourier frame; and it turns out that this condition on $\{a_n\} \subseteq \mathbb{R}$ is not only sufficient but necessary in order that $\{e_{a_n}\}$ be a Fourier frame [33, Lemma 2] (see Theorem 40).

There is the following closely related problem for a given sequence $\{a_n\} \subseteq \mathbb{R}$: determine

$$\Omega_r \equiv \sup\{\Omega \geq 0 : \{e_{a_n}\} \text{ is a Fourier frame for } L^2[-\Omega, \Omega]\}.$$

The quantity $\Omega_r \in [0, \infty]$ is the *frame radius* of $\{a_n\}$, and the problem has recently been resolved by Jaffard [33, Theorem 3], and requires the Duffin-Schaeffer theorem (Theorem 38), as well as an important result by Landau [41] (see Section 9).

Jaffard's solution includes the following formula for the frame radius in the case $\Omega_r \in (0, \infty)$:

$$\Omega_r = \frac{1}{2} \sup_b \Delta(b), \tag{48}$$

where b is a uniformly dense subsequence of $\{a_n\}$ having uniform density $\Delta(b)$.

The proof of (48) uses Theorem 38 to prove the inequality,

$$\Omega_r \geq \frac{1}{2} \sup_b \Delta(b). \tag{49}$$

To verify (49) let us assume for simplicity that $\{a_n\}$ is uniformly dense with uniform density Δ. If $\Omega_r < \frac{1}{2}\Delta$ we can choose $\Omega \in (\Omega_r, \frac{1}{2}\Delta)$ and contradict the definition of Ω_r by applying Theorem 38; therefore $\Omega_r \geq \frac{1}{2}\Delta$. Details to prove the general case, (49), are found in [33] and are not difficult. The opposite inequality not only requires Landau's theorem, mentioned above, but also a more formidable contribution by Jaffard.

§8. Irregular sampling and frames

The results of Section 7 will be used in conjunction with Theorem 46 to formulate sampling formulas such as Equation (50) in terms of given sequences $\{t_n\}$ of sampling points.

Theorem 46 [9]. *Suppose $\Omega > 0$ and $\Omega_1 > \Omega$, and let the sequence $\{t_n\} \subseteq \mathbb{R}$ have the property that $\{e_{-t_n}\}$ is a Fourier frame for $L^2[-\Omega_1, \Omega_1]$ with frame operator S. Further, let $s \in L^2(\mathbb{R})$ have the properties that $\hat{s} \in L^\infty(\hat{\mathbb{R}})$, $supp\, \hat{s} \subseteq [-\Omega_1, \Omega_1]$, and $\hat{s} = 1$ on $[-\Omega, \Omega]$. Then*

$$\forall\ f \in PW_\Omega,\ \ f = \sum c_n(f) \tau_{t_n} s \ \text{ in } \ L^2(\mathbb{R}) \tag{50}$$

where

$$c_n(f) = \left\langle S^{-1}(\hat{f}\mathbf{1}_{(\Omega_1)}), e_{-t_n} \right\rangle_{[-\Omega_1, \Omega_1]}.$$

Proof: Since $\{e_{-t_n}\}$ is a frame for $L^2[-\Omega_1, \Omega_1]$ and $supp\, \hat{f} \subseteq [-\Omega, \Omega]$ we have

$$\hat{f} = \hat{f}\mathbf{1}_{(\Omega_1)} = \sum \left\langle S^{-1}(\hat{f}\mathbf{1}_{(\Omega_1)}), e_{-t_n} \right\rangle_{[-\Omega_1, \Omega_1]} e_{t_n} \ \text{ in } \ L^2[-\Omega_1, \Omega_1]. \tag{51}$$

Equation (51) is a direct consequence of Theorem 14a and the fact that S^{-1}, being a positive operator, is self-adjoint. Using the hypothesis, $f \in PW_\Omega$, we can rewrite (51) as

$$\hat{f} = \sum c_n(f)(e_{-t_n}\mathbf{1}_{(\Omega_1)}) \ \text{ in } \ L^2(\hat{\mathbb{R}}). \tag{52}$$

In fact,

$$\begin{aligned} &\|\hat{f} - \sum_M^N c_n(f)(e_{-t_n}\mathbf{1}_{(\Omega_1)})\|_2^2 \\ &\qquad = \int_{-\Omega_1}^{\Omega_1} |\hat{f}(\gamma) - \sum_M^N c_n(f) e_{-t_n}(\gamma)|^2 d\gamma \\ &\qquad = \|\hat{f} - \sum_M^N c_n(f) e_{-t_n}\|^2_{L^2[-\Omega_1, \Omega_1]}, \end{aligned}$$

and so (52) is valid.

Next, we note that $\hat{f} = \hat{f}\hat{s}$, and, hence,

$$\begin{aligned} &\|\hat{f} - \sum_M^N c_n(f)(e_{-t_n}\hat{s})\|_2^2 \\ &\qquad = \|\hat{f}\hat{s} - \sum_M^N c_n(f)(e_{-t_n}\mathbf{1}_{(\Omega_1)})\hat{s})\|_2^2 \\ &\qquad \le \|\hat{s}\|_\infty^2 \ \|\hat{f} - \sum_M^N c_n(f)(e_{-t_n}\mathbf{1}_{(\Omega_1)})\|_2^2. \end{aligned}$$

Using this estimate, Equation (52), and the hypotheses on s, we obtain

$$\hat{f} = \sum c_n(f)(e_{-t_n}\hat{s}) \text{ in } L^2(\hat{\mathbb{R}}). \tag{53}$$

Finally, we obtain Equation (50) from Equation (53) and the Plancherel theorem. ■

In the following result, an operation must be performed on a given sequence to construct a subsequence $\{t_n\}$ with which we can implement Theorem 46. As such, it doesn't quite fit into the mold of Section 7, so we state it now as a lemma to be used in the proof of Theorem 48.

Lemma 47 [29]. *Let $\{a_k\} \subseteq \mathbb{R}$ be a strictly increasing sequence for which*

$$\lim_{k\to\pm\infty} a_k = \pm\infty$$

and

$$\sup_k \, (a_{k+1} - a_k) \equiv T < \infty.$$

Assume $\Omega_1 > 0$ satisfies the condition

$$2T\Omega_1 < 1.$$

There is a constructible subsequence $\{a_{k_n}\}$ of $\{a_k\}$ such that, setting $-t_n = a_{k_n}, \{e_{-t_n}\}$ is a frame for $L^2[-\Omega_1, \Omega_1]$.

Proof: Since $2T\Omega_1 < 1$ we can choose $\epsilon > 0$ so that

$$T < \frac{1}{2\Omega_1 + \epsilon}.$$

Next, choose $\delta > 0$ such that

$$T + \delta < \frac{1}{2\Omega_1 + \epsilon},$$

and define intervals I_n as

$$\forall\, n \geq 1, \quad I_n = \left[\frac{n}{2\Omega_1 + \epsilon} - \frac{1}{2}\Big(\frac{1}{2\Omega_1 + \epsilon} - \delta\Big), \right.$$
$$\left. \frac{n}{2\Omega_1 + \epsilon} + \frac{1}{2}\Big(\frac{1}{2\Omega_1 + \epsilon} - \delta\Big)\right].$$

Note that $\{I_n\}$ is a disjoint collection since

$$\Big(\frac{n+1}{2\Omega_1 + \epsilon} - \frac{1}{2}\Big(\frac{1}{2\Omega_1 + \epsilon} - \delta\Big)\Big)$$
$$- \Big(\frac{n}{2\Omega_1 + \epsilon} + \frac{1}{2}\Big(\frac{1}{2\Omega_1 + \epsilon} - \delta\Big)\Big) = \delta > 0. \tag{54}$$

The length of each interval I_n is

$$|I_n| = \frac{1}{2\Omega_1 + \epsilon} - \delta > T.$$

Therefore, by the definition of T, each I_n must contain elements of $\{a_k\}$. For each n, choose precisely one such element a_{k_n}, *e.g.*, the smallest or largest element of $\{a_k\} \cap I_n$, and designate it as $-t_n = a_{k_n}$.

Writing $a_n \equiv -t_n$, (54) implies that

$$\forall\, n, \quad (a_{n+1} - a_n) \geq \delta > 0.$$

Further,

$$|a_n - \frac{n}{2\Omega_1 + \epsilon}| \leq \frac{1}{2}\Big(\frac{1}{2\Omega_1 + \epsilon} - \delta\Big) \equiv L,$$

by the definition of I_n. Consequently, $\{a_k\}$ is a uniformly dense sequence with uniform density $\Delta \equiv 2\Omega_1 + \epsilon > 2\Omega_1$. By the Duffin-Schaeffer theorem, Theorem 38, $\{e_{-t_n}\}$ is a frame for $L^2[-\Omega_1, \Omega_1]$. ■

In general, Lemma 47 is false when $2T\Omega_1 = 1$, *e.g.*, [29].

Combining Theorem 46 with Lemma 47 we obtain the following.

Theorem 48. *Let $\{a_k\} \subseteq \mathbb{R}$ be a strictly increasing sequence for which*

$$\lim_{k \to \pm\infty} a_k = \pm\infty$$

and

$$\sup_k \,(a_{k+1} - a_k) \equiv T < \infty.$$

For a given $\Omega > 0$, assume $\Omega_1 > \Omega$ satisfies the condition,

$$2T\Omega_1 < 1,$$

and let $s \in L^2(\mathbb{R})$ have the properties that $\hat{s} \in L^\infty(\hat{\mathbb{R}})$, $\operatorname{supp} \hat{s} \subseteq [-\Omega_1, \Omega_1]$, and $\hat{s} = 1$ on $[-\Omega, \Omega]$. There is a constructible subsequence $\{a_{k_n}\}$ of $\{a_k\}$ such that, setting $-t_n = a_{k_n}, \{e_{-t_n}\}$ is a frame for $L^2[-\Omega_1, \Omega_1]$ with frame operator S, and

$$\forall\, f \in PW_\Omega, \quad f = \sum \Big\langle S^{-1}(\hat{f}\mathbf{1}_{(\Omega_1)}), e_{-t_n} \Big\rangle_{[-\Omega_1, \Omega_1]} \tau_{t_n} s \quad \text{in} \quad L^2(\mathbb{R}).$$

Naturally, the implementation of Theorem 46 depends on the computation of the coefficients in Equation (50). For the case of exact frames these coefficients are computed in the following result in terms of the Lagrange interpolating system introduced in Theorem 32. We shall not now pursue the expansions resulting from a combination of Theorem 46 and Theorem 49, and we refer to [9, Theorem 6.3] for a proof of Theorem 49.

Theorem 49 [9]. *Suppose $\Omega > 0$ and $\Omega_1 > \Omega$, and let the sequence $\{t_n\} \subseteq \mathbb{R}$ have the property that $\{e_{a_n}\}$ is an exact frame for $L^2[-\Omega_1, \Omega_1]$ with frame operator S, where $a_n = -t_n$. Define the function s_n in terms of its involution $\tilde{s}_n(t) = \overline{s}_n(-t)$, where*

$$\forall\, t \in \mathbb{R}, \quad \tilde{s}_n(t) = \int_{-\Omega_1}^{\Omega_1} \overline{h}_n(\gamma) e^{2\pi i t \gamma} d\gamma,$$

and where $\{h_n\} \subseteq L^2[-\Omega_1, \Omega_1]$ is the unique sequence for which $\{e_{a_n}\}$ and $\{h_n\}$ are biorthonormal. The coefficients of the expansion in Equation (50) are

$$\forall\, n, \quad \left\langle S^{-1}(\hat{f}\mathbf{1}_{(\Omega_1)}), e_{-t_n} \right\rangle_{[-\Omega_1, \Omega_1]} = \langle f, s_n \rangle, \tag{55}$$

where $f \in PW_\Omega$.

Of course the real purpose of Theorem 46 is to provide sampling formulas with effectively computed coefficients, where the sampling sequences are not constrained by exactness. This is the role of the following algorithm which is expanded upon in [10].

Algorithm 50 [9]. It is possible to estimate the coefficients in Equation (50), and, in the process, to see to what extent the coefficients contain information from the sampled values $f(t_n)$.

a. Let $\{e_{-t_n}\}$ be a frame for $L^2[-\Omega_1, \Omega_1]$ with frame operator S and frame bounds A and B. Recall that if S_1 and S_2 are operators, then $S_1 \geq S_2$ means that

$$\forall\, f, \quad \langle S_1 f, f \rangle \geq \langle S_2 f, f \rangle.$$

Thus, by our frame hypothesis, we have

$$AI \leq S \leq BI,$$

where I is the identity operator. Consequently, we can compute

$$I - \frac{2}{A+B} S \leq I - \frac{2}{A+B} AI = \left(\frac{A+B-2A}{A+B}\right) I$$

and

$$I - \frac{2}{A+B} S \geq I - \frac{2}{A+B} BI = \left(\frac{A+B-2B}{A+B}\right) I;$$

and, hence,

$$\|I - \frac{2}{A+B} S\| \leq \frac{B-A}{A+B} < 1, \tag{56}$$

where the norm in (56) is the operator norm.

b. Because of (56) and the Neumann expansion we have

$$S^{-1} = \frac{2}{A+B} \sum_{k=0}^{\infty} \Big(I - \frac{2}{A+B} S \Big)^k, \tag{57}$$

where the convergence criterion in (57) is the operator norm topology on the space of continuous linear operators on $L^2[-\Omega_1, \Omega_1]$ (into itself). In fact, (56) and the Neumann expansion really yield

$$\Big(\frac{2}{A+B} S \Big)^{-1} = \sum_{k=0}^{\infty} \Big(I - \frac{2}{A+B} S \Big)^k; \tag{58}$$

and

$$\Big(\frac{2}{A+B} S \Big)^{-1} (g) = f$$

means

$$\frac{A+B}{2} g = Sf,$$

so that since S^{-1} is linear we have

$$\frac{A+B}{2} S^{-1} g = f. \tag{59}$$

Equations (58) and (59) give Equation (57).

c. Substituting (57) into (50), we have

$$c_n(f) = \sum_{k=0}^{\infty} \frac{2}{A+B} \Big\langle \Big(I - \frac{2}{A+B} S \Big)^k \Big(\hat{f} \mathbf{1}_{(\Omega_1)} \Big), e_{-t_n} \Big\rangle_{[-\Omega_1, \Omega_1]} \tag{60}$$

for $f \in PW_\Omega$, $\Omega < \Omega_1$.

We can easily calculate the terms of Equation (60), e.g.,

$$\frac{2}{A+B} \Big\langle \hat{f} \mathbf{1}_{(\Omega_1)}, e_{-t_n} \Big\rangle_{[-\Omega_1, \Omega_1]} = \frac{2}{A+B} f(t_n)$$

and

$$\frac{2}{A+B} \Big\langle \Big(I - \frac{2}{A+B} S \Big) \Big(\hat{f} \mathbf{1}_{(\Omega_1)} \Big), e_{-t_n} \Big\rangle_{[-\Omega_1, \Omega_1]}$$
$$= \frac{2}{A+B} f(t_n) - \Big(\frac{2}{A+B} \Big)^2 \Big[2\Omega_1 f(t_n) + \sum_{n_1 \neq n} f(t_{n_1}) d_{2\pi\Omega_1}(t_n - t_{n_1}) \Big]$$

are the cases $k = 0$ and $k = 1$, respectively. Thus, if we truncate the expansion (60) after the $k = 0$ term, then

$$c_n(f) \approx \frac{2}{A+B} f(t_n),$$

which is a particularly auspicious estimate when $A = B = 1$.

Motivated by Theorems 46 and 48, and noting the importance of the frame bounds in Algorithm 50 we have the following.

Theorem 51. *Suppose $\Omega > 0$, and let the sequence $\{a_n\} \subseteq \mathbb{R}$ have the properties that $\{a_n\}$ is strictly increasing, $\lim_{n\to\pm\infty} a_n = \pm\infty$, and*

$$0 < d \leq \inf(a_{n+1} - a_n) \leq \sup(a_{n+1} - a_n) = T < \infty.$$

Assume $2T\Omega < 1$. Then $\{e_{a_n}\}$ is a Fourier frame for $L^2[-\Omega, \Omega]$ with frame bounds A and B satisfying the inequalities

$$A \geq \frac{(1 - 2T\Omega)^2}{T}$$

and

$$B \leq \frac{4}{\pi^2 \Omega d^2}(e^{\pi\Omega d} - 1).$$

Proof: The upper bound for B is a consequence of Lemma 42b.

The lower bound for A depends on the following result by Gröchenig [27, Theorem 1]. *Let $K : PW_\Omega \to PW_\Omega$ be defined as*

$$Kf = d_{2\pi\Omega} * \sum f(t_n)\mathbf{1}_n,$$

where $t_n = -a_n$ and $\mathbf{1}_n$ is the characteristic function of the interval

$$[(a_{n+1} + a_n)/2, (a_{n+2} + a_{n+1})/2].$$

Then

$$\|Kf\|_2 \leq (1 + 2T\Omega)\|f\|_2 \quad \text{and} \quad \|f - Kf\|_2 \leq 2T\Omega\|f\|_2.$$

Consequently, $\|K^{-1}\| \leq 1/(1 - 2T\Omega)$, so that since $f = K^{-1}Kf$ we have

$$\begin{aligned} \|f\|_2 &\leq \frac{1}{1 - 2T\Omega}\|\sum f(t_n)\mathbf{1}_n\|_2 \\ &\leq \frac{\sqrt{T}}{1 - 2T\Omega}\left(\sum |f(t_n)|^2\right)^{1/2}, \end{aligned}$$

for all $f \in PW_\Omega$, e.g., [29, Corollary 4.4.4]. ■

Remark 52. Our "major frame" sampling formula in this section did not deal with frames for $L^2(\mathbb{R})$, but with Fourier frames for $L^2[-\Omega_1, \Omega_1]$. This provides a gain in simplicity and an opportunity to implement the results from Section 7. On the other hand, we lose the flexibility to consider aliasing as we did in Section 5. Further, we can prove Theorem 46 in terms of Gabor frames for $L^2(\mathbb{R})$. This will allow us to deal with aliasing, and to obtain better implementation of Algorithm 50, *e.g.*, [10].

The following examples are concerned with Gabor frames for $L^2(\mathbb{R})$ and $L^2(\hat{\mathbb{R}})$. Details are found in [9, Section 4], and the material plays a role in [10].

Example 53. Let $\{a_n\}$, $\{b_m\}$ be real sequences and let $\Omega > 0$. Assume that $\{e_{a_n}\}$ is a Fourier frame for $L^2[-\Omega, \Omega]$, and that there exist $d, \Delta > 0$ such that

$$\forall\, m \in \mathbb{Z}, \qquad 0 < d \leq b_{m+1} - b_m \leq \Delta < 2\Omega,$$

where $\lim_{m\to\pm\infty} b_m \pm \infty$. (In particular, $\{b_m\}$ is uniformly discrete.) Suppose $g \in PW_\Omega$ has the properties that $\hat{g} \in L^\infty(\hat{\mathbb{R}})$ and $\inf\{|\hat{g}(\gamma)|^2 : \gamma \in I\} > 0$ for some interval $I \subseteq [-\Omega, \Omega]$ having length $|I| = \Delta$. Then $\{e_{a_n}\tau_{b_m}\hat{g}\}$ is a frame for $L^2(\hat{\mathbb{R}})$ (see Theorem 51). This result follows once we verify the hypotheses of Theorem 30.

Example 54. Suppose we assume the hypotheses of Theorem 30, and that we consider the case $a_n = na$ and $a = \frac{1}{2\Omega}$.

a. Let S be the frame operator for the frame $\{e_{a_n}\tau_{b_m}\hat{g}\}$ (for $L^2(\hat{\mathbb{R}})$) obtained by Theorem 30. A routine calculation allows us to conclude that

$$\forall\, f \in L^2(\mathbb{R}), \qquad S^{-1}\hat{f} = \frac{1}{2\Omega}\frac{\hat{f}}{G},$$

where $G(\gamma) = \sum |\hat{g}(\gamma - b_m)|^2$.

b. Part a allows us to write

$$S^{-1}(e_{na}\tau_{b_m}\hat{g}) = \frac{1}{2\Omega}\frac{e_{na}\tau_{b_m}\hat{g}}{G},$$

whereas

$$e_{na}\tau_{b_m}S^{-1}\hat{g} = \frac{1}{2\Omega}\frac{e_{na}\tau_{b_m}\hat{g}}{\tau_{b_m}G}.$$

Consequently, *the operators S^{-1} and $e_{na}\tau_{b_m}$ are not commutative for irregular sequences.* For perspective, recall that our frame approach to the classical sampling theorem utilized the commutativity of these operators when we invoked Theorem 20 in the proof of Theorem 25.

§9. Stability and uniqueness

For a given sequence $\{a_n\} \subseteq \mathbb{R}$, a major part of the sampling theory we have presented has depended on whether $\{e_{a_n}\}$ is a frame or exact frame for $L^2[-\Omega, \Omega]$. A more basic problem is to determine whether $\{e_{a_n}\}$ is *complete* in $L^2[-\Omega, \Omega]$, *i.e.*, whether the linear span of $\{e_{a_n}\}$ is dense in $L^2[-\Omega, \Omega]$. The ultimate contribution to this problem is due to Beurling and Malliavin [14], [15] (see [36], [54] for superb expositions of this profound material). Our remarks, although elementary, are geared to this work of Beurling and Malliavin, as well as other important contributions by Beurling [12], [13, Volume 2, pages 341–350] and Landau [40], [41].

Definition 55. Let $\{t_n\}$ be a real sequence, set $a_n = -t_n$ for each n, and let $\Omega > 0$.

a. $\{t_n\}$ is a *uniqueness sequence for* PW_Ω if for all $f \in PW_\Omega$ for which $f(t_n) = 0$ for each n, we can conclude that f is the 0-function.

b. $\{t_n\}$ is an *energy stable sequence for* PW_Ω if there is a constant $A > 0$ such that

$$\forall\, f \in PW_\Omega, \quad A\|f\|_2^2 \le \sum |f(t_n)|^2. \tag{61}$$

Thus, for energy stable sequences, small errors in the samples cause a small error in estimating the energy of the original signal.

Theorem 56. *Let* $\{t_n\}$ *be a real sequence, set* $a_n = -t_n$ *for each* n, *and let* $\Omega > 0$.

a. *If* $\{e_{a_n}\}$ *is a Fourier frame for* PW_Ω *then* $\{t_n\}$ *is an energy stable sequence for* PW_Ω; *and if* $\{t_n\}$ *is an energy stable sequence for* PW_Ω *then it is a uniqueness sequence for* PW_Ω.

b. $\{t_n\}$ *is a uniqueness sequence if and only if* $\{e_{a_n}\}$ *is complete in* $L^2[-\Omega, \Omega]$.

c. *If* $\{t_n\}$ *is a uniformly discrete and energy stable sequence for* PW_Ω *then* $\{e_{a_n}\}$ *is a Fourier frame for* PW_Ω. *In general, the converse is false.*

Proof:

a. The first part follows from the lower frame bound inequality. The second part follows since the hypothesis implies $\|f\|_2 = 0$ when $f \in PW_\Omega$ vanishes on $\{t_n\}$, and since $f = 0$ on $\mathbb{R}$ when $\|f\|_2 = 0$.

b. Suppose $\{t_n\}$ is a uniqueness sequence and $\hat{f}$ annihilates $\{e_{a_n}\}$, where $f \in PW_\Omega$. By the Fourier inversion formula, f vanishes on $\{t_n\}$; and so, by uniqueness, we have that $f = 0$ on $\mathbb{R}$. We conclude that $\{e_{a_n}\}$ is complete by the Hahn-Banach theorem.

For the converse, take any $f \in PW_\Omega$ which vanishes on $\{t_n\}$. Then, by the Fourier inversion formula, $\hat{f}$ annihilates $\{e_{a_n}\}$; and so, by completeness, we have that $\hat{f} = 0$ on $[-\Omega, \Omega]$. We conclude that $f = 0$ on $\mathbb{R}$, and, consequently, $\{t_n\}$ is a uniqueness sequence for PW_Ω.

c. The first part follows from Lemma 42b of Section 7 and the definition of an energy stable sequence for PW_Ω. The second part is a consequence of Example 41 and Theorem 44, *i.e.*, there are Fourier frames $\{e_{a_n}\}$, where $\{a_n\}$ is not uniformly discrete. ∎

Discussion 57. *Stability.*

a. The classical sampling formula tells us that each $f \in PW_\Omega$ is uniquely determined by its values on $\{nT\}$, where $2T\Omega = 1$. Suppose $\Omega_1 > \Omega$ and $\{nT_1\}$ is the sampling sequence for $f \in PW_\Omega$, where $2T_1\Omega_1 = 1$. Then $2T_1\Omega < 1$, and knowledge of $\{f(nT_1)\}$ reflects *oversampling*, which is essential information for many applications, *e.g.*, [62]. (*Undersampling* gives rise to numerical *instability* and aliasing, *e.g.*, Section 2, Definition/Remark 6 and Section 11.) On the other hand, if the values of $f \in PW_\Omega$ are known on $\{nT_1 : n < 0\}$ then the Pólya-Carlson theorem uniquely determines f on all of $\mathbb{R}$ (see [3], [7], [11], [25], [54] for discussions of this type of uniqueness theorem and prediction theory). At first glance, this result would seem to imply that knowledge of an oversampled speech signal $f \in PW_\Omega$ allows us to predict the speaker's words in the future. As Pollak [53, pages 74–75] observed, the weakness in this argument is due to the effect of instability. In fact, it can be proved that an arbitrarily small error in $f(-T_1)$ can produce an arbitrarily large error in $f(t)$, for arbitrarily small $t > 0$ (see [19, Chapter 9.2]).

b. Let $\{t_n\}$ be a real sequence. Suppose Y is a metric space of continuous functions on $\mathbb{R}$, and let X be a metric space of sequences $\{f(t_n) : f \in Y\}$; we denote the corresponding metrics by ρ_Y and ρ_X. Assume the mapping,

$$\begin{aligned} L : X &\to Y \\ \{f(t_n)\} &\mapsto f, \end{aligned}$$

is well-defined. Then the sampling process, $f \mapsto \{f(t_n)\}$, is *stable* for the metrics ρ_Y and ρ_X if the linear map L is uniformly continuous, *i.e.*, if

$$\forall\, \epsilon > 0,\ \exists\, \delta > 0 \quad \text{such that} \quad \rho_X(\{f(t_n)\}, \{g(t_n)\}) < \delta \quad \text{implies} \quad \rho_Y(f, g) < \epsilon.$$

Recall that if metric spaces X and Y are also topological vector spaces then linear maps $L : X \to Y$ which are continuous at the origin are uniformly continuous, *i.e.*, for every neighborhood W of $0 \in Y$ there is a neighborhood V of $0 \in X$ such that

$$\forall\, x_f, x_g \in X, \quad x_f - x_g \in V \quad \text{implies} \quad L(x_f - x_g) \in W.$$

In the case of (61), we have the situation that $Y = PW_\Omega$, $X = \{f(t_n) : f \in Y\}$, and ρ_Y and ρ_X are the metrics defined from the usual norms on $L^2(\mathbb{R})$ and $l^2(\mathbb{Z})$, respectively. The mapping L is well-defined if the Bessel map

$$\begin{aligned} B : PW_\Omega &\to l^2(\mathbb{Z}) \\ f &\mapsto \{f(t_n)\} \end{aligned}$$

is well-defined and injective (see [7, Subsection 2.5]).

An aspect of stability concerns conditions to ensure the preservation of expansion properties under small perturbations. A typical problem is to find conditions so that a sequence $\{g_n\}$ of functions which is close to a basis $\{f_n\}$ is itself a basis, *e.g.*, Definition 16. The solution to this problem is due to Paley and Wiener [49, pages 100–106] for orthonormal bases $\{f_n\}$ on PW_Ω, and the generalization to bases is due to Boas (1940) (see the treatment in [64]). "Paley-Wiener stability" in the sense of the following theorem has even become a topic in the theory of locally convex topological vector spaces. We shall state the result at the more down to earth level of Hilbert spaces.

Theorem 58. *Let H be a separable Hilbert space and let $\{f_n\} \subseteq H$ be a basis of H. Suppose that $\{g_n\} \subseteq H$ and $\theta \in [0,1)$ satisfy the condition that*

$$\forall\, F \subseteq \mathbb{Z}, \quad \text{where card } F < \infty, \quad \text{and} \quad \forall\, c_n \in \mathbb{C}, \quad \text{where} \quad n \in F,$$

$$\|\sum_{n\in F} c_n(f_n - g_n)\| \leq \theta \|\sum_{n\in F} c_n f_n\|. \tag{62}$$

Then $\{g_n\}$ is a basis, and, even more, there is a topological isomorphism $L : H \to H$ such that $Lf_n = g_n$ for each n (see Theorem 17c).

Proof: The hypothesis (62) implies the continuity of the linear map $\tilde{L}$ defined by

$$\tilde{L}\left(\sum c_n f_n\right) = \sum c_n(g_n - f_n),$$

where $\sum c_n f_n \in H$; and, further, $\tilde{L}$ is bounded by $\theta < 1$ in the operator norm (see (56)). Thus, $L \equiv I - \tilde{L}$ is a topological isomorphism and $Lf_n = f_n - (f_n - g_n) = g_n$. ■

Example 59. In the case $\{f_n\}$ is an orthonormal basis, or even an exact frame, Theorem 58 allows us to conclude that $\{g_n\}$ is an exact frame when (62) holds.

The following is a well-known calculation for the setting $H = L^2[-\Omega, \Omega]$. It provides an elementary proof for a special case of Kadec's theorem quoted in Definition/Discussion 33b.

Theorem 60. *Let $\{a_n\}$ be a real sequence and let $\Omega > 0$. Suppose*

$$\sup_n |a_n - \frac{n}{2\Omega}| \equiv L < \frac{\log 2}{\pi}\left(\frac{1}{2\Omega}\right). \tag{63}$$

Then $\{e_{a_n}\}$ is an exact frame for $L^2[-\Omega, \Omega]$.

Proof: Let $H = L^2[-\Omega, \Omega]$ and denote the norm on $L^2[-\Omega, \Omega]$ by $\|\ldots\|$. Recall that $\{e_{a_{n/(2\Omega)}}\}$ is an orthonormal basis of H. Clearly,

$$\begin{aligned} e_{a_n}(\gamma) - e_{n/(2\Omega)}(\gamma) &= e^{\pi i n\gamma/\Omega}\left(e^{2\pi i\gamma(a_n - \frac{n}{2\Omega})} - 1\right) \\ &= e^{\pi i n\gamma/\Omega} \sum_{k=1}^{\infty} \frac{[2\pi i\gamma(a_n - \frac{n}{2\Omega})]^k}{k!}, \end{aligned} \tag{64}$$

where the right side is defined on $[-\Omega, \Omega]$ and extended 2Ω-periodically on $\hat{\mathbb{R}}$. Also, we have the estimate

$$\|\gamma^k f(\gamma)\| = \left(\frac{1}{2\Omega}\int_{-\Omega}^{\Omega} |\gamma^k f(\gamma)|^2 d\gamma\right)^{1/2} \leq \Omega^k \|f\| \tag{65}$$

for all $f \in H$ and $k \geq 0$.

For a given finite sequence $\{c_n : n \in F\}$ we use (64), (65), and the orthonormality of $\{e_{n/(2\Omega)}\}$ to make the estimate,

$$\begin{aligned}
&\|\sum_{n\in F} c_n(e_{a_n} - e_{n/(2\Omega)})\| \\
&\quad = \|\sum_{n\in F} c_n e_{n/(2\Omega)}(\gamma) \sum_{k=1}^{\infty} \frac{[2\pi i\gamma(a_n - \frac{n}{2\Omega})]^k}{k!}\| \\
&\quad \leq \|\sum_{k=1}^{\infty} \frac{(2\pi i)^k}{k!}\gamma^k (\sum_{n\in F} c_n(a_n - \frac{n}{2\Omega})^k e_{n/(2\Omega)}(\gamma))\| \\
&\quad \leq \sum_{k=1}^{\infty} \frac{(2\pi\Omega)^k}{k!}\|\sum_{n\in F} c_n(a_n - \frac{n}{2\Omega})^k e_{n/(2\Omega)}\| \\
&\quad = \sum_{k=1}^{\infty} \frac{(2\pi\Omega)}{k!}\Big(\sum_{n\in F} |c_n|^2 |a_n - \frac{n}{2\Omega}|^{2k}\Big)^{1/2} \\
&\quad \leq \sum_{k=1}^{\infty} \frac{(2\pi L\Omega)^k}{k!}\|\sum_{n\in F} c_n e_{n/(2\Omega)}\| \\
&\quad = \big(e^{2\pi L\Omega} - 1\big)\|\sum_{n\in F} c_n e_{n/(2\Omega)}\|.
\end{aligned}$$

This estimate allows us to implement (62) if $e^{2\pi L\Omega} - 1 < 1$, *i.e.*, if $2\pi L\Omega < \log 2$; but this is precisely our hypothesis (63). The result follows from Theorem 58 and our observation in Example 59. ■

In light of the Kadec-Levinson bound of "1/4", we should note that $\frac{\log 2}{\pi} < \frac{1}{4}$ since $\log 16 < \pi = 3.14\ldots$. In fact, $\log 16 = 2.77259\ldots$.

Discussion 61. Two elegant pieces of analysis should be mentioned with regard to the Paley-Wiener stability condition (62). Harry Pollard proved that, when dealing with completeness, (62) can be replaced by

$$\|\sum_{n\in F} c_n(f_n - g_n)\|^2 \leq \theta_f^2 \|\sum_{n\in F} c_n f_n\|^2 + \theta_g^2 \|\sum_{n\in F} c_n g_n\|^2, \tag{66}$$

for some $\theta_f, \theta_g \in [0, 1)$ (Annals of Math., 45 (1944), 738–739). Béla Sz. Nagy proved that, when dealing with exact frames, (62) can be replaced by a condition even weaker than (66) involving the additional mixed term

$$\|\sum_{n\in F} c_n f_n\| \|\sum_{n\in F} c_n g_n\|$$

(*Duke Math. J.* **14** (1947), 975–978).

Example 62.

a. Let $\{a_n\}$ be a real sequence satisfying the density condition (34), viz.,

$$|a_n - \frac{n}{\Delta}| \leq L. \tag{67}$$

Also, suppose $\Omega > 0$ satisfies

$$2\Omega < \Delta. \tag{68}$$

For this calculation, assume $n \in \mathbb{N}$ and $a_n > 0$. From (67) we obtain $\Delta L \geq \Delta a_n - n$ so that $(\frac{L}{n} + \frac{1}{\Delta}) \geq \frac{a_n}{n}$, and hence

$$\frac{n}{a_n}\left(\frac{\Delta L}{n} + 1\right) \geq \Delta.$$

Consequently, (68) allows us to conclude that

$$\overline{\lim_{n\to\infty}} \frac{n}{a_n} > 2\Omega. \tag{69}$$

b. The point of part a is that uniformly discrete sequences satisfying (67) and (68) give rise to Fourier frames $\{e_{a_n}\}$ for $L^2[-\Omega, \Omega]$ by the Duffin-Schaeffer theorem (Theorem 38), and condition (69) is weaker than (67) and (68). The hope is that (69) would lead to completeness which, of course, is weaker than the frame property. Such is the case as we see in the following result (Theorem 63) of Paley and Wiener (1934).

c. This particular theorem of Paley and Wiener is a significant extension of a completeness theorem due to Pólya and Szász (Jahresbericht der Deutschen Mathematiker Vereinigung, 43 (1933), 20), whose density condition is (69) with "$\underline{\lim}$" instead of "$\overline{\lim}$". The added depth of this Paley and Wiener result is due to the use of their fundamental theorem on quasi-analytic functions, *viz.* [49, Theorem XII], which, in turn, is closely related to the Beurling-Malliavin theory and has many other applications, *e.g.*, [3], [4].

Theorem 63. (Paley-Wiener, [49, Theorem XXVIII]). *Let $\{a_n\} \subseteq \mathbb{R}$ be a strictly increasing sequence of positive numbers, and assume the density condition*

$$\overline{\lim_{n\to\infty}} \frac{n}{a_n} > 2\Omega.$$

Then $\{e_{\pm a_n}\}$ is complete in $L^2[-\Omega, \Omega]$.

Discussion 64. Theorems 60, 38 (and 44), and 63 have density hypotheses and conclude with exact frame, frame, and completeness properties, respectively. Theorem 60, depending as it does on Theorem 58, shows how its density hypothesis is tied-in with Paley-Wiener stability.

At the completeness level, a celebrated example by Kahane [35] exhibits sparse non-uniformly distributed sequences $\{a_n\} \subseteq \mathbb{R}$ for which $\{e_{a_n}\}$ is complete in $L^2[-\Omega, \Omega]$, for arbitrarily large Ω. The Beurling-Malliavin theory referenced at the beginning of this section places such examples in the same theory as the seemingly more regular sequences of Theorem 63 (or Theorem 60).

We shall not go into the Beurling-Malliavin completeness theory. Instead, we shall define the lower and upper uniform Beurling densities mentioned in Section 7, Definition 36b. The lower uniform Beurling density is the correct condition to formulate a converse of the frame result, Theorem 38.

Definition 65.

a. Let $\{a_n\} \subseteq \mathbb{R}$ be a strictly increasing sequence for which $\lim_{n\to\pm\infty} a_n = \pm\infty$, and define

$$n^-(r) = \inf_I n_I(r) \quad \text{and} \quad n^+(r) = \sup_I n_I(r),$$

where $I \subseteq \mathbb{R}$ is an interval, $r > 0$, and

$$n_I(r) = \text{card}\{a_n \in I : |I| = r\}.$$

Clearly, there are sequences $\{a_n\}$ for which

$$\forall\, r > 0, \quad n^+(r) = \infty.$$

The *lower* and *upper uniform Beurling densities* of $\{a_n\}$ are

$$\Delta^- \equiv \Delta^-(\{a_n\}) = \lim_{r\to\infty} \frac{n^-(r)}{r}$$

and

$$\Delta^+ \equiv \Delta^+(\{a_n\}) = \lim_{r\to\infty} \frac{n^+(r)}{r},$$

respectively. These limits exist since n^- is superadditive and n^+ is subadditive (see [37, Subsections 10.6–10.11]).

b. Suppose $\Delta^-(\{a_n\}) = \Delta^+(\{a_n\}) \equiv \Delta \in [0,\infty)$. Then the natural density of $\{a_n\}$ (Definition 36b),

$$\lim_{r\to\infty} \frac{n_0(r)}{r},$$

exists and equals Δ since

$$n^-(r) \leq \text{card}\{a_n \in [\frac{-r}{2}, \frac{r}{2}]\} \leq n^+(r).$$

c. Uniform Beurling density is closely related to the more classical notion defined by the right side of (69). For example, $\text{card}\{a_j \in I : |I| = a_n\} = n$ if $I = [0, a_n]$ in the case $a_0 = 0$ and $a_j > 0$ if and only if $j \geq 1$. Thus,

$$\frac{n^-(a_n)}{a_n} \leq \frac{n}{a_n} \leq \frac{n^+(a_n)}{a_n}.$$

In fact, the Duffin-Schaeffer theorem (Theorem 38) is valid if the uniformly discrete sequence $\{a_n\}$ satisfies the condition, $\Delta^-(\{a_n\}) > 2\Omega$.

We can now state the converse of the Duffin-Schaeffer theorem alluded to before Definition 65 (see Section 7, Formula 45).

Theorem 66. (Landau [41]). *Let $\Omega > 0$ and let $\{a_n\} \subseteq \mathbb{R}$ be a uniformly discrete energy stable sequence for PW_Ω. (Thus, $\{e_{a_n}\}$ is a Fourier frame for $L^2[-\Omega, \Omega]$.) Then*

$$\Delta^-(\{a_n\}) \geq 2\Omega.$$

Discussion 67. *Frames and completeness.*

Theorem 66 is a special case of a deep contribution by Landau [41]. Landau's results extend to higher dimensions and multiband signals. This latter notion means that N-frequency bands are analyzed instead of $[-\Omega, \Omega]$ (see the subband coding inherent in the analysis of decimation and aliasing in Section 10).

A critical insight, which one can extract from Landau's work, is that irregular sampling reconstruction formulas such as those in Section 8 necessarily depend on a close relation between the bandwidth and the lower uniform Beurling density of the sampling set. This relationship is not fundamental in dealing with the analogous problem for obtaining completeness, thereby adding a level of complexity to this latter problem, *e.g.*, Discussion 64.

Definition 68.

a. Let $D \subseteq \mathbb{R}$ and $E \subseteq \hat{\mathbb{R}}$ be closed sets. *Balayage is possible for* (D, E) if

$$\forall\, \mu \in M_b(\mathbb{R}),\ \exists\, \nu_\mu \in M_b(D) \quad \text{such that} \quad \forall\, \gamma \in E,$$

$$\int e^{2\pi i t\gamma} d\mu(t) = \int e^{2\pi i t\gamma} d\nu_\mu(t).$$

b. This notion of balayage stems from Poincaré's balayage process in potential theory; and, historically, E was a collection of potential theoretic kernels instead of group characters or frequencies.

A dual formulation of studying balayage problems has the flavor of classical spectral synthesis, *e.g.*, [1, Equation (3.2.52)]. Further, Beurling's theorem, Theorem 69, is a special case of his balayage theory which depends on specific systhesizable sets [12], [13, Volume 2, pages 341–350].

Theorem 69. (Beurling 1959–1960). *Let $D \subseteq \mathbb{R}$ be a closed set and let $E = [-\Omega, \Omega]$ for some $\Omega > 0$. Balayage is possible for (D, E) if and only if D contains a uniformly discrete sequence $\{a_n\}$ whose lower uniform Beurling density satisfies the inequality,*

$$\Delta^-(\{a_n\}) > 2\Omega.$$

Problem 70. Discussion 43b and the definition of (energy) stable sequences lead us to the following problem: *find those sequences* $\{t_n\}$ *such that the formula,*

$$\int f(t)dt = \sum f(t_n), \tag{70}$$

or similar "continuous analogues of series" formulas are valid for all $f \in PW_\Omega$ *for which* $\int f(t)dt$ *exists.* The existence of the integral can be considered either in the ordinary sense or as a principal value integral. The general question is quite difficult, and, in light of the weights mentioned in Discussion 43b, there are a number of related questions.

In the case $t_n = nT$, we shall see that the analogue of (70) itself is

$$\int f(t)dt = T\sum f(nT). \tag{71}$$

Even in this case there are questions to be answered if $\{f(nT)\} \in l^2(\mathbb{Z})\backslash l^1(\mathbb{Z})$.

The following elementary result gives some positive information in the case of regular sampling sequences.

Theorem 71. *Let* $T, \Omega > 0$ *and let* $f \in PW_\Omega$.

a. *If* $\{f(nT)\} \in l^1(\mathbb{Z})$ *and* $2T\Omega \leq 1$ *then* $\int f(t)dt$ *exists as a principal value integral and (71) is valid.*

b. *Assume* $\hat{f} \in C_c^\infty(\hat{\mathbb{R}})$ *or that*

$$\exists\, C \quad \text{such that} \quad \forall\, t \in \mathbb{R},\ |f(t)| + |f^{(1)}(t)| + |f^{(2)}(t)| \leq \frac{C}{1+t^2}. \tag{72}$$

If $T\Omega \leq 1$ *then (71) is valid.*

c. *If* $2T\Omega = 1$ *then* $\{f(nT)\} \in l^2(\mathbb{Z})$ *and*

$$\int |f(t)|^2 dt = T\sum |f(nT)|^2.$$

Proof:

a. By the classical sampling theorem, the right side of

$$f = T\sum f(nT)\tau_{nT}d_{2\pi\Omega}$$

converges uniformly on $\mathbb{R}$. Thus,

$$\begin{aligned}\forall\, S > 0, \quad \int_{-S}^{S} f(t)dt &= T\sum f(nT)\int_{-S}^{S} \tau_{nT}d_{2\pi\Omega}(t)dt \\ &\equiv T\sum f(nT) - \sum f(nT)\epsilon(S,n);\end{aligned} \tag{73}$$

and, now, a standard argument, first recorded in [34] (as far as I know) and using the hypothesis $\{f(nT)\} \in l^1(\mathbb{Z})$, allows us to compute

$$\lim_{S\to\infty} \sum f(nT)\epsilon(S,n) = 0.$$

Hence, the principal value integral $\int f(t)dt$ exists and (71) is valid by (73).

b. The function-form of the Poisson summation formula recorded in Section 2, Equation (7) is

$$\sum \hat{f}\left(\frac{n}{T}\right) = T \sum f(nT). \tag{74}$$

If $f \in PW_\Omega$, $\hat{f}$ is continuous, and $\Omega \leq 1/T$ then (74) becomes

$$\hat{f}(0) = T \sum f(nT)$$

and this is (71). Our hypothesis, $\hat{f} \in C_c^\infty(\hat{\mathbb{R}})$ or (72), implies the continuity of $\hat{f}$ and the validity of (74), *e.g.*, [38].

c. By hypothesis, even for $2T\Omega \leq 1$, $f = T\sum f(nT)\tau_{nT}d_{2\pi\Omega}$ in $L^2(\mathbb{R})$, and so $\hat{f} = T\sum f(nT)e_{-nT}\mathbf{1}_\Omega$ in $L^2(\hat{\mathbb{R}})$. If $2T\Omega = 1$ and we extend $\hat{f}$ $\frac{1}{T}$-periodically on $\hat{\mathbb{R}}$, then we can conclude that $\{f(nT)\} \in l^2(\mathbb{Z})$ by the square integrability of $\hat{f}$ on $[-\Omega,\Omega]$.

Since

$$\int_{-\Omega}^{\Omega} e^{-2\pi i(m-n)T\gamma}d\gamma = \frac{\sin(2\pi(m-n)T\Omega)}{\pi(m-n)T}$$

and $\{f(nT)\} \in l^2(\mathbb{Z})$, we have

$$\|f\|_2 = T\left(\sum_{m,n} f(mT)\overline{f}(nT)\frac{\sin(2\pi(m-n)T\Omega)}{\pi(m-n)T}\right)^{1/2}$$

by the Plancherel theorem, and the result follows. ■

Discussion 72. There are several interesting historical remarks concerning (71) in [31, page 63] and some wonderful examples in [20].

In the proof of Theorem 71a, we cannot integrate the classical sampling formula term by term over $\mathbb{R}$, even though the series converges uniformly. There are counterexamples, and we mention the point since there has been a little confusion on this issue in some of the literature. Of course, formally, (71) is an immediate consequence of the classical sampling formula.

In the proof of Theorem 71b, we only required $T\Omega \leq 1$ instead of $2T\Omega \leq 1$. Also, our hypotheses for the validity of the Poisson summation formula can be weakened, *but* the formula cannot be used indiscriminately. One of the difficulties in Problem 70 is that "Poisson summation formulas" require some regularity, and there are deep open questions here concerning fundamental problems in number theory.

Because of the proof of Theorem 71c, we note that *if* $2T\Omega \leq 1$, *then the mapping,*

$$\begin{aligned} PW_\Omega &\to l^2(\mathbb{Z}) \\ f &\mapsto \{Tf(nT)\}, \end{aligned}$$

is a well-defined, continuous linear injection (see the discussion on Bessel sequences in [7, Section 2]). To verify this claim, consider the function G and sequence $\{c_n\}$ of Fourier coefficients defined in Theorem 1. By the Plancherel theorem for Fourier series, we can assert that $\sum |c_n|^2 < \infty$; and each $c_n = Tf(nT)$ by Proposition 9. Thus, the mapping is well-defined, and the fact it is an injection is clear. The continuity is a consequence of the inequality,

$$\left(\sum |Tf(nT)|^2\right)^{1/2} = \sqrt{T}\|f\|_2,$$

or (for more general situations) of the uniform boundedness principle. Similar easy calculations yield the *surjectivity* of the above mapping in the case $2T\Omega = 1$.

Example 73.

a. Our proof of Theorem 71b combined the Poisson summation formula and bandlimitedness to obtain (71). Although this is an entirely standard procedure, we do mention that Wiener used it in a calculation for a work by Bhatia and Krishnan on light scattering (Proc. Royal Soc., London, Ser. A, 192 (1948), 181–194). Wiener verified that if $2T\Omega \leq 1$ then

$$\forall\, t \in \mathbb{R}, \quad \frac{1}{2T\Omega} = \sum \left(\frac{\sin 2\pi\Omega(t - nT)}{2\pi\Omega(t - nT)}\right)^2, \tag{75}$$

(see the classical sampling formula). Equation (75) is an immediate consequence of the Poisson summation formula applied to

$$f(u) = 4\pi\Omega w(2\Omega u - 4\Omega\pi t)$$

where $t \in \mathbb{R}$ is fixed and w is the *Fejér kernel*

$$w(u) = \frac{1}{2\pi}\left(\frac{\sin u/2}{u/2}\right)^2,$$

whose Fourier transform is $\max(1 - 2\pi|\gamma|, 0)$. (We use "$w$" for Fejér since we cannot use "f", and Fejér's original surname was Weiss.) (75) is not unexpected since $f(t) \equiv 1$, although not in $L^2(\mathbb{R})$, has distributional Fourier transform δ which has the required support property, viz., $supp\, \delta = \{0\} \subseteq [-\Omega, \Omega]$, and the sampling function $\hat{w}$ is 1 on the support of $\hat{f}$, *e.g.*, Theorem 46.

b. The proof of the Poisson summation formula involves a periodization procedure. Thus, if $f \in L^1(\mathbb{R})$ then

$$F(t) \equiv \sum f(t+nT), \quad t \in \mathbb{R},$$

is T-periodic and $F \in L^1(\mathbb{R}/T\mathbb{Z})$, *i.e.*, F defined on $\mathbb{R}$ is Lebesgue integrable on every interval of length nT. With this in mind we shall now take another look at part a but with some of the material of Section 10 in mind. For convenience, we let $T = 1$ and, because of Section 10, we switch to the frequency axis.

First, let $g \in L^2(\mathbb{R})$, and note that $\{\tau_n g\}$ is orthonormal if and only if

$$\sum |\hat{g}(\gamma+n)|^2 = 1 \text{ a.e.} \tag{76}$$

Now let $g \equiv \mathbf{1}_{[0,1)}$, so that $\{\tau_n g\}$ is an orthonormal basis of some subspace V_0 of $L^2(\mathbb{R})$. For this function, the method of multiresolution analysis in wavelet theory generates the classical Haar orthonormal basis of $L^2(\mathbb{R})$, *e.g.*, [47].

Note that

$$\hat{g}(\gamma) = e^{-\pi i \gamma} d_\pi(\gamma), \tag{77}$$

and, by (76),

$$G(\gamma) \equiv \sum |\hat{g}(\gamma+n)|^2 = 1 \text{ a.e.} \tag{78}$$

$G \in L^1(\mathbb{R}/\mathbb{Z})$ since $\hat{g} \in L^2(\mathbb{R})$; and, because of (78), the nth Fourier coefficient of G is δ_{0n}. Lest the equivalence of orthonormality and (76) cause trepidation, the Fourier coefficients of G can be computed directly as

$$g * \tilde{g}(-n) = \int_{-n}^{1-n} g(t)dt = \delta_{0n}.$$

Combining (77) and (78), we obtain

$$1 = \sum \left(\frac{\sin \pi(\gamma - n)}{\pi(\gamma - n)} \right)^2.$$

In particular,

$$\frac{\pi^2}{\sin^2 \pi\gamma} = \sum \frac{1}{(\gamma+n)^2}, \quad \gamma \notin \mathbb{Z}, \tag{79}$$

so that

$$\pi^2 = \sum \frac{1}{(n+\frac{1}{2})^2};$$

and we can even obtain Euler's formula,

$$\frac{\pi^2}{6} = \sum_{n=1}^{\infty} \frac{1}{n^2},$$

by taking limits correctly in (79).

§10. Irregular sampling - approaches and topics

Recent Contributions 74. We begin our final section by listing contributions by others on the topic of irregular sampling, including new closely related methods. Our caveat about this list was made in the Introduction; and, generally, we shall only give a "sampling" of works by some authors who publish "regularly" in the field. We shall not list work referenced in [21], [31], except to point out that Kluvánek's fundamental article [39] gets better with time. Further, there is a new book by Marks [45] and a forthcoming venture by the authors of [21] and [31]. We emphasize again that many contributions are not cited.

Cenker, C., H. Feichtinger, and M. Herrmann, Iterative algorithms in irregular sampling, a first comparison of methods, 1991, preprint.

Cetin, A., Reconstruction of signals from Fourier transform samples, *Signal Processing* **16** (1989), 129–148.

Donoho, D. and B. Logan, Signal recovery and the large sieve, preprint.

Feichtinger, H. and K. Gröchenig, Error analysis in regular and irregular sampling theory, April 1990, preprint.

Feichtinger, H., K. Gröchenig, and M. Herrmann, Iterative methods in irregular sampling theory, numerical results, *ASST* 1990, *Proc. Conf.*, Aachen, September 1990.

Frazier, M. and R. Torres, The sampling theorem, ϕ– transform, and Shannon wavelets for R, Z, T, and Z_N, 1991, preprint.

Gröchenig, K., Reconstruction algorithms in irregular sampling, 1991, preprint.

Houdré, C., On path reconstruction of processes from irregular samples, in *Probabilistic and stochastic advances in analysis, with applications*, J. S. and J. L. Byrnes (eds.), NATO-ASI, 1991, Series C, Kluwer Academic Publishers, to appear.

Luthra, A. Extension of Parseval's relation to nonuniform sampling, *IEEE Trans. ASSP* **36** (1988), 1909–1911.

Marvasti, F. and M. Analoui, Recovery of signals and nonuniform samples using iterative methods, *Proc. Int. Symp. CAS*, Portland, OR, 1989.

Rawn, M., A stable nonuniform sampling expansion involving derivatives, *IEEE Trans. IT* **35** (1989), 1223–1227.

Sauer, K. and J. Allebach, Iterative reconstruction of band– limited images from nonuniformly spaced samples, *IEEE Trans. CAS* **34** (1987), 1497–1506.

Seip, K., An irregular sampling theorem for functions bandlimited in a generalized sense, *SIAM J. Appl. Math.* **47** (1987), 1112–1116.

Torres, R., Spaces of sequences, sampling theorem, and functions of exponential type, 1990, preprint.

Walter, G., Some wavelet sampling theorems, 1990, preprint.

Walter, G., Nonuniform sampling of bandlimited functions of polynomial growth, 1990, preprint.

Zeevi, Y. and E. Shlomot, Pyramidal image representation in nonuniform systems, *SPIE, Vol. 1001, Visual Communications and Image Processing* (1988), 563–570.

Now let's turn to the second part of Section 10, where we shall discuss aliasing and the transition from coherent states, which has been the topic of this chapter, to formulating a backdrop for a similar treatment of wavelets and wavelet packets.

Background 75. Let $T > 0$ be a sampling period. The measure (on $\mathbb{R}$),

$$p = \sum \delta_{nT},$$

is the *sampling impulse train* where δ_{nT} is the Dirac measure at nT. The measure p is tempered (but not bounded); and, by the Poisson summation formula (Equation (7)), $\hat{p}$ is the tempered measure,

$$\hat{p} = \frac{1}{T} \sum \delta_{n/T}.$$

If f is continuous on $\mathbb{R}$ then the product fp exists as a measure, and

$$fp = \sum f(nT)\delta_{nT} \tag{80}$$

is the *sampled signal.* The "distributional" verification of (80) is immediate:

$$\forall\ g \in C_c(\mathbb{R}), \quad fp(g) = p(fg) = \sum f(nT)g(nT),$$

and

$$\left(\sum f(nT)\delta_{nT}\right)(g) = \sum f(nT)g(nT).$$

We write $f_p = fp$, and assume f is well-behaved enough to ensure that the exchange formula,

$$\hat{f}_p = \hat{f} * \hat{p},$$

is valid, *e.g.*, [57] provides conditions for validity. Thus

$$\hat{f}_p = \frac{1}{T} \sum \tau_{n/T} \hat{f}.$$

Let us consider the sampling frequency $\frac{1}{T}$ and the low pass filter $h = Td_{\pi/T}$ having frequency response $\hat{h} = T\mathbf{1}_{1/(2T)}$. The formal output of the sampled signal f_p in this linear time- invariant system is

$$h * f_p = T \sum f(nT)\tau_{nT} d_{\pi/T}. \tag{81}$$

(If we define a frequency bandwidth 2Ω by $2T\Omega = 1$, then the right side of (81) is the classical sampling formula.) Formally, then,

$$(h * f_p)^\wedge = \mathbf{1}_{(\frac{1}{2T})} \sum \tau_{n/T} \hat{f}. \tag{82}$$

Example 76. Let $f(t) = \sin 2\pi t\gamma_0$, for some $\gamma_0 > 0$. Distributionally, we calculate

$$\hat{f} = \frac{1}{2i}(\tau_{\gamma_0}\delta - \tau_{-\gamma_0}\delta), \tag{83}$$

since, for example,

$$(\tau_{\gamma_0}\delta)^\wedge(t) = (\tau_{\gamma_0}\delta(\gamma))(e^{2\pi it\gamma}) = \delta_{\gamma_0}(\gamma)(e^{2\pi it\gamma}) = e^{2\pi it\gamma_0}.$$

Now let us choose a sampling period T, with corresponding sampling frequency $1/T$, so that

$$\gamma_0 \in \left(\frac{1}{2T}, \frac{1}{T} \right).$$

Consider the linear time-invariant system L defined by (81) where $h = Td_{\pi/T}$. Because of (83), the right side of (82) consists of the two nonzero terms,

$$\mathbf{1}_{(\frac{1}{2T})}\tau_{1/T}\hat{f} = -\frac{1}{2i}\,\delta_{-\gamma_0+\frac{1}{T}}$$

and

$$\mathbf{1}_{(\frac{1}{2T})}\tau_{-1/T}\hat{f} = -\frac{1}{2i}\,\delta_{\gamma_0-\frac{1}{T}}.$$

Note that $-\gamma_0 + \frac{1}{T} > 0$ and $\gamma_0 - \frac{1}{T} < 0$. In any case,

$$(h * f_p)^\wedge = -\frac{1}{2i}(\delta_{\frac{1}{T}-\gamma_0} - \delta_{-(\frac{1}{T}-\gamma_0)}), \quad \frac{1}{T} - \gamma_0 > 0.$$

Thus, if $\gamma_0 \in (\frac{1}{2T}, \frac{1}{T})$, the original frequency γ_0 takes on the identity or *alias* of the lower (and, incidentally, negative in this example) frequency $\gamma_0 - \frac{1}{T}$, *i.e.*, the output of the sampled signal is

$$h * f_p(t) = \sin\Big(2\pi t(\gamma_0 - \frac{1}{T})\Big)$$

instead of $\sin 2\pi t\gamma_0$. We *undersampled* the original signal f, and consequently were *not* able to reconstruct it by the right side of (81). Undersampling means that the sampling period is big so that the sampling frequency is small; in our case, this meant that $\frac{1}{2T} < \gamma_0$ (see Definition/Remark 6). (The condition $\gamma_0 < \frac{1}{T}$ was not essential to make this point about undersampling and aliasing, but serves to pick out certain elements on the right side of (82).)

Definition 77.

a. Let f be a discrete signal, *i.e.*, f is a complex-valued function defined on $\mathbb{Z}$. For example, f could be the sampled signal for some continuous function in the case $T = 1$. The *decimation* or *downsampling* of f is the discrete signal $f_2 : \mathbb{Z} \to \mathbb{C}$ defined as

$$\forall\, n \in \mathbb{Z}, \quad f_2[n] = f[2n].$$

We hasten to point out that this terminology is in the spirit but not the letter of the word "decimation". Also, we could just as easily have defined f_k as $f_k[n] = f[kn]$.

b. Formally, the *Fourier series* of $f : \mathbb{Z} \to \mathbb{C}$ is the 1- periodic function,

$$F(\gamma) \equiv \sum f[n] e^{-2\pi i n \gamma}, \tag{84}$$

where the *Fourier coefficients* are

$$f[n] = \int_0^1 F(\gamma) e^{2\pi i n \gamma}\, d\gamma, \tag{85}$$

(see the proof of Theorem 1). A rigorous treatment of the theory of Fourier series is contained in [38]. The most natural and difficult (to prove) convergence result in the theory is Carleson's theorem (1966): *if* $F \in L^2(\mathbb{R}/\mathbb{Z})$ *and* f *is defined by* (85) *then the series in (84) converges to* F a.e.

c. Let the discrete signals f and f_2 have Fourier series F and F_2, respectively. Formally, it is easy to see that F_2 is a 1-periodic function and

$$\forall\, \gamma \in \hat{\mathbb{R}}, \quad F_2(\gamma) = \frac{1}{2}\Big(F(\frac{\gamma}{2}) + F(\frac{\gamma}{2} + \frac{1}{2})\Big). \tag{86}$$

d. In Equation (80) we defined the sampled signal of a given continuous function. We shall now define the *sampled signal* of a given discrete signal f. With Equation (80) in mind, we define

$$\delta[m - n] = \delta_{mn},$$

and consider the *sampling period* $T = 2$. Thus, the *sampled signal* f_p of f is

$$\forall\, j \in \mathbb{Z}, \quad f_p[j] = \sum_n f[2n]\delta[j - 2n].$$

If F_p is the Fourier series of f_p, then F_p is a $\frac{1}{2}$- periodic function and

$$\forall\, \gamma \in \hat{\mathbb{R}}, \quad F_p(\gamma) = \frac{1}{2}\Big(F(\gamma) + F(\gamma + \frac{1}{2})\Big).$$

e. The process of decimation can be viewed as taking place in two steps, viz., forming f_p from f and then forming f_2 from f_p. In terms of Fourier series, this process is reflected by the two equations,

$$F_p(\gamma) = \frac{1}{2}\Big(F(\gamma) + F(\gamma + \frac{1}{2})\Big)$$

and

$$F_2(\gamma) = F_p(\frac{\gamma}{2}).$$

Definition/Discussion 78. *Aliasing.*

In characterizing aliasing in terms of undersampling, we say that *aliasing occurs in the decimation process* if

$$supp\, F \cap supp\, \tau_{-1/2}F \neq \phi \tag{87}$$

in the first step of decimation, when forming f_p from f. (Technically, we should deal with the interiors of the supports in (87).) In this case, F is the *unaliased component* and $\tau_{-1/2}F$ is the *aliased component*. The reason for this terminology is that there is γ_0 for which $F(\gamma_0) = F(\gamma_0 + \frac{1}{2})$. Consequently, we have the expected expansion,

$$F(\gamma_0) = \sum f[n]e^{-2\pi i n\gamma_0},$$

and the "aliased" expansion,

$$F(\gamma_0) = \sum (-1)^n f[n]e^{-2\pi i n\gamma_0},$$

(see Definition/Discussion 6).

The decimation process and accompanying aliasing phenomenon logically lead us to the method of subband coding, to the notion of a *quadrature mirror filter* (QMF), and to our closing material, which can be viewed as an introduction to wavelets and wavelet packets.

Before defining a QMF (Definition 79), we shall make the following remarks about *subband coding*. First, a subband coding procedure filters a spectrum, *e.g.*, a speech spectrum, into separate frequency subbands, and downsamples the signals corresponding to the given subbands. In the case of two subbands we have the "2-decimations" of Definition 77a and Theorem 80; and, hence, we preserve, and do not increase, the number of data points from the original signal. The resulting signals are encoded for transmission and transmitted. Finally, they are decoded, interpolated, *e.g.*, by inserting zeros, and reconstructed as a single message at the receiver. There are important reasons for this procedure. These include limiting noise from encoding and decoding to the relevant subbands, allocating bits based on perceptual criteria for a given subband, providing a structure to deal with compression problems, and sending "images" at various levels of resolution to control "expense". The early work on subband coding is well documented in the IEEE ICASSP and information sciences publications from the 1970s and early 1980s.

QMFs are used in subband coding to remove aliasing, *e.g.*, Theorem 81.

Definition 79. Let $H \in L^2(\mathbb{R}/\mathbb{Z})$ be the Fourier series of $h \in l^2(\mathbb{Z})$. The discrete signal h is a *quadrature mirror filter* with frequency response H if

$$|H(\gamma)|^2 + |H(\gamma + \frac{1}{2})|^2 = 2 \qquad a.e. \tag{88}$$

In particular, $\|H\|_{L^2(\mathbf{T})} = 1$.

Theorem 80. *Let g_i, h_i, g_o, h_o be filters corresponding to linear time-invariant systems, and let f be a discrete signal. ("i" is for "in" and "o" is for "out".) Consider the filterings/decimations*

$$f_{g_i}[n] \equiv \sum \overline{g}_i[k-2n]f[k]$$

$$f_{h_i}[n] \equiv \sum \overline{h}_i[k-2n]f[k]$$

and the inserting zeros/filterings

$$g_o * f_{0,g_i} \qquad \text{and} \qquad h_o * f_{0,h_i},$$

where

$$f_{0,g_i}[2n] \equiv f_{g_i}[n] \qquad \text{and} \qquad f_{0,g_i}[2n+1] \equiv 0.$$

Define

$$f_{g,h} \equiv g_o * f_{0,g_i} + h_o * f_{0,h_i},$$

and let the Fourier series of the various sequences be denoted by the corresponding capitalized letter. Then

$$\begin{aligned} F_{g,h} = & \frac{1}{2}F(G_o\overline{G}_i + H_o\overline{H}_i) \\ & + \frac{1}{2}\, \tau_{-1/2}F(G_o\tau_{-1/2}\overline{G}_i + H_o\tau_{-1/2}\overline{H}_i). \end{aligned} \tag{89}$$

By Definition/Discussion 78, the second term on the right side of (89) is the *aliased component.* The following result shows how QMFs give rise to alias-free reconstruction; and, together with Theorem 80, it is an essential component of the wavelet packet point of view.

Theorem 81. *Consider the setup of Theorem 80, and suppose $g = g_i = g_o$, $h = h_i = h_o$, and*

$$G(\gamma) = -e^{-2\pi i\gamma}\, \overline{H}(\gamma + \frac{1}{2}). \tag{90}$$

Assume h is a QMF. Then the aliased component of (89) vanishes, and

$$F_{g,h} = F.$$

If h is a low-pass filter, then g defined by (90) is a high-pass filter; and the filterings/decimations f_g and f_h defined in Theorem 80 give rise to relatively disjoint subbands. In the case of (90),

$$g[n] = (-1)^n \, \overline{h}[-n+1]. \tag{91}$$

We mention this because of Theorem 83, where Equation (91) appears again (see [47] for the development of Definition 82, the proof of Theorem 83, and historical comments).

Definition 82. A *multiresolution analysis* $\{V_j, \theta\}$ of $L^2(\mathbb{R})$ is an increasing sequence of closed linear subspaces $V_j \subseteq L^2(\mathbb{R})$ and an element $\theta \in V_0$ for which the following properties hold:

i. $\bigcap V_j = \{0\}$ and $\overline{\bigcup V_j} = L^2(\mathbb{R})$,
ii. $f(t) \in V_j$ if and only if $f(2t) \in V_{j+1}$,
iii. $f \in V_0$ and $k \in \mathbb{Z}$ imply $\tau_k f \in V_0$,
iv. $\{\tau_k \theta\}$ is an exact frame for the Hilbert space V_0.

Let $\{V_j, \theta\}$ be a multiresolution analysis of $L^2(\mathbb{R})$. If $\varphi \in V_0$ is defined by

$$\hat{\varphi}(\gamma) = \theta(\gamma)\left(\sum |\hat{\theta}(\gamma+n)|^2\right)^{-1/2},$$

then $\{\tau_k \varphi\}$ is an orthonormal basis of V_0 (see Equation (76)).

Theorem 83. *Let $\{V_j, \varphi\}$ be a multiresolution analysis of $L^2(\mathbb{R})$, where $\{\tau_k \varphi\}$ is an orthonormal basis of V_0.*

a. *Let W_j be the orthogonal complement of V_j in V_{j+1}, i.e.,*

$$V_{j+1} = V_j \oplus W_j.$$

There is $\psi \in W_0$ such that $\{\tau_k \psi\}$ is an orthonormal basis of W_0, and $\{\psi_{m,n}\}$ is an orthonormal basis of $L^2(\mathbb{R})$, where

$$\psi_{m,n}(t) = 2^{m/2} \, \psi(2^m t - n).$$

$\{\psi_{m,n}\}$ is called a wavelet basis.

b. *The functions ϕ and ψ satisfy the following properties:*

i. $\varphi(t) = \sqrt{2} \sum h[n] \varphi(2t-n)$

and

$$\hat{\varphi}(2\gamma) = \tfrac{1}{\sqrt{2}} H(\gamma) \hat{\varphi}(\gamma),$$

where h is a QMF having Fourier series H;

ii. $\psi(t) = \sqrt{2} \sum g[n] \varphi(2t-n)$

and

$$\hat{\psi}(2\gamma) = -e^{-2\pi i \gamma}\left(\tfrac{1}{\sqrt{2}}\right) \overline{H}(\gamma + \tfrac{1}{2}) \hat{\varphi}(\gamma),$$

where

$$g[n] = (-1)^n \overline{h}[-n+1].$$

§11. Notation

$\mathbb{R}$ is the real line and $\mathbb{C}$ is the set of complex numbers. $L^p(\mathbb{R})$, $1 \le p < \infty$, is the space of Lebesgue measurable functions $f : \mathbb{R} \to \mathbb{C}$ for which

$$\|f\|_p \equiv \left(\int |f(t)|^p dt \right)^{1/p} < \infty,$$

where "$\int$" indicates integration over $\mathbb{R}$. The usual adjustment is made to define $L^\infty(\mathbb{R})$ and its norm $\| \cdots \|_\infty$. The Fourier transform of $f \in L^1(\mathbb{R})$ is the function $\hat{f}$ defined by

$$\hat{f}(\gamma) \equiv \int f(t) e^{-2\pi i t \gamma} dt,$$

where $\gamma \in \hat{\mathbb{R}}(= \mathbb{R})$; and a similar definition is made for $f \in L^2(\mathbb{R})$. If F is defined on $\hat{\mathbb{R}}$ then

$$F^\vee(t) \equiv \int F(\gamma) e^{2\pi i t \gamma} d\gamma.$$

The pairing between f and $\hat{f}$ is designated by $f \leftrightarrow \hat{f}$, and the space of absolutely convergent Fourier transforms is $A(\hat{\mathbb{R}})$ with norm $\|\hat{f}\|_A \equiv \|f\|_1$.

$C^\infty(\mathbb{R})$ is the space of infinitely differentiable functions on $\mathbb{R}$, $\mathcal{S}(\mathbb{R})$ is the Schwartz space of "rapidly decreasing" elements of $C^\infty(\mathbb{R})$, and $C_c^\infty(\mathbb{R})$ is the space of compactly supported elements of $C^\infty(\mathbb{R})$. $C(\mathbb{R})$ is the space of continuous functions on $\mathbb{R}$, and $C_c(\mathbb{R})$ is the space of compactly supported elements of $C(\mathbb{R})$. $C^m[a,b]$ is the space of m-times continuously differentiable functions on $[a,b]$. The space of pseudo-measures is the Banach space dual of $A(\mathbb{R})$ with norm $\| \cdots \|_A$; and the space $M_b(\mathbb{R})$ of bounded Radon measures is the Banach space dual of $C_c(\mathbb{R})$ (or $C_c^\infty(\mathbb{R})$ or $\mathcal{S}(\mathbb{R})$) taken with the norm $\| \cdots \|_\infty$. δ_a is the Dirac measure supported by the point $a \in \mathbb{R}$, and $\delta_0 \equiv \delta$ is the unit under convoluion in the Banach algebra $M_b(\mathbb{R})$.

$\mathbb{Z}$ is the set of integers, $\mathbb{N}$ is the set of positive integers and $\mathbf{T} = \hat{\mathbb{R}}/\mathbb{Z}$ is the circle group identified with any interval $[\alpha, \alpha+1) \subseteq \hat{\mathbb{R}}$. $l^p(\mathbb{Z})$, $1 \le p < \infty$, is the space of sequences $f : \mathbb{Z} \to \mathbb{C}$ for which $\left(\sum |f(n)|^p\right)^{1/p} < \infty$, where "$\sum$" indicates summation over $\mathbb{Z}$. Just as $\hat{\mathbb{R}}$ is the dual group of $\mathbb{R}$, so $\mathbf{T}$ is the dual group of $\mathbb{Z}$. $L^p(\mathbf{T})$, $1 \le p < \infty$, is the space of 1-periodic Lebesgue measurable functions F on $\hat{\mathbb{R}}$ for which

$$\|F\|_{L^p(\mathbf{T})} \equiv \left(\int_0^1 |F(\gamma)|^p d\gamma \right)^{1/p} < \infty.$$

If f is defined on $\mathbb{R}$ and μ is a positive measure then

$$\|f\|_{2,\mu} \equiv \left(\int |f(t)|^2 d\mu(t) \right)^{1/2}.$$

The following special symbols are used throughout the chapter.

$$(\tau_a g)(t) \equiv g(t-a),$$

$$e_b(t) \equiv e^{2\pi i t b},$$

$$\mathbf{1}_S(t) \equiv \begin{cases} 1, & \text{if} \quad t \in S \\ 0, & \text{if} \quad t \notin S, \end{cases}$$

$$\mathbf{1}_{(\Omega)}(t) \equiv \mathbf{1}_{[-\Omega,\Omega]}(t),$$

$$d(t) \equiv \frac{\sin t}{\pi t}, \quad \text{“}d\text{” for Dirichlet},$$

$$f_\lambda(t) \equiv \lambda f(t\lambda),$$

$$\delta_{mn} = \begin{cases} 1, & \text{if} \quad m = n \\ 0, & \text{if} \quad m \neq n, \end{cases}$$

card S is the cardinality of S, $|S|$ is the Lebesgue measure of S, ϕ is the empty set, and *supp* F is the support of F.

References

1. Benedetto, J., *Spectral Synthesis*, Academic Press, NY, 1975.
2. Benedetto, J., *Real Variable and Integration*, B. G. Teubner, Stuttgart, 1976.
3. Benedetto, J., Fourier uniqueness criteria and spectrum estimation theorems, in *Fourier Techniques and Applications*, J. F. Price (ed.), Plenum Plublishing, 1985, 149–170.
4. Benedetto, J., A quantitative maximum entropy theorem for the real line, *Integral Equations and Operator Theory* **10** (1987), 761–779.
5. Benedetto, J., Gabor representations and wavelets, in *Commutative Harmonic Analysis*, D. Colella (ed.), *Contemp. Math.* **19**, Am. Math. Soc., Providence, 1989, 9–27.
6. Benedetto, J., Uncertainty principle inequalities and spectrum estimation, in *Recent Advances in Fourier Analysis and its Applications*, J. S. and J. L. Byrnes (eds.), NATO-ASI Series C, **315**, Kluwer Academic Publishers, 1990, 143–182.
7. Benedetto, J., Stationary frames and spectral estimation, in *Probabilistic and Stochastic Advances in Analysis, with Applications*, J. S. and J. L. Byrnes (eds.), NATO-ASI, Series C, Kluwer Academic Publishers, 1991, to appear.
8. Benedetto, J., C. Heil, and D. Walnut, Remarks on the proof of the Balian-Low theorem, *Canad. J. Math.*, to appear.
9. Benedetto, J. and W. Heller, Irregular sampling and the theory of frames, I, *Note Math*, 1991, to appear.

10. Benedetto, J. and W. Heller, Irregular sampling and the theory of frames, II, preprint.
11. Benedicks, M., Supports of Fourier transform pairs and related problems on positive harmonic functions, *TRITA-MAT* **10**, Stockholm, 1980.
12. Beurling, A., Local harmonic analysis with some applications to differential operators, Belfer Graduate School of Science, *Annual Science Conference Proceedings*, 1966, 109–125.
13. Beurling, A., *The Collected Works of Arne Beurling*, L. Carleson, P. Malliavin, J. Neuberger, and J. Wermer (eds.), Birkhäuser, Boston, MA, 1989.
14. Beurling, A. and P. Malliavin, On Fourier transforms of measures with compact support, *Acta Math.* **107** (1962), 291–309.
15. Beurling, A. and P. Malliavin, On the closure of characters and the zeros of entire functions, *Acta Math.* **118** (1967), 79–93.
16. Beutler, F., Sampling theorems and bases in Hilbert space, *Inform. Contr.* **4** (1961), 97–117.
17. Beutler, F., Error-free recovery of signals from irregularly spaced samples, *SIAM Review* **8** (1966), 328–335.
18. Boas, R., Entire functions bounded on a line, *Duke J. Math.* **6** (1940), 148–169.
19. Boas, R., *Entire Functions*, Academic Press, NY, 1954.
20. Boas, R. and H. Pollard, Continuous analogues of series, *Amer. Math. Monthly* **80** (1973), 18–25.
21. Butzer, P., W. Splettstösser, and R. Stens, The sampling theorem and linear prediction in signal analysis, *Jber. d. Dt. Math.-Verein* **90** (1988), 1–70.
22. Chui, C. K. and J. Z. Wang, A cardinal spline approach to wavelets, *Proc. Amer. Math. Soc.*, 1991, to appear.
23. Daubechies, I., The wavelet transform, time-frequency localization and signal analysis, *IEEE Trans. Inform. Theory* **36** (1990), 961–1005.
24. Duffin, R. and A. Schaeffer, A class of nonharmonic Fourier series, *Trans. Amer. Math. Soc.* **72** (1952), 341–366.
25. Gabardo, J.-P., Tempered distributions with spectral gaps, *Math. Proc. Camb. Phil. Soc.* **106** (1989), 143–162.
26. Gabor, D., Theory of communication, *J. IEE (London)* **93** (1946), 429–457.
27. Gröchenig, K., Reconstruction algorithms in irregular sampling, preprint.
28. Heil, C. and D. Walnut, Continuous and discrete wavelet transforms, *SIAM Review* **31** (1989), 628–666.
29. Heller, W., Frames of exponentials and applications, Doctoral Thesis, University of Maryland, College Park, MD, 1991.
30. Higgins, J., *Completeness and Basis Properties of Sets of Special Functions*, Cambridge University Press, 1977.
31. Higgins, J., Five short stories about the cardinal series, *Bull. Amer. Math. Soc.* **12** (1985), 45–89.

32. Hörmander, L., *The Analysis of Linear Partial Differential Operators*, I and II, Springer-Verlag, NY, 1983.
33. Jaffard, S., A density criterion for frames of complex exponentials, preprint.
34. Jagerman, D., Bounds for truncation error of the sampling expansion, *SIAM J. Appl. Math.* **14** (1966), 714–723.
35. Kahane, J.-P., Sur la totalité des suites d'exponentielles imaginaires, *Ann. Inst. Fourier* **8** (1958), 273–275.
36. Kahane, J.-P., *Travaux de Beurling et Malliavin*, *Séminaire Bourbaki* No. 255, 1961/62, reprinted and updated by Benjamin Publishers, NY, 1966.
37. Kahane, J.-P., *Séries de Fourier Absolument Convergentes*, Springer-Verlag, NY, 1970.
38. Katznelson, Y., *An Introduction to Harmonic Analysis*, John Wiley and Sons, 1968, reprinted by Dover, NY, 1976.
39. Kluvánek, I., Sampling theorem in abstract harmonic analysis, *Mat. Fyz. Casopsis Sloven. Akad. Vied.* **15** (1965), 43–48.
40. Landau, H., Sampling, data transmission, and the Nyquist rate, *Proc. IEEE* **55** (1967), 1701–1706.
41. Landau, H., Necessary density condition for sampling and interpolation of certain entire functions, *Acta Math.* **117** (1967), 37–52.
42. Landau, H., An overview of time and frequency limiting, in *Fourier Techniques and Applications*, J. Price (ed.), Plenum Press, NY, 1985, 201–220.
43. Landau, H., On the density of phase-space expansions, preprint.
44. Levinson, N., Gap and density theorems, *Am. Math. Soc. Colloq. Publ.* **26**, Am. Math. Soc., 1940.
45. Marks, R., *Introduction to Shannon Sampling and Interpolation Theory*, Springer-Verlag, NY, 1991.
46. Meyer, Y., Wavelets and operators, *CEREMADE*, Université de Paris-Dauphine, 1989.
47. Meyer, Y., *Ondelettes et Opérateurs*, I, Hermann, Paris, 1990.
48. Oppenheim, A. and A. Willsky, *Signals and Systems*, Prentice Hall, NJ, 1983.
49. Paley, R. and N. Wiener, Fourier transforms in the complex domain, *Am. Math. Soc. Colloq. Publ.* **19**, Am. Math. Soc., 1934.
50. Papoulis, A., Error analysis in sampling theory, *Proc. IEEE* **54** (1966), 947–955.
51. Papoulis, A., *Signal Analysis*, McGraw-Hill, NY, 1977.
52. Plancherel, M., Intégrales de Fourier et fonctions entières, *Colloque sur l'analyse harmonique*, Nancy (1947), **15** (1949), 32–43.
53. Pollak, H., *New Directions in Mathematics*, Dartmouth Mathematics Conference, J. Kemeny, R. Robinson, and R. Ritchie (eds.), Prentice Hall, NJ, 1963.
54. Redheffer, R., Completeness of sets of complex exponentials, *Advances in Math.* **24** (1977), 1–63.

55. Rieffel, M., Von Neumann algebras associated with pairs of lattices in Lie groups, *Math. Ann.* **257** (1981), 403–418.
56. Rudin, W., *Real and Complex Analysis*, McGraw-Hill, NY, Third Edition, 1987.
57. Schwartz, L., *Théorie des Distributions*, Hermann, Paris, 1966.
58. Seip, K. and R. Wallstén, Sampling and interpolation in the Bargmann-Fock space, 1990–1991, preprint.
59. Stein, E. and G. Weiss, *An Introduction to Fourier Analysis on Euclidean Spaces*, Princeton University Press, 1971.
60. von Neumann, J., *Mathematical Foundations of Quantum Mechanics*, Princeton University Press, 1932, 1949, and 1955.
61. Weinert, H., (ed.), *Reproducing Kernel Hilbert Spaces*, Hutchinson Ross Publishing.
62. Wexler, J. and S. Raz, Discrete Gabor expansions, *Signal Processing* **21** (1990), 207–220.
63. Yao, K. and J. Thomas, On some stability and interpolatory properties of nonuniform sampling expansions, *IEEE Trans. Circuit Theory* **14** (1967), 404–408.
64. Young, R., *An Introduction to Nonharmonic Fourier Series*, Academic Press, NY, 1980.

This research was suppported by Prometheus Inc. under DARPA Contract DAA-H01-91-CR212 and by NSF Contract DMS-90-02420. I would also like to thank Professor Louis Raphael for an informative conversation about splines.

John J. Benedetto
Department of Mathematics
University of Maryland
College Park, MD 20742
jjb@math.umd.edu

Families of Wavelet Transforms in Connection with Shannon's Sampling Theory and the Gabor Transform

Akram Aldroubi and Michael Unser

Abstract. In this chapter, we look at the algebraic structure of nonorthogonal scaling functions, the multiresolutions they generate, and the wavelets associated with them. By taking advantage of this algebraic structure, it is possible to create families of multiresolution representations and wavelet transforms with increasing regularity that satisfy some desired properties. In particular, we concentrate on two important aspects. First, we show how to generate sequences of scaling functions that tend to the ideal lowpass filter and for which the corresponding wavelets converge to the ideal bandpass filter. We give the conditions under which this convergence occurs and provide the link between Mallat's theory of multiresolution approximations and the classical Shannon Sampling Theory. This offers a framework for generating generalized sampling theories. Second, we construct families of nonorthogonal wavelets that converge to Gabor functions (modulated Gaussians). These latter functions are optimally concentrated in both time and frequency and are therefore of great interest for signal and image processing. We obtain the conditions under which this convergence occurs; thus allowing us to create whole classes of wavelets with asymptotically optimal time-frequency localization. We illustrate the theory using polynomial splines.

§1. Introduction

A recurrent question in signal processing is how to best represent signals. This issue arises naturally in applications such as detection, data storage, data compression, signal analysis, signal transmission, etc. When the signals are continuous, it is usually desired to represent them by discrete sequences. One approach is simply to sample the functions on uniform grids. However, there are infinitely many continuous functions having the same sample values. For this reason, a signal is first preprocessed and sampled so that, after the whole process, the sample values are the "best" description of the original signal.

Wavelets–A Tutorial in Theory and Applications
C. K. Chui (ed.), pp. 509–528.

ISBN 0-12-174590-2

This procedure is contained in the well known Shannon Sampling Theorem ([15] and [21]). The theory as well as the class of functions that are completely characterized by their sample values will be reviewed in Section 6.

Another widely used approach is to discretize a signal by describing it as a weighted sum of elementary signal packets (in fact Shannon's sampling theory can also be described in this way). The Gabor representation uses modulated Gaussian functions as the elementary packets [14]. This representation has been praised because modulated Gaussian functions are optimal in terms of their time-frequency localization. This means that the expansion coefficients describe, in an optimal sense, the frequency components at their occurrence in time (or space).

Recently, the concepts of multiresolutions and wavelet transforms have offered an infinite variety of discrete representations to choose from ([13], [17], and [19]). Moreover, the wavelet representations have the desirable property of being localized in both the time (space) and the frequency domains. However, no wavelets can achieve the optimal time-frequency localization of Gabor functions. Still, we will show how to construct families of wavelets that tend to Gabor functions.

In this chapter, we present a unified view of these various signal representation theories. To do this, we identify their essential properties. We use these properties to extend some results from multiresolution analysis and to generalize Shannon's Sampling Theory. We then introduce the new concept of wavelet sequences and apply it to create "Shannon's-type" and "Gabor's-type" scaling and wavelets sequences.

The presentation is organized as follows: in Section 3, we briefly review Stephane Mallat's theory of multiresolution analysis and orthogonal wavelet transforms. In Section 4, we show how to extend some of his results to nonorthogonal scaling functions. We define equivalent classes of scaling and wavelet functions and construct basis functions with certain special properties. In Section 5, we define related sequences of scaling and wavelet functions with an increasing regularity index n. We concentrate on three particular sequences: the basic, the cardinal, and the orthogonal sequences. The connection with Shannon's Theory is provided in Section 6. In particular, we show that the cardinal and orthogonal sequences tends to ideal lowpass and bandpass filters as the regularity index n tends to infinity. Finally, in Section 7, we discuss the Gabor transform which is optimal in terms of its time-frequency localization. We then show that the basic wavelet sequence constructed in Section 5 tends to Gabor functions. This implies that we can obtain wavelets that are nearly time-frequency optimal. The various aspects of the theory are illustrated using polynomial splines. Many results in this chapter are new; their proofs can be found in [1] and will be published elsewhere.

§2. Definitions and notations

The signals considered here are defined on $\mathbb{R}$ and belong to the space of measurable, square-integrable functions: L_2. The space of square summable sequences (discrete functions) is denoted by l_2.

The symbol "$*$" will be used for three slightly different binary operations that are defined below: the convolution, the mixed convolution, and the discrete convolution . The ambiguity should be easily resolved from the context.

For two functions f and g defined on $\mathbb{R}$, $*$ denotes the usual convolution:

$$(f * g)(x) = \int_{-\infty}^{+\infty} f(\xi)g(x-\xi)d\xi, \quad x \in \mathbb{R}. \tag{1}$$

The mixed convolution between a sequence $b(k), k \in \mathbb{Z}$ and a function f on $\mathbb{R}$ is the function $b * f$ on $\mathbb{R}$ given by:

$$(b * f)(x) = \sum_{k=-\infty} b(k)f(x-k), \quad x \in \mathbb{R}. \tag{2}$$

The discrete convolution between two sequences a and b is the sequence $a * b$:

$$(a * b)(k) = \sum_{i=-\infty}^{i=+\infty} a(i)b(k-i), \quad k \in \mathbb{Z}. \tag{3}$$

Whenever it exists, the convolution inverse $(b)^{-1}$ of a sequence b is defined to be:

$$\left((b)^{-1} * b\right)(k) = \delta_0(k), \tag{4}$$

where δ_0 is the unit impulse; *i.e.*, $\delta_0(0) = 1$ and $\delta_0(k) = 0$ for $k \neq 0$.

The reflection of a function f (resp., a sequence b) is the function f' (resp., the sequence b') given by:

$$f'(x) = f(-x), \quad \forall\, x \in \mathbb{R}, \tag{5}$$

$$b'(k) = b(-k), \quad \forall\, k \in \mathbb{Z}. \tag{6}$$

The modulation $\tilde{b}(k)$ of a sequence b is obtained by changing the signs of the odd components of b:

$$\tilde{b}(k) = (-1)^k b(k). \tag{7}$$

The decimation (or down-sampling) operator by a factor of two $[.]_{\downarrow 2}$ assigns to a sequence b the sequence $[b]_{\downarrow 2}$ which consists of the even components of b only:

$$[b]_{\downarrow 2}(k) = b(2k), \quad \forall\, k \in \mathbb{Z}. \tag{8}$$

A pseudo-inverse to the decimation operator is the up-sampling operator $[.]_{\uparrow 2}$. This operator expands a sequence b by inserting zeros between its components:

$$[b]_{\uparrow 2}(k) = \begin{cases} b(k/2), & k \text{ even} \\ 0, & k \text{ odd} \end{cases} . \tag{9}$$

§3. Multiresolution approximations

As defined by Stephane Mallat, a multiresolution structure of L_2 is a sequence of spaces $V_{(j)}$ satisfying:

(i) $\operatorname{clos}\left(V_{(0)}\right) = V_{(0)} \subset L_2$;

(ii) $V_{(j)} \subset V_{(j-1)}$;

(iii) $\operatorname{clos}\left(\bigcup_{j=-\infty}^{j=+\infty} V_{(j)}\right) = L_2$;

(iv) $\bigcap_{j=-\infty}^{j=+\infty} V_{(j)} = \{0\}$;

(v) $f(x) \in V_{(j)} \Leftrightarrow f(2x) \in V_{(j-1)}$;

(vi) there exists an isomorphism I *from* $V_{(0)}$ onto l_2 which commutes with the action of $\mathbb{Z}$.

Relative to such a structure, a multiresolution approximation $\{..., s_{(-1)}, s_{(0)}, s_{(1)}, ...\}$ of a signal s consists of the L_2 least square approximations $s_{(j)}$ of s on the spaces $V_{(j)}$. Because of property (ii), if $i < j$ then $s_{(i)}$ can be viewed as a finer approximation than $s_{(j)}$. In signal processing, $\{..., s_{(-1)}, s_{(0)}, s_{(1)}, ...\}$ is called a fine-to-coarse pyramid representation. It is used for finding efficient algorithms in applications such as coding, edge detection, texture discrimination, fractal analysis, etc. ([6], [17], [23], and [29]).

Properties (i), (ii), and (iii) allow us to approximate a function to any desired accuracy by taking a sufficiently large approximation space $V_{(j)}$ while property (iv) states that coarser and coarser approximations will eventually contain no information about the original function. The similarity property (v) and property (vi) are essential for numerical computations. An important result due to S. Mallat is that, in a multiresolution structure, each subspace $V_{(j)}$ can be induced by an orthonormal basis $\{2^{-j/2}\phi(2^{-j}x - k)\}_{k\in\mathbb{Z}}$. The basis vectors are generated by dilating and translating a unique function ϕ called the orthogonal scaling function. Formally, this can be stated as follows:

$$V_{(j)}(\phi) = \left\{ v : v(x) = \sum_{k=-\infty}^{\infty} c_{(j)}(k)\phi_{2^j}(x - 2^j k), \quad c_{(j)} \in l_2 \right\}, \tag{10}$$

where

$$\phi_{2^j}(x) = 2^{-j/2}\phi(2^{-j}x). \tag{11}$$

Associated with the space $V_{(j)}$, is the wavelet space $W_{(j)}$ which is defined to be the orthogonal complement of $V_{(j)}$ relative to $V_{(j-1)}$:

$$W_{(j)} \oplus V_{(j)} = V_{(j-1)}. \tag{12}$$

Similarly to $V_{(j)}$, the wavelet space $W_{(j)}$ can also be generated by an orthonormal basis $\{2^{-j/2}\psi(2^{-j}x - k)\}_{k\in\mathbb{Z}}$ induced by the orthogonal wavelet function

ψ ([13], [17], and [19]):

$$W_{(j)}(\psi) = \left\{ w : w(x) = \sum_{k=-\infty}^{\infty} d_{(j)}(k)\psi_{2^j}(x - 2^j k), \quad d_{(j)} \in l_2 \right\}, \tag{13}$$

where

$$\psi_{2^j}(x) = 2^{-j/2}\psi(2^{-j}x). \tag{14}$$

Equation (12) implies that the difference between the two approximations $s_{(j)}$ and $s_{(j-1)}$ of a function s is equal to the orthogonal projection of s onto $W_{(j)}$. For this reason, the spaces $W_{(j)}$ are also called the detail spaces, or sometimes the residual spaces. They constitute a direct sum of the square integrable function space L_2:

$$L_2 = \bigoplus_j W_{(j)} := \ldots \oplus W_{(1)} \oplus W_{(0)} \oplus W_{(-1)} \oplus \ldots. \tag{15}$$

Thus, a function s can be represented by the coefficients $\{\ldots \mathbf{d}_{(-1)}, \mathbf{d}_{(0)}, \mathbf{d}_{(1)}, \ldots\}$ (in Equation (13)) of the orthogonal projection of s onto $W_{(j)}$. This representation is what is called the discrete wavelet transform.

The simplest example of orthogonal scaling function is the rectangular or indicator function $\chi_{[0,1)}$ which takes the value 1 in the interval $[0, 1)$ and the value 0 elsewhere. The corresponding wavelet is the Haar function which takes the value 1 in $[0, \frac{1}{2})$ and -1 in the interval $[\frac{1}{2}, 1)$. The associated multiresolution spaces $V_{(j)}$ are the piecewise constant functions or splines of order zero. Other examples are the orthogonal spline scaling functions and their corresponding orthogonal spline wavelets which have been found independently by Lemarié and Battle ([5] and [16]).

Another important scaling function is the sinc function (*i.e.*, the impulse response of the ideal lowpass filter) used in the Shannon's Sampling Theory. The associated wavelet is a modulated sinc (the ideal bandpass filter) ([15], [21]). The link between Shannon Sampling Theory and polynomial splines approximations can be found in some of our previous work ([2] and [27]). The theory for discrete signals has also been considered by Aldroubi *et al.* ([3] and [4]).

§4. Nonorthogonal scaling functions and wavelets

A useful and constructive method to build a multiresolution structure is to start with a scaling function ϕ and use Equations (10) and (11) to define the subspaces $V_{(j)}$. Clearly, the scaling function cannot be chosen arbitrarily. In fact, because $V_{(1)} \subset V_{(0)}$, there must be a sequence $u(k)$ that relates the dilated function ϕ_2 and the scaling function ϕ:

$$\phi_2(x) = (u * \phi)(x), \tag{16}$$

where the $*$ is the mixed convolution defined by Equation (2). The sequence $u(k)$ in the relation will be called the generating sequence and the relation is usually referred to as the two scale relation.

Sufficient conditions on $u(k)$ and its Fourier transform $U(f)$ as well as a method of constructing the orthogonal scaling function can be found in [17]. Fast algorithms for computing the projections onto the corresponding spaces $V_{(j)}$ and $W_{(j)}$ are described in [18]. The construction method as well as the computational algorithms are closely related to the multirate filter banks developed in signal processing ([12] and [30]).

If we drop the orthogonality constraint and require only that $\{2^{-j/2}\phi(2^{-j}x-k) : k \in \mathbb{Z}\}$ be an unconditional basis of $V_{(j)}$ then we can start from a sequence $u(k)$ with Fourier transform $U(f)$ and, similarly to [17], construct a scaling function λ with Fourier transform $L(f)$ as described in the following proposition:

Proposition 1. *Let* $U(f) = \sum_{k=-\infty}^{\infty} u(k)e^{-i2\pi kf}$ *be such that:*

$$2^{-1/2}\,|U(0)| = 1, \tag{17}$$

$$2^{-1/2}\,|U(f)| \leq 1, \tag{18}$$

$$U(f) \neq 0, \quad \forall\, f \in [-1/4, 1/4], \tag{19}$$

$$\prod_{i=1}^{n} 2^{-1/2} U(\frac{f}{2^i}) \overset{L_2}{\rightarrow} L(f), \tag{20}$$

$$L(f) = O(|f|^{-1-\varepsilon}), \tag{21}$$

then $L(f)$ *is the Fourier transform of a scaling function* λ *which generates a multiresolution* $V_{(j)}(\lambda)$ *as in Equations (10) and (11).*

The conditions in Proposition 1 are the same as the one given by Mallat in [17] except that we have droped the Quadrature Mirror Filter requirement and added Equation (21). Except for the technical condition (19), all the others can be inferred by taking the Fourier transform of Equation (16) and solving for $L(f)$. Condition (21) is a regularity requirement which implies the continuity of the function λ.

The set $\{2^{-j/2}\lambda(2^{-j}x-k) : k \in \mathbb{Z}\}$ generated from a scaling function obtained by the procedure of Proposition 1 are not necessarily orthogonal. If $U(f)$ is chosen to be real and symmetric then λ will also be real and symmetric. Moreover, if $u(k)$ has finitely many nonzero values then λ will have compact support; *i.e.*, $\lambda(x)$ will be zero outside a closed bounded set [13]. Nonorthogonal scaling functions and wavelets using splines were proposed by Chui and Wang ([9], [10], and [11]) and, in the context of signal processing, by Unser *et al.* ([24], [27], and [28]).

4.1. Law of composition of scaling functions

It should be noted that, if $U_1(f)$ and $U_2(f)$ are the Fourier transforms of two sequences satisfying the conditions of Proposition 1 then so does their weighted product :

$$U = U_1 \bullet U_2 = 2^{-1/2} U_1 U_2. \tag{22}$$

This remark depicts an obvious algebraic structure which we state in the following proposition.

Proposition 2. *The product (22) defines an internal law of composition for the set G which consists of all sequences satisfying the conditions of Proposition 1.*

Using the property that the Fourier transform converts a convolution product into a multiplication product, this last proposition allows us to use convolution to generate new scaling functions. Thus, the convolution $\lambda * \zeta$ of two scaling functions λ and ζ generated by the sequences $u_1(k)$ and $u_2(k)$ is also a scaling function generated by the discrete convolution $u_1 * u_2$. An interesting property is that the new function $\lambda * \zeta$ is more regular than its individual constitutive atoms λ and ζ. In fact, if λ is r_1 regular and ζ is r_2 regular, then $\lambda * \zeta$ has $r_1 r_2$ regularity. Another interesting property is that if $U_1(f)$ has a zero of order p_1 at $f = 1/2$ and $U_2(f)$ has a zero of order p_2 at $f = 1/2$ then $U_1(f)U_2(f)$ has zero of order $p_1 p_2$ at $f = 1/2$. This implies that the degree of the polynomials that can be exactly represented in terms of $\lambda * \zeta$ is $p_1 p_2$ instead of p_1 if the representation λ is used or p_2 if ζ is used [22].

Also, there are useful properties that are invariant under convolution. For instance, symmetry is preserved; *i.e.*, if λ and ζ are symmetric then $\lambda * \zeta$ is also symmetric. Moreover, if λ and ζ have compact support, then $\lambda * \zeta$ have compact support as well.

4.2. Equivalent scaling functions

It should be noted that two non-identical scaling functions may generate the same multiresolution $V_{(j)}$. In fact, given a scaling function λ that generates the multiresolution $V_{(j)}(\lambda)$, we can use the mixed convolution to construct another function φ generating the same multiresolution $V_{(j)}(\lambda) = V_{(j)}(\varphi)$. This is done by convolving λ with an invertible convolution operator $p(k)$ on l_2 (mixed convolution as in Equation (2)):

$$\varphi(x) = (p * \lambda)(x). \tag{23}$$

Thus, the set of scaling functions can be partitioned into equivalence classes by the relation that associates two scaling functions whenever they generate the same multiresolution. Moreover, two scaling functions belonging to the same equivalence class are always related by Equation (23).

As an example, we can choose p to be the inverse of the operator-square-root of the autocorrelation function $a(k)$ of λ:

$$p(k) = (a)^{-1/2}(k), \tag{24}$$

$$a(k) = (\lambda * \lambda')(k), \quad k \in \mathbb{Z}, \tag{25}$$

where the operator-square-root $a^{1/2}$ is such that $a(k) = (a^{1/2} * a^{1/2})(k)$. For this choice of p we obtain the orthogonal scaling function ϕ of Stéphane Mallat. To verify this claim, a simple calculation will show that:

$$(\phi * \phi')(k) = \begin{cases} 1, & k = 0 \\ 0, & k = \pm 1, \pm 2, \ldots \end{cases} \tag{26}$$

which is equivalent to the orthogonality condition.

4.3. Nonorthogonal wavelets

Similarly to the case of scaling functions, there are infinitely many ways to choose a wavelet ω so that $\{2^{-j/2}\omega(2^{-j}x - k) : k \in \mathbb{Z}\}$ is a basis of $W_{(j)}$. However, although $\{2^{-j/2}\omega(2^{-j}x - k) : k \in \mathbb{Z}\}$ are not necessarily orthogonal at a given level j, they retain the orthogonality between two wavelet spaces at different resolutions. The reason for this is that $W_{(j)}$ is still the orthogonal complement of $V_{(j)}$ relative to $V_{(j-1)}$.

Since $W_{(j)}$ is included in the space $V_{(j-1)}$, any wavelet function ω generating $W_{(j)}$ can be expressed in terms of the scaling function λ generating $V_{(j)}$. For our purpose, we begin by choosing the basic wavelet μ defined by:

$$2^{-1/2}\mu(x/2) = \mu_2(x) = \sum_{k=-\infty}^{\infty} (\delta_1 * \tilde{u}' * \tilde{a})(k)\lambda(x - k), \tag{27}$$

where δ_1 is the unit pulse centered at 1 (*i.e.*, $\delta_1(1) = 1$ and $\delta_1(k) = 0$ for $k \neq 1$), where $a(k)$ is the sampled autocorrelation of λ as in (25), and where $u(k)$ is its generating sequence. An interesting property of the construction (27) is that if the generating sequence $u(k)$ has finitely many nonzero values, then the wavelet μ has compact support.

To obtain other wavelet functions ω generating the same spaces $W_{(j)}$, we use an invertible convolution operator $q(k)$ on l_2 as in (23) to get:

$$\omega(x) = (q * \mu)(x). \tag{28}$$

For example, we can choose q to obtain the orthonormal wavelet function ψ of Mallat:

$$q(k) = \left(a * [\tilde{a} * a]_{\downarrow 2}\right)^{-1/2}(k), \tag{29}$$

where a is the sampled autocorrelation function defined by (25) where $\tilde{a}$ is the modulation of a as defined by Equation (7) and where the decimation operator $[.]_{\downarrow 2}$ is defined by Equation (8).

§5. Families of scaling and wavelet functions

As we have mentioned in Subsection 4.1, we can generate new multiresolution structures simply by convolution of known scaling functions. We can even start with a single function and generate, by repeated convolution, an infinite sequence of scaling functions. Well known cases that can be obtained in this fashion are the B-spline functions $\beta^n(x)$ of order n, which are used as basis for generating certain polynomial spline function spaces. In fact, it is the structure and properties of B-spline functions that motivated part of this work. In this sense, the constructions that follow can be viewed as a generalization of polynomial splines.

5.1. Basic scaling functions and wavelets

We start with a scaling function λ and generate, by convolution, a sequence of increasingly regular scaling functions λ^n as follows:

$$\lambda^n(x) = \lambda * \lambda * \lambda * \ldots * \lambda. \quad (n-1 \text{ convolution}). \tag{30}$$

These functions will be called the basic scaling functions of order n. For each λ^n, there is a corresponding basic wavelet function μ^n obtained from the construction (27):

$$2^{-1/2}\mu^n(x/2) = \mu_2^n(x) = \sum_{k=-\infty}^{\infty} (\delta_1 * \tilde{u}^{n\prime} * \tilde{a}^n)(k)\lambda^n(x-k), \tag{31}$$

$$a^n(k) = (\lambda^n * \lambda^{n\prime})(k), \quad \forall\, k \in \mathbb{Z}, \tag{32}$$

$$u^n = 2^{-(n-1)/2} u * u * \ldots * u, \quad (n-1 \text{ discrete convolutions}). \tag{33}$$

The remarks of Subsection of 4.1 imply that the regularity of the scaling function λ^n and the wavelet μ^n increases with increasing values of the order n. Also, if the starting generating sequence u is finite then the basic scaling and wavelet functions λ^n and μ^n have compact support. Moreover, λ^n and μ^n will be "linear phase" (have a vertical axe of symmetry) if u has a vertical axe of symmetry. This property is relevant in signal processing for obtaining representations that have no phase distortions.

5.2. Cardinal families of scaling functions and wavelets

The basic scaling functions λ^n and the corresponding wavelets μ^n generate the multiresolution spaces $V_{(j)}^n$ and the wavelet spaces $W_{(j)}^n$. We can use Equations (23) and (28) to create other scaling functions and wavelets associated with the same multiresolution spaces $V_{(j)}^n$. A particular scaling function of interest to us is the cardinal scaling function η^n which has zeros at all the integers except at the origin where its value is 1:

$$\eta^n(k) = \delta_0(k) = \begin{cases} 1 & k = 0 \\ 0 & k = \pm 1, \pm 2, \ldots \end{cases} \tag{34}$$

The expression of the cardinal scaling function in terms of λ^n is given by:

$$\eta^n(x) = \left((b^n)^{-1} * \lambda^n\right)(x), \tag{35}$$

$$b^n(k) = \lambda^n(k), \quad \forall\, k \in \mathbb{Z}. \tag{36}$$

An associated wavelet which has the same property (34) as the cardinal scaling function is the cardinal wavelet given by:

$$2^{-1/2}\nu^n(x/2) = \nu_2^n(x) = \left([c^n]_{\uparrow 2} * \mu_2^n\right)(x), \tag{37}$$

$$c^n(k) = 2^{-1/2}\left([\delta_1 * b^n * \tilde{u}^{n\prime} * \tilde{a}^n]_{\downarrow 2}\right)^{-1}(k) \quad \forall\, k \in \mathbb{Z}, \tag{38}$$

where μ_2^n is the basic wavelet in Equation (31).

An important property of a cardinal representation of a signal s is that the expansion coefficients in the representation are precisely equal to its sampled values. This implies that the expansion coefficients can themselves be viewed as a faithful representation of the sampled signal $s(k), k \in \mathbb{Z}$. This property will be further discussed in Section 6.

5.3. Orthogonal families of scaling functions and wavelets

Another family of interest is the orthogonal scaling sequence ϕ^n. It is obtained from the basic sequence λ^n using Equations (23)-(25) as follows:

$$\phi^n(x) = \left((a^n)^{-1/2} * \lambda^n\right)(x), \tag{39}$$

$$a^n(k) = (\lambda^n * \lambda^{n\prime})(k), \quad \forall\, k \in \mathbb{Z}. \tag{40}$$

The corresponding orthogonal wavelet ψ^n is obtained from the basic wavelet μ^n using Equations (28) and (29):

$$\psi_2^n(x) = \left([o^n]_{\uparrow 2} * \mu_2^n\right)(x), \tag{41}$$

where

$$o^n(k) = \left(a^n * [\tilde{a}^n * a^n]_{\downarrow 2}\right)^{-1/2}(k), \quad \forall\, k \in \mathbb{Z}. \tag{42}$$

5.4. Examples using polynomial splines

Polynomial spline functions of order m are C^{m-1} functions that are formed by patching together polynomials of degree m at grid points called the knot points. Here, we will only consider uniformly spaced knot points. In this case, Schoenberg in a remarkable paper shows that any polynomial spline function $s(x)$ of order m can be represented in terms of a unique bell shaped function $\beta^m(x)$; the B-spline function of order m [20]. As an example, when the grid is $\mathbb{Z}$, $s(x)$ can be written as :

$$s(x) = \sum_{k=-\infty}^{\infty} c(k)\beta^m(x-k), \tag{43}$$

where the symmetric B-spline $\beta^m(x)$ is given by:

$$\beta^m(x) = (\beta^0 * \beta^0 * \cdots * \beta^0)(x), \quad (m \text{ convolution}), \tag{44}$$

and where $\beta^0(x)$ is the rectangular function rect(x) defined by Equation (47).

The fact that spline functions are constructed from pieces of polynomials and that they can be easily represented by compactly supported functions $\beta^m(x)$ makes them computationally, as well as theoretically, very attractive. Moreover, it can be easily shown that, for m odd, $\beta^m(x)$ are scaling functions. For m even, the shifted functions $\beta^m(x-1/2)$ are also scaling functions. In fact, there are numerous examples of spline scaling functions and spline wavelets. The simplest of all, the Haar basis function used in the Haar transform, is a spline wavelet of order zero. The Lemarié and Battle orthogonal scaling and wavelet functions are splines of order m ([5] and [16]). For $m = 2n+1$, they can be obtained from the tent function $\tau = \beta^1$ by substituting τ for λ in Equations (39)-(42). The generating sequence u for the tent function τ (see Equation (16)) is given by:

$$u(k) = \begin{cases} 2^{-1/2}, & k = 0 \\ 2^{-3/2}, & k = \pm 1 \\ 0, & \text{otherwise} \end{cases} . \tag{45}$$

Nonorthogonal spline scaling and wavelet functions with various desired properties have also been proposed ([7-11], [24-25], and [27]). For example the cardinal spline functions and wavelets are obtained by replacing λ^n in Equations (35)-(38) by τ^n. The basic spline wavelets proposed by Chui and Wang [11] and by Unser *et al.* [24] are obtained from Equation (31) by replacing λ^n by τ^n.

§6. Multiresolution families and Shannon's Sampling Theory

Shannon's Sampling Theory describes how to process a function before a uniform sampling. The goal is to "best" reconstruct the original function from the sample values. Clearly, there are many different functions having the same sample values. Thus, it is important to describe the class of functions that are fully characterized by a uniform sampling. This class is the set of bandlimited functions (*i.e.*, functions that have a compactly supported Fourier transforms). In particular, a function $s(x)$ with Fourier transform $S(f)$ supported in $[-\frac{1}{2}, \frac{1}{2}]$ (*i.e.*, $S(f) = 0$, $\forall f \notin [-\frac{1}{2}, \frac{1}{2}]$) can be recovered from its samples $s(k)$, $k \in \mathbb{Z}$, using the sinc-interpolation:

$$s(x) = \sum_{k=-\infty}^{\infty} s(k)\text{sinc}(x-k). \tag{46}$$

where $\text{sinc}(x) = \sin(\pi x)/\pi x$. Equivalently, $S(f)$ can be recovered by multiplying the Fourier transform of the sampled sequence with the ideal lowpass filter. The ideal lowpass filter is the rect function :

$$\text{rect}(f) = \begin{cases} 1, & |f| < 1/2 \\ 0, & \text{elsewhere.} \end{cases} \tag{47}$$

In general, however, the functions to be represented by their samples are not band-limited. To address this problem, the functions are first preprocessed. This is done by first applying an ideal lowpass filter, thus forcing them to be bandlimited before the sampling. The whole process, which is depicted in Figure 1, is what is known as Shannon's Sampling Theory.

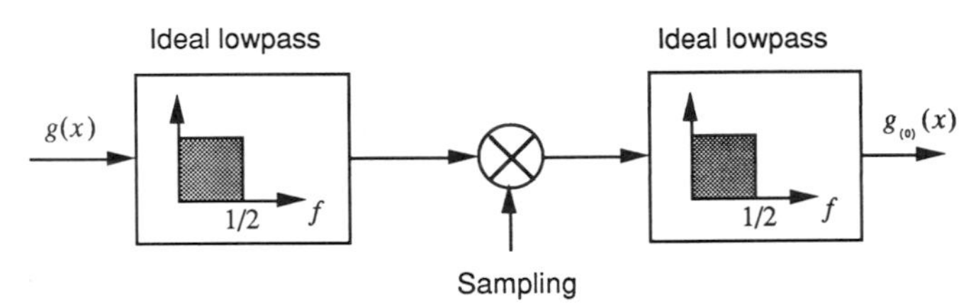

Figure 1. Block diagram for Shannon's sampling procedure. The function g is prefilter by the ideal lowpass filter before sampling. The reconstruction $g_{(0)}$ is obtained from the samples by a second ideal lowpass filtering.

6.1. Multiresolution interpretation of Shannon' Sampling Theory

If we consider the subspace of L_2 generated by the span of all the shifted sinc functions:

$$V_{(0)} = \left\{ v : v(x) = \sum_{k=-\infty}^{\infty} c(k)\text{sinc}(x-k), \quad c \in l_2 \right\}, \tag{48}$$

then, Shannon's Sampling Theory is equivalent to the least square approximation of L_2 in $V_{(0)}$. Taking this point of view, the whole procedure in Figure 1 applied to a function g is expressed as:

$$g_{(0)}(x) = \sum_{k=-\infty}^{\infty} (g * \text{sinc})(k)\text{sinc}(x). \tag{49}$$

In fact, this interpretation was suggested by Shannon in his original paper [21].

It is easy to show that the sinc function is a scaling function. Thus, by defining $\phi(x) := \text{sinc}(x)$ in Equations (10) and (11), we generate a multiresolution structure $V_{(j)}$ of L_2. Since the sinc function vanishes at all integers except at the origin, the coefficients $(g * \text{sinc})(k)$ in Equation (49) are exactly equal to the samples of the approximation $g_{(0)}(k)$. Because of this property, the sinc function is a cardinal scaling function and the resolution associated with it is a cardinal multiresolution. Interestingly enough, sinc is also an orthogonal scaling function.

6.2. Generalized sampling theory

By adopting the previous interpretations, we can generalize Shannon's Sampling Theory. This is achieved simply by using an arbitrary scaling function φ. The whole sampling procedure applied to a function g is reduced to finding the L_2 projection of g on the space $V_{(0)}(\varphi)$:

$$V_{(0)}(\varphi) = \left\{ v : v(x) = \sum_{k=-\infty}^{\infty} c(k)\varphi(x-k), \quad c \in l_2 \right\}. \tag{50}$$

By computing the projection of g and by making the appropriate identification we get:

$$g_{(0)}(x) = \sum_{k=-\infty}^{\infty} (g * \overset{o}{\varphi})(k)\varphi(x), \tag{51}$$

where $\overset{o}{\varphi}$ is also called the dual of φ and is given by:

$$\overset{o}{\varphi}(x) = ((a)^{-1} * \varphi')(x), \tag{52}$$

$$a(k) = (\varphi * \varphi')(k), \quad \forall\, k \in \mathbb{Z}. \tag{53}$$

In this context, the function g is first preprocessed using the prefilter with impulse response $\overset{o}{\varphi}$ and then sampled. The sample values $(g * \overset{o}{\varphi})(k)$ can then be used to reconstruct the "best" (least square) approximation $g_{(0)}$ of g given by Equation (51).

By analogy with Shannon's Sampling Theory, we can say that the functions that are fully characterized by their sample values are the φ-limited functions (*i.e.*, the class that is described by Equation (50)).

Finally, the assumption that φ be a scaling function is in no way crucial to the generalization. The only important requirement is that the space $V_0(\varphi)$ be a closed subspace of L_2.

6.3. Sampling with cardinal and orthogonal functions

It should be noted that, in general, the samples $(g * \overset{\circ}{\varphi})(k)$ are not equal to the sampled approximation $g_{(0)}(k)$. If we require equality, then we need to replace φ in Equations (50)-(53) by the cardinal scaling function η. The function η vanishes at all integers except at 0 where it takes the value 1. It can be obtained from φ as described by Equations (35) and (36):

$$\eta(x) = \left((b)^{-1} * \varphi\right)(x), \tag{54}$$

$$b(k) = \varphi(k), \quad \forall\, k \in \mathbb{Z}. \tag{55}$$

More generally, starting from a scaling function λ, we obtain a sequence of cardinal scaling functions η^n by using Equations (30), (35), and (36). If we replace φ in Equations (50)-(53) by η^n, we obtain a cardinal sampling procedure for each member of the family. In fact, under the appropriate conditions on λ, the sequence of cardinal sampling procedures tends to the scheme of Shannon as n goes to infinity. This result, which further emphasizes the link with Shannon's Sampling Theory is stated in the following theorem.

Theorem 3. *Let $\lambda(x)$ be a scaling function and $L(f)$ its Fourier transform. Let η^n be the sequence of cardinal scaling functions generated by Equations (35) and (36) and let $\overset{\circ}{\eta}{}^n$ be the corresponding duals. If the Fourier transform $L(f)$ of $\lambda(x)$ satisfies:*

$$|L(f)| > |L(f-i)|, \quad \forall\, f \in (-1/2, 1/2), \forall\, i \in \mathbb{Z} \quad i \neq 0, \tag{56}$$

$$\sum_i \frac{|L(f-i)|}{|L(f)|} < \infty, \quad f \in (-1/2, 1/2), \tag{57}$$

then the filters $H^n(f)$ and $\overset{\circ}{H}{}^n(f)$ corresponding to η^n and $\overset{\circ}{\eta}{}^n$ converge to the ideal lowpass filter as the order n tends to infinity:

$$\lim_{n\to\infty} H^n(f) = rect(f), \tag{58}$$

$$\lim_{n\to\infty} \overset{\circ}{H}{}^n(f) = rect(f). \tag{59}$$

With a few more conditions on the scaling function, L_p convergence can also be obtained as in the case of polynomial splines ([2] and [28]). Conditions (56) and (57) essentially mean that $L(f)$ is a non-ideal lowpass filter in the frequency band $(-\frac{1}{2}, \frac{1}{2})$. They can be satisfied by appropriately choosing $U(f)$ (*e.g.*, $U(f)$ smooth and $U(f)$ monotonically decreasing). The theorem can be viewed as stating that the ideal lowpass filter can be approximated as closely as needed by the sequences η^n and $\overset{\circ}{\eta}{}^n$. These are obtained by repeated convolutions and a simple correction of a single non-ideal lowpass filter. Moreover, by the remarks in Subsection 4.1, η^n and $\overset{\circ}{\eta}{}^n$ increase their regularity (smoothness) with increasing values of n.

Instead of cardinal scaling functions η^n, we can select orthogonal scaling functions ϕ^n (*i.e.*, $\{2^{-j/2}\phi(2^{-j}x-k)\}_{k\in\mathbb{Z}}$ is an orthogonal set). They are obtained from λ using Equations (39) and (40). As previously noted, this orthogonality property also holds for the sinc scaling function. An interesting property of orthogonal functions is that they are equal to the reflection of their duals. This means that the preprocessing and the reconstruction filters in the sampling procedure are the same. Similarly to the cardinal scaling family, the Fourier transforms of the sequence ϕ^n also tend to the ideal lowpass filter as n goes to infinity. It can also be shown that the Fourier transforms of the cardinal and orthogonal wavelets (37) and (41) tend to the ideal bandpass filter as the order n goes to infinity.

In Figure 2, spline scaling functions are used to illustrate the extension to Shannon's theory and the above theorem. This figure shows the totality of the sampling and representation procedure; it consists of a prefiltering followed by a sampling and finally a reconstruction using a postfilter. As can be seen from the graphs, the cardinal and orthogonal spline filters of order 3 are already a good approximations to the ideal filter. The graphs also show that the orthogonal filters and their duals are equal.

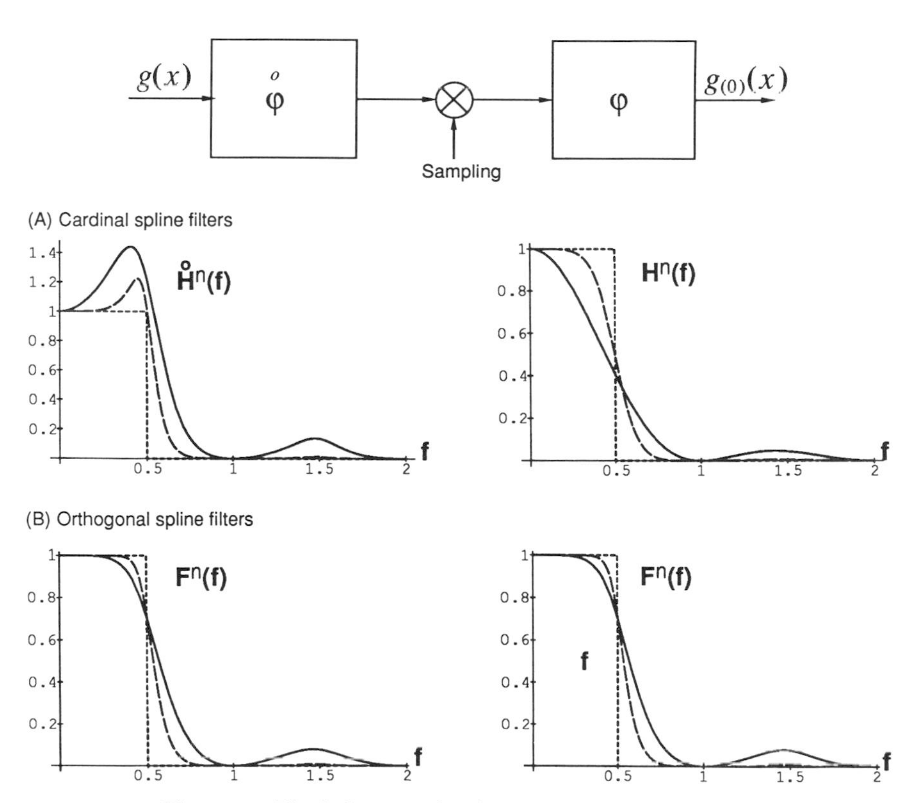

Figure 2. Block diagram for the general sampling theory.

Graph (A) corresponds to the cardinal spline filters and Graph (B) corresponds to the Orthogonal spline filters. The filters of order $n = 1$ are drawn with a continuous lines, the ones of order $n = 3$ are in dashed lines (- - - -), and the ideal lowpass filter is in dotted line $(\cdots)$.

§7. Multiresolution families in connection with Gabor transform

An important property of wavelets is that they are localized in both space (or time) and in frequency. This property is useful because a decomposition of a signal in terms of a wavelet ω determines the signal frequency content at "each" location in space (or time). A measure of the time and frequency localization of a function is given by the product of the standard deviations of its squared modulus and the squared modulus of its Fourier transform. This measure is bounded below by the constant $(4\pi)^{-2}$ [14]. The lower bound cannot be achieved by any wavelet. The only time-frequency optimal functions that achieve the lower bound are the Gabor functions:

$$g(x) = \exp\left(i\Omega(x - x_0) - i\theta\right) \frac{1}{\sqrt{2\pi}\sigma} \exp\left(-\frac{(x - x_0)^2}{2\sigma^2}\right). \tag{60}$$

These are modulated Gaussian functions with four parameters: the offset x_0, the standard deviation σ, the modulation frequency Ω, and the phase shift θ. All the parameters can be chosen arbitrarily.

It can be shown that there are no values of the parameters that can force the corresponding function to form a wavelet basis of L_2. Moreover, this also holds for the real and complex parts of Gabor functions, even though it is still possible to create a sequence of wavelets that approaches the real part of Gabor functions as close as we choose. One way of doing this is by choosing the nonorthogonal family of wavelets given by Equation (31). This family is easy to construct since the only operations that are involved are: convolutions, discrete modulations and a single reflection. The way in which this family tends to the real part of a Gabor function as the order n tends to infinity is stated in the following theorem.

Theorem 4. *Let λ be a scaling function, $L(f)$ its Fourier transform and $u(k)$ the corresponding generating function. Let μ_2^n be the sequence of nonorthogonal scaling function generated by Equations (31), (32), and (33). Let $U(f)$ be symmetric and $U(-1/2) = 0$. Furthermore assume that :*

$$|L(f)| > |L(f - i)|, \quad \forall\, f \in (-1/2, 1/2), \forall\, i \in \mathbb{Z} \quad i \neq 0, \tag{61}$$

$$\sum_i |L(f - i - 1/2)| < \infty, \tag{62}$$

then the Fourier transforms $G_2^n = e^{i2\pi f} M_2^n(F)$ of $\delta_{-1} * \mu_2^n$ have the convergence property given by

$$\lim_{n\to+\infty}\left\{\frac{1}{(Z(f_0))^n}\left(G_2^n\left(\frac{f}{2\pi\sigma_0\sqrt{n}}\pm f_0\right)\right)\right\}=\exp\left(-f^2/2\right), \tag{63}$$

where $Z(f)$, $f_0 \in (-\frac{1}{2},\frac{1}{2})$, *and* σ_0 *are given by*

$$Z(f)=2^{-1/2}U(f-1/2)L(f)|L|^2(f-1/2), \tag{64}$$

$$Z(f_0)>|Z(f)|,\quad \forall\, f\in R, f\neq f_0, \tag{65}$$

$$\left.\frac{d^2Z}{df^2}\right|_{f=f_0}=-(2\pi\sigma_0)^2Z(f_0). \tag{66}$$

The conditions that $U(f)$ be symmetric and $U(1/2)=0$ are not restrictive. The first is simply the requirement that λ be real symmetric. The second is a standard condition that is usually desirable as mentioned in Subsection 4.1. Conditions (61) and (62) constitute the requirement that $L(f)$ essentially be a lowpass filter in the frequency band $(-\frac{1}{2},\frac{1}{2})$. They can be met by choosing $U(f)$ appropriately (*e.g.*, $U(f)$ smooth and decreasing). Roughly speaking, the theorem states that the basic wavelet of order n is essentially the real part of a Gabor function. It is centered at the origin, its standard deviation is $n^{1/2}\sigma_0$ and its modulation frequency is $\Omega=2\pi f_0\sigma n^{1/2}$. If (63) also holds in L_2 then Parseval identity gives us that

$$\frac{\sigma_0\sqrt{n}}{(Z(f_0))^n}\delta_{-1}*\mu_2^n\left((\sigma_0\sqrt{n})x\right)\overset{L_2}{\to}\frac{\cos\left(2\pi f_0\sigma_0\sqrt{n}x\right)}{\sqrt{2\pi}}\exp\left(-\frac{x^2}{2}\right). \tag{67}$$

It should be noted that the sequence of wavelets satisfying the conditions of the theorem above is symmetric. They also have linear phase, and their regularity increases with increasing values of the order n. Moreover, if the generating sequence $u(k)$ is finite, then the wavelets have compact support (see Subsection 4.1).

As a particular case, the B-spline wavelets proposed in ([11] and [26]) can be obtained by replacing λ^n in Equations (30)-(33) by the function τ^n defined in Subsection 5.4. They satisfy all the properties of Theorem 4 and converge to the real part of the Gabor function with frequency $f_0 \cong 0.41$. Moreover, the convergence (63) is in all L_p norms and (67) holds in L_q for $q\in[2,\infty)$ [26].

Figure 3 illustrates the basic B-spline wavelets and their relation to the real part of Gabor functions. Even though the theorem is only an asymptotic result, the graph already shows a very good agreement between the B-spline wavelet of order 3 and its Gabor limit. In fact, the time-frequency localization measure defined at the beginning of this section is found to be within 2% of the optimal number $(2\pi)^{-2}$.

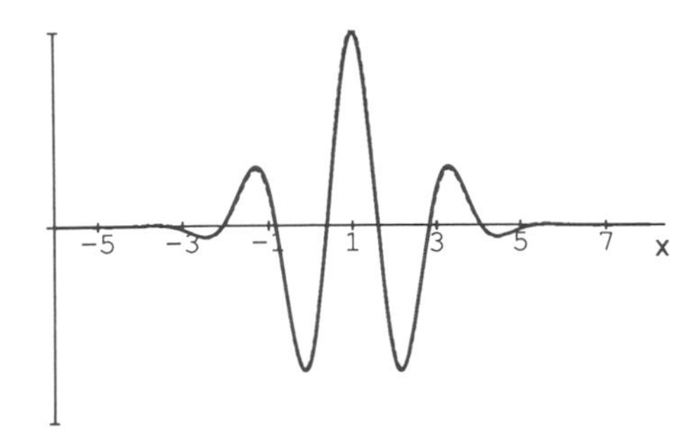

Figure 3. Gabor function (dashed lines) and the Basic spline function of order $n = 3$ (continuous line).

§8. Concluding remarks

Multiresolutions and the related wavelet representations offer an infinite variety of signal representations. Which one to choose depends on the application. As we have shown here, we have an easy procedure to construct scaling and wavelet functions that have certain desired properties. We only need to start from a generating sequence or a scaling function. Then, by using simple convolutions and appropriate corrections, we obtain multiresolution and wavelet transforms with some special chosen features. For example, we can construct arbitrarily regular cardinal and orthogonal functions that are linear phase. Also, we can construct smooth wavelets that tend to a Gabor functions and cardinal sequences that tend to the ideal filters of Shannon. The techniques that we have developed here are general. They can be easily adapted to obtain scaling and wavelets functions with some special features other than the ones presented in this chapter.

References

1. Aldroubi, A. and M. Unser, Multiresolution structure and wavelets : algebraic properties and relation with Shannon's Sampling Theory and the Gabor transform, NCRR Report, National Institutes of Health, 1991, preprint.
2. Aldroubi, A., M. Unser, and M. Eden., Cardinal spline filters: stability and convergence to the ideal sinc interpolator, NCRR Report, National Institutes of Health, 1991, preprint.
3. Aldroubi, A., M. Unser, and M. Eden., Discrete least squares spline filters: convergence properties and application to multiresolution representation of signals and discrete wavelet decomposition, NCRR Report, National Institutes of Health, 1991, preprint.
4. Aldroubi, A., M. Unser, and M. Eden., Asymptotic properties of least square spline filters and application to multi-scale decomposition of sig-

nals, in *Proc. International Symposium on Information Theory and its Applications*, Waikiki, Hawaii, 1990, 271–274.

5. Battle, G., A block spin construction of ondelettes, Part I: Lemarié functions, *Comm. Math. Phys.* **110** (1987), 601–615.
6. Burt, P. J. and E. H. Adelson, The Laplacian pyramid as a compact code, *IEEE Trans. Comm.* **31** (4) (1983), 337–345.
7. Chui, C. K., An overview of wavelets, in *Approximation Theory and Functional Analysis*, C. K. Chui (ed.), Academic Press, Boston, MA, 1991, 47–72.
8. Chui, C. K., J. Stöckler, and J. D. Ward, Compactly supported box-spline wavelets, CAT Report # 230, Texas A&M University, 1990.
9. Chui, C. K. and J. Z. Wang, A general framework of compactly supported splines and wavelets., CAT Report #219, Texas A&M University, 1990.
10. Chui, C. K. and J. Z. Wang, A cardinal spline approach to wavelets, *Proc. Amer. Math. Soc.*, 1991, to appear.
11. Chui, C. K. and J. Z. Wang, On compactly supported spline wavelets and a duality principle, *Trans. Amer. Math. Soc.*, 1991, to appear.
12. Croisier, A., D. Esteban, and C. Galand, Perfect channel splitting by use of interpolation, decimation, tree decomposition techniques, in *Proc. Int. Conf. on Information Sciences/Systems*, Patras, 1976, 443–446.
13. Daubechies, I., Orthonormal bases of compactly supported wavelets, *Comm. Pure and Appl. Math.* **41** (1988), 909–996.
14. Gabor, D., Theory of communication, *J. IEE* **93** (1946), 429–457.
15. Jerri, A. J., The Shannon Sampling Theorem-its various extensions and applications: a tutorial review, *Proc. IEEE* **65** (1977), 1565–1596.
16. Lemarié, P. G., Ondelettes à localisation exponentielles, *J. Math. Pure et Appl.* **67** (1988), 227–236.
17. Mallat, S., Multiresolution approximations and wavelet orthogonal bases of $L_2(R)$, *Trans. Amer. Math. Soc.* **315** (1989), 69–87.
18. Mallat, S., A theory of multiresolution signal decomposition: the wavelet representation, *IEEE Trans. Pattern Anal. Machine Intell.* **11** (1989), 674–693.
19. Meyer, Y., *Ondelettes*, Hermann, Editeurs des Sciences et des Arts, Paris, 1990.
20. Schoenberg, I. J., Contribution to the problem of approximation of equidistant data by analytic functions, *Quart. Appl. Math.* **4** (1946), 45–99, 112–141.
21. Shannon, C. E., Communications in the presence of noise, *Proc. IRE* **37** (1949), 10–21.
22. Strang, G., Wavelets and dilation equations: a brief introduction, *SIAM Review* **31** (1989), 614–627.
23. Terzopoulos, D., Image analysis using multigrid relaxation methods, *IEEE Trans. Pattern Anal. Mach. Intell.* **8** (1986), 129–139.
24. Unser, M., A. Aldroubi, and M. Eden, A family of polynomial spline wavelet transforms, NCRR Report, National Institutes of Health, 1990,

preprint.
25. Unser, M., A. Aldroubi, and M. Eden, The L_2 polynomial spline pyramid: a discrete representation of continuous signals in scale-space, *IEEE Trans. Inform. Theory*, to appear.
26. Unser, M., A. Aldroubi, and M. Eden, On the asymptotic convergence of B-spline wavelets to Gabor functions, *IEEE Trans. Inform. Theory* (special issue on wavelet transform), to appear.
27. Unser, M., A. Aldroubi, and M. Eden., Polynomial spline signal approximations: filter design and asymptotic equivalence with Shannon's Sampling Theorem, *IEEE Trans. Inform. Theory*, to appear.
28. Unser, M., A. Aldroubi, and M. Eden., A sampling theory for polynomial splines, in *Proc. International Symposium on Information Theory and its Applications*, Waikiki, Hawaii, 1990, 279–282.
29. Unser, M. and M. Eden, Multiresolution feature extraction and selection for texture segmentation, *IEEE Trans. Pattern Anal. Machine Intell.* **11** (1989), 717–728.
30. Vetterli, M., Multidimensional sub-band coding: some theory and algorithms, *Signal Processing* **6** (1984), 97–112.

Akram Aldroubi
Mathematics and Image Processing Group
BEIP/NCRR
National Center for Research Resources
Building 13, Room 3W13
National Institutes of Health
Bethesda, MD 20892
aldroubi%aamac.dnet@dxi.nih.gov

Michael Unser
Mathematics and Image Processing Group
BEIP/NCRR
National Center for Research Resources
Building 13, Room 3W13
National Institutes of Health
Bethesda, MD 20892
unser@helix.nih.gov

Wavelets in $H^2(\mathbf{R})$:

Sampling, Interpolation, and Phase Space Density

Kristian Seip

Abstract. A density concept of Beurling yields a complete description of sampling and interpolation in the Bargmann-Fock space. Similarly, regular sets of sampling and interpolation for weighted Bergman spaces can be described in terms of a certain critical density. Key words for problems arising from this insight are phase space density in general, critical sampling densities and a possible Balian-Low type theorem for arbitrary progressive wavelets.

§1. Introduction

One of the fundaments of modern communication theory is Whittaker's cardinal series [37], or Shannon's sampling theorem [34]:

$$f(t) = \sum_n f(n) \frac{\sin \pi(t-n)}{\pi(t-n)},$$

valid for every f in the Paley-Wiener space PW consisting of all entire functions of exponential type at most π that are square-integrable on the real line. A function in PW can thus be reconstructed after sampling its values on a discrete set, in this case the integers. Moreover, since the functions $e_n(t) = \frac{\sin \pi(t-n)}{\pi(t-n)}$ constitute an orthonormal basis for PW, the cardinal series provides a solution to an interpolation problem as well: For every sequence $\{a_n\}$ in l^2, there exists an f in PW such that $f(n) = a_n$ for all n.

It is important for applications and intrinsically interesting to have a characterization of the discrete sets $\{t_n\}$ of real numbers for which every f in PW can be retrieved from its sampled values $\{f(t_n)\}$. When thinking of the elements of PW as bandlimited signals, it is not sufficient, as pointed out by Landau [21], to require $\{t_n\}$ to be a set of uniqueness only; *i.e.*, $f(t_n) = 0$ for all n implies $f \equiv 0$. It is then also essential that the reconstruction of a

Wavelets–A Tutorial in Theory and Applications
C. K. Chui (ed.), pp. 529–540.

ISBN 0-12-174590-2

function f from its sampled values $\{f(t_n)\}$ be stable, conveniently stated by requiring the existence of positive numbers A and B such that

$$A \int |f(x)|^2 dx \leq \sum_n |f(t_n)|^2 \leq B \int |f(x)|^2 dx$$

for every f in PW. The set $\{t_n\}$ is said to be a *sampling set* for PW if this holds. Similarly, $\{t_n\}$ is said to be an *interpolating set* for PW if for every sequence $\{a_n\}$ in l^2 there is an f in PW such that $f(t_n) = a_n$ for all n. A set which is both sampling and interpolating (like the integers) corresponds to an ideal situation with stability and no redundant information; *i.e.*, the set fails to be one of uniqueness after the removal of any of its elements.

Roughly speaking, we have that a sequence of points is a sampling set if its density in every part of the line is larger than 1, and only if this density is larger than or equal to 1. Similarly, a sequence is an interpolating set if its density in every part of the line is smaller than 1, and only if the same density is smaller than or equal to 1. These statements can be made precise with the help of a density concept of Beurling [4], and they can be extracted from the papers [13, 36, 22]. The critical density appearing here is the Nyquist rate; indeed, these results give a precise mathematical meaning to this phrase.

The ideas involved in this classical situation can be transferred to other contexts. For instance, a wavelet basis, or indeed any basis of "coherent states", is generated by a sequence of points in phase space that is both a sampling and an interpolating set. In general, sampling sets are "dense", while interpolating sets are "sparse". We must therefore satisfy competing density requirements in order to construct a basis, and we might expect that this concept of basis decoupling would lead us to detect critical sampling densities.

In this chapter, we will describe a line of research whose goal has been to develop analogues of the indicated results for bandlimited functions within the context of certain time-frequency representations; more specifically, this has taken place in the Bargmann-Fock space of entire functions and in weighted Bergman spaces on a half-plane. Our objective has been to give a general characterization, in terms of suitable phase space density concepts, of the information needed to represent signals as vectors in these function spaces.

This note deserves its place in a book on wavelets because the continuous wavelet transform takes us into weighted Bergman spaces for a particular family of wavelets. Although we have definitive results only for Bergman spaces, our theorems suggest some general properties of the continuous wavelet transform that do not seem to have been recognized previously.

It is instructive first to consider the situation in the Bargmann-Fock space, which is closely related, but easier to handle than the Bergman spaces. The complete and strikingly simple solution that we have obtained here, serves, we believe, as a motivation for our interest in the still unsolved problems for the Bergman spaces.

§2. Density theorems for the Bargmann-Fock space

The Bargmann-Fock space is an instance of a coherent state representation. A set of (Weyl-Heisenberg) coherent states is a family of functions $g^{(x,y)}$, generated from one function $g \in L^2(\mathbf{R})$. $g^{(x,y)}$ is obtained from $g(t)$ by translations on both sides of the Fourier transform; *i.e.*,

$$g^{(x,y)}(t) = e^{iyt}g(t-x).$$

If g is well-localized both in time and frequency, we can think of the inner-product

$$Tf(x,y) = \langle f, g^{(x,y)} \rangle$$

as a "snapshot" of the signal f at time x and frequency y. By letting x and y vary in a continuous fashion, $Tf(x,y)$ then gives a picture of f in a time-frequency space (like a musical score; see the introduction of [8]). In fact, T sets up an isometry from $L^2(\mathbf{R})$ into $L^2(\mathbf{R}^2)$, so that no information on our signals is lost in this representation.

A very natural choice for g is a Gaussian, since the $g^{(x,y)}$ are then extremals of the Heisenberg relation. For the Gaussian g, the functions $g^{(x,y)}$ are known as the *canonical coherent states* of quantum mechanics and *Gabor wavelets* in signal analysis. The main reason for the importance of the Bargmann-Fock space is that for such g, the map T brings us into the space. More precisely, let

$$g(t) = \pi^{-\frac{1}{4}}e^{-\frac{1}{2}t^2},$$

and define the Bargmann transform,

$$Bs(z) = e^{\frac{1}{4}|z|^2 - \frac{1}{2}ixy}\langle s, g^{(x,y)} \rangle,$$

$z = x + iy$. B is a unitary map from $L^2(\mathbf{R})$ onto $F_{\frac{1}{2}}$, where the F_α is the Bargmann-Fock space,

$$F_\alpha = \{f \text{ entire} : \ \|f\|_\alpha^2 = \int_{\mathbf{C}} |f(z)|^2 d\mu_\alpha(z) < \infty\},$$

$\alpha > 0$ and $d\mu_\alpha(z) = \frac{\alpha}{\pi}e^{-\alpha|z|^2}dxdy$. We refer the reader to [15] for general information, and [9] for more background on the problems treated here, as well as a brief account of the importance of the Bargmann-Fock space in quantum mechanics, signal analysis and mathematics.

For applications, it is necessary to restrict the evaluation of the functions in F_α to a discrete set of points. We say that a discrete set Γ of complex numbers is a *sampling set* for F_α if there exist positive numbers A and B such that

$$A\|f\|_\alpha^2 \leq \sum_{z\in\Gamma} e^{-\alpha|z|^2}|f(z)|^2 \leq B\|f\|_\alpha^2 \tag{1}$$

for all $f \in F_\alpha$. This implies that if Γ is a sampling set, then stable recovery of f is possible from a knowledge of its values on Γ; again, mere uniqueness of

f from its sampled values is not sufficient in practice. If to every l^2-sequence $\{a_j\}$ of complex numbers there exists an $f \in F^2_\alpha$ such that $e^{-\frac{\alpha}{2}|z_j|^2} f(z_j) = a_j$ for all j, then $\Gamma = \{z_j\}$ is said to be an *interpolating set* for F_α. A sampling set corresponds, in the terminology of [13], to a *frame* of coherent states. A set which is both sampling and interpolating (an impossibility) would correspond to a *Riesz basis* of coherent states.

Frames were introduced by Duffin and Schaeffer in their study of nonharmonic Fourier series [13]. The general importance of frames in applied mathematics was stressed by Daubechies, Grossmann and Meyer in [10]. An extensive general investigation of frames in the context of short-time Fourier and wavelet transforms was made by Daubechies in [8].

Daubechies and Grossmann posed the problem of finding the lattices $z_{mn} = ma + inb$, $m, n \in \mathbf{Z}$, that are sampling sets for F_α [9]. They proved that a lattice could be a sampling set only if $ab < \pi/\alpha$ and conjectured this condition also to be sufficient. For $ab = \pi/(\alpha N)$, N an integer ≥ 2, they found (1) to hold by providing explicit expressions for the optimal constants A, B. Daubechies was later able to show that a lattice is a set of sampling whenever $N^{-1} < 0.996$ [8].

We have found that the density criterion of the Daubechies-Grossmann conjecture applies not only to lattices, but also to *arbitrary* discrete sets. In order to state this result, we need Beurling's density concept as generalized by Landau [22]. We consider then *uniformly discrete sets*; i.e., discrete sets $\Gamma = \{z_j\}$ for which $\inf_{j \neq k} |z_j - z_k| > 0$. We fix a compact set I of measure 1 in the complex plane, whose boundary has measure 0. Let $n^-(r)$ and $n^+(r)$ denote, respectively, the smallest and largest number of points from Γ to be found in a translate of rI. We define the lower and upper uniform densities of Γ to be

$$D^-(\Gamma) = \liminf_{r \to \infty} \frac{n^-(r)}{r^2}; \qquad D^+(\Gamma) = \limsup_{r \to \infty} \frac{n^+(r)}{r^2},$$

respectively. It was proved by Landau that these limits are independent of I.

Inspired by the density theorems obtained by Beurling in [4], we have established the following results.

Theorem 1. *A discrete set Γ is a sampling set for F_α if and only if it can be expressed as a finite union of uniformly discrete sets and contains a uniformly discrete subset Γ' for which $D^-(\Gamma') > \alpha/\pi$.*

Theorem 2. *A discrete set Γ is an interpolating set for F_α if and only if it is uniformly discrete and $D^+(\Gamma) < \alpha/\pi$.*

The sufficiency of the conditions in the theorems was proved in [33] (the Daubechies-Grossmann conjecture was also proved independently by Lyubarskii [23]), and the necessity in [32].

The simplicity of these results is quite remarkable when compared to the situation in the Paley-Wiener space, where, as indicated above, the necessary and sufficient density criteria do not coincide. As illustrated by the cardinal series, we may have sets which are both sampling and interpolating, and there is an extensive literature on Riesz bases of complex exponentials [38].

The lattice with $a = b = \sqrt{\pi/\alpha}$ is called the *von Neumann lattice*, since von Neumann claimed (without proof) that it is a set of uniqueness [24]; many proofs have later been given in [2], [27], [1], and [33]. (See [17] for an attempted repair of the "defect" of the von Neumann lattice that it is neither a sampling set nor an interpolating set.)

An interesting question is if the density conditions apply to transforms associated to other families of (Weyl-Heisenberg) coherent states; see the discussion in [8, pp. 977-981] for some information on this problem.

§3. Continuous wavelet transforms and Bergman spaces

We refer to [18] or Grossmann and Kronland-Martinet's chapter in [5] for a general introduction to the continuous wavelet transform and its usage in signal analysis. A main point, as stressed in [5], is that the continuous wavelet transform provides an easily interpretable visual representation of signals. To obtain the desired display of real signals, one considers only positive frequencies; *i.e.*, one restricts to the Hardy space $H^2(\mathbf{R})$ and work with the wavelet transform associated with a *progressive wavelet.*

Briefly speaking, the general situation is as follows. We fix an admissible wavelet $g \in H^2(\mathbf{R})$; *i.e.*,

$$c_g = 2\pi \int_0^\infty |\hat{g}(\xi)|^2 \, \xi^{-1} d\xi < \infty.$$

For a dilation parameter $y > 0$ and a translation parameter x we put $u = (x, y)$, and define

$$\tau(u)g(t) = y^{-\frac{1}{2}} g\left(\frac{t-x}{y}\right).$$

The continuous wavelet transform $S(u)$ of a function $s(t) \in H^2(\mathbf{R})$ is obtained from this family of coherent states by forming the inner-products

$$S(u) = \langle s, \tau(u)g \rangle.$$

As probably well known to the readers of this book, this map is isometric (up to a constant factor) from $H^2(\mathbf{R})$ into the space of functions square integrable with respect to the area measure $y^{-2}dxdy$.

We may interpret the time-scale half-plane as the upper half-plane of $\mathbf{C}$. We write $U = \{z = x + iy : \; y > 0\}$ and note that for

$$\hat{g}(\xi) = \begin{cases} \xi^{\frac{\alpha-1}{2}} e^{-\xi}, & \xi > 0 \\ 0, & \xi \le 0, \end{cases} \tag{2}$$

$\alpha > 1$, the wavelet transform maps $H^2(\mathbf{R})$ unitarily (modulo a constant factor) onto the Bergman space

$$A_\alpha(U) = \{f \text{ analytic in } U : \; \|f\|_\alpha = \int\int_U |f(x+iy)|^2 y^{\alpha-2} dxdy < \infty\};$$

that is, to be precise, after we have pulled the factor $y^{-\frac{\alpha}{2}}$ appearing in the transform, into the weighted measure.

Bergman spaces are interesting in their own right, and they have been the subject of quite an extensive mathematical activity in recent years (see the recent monograph [39] and the references therein). Still, many fundamental and interesting problems for these spaces remain, one of which will be mentioned below. When we work with Bergman spaces instead of general continuous wavelet transforms, we have, of course, more powerful tools at our disposal, and consequently, we can in general expect to be able to prove stronger results. Such results can be of considerable intrinsic interest (see, *e.g.*, [12], and [30]). Sometimes they may also illuminate the general picture; we will indicate that the results discussed below could have such an impact.

The most notable fact about the wavelet transform associated with (2) is that it yields a phase space representation of quantum mechanics on the half-plane, relevant in particular for the hydrogen atom. Investigations of this relation have been made by Paul. We refer the interested reader to his part of [26], and the references therein.

§4. A von Neumann type lattice for weighted Bergman spaces

Discrete representations of continuous wavelet transforms were extensively studied by Daubechies in [8]. She considered regular lattices $\Gamma(a,b) = \{u_{mn} = (a^m, bna^m);\ m,n \in \mathbf{Z}\}$, defined for $a > 1$, $b > 0$. As already been stressed, one requires both uniqueness and numerical stability, which in this situation mean that there exist positive numbers A, B such that

$$A\|s\|^2 \leq \sum_{m,n} |S(u_{mn})|^2 \leq B\|s\|^2$$

for all $s \in H^2(\mathbf{R})$, where now $\|\cdot\|$ denotes the H^2 norm. If this holds, we say that $\Gamma(a,b)$ is a *sampling set* for the wavelet transform associated with g, or equivalently, that the sequence of functions $\{\tau(u_{mn})g\}$ is a *frame* in $H^2(\mathbf{R})$. [8] contains excellent numerical estimates of the range of the lattice parameters a and b, as well as of the optimal values of A and B for fairly arbitrary wavelets (as discussed in [8], such estimates ensure efficient reconstruction methods).

The question about a possible critical "sampling density" was touched upon in [8] with a a striking example showing that we might have bases corresponding to lattices of different "densities". The results presented below provide a good opportunity to take up the thread from this example and to some extent clarify the matter. We will return to this argument in the last section.

Daubechies' estimates are not completely sharp. For the Bergman spaces, however, optimal results have recently been obtained. By also describing interpolating sets, we have been led to discover a von Neumann type lattice. Thus, for the Bergman spaces, a critical sampling density does in fact exist.

Before stating these results, let us, for the sake of clarity, specify the notions involved. Quite generally, a sequence of points $\{z_j\}$ is said to be a sampling set for $A_\alpha(U)$ if there exist positive numbers A, B such that

$$A\|f\|_\alpha \leq \sum_j |f(z_j)|^2 y_j^\alpha \leq B\|f\|_\alpha$$

for all $f \in A_\alpha(U)$. $\{z_j\}$ is an interpolating set for $A_\alpha(U)$ if to every sequence $\{a_j\}$ for which $\{a_j y_j^{\frac{\alpha}{2}}\} \in l^2$, we can find an $f \in A_\alpha(U)$ such that $f(z_j) = a_j$ for all j. With a slight abuse of notation, now $\Gamma(a,b) = \{z_{mn}\}_{m,n\in\mathbf{Z}}$, where $z_{mn} = a^m(bn + i)$.

Theorem 3. *$\Gamma(a,b)$ is a sampling set for $A_\alpha(U)$ if and only if $b\ln a < 4\pi/(\alpha-1)$.*

Theorem 4. *$\Gamma(a,b)$ is an interpolating set for $A_\alpha(U)$ if and only if $b\ln a > 4\pi/(\alpha-1)$.*

For the critical value $b\ln a = 4\pi/(\alpha-1)$, $\Gamma(a,b)$ is a set of uniqueness for $A_\alpha(U)$. We remark that a simple argument shows that there is in fact *no* sequence of points in U that is both a sampling and an interpolating set for $A_\alpha(U)$.

We refer to [31] for proofs. The main point is that $\Gamma(a,b)$ is invariant under the Fuchsian group of dilations generated by the transformation $z \mapsto az$, and that we are able to construct explicitly an automorphic form with respect to this group, whose zero-set is precisely $\Gamma(a,b)$. The multiplicative behavior of the automorphic form yields sharp lower and upper bounds on its growth. The resulting estimates enable us to use a technique that was applied to the corresponding problem in the Bargmann-Fock space, and that also works for entire functions of exponential type (or bandlimited functions).

More specifically, when we deal with entire functions of exponential type, we may exploit an automorphic function associated with the periodic group, namely: the sine function. When we work with lattices in the Bargmann-Fock space, we consider the associated double-periodic group and its Weierstrass σ-function. In both cases, the key estimates are consequences of the invariance properties of the special functions.

In fact, the hard estimates needed to prove the sufficiency of the *general* density conditions discussed above, are also consequences of these invariance properties. For instance, in the Bargmann-Fock space, we obtain, via the right σ-function, sufficiently sharp estimates on the growth of a certain infinite product associated to a discrete set which is "close" to a lattice (see [33]).

For the Bergman spaces, this approach fails spectacularly. By the kind of argument used in the proof of Theorem 5 below, this may be realized without any attempt to perform estimates. We are left with a hard and interesting problem: What are the analogues of the Beurling type density theorems for the Bergman spaces?

We mention that "arbitrary" sets of sampling and interpolation have been studied in the mathematical literature (see [7], [28], and [20]), more recently even for arbitrary wavelet transforms (see [14] and [16]; see also [25] for a short and simple exposi- tion). The density concept used in these papers is, however, not adequate for a complete characterization of sampling or interpolating sets.

§5. Arbitrary progressive wavelets and phase space density

We now declare the phase space density of a regular lattice $\Gamma(a,b)$ to be the number $(b \ln a)^{-1}$. We mention also that when we imprecisely refer to "nice" wavelets in the sequel, we think of wavelets that are reasonably well localized both in time and frequency. A mild requirement that probably suffices for our discussion is that the function

$$G(u) = \langle g, \tau(u)g \rangle$$

be integrable with respect to the invariant measure $y^{-2}dxdy$.

We will now look at the argument about phase space density that we alluded to in the previous section. Let g be a wavelet either in $H^2(\mathbf{R})$ or in $L^2(\mathbf{R})$, and put $u_{mn} = (a^m, a^m bn)$ and $u_{mn}^{(\epsilon)} = (a^m, a^m b(1+\epsilon)n)$. In [8, Theorem 2.10, p. 985], it was shown that for a certain g, with $\{\tau(u_{mn})g\}$ $(a=2,\ b=1)$ an orthonormal basis for $L^2(\mathbf{R})$, $\{\tau(u_{mn}^{(\epsilon)})g\}$ is a Riesz basis for $L^2(\mathbf{R})$ for all sufficiently small ϵ. In fact, we have the following general theorem.

Theorem 5. *Suppose $\{\tau(u_{mn})g\}$ is a Riesz basis for $H^2(\mathbf{R})$ $(L^2(\mathbf{R}))$ and that $G(u)$ is integrable. Then $\{\tau(u_{mn}^{(\epsilon)})g\}$ is a Riesz basis for $H^2(\mathbf{R})$ $(L^2(\mathbf{R}))$ for sufficiently small ϵ.*

Thus, whenever we have a nice wavelet basis, the wavelet generates a scale of bases of varying phase space densities. This is indeed a striking feature of wavelet bases.

We sketch the proof of the theorem, since it provides a nice "explanation" of this phenomenon.

The first part of the argument is quite general: Suppose $\{f_j\}$ is a Riesz basis for some Hilbert space H, and let $\{e_j\}$ be some other sequence in H. We assume the existence of a positive constant B such that

$$\sum |\langle f, e_j \rangle|^2 \le B\|f\|^2 \tag{3}$$

for all $f \in H$.

We want to derive a condition under which $\{e_j\}$ is also a Riesz basis. According to the Paley-Wiener stability criterion [38, p. 38], we may then consider the vector

$$f = \sum_j c_j (f_j - e_j),$$

where $\{c_j\}$ is an arbitrary l^2 sequence; $f \in H$ by (3). By the Cauchy-Schwarz inequality, we obtain

$$\|f\|^4 = \Big(\sum c_k \langle f, f_k - e_k \rangle\Big)^2 \le \sum_j |c_k|^2 \sum_k |\langle f, f_k - e_k \rangle|^2.$$

By the assumptions on $\{f_j\}$ and $\{e_j\}$, it is easy to see that we have the estimate

$$\|f\|^2 \leq C \sum |c_j|^2 \leq C' \| \sum c_j f_j \|^2.$$

Letting λ be the smallest number L such that

$$\sum |\langle f, f_k - e_k \rangle|^2 \leq L^2 \|f\|^2 \tag{4}$$

for arbitrary $f \in H$, we obtain

$$\| \sum c_k (f_k - e_k) \| \leq C\lambda \| \sum c_k f_k \|.$$

The problem is thus to determine when is $C\lambda < 1$.

We employ this scheme to the sequences $\{\tau(x_{mn})g\}$ and $\{\tau(u'_{mn})g\}$, with $u'_{mn} = ((1+\epsilon)^{-1}a^m, a^m nb)$. Using the techniques of [25], it is a relatively simple matter to show that $C\lambda < 1$ for all sufficiently small ϵ. To finish the proof, we need only remark that $\{\tau(x'_{mn})g\}$ is a Riesz basis if and only if $\{\tau(x^{(\epsilon)}_{mn})g\}$ is one.

We see that the theorem rests on the following remarkable fact. If we perturb each point of the discrete set slightly (in a uniform way), we may change the density of the set. This stands in marked contrast to the situation in the Bargmann-Fock space.

We can deduce similar perturbation theorems both for sampling and for interpolation. As our results for the Bergman spaces show, this does not contradict the existence of a critical density. Arguments of the above type merely imply that if there is a critical value for $b \ln a$, we cannot have a Riesz basis when the wavelet is reasonably nice.

This observation is interesting in conjunction with a problem of long standing in wavelet theory. The question is if some precise version of the following statement is true : There is no nice g so that $\{\tau(x_{mn})g\}$ is an orthonormal basis for $H^2(\mathbf{R})$. This would be the analogue of the Balian-Low theorem [3] in the wavelet situation. The Balian-Low theorem states that if the set of functions $\{e^{i2\pi mt}g(t-n)\}_{m,n\in\mathbf{Z}}$ is an orthonormal basis for $L^2(\mathbf{R})$, then $tg(t)$ and $g'(t)$ cannot both belong to $L^2(\mathbf{R})$. Actually, the first paper about wavelet frames was written partly because the authors expected that a similar no-go-theorem would exist in the wavelet situation [10] (they were then unaware of the fact that an orthonormal wavelet basis had already been constructed [35]). As an interesting digression, we mention that later on, the theory of orthonormal wavelet bases inspired constructions of nice orthonormal bases in which one gets around the Balian-Low "obstacle" by making some simple modifications of the structure of the sequence (see [11] and [6]).

It is now easy to see that when we use the formalism of multiscale analysis to construct orthonormal wavelet bases, we need wavelets containing both negative and positive frequency components. However, it is not known whether the multiresolution analysis approach yields all "good" wavelet bases.

At an early stage in our study of sampling and interpolation, we made some speculation about possible critical densities [29]. Based on a rather hand-waving argument, we suggested $b \ln a = c_g/\|g\|^2$ to be a critical quantity for regular lattices. It is interesting to note that for the weighted Bergman spaces, we have in fact $c_g/\|g\|^2 = 4\pi/(\alpha - 1)$, so our speculation was at least partly justified. The unsettled question about a possible Balian-Low type theorem thus leads us to ask the following question: For a general nice and progressive g, is it necessary for $\{x_{mn}\}$ to be a sampling (interpolating) set that $b \ln a < c_g/\|g\|^2$ ($b \ln a > c_g/\|g\|^2$)? Moreover, for which progressive wavelets are these conditions sufficient?

References

1. Bacry, H., A. Grossmann, and J. Zak, Proof of the completeness of lattice states in the kq-representation, *Phys. Rev. B* **12** (1975), 1118–1120.
2. Bargmann, V., P. Butero, L. Girardello, and J. R. Klauder, On the completeness of coherent states, *Rep. Mod. Phys.* **2** (1971), 221–228.
3. Battle, G., Heisenberg proof of the Balian-Low theorem, *Lett. Math. Phys.* **15** (1988), 175–177.
4. Beurling, A., *The Collected Works of Arne Beurling, Vol. 2 Harmonic Analysis*, L. Carleson, P. Malliavin, J. Neuberger, and J. Wermer (eds.), Birkhäuser, Boston, MA, 1989, 341–365.
5. Beylkin, G., R. R. Coifman, I. Daubechies, S. Mallat, Y. Meyer, L. A. Raphael, and M. B. Ruskai, eds., *Wavelets and Their Applications*, Jones and Bartlett, Cambridge, MA, 1992, to appear.
6. Coifman, R. R. and Y. Meyer, Remarques sur l'analyse de Fourier à fenêtre, *C. R. Acad. Sci. Paris* Série I **312** (1991), 259–261.
7. Coifman, R. R. and R. Rochberg, Representation theorems for holomorphic and harmonic functions in L^p, *Astérisque* **77** (1980), 11–66.
8. Daubechies, I., The wavelet transform, time-frequency localization and signal analysis, *IEEE Trans. Inform. Theory* **36** (1990), 961–1005.
9. Daubechies, I. and A. Grossmann, Frames in the Bargmann space of entire functions, *Comm. Pure and Appl. Math.* **41** (1988), 151–164.
10. Daubechies, I., A. Grossmann, and Y. Meyer, Painless nonorthogonal expansions, *J. Math. Phys.* **27** (1986), 1271–1283.
11. Daubechies, I., S. Jaffard, and J. L. Journé, A simple Wilson basis with exponential decay, *SIAM J. Math. Anal.* **22** (1991), 549–568.
12. Daubechies, I. and T. Paul, Time-frequency localization operators - a geometric phase space approach II, *Inverse Problems* **4** (1988), 661–680.
13. Duffin, R. J. and A. C. Schaeffer, A class of nonharmonic Fourier series, *Trans. Amer. Math. Soc.* **72** (1952), 341–366.
14. Feichtinger, H. and K. Gröchenig, Banach spaces related to integrable group representations I, *J. Functional Anal.* **86** (1989), 307–340. II, *Mh. Math.* **108** (1989), 129–148.
15. Folland, G. B., *Harmonic Analysis in Phase Space*, Princeton University Press, Princeton, NJ, 1989.

16. Gröchenig, K., Describing functions: Atomic decompositions versus frames, *Mh. Math.*, to appear.
17. Gröchenig, K. and D. Walnut, A Riesz basis for the Bargmann-Fock space related to sampling and interpolation, 1991, in manuscript.
18. Grossmann, A., R. Kronland-Martinet, and J. Morlet, Reading and understanding continuous wavelet transforms, in *Wavelets*, J. M. Combes, A. Grossmann, and Ph. Tchamitchian (eds.) Springer-Verlag, NY, 1989, 2–20.
19. Grossmann, A., J. Morlet, and T. Paul, Transforms associated to square integrable group representations I. *J. Math. Phys.* **26** (1985), 2473–2479. II, Ann. l'Inst. Henri Poincaré, *Phys. Theor.* **45** (1986), 293–309.
20. Janson, S., J. Peetre, and R. Rochberg, Hankel forms and the Fock space, *Revista Mat. Iberoamer.* **3** (1987), 61–138.
21. Landau, H. J., Sampling, data transmission, and the Nyquist rate, *Proc. IEEE* **55** (1967), 1701–1706.
22. Landau, H. J., Necessary density conditions for sampling and interpolation of certain entire functions, *Acta Math.* **117** (1967), 37–52.
23. Lyubarskii, Y., Frames in the Bargmann Space of Entire Functions, *Proceedings of Seminar on Complex Analysis*, Lecture Notes in Math. #15, Springer-Verlag, NY, to appear.
24. von Neumann, J., *Foundations of Quantum Mechanics*, Princeton University Press, Princeton, NJ, 1955.
25. Olsen, P. A. and K. Seip, A note on irregular discrete wavelet transforms, *IEEE Trans. Inform. Theory*, Special issue on wavelets, to appear.
26. Paul, P. and K. Seip, Wavelets and quantum mechanics, in *Wavelets and Their Applications*, G. Beylkin, R. R. Coifman, I. Daubechies, S. Mallat, Y. Meyer, L. A. Raphael, and M. B. Ruskai (eds.), Jones and Bartlett, Cambridge, MA, 1992, to appear.
27. Perelomov, A. M., On the completeness of a system of coherent states, *Theor. Math. Phys.* **6** (1971), 156–164.
28. Rochberg, R., Interpolation by functions in Bergman spaces, *Mich. Math. J.* **29** (1982), 229–236.
29. Seip, K., Mean value theorems and concentration operators in Bargmann and Bergman spaces, in *Wavelets*, J. M. Combes, A. Grossmann, and Ph. Tchamitchian (eds.), Springer-Verlag, NY, 1989, 209–215.
30. Seip, K., Reproducing formulas and double orthogonality in Bargmann and Bergman spaces, *SIAM J. Math. Anal.* **22** (1991), 856–876.
31. Seip, K., Regular sets of sampling and interpolation for weighted Bergman spaces, *Proc. Amer. Math. Soc.*, to appear.
32. Seip, K., Density theorems for sampling and interpolation in the Bargmann-Fock space, Report Institut Mittag-Leffler #27 1990/91, Institut Mittag-Leffler, Djursholm, 1991.
33. Seip, K. and R. Wallstén, Sampling and interpolation in the Bargmann-Fock space, Report Institut Mittag-Leffler #4 1990/91, Institut Mittag-Leffler, Djursholm, 1990.

34. Shannon, C. E., Communication in the presence of noise, *Proc. IRE* **37** (1949), 10–21.
35. Strömberg, J.-O., A modified Franklin system and higher order spline systems on $\mathbf{R}^n$ as unconditional bases, in *Conference in Honor of Antoni Zygmund, II*, W. Beckner, A. P. Calderón, R. Fefferman, and P. W. Jones (eds.), Wadsworth, Belmont, 1981, 475–493.
36. Valiron, G., Sur la formule d'interpolation de Lagrange, *Bull. Sci. Math.* **49** (1925), 181–192.
37. Whittaker, E. T., On the functions which are represented by the expansions of the interpolation theory, *Proc. R. Soc. Edinburgh* **35** (1915), 181–194.
38. Young, R. M., *An Introduction to Nonharmonic Fourier Series*, Academic Press, NY, 1980.
39. Zhu, K., *Operator Theory in Function Spaces*, Marcel Dekker, NY, 1990.

Kristian Seip
Division of Mathematical Sciences
University of Trondheim
N-7034 Trondheim-NTH
Norway
seip@imf.unit.no

Part VII

Applications to Numerical Analysis and Signal Processing

Orthonormal Wavelets, Analysis of Operators, and Applications to Numerical Analysis

S. Jaffard and Ph. Laurençot

Abstract. We give a review of the algorithms of construction of orthonormal bases of wavelets, and of the main properties of the wavelet decomposition thus obtained. We show how to use these bases in order to study large classes of operators. Finally, we give theoretical and numerical applications of these properties for the resolution of Partial Differential Equations.

§1. Introduction

In this paper, our purpose is to show how orthonormal bases of wavelets can be used to obtain theoretical and numerical results for the resolution of P.D.E.'s. This field has only very recently started to be investigated, and there are now few results and many open problems. Therefore we shall concentrate on the prerequisites of this goal, namely the constructions of orthonormal bases of wavelets, the properties of the decomposition of a function on these bases, and the way some classical operators act on wavelets. The two directions we shall consider with more care, since they appear to be a key in the study of P.D.E.'s, are the local analysis performed by the wavelet decomposition and the *almost diagonalization* of differential operators. We shall also mention many open problems related to this study. This paper is essentially self contained. Many results however are given in a simplified form, or their proofs are just sketched. Therefore it cannot of course replace the study of the fundamental reference in this area, namely Yves Meyer's *Ondelettes et Opérateurs* [42], nor the many research papers quoted in the bibliography. It should really be seen as an introduction to the subject of wavelets and P.D.E.'s.

The term *wavelet bases* tends now to cover more and more things. The initial definition is perhaps the following. An orthonormal wavelet basis of $L^2(\mathbb{R})$ is of the form $\psi_{j,k} = 2^{j/2}\psi(2^j x - k)$ where j and k belong to $\mathbb{Z}$, and ψ is a smooth and well localized function. This last condition needs to be made more precise. For instance, we can suppose that

$$| \psi^{(\alpha)}(x) | \leq C \exp(-\gamma | x |)$$

Wavelets–A Tutorial in Theory and Applications
C. K. Chui (ed.), pp. 543–601.

ISBN 0-12-174590-2

for $\alpha \leq N$ and a positive γ. The decomposition of a function f in this basis is then local at high frequencies (that is for large j's), since

$$f = \sum_{j,k} c_{j,k} \psi_{j,k}$$

where

$$c_{j,k} = \int f \psi_{j,k}$$

and the function $\psi_{j,k}$ has a numerical support essentially of size 2^{-j}, so that the wavelet coefficients for j large yield a local information on f around $k2^{-j}$. An extremely important feature of this decomposition is that it is performed on functions that have cancellation. Actually, for $l < N$,

$$\int x^l \psi(x) dx = 0.$$

Let us sketch the proof when $l = 0$. We choose $a \in \mathbb{R}$ such that $\psi(a) \neq 0$. Let $j > 0$ and $k = [2^j a]$.

$$0 = \int \psi(x) \psi_{j,k}(x) dx = \int \psi(a) \psi_{j,k}(x) dx + O(\int \mid x - a \mid\mid \psi_{j,k}(x) \mid dx).$$

Using the localization estimates of ψ, we get

$$\int \mid x - a \mid\mid \psi_{j,k}(x) \mid dx \leq C 2^{-3j/2}$$

so that

$$\int \psi(a) \psi_{j,k}(x) dx \left(= 2^{-j/2} \psi(a) \int \psi \right) \leq C 2^{-3j/2}.$$

Choosing j large, we get $\int \psi = 0$. To prove that moments of higher order vanish, the same proof works, but one has to substract the Taylor expansion of ψ in a, hence the link between the smoothness of the wavelet and the number of vanishing moments. This last property implies that $\hat{\psi}_{j,k}$ is well localized near the frequencies of order of magnitude 2^j and -2^j.

The fact that wavelets have vanishing moments allows us to estimate the smoothness of the function f in terms of decay estimates of its wavelet coefficients for large j's. Let us give a precise result.

Let $0 < \alpha < 1$. A function f is uniformly C^α if

$$\mid f(x) - f(y) \mid \leq C \mid x - y \mid^\alpha .$$

Then, f is uniformly C^α if and only if

$$\mid c_{j,k} \mid \leq C 2^{-(\frac{1}{2}+\alpha)j}.$$

Let us prove this assertion. First, suppose that f is uniformly C^α. Then

$$\begin{aligned}
| c_{j,k} | &= \left| \int f(x)\psi_{j,k}(x)dx \right| \\
&= \left| \int (f(x) - f(k2^{-j}))\psi_{j,k}(x)dx \right| \\
&\leq C \int | x - k2^{-j} |^\alpha \frac{2^{j/2}}{(1+ | x - k2^{-j} |)^2} dx \\
&\leq C2^{-(\frac{1}{2}+\alpha)j}.
\end{aligned}$$

Conversely, if $| c_{j,k} | \leq C2^{-(\frac{1}{2}+\alpha)j}$,

$$| f(x) - f(y) | \leq C \sum_j \sum_k 2^{-(\frac{1}{2}+\alpha)j} | \psi_{j,k}(x) - \psi_{j,k}(y) | .$$

Let J be such that $2^{-J} \leq | x - y | < 2^{-J+1}$. Using the mean value theorem,

$$\sum_{j \leq J} \sum_k 2^{-(\frac{1}{2}+\alpha)j} | \psi_{j,k}(x) - \psi_{j,k}(y) | \leq$$

$$\sum_{j \leq J} \sum_k C2^{-\alpha j} 2^j | x - y | \sup\left(\frac{1}{(1+ | x - k2^{-j} |)^2}, \frac{1}{(1+ | y - k2^{-j} |)^2}\right).$$

The sum in k is bounded by a constant independant of j, and the sum in j is bounded by $C2^{(1-\alpha)J} | x - y | \leq C | x - y |^\alpha$. Furthermore

$$\begin{aligned}
&\sum_{j > J} \sum_k 2^{-(\frac{1}{2}+\alpha)j} | \psi_{j,k}(x) - \psi_{j,k}(y) | \\
&\leq \sum_{j > J} \sum_k 2^{-(\frac{1}{2}+\alpha)j} (| \psi_{j,k}(x) | + | \psi_{j,k}(y) |) \\
&\leq 2C \sum_{j > J} \sum_k \frac{2^{-\alpha j}}{(1+ | y - k2^{-j} |)^2} \leq C2^{-\alpha J} \leq C | x - y |^\alpha .
\end{aligned}$$

Here, we have a result about global smoothness. Actually, using the localization of the wavelets, we shall prove in Subsection 3.1 a similar result concerning the pointwise smoothness of f. Notice that in this proof we just use the cancellation of the wavelets and their localization, but not their exact form. We usually will not need it to obtain estimates. We shall see that it is important only for the fast algorithms of decomposition.

§2. Construction of orthonormal bases of wavelets

2.1. Wavelets on $\mathbb{R}$

2.1.1. Multiresolution analysis

We shall first describe the standard way, due to S. Mallat (see [38] or [42]), to construct orthonormal bases of wavelets. Let us mention that all known orthonormal (smooth and well localized) wavelet bases are obtained through this procedure. We shall first stick to the dimension 1 for the sake of simplicity. A multiresolution analysis is an increasing sequence $(V_j)_{j\in\mathbb{Z}}$ of closed subspaces of L^2 such that

1. $\bigcap V_j = \{0\}$
2. $\bigcup V_j$ is dense in L^2
3. $f(x) \in V_j \Leftrightarrow f(2x) \in V_{j+1}$
4. $f(x) \in V_0 \Leftrightarrow f(x+1) \in V_0$
5. There is a function g in V_0 such that the $g(x-k)_{k\in\mathbb{Z}}$ form a Riesz basis of V_0.

Recall that, for a Hilbert space H, (e_n) is a Riesz basis of H, if it is a basis and there exist positive constants C_1 and C_2 such that

$$C_1 \sum \mid a_n \mid^2 \leq \| \sum a_n e_n \|^2 \leq C_2 \sum \mid a_n \mid^2 \tag{1}$$

for any sequence (a_n) in l^2.

We also require g to be smooth and well localized. This last condition will be made more precise afterwards. Let us mention now that localization and smoothness are mutually exclusive. For instance, a wavelet with exponential decay cannot be C^∞. If a wavelet is compactly supported, the size of its support is essentially proportional to its degree of smoothness (see [13]). We shall say that a multiresolution analysis is m-regular, if g and its m first derivatives have fast decay. In the following we often will not be precise on the regularity needed on the wavelets, supposing they are *smooth* enough. However the smoothness actually needed will be clear in the proofs.

The sequence of spaces V_j can be interpreted as follows: if f_j is the projection of f on V_j, we *see* on f_j the details of f of size larger than 2^{-j}.

A simple example of multiresolution analysis is obtained by taking for V_j the space of continuous and piecewise linear functions on the intervals $[k2^{-j}, (k+1)2^{-j}]$, $j, k \in \mathbb{Z}$. A possible choice for g is the *hat* function which is the function of V_0 taking the value 1 for $x = 0$ and vanishing at the other integers.

Of course, such an example does not provide spaces of smooth functions. This can be improved using splines. Let V_0 be the subspace of L^2 composed of functions which are piecewise polynomials of degree m on each interval $[k, k+1]$ and are C^{m-1}. A possible choice for g is the B-spline σ such that

$$\hat{\sigma}(\xi) = \left(\frac{\sin \xi/2}{\xi/2}\right)^{m+1}.$$

Once a multiresolution analysis is given, it is easy to obtain an orthonormal basis of V_0. Let, for instance

$$\hat{\varphi}(\xi) = \hat{g}(\xi)(\sum | \hat{g}(\xi + 2k\pi) |^2)^{-1/2}.$$

Then, the $\varphi(x-k)$ forms an orthonormal basis of V_0 and, by dilation,

$$\varphi_{j,k}(x) = 2^{j/2}\varphi(2^j x - k)$$

forms an orthonormal basis of V_j. Remark that other choices are possible for $\hat{\varphi}$: one can multiply the one defined above by a 2π periodic function of modulus 1.

An important property of the function φ is that $| \int \varphi |= 1$. It can be proved, for instance, by computing the projection of the characteristic function of $[0,1]$ on the V_j's for $j \to +\infty$. We can thus always choose φ such that $\int \varphi = 1$.

Define W_j as the orthogonal complement of V_j in V_{j+1}. This space is sometimes called the *innovation space* for the following reason. The projection of f on W_j tells what one has to add to f approximated with a precision of size 2^{-j} to obtain details twice smaller. An easy consequence of the definition of a multiresolution analysis is that the W_j are mutually orthogonal, and their direct sum is equal to L^2.

Let us show how to obtain an orthonormal basis of W_0 of the form $\psi(x-k)$.

Since $V_{-1} \subset V_0$, there exists a sequence h_k in l^2 such that

$$\frac{1}{2}\varphi(\frac{x}{2}) = \sum h_k \varphi(x-k), \tag{2}$$

that is $\hat{\varphi}(2\xi) = m_0(\xi)\hat{\varphi}(\xi)$ where

$$m_0(\xi) = \sum h_k e^{-ik\xi}.$$

Then, a possible choice for ψ is $\hat{\psi}(2\xi) = m_1(\xi)\hat{\varphi}(\xi)$ where

$$m_1(\xi) = e^{i\xi}\overline{m_0(\xi+\pi)}.$$

Since the W_j are obtained from each other by dilation, and are mutually orthogonal, the functions $2^{j/2}\psi(2^j x - k)$ form an orthonormal basis of $L^2(\mathbb{R})$.

Let us now come back to the orthogonal decomposition $V_0 = V_{-1} \oplus W_{-1}$. We have two orthonormal bases of V_0: the first one is the $\varphi(x-k)$, $k \in \mathbb{Z}$, and the second one is the union of the $\frac{1}{\sqrt{2}}\varphi(\frac{x}{2}-k)$ and $\frac{1}{\sqrt{2}}\psi(\frac{x}{2}-k)$. The existence of two bases implies the existence of an isometry mapping the coordinates of a function in the first basis on its coordinates in the second basis. Let us describe precisely this isometry. Let

$$h_k = \frac{1}{2}\int \varphi(\frac{x}{2})\bar{\varphi}(x-k)dx$$

$$g_k = \frac{1}{2}\int \psi(\frac{x}{2})\bar{\varphi}(x-k)dx.$$

Let $f \in V_0$ and let us write its decomposition on the two bases of V_0

$$f(x) = \sum_k c_k^0 \varphi(x-k)$$

$$= \sum_l c_l^1 \frac{1}{\sqrt{2}}\varphi(\frac{x}{2}-l) + \sum_l d_l^1 \frac{1}{\sqrt{2}}\psi(\frac{x}{2}-l).$$

Since this decomposition is orthonormal, we obtain

$$c_l^1 = \langle f \mid \frac{1}{\sqrt{2}}\bar{\varphi}(\frac{x}{2}-l)\rangle = \sum_k c_k^0 \langle \varphi(x-k) \mid \frac{1}{\sqrt{2}}\bar{\varphi}(\frac{x}{2}-l)\rangle,$$

so that

$$c_l^1 = \frac{1}{\sqrt{2}}\sum_k c_k^0 \bar{h}_{k-2l}, \tag{3}$$

and, similarly

$$d_l^1 = \frac{1}{\sqrt{2}}\sum_k c_k^0 \bar{g}_{k-2l}. \tag{4}$$

Suppose now that a signal is sampled and thus known by a sequence of discrete values $(c_k^0)_{k\in\mathbb{Z}}$. We can consider it to be the coefficients of a function of V_0 on the $\varphi(x-k)$. The isometry transforming the sequence (c_k^0) into (c_k^1, d_k^1) can be written $F = (F_0, F_1)$ where F_0 and F_1 are commuting with even translations: Formulas (3) and (4) show that they are discrete convolutions where we only keep each other term. In the terminology of signal analysis, F_0 and F_1 are said to be quadrature mirror filters. This notion has been introduced in 1977 by D. Esteban and C. Galand for improving the quality of digital transmission of sound. Iterating $|j|-1$ times the filter F_0 and then applying once the filter F_1, we obtain the coefficients on the $\psi_{j,k}$ for $j \leq 0$. Each level requires only a discrete convolution. This algorithm constitutes the Fast Wavelet Transform (see [39]).

We remark that, if there exist only a finite number of nonvanishing entries h_k and g_k, the whole procedure is performed in $O(N)$ operations.

Remark also that, in order to compute the wavelet decomposition of a function, we do not use the values of the wavelet (which sometimes is a very complicated function), but just the sequences (h_k) and (g_k).

Let us give a few examples of wavelets, which can be thus constructed. The example of the spline multiresolution analysis leads to the following formulas for φ and ψ

$$\hat{\varphi}(\xi) = \left(\frac{\sin \xi/2}{\xi/2}\right)^{m+1} \left(P_m(\sin^2(\xi/2))\right)^{-1/2}$$

where P_m is the polynomial of degree m defined by

$$P_m(\sin^2 \xi) = \sin^{2m+2}(x) \sum_k \frac{1}{(\xi + k\pi)^{2m+2}} = -\frac{1}{(2m+1)!} \frac{d^{2m+1}}{dx^{2m+1}} (\cot \xi),$$

and

$$\hat{\psi}(2\xi) = e^{-i\xi} \left(\frac{P_m(\cos^2 \xi/2)}{P_m(\sin^2 \xi)} \right)^{1/2} \hat{\varphi}(\xi).$$

These wavelets have been independantly discovered by G. Battle and P. G. Lemarié (see [34]). A slightly different wavelet (using the same multiresolution analysis) had been previously discovered by J. O. Stromberg [50]. Let us describe another wavelet that we shall use in the following. It is a C^∞ wavelet, which is not very well localized (it has *only* fast decay, while the other numerically used wavelet bases have exponential decay, or a compact support). Thus, we shall see that this basis is extremely useful for theoretical computations, but not much used for numerical purposes.

Let $\hat{\varphi}$ be a C^∞, even function supported by $[-4\pi/3, 4\pi/3]$ such that

$$\begin{cases} 0 \leq \hat{\varphi} \leq 1 \\ \hat{\varphi}(\xi) = 1 \quad \text{if } |\xi| \leq 2\pi/3 \\ \hat{\varphi}(\xi)^2 + \hat{\varphi}(2\pi - \xi)^2 = 1 \quad \text{if } 0 \leq \xi \leq 2\pi. \end{cases}$$

The function φ is in the Schwartz class. The corresponding wavelet ψ is also in the Schwartz class. Its Fourier transform is C^∞ and supported by the set

$$[-8\pi/3, -2\pi/3] \cup [2\pi/3, 8\pi/3].$$

It is sometimes called the *Littlewood-Paley* wavelet for the following reason. Let $f_j = \sum_k c_{j,k}\psi_{j,k}$. The decomposition $f = \sum_{j \in \mathbb{Z}} f_j$ is similar to the Littlewood-Paley decomposition, which is obtained by splitting f as a sum of functions whose supports in the Fourier domain are essentially located on the dyadic intervals $[-2^{j+1}, -2^j] \cup [2^j, 2^{j+1}]$. Many properties of the wavelet decomposition will be a consequence of this dyadic localization in the Fourier domain (see Subsection 4.2).

2.1.2. Wavelets in several dimensions

There are two possibilities for constructing wavelets in several dimensions. One consists in generalizing the concept of multiresolution analysis in several dimensions and essentially following the one-dimensional construction; this allows, for instance, to construct wavelets adapted to certain tesselations of the space (see [42], [28], and [13]). Another method consists in obtaining the wavelets using the one-dimensional construction. Let us explain it in two dimensions. We define the following multiresolution analysis.

Let $\mathcal{V}_j$ be the subspace of $L^2(\mathbb{R}^2)$ defined by $\mathcal{V}_j = V_j \otimes V_j$. Clearly, we define thus a multiresolution analysis in two dimensions (here $\mathcal{V}_0$ is invariant under translations in $\mathbb{Z} \times \mathbb{Z}$). An orthonormal basis of $\mathcal{V}_j$ is given by the functions

$$\Phi_{j,(k_1,k_2)}(x,y) = \varphi_{j,k_1}(x)\varphi_{j,k_2}(y).$$

Define now $\mathcal{W}_j$ as the orthogonal complement of $\mathcal{V}_j$ in $\mathcal{V}_{j+1}$. We have

$$\begin{aligned}\mathcal{V}_{j+1} &= V_{j+1} \otimes V_{j+1} \\ &= (V_j \oplus W_j) \otimes (V_j \oplus W_j) \\ &= (V_j \otimes V_j) \oplus (V_j \otimes W_j) \oplus (W_j \otimes V_j) \oplus (W_j \otimes W_j),\end{aligned}$$

so that

$$\mathcal{W}_j = (V_j \otimes W_j) \oplus (W_j \otimes V_j) \oplus (W_j \otimes W_j)$$

and a basis of $\mathcal{W}_j$ is given by the functions $\psi^{(i)}_{j,k} = 2^j \psi^{(i)}(2^j x - k)$ where

$$\begin{cases} \psi^{(1)}(x,y) = \psi(x)\varphi(y) \\ \psi^{(2)}(x,y) = \varphi(x)\psi(y) \\ \psi^{(3)}(x,y) = \psi(x)\psi(y). \end{cases}$$

We see that, in two dimensions, we have not just one but three wavelets. This construction immediately generalizes to the n-dimensional case; in that case, we obtain $2^n - 1$ wavelets.

Remark that another possible basis of $L^2(\mathbb{R}^2)$ is given by the functions $\psi_{j,k}(x)\psi_{j',k'}(y)$. This basis, unlike the one we have just constructed, has unrelated localizations in the x and y directions. These directions play a very priviledged role; therefore it seems less natural for decomposing functions. But this drawback can be compensated by the fact that the analysing functions have cancellation in both directions (integrals in x **and** in y vanish, which is not the case for $\psi^{(1)}$ and $\psi^{(2)}$); therefore, the coefficients of a smooth function may be encoded using less nonvanishing coefficients. This point is discussed in [37]; we shall return to it in Subsection 4.1 since these two ways for decomposing functions give rise to two ways for decomposing distribution kernels of operators, and thus yield two ways for analysing operators.

2.1.3. Compactly supported wavelets

Let us come back to the formula $\hat{\varphi}(2\xi) = m_0(\xi)\hat{\varphi}(\xi)$. Iterating it, we obtain, at least formally,

$$\hat{\varphi}(\xi) = \prod_{j=1}^{\infty} m_0\left(\frac{\xi}{2^j}\right).$$

This formula can allow one to construct the function φ, and thus the multiresolution analysis from the function m_0. One may wonder under which conditions on m_0 this product will converge and define a function φ which will have the right properties for a multiresolution analysis. This problem has been thoroughly investigated by A. Cohen in [13]. We shall not describe these results here but just give examples found by I. Daubechies (see [15]) where there are only a finite number of non-vanishing entries h_k (in which case the g_k will also have a finite number of non-vanishing entries). Then φ and ψ will be compactly supported. This requirement means that m_0 will be a trigonometric polynomial.

The function m_0 is constructed as follows. We shall look for m_0 of the following form

$$m_0(\xi) = (1 + e^{i\xi})^N P(\xi)$$

where P is a trigonometric polynomial and we require $m_0(0) = 1$, $\mid m_0(\xi) \mid^2 + \mid m_0(\xi + \pi) \mid^2 = 1$ and $m_0(\xi) \neq 0$ if $\mid \xi \mid \leq \pi/2$. Let

$$Q(x) = \left(\frac{1+x}{2}\right)^N \sum_{k=0}^{N-1} C_{k+N-1}^k \left(\frac{1-x}{2}\right)^k .$$

The function $Q(\cos\xi)$ is a trigonometric polynomial, and a theorem of Riesz ensures the existence of a trigonometric polynomial $P(\cos\xi)$ such that $\mid P \mid^2 = Q$. We thus obtain wavelets supported by $[0, 2N-1]$.

2.1.4. Open problems

We have just seen that there exist many orthonormal wavelet bases of L^2. Nonetheless, if L^2 is replaced by the Hardy space H^2 which is the subspace of L^2 composed of functions whose Fourier transform vanishes for $\xi \leq 0$, then an open problem is to find an orthonormal basis of H^2 of the form $2^{j/2}\psi(2^j x - k)$ where ψ would be smooth and well localized (the maximal localization and smoothness possible would have to be also determined). Such a construction would be useful in mathematical analysis and signal analysis. One of the difficulties lies in the following result (see [28, Chapter 3]):

Theorem 1. *There exists no multiresolution analysis on H^2 which would give a wavelet ψ such that $\hat{\psi}$ is continuous and*

$$\mid \hat{\psi} \mid \leq C \mid \xi \mid^{-\alpha} \quad \text{for any } \alpha > 1/2.$$

Thus one of the problems, if such a basis exists, lies in constructing a wavelet basis which would not be obtained through a multiresolution analysis. Such examples, even for L^2, have not been found: all known orthonormal (smooth and well localized) wavelet bases are obtained through this procedure, thus it is also an open problem to determine if there exist smooth and well localized wavelet bases of L^2 which would not be obtained through a multiresolution analysis.

2.2. Construction on domains

If we keep in mind that we want to use wavelets in order to study operators or to solve Partial Differential Equations, we clearly cannot be satisfied with constructions on $\mathbb{R}$ or $\mathbb{R}^n$. However, as we shall see, it is largely an open problem to construct an orthonormal wavelet basis on a domain Ω which would keep the numerical simplicity, the fast algorithms, and would simultaneously be bases for certain Sobolev spaces (like H_0^s, H^s,...) . First, in Subsection 2.2.1., we shall construct wavelets adapted to periodic functions.

As regards to wavelets on a domain Ω which would be bases of the Sobolev space H_0^s, an intuitive idea is to start with a usual basis on $\mathbb{R}^n$ of the form

$$2^{nj/2}\psi^{(i)}(2^j x - k),$$

to keep all the wavelets which are *essentially* localized inside Ω and to *modify* them so that they vanish at the boundary (so that they can belong to H_0^s). In the case where Ω is a segment, this program can be precisely achieved using compactly supported wavelets. Thus, for each j, we just have to modify a given number of wavelets, namely all those whose support intersects the endpoints of the segment (see Subsection 2.2.2.). When Ω is more general, the corresponding construction is a modification of exponentially decreasing wavelets. Thus, the supports of all wavelets *essentially localized* inside the domain intersect the boundary, and thus, all wavelets have to be modified, but this modification becomes numerically negligeable when $2^j \mathrm{dist}(k2^{-j}, \partial\Omega)$ is large (see Subsection 2.2.3.).

2.2.1. Periodized wavelets

We shall construct a wavelet basis of the periodized functions on $\mathbb{R}$ of period 1 (following [42]). This basis will be obtained by periodizing the wavelets. We shall prove the following result.

Proposition 2. *Let* $\bar{\psi}_{j,k} = \sum_{l\in\mathbb{Z}} \psi_{j,k}(x+l)$. *The constant function and* $\bar{\psi}_{j,k}$ *for* $j \geq 0$ *and* $k = 0, \ldots 2^j - 1$ *form an orthonormal basis of* $L^2([0,1])$.

In order to prove this proposition, let us first introduce the following notations. Let V_j^∞ be the closure of V_j in L^∞, and P_j be the set of functions of V_j^∞ of period 1.

Lemma 3. *If* $j \leq 0$, P_j *is the set of constant functions. If* $j > 0$, $\dim P_j = 2^j$.

Since $P_j \subset P_{j+1}$, We only have to prove the first assertion for $j = 0$.

Let us first prove that $\sum \varphi(x-k) = 1$. Since this function is 1-periodic, we can do that by checking its Fourier coefficients a_l.

$$a_l = \int_0^1 \sum_k \varphi(x-k)e^{2i\pi lx}dx = \sum_k \int_k^{k+1} \varphi(x)e^{2i\pi lx}dx = \hat{\varphi}(2\pi l).$$

But $\hat{\varphi}(0) = \int \phi = 1$ and, since $\sum \mid \hat{\varphi}(\xi + 2k\pi) \mid^2 = 1$, this implies that $\hat{\varphi}(2\pi l) = 0$ if $l \neq 0$. Thus $\sum \varphi(x-k) = 1$.

If $f \in P_0$, $f = \sum c_k \varphi(x-k)$. Since f is 1-periodic,

$$c_k = \int f(x)\varphi(x-k)dx = \int f(x)\varphi(x)dx = c_0,$$

thus f is a multiple of $\sum \varphi(x-k)$ so that it is a constant.

Let us prove the second assertion of the lemma. If $f \in P_j$,

$$f = \sum c_{j,k}\varphi_{j,k}$$

The periodicity of f implies that $c_{j,k} = c_{j,k+2^j l}$. Thus

$$f = \sum_0^{2^j-1} c_{j,k}(\sum_{l\in\mathbb{Z}} \varphi_{j,k+2^j l}).$$

The functions $\sum\limits_{l\in Z} \varphi_{j,k+2^j l}$ are orthogonal, hence the second part of the lemma.

If Q_j is the orthogonal complement of P_j in P_{j+1}, an orthonormal basis of Q_j is supplied by the functions $\sum\limits_{l\in Z} \psi_{j,k+2^j l}$.

The fast decomposition algorithms are immediately deduced from the usual algorithm on $\mathbb{R}$ by periodization . We obtain

$$c_l^1 = \frac{1}{\sqrt{2}} \sum_{k=0}^{2^{j+1}-1} c_k^0(\sum_m h_{k-2l-2^{j+1}m}),$$

and, similarly

$$d_l^1 = \frac{1}{\sqrt{2}} \sum_{k=0}^{2^{j+1}-1} c_k^0(\sum_m g_{k-2l-2^{j+1}m}).$$

These wavelets are a basis of the space H^s-periodic (the space of periodic functions locally in H^s).

This construction immediately generalizes to functions in two dimensions, periodized in two directions, and similarly in higher dimensions.

2.2.2. Wavelets on a segment

These wavelets are obtained from the compactly supported wavelets. We shall follow [45]. As in Subsection 2.1.3, we start with $2N$ coefficients $h_0, ...,$ h_{2N-1} and $2N$ coefficients $g_0, ..., g_{2N-1}$, and Equation (2) becomes

$$\frac{1}{2}\varphi(x) = h_0\varphi(2x) + ... + h_{2N-1}\varphi(2x-(2N-1)),$$

and, similarly, from the definition of ψ, we obtain

$$\frac{1}{2}\psi(x) = g_0\varphi(2x) + ... + g_{2N-1}\varphi(2x - (2N-1)),$$

which implies that

$$\begin{aligned}\varphi(2x) =& \bar{h}_0\varphi(x) + \bar{h}_2\varphi(x+1) + ... + \bar{h}_{2N-2}\varphi(x+N-1) \\ &+ \bar{g}_0\psi(x) + \bar{g}_2\psi(x+1) + ... + \bar{g}_{2N-2}\psi(x+N-1),\end{aligned} \tag{5}$$

and

$$\begin{aligned}\varphi(2x-1) =& \bar{h}_1\varphi(x) + \bar{h}_3\varphi(x+1) + ... + \bar{h}_{2N-1}\varphi(x+N-1) \\ &+ \bar{g}_1\psi(x) + \bar{g}_3\psi(x+1) + ... + \bar{g}_{2N-1}\psi(x+N-1).\end{aligned} \tag{6}$$

Let S_j be the set of k's such that $-2N+2 \le k \le 2^j - 1$, which is the set of k such that $\varphi_{j,k}$ has a nonvanishing intersection with $(0,1)$, and let $V_j^{[0,1]}$ be the restriction to $[0,1]$ of the functions in V_j.

Proposition 4. *If $2^j \ge 4N-4$, the $(\varphi_{j,k})$ for $k \in S_j$ are a basis of $V_j^{[0,1]}$, and*

$$\| \sum_{k\in S_j} c_{j,k}\varphi_{j,k} \|^2 \equiv \sum_{k\in S_j} | c_{j,k} |^2$$

where the constants in the equivalence are independant of j.

Clearly, the $(\varphi_{j,k})$ span $V_j^{[0,1]}$. Let us prove their independence. Divide the set S_j into three disjoint subsets S_j^1, S_j^2, and S_j^3, depending when the interior of the support of $\varphi_{j,k}$ contains 0, is included in $[0,1]$, or contains 1.

Suppose that $f = \sum_{k\in S_j} c_{j,k}\varphi_{j,k}$ vanishes on $[0,1]$.

If $k \in S_j^2$, $c_{j,k} = \int_0^1 f\varphi_{j,k} = 0$. Let l be the largest integer in S_j^1 such that $c_{j,k} \neq 0$. On $[\frac{l+2N-2}{2^j}, \frac{l+2N-1}{2^j}]$, we have $f = c_{j,l}\varphi_{j,l}$, thus $c_{j,l} = 0$, so that $c_{j,l} = 0$ if $k \in S_j^1$, and the same proof gives the same result for $k \in S_j^3$.

Let us now prove the equivalence in Proposition 4. If $f = \sum_{k\in S_j} c_{j,k}\varphi_{j,k}$, let $f^1 = \sum_{k\in S_j^1} c_{j,k}\varphi_{j,k}$ and define similarly f^2 and f^3. These three functions are orthogonal, so that, by Bessel's inequality,

$$\sum_{k\in S_j^2} | c_{j,k} |^2 = \int_0^1 | f^2(x) |^2 \, dx \le \int_0^1 | f(x) |^2 \, dx,$$

hence,

$$\| f^1 + f^3 \|_{L^2[0,1]} \leq 2 \| f \|_{L^2[0,1]}.$$

Since the supports of f^1 and f^3 are disjoint,

$$\| f^1 \| + \| f^3 \| \leq 2\sqrt{2} \| f \| .$$

The norms of f^1 and f^3 are evaluated by a rescaling and the estimates do not depend on j, hence Proposition 4.

Let us show how to obtain an orthonormal basis of $V_j^{[0,1]}$. For a given j, we orthonormalize the functions $\varphi_{j,k}$, $k \in S_j^1$ by the Gram-Schmidt algorithm, and we do the same for S_j^3. By rescaling, for each j, we obtain the same set of functions, and we have thus obtained an orthonormal basis of $V_j^{[0,1]}$ composed of the three sets of functions:

$2^{j/2}\varphi(2^j x - k)$ for $k \in S_j^2$; $2^{j/2}\varphi_k^0(2^j x)$ for $k \in S_j^1$; and $2^{j/2}\varphi_k^1(2^j(1-x))$ for $k \in S_j^3$.

Let us now construct the wavelets. We shall first prove the following proposition

Proposition 5. *For any $j \geq 4N - 4$, the union of the above basis of $V_j^{[0,1]}$ and of the $\psi_{j,k}$ such that $-N+1 \leq k \leq 2^j - N$ is a basis of $V_{j+1}^{[0,1]}$.*

This proposition will be an easy consequence of the following lemma.

Lemma 6. *The functions $\psi(2^j x - k)$, $-2N+2 \leq k \leq -N$, once restricted to $[0,1]$ belong to $V_j^{[0,1]}$.*

After a rescaling, we only have to prove that the functions $\psi(x-k)$, $-2N+2 \leq k \leq -N$, once restricted to $[0,\infty)$ belong to $V_0^{[0,\infty)}$.

This means that, if more than half of the support of $\psi(x-k)$ is outside $[0,\infty)$, the restriction of $\psi(x-k)$ to $[0,\infty)$ belongs to V_0. In order to prove the lemma, let us come back to Equations (5) and (6). Replacing x by $x+2N-2$, then by $x+2N-3$,..., we get for $x \geq 0$, since $\varphi(2x+2N-1) = 0$ if $x \geq 0$

$$\bar{g}_0\psi(x+2N-2) + \bar{h}_0\varphi(x+2N-2) = 0,$$

$$\bar{g}_0\psi(x+2N-3)\bar{g}_2\psi(x+2N-2) + \bar{h}_0\varphi(x+2N-3)\bar{h}_2\varphi(x+2N-2) = 0,$$

and so on. One proves then by induction that $\psi(x+2N-2)$ restricted to $[0,+\infty)$ belongs to $V_0^{[0,\infty)}$, then the same holds for $\psi(x+2N-3)$,....

In order to prove Proposition 5, we remark that we have the right number of functions for a basis of $V_{j+1}^{[0,1]}$ (namely $2^{j+1} + 2N - 2$), so that we only have to prove that they span $V_{j+1}^{[0,1]}$, which is a consequence of Lemma 3.

Let us now introduce $W_j^{[0,1]}$, the orthogonal complement of $V_j^{[0,1]}$ in $V_{j+1}^{[0,1]}$. An orthonormal basis of $W_j^{[0,1]}$ is constructed as follows:

First, we take the $\psi_{j,k}$ such that $0 \le k \le 2^j - 2N + 1$ without any change. Consider now the $\psi_{j,k}$ such that $-N+1 \le k \le -1$; they are orthonormal to the previous set; we orthonormalize them, via the Gram-Schmidt algorithm, so that they become orthogonal to the $2^{j/2}\varphi_k^0(2^j x)$, $k \in S_j^1$. We do the same around 1, and, since this orthogonalization is invariant by rescaling, it yields, up to a dilation, the same functions at all scales, and we obtain an orthonormal basis of $W_j^{[0,1]}$ composed of functions $2^{j/2}\psi(2^j - k)$, $k \in S_j^2$; $2^{j/2}\psi_k^0(2^j x)$, $k \in S_j^1$; $2^{j/2}\psi_k^1(2^j(1-x))$, $k \in S_j^3$.

Actually we have thus constructed a basis of the $H^s([0,1])$ spaces, and the following characterization holds

$$f \in H^s \Leftrightarrow \sum | c_{j,k} |^2 \, 2^{2sj} < \infty.$$

The proof of this result is very similar to the proof of Proposition 23 in Subsection 4.2.

2.2.3. Some results on orthonormalization procedures

The technique to construct wavelet bases on domains is basically the following. One starts with spaces of spline functions, or of finite elements defined on the domain, and then, one applies an orthonormalization procedure. We can no more use the one that gave φ from g in Subsection 2.1, since the explicit formula on the Fourier transform was made possible only because V_0 was invariant by translation, which won't be the case when we work on a domain. Thus, we now describe the one we shall use since, though classical (see [48]), it is not as widely known as the Gram-Schmidt algorithm.

Let H be an Hilbert space, and (e_n) a Riesz basis of H. Let G be the operator defined on H by

$$G(f) = \sum \langle f \mid e_n \rangle e_n.$$

Lemma 7. *The operator G is a self-adjoint positive definite operator, if (h_n), defined by*

$$h_n = G^{-1/2}(e_n),$$

forms an orthonormal basis of H.

Proof: The operator G is self-adjoint positive because

$$\langle G(f) \mid g \rangle \; = \sum \langle f \mid e_n \rangle \langle g \mid e_n \rangle.$$

Since the (e_n) are a Riesz basis,

$$C_1 \sum \langle f \mid e_n \rangle^2 \le \| \sum \langle f \mid e_n \rangle e_n \|^2 \le C_2 \sum \langle f \mid e_n \rangle^2$$

so that

$$C_1\langle G(f) \mid f\rangle \ \leq \| \ G(f) \ \|^2 \leq C_2\langle G(f) \mid f\rangle;$$

hence G is positive definite and

$$C_1 \mathbf{Id} \leq G \leq C_2 \mathbf{Id}$$

where C_1 and C_2 are positive.

Let us prove that the (h_n) forms an orthonormal basis of H. Since G is positive definite, $G^{-1/2}$ can be defined (as the only positive operator T such that $T^2 = G^{-1}$), and it is an isomorphism on H. Hence, the $G^{-1/2}(e_n)$ are a basis of H. We first check that

$$\forall \ x \in H, \ \sum \langle G^{-1/2}(e_n) \mid x\rangle G^{-1/2}(e_n) = x. \tag{7}$$

We have

$$G(y) = \sum \langle e_n \mid y\rangle e_n$$

so that

$$G^{1/2}(G^{1/2}(y)) = \sum \langle G^{-1/2}(e_n) \mid G^{1/2}(y)\rangle e_n$$

and

$$G^{1/2}(y) = \sum \langle G^{-1/2}(e_n) \mid G^{1/2}(y)\rangle G^{-1/2}(e_n).$$

Hence Equation (7) if we take $x = G^{1/2}(y)$ (which is possible since $G^{1/2}$ is an isomorphism). The $G^{-1/2}(e_n)$ are a basis of H. If we take $x = G^{-1/2}(e_m)$ in Equation (7), we obtain

$$\sum_{n \neq m} \langle G^{-1/2}(e_n) \mid G^{-1/2}(e_m)\rangle G^{-1/2}(e_n) +$$

$$(\| \ G^{-1/2}(e_m) \ \|^2 - 1) G^{-1/2}(e_m) = 0.$$

Since the (e_n) are a basis of H, this implies

$$\langle G^{-1/2}(e_n) \mid G^{-1/2}(e_m)\rangle \ = \delta_{n,m}$$

and the (h_n) are an orthonormal basis of H.

We shall apply this construction in cases in which the (e_n) will be localized functions, and we shall show that the orthonormal basis thus constructed has a similar localization. It will be a consequence of the following lemmas.

Let T be a discrete subset of $\mathbb{R}^n$ such that

$$\forall \ k, l \in T, \quad \inf_{k \neq l} \mid k - l \mid \geq C > 0.$$

Define A^γ as the space of matrices $B(k,l)$ indexed by $T \times T$ such that

$$\mid B(k,l) \mid \leq C \exp(-\gamma \mid k - l \mid).$$

Lemma 8. *Let* $0 < \gamma' < \gamma$. *Then*

$$G \in A^{\gamma} \text{ and } H \in A^{\gamma'} \Rightarrow GH \in A^{\gamma'} \text{ and } HG \in A^{\gamma'}.$$

This holds because

$$\begin{aligned} |\,(GH)(k,l)\,| &\leq C \sum_m \exp(-\gamma \mid k-m \mid) \exp(-\gamma' \mid m-l \mid) \\ &\leq C \exp(-\gamma' \mid k-l \mid) \sum_m \exp(-(\gamma-\gamma') \mid k-m \mid) \\ &\leq C \exp(-\gamma' \mid k-l \mid). \end{aligned}$$

Lemma 9. *Let* G *be a positive matrix in* A^{γ} *such that*

$$C_1 \mathbf{Id} \leq G \leq C_2 \mathbf{Id}.$$

Then $\exists\, \gamma' > 0$ *such that* $G^{-1/2} \in A^{\gamma'}$, *where* γ' *depends only on* γ, C, C', C_1 *and* C_2.

Proof: We have

$$G = \| G \| \,(Id - R)$$

with $\| R \|_{l^2} = r < 1$. So that

$$G^{-1/2} = \| G \|^{-1/2} \sum_{n=0}^{\infty} C_n^{-1/2} R^n.$$

Using Lemma 8 iteratively, we get

$$R^n(k,l) \leq C^n \exp(-\gamma' \mid k-l \mid).$$

But, since $\| R \|_{l^2} \leq r$, we also have $\mid R^n(k,l) \mid \leq r^n$. Taking the best of the two estimates, we obtain Lemma 9.

We remark that the same result holds for G^{-1} or for any other power of G.

2.2.4. Wavelets on arbitrary domains

We shall construct in this section wavelets which are an orthonormal basis of $L^2(\Omega)$ and are bases of the Sobolev spaces $H_0^s(\Omega)$. For that, following [30], we shall construct a multiresolution analysis adapted to Ω. We will not be able to use the Fourier transform as in $\mathbb{R}$, and instead, we shall use the orthonormalization procedure described in the preceding section.

Consider the space V_j of functions that are C^{2m-2}, vanish outside Ω, and are polynomials of degree $2m-1$ in each variable in the cubes

$$k2^{-j} + 2^{-j}[0,1]^n, \quad \text{with } k \in \mathbb{Z}^n.$$

Let $\sigma_m(x_1 - m, ..., x_n - m) = \xi * ... * \xi$ ($2m$ times), where ξ is the characteristic function of $[0,1]^n$. The Fourier transform of σ_m is

$$\left(\frac{\sin \xi_1/2}{\xi_1/2}\right)^{2m} \cdots \left(\frac{\sin \xi_n/2}{\xi_n/2}\right)^{2m}.$$

These functions are the classical B-splines in dimension n. Let

$$\sigma_{j,k} = 2^{nj/2}\sigma(2^j x - k),$$

and let Λ_j be the set of points $k2^{-j}$ such that $\text{supp}(\sigma_{j,k}) \subset \Omega$. First, we have

Proposition 10. *The* $(\sigma_{j,k})_{k2^{-j}\in\Lambda_j}$ *are a Riesz basis of* V_j.

The equivalence of norms in the definition of a Riesz basis holds because it holds on $\mathbb{R}^n$, if we take all the $\sigma_{j,k}$, $k \in \mathbb{Z}^n$ (this is easily checked using the Fourier transform). Let us sketch the proof of the completeness. If f is a function compactly supported in V_j, its support has a corner. One easily checks that, on the cube of size 2^{-j} located in that corner, f coincides, up to a multiplicative constant, with a function $\sigma_{j,k}$. We substract this function and we can make the support of f shrink until ultimately it has vanished.

In order to obtain an orthonormal basis of V_j, we use the orthonormalization procedure described in the preceding section. Since the Gram matrix of the $\sigma_{j,k}$ has exponential decay, using Lemma 9, we obtain the following localization for the functions $\varphi_{j,k}$

$$| \partial^\alpha \varphi_{j,k}(x) | \leq C 2^{j\alpha} 2^{nj/2} \exp(-\gamma 2^j \mid x - k2^{-j} \mid)$$

for $| \alpha | \leq 2m - 2$, and a positive γ.

Let us now describe how to obtain the wavelets. We first remark that a function in V_j is determined by its values on the $k2^{-j}$. More precisely, there exist C_1 and C_2 positive such that

$$C_1 \| f \|_2 \leq (\sum_{\lambda\in\Lambda_j} | f(\lambda) |^2)^{1/2} \leq C_2 \| f \|_2 .$$

This holds because it is equivalent to the fact that the positive self adjoint matrix $G_{k,l} = \sigma_m(k - l)$ is invertible. As before, we first check it for $k \in \mathbb{Z}^n$, where it is straightforward using the Fourier transform, and we then use the fact that, if

$$C_1 Id \leq G \leq C_2 Id,$$

this property is stable under restriction, with the same constants. Thus the constants in the equivalence are independant of j. Let $(L_{j,k}) = G^{-1}(\sigma_{j,k})$. The $L_{j,k}$ are called cardinal splines because of the following property which makes them useful for interpolation

$$L_{j,k}(l2^{-j}) = \delta_{k,l}.$$

The orthonormal basis of W_j is obtained by projecting on W_j the functions $L_{j,k}$ such that $k2^{-j} \in \Lambda_{j+1} \backslash \Lambda_j$ and then by orthonormalizing (by the procedure described in Subsection 2.2.3) the set thus obtained. Here again

$$| \partial^\alpha \psi_{j,k}(x) | \leq C 2^{j\alpha} 2^{nj/2} \exp(-\gamma 2^j | x - k2^{-j} |)$$

for $| \alpha | \leq 2m-2$, and a positive γ. This estimate holds because all the matrices of the transforms performed on the bases in order to obtain the $\psi_{j,k}$ have entries exponentially decreasing away from the diagonal, so that we can apply several times Lemma 9.

Actually, though these wavelets are not the same as in the case $\Omega = \mathbb{R}^n$, they are *almost* the same; that is, numerically, only the wavelets that are close to the boundary are modified. Thus, we can essentially keep the fast decomposition algorithms, with only small modifications of the values of the associated filters near the boundary. More precisely, there exists *classical* wavelets (of the type described in Subsection 2.1.2.) $\bar{\psi}_{j,k}$ on $\mathbb{R}^n$ such that (see[27])

$$\| \partial^\alpha (\psi_{j,k} - \bar{\psi}_{j,k}) \|_\infty \leq C 2^{(\frac{n}{2}+\alpha)j} \exp(-\gamma 2^j d(k2^{-j}, \partial\Omega)).$$

We shall prove in Subsection 4.2 that the wavelets we have constructed in this section are bases of the $H_0^s(\Omega)$ spaces and that the following characterization holds

$$\| f \|^2_{H_0^s(\Omega)} \equiv \sum | c_{j,k} |^2 \, 2^{2js}.$$

2.2.5. Open problems

A key point for the resolution of PDE's on a domain Ω using wavelets is to construct wavelets which are an orthonormal basis of $L^2(\Omega)$ and bases of Sobolev spaces on the domain. Let us sumarize the existing results.

> On a segment (or a square, a cube,...) there exists an orthonormal basis of compactly supported wavelets of the form $2^{j/2}\psi(2^j x - k)$, except for N functions at each level j localized near the end points of the segment, where the function ψ has to be modified. There exist such bases which are also bases for the H^s spaces or for the H_0^s spaces; but for no other spaces (such as, for instance, $H_0^1 \cap H^s$, which would be of importance for the resolution of elliptic problems with Dirichlet boundary values).
>
> For a general domain, there exists a similar basis with exponential decay, but all the wavelets have to be modified (though the correction exponentially decreases as a function of $2^j \mathrm{dist}(k2^{-j}, \partial\Omega)$), and they are bases of the H_0^s. Non-smooth bases (their elements are not C^1) which work for H_0^1 also exist (see [34] and [23]).

Bases of the first kind which would work for more general domains, and bases, even of the second kind, but which would be bases of other Sobolev space, or could be adapted to mixed boundary conditions would be extremely useful.

§ 3. Local analysis by wavelets

One of the most attractive feature of the wavelets is that they give completely local information on the functions analysed. We shall study two examples of this property. First, we shall show that we can almost characterize the pointwise smoothness of a function (that is the existence of a Taylor expansion of a given order at a given point) by conditions on the moduli of the coefficients on the wavelets localized near the considered point. Then we shall show that, if f is a function whose smoothness is not everywhere the same (think of smooth functions with discontinuities), there is an optimal way to approximate it using low frequency wavelets everywhere and adding high frequency wavelets near the singularities. We shall see in Section 5 applications of this result for the resolution of evolution equations, since their solutions often have this feature of being smooth with some localized discontinuities (think of Burgers' equation for instance).

3.1. Analysis of pointwise smoothness

In this section, following [25], we shall compare the usual definition of pointwise smoothness and a corresponding condition on the wavelet coefficients. The purpose is to show that this last condition provides a "good substitute" for the pointwise Hölder regularity condition: it is easily checkable from the knowledge of the wavelet coefficients; it can be very precisely compared with the Hölder condition; and it has more functional properties. We also give applications of these properties.

Let s be a strictly positive real number. Let us recall the usual definition of the Hölder criterion at x_0. A function f belongs to $C^s_{x_0}$ if there exists a polynomial $P(x)$ of degree equal to the integral part of s such that

$$f(x) = P(x - x_0) + O(| x - x_0 |^s). \tag{8}$$

The following properties are classical. If f belongs to $C^s_{x_0}$, nothing is implied on the derivatives of f; actually, some insight about the functions for which this loss of regularity holds is given in [40]. In dimension 1, the primitive of f belongs to $C^{s+1}_{x_0}$. In dimension larger than 1, the fractional integration of order 1 maps $C^s_{x_0}$ into $C^{s+1}_{x_0}$.

3.1.1. The wavelet criterion of pointwise smoothness

Theorem 11. *If f belongs to $C^s_{x_0}$,*

$$| c_{j,k} | \leq C 2^{-(\frac{n}{2}+s)j} (1 + | 2^j x_0 - k |^s). \tag{9}$$

Conversely, if Equation (9) holds and f is uniformly C^β for a positive number β, there exists a polynomial P (depending on x_0) of degree less than s such that, if $| x - x_0 | \leq 1$,

$$| f(x) - P(x - x_0) | \leq C | x - x_0 |^s \log \frac{2}{| x - x_0 |}, \tag{10}$$

and this result is optimal.

We remark that, in this theorem, we are looking for regular points in an irregular background, which is more subtle than the usual approach that consists in finding irregular points in a C^∞ or analytical background (search of the different singular supports).

This theorem can be interpreted as a tauberian theorem. We have information on the behavior of the averages of f (its Littlewood-Paley decomposition) and a tauberian condition of minimal global regularity, which allow to obtain a pointwise result.

Proof: Suppose that f belongs to $C^s_{x_0}$. Then

$$\begin{aligned}
| c_{j,k} | &= | \int f(x)\psi_{j,k}(x)dx | \\
&= | \int (f(x) - P(x-x_0))\psi_{j,k}(x)dx | \\
&\text{(because the wavelets have vanishing moments)} \\
&\leq C \int | x - x_0 |^s \frac{2^{nj/2}}{(1+2^j \mid x - k2^{-j} \mid)^N} dx \\
&\leq C 2^{nj/2} \int \frac{| x - k2^{-j} |^s + | k2^{-j} - x_0 |^s}{(1+2^j \mid x - k2^{-j} \mid)^N} dx \\
&\leq C 2^{-(\frac{n}{2}+s)j} (1 + | 2^j x_0 - k |^s).
\end{aligned}$$

Let us now prove the converse result.
Define j_0 and j_1 by

$$2^{-j_0-1} \leq | x - x_0 | < 2^{-j_0} \quad \text{and} \quad j_1 = \frac{s}{\beta} j_0.$$

Let

$$f_j(x) = \sum_k c_{j,k}\psi_{j,k}.$$

From Equation (9), we deduce

$$| f_j(x) | \leq C 2^{-sj} (1 + 2^j \mid x - x_0 \mid)^s$$

and, for any l,

$$| f_j^{(l)}(x) | \leq C 2^{(l-s)j} (1 + 2^j \mid x - x_0 \mid)^s.$$

If g is a function, let $T(g)(x_0)$ be the Taylor expansion of g at the order $[s]$ at x_0. Then

$$| f(x) - T(f)(x_0) | \leq$$
$$\sum_{j \leq j_0} | f_j(x) - T(f_j)(x_0) | + \sum_{j \geq j_0} | f_j(x) | + \sum_{j \geq j_0} | T(f_j)(x_0) | .$$

Let $l = [s] + 1$. The first term is bounded by

$$C \mid x - x_0 \mid^l \sum_{j \leq j_0} \sup_{[x,x_0]} \mid f_j^l(x_0) \mid \leq C \mid x - x_0 \mid^l \sum_{j \leq j_0} 2^{(l-s)j} \leq C \mid x - x_0 \mid^s,$$

as regards the second term,

$$\sum_{j_0 \leq j < j_1} \mid f_j(x) \mid \leq \sum_{j_0 \leq j < j_1} \mid x - x_0 \mid^s \leq C(j_1 - j_0) \mid x - x_0 \mid^s$$

and

$$\sum_{j \geq j_1} \mid f_j(x) \mid \leq \sum_{j \geq j_1} 2^{-\beta j} \leq C \mid x - x_0 \mid^s$$

(because f is uniformly C^β).

The third term is bounded by

$$\sum_{j \geq j_0} \sum_{|m|} \mid x - x_0 \mid^m 2^{(m-s)j} \leq C \mid x - x_0 \mid^s .$$

Hence the converse part of the theorem, since $j_0 - j_1 \leq C \log \dfrac{2}{\mid x - x_0 \mid}$. We now give Y. Meyer's counter-examples that show its optimality.

Assume that ψ is a compactly supported wavelet, as constructed in Subsection 2.1.3. One easily checks that it is possible to suppose

$$\psi(0) \neq 0.$$

We first prove that the global C^β assumption is needed in Theorem 11, and cannot be replaced by the uniform continuity.

Let m be a positive integer and ϵ_m a real number such that $2^m \epsilon_m$ is an integer, and $\epsilon_m \to 0$ when $m \to \infty$. The precise value of ϵ_m will be given later. Let α be such that $0 < \alpha < 1$. The wavelet coefficients of the counter-example f are defined by:

if $2^m \leq j \leq 2^{m+1}$ and $k = \epsilon_m 2^j$, $C_{j,k} = 2^{-j/2} \epsilon_m^\alpha$;
else, $C_{j,k} = 0$.

Then define

$$f(x) = \sum_{m=0}^{\infty} f_m(x)$$

with

$$f_m(x) = \epsilon_m^\alpha \sum_{2^m \leq j \leq 2^{m+1}} \psi(2^j(x - \epsilon_m)).$$

Choose then ϵ_m such that

$$\frac{1}{2m} \leq 2^m \epsilon_m^\alpha \leq \frac{1}{m}.$$

The supports of the f_m are disjoint, $| f_m(x) | \leq C2^m \epsilon_m^\alpha$, and $f(0) = 0$. Then f is continuous (because $2^m \epsilon_m^\alpha \to 0$). But

$$f_m(\epsilon_m) = C2^m \epsilon_m^\alpha$$

so that

$$\limsup \frac{| f(x) - f(x_0) |}{x^\gamma} \geq \limsup C2^m \epsilon_m^{\alpha-\gamma} = +\infty \quad \forall \, \gamma > 0.$$

Hence f is not in C_0^γ for any value of γ, although condition (9) holds at 0.

The following counter-example shows that the logarithmic term is needed in Equation (10).

Take the same construction as before, but with

$$\epsilon_m = 2^{-\beta 2^m} \text{ for a given } \beta > 0.$$

Then

$$\frac{| f(\epsilon_m) - f(0) |}{\epsilon_m^\alpha} \geq C2^m \geq C' \log | \epsilon_m | .$$

Hence the optimality of the logarithmic term.

It should be noticed that other conditions similar to condition (9) can be introduced in order to be compared with other types of pointwise regularity conditions. For example, a comparison with pointwise differentiability is given by the following proposition, the proof of which is similar to the one of Theorem 11.

Proposition 12. *Let f be a function differentiable at x_0 with wavelet coefficients $c_{j,k}$. Let λ be the point $(k2^{-j}, 2^{-j})$ in the upper half-plane. Then, the following estimate holds*

$$| c_{j,k} | \leq C\eta(\lambda) 2^{-(\frac{n}{2}+1)j} (1 + | k - 2^j x_0 |)$$

where $\eta(\lambda) \leq 1$ and $\eta(\lambda) = o(1)$ when λ tends to $(x_0, 0)$. Conversely, if f is in $C^\beta(\mathbb{R}^n)$ for a strictly positive β and if there exists a positive function θ defined for positive values of j such that $\sum \theta(j) < \infty$ and

$$| c_{j,k} | \leq C\eta(\lambda)\theta(j) 2^{-(\frac{n}{2}+1)j} (1 + | k - 2^j x_0 |),$$

then f is differentiable at x_0.

3.1.2. Remarks and applications

Let x_0 be given. We have obtained conditions under which the following expansion holds at x_0

$$f(x) = P(x - x_0) + O(| x - x_0 |^s). \tag{11}$$

This can be interpreted as a generalized Taylor expansion : If f were C^n at x_0, Taylor's formula would prove that this expansion holds for $s = n$ and

$$P(x - x_0) = \sum_{|\alpha| \leq n} \frac{f^{(\alpha)}(x_0)}{\alpha!}(x - x_0)^\alpha.$$

The existence of Equation (11) with $s > 1$ implies that f is differentiable at x_0, which can be seen by rewriting Equation (10) as

$$\frac{| f(x) - P(x - x_0) |}{| x - x_0 |} \leq C | x - x_0 |^{s-1},$$

where P is a polynomial of degree 1. But, even if s is large, (11) does not imply more than the differentiability of f at x_0; it may hold while f is not differentiable in a neighborhood of x_0, in which case f is certainly not two times differentiable.

It is interesting to notice that, though some explicit conditions on the wavelet coefficients guarantee the existence of such an expansion, the coefficients of the expansion (*i.e.*, the derivatives of f at x_0) cannot be obtained from the wavelet coefficients. This is so because wavelet have vanishing moments and thus cannot "see" polynomials.

"Good substitutes" for the $C^s_{x_0}$ condition were introduced already in 1961 by A. P. Calderón and A. Zygmund [10] as follows.

A function f belongs to $T^p_u(x_0)$ if there exists a polynomial P of degree less than u such that

$$(\frac{1}{\rho^n} \int_{|x-x_0| \leq \rho} | f(x) - P(x - x_0) |^p \, dx)^{1/p} \leq C\rho^u,$$

for ρ small enough. The case $p = \infty$ corresponds to the usual Hölder criterion. If f belongs to $C^u(x_0)$, then f belongs to $T^p_u(x_0)$ for any p. This weak form of pointwise regularity is preserved under fractional integration and singular integral transformations. However, it is not as closely related to the $C^u_{x_0}$ spaces as condition (9) is, and a result such as Theorem 11 cannot hold for this condition.

However, using condition (9), the ideas developped in [10] give the following corollary, which is an optimal result of pointwise Hölder regularity for elliptic operators and follows from Theorem 11.

Corollary 13. *Let Λ be a partial differential operator of order m, with smooth coefficients and elliptic at x_0. If $\Lambda f = g$, and if g is a function that belongs to*

$C^s_{x_0}$, *then there exists a polynomial* P *of degree less than* $s+m$ *such that, for* $\mid x - x_0 \mid \leq 1$,

$$\mid f(x) - P(x) \mid \leq C \mid x - x_0 \mid^{s+m} \log \frac{2}{\mid x - x_0 \mid}.$$

The proof of this corollary makes use of the decomposition discovered by Calderón and Zygmund of the inverse of an elliptic operator as a product of a fractional integration and a pseudo-differential operator, which is the sum of a regularizing operator and of an operator of order 0; this result holds because condition (9) is preserved under the action of such operators (see [25] for details).

The "almost characterization" of pointwise smoothness has allowed M. Holschneider and Ph. Tchamitchian (see [21]) to obtain a new easy proof of the exact pointwise regularity of the Riemann-Weierstrass function

$$\sum_{n \geq 1} \frac{1}{n^2} \cos(n^2 x).$$

More recently, it has allowed to compute exactly the maximal Hausdorff dimension of the set on which a function that belongs to a given Sobolev space is not C^α (see [29]).

3.2. Local approximation properties

Let f be a smooth function defined on $\mathbb{R}^n$. Let f_ϵ be an approximation of f *at the scale* ϵ. The L^p order of the method of approximation is by definition the exponant α (if any) such that

$$\| f - f_\epsilon \|_{L^p} \leq C\epsilon^\alpha.$$

We now want to estimate this order of approximation when f_ϵ is an approximation of f given by its projection on the subspace generated by a subbase of an orthonormal basis of wavelets (see [26]).

First, we suppose that f belongs to $W^{s,p}$ (which, when s is an integer, is the space of all functions in L^p whose derivatives of order at most s also belong to L^p). Since we do not make some specific local hypotheses on the function f, we do not have reasons to approximate f more accurately at some places than others, and we choose its projection on the space V_J generated by all the wavelets $\psi_{j,k}$ for $j < J$. We shall note this projection $E_J(f)$. Thus

$$E_J(f) = \sum_{j<J} c_{j,k}\psi_{j,k}.$$

In this case, estimating the order of approximation is a straightforward consequence of the following characterization (see [42]).

Proposition 14. *Let $\chi_{j,k}$ be the characteristic function of the dyadic cube centered on $k2^{-j}$ and of size 2^{-j}, and $c_{j,k}$ the wavelet coefficients of f. If $0 \leq s < r$,*

$$f \in W^{s,p} \Leftrightarrow \left(\sum \mid c_{j,k} \mid^2 (1 + 4^{js}) 2^{nj} \chi_{j,k}(x)\right)^{1/2} \in L^p.$$

We shall actually prove this characterization when $p = 2$ in Subsection 4.2.

Corollary 15. *Under the preceding hypotheses,*

$$\parallel f - E_J(f) \parallel_{L^p} \leq C 2^{-Js} \parallel f \parallel_{W^{s,p}} . \tag{12}$$

Proof: If f is in $W^{s,p}$,

$$\parallel f - E_J(f) \parallel_{L^p} \leq C \parallel (\sum_{j \geq J} \mid c_{j,k} \mid^2 2^{nj} \chi_{j,k}(x))^{1/2} \parallel_{L^p}$$

$$\leq C \parallel (\sum_{j \geq J} \mid c_{j,k} \mid^2 (1 + 4^{(j-J)s}) 2^{nj} \chi_{j,k}(x))^{1/2} \parallel_{L^p}$$

$$\leq CC' 2^{-Js} \parallel f \parallel_{W^{s,p}} .$$

Suppose now that f is not smooth everywhere, but perhaps belongs only to $W^{s',p}$ for an $s' < s$ in some regions. Then the order of approximation given by Equation (12) will of course no longer hold. But it might still hold if we allow f to be approximated more precisely, by more wavelets, in the region where it is not smooth. We shall show precisely how this local refinement should be performed in order to recover the optimal order of approximation given by Equation (12). We shall have to add higher frequency wavelets in the *non-smoothness* region and also in a certain neighborhood, this neighborhood depending on the scale index j of the wavelet. Not surprisingly, the size of this *security zone* will depend on the localization of the wavelets; the more localized they are, the smaller this zone will be.

3.2.1. Notations and statement of results

We shall consider the following cases for the localization of the wavelets. The first one corresponds to the compactly supported wavelets, and the second one to the spline wavelets.

Case A: The wavelets are compactly supported in a ball centered on 0, of radius A.

Case B: The following estimates holds

$$\mid \partial^\alpha \psi(x) \mid \leq C \exp(-\gamma \mid x \mid)$$

for $| \alpha | \leq r$ and a $\gamma > 0$.

We shall make the following assumption on f. We suppose that f belongs to $W^{s',p}(\mathbb{R}^n)$, and f belongs to $W^{s,p}(\Omega)$, for $s' < s$ and a certain open set $\Omega \in \mathbb{R}^n$. If s is an integer, this will mean

$$| \alpha | \leq s \Rightarrow \int_{\Omega} | \partial^{\alpha} f |^p \, dx < \infty.$$

Recall that we want to choose a certain set A of indexes (j, k) such that

$$\| f - \sum_{(j,k) \in A} c_{j,k} \psi_{j,k} \|_{L^p} \leq C 2^{-Js} \| f \|_{W^{s,p}} .$$

Using Corollary 15, we know that, if f belonged to $W^{s,p}(\mathbb{R}^n)$, we would have to use all wavelets $\psi_{j,k}$ such that $j \leq J$; if f belonged only to $W^{s',p}(\mathbb{R}^n)$, using again Corollary 15, we would have to use more wavelets, namely all wavelets such that $j \leq sJ/s'$. Thus, because of the localization of the wavelets, we can guess that, roughly, we shall have to take all wavelets localized outside Ω such that $j \leq sJ/s'$ and, outside that region, all wavelets such that $j \leq J$. Which wavelets to take near the boundary of Ω is being made clear by the following theorem.

Theorem 16. *Define the set of indexes $A(J, s, s', \Omega)$ as follows*

(1) *If $j \leq J$, $(j, k) \in A(J, s, s', \Omega)$*
(2) *If $j \geq sJ/s'$, $(j, k) \notin A(J, s, s', \Omega)$*
(3) *Else*
  (3.1) *In Case A, $(j, k) \in A(J, s, s', \Omega)$ if $d(k2^{-j}, \mathbb{R}^n - \Omega) \leq C_1 2^{-j}$.*
  (3.2) *In Case B, $(j, k) \in A(J, s, s', \Omega)$ if $d(k2^{-j}, \mathbb{R}^n - \Omega) \leq C_2 j 2^{-j}$, where C_1 is the diameter of the largest compactly supported wavelet used, and C_2 will be specified later.*

Then, under the preceding hypotheses

$$\| f - \sum_{(j,k) \in A(J,s,s',\Omega)} c_{j,k} \psi_{j,k} \|_{L^p} \leq C 2^{-Js} \| f \|_{W^{s,p}} . \tag{13}$$

3.2.2. Proof of Theorem 16

By Corollary 15, since $f \in W^{s',p}(\mathbb{R}^n)$,

$$\| f - \sum_{j \leq Js/s'} c_{j,k} \psi_{j,k} \|_{L^p} \leq C 2^{-Js} \| f \|_{W^{s,p}} .$$

Thus, we only have to prove that

$$\| \sum c_{j,k} \psi_{j,k} \|_{L^p} \leq C 2^{-Js} \| f \|_{W^{s,p}}$$

where the sum is restricted to the indexes such that

$$j \geq Js \text{ and } (j,k) \in A(J,s,s',\Omega).$$

For these indexes we shall first obtain an upper bound for the wavelet coefficients $c_{j,k}$ which will be a consequence of the following lemma.

Lemma 17. *Let $F = \mathbb{R}^n \backslash \Omega$. The wavelets $\psi_{j,k}$ can be written as the sum of two functions $\theta_{j,k}$ and $\omega_{j,k}$ such that: $\theta_{j,k}$ and $\omega_{j,k}$ have the same uniform decay estimates as $\psi_{j,k}$; $\theta_{j,k}$ is supported by the square $S^1_{j,k}$ centered on $k2^{-j}$ and of size $n^{-1/2}d(k2^{-j},F)$, while $\omega_{j,k}$ vanishes on the square $S^2_{j,k}$ centered on $k2^{-j}$ and of size $n^{-1/2}d(k2^{-j},F)/2$; the moments of order less than r of $\theta_{j,k}$ all vanish, and thus the function $\theta_{j,k}$ can be written $\theta_{j,k} = \sum_{|\alpha|=s} \partial^\alpha \phi^\alpha_{j,k}$ where the $2^{sj}\phi^\alpha_{j,k}$ have a vanishing integral and have the same uniform estimates as $\psi_{j,k}$ and are supported by the cube centered on $k2^{-j}$ and of size $n^{-1/2}d(k2^{-j},F)$.*

The proof of this lemma is quite technical and can be found in [26].

Let us now come back to the proof of Theorem 16. We shall prove that

$$\| \left(\sum | \langle f \mid \theta_{j,k} \rangle |^2 (1+4^{js}) 2^{nj} \chi_{j,k}(x) \right)^{1/2} \|_{L^p} < \infty \tag{14}$$

and

$$\| \left(\sum | \langle f \mid \omega_{j,k} \rangle |^2 (1+4^{js}) 2^{nj} \chi_{j,k}(x) \right)^{1/2} \|_{L^p} < \infty \tag{15}$$

which, using the same proof as in Corollary 15, will imply the theorem.

Estimate Equation (14) holds because

$$\langle f \mid \theta_{j,k} \rangle = \sum_{j,k,|\alpha|=s} (-1)^\alpha \langle \partial^\alpha f \mid 2^{sj} \phi_{j,k} \rangle 2^{-sj},$$

$\partial^\alpha f \in L^p$, and the $2^{sj}\phi_{j,k}$ are *vaguelettes*, which means that they have the same decay estimates as the wavelets and a vanishing integral; the L^p estimate Equation (15) is thus a consequence of the following lemma (see [42]).

Lemma 18. *If the $g_{j,k}$ are vaguelettes,*

$$\| \left(\sum | \langle f \mid g_{j,k} \rangle |^2 2^{nj} \chi_{j,k}(x) \right)^{1/2} \|_{L^p} \leq C \| f \|_{L^p}.$$

Actually, when $p = 2$, this Lemma will be proved in Subsection 4.2.

Let us now prove Equation (15).

$$| \langle f \mid \omega_{j,k} \rangle | \leq \| f \|_{L^1} \sup_{|x| \geq n^{-1/2}d/2} | \omega_{j,k} |$$

where $d = \text{dist}(k2^{-j}, F)$. Thus, if $\gamma' = \gamma n^{-1/2}/2$,

$$| \langle f \mid \omega_{j,k} \rangle | \leq C 2^{nj/2} \exp(-\gamma' d).$$

Since the result is clearly local, we can restrict to a square of side 1. Thus at each level j, we shall handle with at most 2^{nj} wavelets, and for each j

$$\int \left(\sum_k | \langle f \mid \omega_{j,k} \rangle |^2 (1 + 4^{js}) 2^{nj} \chi_{j,k}(x) \right)^{p/2} dx \leq 2^{2nj} \exp(-2\gamma' d)(1 + 4^{js}).$$

Since $d2^j \geq C_2 j$, if we choose C_2 such that

$$\frac{C_2 \gamma'}{\log(2)} > 2n + 2s$$

the series (in j) of the L^p norms will converge and Equation (15) is proved.

3.2.3. Conclusion

Theorem 1 shows that if f is a smooth function (in H^s) defined on $\mathbb{R}$ with a finite number of discontinuities, approximated first using wavelets up to a certain scale, the order of approximation is improved by adding a few small scale wavelets near the discontinuities. In the uniform case considered in Corollary 15, we use 2^J wavelets for an order of approximation less than $2^{-J/2}$ (this is because such a function belongs to $H^{s'}(\mathbb{R})$ for any $s' < 1/2$ but does not belong to $H^{1/2}$). In the case of spline wavelets, an immediate calculation shows that we have to add only a number of wavelets of the order of $S^2 J^2$ to make the order of approximation become 2^{-Js}. Thus the order of approximation is drastically improved by a (proportionally) very small increase of the number of wavelets used.

Of course, we have here a theoretical justification to the good performances of the numerical methods used, for instance, for solving adaptatively Burgers' equation (see Section 5). In these algorithms, a local refinement is used where the discontinuities appear. Theorem 11 can be used there to calculate precisely the order of approximation of these methods. If the Littlewood-Paley wavelets were used, they would give an infinite order of approximation method even after the apparition of the discontinuities, and this by adding just an arbitrarily small proportion of the number of wavelets used to obtain the 1/2 order of approximation!

As was shown in [23], this method of local refinement can also be used to solve equations of the form $A(f) = g$ where A is a smooth elliptic operator and g is a smooth function with some singularities. This may be of interest in the context of adaptative implicit schemes for solving evolution equations.

§4. Analysis of linear operators

4.1. Standard and nonstandard decomposition

We follow in this section the descriptions of [8] and [46]. In this section, once for all, let m be a positive integer and $(V_j)_j$ an m-regular multiresolution analysis of $L^2(\mathbb{R})$. As in the previous sections, we denote by ϕ and ψ the scaling function and the wavelet associated with this multiresolution analysis. Then $(\phi_{j,k})_k$ is an orthonormal basis of V_j and $(\psi_{j,k})_k$ is an orthonormal basis of W_j, the orthogonal complement of V_j in V_{j+1}. For a given f in $L^2(\mathbb{R})$, we consider the sequence $(P_j f)_j$ of the orthogonal projections of f on the subspaces V_j. This sequence satisfies the following property

$$\lim_{j\to+\infty} \|P_j f - f\|_{L^2} = 0,$$

so $P_j f$ is an approximation of f but with special properties. In fact, $P_j f$ can be seen as a copy of f where all details of scale lower than 2^{-j} would have been removed. Moreover, going from $P_j f$ to $P_{j+1} f$ consists in adding to $P_j f$ the details of scale $2^{-(j+1)}$ of f. Thus, these properties allow an accurate decomposition of f and a good understanding of its behavior, either in space and phase.

The next idea is to try to achieve the same kind of decomposition on operators. For that purpose, let T be a continuous linear operator on $L^2(\mathbb{R})$ and define, for all integer j the operator T_j by

$$T_j = P_j T P_j \tag{16}$$

which is a continuous linear operator over V_j. The first result we can state is the following

$$\forall\ f \in L^2(\mathbb{R})\ \lim_{j\to+\infty} \|P_j T P_j f - Tf\|_{L^2} = 0.$$

Following what is done to functions in $L^2(\mathbb{R})$, informations on the operator T will be gathered from the properties of T_j. The remaining work is then to develop algorithms to study the sequence (T_j). But, before presenting precisely the methods involved, let us make a few remarks to introduce them.

A1. Let $(H, (.|.))$ be an Hilbert space and $(\epsilon_\lambda)_\lambda$ be an orthonormal basis of H. We then consider a linear operator A over H. For x in H, Ax is computed by

$$\begin{aligned} Ax &= \sum_\lambda (x|\epsilon_\lambda)\, A\epsilon_\lambda \\ &= \sum_\lambda \sum_{\lambda'} (A\epsilon_\lambda|\epsilon_{\lambda'})\, (x|\epsilon_\lambda)\, \epsilon_{\lambda'} \end{aligned}$$

so that A can be depicted by an '*infinite matrix*' whose coefficients are

$$(A\epsilon_\lambda|\epsilon_{\lambda'}) .$$

Depending on the chosen basis, the study of these coefficients may easily reveal certain properties of A.

A2. Suppose now that H can be written as the orthogonal sum of two nontrivial subspaces, say E_1 and H_1. Introducing P_{E_1} and P_{H_1}, the orthogonal projections on E_1 and H_1, A can be written as below

$$A = P_{E_1}AP_{E_1} + P_{E_1}AP_{H_1} + P_{H_1}AP_{E_1} + P_{H_1}AP_{H_1}$$

which enables us to depict A by an '*infinite matrix*' divided in four blocks

$$\begin{pmatrix} B & C \\ D & A_1 \end{pmatrix}$$

where B represents $P_{E_1}AP_{E_1}$, C represents $P_{E_1}AP_{H_1}$, D represents $P_{H_1}AP_{E_1}$, and A_1 represents $P_{H_1}AP_{H_1}$. Then, provided that H_1 can be written as the orthogonal sum of two nontrivial subspaces, we can achieve the same decomposition on A_1 and iterate the algorithm. The '*infinite matrix*' thus obtained, we may find through the study of its coefficients, the properties of A.

Coming back to wavelets, the two decompositions we mentionned above are the ones currently used. We will describe more precisely both of them in the next two subsections, beginning by the so-called standard decomposition (which relies on the algorithm **A1**) and ending with the nonstandard decomposition (which relies on the algorithm **A2**).

4.1.1. The standard decomposition

Let J be any positive integer and consider T_J. To achieve what we planned above, we need a basis of V_J. At this point, we may choose either

$$\mathcal{B}_1 = \{(\phi_{j,k})_k\}$$

or

$$\mathcal{B}_2 = \{(\psi_{j,k})_k, 0 \leq j \leq J-1\} \cup \{(\phi_{0,k})_k\} .$$

But, the elements of $\mathcal{B}_1$ have no cancellation properties so the analysis of T_J with it would not lead to interesting results. On the opposite, useful estimates on the coefficients of the *infinite matrix* depicting T_J will be obtained with $\mathcal{B}_2$ because of the cancellation and localization properties of the wavelets. As examples of this algorithm will be used extensively in the next sections, no more details about it will be given in this section.

4.1.2. The nonstandard decomposition or BCR algorithm

Let still J be any positive integer.

$$V_J = V_{J-1} \oplus W_{J-1}$$

$\forall\, v_J \in V_J \,\exists\, (v_{J-1}, w_{J-1}) \in V_{J-1} \times W_{J-1}$ so that $v_J = v_{J-1} + w_{J-1}$.

Then,

$$T_J v_J = B_{J-1} v_{J-1} + C_{J-1} w_{J-1} + A_{J-1} w_{J-1} + T_{J-1} v_{J-1}$$

where

$B_{J-1} = Q_{J-1} T P_{J-1}$ is a linear operator from V_{J-1} to W_{J-1},
$C_{J-1} = P_{J-1} T Q_{J-1}$ is a linear operator from W_{J-1} to V_{J-1},
$A_{J-1} = Q_{J-1} T Q_{J-1}$ is a linear operator from W_{J-1} to W_{J-1},

and $Q_{J-1} = P_J - P_{J-1}$ is the orthogonal projection on W_{J-1}. We then iterate this decomposition on T_{J-1}, T_{J-2}, ...,T_1. At each step, each T_j can be depicted by an *infinite matrix* which writes

$$T_j = \begin{pmatrix} A_{j-1} & B_{j-1} \\ C_{j-1} & T_{j-1} \end{pmatrix}.$$

Finally, we have for T_J

$$T_J = \begin{pmatrix} A_{J-1} & & & B_{J-1} & \\ & & & & \\ & A_{J-2} & & B_{J-2} & \\ C_{J-1} & & \ddots & \vdots & \\ & C_{J-2} & \cdots & A_0 & B_0 \\ & & & C_0 & T_0 \end{pmatrix}.$$

In order to get an efficient algorithm, Beylkin et al. in [8] introduce a stretched matrix M and a stretched vector $\tilde{v_J}$ defined by

$$M = \begin{pmatrix} A_{J-1} & B_{J-1} & & 0 & & \\ C_{J-1} & 0 & & & & \\ & & A_{J-2} & B_{J-2} & & 0 \\ & 0 & C_{J-2} & 0 & & \\ & & & & \cdots & 0 \\ & & & 0 & 0 & T_0 \end{pmatrix}$$

and by

$$\tilde{v_J} = (w_{J-1}, v_{J-1}, w_{J-2}, v_{J-2}, \ldots, w_0, v_0, v_0)$$

where $w_j \in W_j$, $v_j \in V_j$ and $v_{j+1} = v_j + w_j$ for all $j \in \{0, \ldots, J-1\}$.

Moreover, it is easy to see that $T_J v_j$ can be computed from $M\tilde{v_J}$.

Before stating a few results about standard and nonstandard decomposition, we should notice that these two algorithms may be easily explained using Schwartz kernels' theorem. In fact, if T is a continuous linear operator on $L^2(\mathbf{R})$, there exists $K \in \mathcal{S}'(\mathbb{R} \times \mathbb{R})$ so that, formally,

$$\forall\, f \in L^2(\mathbb{R}), \qquad Tf(x) = \int K(x,y) f(y)dy.$$

If K belongs to $L^2(\mathbb{R}^2)$, we may analyze it using wavelets: at this point, two wavelet bases of $L^2(\mathbb{R}^2)$ are available: the standard one built from a multiresolution analysis of $L^2(\mathbb{R})$

$$(\phi_{j,k}(x)\psi_{j,k'}(y), \psi_{j,k}(x)\phi_{j,k'}(y), \psi_{j,k}(x)\psi_{j,k'}(y))_{j,k,k'}$$

which leads to the nonstandard decomposition we describe above and the wavelet basis obtained by the tensorial product of a wavelet basis of $L^2(\mathbb{R})$

$$(\psi_{j,k}(x)\psi_{j',k'}(y))_{j,j',k,k'}$$

which leads to the standard decomposition.

4.1.3. Some numerical results

In this subsection, we are going to restrict ourselves to a wide class of operators (including the pseudo-differential operators of order 0 and the Hilbert transform), so called the Calderon-Zygmund operators. The reason for this restriction is that, if we denote by T such an operator, the matrices of the $(T_j)_j$ defined by Equation (16) obtained by the BCR algorithm are sparse. Roughly speaking, the pair of functions arising in the nonstandard decomposition, namely

$$\psi_{j,k}(x)\phi_{j,k'}(y)$$

$$\phi_{j,k}(x)\psi_{j,k'}(y)$$

and

$$\psi_{j,k}(x)\psi_{j,k'}(y)$$

have cancellation either in one or two directions, but *live* at the same scale: this last property prevents the interaction of phenomenons of different scales and leads to small coefficients (in absolute value) in the matrix, provided they are far enough from the diagonal. On the opposite, the pair of functions arising in the standard decomposition, namely

$$\psi_{j,k}(x)\psi_{j',k'}(y)$$

live at different scales, but all of them have cancellation in both directions: this property may also lead to sparse matrices.

The aim of our numerical experiments was to compare the sparcity of the matrices resulting from both standard and nonstandard decomposition for some operators. Before going further, we must recall that the results of these decompositions depend strongly on the wavelet basis chosen and may vary a lot from one wavelet basis to another. Some discussion about this point may be found in [37], [8], and [46].

For the sake of simplicity, we consider the 1-periodic multiresolution analysis of $L^2([0,1])$ generated by the 1-periodic splines of degree 5. For a given operator T, we choose a positive integer J and compute both

$$\text{(i)}\quad \begin{cases} (T\psi_{j,k}|\psi_{j\prime,k\prime}) \\ (T\psi_{j,k}|\phi_{0,0}) \\ (T\phi_{0,0}|\psi_{j,k}) \\ 0 \le j \le J-1, \quad 0 \le j\prime \le J-1 \\ 0 \le k \le 2^{j-1}, \quad 0 \le k\prime \le 2^{j\prime-1} \end{cases}$$

and

$$\text{(ii)}\quad \begin{cases} (T\phi_{j,k}|\psi_{j,k\prime}) \\ (T\psi_{j,k}|\phi_{j,k\prime}) \\ (T\psi_{j,k}|\psi_{j,k\prime}) \\ 0 \le j \le J-1 \\ 0 \le k \le 2^{j-1}, \quad 0 \le k\prime \le 2^{j-1}. \end{cases}$$

We then evaluate the number of coefficients which are in $[10^{-n+1}, 10^{-n}]$ in the two matrices (after a renormalization of the coefficients so that the largest becomes 1). The histograms below gather such results. On the left is the standard algorithm, and on the right the BCR algorithm.

In fact, the first experiments we made seem to show that none of this two decomposition is always the best. We obtain better results for the standard decomposition in the cases of the first derivative and of the Hilbert Transform, and better results for the BCR decomposition for the pointwise multiplication. These results are partial and may depend on the wavelet used.

The problem of comparing the two methods of decomposition we described is very similar to the problem raised at the end of Subsection 2.1.2. Consider for instance the distribution kernel of the Hilbert transform, which is $\dfrac{1}{x-y}$. It is a function which is smooth except on the diagonal, and the remarks made at the end of Subsection 2.1.2 about the best way to store such a function on the two possible *wavelet-type* bases in two dimensions apply here.

Considering the partial results of [8], [46], and [37], this section, and the next, a likely conclusion is that the standard algorithm fits well operators that have *cancellation properties* (for instance, following the terminology used in the study of Calderon-Zygmund operators, operators such that $T(1) = T^*(1) = 0$); and the nonstandard decomposition fits the operators which do not have these

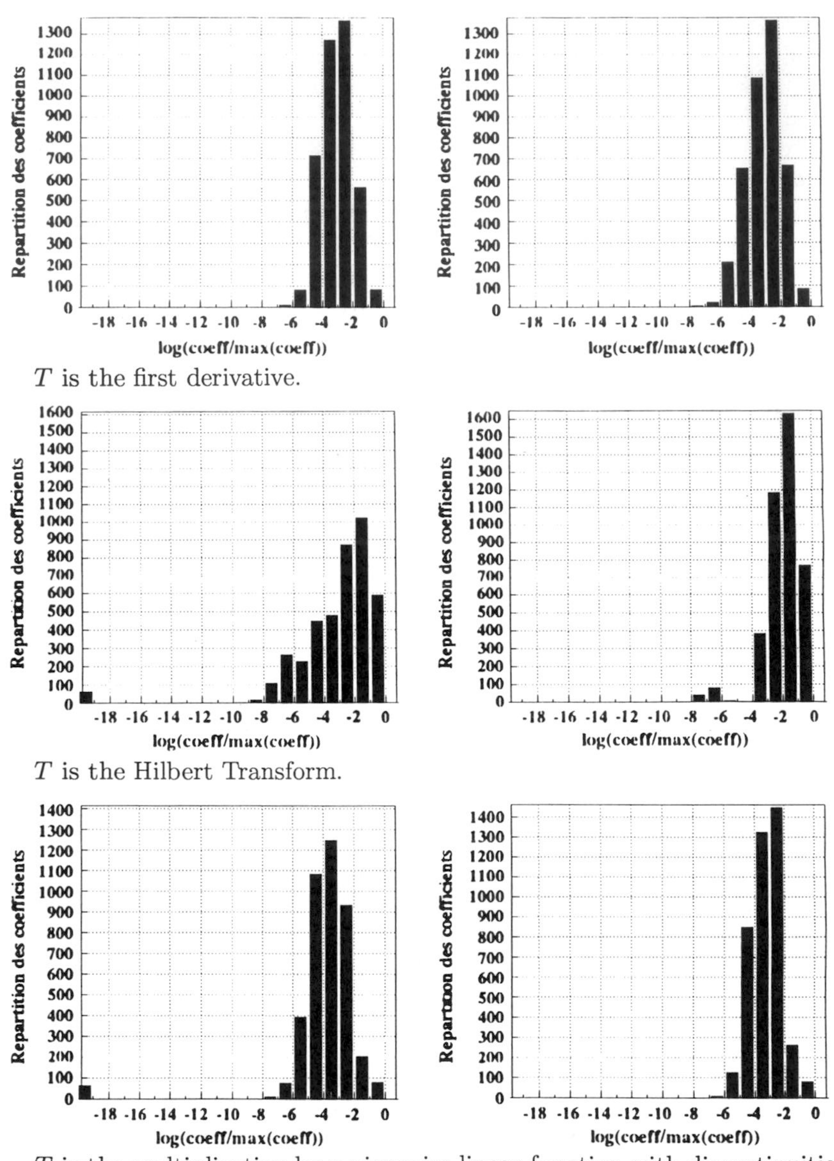

T is the first derivative.

T is the Hilbert Transform.

T is the multiplication by a piecewise linear function with discontinuities.

properties (such as for instance pointwise multiplication by a non-smooth function).

In the following two sections, we shall analyse operators using the standard decomposition for the following reasons: They will have *enough cancelletion* so that their boundedness properties will be checkable using simple conditions on the entries of the corresponding matrices, and this decomposition will have additional properties that we shall describe.

4.2. An example: analysis of the Laplacian in $\mathbb{R}^n$

As a prerequisite, we first have to define some classes of matrices which we shall use often in the following. Their usefulness will be a consequence of the three following properties.

1. They will be algebras.
2. They will be large enough so that a large class of operators we shall consider (for instance differential operators or their inverses) will be the products of a matrix in this space with diagonal matrices.
3. Their properties (as boundedness on certain functional spaces or estimates on the Green functions) will be straightforward to check.

Definition 19. *We say that an infinite matrix $T(\lambda, \lambda')$ belongs to $\mathcal{M}^\gamma$ if*

$$| T(\lambda, \lambda') | \leq C\omega_\gamma(\lambda, \lambda')$$

where

$$\omega_\gamma(\lambda, \lambda') = \frac{2^{-|j-j'|(\frac{n}{2}+\gamma)}}{1+(j-j')^2}\left(\frac{2^{-j}+2^{-j'}}{2^{-j}+2^{-j'}+ | \lambda-\lambda' |}\right)^{n+\gamma} . \tag{17}$$

An operator U belongs to $Op(\mathcal{M}^\gamma)$ if its matrix in a wavelet basis belongs to $\mathcal{M}^\gamma$, that is if

$$| \langle U(\psi_\lambda) \mid \psi_{\lambda'}\rangle | \leq C\omega_\gamma(\lambda, \lambda').$$

The space $\mathcal{M}^\gamma$ is an algebra, that is if $T \in \mathcal{M}^\gamma$ and $T' \in \mathcal{M}^\gamma \Longrightarrow TT' \in \mathcal{M}^\gamma$. The verification of this property is a straightforward computation left to the reader. Thus, we define

$$\| T \|_\gamma = \frac{1}{\alpha} \sup_{\lambda,\lambda'} \frac{|T(\lambda, \lambda')|}{\omega_\gamma(\lambda, \lambda')},$$

where α is such that, for all $T, T' \in \mathcal{M}^\gamma$,

$$\| TT' \|_\gamma \leq \| T \|_\gamma \| T' \|_\gamma . \tag{18}$$

An operator in $\mathcal{M}^\gamma$ is continuous on l^2. It will be a consequence of Schur's Lemma which we now state.

Lemma 20. *Let $A(k, l)$ be an infinite matrix, and $c_l > 0$ weights such that*

$$\sum_l | a(k,l) | c_l \leq c_k$$

and

$$\sum_k | a(k,l) | c_k \leq c_l;$$

then A is bounded on l^2.

This lemma is a consequence of Cauchy-Schwartz inequality. The continuity of the matrices of $\mathcal{M}^\gamma$ on l^2 is obtained by applying this lemma with weights $2^{-nj/2}$.

As a first example, we shall now analyse in the Littlewood-Paley wavelet basis the fractional powers of the Laplacian.

Recall that, if $s \in \mathbb{R}$, Δ^s is the operator defined in the Fourier transform as follows

$$\widehat{\Delta^s(f)}(\xi) = \mid \xi \mid^{2s} \hat{f}(\xi).$$

Let G be the matrix of this operator in the wavelet basis.

Proposition 21. *The following decomposition holds: $G = DAD$, where D is the operator diagonal in a wavelet basis such that $D(j,k,j,k) = 2^{sj}$, and A and A^{-1} belong to $\mathcal{M}^\gamma$ for any $\gamma > 0$.*

Proof: If $\mid j-j' \mid > 1$, $G(j,k,j',k') = 0$, because the Fourier transforms of $\widehat{\psi}_{j,k}$ and $\widehat{\psi}_{j',k'}$ have disjoint supports.

If $j = j'$,

$$G(j,k,j,k') = 2^{-nj} \int e^{i(k-k')\xi/2^j} \mid \xi \mid^{2s} \mid \hat{\psi} \mid^2 (\xi/2^j) d\xi$$

$$= 2^{2sj} \int e^{i(k-k')\xi} \mid \xi \mid^{2s} \mid \hat{\psi} \mid^2 (\xi) d\xi.$$

Since $\mid \hat{\psi} \mid^2$ is in the Schwartz class, its Fourier transform has fast decay, thus, for any γ,

$$\mid G(j,k,j,k') \mid \leq \frac{C_\gamma}{(1+ \mid k-k' \mid)^\gamma}.$$

The proof when $\mid j - j' \mid = 1$ is similar. Thus $A \in \mathcal{M}^\gamma$. The estimate for A^{-1} is obtained by remarking that it is the estimate on A when changing s into $-s$.

As a corollary, let us prove the characterization of Sobolev spaces.

Proposition 22. *Let f be in $H^s(\mathbb{R}^n)$. Then the following equivalence holds*

$$\| f \|_{H^s}^2 \equiv \sum \mid c_{j,k} \mid^2 (1 + 2^{2j})^s.$$

Proof: Let $s \geq 0$. $\| f \|_{H^s}^2 = \| f \|_{L^2}^2 + \| \Delta^{s/2} f \|_{H^s}^2$.

Let C be the vector of the wavelet coefficients of f.

$$\| \Delta^{s/2} f \|_{H^s}^2 \equiv \langle A(DC) \mid DC \rangle \equiv \| (DC) \|^2 = \sum \mid c_{j,k} \mid^2 2^{2js}.$$

Hence the proposition when $s \geq 0$. The case $s < 0$ follows by duality.

Let us now introduce the notion of *vaguelettes* (introduced in [42]).

Functions $(f_{j,k})$ are vaguelettes (of order γ) if their matrix in a wavelet basis (of smoothness larger than γ) belongs to $\mathcal{M}^\gamma$.

Then, for all $l = (l_1, ..., l_n)$ such that $| l | \leq \gamma$,

$$\int (x - k2^{-j})^l f_{j,k}(x)dx = 0$$

and

$$| \partial^l f_{j,k}(x) | \leq \frac{c(p)2^{(n/2+|l|)j}}{(1+ | 2^j x - k |)^\gamma}.$$

As an example of the use of vaguelettes, let us sketch the proof of the following characterization of Sobolev spaces $H_0^s(\Omega)$ using the wavelets defined in Subsection 2.2.5.

Proposition 23. *Let f be in $H_0^s(\Omega)$. Then*

$$\| f \|_{H_0^s}^2 \equiv \sum | c_{j,k} |^2 \, 2^{2js}.$$

The idea of the proof is the following. We extend the wavelets outside Ω so that the size estimates remain the same, but so that the moments of the new wavelets vanish. This is possible if, for instance, the domain is Lipschitz. The new functions are vaguelettes on $\mathbb{R}^n$, and f extended by 0 outside Ω has the same coefficients on this system. The proposition is then, a consequence of Proposition 21.

4.3. Analysis of elliptic operators

4.3.1. Introduction

Let us recall at first some properties of Galerkin methods based on finite elements or of finite difference methods for the resolution of elliptic problems in a bounded domain.

In all these methods, once the problem has been properly discretized, one is led to solve a system which is ill-conditioned. Typically, for a second order elliptic problem in two dimensions, one obtains a matrix M such that

$$\kappa = \| M \| \, \| M^{-1} \| = O(1/h^2)$$

where h is the size of the discretization. Such ill-conditioning has two drawbacks; it leads to numerical instabilities and to slow convergence for iterative resolution algorithms. In order to avoid this problem, one usually uses a preconditioning, which amounts to finding an easily invertible matrix D such that $D^{-1}MD^{-1}$ (or $D^{-1}M$, depending on the method used) will have a better condition number κ. For the example we considered, the usual preconditioning methods on general domains (SSOR or DKR on a conjugate gradient method, for instance) make κ become $O(1/h)$. We are going to show that, if one uses a wavelet method, the simplest conceivable preconditioning, namely when D is a diagonal matrix, yields a $\kappa = O(1)$ in any dimension (provided that the domain has a Lipschitz boundary). This property will be a direct consequence of the characterization of Sobolev spaces on the wavelet coefficients.

We shall consider the following Dirichlet type boundary value problem

$$(A) \quad \begin{cases} -\nabla.(A\nabla u) + bu = g, & \text{with } g \in H^{-1}(\Omega) \\ u(x) = h \; for \; x \in \partial\Omega, & \text{with } h \in H^{1/2}(\partial\Omega). \end{cases}$$

We suppose that Ω is a bounded domain of $\mathbb{R}^n$ with a Lipschitz boundary and that

$$\begin{cases} \forall\, X \in \mathbb{R}^n,\; C_1 \parallel X \parallel \leq \langle AX \mid X\rangle \leq C_2 \parallel X \parallel \\ 0 \leq b(x) \leq C_3 < +\infty. \end{cases} \tag{19}$$

4.3.2. The wavelet method for a Dirichlet boundary problem

In order to solve (A), we use the wavelet basis constructed in Subsection 2.2.4. Recall that, if a function f belongs to $H_0^1(\Omega)$, the following condition on its wavelet coefficients $c_{j,k}$ $(= \langle f, \psi_{j,k}\rangle)$ holds

$$C \sum \mid 2^j c_{j,k} \mid^2 \leq \parallel f \parallel_{H_0^1}^2 \leq C' \sum \mid 2^j c_{j,k} \mid^2 . \tag{20}$$

We now use the Galerkin method to solve (A) (a related method has also been proposed in [51] when $\Omega = \mathbb{R}^n$). As usual, we first reduce (A) to an homogeneous problem by supposing that we can find a smooth function h' which extends h inside Ω. Then $u' = u - h'$ will be the solution of the following problem (A'), if we replace in (A') g by $g' = g - \nabla.(A\nabla h') + bh'$.

$$(A') \quad \begin{cases} -\nabla.(A\nabla u) + bu = g & \text{with } g \in L^2(\Omega) \\ u(x) = 0 & \text{for } x \in \partial\Omega. \end{cases}$$

The variational form of this problem is

$$\forall\, v \in H_0^1, \quad \int_\Omega A(\nabla u).\nabla v + \int_\Omega buv = \int_\Omega gv.$$

The Galerkin approximation will consist in finding u in V_j such that

$$\forall\, v \in V_j, \quad \int_\Omega A\nabla u.\nabla v + \int_\Omega buv = \int_\Omega gv;$$

which, once the functions are expressed by their coordinates in the wavelet basis, amounts to solving the problem $MX = Y$, where the matrix M is given by

$$M_{(j,k),(j',k')} = \langle A\nabla\psi_{j,k} \mid \nabla\psi_{j',k'}\rangle + \langle b\psi_{j,k} \mid \psi_{j',k'}\rangle,$$

$\langle | \rangle$ denotes the L^2 scalar product, and the vectors X and Y are

$$X = (\langle u \mid \psi_{j,k} \rangle)$$

and

$$Y = (\langle g \mid \psi_{j,k} \rangle).$$

Let X be the vector $(c_{j,k})$, and $u = \sum c_{j,k} \psi_{j,k}$. Then

$$X^t M X = \langle A \bigtriangledown u \mid \bigtriangledown u \rangle + \langle bu \mid u \rangle.$$

Thus, because of Equation (19),

$$C_1 \parallel u \parallel_{H_0^1}^2 \leq X^t M X \leq (C_2 + C_3 C") \parallel u \parallel_{H_0^1}^2,$$

where $C"$ is the smallest number such that

$$\parallel f \parallel_{L^2}^2 \leq C" \parallel \bigtriangledown f \parallel_{L^2}^2 \ (= \parallel f \parallel_{H_0^1}^2).$$

Using Equation (20), we get

$$CC_1 \sum \mid 2^j c_{j,k} \mid^2 \leq X^t M X \leq C'(C_2 + C" C_3) \sum \mid 2^j c_{j,k} \mid^2 .$$

Hence, if D is the diagonal matrix defined by

$$D_{(j,k),(j',k')} = 2^j \delta_{(j,j')} \delta_{(k,k')},$$

for any vector X,

$$CC_1 \parallel X \parallel^2 \leq X^t D^{-1} M D^{-1} X \leq C'(C_2 + C_3) \parallel X \parallel^2 .$$

We have thus proved the following Theorem.

Theorem 24. *For a Galerkin method using wavelets to approximate the solution of (A), the condition number of $D^{-1} M D^{-1}$ is bounded by $\dfrac{C'(C_2 + C" C_3)}{CC_1}$ (and is thus independant of the mesh size).*

Remarks. The coefficients in D are known a priori and do not depend on the specific second order elliptic operator considered, and κ depends only on the constants appearing in the Sobolev equivalences and on the matrix A and the function b.

The regularity of the wavelet basis used has not been used up to now, but will be important to prove the sparcity of the matrix $D^{-1} M D^{-1}$.

From a *pseudo-differential* point of view, the method described amounts to writing the inverse of an elliptic operator of order 2 on a domain as the product of a pseudo-differential operator of order 0, which is bounded and invertible on L^2, and some kind of fractional integration diagonal in the wavelet basis, which consists in multiplying the wavelet $\psi_{j,k}$ by 2^{-2j}.

The wavelets that are bases of $H^1(\Omega)$ can be used to solve the same type of problems with Neumann boundary conditions (see [23]). One obtains here also a bounded condition number. The drawback is that the wavelets are not very smooth, so that the corresponding matrix that we have to invert will not be very sparce.

4.3.3. Sparcity of the matrices in the wavelet method

An advantage of the finite elements or finite difference methods is the sparcity of the matrices associated to the discretized problems.

The wavelets do not have a localization as sharp as, say, finite elements (in fact it is a key-point in the improvement of the conditioning, see [23]). Hence, there are more entries in the matrix G to store and more computations (for a given step in an inversion procedure). Because this seems to be an important drawback, we shall study it in details and give precise decay estimates for the entries of G.

The decay estimates will depend on the smoothness of the coefficients of the matrix A. We recall that the wavelets are indexed by $j \geq 1$ and $k \in \mathbb{Z}^n$, and are such that $k2^{-j} \in \Omega$; in the following, we note $\lambda = k2^{-j}$ and $\lambda' = k2^{-j'}$. Then

Proposition 25. *Suppose that A and b belong to $\mathbf{W}^{\mathbf{m}}_{\infty}$, then, there exist C and γ positive such that*

$$\mid G_{(j,k),(j',k')} \mid \leq C 2^{-|j-j'|(m+n/2)} \exp(-\gamma 2^{\inf(j,j')} \mid \lambda - \lambda' \mid).$$

This estimate shows, for a given accuracy, which entries of the matrix G do not need to be calculated. In this proposition, the wavelets are supposed to be at least C^m. Recall that W^m_∞ is the subspace of L^∞ composed of functions which have their derivatives of order at most m in L^∞.

Proof: We have to estimate

$$M_{(j,k),(j',k')} = \int A \nabla \psi_{j,k} \nabla \psi_{j',k'}$$

(the estimate for $\int b\psi_{j,k}\psi_{j',k'}$ is similar). We can suppose that $j \geq j'$. By Taylor's formula, there exist α_k's such that

$$\mid A \nabla \psi_{j',k'} - \sum_{k<m} \alpha_k (x-\lambda')^k \mid \leq C \mid x - \lambda \mid^m \sup_{B(\lambda, |x-\lambda|)} \mid \partial^m (A \nabla \psi_{j',k'}) \mid .$$

For the wavelets used in the Dirichlet problem, the moments of $\psi_{j,k}$ of order at most m do not exactly vanish. However, following a trick already mentioned in Subsection 4.2, we can suppose that they do when computing $M_{(j,k),(j',k')}$, because we can extend $\psi_{j,k}$ outside Ω so that its moments vanish but without changing the uniform estimates. This does not change $M_{(j,k),(j',k')}$, since $\psi_{j',k'}$ vanishes outside Ω. Hence

$$\mid M_{(j,k),(j',k')} \mid \leq C 2^{(n/2+1)j} 2^{(n/2+m+1)j'} \times$$

$$\int_\Omega \mid x - \lambda \mid^m \exp(-\gamma 2^j \mid x - \lambda \mid) \inf(1, \exp(-\gamma 2^{j'} (\mid \lambda - \lambda' \mid - \mid \lambda - x \mid))).$$

If $x \in B(\lambda, \mid \lambda - \lambda' \mid /2)$,

$$| M_{(j,k),(j',k')} | \leq C2^{(m+n/2)(j'-j)}2^{j+j'} \times$$

$$\exp(-(\gamma/2)2^{j'} \mid \lambda - \lambda' \mid) \int_\Omega 2^{nj}(2^j \mid x - \lambda \mid)^m \exp(-\gamma 2^j \mid x - \lambda \mid)$$

$$\leq C2^{j+j'}2^{-(m+n/2)|j-j'|} \exp(-\gamma' 2^{j'} \mid \lambda - \lambda' \mid).$$

If $x \notin B(\lambda, \mid \lambda - \lambda' \mid /2)$, $\mid M_{(j,k),(j',k')} \mid \leq$

$$\leq C \int_\Omega \mid x - \lambda \mid^m 2^{(\frac{n}{2}+1)j} \exp(-\frac{\gamma}{2}2^j(\mid x - \lambda \mid + \mid \lambda - \lambda' \mid)2^{(\frac{n}{2}+m+1)j'})$$

$$\leq C2^{j+j'}2^{-(m+n/2)|j-j'|} \exp(-\gamma' 2^j \mid \lambda - \lambda' \mid).$$

Hence Proposition 25, since

$$G_{(j,k),(j',k')} = 2^{-j}2^{-j'} M_{(j,k),(j',k')}.$$

We remark that, if V_p is a space of dimension N (hence, if G is a $N \times N$ matrix), then for a given accuracy ϵ, there are no more than $CN \mid \log \epsilon \mid$ entries larger than ϵ; which is the same order of magnitude as the G with entries decreasing exponentionally fast away from the diagonal. In fact these entries can be calculated using fast wavelet algorithms, starting with the entries $\langle A\phi_{j,k}(x) \mid \phi_{j,l}(x)\rangle$, and adapting the cascade algorithms introduced by Mallat in [38].

Actually, using smooth wavelets improves the sparcity of the matrix G (as was just proved), but it does not improve the order of convergence of the method. This is so because the derivatives of smooth wavelets vanish on $\partial\Omega$, which is, in general, not the case of the solutions of (A). Hence, the approximation cannot hold for a topology stronger than H_0^1 (at least near the boundary).

Here we see again the necessity to find wavelets which would be unconditional bases of $H_0^1 \cap H^s$. For such wavelets, the order of approximation of the method would be optimal up to the boundary.

4.3.4. Remarks

The main purpose of the first numerical experiments is to check if, numerically, the condition number of the matrix to be inverted is effectively bounded and if this bound is reasonable.

We have so far emphasized the very good condition number that can be obtained using wavelets. This is important for two reasons. First because ill-conditioning leads to numerical instabilities (a small error in the entries of the matrix or in the second term can lead to large errors in the solution).

The second reason is that, when using an iterative method to solve a linear system, ill-conditioning can make the convergence become very slow. For example, in the conjugate gradient method, the number of iterations needed for a given accuracy is proportionnal to $\sqrt{\kappa}$ (see [49]). This actually implies that, if one uses a conjugate gradient method after discretization in a wavelet basis, the number of iterations needed is bounded by a constant which depends neither on the dimension of space nor on the fineness of the discretization, in sharp contrast with other methods. As was already mentioned, the usual preconditionings (which are usually not diagonal) yield a condition number which is $O(1/h)$.

We present results for the second derivative in the 1-Dimensional case. Our example is piecewise linear wavelets and Dirichlet boundary conditions. In the first column, we give the number of points of discretisation, in the second the condition number of the matrix after preconditioning, in the two last columns, we check the sparceness of the matrix, giving the percentage of coefficients smaller than 10^{-2} and 10^{-3}.

	Cond.	$< 10^{-2}$	$< 10^{-3}$
8	4.12	0	0
16	5.73	0.27	0.1
32	6.56	0.53	0.31
64	7.25	0.73	0.54
128	7.70	0.84	0.72
256	7.97	0.92	0.84

These results show that the condition number increases very slowly when N is large. The percentage of coefficients that are not numerically vanishing is quite high, and an important open question now is to find out wavelets for which the differential operators would be *as diagonal as possible.*

We shall now go a little further in the comparison between classical and wavelet methods.

Complexity of calculations:

A drawback of the wavelet method is the following. The wavelets are not as easy to calculate as finite elements, and thus the entries of the matrix G are also more difficult to calculate. However, for the Dirichlet problem, it should be noted that the asymptotic behavior of the wavelets makes this computation not as bad as it might appear at first sight; and also that computing the wavelets for a given domain is done once and for all. Hence, the method may be costly for the resolution of one problem, but, if we keep in mind that the numerical resolution of many problems uses an iterative procedure involving the resolution of a large number of linear elliptic problems *on the same domain*, the method will in such cases become numerically efficient.

In contrast with other methods, a priori knowledge on the smoothness of the righthand side of the equation indicates which wavelet coefficients we actually have to calculate, and which can be disregarded for a given accuracy (as was noticed in Section 4).

Generalizations:

Another drawback of the method is that the functions u and v that appear in the variational formulation need to be in the same space. This prevents the direct use of this method for problems which do not have such a variational formulation. In some cases this drawback can be overcome. It was for instance the case in the nonhomogeneous Dirichlet problem (A), the direct variational formulation of which yields u and v that are not in the same space. The transformation of this problem into an homogeneous one overcomes this difficulty.

Another way to overcome the boundary problem would be to use wavelets in conjunction with penalization methods, because, in these methods, the boundary conditions do not appear in the definition of the space where the variational problem takes place.

It should be noticed that the only property of the differential operator that we use is the equivalence between the norm associated with this operator and the H^1 or the H_0^1 norm. Thus the wavelet method that we described can be used without any change for more general second order elliptic operators. This is especially the case for the Neumann boundary problem. In that case, we have to use wavelets that are bases of $H^1(\Omega)$. We already mentioned that there exist no such constructions yielding smooth and numerically simple wavelets except for special geometries (such as cubes). In other problems where this condition does not appear, one has to use the whole collection of wavelets, including the first constant one, and the technique is the same as in our model problem.

In order to solve higher order problems, one requires smooth wavelet bases. Those constructed in Subsection 2.2.5 can only be used if the variational problem is set in H_0^m. An example of this setting is given by the following plane clamped plate problem

$$\begin{cases} -\Delta^2 u = f \\ u(x) = \dfrac{\partial u}{\partial n}(x) = 0 \quad \text{for } x \in \partial\Omega. \end{cases}$$

A variational formulation is

$$\forall\, v \in H_0^2, \quad \int \Delta u \Delta v = \int fv.$$

The method described above can be applied without any change, since, provided that the wavelets are smooth enough, we have

$$C \sum \mid 2^{2j} c_{j,k} \mid^2 \leq \parallel f \parallel_{H_0^2}^2 \leq C' \sum \mid 2^{2j} c_{j,k} \mid^2 .$$

The Galerkin method described above if using as preconditioning the diagonal matrix

$$D_{(j,k),(j',k')} = 2^{2j} \delta_{(j,j')} \delta_{(k,k')},$$

will provide a discretization matrix, the condition number of which is bounded by a constant.

4.3.5. Recent extensions

We have seen in Subsection 4.2 that the decomposition of an operator A as $D\bar{A}D$ with D diagonal in a wavelet basis and $\bar{A}$ continuous and invertible in $\mathcal{M}^\gamma$ gives very precise informations on this operator, such as the spaces on which it is continuous and invertible, or estimates on its Green function and its derivatives. We also saw in Subsection 4.3 that it can yield numerical methods of resolution. Such a decomposition clearly holds for elliptic differential or pseudo-differential operators, but it is largely an open problem to understand more generally which operators can be decomposed in such a way; however, let us now describe an example of a singular elliptic operator for which such a decomposition holds (see [24]).

Let A be the operator

$$A = -\nabla(a\nabla) + Id \tag{21}$$

on $\mathbb{R}^n$, where a is a positive function which can vanish, which is uniformly C^k for k larger than 2, and which satisfies the following condition

$$|a(x) - a(y)| \leq c[a(x)^{\alpha+1/2}|x-y| + c|x-y|^{\alpha+2}]$$

which essentially means that a has zeros of order strictly larger than 2 at the points where it vanishes.

Theorem 26. *Under the preceding hypotheses, for $\gamma < k-2$, we have $A = D\bar{A}D$ where D is diagonal in a wavelet basis, $\bar{A} \in Op(\mathcal{M}^\gamma)$, and $\bar{A}^{-1} \in Op(\mathcal{M}^\gamma)$.*

Theorem 26 is obtained by remarking that for high j's the operator A acts on wavelets almost as if it had constant coefficients (that is we can replace the function $a(x)$ by $a(\lambda)$). Thus we *almost* invert A by simply inverting this new operator with *locally constant* coefficients.

The following corollary is an easy consequence of Theorem 26.

Corollary 27. *The Green function of A satisfies the following estimates if $\alpha + \beta < k - 2$,*

$$|\partial_x^\alpha \partial_y^\beta G(x,y)| \leq \frac{C}{|x-y|^{n-2+|\alpha|+|\beta|} \sup(\sqrt{a(x)a(y)}, |x-y|^2)}$$

except if $n = 2$, $\alpha = 0$, $\beta = 0$, in which case

$$|G(x,y)| \leq C \inf(|\log|x-y||, \frac{1}{|x-y|^2\sqrt{a(x)a(y)}}).$$

4.4. A comparison with multigrid algorithms

We shall now draw a comparison between wavelet and multigrid methods. They are closely related since both use the same ingredients: *hierarchical bases* and *smoothing resolution algorithms* . The difference lies in the way of using them. The multigrid methods mix decomposition on hierarchical bases and resolution, while for wavelets, these two parts are performed once for all successively. As a case study, we shall consider again the Poisson problem with Dirichlet boundary conditions.

4.4.1. Hierarchical bases and inversion algorithms

Both methods use a decomposition on hierarchical bases adapted to the domain Ω we consider. Suppose that we are given an increasing sequence $(V_n)_{n\geq 0}$ of subspaces of $H_0^1(\Omega)$, each space V_n having a basis (e_n^j). The (e_n^j) will be finite elements or spline functions associated to an increasing sequence of finer and finer meshes. Remark that the geometry of the triangulations must satisfy some compatibility and non-degenerescence properties that we will not detail here. Let h_n be the *average size* of the triangles or of the cubes used in the construction of V_n; h_n gives the order of magnitude of the fineness of discretization (for wavelets, we shall have $h_n = 2^{-n}$).

Consider the Poisson problem

$$\begin{cases} -\Delta u = f \quad \text{on } \Omega \\ u = 0 \quad \text{on } \partial\Omega \end{cases} \tag{22}$$

discretized on the largest V_N by a Galerkin method. The discrete problem, in variational form is

$$\begin{cases} \text{find } u \in V_N \ \text{ such that} \\ \langle \nabla u \mid \nabla v \rangle = \langle f \mid v \rangle \ \forall\ v \in V_N. \end{cases} \tag{23}$$

Writing this problem on the basis (e_N^j), we obtain

$$A_N X = Y \tag{24}$$

where A_N is the matrix of the scalar product $\int \nabla f \nabla g$ in the basis (e_N^j); that is

$$A_N(i,j) = \int \nabla e_N^i(x) \nabla e_N^j(x)\, dx.$$

We shall now consider the iterative methods of resolution, more precisely the so-called smoothing methods. Let us first justify this term.

This category contains the Jacobi, Gauss-Seidel, gradient, or conjugate gradient methods, which all have the property of converging very quickly on the*high frequency* part of the solution (that is for the components of the solution on the eigenfunctions associated to the highest eigenvalues) and much more slowly on the *low frequency* components. After a few iterations, the approximate solution X_{app} will have a high frequency component very close to the one of the exact solution, and the remaining problem is $A\tilde{X} = Y - AX_{app}$ (if $\tilde{X} = X - X_{app}$), where $\tilde{X}$ will have only low frequency components. Thus, the problem has been *smoothed*. The algorithms mentioned above converging slowly, their use without modification is not efficient. The idea of multigrid and wavelet methods is to use the smoothing property, that is the quick convergence on the high frequency part in order to improve this speed and, in both cases, obtain fast resolution methods.

4.4.2. Description of the multigrid methods

The multigrid method consists in the following algorithm that we shall write **A(N)**.

1. One performs ν iterations of the smoothing method, thus obtaining an approximation X_{app} of the discretized problem (22).
   The remaining problem is $A_N\tilde{X} = Y - AX_{app}$. We saw that the right handside is smooth. One can thus, without an appreciable loss of precision, project it on V_{N-1}, and calculate $P_{N-1}(Y - AX_{app})$ (where P_N is the projection on V_N).
   One *solves* then (we shall be precise what we mean by that) the problem projected on the rough grid: $A_{N-1}\bar{X} = P_{N-1}(Y_A X_{app})$.
2. One goes back to the finer grid, that is $\bar{X}$ is the vector of coordinates of a function in V_{N-1}, and one calculates its coordinates X' on the basis of V_N. X' is an approximation of $\tilde{X}$ appearing at Step 2.
3. The approximate solution is $X_{app} + X'$.

If Step 3 is an exact resolution (by any method), one obtains the two grids algorithm. Remark that **A(N)** is only an approximate solution of the discretized problem, even if Step 3 is an exact resolution. This is so because $\bar{X}$ is solution of

$$A_{N-1}\bar{X} = P_{N-1}(Y - AX_{app}),$$

which does not mean that X' is solution of

$$A_N X' = Y - AX_{app}.$$

In order to analyse this point more precisely, we can write V_N as a sum of two orthogonal subspaces V_{N-1} and W_{N-1}. The matrix of the discretized Laplacian can be written (supposing that we have a basis of W_{N-1})

$$\begin{matrix} & V_{N-1} & W_{N-1} \\ V_{N-1} & \left(\begin{matrix} A_{N-1} \\ C_{N-1} \end{matrix}\right. & \left.\begin{matrix} B_{N-1} \\ D_{N-1} \end{matrix}\right) \\ W_{N-1} & & \end{matrix}.$$

We have

$$A_N X' = A_{N-1}X' + C_{N-1}X'.$$

The method would be exact if $C_{N-1}X'$ vanished. The resolution is approximative because we disregard this correlation term between V_{N-1} and W_{N-1}. Since the algorithm is approximative, it is not necessary to perform an exact resolution at Step 3. One could for instance use the approximate resolution supplied by the algorithm **A(N-1)**; **A(N)** is thus defined in a recursive way; and one obtains the multigrid scheme *in V*. One could also wish a more precise resolution at Step 3 by iterating two times **A(N-1)**; one obtains for **A(N)** the algorithm called *in W*. The solution of the initial problem is obtained through a certain number of iterations of **A(N)**.

Let us come back to the error commited at Step 4 in order to see how to correct it efficiently. This correction will precisely lead us to the wavelet resolution. Let us consider again the matrix of the Laplacian above.

The multigrid method corresponds to a sequence of partial resolutions associated to this matrix and its sub-matrices. More precisely, the spaces W_i do not appear explicitly in the resolution algorithm; nonetheless, we can say that the resolution is performed accurately and precisely on these spaces at each step since the improvement on lower frequencies is very small.

The wavelet method takes into account the fact that the above matricial form is already a frequential decomposition of the Laplacian, each space W_i *occupying* on the matrix the place which it occupies effectively in the frequency domain.

Thus, the wavelet method does not calculate iteratively the component of the solution on each W_i. Such a method , as we saw, has problems rising from the interactions between scales. Here, we *move* the frequencies to group them in one zone of rather small width, which allows then to use fully the power of the chosen iterative method. For that, we first choose a basis (f_n^j) of W_j and we write the matrix of the Laplacian discretized on the $(2^{-n} f_n^j)$. This matrix is

$$M_{(n,j),(n',j')} = 2^{-n} 2^{-n'} \langle \nabla f_n^j \mid \nabla f_{n'}^{j'} \rangle.$$

(The renormalization of the elements of the basis can be interpreted as a diagonal preconditioning of the initial matrix). Because of the dyadic localization of the spaces W_j in frequency, this transformation of the Laplacian amounts to grouping its spectrum between two frequencies C_1 and C_2 which do not depend on the number of W_n we take; namely, the condition number of the method is independant of the size of discretization. The problem is then solved by applying a given number of times the chosen iterative resolution procedure.

The wavelet method we presented is thus closely related to the multigrid method. Nonetheless, the wavelet method uses more information (we need to know a basis of W_n which is not requested in multigrid methods); but the iterative method is then more efficiently used. There is also a certain conceptual difference. Multigrid methods mix the decomposition on hierarchical bases and the resolution steps, while these two parts are performed successively in the multigrid method. In both cases, the resolution is performed in $O(\mathrm{Dim}(V_N))$ operations.

§5. Resolution of evolution equations

5.1. Introduction

In this section, we will discuss numerical methods using wavelets for the resolution of evolution partial differential equations, that is equations depending on time. This field remains fairly open today. This may be explained by the recent appearance of the theory of wavelets but also by the problems occurring, even in the time-independent case, when constructing an orthonormal wavelet basis on any open set of $\mathbb{R}^n$. As we have said in the previous section, numerical algorithms to compute the solutions of $\Delta u = f$ over an open set of $\mathbb{R}^n$ using wavelets are currently developed [23]. In comparison with the methods presented above, algorithms for the resolution of partial differential equations need to take account of two more problems:

- the discretization in time
- the stability of the method.

To our knowledge, people have started to work on evolution partial differential equations in which occurs a single first order time derivative. The time behavior is then treated by classical finite difference schemes such as Crank-Nicholson or Adams-Bashforth ones; however, a particular time scheme can be found in the work of Bacry et al. [3]. But today, several works have already been achieved on the subject, particularly for evolution equations in space dimension 1 with periodic boundary conditions ([3],[20],[31],[32],[36],[37], and [47]) or with Dirichlet boundary conditions ([20], and [37]). Evolution equations in space dimension 2 are currently studied.

We will now try to explain why wavelets can be useful to this field of mathematics. Roughly speaking, the continuous wavelet transform can be seen as a mathematical microscope which can be moved over $\mathbb{R}^n$ with different resolutions. Thus it seems to be the good tool to study phenomenons which occurs all over an open set of $\mathbb{R}^n$ at different scales. It was first used to analyze complex signals and it brought more tractable informations on the signal than the Fourier transform and its derivatives. Some evolution partial differential equations, particularly in fluid mechanics, have also solutions with complex structures. So it seems to be a good idea to try to solve these equations with the wavelet transform. It was both the discovery of multiresolution analysis and the construction of orthonormal wavelet bases which really made it possible. Such equations are the équations of fluid mechanics, either with or without turbulence models. Their solutions are hard to compute because of the large number of unknown quantities to determine and the complexity of the phenomenons occurring during the evolution in time. In some cases, even well-understood numerical methods like finite elements, finite volumes, or finite differences fail, both because of power shortage of today computers (which lead to coarse meshes and poor precision) and then of problems of modelization. It is hoped that numerical algorithms using wavelets will permit a reduction of the number of degrees of freedom to be determined and a better computation of the complex phenomenons.

Then, in order to study the efficiency of wavelet numerical methods, a test problem needs to be chosen; people chose the viscous Burgers' equation in space dimension 1 for two main reasons: first, this equation has been intensely studied from a numerical point of view ([5]) which is very useful to make comparison with other methods. Then, during the time evolution, the solution presents a localized sharp gradients region and the ability of computing accuratly this phenomenon measures the performance of a numerical method. Moreover, this equation can be considered as an one space dimension Navier-Stokes-like equation, a more difficult problem to be faced later. Then, the choice of periodic boundary conditions was made for simplicity as the main goal of these studies was to prove the efficiency of numerical algorithms using wavelets. Thus the test problem writes

$$\begin{aligned} \frac{\partial u}{\partial t} + u\frac{\partial u}{\partial x} &= \nu\frac{\partial^2 u}{\partial x^2} \\ u(x+1,t) &= u(x,t) \\ u(x,0) &= u_0(x) \end{aligned}$$

where ν is a small positive real number.

Some references on computing solutions of Burgers' equation using wavelets are the papers of Maday et al., of Liandrat et al., of Glowinski et al., and of Latto & Tenenbaum.

5.2. Numerical algorithms and wavelets

We now try to explain what people expect from an efficient algorithm to solve partial differential equations and then, we will try to show what answers may be brought by wavelets.

What is first expected from a numerical algorithm is the quality of the approximation, that is the set $\mathcal{E}$ of approximation to which belongs the computed solution must be close to the exact solution u. That is

$$\inf_{v\in\mathcal{E}} \; d(u,v) << 1.$$

And $\mathcal{E}$ needs also to be small enough to allow the computation of the numerical solution.

Then, of course, the algorithm needs to be fast.

Finally, in order to reduce the time of computation, the algorithm needs to select the minimal set of approximation at each time step so that the computed solution remains close to the exact one. This last point is called adaptivity and ensures that no unnecessary quantity will be computed. But, to achieve this point, we need to find criterions that automatically choose at best this minimal set from quantities given by the computed solution at the previous time. Moreover, these criterions must be easy to implement.

Now, let us give a few reasons which allow us to think that numerical algorithms using wavelets may supply attractive solutions to the three requests stated above.

The first point and especially the order of approximation for adaptative methods has been studied precisely in Section 3. Coming back to the choice of the generating wavelet, we just recall that the quality of the approximation is given by the regularity of the wavelet. We must also keep in mind that, in the case of splines or Daubechies' wavelets, as the degree of the splines or the order of Daubechies wavelets grows, regularity and cancellation properties are improved while the (numerical) support of the wavelet becomes wider and wider. Hence, there is a tradeoff between the order of approximation and the numerical cost. It is not clear how to determine an *optimal wavelet* under this respect.

Then, in the periodic case or with compactly supported wavelets, fast numerical algorithms using the properties of multiresolution analysis are available for the fast computation of the wavelet coefficients.

Finally, some important properties of wavelets provide the ideas to build adaptative algorithms. Thus, if the solution of the partial differential equation we wish to compute is smooth in some regions, only a few wavelet coefficients will be needed to get a good approximation of the solution in those regions (see Section 3 above). Practically, only the wavelet coefficients of the low-frequencies wavelets whose support are in these regions will be necessary. On the other hand, the greatest coefficients (in absolute value) will be localized near the singularities; this allows to define and implement easily criterions of adaptivity through time evolution. We will give more details about adaptivity in the next section.

At last, it must not be forgotten that these criterions depend also on the properties of the linear and nonlinear operators in the equation. So good criterions of adaptivity are not straightforward and must rely on theoretical properties of the equation.

5.3. Numerical results

We are now going to present a few numerical algorithms using wavelets that have been successfully implemented. We will first turn towards the works of Maday et al. [37] and of Liandrat et al. [36].

To our knowledge, all the wavelets currently used for computing numerical solutions of evolution partial differential equations come from a multiresolution analysis (except in [6] where Gaussian derivatives are used). Thus, a multiresolution analysis is a set of finite dimension subspaces (V_j) of L^2 which satisfies

(i) $V_j \subset V_{j+1}$

(ii) Let P_j be the orthogonal projection on V_j and $u \in L^2$, then

$$\lim_{j \to +\infty} \|P_j u - u\|_{L^2} = 0.$$

These two properties fit well to the frame of Galerkin methods.

Let us first recall a few things about these methods. We consider an evolution partial differential equation which we write

$$\frac{\partial u}{\partial t} + Au = 0 \tag{25}$$

with periodic boundary conditions

$$u(x+1, t) = u(x, t) \tag{26}$$

and initial data

$$u(x, 0) = u_0(x). \tag{27}$$

The operator A is a linear or nonlinear operator of space variables (for example, for the Burgers' equation in space dimension 1, $Au = uu_x - \nu u_{xx}$ where ν is a small positive real number).

Then, let V be an Hilbert space with good properties related to A. We say that u is a weak solution of (25)-(26)-(27) in V if it satisfies $(\mathcal{P})$.

$$(\mathcal{P}) \left\{ \begin{array}{r} \text{Find } u \in V \text{ so that } \forall\, v \in V, (u_t|v)_V + (Au|v)_V = 0 \\ (u(.,0) - u_0|v)_V = 0 \end{array} \right.$$

where $(.|.)_V$ denotes the scalar product in V. Provided Equations (25), (26), and (27) has a single solution in V, the solution of $(\mathcal{P})$ satisfies Equations (25), (26), and (27) in the sense of distributions. We are then given a set of finite dimension subspaces (V_h) of V such that

(i) $h < h' \Rightarrow V_h \subset V_{h'}$

(ii) Let P_h be the orthogonal projection on V_h and v be in V, then

$$\lim_h \|P_h u - u\|_V = 0$$

and we consider the approximate problem

$$(\mathcal{P}_h) \left\{ \begin{array}{r} \text{Find } u_h \in V_h \text{ so that } \forall\, v_h \in V_h, (u_{ht}|v_h)_{V_h} + (Au_h|v_h)_{V_h} = 0 \\ (u_h(.,0) - u_{h0}|v_h)_{V_h} = 0. \end{array} \right.$$

Under nice hypotheses on the equation and on V, we claim that u_h, solution of $(\mathcal{P}_h)$, converges to u, solution of $(\mathcal{P})$ in a certain sense. Taking a basis of V_h, $(\mathcal{P}_h)$ becomes a linear or nonlinear system of a finite number of equations. The solution of such a problem may be then computed by usual techniques.

Now let us give a few details on the algorithms described in [37] and in [36] and of their adaptative procedures.

Both of them are Galerkin methods. We take for the set (V_h) the subspaces (V_j) of the multiresolution analysis defined by the splines functions of degree m. Then, let J be a positive integer and we write the approximate problem $(\mathcal{P}_J)$

$$(\mathcal{P}_J)\begin{cases}\text{Find } u_J \in V_J \text{ so that } \forall\ v \in V_J, (u_{Jt}|v) + (Au_J|v) = 0\\ \qquad\qquad\qquad\qquad (u_J(x,0) - P_J u_0|v) = 0.\end{cases}$$

Then, we have to choose a basis of V_J and we take the family

$$\{\psi_{j,k}, 0 \le j \le J-1, 0 \le k \le 2^j - 1\} \cup \{\phi_{0,0}\}.$$

Still, a difficult part of the work remains to be done: the time discretization. Actually, in both algorithms, the linear terms are treated with an implicit finite difference scheme and the nonlinear ones with an explicit or semi-implicit finite difference scheme so that the resulting system of equations is linear. Let us sketch the algorithms presented in [36] and [37]. The equation studied in these papers is the Burgers' equation with the following periodic boundary conditions:

$$u_t + uu_x - \nu u_{xx} = 0 \tag{28}$$

$$u(x+1,t) = u(x,t) \tag{29}$$

$$u(x,0) = u_0(x) \tag{30}$$

where ν is a small positive real number. Let Δt be the time step and $U^n(x) = u(x, n\Delta t)$ where u is the solution of Equations (28), (29), and (30). We may write the following time discretization

$$\frac{U^{n+1} - U^n}{\Delta t} + U^n \frac{\partial U^n}{\partial x} - \nu \frac{\partial^2 U^{n+1}}{\partial x^2} = 0$$

which can also be written

$$(Id - \nu \Delta t \frac{\partial^2}{\partial x^2}) U^{n+1} = U^n - \Delta t U^n \frac{\partial U^n}{\partial x}. \tag{31}$$

In [36], a family of functions $\theta_{j,k}, 0 \le j \le J-1, 0 \le k \le 2^j - 1$ defined by

$$(Id - \nu \Delta t \frac{\partial^2}{\partial x^2}) \theta_{j,k} = \psi_{j,k}$$

is then introduced and the wavelet coefficients of U^{n+1} are computed by

$$(U^{n+1}|\psi_{j,k}) = (U^n|\theta_{j,k}) + \frac{\Delta t}{2}(U^{n2}|\frac{\partial \theta_{j,k}}{\partial x}).$$

The knowledge of the wavelet coefficients of U^n allows the computation of the right term of the last equality and gives the expected result. In [37], we take the scalar product of Equation (31) with any wavelet $\psi_{j,k}, 0 \leq j \leq J-1, 0 \leq k \leq 2^j - 1$ which leads to the following linear system

$$(I - \nu \Delta t M)(U^{n+1}|\psi_{j,k})_{j,k} = B^n$$

where M is the matrix of the operator $P_J \frac{\partial^2}{\partial x^2} P_J$ in the wavelet basis of V_J and B^n is computed from the wavelet coefficients of U^n.

Thus, the main difference between the two methods lies in the computation of the wavelet coefficients of U^{n+1}. In [37], they are the solution of a linear system which is to be solved by classical techniques. On the other hand, in [36], they are given explicitly by scalar products concerning U^n and a set of wavelet-like functions. In both cases, either the matrix M or the set of functions $(\theta_{j,k})$ are computed once at the beginning of the computation. Moreover, the nonlinear terms are actually computed by a pseudospectral method which is very expensive in time. To avoid this, a method to compute directly the wavelet coefficients of the nonlinear term should be developped. An idea to achieve it relies on estimates of the scalar products concerned using the localization properties of the wavelets and is currently developped. But such estimates may fail in the case of a more general nonlinear term.

We will now give a few information about adaptative procedures. We are still given an integer J and we choose ϵ, a small positive real number. We call C_J^n, the set of all integer pairs (j,k) such that $0 \leq j \leq J-1$ and $0 \leq k \leq 2^j - 1$. Then we define R_J^n by:

$$R_J^n = \{(j,k) \in C_J^n \text{ so that } |(u(n\Delta t,.)|\psi_{j,k})| \geq \epsilon\}$$

to which we add a set of other wavelet coefficients to anticipate the behavior of the solution. Finally, we compute $(u((n+1)\Delta t,.)|\psi_{j,k})$ for all integer pairs in R_J^n.

These algorithms have both been successfully tested on the Burgers' equation in space dimension 1 with periodic boundary conditions. In this case, starting from a smooth initial condition, it appears a localized sharp gradients region which can travel over the interval $[0,1]$. Elsewhere the solution is smooth. The results were compared to those quoted in [5] and the adaptative procedures prove their efficiency with the comparison of the number of the wavelet coefficients and the number of Fourier coefficients necessary to get an accurate approximation of the exact solution. For more details, the reader should look in [36] and [37].

We will then quote another result obtained with the algorithms presented in [36]. They have also been successfully tested on the Korteweg-De Vries equation ([32]). This equation includes the same nonlinear term as the Burgers' eqaution but a third order space derivative occurs instead of a second order space derivative; the problem studied writes

$$v_t + vv_x + \nu v_{xxx} = 0 \tag{32}$$

$$v(x+1,t) = v(x,t) \tag{33}$$

$$v(x,0) = v_0(x). \tag{34}$$

Let v be a solution of Equations (32), (33), and (34). Then it satisfies the following properties:

First, provided v_0 is smooth, v is smooth and

$$\forall\, t \in \mathbf{R}^+, \quad \int_0^1 |v(x,t)|^2 dx = \int_0^1 |v_0(x,t)|^2 dx. \tag{35}$$

Secondly, *solitary waves* phenomenons may occur during the time evolution. Using the algorithms presented in [36], we were able to observe such phenomenons. But the time scheme used reduces the L^2 norm so Equation (35) is not true during the computation. This prevents us from computing a good approximation of the exact solution for large times. We must also notice that, as the solutions of the Korteweg-De Vries equations are smooth for smooth initial data, the adaptative procedures are of less interest than for the Burgers' equation.

In the works of Glowinski et al. [20] and of Latto & Tenenbaum [31], the algorithm is also built on a Galerkin method but the basis taken for V_J is not the set $\{\psi_{j,k}, 0 \le j \le J-1, 0 \le k \le 2^j - 1\}$ but the family $\{\phi_{J,k}, 0 \le k \le 2^J - 1\}$ of scaling functions defined in [15]. Thus, no use is made of the cancellation properties of the wavelets, which is the keystone of the adaptative methods we mentionned, so this method seems to be closer to splines methods rather than to wavelet algorithms.

At last, a totally different method, called the travelling wavelets method, is proposed in [6]. It uses the continuous wavelet transform rather than the discrete one and the wavelets used are Gaussian derivatives. Thus, its spirit is close to particles' method.

5.4. Forthcoming prospects

The good results actually given by the numerical methods using wavelets bring hope for further developments. In particular, the use of adaptative procedures reduces the number of degrees of freedom and save computation time for the examples already studied. However, some points require improvements and there is still a great deal of work to be done.

First, the computation of the scalar products concerning the nonlinear terms must be improved, not only in the case of the Burgers' equation but for any usual nonlinear terms. We then turn to the two main improvements to be realized: the boundary conditions and the dimensions greater than one. As we have already said in Subsection 5.1, most examples actually studied had periodic boundary conditions. In order to allow a wider field of applications, we need to be able to compute solutions of partial differential equations over any open set of $\mathbb{R}^n$, either bounded or unbounded with any boundary conditions. Some preliminary tests, using a change of variables, have already been done [37]. But, as the change of variables is nonlinear, the resulting wavelets lose their cancellation properties and the fact that the wavelet coefficients are small where the solution is smooth is no longer true. Another idea lies in the domain decomposition methods, already used for spectral methods. But here comes back the problem of the construction of orthonormal wavelet basis over an open set of $\mathbb{R}^n$ which we already discussed in Section 2.

Another crucial step to be crossed is that of space dimension 2. In fact, in that case, two choices are to be made: first, as mentionned in Subsection 5.2, a family of wavelets has to be chosen, regarding their localization, regularity, and cancellation properties. Then, orthonormal bases of wavelets of $L^2(\mathbb{R}^2)$ may be built from orthonormal bases of wavelets of $L^2(\mathbb{R})$ by tensorial products. However, two constructions may occur: one may choose either tensor products of one dimensionnal wavelets or the standard two-dimensionnal wavelets. The *phase space* localization is better in the second one but the first one often gives better compression rates [37]. This problem is quite the same as the comparison between standard and nonstandard methods for operators and we refer to Section 4 for a more extensive discussion on this topic. In the case of partial differential equations in space dimension 2, one of these constructions has to be selected in order to optimize the adaptative procedure, but to our knowledge the question of which is the best remains open. In particular, it seems to depend very much on the family of wavelets used.

At last, we have seen that adaptative procedures are very efficient to get good approximations of solutions of equations where only a few localized singularities occur. But, in case of very turbulent signals where singularities take place everywhere and at each scale, they may be inefficient. So people look towards the recent development of the wavelet packets [11] or the Malvar's wavelets [41]. There are orthonormal bases of wavelet-like functions. For a given signal, one wavelet packet or one Malvar's wavelet is constructed, that fits perfectly to it. These methods actually give very good results in signal processing. Using them may lead to more efficient adaptative procedures. But, in the particular case of solving partial differential equations, the solution evolves through time, the signal changes its shape, and one can wonder if such methods can be used. Moreover, both wavelet packets and Malvar's wavelets have bad localization properties in the phase space which may prevent them from being as efficient in this field as they are in signal processing.

§ 6. Conclusion

Let us mention two recent directions of research around the topics of wavelets and P.D.E.'s. An intriguing problem was to know if there can exist wavelets orthonormal for other scalar products than L^2. Examples of interest are supplied by the scalar products induced by differential operators, that is $\int A(f)\bar{g}$, where A is a self-adjoint elliptic operator. A natural idea is to try to construct a multiresolution analysis for this norm. However, such a construction cannot lead to smooth, well localized wavelets [44]. Another idea is to start with an L^2 orthonormal basis of wavelets and to apply the Green function of the square root of A. We thus obtain an orthonormal basis for the right scalar product, and it remains to prove its localization. Such wavelets have been constructed in [7]. An extremely simple case is when the operator is homogeneous with constant coefficients. Suppose for instance that $A = \Delta$. Let

$$\hat{\omega}(\xi) = \frac{1}{\mid \xi \mid}\hat{\psi}(\xi).$$

The $2^{(\frac{n}{2}-1)j}\omega(2^j x - k)$ are an orthonormal basis for the norm $\int \mid \nabla f \mid^2 dx$.

Paul Federbush has also very recently used wavelets to obtain new estimates on the strong solutions of Navier-Stokes equations [19].

References

1. Adelson, E. H., E. Simoncelli, and R. Hingorani, Orthogonal pyramid transforms for image coding, *SPIE Vision. Comm. and Image Proc.* **845** (1987).
2. Argoul, F., A. Arnéodo, G. Grasseau, Y. Gagne, E. J. Hopfinger, and U. Frisch, Wavelet analysis of turbulence data reveals the multifractal nature of the Richardson cascade, *Nature* **338** (6210) (1989).
3. Bacry, E., S. Mallat, G. Papanicolaou, Time and space wavelet adaptative scheme for partial differential equations, Communication at ' Wavelets and Turbulence', Princeton, June 1991.
4. Bank, R. E. and L. R.Scott, On the conditioning of finite element equations with highly refined meshes, *SIAM J. Numer. Anal.* **26** (1989).
5. Basdevant, C., M. Deville, P. Haldenwang, J. M. Lacroix, J. Ouazzani, R. Peyret, P. Orlandi, and A. T. Patera, Spectral and finite difference solutions of the Burgers equation, *Computers & Fluids* **14** (1) (1986), 23–41.
6. Basdevant, C., M. Holschneider, V. Perrier, Méthode des ondelettes mobiles, *C.R. Acad. Sci. Paris*, Série I (1990), 647–652.
7. Benassi, A., S. Jaffard, and D. Roux, Bases d'ondelettes orthonormées pour un opérateur différentiel, Preprint, 1991.
8. Beylkin, G., R. R. Coifman, and V. Rokhlin, Fast wavelet transform, preprint,1989.
9. Bourgain, J., A remark on the uncertainty principle of Hilbertian basis, *J. of Funct. Anal.* **79** (1988), 136–143.

10. Calderón, A. P. and A. Zygmund, Singular integral operators and differential equations, *Amer. J. Math.* **79** (1957), 901–921.
11. Coifman, R. R. and Y. Meyer, Remarques sur l'analyse de Fourier á fenêtres, *C.R.A.S.* **312** (1991), 259–261.
12. Coifman, R. R., Y. Meyer, and V. Wickerhauser, Wavelet analysis and signal processing, preprint, 1991.
13. Cohen, A., Ondelettes, Analyses multirésolutions et traitement numerique du signal, Doctoral Thesis, Univ. Paris-Dauphine, 1990.
14. Daubechies, I., The wavelet transform, time-frequency localization and signal analysis, *IEEE Trans. Inform. Theory* **36** (1990), 961–1005.
15. Daubechies, I., Orthonormal bases of compactly supported wavelets, *Comm. Pure and Appl. Math.* **41** (1988), 909–996.
16. Daubechies, I., A. Grossmann, and Y. Meyer, Painless nonorthogonal expansions, *J. Math. Phys.* **27** (1986), 1271–1283.
17. Daubechies, I., S. Jaffard and J. L. Journé, A simple Wilson orthonormal basis with exponential decay, *SIAM J. Math. Anal.* **22** (2) (1991), 554–572.
18. Farge, M. and D. Rabreau, Transformée en ondelettes pour détecter et analyser les structures cohérentes dans les écoulements turbulents bidimensionnels, *C.R.A.S.* **307**, Série 2, (1988), 1479–1486.
19. Federbush, P., Navier and Stokes meet the wavelet, preprint, 1991.
20. Glowinski, R., W. Lawton, M. Ravachol, and E. Tenenbaum, Wavelet solution of linear and nonlinear elliptic, parabolic, and hyperbolic problems in one space dimension, in *Proc. Ninth International Conf. on Computing Methods in Appl. Sciences and Engineering*, R. Glowinski and A. Lichnewsky (eds.), SIAM Publ., Philadelphia, PA, 1990.
21. Holschneider, M. and Ph. Tchamitchian, Pointwise analysis of Riemann's nondifferentiable function, *Inventiones Mathematicae*, to appear.
22. Jaffard, S., Exposants de Holder en des points donnés et coefficients d'ondelettes, *C.R.A.S.* **308** Série 1 (1989), 79–81.
23. Jaffard, S., Wavelet methods for fast resolution of elliptic problems, *SIAM J. Numer. Anal.*, to appear.
24. Jaffard, S., Analyse par ondelettes d'un problème elliptique singulier, *Journal de Mathématiques Pures et Appliquées*, to appear.
25. Jaffard, S., Pointwise smoothness, two-microlocalization and wavelet coefficients, *Publicacions Matematiques* **35** (1991), 155–168.
26. Jaffard, S., Local order of approximation by wavelets, preprint, 1991.
27. Jaffard, S., Propriétés des matrices "bien localisées" près de leur diagonale et quelques applications, *Annales de l'IHP*, série Problèmes non linéaires **7** (5) (1990), 461–476.
28. Jaffard, S., Thèse de l'école Polytechnique, 1989.
29. Jaffard, S., Sur la dimension de Hausdorff des points singuliers d'une fonction, preprint, 1991.
30. Jaffard, S. and Y. Meyer, Bases d'ondelettes dans des ouverts de $\mathbb{R}^n$, *J. Math. Pures et Appl.* **68** (1989), 95–108.

31. Latto, A., and E. Tenenbaum, Compactly supported wavelets and the numerical solution of the Burgers' equation, *C.R. Acad. Sci. Paris* **t.311** Série I (1990), 903-909.
32. Laurençot, Ph., Résolution par ondelettes 1-D de l'équation de Burgers avec viscosité et de l'équation de Korteweg-De Vries avec conditions limites périodiques, Rapport de DEA, Université Paris XI, 1990.
33. Lemarié, P. G., Ondelettes à localisation exponentielle, *J. Math. Pures et Appl.* **67** (1988), 227–236.
34. Lemarié, P. G., Deux bases d'ondelettes sur un polygone de $\mathbb{R}^2$, Preprint Université Paris XI, 1990.
35. Lemarié, P. G. and Y. Meyer, Ondelettes et bases Hilbertiennes, *Rev. Math. Iberoamericana* **1** (1986).
36. Liandrat, J., V. Perrier, and Ph. Tchamitchian, Numerical resolution of the regularized Burgers equation using the wavelet transform, preprint, 1990.
37. Maday, Y., V. Perrier, and J. C. Ravel, Adaptativité dynamique sur bases d'ondelettes pour l'approximation d'équations aux dérivées partielles, *C.R.A.S.* (1991).
38. Mallat, S., Multiresolution approximations and wavelet orthonormal bases of $L^2(\mathbb{R})$, *Trans. Amer. Math. Soc.* (1989).
39. Mallat, S., A theory for Multiresolution signal decomposition: the wavelet representation, *IEEE Trans. Pattern Anal. and Machine Intell.* **11** (1989), 674-693.
40. Mallat, S. and W. L. Hwang, Singularity detection and processing with wavelets, preprint, 1991.
41. Malvar, H., Lapped transforms for efficient transform/subband coding, *IEEE on T.A.S.S.P.* **38** (6) (1990), 969–978.
42. Meyer, Y., *Ondelettes et Opérateurs*, in two volumes, Hermann, Paris, 1990.
43. Meyer, Y., Wavelets and applications, in *Proceedings of the International Congress of Mathematics*, Kyoto, 1990.
44. Meyer, Y., Private communication.
45. Meyer, Y., Ondelettes sur l'intervalle, *Rev. Math. Iberoamericana*, to appear.
46. Meyer, Y., Ondelettes et calcul scientifique performant, INRIA, 1989.
47. Perrier, V., Ondelettes et simulation numérique, Thèse de l'Université Paris VI, Février 1991.
48. Schweinler, H. C. and E. P. Wigner, Orthogonalization methods, *J. Math. Phys.* **11** (1970), 1693–1694.
49. Strang, G. and G. J. Fix, *An Analysis of the Finite Element Method*, Prentice hall.
50. Strömberg, J. O., A modified Franklin system and higher order spline systems on $\mathbb{R}^n$ as unconditionnal bases for Hardy spaces, *Conference in honor of Antoni Zygmund* **2**, W. Beckner, A. P. Calderón, R. Fefferman,

and P. W. Jones (eds.), Wadsworth, NY, 1981, 475–493. (eds.), Wadsworth Math. Series, 475–493.

51. Tchamitchian, Ph., Bases d'ondelettes et intégrales singulières: Analyse des fonctions et calcul sur les opérateurs, Habilitation, Université d'Aix-Marseille 2, 1989.

S. Jaffard
CERMA
Ecole Nationale des Ponts et Chaussics
La Courtine
93167 Noisy le Grand
France

Ph. Laurençot
Universite de Besagon
Centere de Mathématiques
16 route de Gray
25030 Besagon Cedex
France

Wavelet Transforms and Filter Banks

R. A. Gopinath and C. S. Burrus

Abstract. Wavelet and short-time Fourier analysis is introduced in the context of frequency decompositions. Wavelet type frequency decompositions are associated with filter banks, and using this fact, filter bank theory is used to construct multiplicity M wavelet frames and tight frames.

§1. Introduction

Wavelet theory deals with the study of the time-scale behavior of functions just as the Short-Time Fourier analysis deals with the time-frequency behavior of functions. Typically, in wavelet analysis a *generating* function, the *wavelet*, is chosen, and an associated transform gives a time-scale representation of functions (typically finite-energy) called the wavelet representation. In a wavelet representation, the one, or both, of the time and scale parameters may be discrete or continuous and, correspondingly, we have the Discrete Wavelet transform (DWT- both parameters are discrete), the Continuous Wavelet Transform (CWT- both parameters are continuous) the Discrete-Scale Wavelet Transform (DSWT - only the time parameter is continuous) and the Discrete-Time Wavelet transform (DTWT - only the scale parameter is continuous) ([19], [6], and [22]). In this chapter we discuss DWTs only. For example, the orthonormal wavelet bases constructed by Daubechies [5], gives rise to a discrete, time-scale representation of finite-energy signals. If the ON wavelet is compactly supported, then finite-power signals can also be localized in time and scale, thus facilitating the study of the time-scale behavior of periodic and non-stationary signals.

The enormous flexibility in the choice of the wavelet (reminiscent of the choice of the window function in Short-Time Fourier analysis), permits the use of optimal wavelets for specific applications. Since compactly supported wavelets are determined by a finite sequence of numbers, one can optimize over these to obtain best suited wavelets for specific applications. Even though, wavelet analysis can be introduced from a viewpoint of a multiscale analysis framework, we try to introduce it from a frequency decomposition framework

Wavelets–A Tutorial in Theory and Applications
C. K. Chui (ed.), pp. 603–654.

ISBN 0-12-174590-2

which includes the familiar STFT and the wavelet analysis as limiting cases, and also admits of generalizations that cannot be easily seen from a multiscale framework.

In general, wavelet analysis is associated with fast computational techniques. This is because of certain efficient computational structures called *filter banks* associated with them. Filter bank theory not only gives computational algorithms for wavelet analysis, but moreover, acts as a starting point for some of the interesting aspects of wavelet theory. Using the frequency decomposition framework we introduce filter banks and then construct multiplicity M wavelet frames and tight frames. We finally, describe how filter banks give a general decomposition of separable Hilbert spaces, and using these results to argue that, regularity of the wavelets may not be as critical as often presupposed (in practical applications).

§2. Frequency decompositions

The essential ideas in Wavelet and Short-Time-Fourier decompositions can be understood in the context of frequency decompositions. Frequency decompositions, overcome some of the shortcomings of Fourier analysis. The Fourier transform decomposes a signal into individual frequency components, but does not tell *when* the frequencies occurred. This is because the complex exponentials, the Fourier basis functions, are infinite in extent. Frequency decompositions generally decompose a signal into a countable number of frequency channels, so that temporal information can be obtained in each frequency channel.

For a function $f(t) \in L^2(\mathbb{R})$, its Fourier transform $\hat{f}(\omega)$ uniquely determines it. Since a frequency decomposition splits it into a number of frequency channels, we have, $\hat{f}_j(\omega)$, such that

$$\hat{f}(\omega) = \sum_j \hat{f}_j(\omega) \quad \text{and} \quad f(t) = \sum_j f_j(t). \tag{1}$$

If the channels are mutually orthogonal, then we call it an orthogonal frequency decomposition.

$$\int f_i(t) f_j^*(t)\, dt = \frac{1}{2\pi} \int \hat{f}_i(\omega) \hat{f}_j^*(t)\, d\omega = \delta_{i-j}. \tag{2}$$

Some orthogonal frequency decompositions can be constructed easily. For example, the frequency axis could split into *non-overlapping* bins and $\hat{f}_j(\omega)$ could be the restriction of $\hat{f}(\omega)$ to the jth bin. Since each of the channels then, becomes ideal "bandpass" channels, the decomposition is not realizable practically. Thus the problem is to come up with realizable frequency decompositions with useful properties. If some structure is introduced into the frequency decomposition, then it is possible to construct orthogonal and non-orthogonal frequency decompositions. This section will examine structured frequency decompositions such as the the Discrete Short Time Fourier Transform (DSTFT) type frequency decomposition and the Discrete Wavelet Transform (DWT) type frequency decomposition. Since the essential features of these frequency decompositions are captured in the non-overlapping frequency channels case, we examine this case only.

Discrete data are easily stored and manipulated; hence,it is desirable to have discrete representations of all signals in a frequency decomposition. Discrete representations could be obtained by sampling, or by expansion in bases. Given an orthogonal frequency decomposition, we can construct orthonormal (ON) bases for $L^2(\mathbb{R})$ that would give a discrete representation. For example, for fixed j, let $\{\psi_{j,k}|j,k \in \mathbb{Z}\}$, be an orthonormal family that spans the jth channel. This means that there exist numbers $a_{j,k}$ such that

$$f_j(t) = \sum_k a_{j,k}\psi_{j,k}(t) \quad \text{and} \quad \hat{f}_j(\omega) = \sum_k a_{j,k}\hat{\psi}_{j,k}(\omega). \tag{3}$$

Then $\{\psi_{j,k}|j,k \in \mathbb{Z}\}$ is an orthonormal basis for $L^2(\mathbb{R})$. It is not obvious that non-orthogonal frequency decompositions with discrete representations can be constructed. Shortly, while discussing DWT type frequency decompositions we will identify a certain *filter bank* structure, that provides the key to the construction of a rich class of computationally efficient frequency decompositions with discrete representations.

Notice that in Equations 1-3, both time and frequency domains have been considered. This is to highlight that frequency decompositions could also be thought of as time decompositions (decomposition of a signal into "time" channels). For example, one could split the time-axis into non-overlapping bins, and one would then have an orthogonal overlapping frequency decomposition and a non-overlapping time decomposition. In both cases, $L^2(\mathbb{R})$, a separable Hilbert space is being decomposed into a possibly infinite sequence of subspaces. This point will be further explored in Section 8.

2.1. The DSTFT decomposition

Short-Time Fourier analysis began essentially to study the non-stationary behavior of speech signals [11] and since then has been used successfully for the analysis of non-stationary signals ([13], [36], and [17]). The DSTFT decomposition is associated with a basis made up of translates and modulates of a single function called the *window* function. All translates of a fixed modulation of the window, span a specific channel. The idealized choice of frequency bins for the DSTFT type frequency decomposition is shown in Figure 1.

For this choice of frequency bins, we can construct an orthonormal basis for each channel, and thus an orthonormal basis for $L^2(\mathbb{R})$ as follows. For an arbitrary $f(t)$, let its Fourier transform restricted to the jth be represented as a Fourier series with coefficients $a_{j,k}$. Then,

$$\hat{f}_j(\omega) = \left[\sum_k a_{j,k}e^{i\omega k}\right] \sqcap (\frac{\omega - j2\pi}{2\pi}) \tag{4}$$

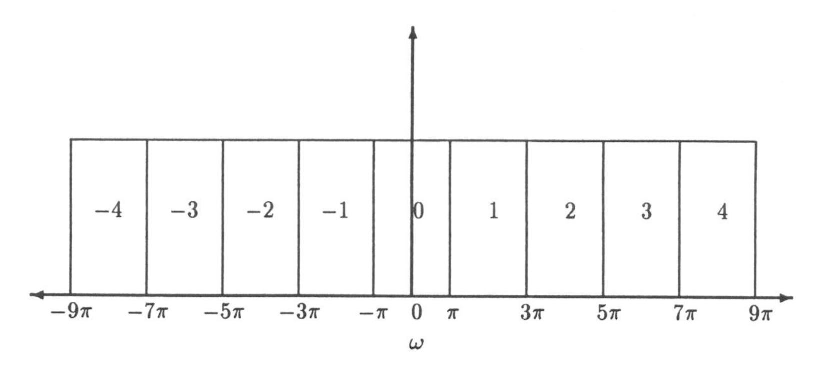

Figure 1. Frequency decomposition for short time Fourier transform.

where

$$\sqcap(\omega) = \begin{cases} 1 & \text{for } -\frac{1}{2} \le \omega \le \frac{1}{2} \\ 0 & \text{otherwise.} \end{cases} \tag{5}$$

Then,

$$\hat{f}(\omega) = \sum_{j,k} a_{j,k} e^{i\omega k} \sqcap \left(\frac{\omega - j2\pi}{2\pi} \right). \tag{6}$$

Taking the inverse Fourier transform on both sides

$$f(t) = \sum_{j,k} a_{j,k} w_{j,k}(t) \tag{7}$$

where

$$w(t) = \frac{\sin(\pi t)}{\pi t} = \operatorname{sinc}(t) \tag{8}$$

$$w_{j,k}(t) = w(t+k) e^{i2\pi jt} \tag{9}$$

Equation (7) is precisely the DSTFT expansion of $f(t)$ with respect to the window $w(t)$, sometimes called the *sinc* window. Then sequence $a_{j,k}$ is referred to as the DSTFT of $f(t)$ with respect to $w(t)$ ([15] and [17]). Since the Fourier basis is ON, by choice of non-overlapping bins the functions $w_{j,k}$ are ON. Thus the DSTFT can be computed by taking inner-products with the basis functions.

$$a_{j,k} = \langle f(t) w_{j,k} \rangle \tag{10}$$

In general orthogonal DSTFT type frequency decompositions can be viewed as generalizations of this decomposition where the frequency channels are non-overlapping, yet mutually orthogonal. The essential idea is the window function $w(t)$, such that modulates and translates $(w(t+k)e^{iw\pi jt})$ form a basis. For fixed j, $\{w_{j,k} | k \in \mathbb{Z}\}$ spans the jth frequency channel which is a frequency shift of the 0th channel. This is a distinguishing feature of the DSTFT. It is quite conceivable that one can have an orthogonal frequency decomposition where the bins are approximations to that in Figure 1, and yet the channels are *not* frequency shifts of each other. In other words, there is no window $w(t)$ such that $w_{j,k}$ spans $L^2(\mathbb{R})$. Clearly such a frequency decomposition is very similar to the STFT, in that the channels are of uniform effective bandwidth and hence useful. Later we will see how to construct many such frequency decompositions. Even though orthonormal DSTFT bases can be constructed ([6], [32], and [8]), there is no general technique to obtain such decompositions with windows that have both desirable localization properties and good numerical algorithms associated with them. Usually one starts with a desirable $w(t)$ ($w_{j,k}$ not necessarily orthonormal) (see [4] and [15]), and hopes that expansions like Equation (7) hold. Since the basis functions $w_{j,k}$ are no longer orthonormal, Equation (10) does not hold and hence cannot be used to compute the DSTFT coefficients$a_{j,k}$. However, for most choices of $w(t)$ there exists a function $\tilde{w}(t)$, the dual window, such that,

$$a_{j,k} = \langle f(t), w_{j,k} \rangle. \tag{11}$$

The set of functions $w_{j,k}$ and $\tilde{w}_{j,k}$ are referred to as a *biorthogonal* basis, which means no more than that a function can be projected onto one set and expanded in the other to recover it.

Incidentally, in the sinc window case, the coefficients, $a_{j,k}$ are seen to be the samples of jth frequency channel and hence there is no need to compute inner products.

$$a_{j,k} = 2^{-j/2} f_j(k). \tag{12}$$

Moreover, for general ON orthogonal or biorthogonal DSTFT type frequency decompositions, the numbers $a_{j,k}$ can be interpreted as generalized samples of the jth frequency channel (up to a constant).

In summary, the DSTFT arises from a frequency decomposition, where the frequency channels are of the same bandwith and are shifts of each other.

2.2. The DWT decomposition

The idealized choice of frequency bins for the DWT type frequency decomposition is shown in Figure 2.

An important feature of this decomposition is that if $f(t)$ is in the jth bin (that is $\hat{f}(\omega)$ is zero outside the jth bin), then $f(2t)$ is in the $j+1^{st}$ bin. Just as channels in the DSTFT case were related by modulates, here the channels are related by *scaling* by factors of 2, (called the multiplicity of the DWT frequency decomposition). Also notice that the bandwith of the jth bin is twice the bandwidth of the $(j-1)^{st}$ bin, and hence bandwidths are constant on a logarithmic scale. Why are such frequency decompositions interesting? The physiology of early mammalian vision indicates that the retinal image is decomposed spatially-oriented frequency channels each having the same bandwidth on a log scale ([31] and [30]). That is, the retinal system, splits the

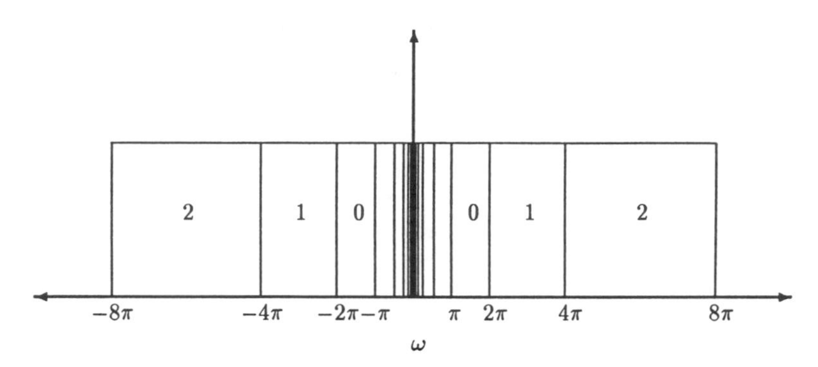

Figure 2. Frequency channels in a wavelet decomposition.

visual input into information at different scales. Furthermore, in many naturally occurring classes of signals like transients, it has been observed that the larger the center frequency of the signal, the smaller its duration, and hence greater its bandwidth. Such a frequency behavior has been found to be useful in the modeling of nearly $\frac{1}{f}$ noise processes [48]. And moreover, from a practical point of view, unlike the DSTFT type of frequency decomposition, this *wavelet* type frequency decomposition, allows of non-trivial and interesting examples, orthogonal and otherwise. The scaling structure is associated with efficient computational structures. Notice also that the channels are symmetric with respect to the origin, and this will make the DWT representation real for real signals.

As in the STFT case, the restrictions of $\hat{f}$ to each of the bins can be expanded as a Fourier series. But then, the coefficients $a_{j,k}$ obtained would in general be complex (for real $f(t)$) and it turns out that one can do better. Let $b_{j,k}$ be the Fourier series coefficients of $B_j(\omega)$, the periodization of $f_j(\omega)$ by period $2^j\pi$ (see Figure 3)

$$B_j(\omega) = \sum_k f_j(\omega + 2\pi 2^j k). \tag{13}$$

Notice that by the sampling theorem $b_{j,k} = f_j(2^{-j}k)$. Then,

$$f_j(\omega) = B_j(\omega)\left(\sqcap\left(\frac{\omega}{2^{j+1}\pi}\right) - \sqcap\left(\frac{\omega}{2^j\pi}\right)\right)$$

$$f_j(t) = \sum_k b_{j,k}(2^{j+1}\phi(2^{j+1}(t-k)) - 2^j\phi(2^j(t-k))) \tag{15}$$

$$f(t) = \sum_j f_j(t) = \sum_{j,k} a_{j,k}\psi_{j,k}(t) \tag{16}$$

where

$$\psi_{j,k}(t) = 2^{j/2}\psi(2^j t - k) \tag{17}$$

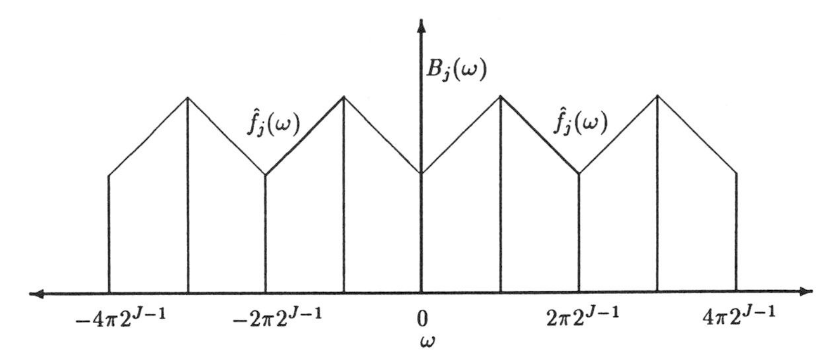

Figure 3. Periodization of $\hat{f}_j(\omega)$.

$$\psi(t) = 2\phi(2t) - \phi(t) \tag{18}$$

$$\phi(t) = \frac{\sin(\pi t)}{\pi t} \tag{19}$$

$$a_{j,k} = \sqrt{2^{-j}} b_{j,k} \tag{20}$$

Equation (16) is the *wavelet* expansion of $f(t)$ with respect to the *wavelet* ψ. The numbers $a_{j,k}$ are the DWT coefficients. Notice that Equation (16) is exactly in the form of Equation (7) and that the functions form an $\psi_{j,k}$ ON family. The essential difference is that now, the functions $w_{j,k}$ are translates and dilates of the wavelet w. From orthonormality, the DWT coefficients are given by

$$a_{j,k} = \langle f(t), \psi_{j,k} \rangle. \tag{21}$$

For this simple wavelet $a_{j,k} = f_j(2^{-j}k)$, and hence the numbers can also be interpreted as the "Nyquist rate" samples of each of the frequency channels. This interpretation is generic for all DWT type frequency decompositions.

So far we have seen that the STFT and the DWT decompositions are very similar in the non-overlapping frequency decomposition case. We now show that, the *scaling* property implies existence of discrete sequences associated with the frequency decomposition that contain all the information about the decomposition. Furthermore, this scaling structure also gives an efficient means to compute $a_{j,k}$.

Notice that the wavelet $\psi(t)$ is the difference of the function ϕ in Equation (19) and a scaled version of it. The function ϕ is bandlimited to $[0, \pi]$, and $\psi(t)$ is the difference of ϕ and a scaled version of ϕ. Let V_j denote the space of signals band-limited to $[0, 2^j\pi]$. Define

$$\phi_{j,k}(t) = 2^{j/2}\phi(2^j t - k). \tag{22}$$

Then for fixed j, $\{\phi_{j,k}\}$ is an orthonormal basis for V_j (the sinc basis). Also for fixed j, $\{\psi_{j,k}\}$ is an orthonormal basis for W_j the space of bandlimited functions in $[2^j\pi, 2^{j+1}\pi]$. We have already seen that $\psi_{j,k}$ forms an orthonormal basis for W_j. Now from Figure 2, it is clear that any function in V_{j+1} can be decomposed into a function in V_j and a function in W_j. Also any function in W_j or V_j is trivially contained in V_{j+1}. In particular, $\phi(t) \in V_0 \subset V_1$ and therefore there exists a sequence $h_0(k)$ such that

$$\phi(t) = \sum_k h_0(k)\phi_{1,k}(t) = \sqrt{2}\sum_k h_0(k)\phi(2t-k). \tag{23}$$

Similarly, $\psi(t) \in W_0 \subset V_1$ and therefore there exists a sequence $h_1(k)$ such that

$$\psi(t) = \sum_k h_1(k)\phi_{1,k}(t) = \sqrt{2}\sum_k h_1(k)\phi(2t-k). \tag{24}$$

Because the function $\phi(t)$ satisfies a *dilation* equation (Equation (23)), it is called the *scaling* function. The sequences h_0 and h_1 are called the *scaling vector* and *wavelet vector* respectively. From the sampling theorem it is easy to show that

$$h_0(k) = \frac{1}{\sqrt{2}}\phi(\frac{k}{2}) = \sqrt{2}\frac{\sin(\frac{k}{2}\pi)}{k\pi} \tag{25}$$

$$h_1(k) = \frac{1}{\sqrt{2}}\psi\left(\frac{k}{2}\right) = \left[\sqrt{2}\phi(k) - \frac{1}{\sqrt{2}}\phi\left(\frac{k}{2}\right)\right] = \sqrt{2}\delta(k) - h_0(k) \tag{26}$$

showing that the wavelet vector is determined by the scaling vector. Since $\{\phi(t-k)\} = \phi_{0,k}(t)$ forms an orthonormal basis for V_0, $\{\psi(t-k)\} = \phi_{0,k}(t)$ forms an orthonormal basis for W_0, and $V_0 \perp W_0$, from Equation (23), and Equation (24)

$$\int \phi(t)\phi(t-l)\,dt = \delta(l) = \sum_k h_0(k)h_0(k+2l), \tag{27}$$

$$\int \psi(t)\psi(t-l)\,dt = \delta(l) = \sum_k h_1(k)h_1(k+2l), \tag{28}$$

$$\int \phi(t)\psi(t-l)\,dt = 0 = \sum_k h_0(k)h_1(k+2l). \tag{29}$$

Thus all the orthogonality properties of the scaling function and wavelet are reflected in the scaling and wavelet vectors. In fact more is true. The scaling vector $h_0(n)$, by itself, completely determines the DWT frequency decomposition. To see this take the Fourier transform of Equation (23) and let $H_0(\omega)$ denote the Fourier transform of $h_0(n)$.

$$\hat{\phi}(\omega) = \left[\frac{1}{\sqrt{2}}H_0(\frac{\omega}{2})\right]\hat{\phi}(\frac{\omega}{2}) \tag{30}$$

or equivalently,

$$\hat{\phi}(\omega) = \hat{\phi}(0) \prod_{j=1}^{\infty} \left[\frac{1}{\sqrt{2}} H_0(\omega/2^j) \right] \tag{31}$$

Thus if the infinite product converges, by the uniqueness of the Fourier transform ϕ and hence ψ and therefore the DWT frequency decomposition is determined completely by $h_0(k)$. A necessary condition for the convergence of the infinite product is that $H_0(0) = \sqrt{2}$, or equivalently

$$\sum_k h_0(k) = \sqrt{2}. \tag{32}$$

Notice also that

$$\int \psi(t)\, dt = 0 \tag{33}$$

$$\int \phi(t)\, dt = 1. \tag{34}$$

The big breakthrough for DWT type orthonormal frequency decompositions came about when Ingrid Daubechies parameterized all finite length sequences $h_0(k)$ satisfying Equation (27) and Equation (32), and showed that many of them give rise to overlapping DWT type orthonormal frequency decompositions. Independently of her efforts, in the theory of Filter Banks (where sequences satisfying Equation (85) are called Conjugate Quadrature Filters (CQFs) or Quadrature Mirror Filters (QMFs)), all such sequences had also been parameterized [45].

We will now see how the DWT can be computed using a "filter bank". Thinking of the channel W_j as the orthogonal complement of V_j in V_{j+1} will be useful. Let $c_{j,k}$ denote expansion coefficients of $f(t) \in V_j$,

$$c_{j,k} = \langle \phi_{j,k}, f(t) \rangle \tag{35}$$

Then $c_{j,k}$ is related to $c_{j-1,k}$ and $b_{j,k}$ in terms of h_0 and h_1 as,

$$c_{j-1,k} = \sum_n h_0(k-2n) c_{j,n} \tag{36}$$

$$b_{j-1,k} = \sum_n h_1(k-2n) c_{j,n} \tag{37}$$

$$c_{j,n} = \sum_k h_0(k-2n) c_{j-1,k} + \sum_k h_1(k-2n) b_{j-1,k}. \tag{38}$$

These equations are obtained by invoking Equation (23) and Equation (24). The computations corresponding to these equations can be represented as in

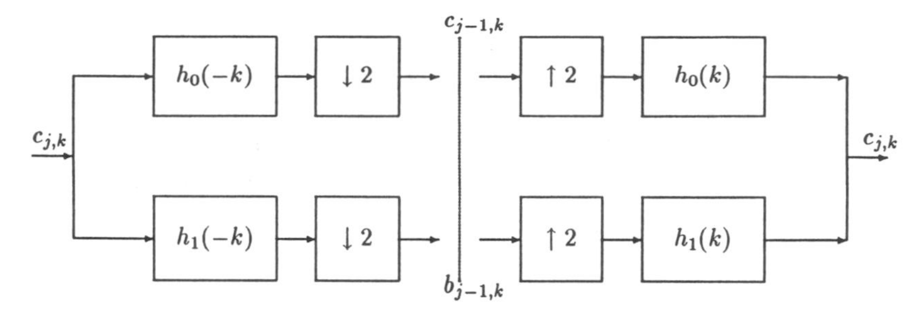

Figure 4. DWT decomposition and the filter bank structure.

Figure 4, known as a two channel filter bank. The symbols $\downarrow$ and $\uparrow$ correspond to the dual operations of sub-sampling and upsampling. That is,

$$[\downarrow 2]x(n) = x(2n) \tag{39}$$

$$[\uparrow 2]x(n) = \begin{cases} x(\frac{n}{2}) & \text{for } n \text{ even} \\ 0 & \text{otherwise.} \end{cases} \tag{40}$$

Given $c_{j,k}$ for fixed j, the filter bank can be used to compute $c_{j-1,k}$ and $b_{j-1,k}$. and hence given $c_{J,k}$, the DWT $b_{j,k}$ can be computed for all j less than J. Now in practical signal analysis, a reasonable assumption to make is that the signals are approximately bandlimited. That is, f can be assumed to reside in some V_J for large enough J. All information about such signals are in $c_{J,k}$ and the DWT coefficients satisfy

$$b_{j,k} = 0 \quad \text{for } j \geq J . \tag{41}$$

Once $c_{J,k}$ is obtained (by taking inner-products or one of the methods discussed in Section. 7), $b_{j,k}$ can be obtained by using the filter bank equations. The filter bank structure is more that just a computational tool. Orthonormal and biorthogonal wavelet bases can be developed with filter bank theory as a starting point and this viewpoint is very useful.

In summary, compared to the DTSTFT, the DWT above has a number of attractive features

1. The coefficients $a_{j,k}$ are real for real signals.
2. There is a natural recursive structure associated with the DWT decomposition that can be used to compute the DWT coefficients recursively. The reason for this is the scaling property of the basis functions.
3. In the example above the DWT coefficients $b_{j,k}$ can be interpreted as the Nyquist rate samples of $f_j(t)$. Though the general DWT decompositions discussed later do not satisfy this, this intuition is useful.

At this point, an apparent disadvantage with the filter bank structure is that since the sequences $h_i(n)$ are of infinite extent they cannot be realized on a digital computer. We will see how filter bank theory gives a solution to this problem.

A natural question at this point is whether STFT type frequency decompositions can be obtained that have the nice computational properties of DWT type decompositions. To answer this question consider Figure 2. The frequency bin $[0, 2^j\pi]$ corresponds to V_j and is split into two bins corresponding to V_{j-1} and W_{j-1} with V_{j-1} corresponding to the low frequency part of V_j and W_{j-1} the high frequency part. The scaling property of the DWT derives from the fact that V_{j-1} is then split into V_{j-2} and W_{j-2}. In order to get uniform bins as in the STFT type frequency decomposition we only need to also split the high frequency part of V_j (namely W_j) into two uniform width bins. Corresponding to these bins there would be two subspaces of W_j with orthonormal bases for each one of them. Clearly there would be sequences relating these bases to the basis for W_j. It turns out that these sequences are precisely the scaling and wavelet vectors. Thus expansion coefficients in these bases can be obtained by starting with $b_{j,k}$, the expansion coefficients in W_j and then using Equation (36) and Equation (37), with $c_{j,k}$ replaced by $b_{j,k}$ and the left hand side interpreted as the expansion coefficients in the new bases. Now each of the new subspaces can further be decomposed into two bins, and this process can be continued any number of times. It turns out in all these cases the bases are related by the same sequences h_0 and h_1. The bevy of functions thus obtained are sometimes called *coiflets* [27], after Coifman who first discovered them. They are also sometimes referred to as *wavelet packets* [37]. In some cases, these decompositions could be carried out ad infinitum. For example, in the DWT decomposition, the low-frequency channel V_j indeed decomposed infinitely many times.

In Section 8 we will show that when bins are split recursively, as described above, one can do so using other sequences h_0 and h_1 that form CQFs. In fact, we will have a variety of efficient decompositions that have the computational advantages of the wavelet type decomposition, but there will be no underlying *scaling function* and *wavelet.* While the recursive decomposition or *scaling* described above that is at the heart of things, and the existence of a scaling function and wavelet is wavelet is intimately related to whether these decompositions can be carried out ad infinitum.

Another feature in the DWT decomposition in Figure 2 is that V_j is split into *two* subspaces corresponding to the split of $[0, 2^j\pi]$, into $[0, 2^{j-1}\pi]$ and $[2^{j-1}\pi, 2^j\pi]$. At each stage the bin corresponding to V_j could have been split into M sub-bins and then we would have a decomposition as shown in Figure 5. The low-frequency sub-bin is further split into M bins and so on, thus introducing the concept of *scale.* Such a decomposition is called a multiplicity M wavelet decomposition. If $W_{i,j}$ denotes the space of functions bandlimited to $[iM^j, (i+1)M^j]$, then it is clear that for all j, we have a decomposition of $W_{0,j}$ into M sub-channels $W_{0,j-1}$ through $W_{M-1,j-1}$. As in the two channel case there would be sequences $h_i(n)$, $i = 0, \ldots, M-1$, relating the basis for $W_{0,j}$ and the basis for $W_{i,j-1}$. These sequences also relate the expansion coefficients as in

Equation (36) and Equation (37). The corresponding *computational* structure relating the expansion coefficients in these bases is an M-channel filter bank associated with the sequences $h_i(n)$.

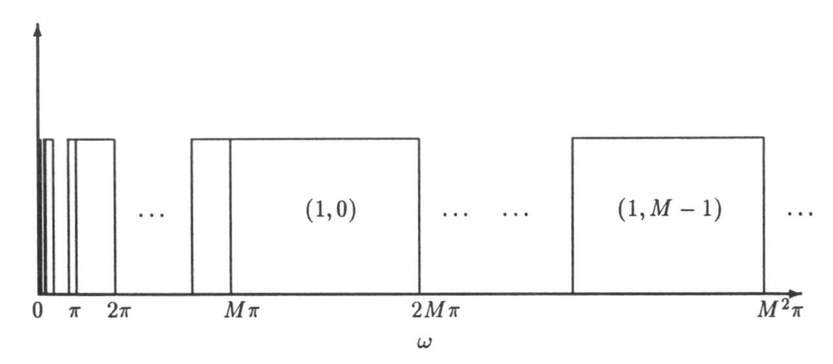

Figure 5. Multiplicity M-DWT decomposition.

The wavelet packets idea can be combined with multiplicity M DWT decompositions in order to give a very rich family of computationally efficient decompositions. In the next two sections we describe the essence of filter bank theory in relation to wavelet theory, and describe how computations can be performed in filter banks. The theory developed will be used to introduce, orthonormal and biorthogonal wavelet bases, far-reaching generalizations of DWT type frequency decompositions.

§3. Filter bank theory

Filter banks are efficient convolution structures that have been extensively used in sub-band coders for speech, and in transmultiplexers (devices to convert time-division multiplexed data to frequency-division multiplexed data and vice-versa). In a filter bank, a data sequence $x(n)$ (see Figure 6), is decomposed into M channels by convolution with sequences $h_i(n), i = 1, 2, \ldots, M-1$, called the *analysis* filters, down-sampled by M on each channel, upsampled by M on each channel, and then convolved with the sequences $g_i(n), i = 1, 2, \ldots, M-1$, called the *synthesis* filters, and recombined to give $y(n)$. A potential application is in speech coding, where the sequences $d_i(n)$ are encoded with different number of bits depending on the sensitivity of the ear to channel i. A fundamental question is whether there exist perfect reconstruction filter banks (*i.e.*, $x(n) = y(n)$). Recall that in Section 2, the wavelet and scaling vectors give rise to a two channel ($M = 2$) perfect reconstruction filter bank! In that case, we also had that the sequences h_i and g_i were time reverses of each other,

$$h_i(k) = g_i(-k) \qquad H_i(z) = G_i(z^{-1}). \tag{42}$$

In order to analyze the filter bank we introduce some standard signal processing definitions and notation. For a sequence $h(k)$ (also called a filter) the bilateral $\mathcal{Z}$-transform (in its region of convergence) is defined to be

$$H(z) = \sum_k h(k) z^{-k}. \tag{43}$$

Usually the domain of convergence is assumed to include the unit circle so that the Discrete Time Fourier Transform of the $h(n)$ is defined (see [42[and [34]). If $h(n) \in \ell^1(\mathbb{Z})$, then

$$\sum_k |h(k)| < \infty, \tag{44}$$

and the unit circle is in the region of convergence. Most of the filters we consider will be in $\ell^1(\mathbb{Z})$. A filter $h(k)$ is said to be *causal* if $h(n) = 0$ for negative n. This means that $H(z)$ is analytic in the exterior of the closed unit disc since $H(z)$ is a power series in z^{-1}. Every filter has a *causal-anticausal* decomposition

$$H(z) = H_c(z) + zH_a(z^{-1}) \tag{45}$$

where $H_c(z)$ and $H_a(z)$ are causal (that is the corresponding sequences $h_c(n)$ and $h_a(n)$ are causal). A non-causal filter is said to be *finitely non-causal*, if there exists an N such that $h(n - N)$ is causal. An equivalent characterization for finite non-causality is that $H_a(z)$ is a polynomial in z^{-1}. A filter $h(k)$ is said to be FIR, if $H_a(z)$ and $H_c(z)$ are polynomials in z^{-1}. In other words, $H(z)$ is a Laurent polynomial (sum of a polynomial in z and z^{-1}). Clearly all FIR filters are finitely non-causal. A filter $h(k)$ is said to be IIR if it is not FIR. The $\mathcal{Z}$-transform of an IIR filter is a Laurent series that is not a Laurent polynomial. A filter is said to be *rational* if $H_a(z)$ and $H_c(z)$ are rational functions of z. A filter is said to be *realizable* if convolution with the sequence $h(n)$ can be implemented. Only *finitely non-causal, rational* filters are realizable. Thus for a realizable filter, $H_a(z)$ is a polynomial, and $H_c(z)$ is a rational. The scaling and wavelet vectors in Section 2 are neither finitely causal, nor rational and hence are not realizable.

The standard problem in filter bank theory is to start with a realizable set of filters $h_i(n)$, with desired frequency behavior, and then to determine $g_i(n)$ (preferably realizable) such that $y(n) = x(n)$. Clearly for arbitrary realizable $h_i(n)$, there may not exist $g_i(n)$, such that $y(n) = x(n)$. And even if $g_i(n)$ did exist, they may not be realizable. As far as filter banks in wavelet theory are concerned, from Section 2, it should be clear that realizability of the filters is not an issue in the construction of wavelet bases. Hence for wavelet theory it is perhaps necessary, even to consider non-realizable filter banks. In any case, from a computational perspective, exact signal analysis is not possible in wavelet bases associated with non-realizable filter banks, and hence it is definitely important to know the theory of realizable filter banks. Not surprisingly, most of filter bank theory has dealt with only a special class of *realizable* filters, namely FIR filters.

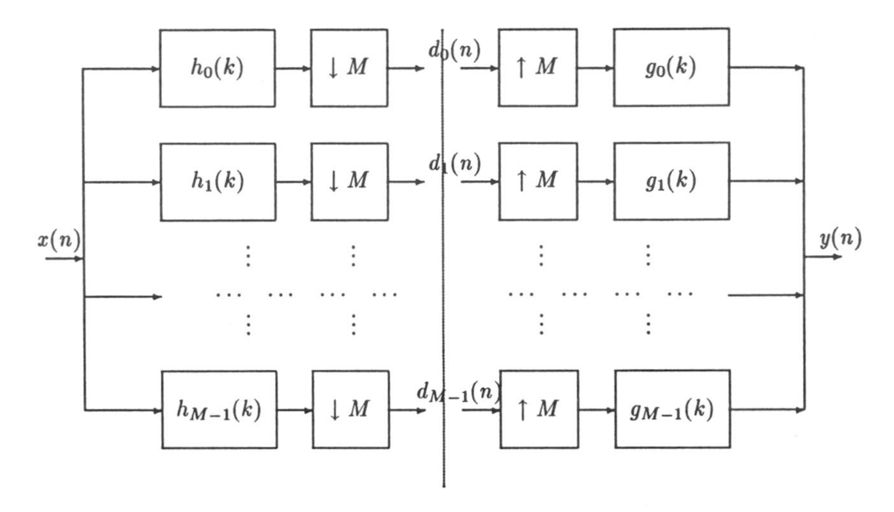

Figure 6. An M channel filter bank.

Most of the literature in filter bank theory specializes to FIR $h_i(n)$ and $g_i(n)$ partly because FIR filters can be easily implemented, and partly because stability issues do not have to be addressed. Surprisingly, FIR filter bank theory includes all orthogonal and biorthogonal wavelet bases with compactly supported wavelets as a special case.

We now review the main results in filter bank theory introducing some new approaches to filter bank theory on the way. The question of existence of perfect reconstruction filter banks can be reduced to the question of invertibility of multi-input multi-output (MIMO) linear shift-invariant system. If we block $x(n)$ and $y(n)$ into M-vectors, then from Figure 6 it is obvious that the analysis bank maps an M-vector signal to another M-vector signal $(d_i(n))$, and the synthesis bank maps this M-vector signal yet another M-vector signal. More precisely, let the input $X(z)$ and output $Y(z)$ be rewritten as

$$X(z) = \sum_{k=0}^{M-1} z^k X_k(z^M), \tag{46}$$

$$Y(z) = \sum_{k=0}^{M-1} z^k Y_k(z^M), \tag{47}$$

Let $H_i(z)$ and $G_i(z)$ denote the $\mathcal{Z}$-transforms of the analysis and synthesis filters. Then $H_i(z)$ and $G_i(z)$ (possibly non-causal and irrational) can be written as

$$H_i(z) = \sum_{k=0}^{M-1} z^{-k} H_{i,k}(z^M), \tag{48}$$

$$G_i(z) = \sum_{k=0}^{M-1} z^{k} G_{i,k}(z^M), \tag{49}$$

where

$$H_{i,k}(z) = \sum_n h_i(Mn+k) z^{-n} \tag{50}$$

$$G_{i,k}(z) = \sum_n g_i(Mn-k) z^{-n}. \tag{51}$$

Now from Figure 6, we have

$$\begin{aligned} D_i(z) &= [\downarrow M] H_i(z) X(z) \\ &= [\downarrow M] \left[\sum_{k=0}^{M-1} z^{-k} H_{i,k}(z^M) \right] \left[\sum_{j=0}^{M-1} z^{j} X_j(z^M) \right] \\ &= \sum_{k=0}^{M-1} H_{i,k} X_k. \end{aligned} \tag{52}$$

The last step follows from the fact that

$$[\downarrow M] z^k = \begin{cases} z^m & \text{for } k = mM \\ 0 & \text{otherwise.} \end{cases}$$

Similarly, for the synthesis bank,

$$\begin{aligned} \sum_i z^k Y_k(z^M) = Y(z) &= \sum_i \{[\uparrow M] D_i(z)\} G_i(z) \\ &= \sum_i D_i(z^M) G_i(z) \\ &= \sum_{i,k} z^k D_i(z^M) G_{i,k}(z^M) \\ &= \sum_k z^k \left[\sum_i D_i(z^M) G_{i,k}(z^M) \right]. \end{aligned} \tag{53}$$

Equating like powers of z we have

$$Y_k(z) = \sum_i G_{i,k}(z) D_i(z). \tag{54}$$

To simplify notation and make things transparent, define the *polyphase* component matrices

$$H_p(z) = \begin{bmatrix} H_{0,0}(z) & H_{0,1}(z) & \cdots & H_{0,M-1}(z) \\ H_{1,0}(z) & H_{1,1}(z) & \cdots & H_{1,M-1}(z) \\ \vdots & \vdots & \cdots & \vdots \\ H_{M-1,0}(z) & H_{M-1,1}(z) & \cdots & H_{M-1,M-1}(z) \end{bmatrix} \tag{55}$$

$$G_p(z) = \begin{bmatrix} G_{0,0}(z) & G_{0,1}(z) & \cdots & G_{0,M-1}(z) \\ G_{1,0}(z) & G_{1,1}(z) & \cdots & G_{1,M-1}(z) \\ \vdots & \vdots & \cdots & \vdots \\ G_{M-1,0}(z) & G_{M-1,1}(z) & \cdots & G_{M-1,M-1}(z) \end{bmatrix}. \tag{56}$$

Let

$$\begin{aligned} \mathbf{x}_p(z) &= [X_0(z), X_1(z), \ldots, X_{M-1}(z)]^T \\ \mathbf{y}_p(z) &= [Y_0(z), Y_1(z), \ldots, Y_{M-1}(z)]^T \\ \mathbf{d}(z) &= [D_0(z), D_1(z), \ldots, D_{M-1}(z)]^T. \end{aligned}$$

Then Equation (52) and Equation (53) can be written compactly as

$$\mathbf{d}(z) = H_p(z)\mathbf{x}_p(z) \tag{57}$$

$$\mathbf{y}_p(z) = G_p^T(z)\mathbf{d}(z) \tag{58}$$

$$\mathbf{y}_p(z) = G_p^T(z)H_p(z)\mathbf{x}_p(z). \tag{59}$$

These equations are perhaps the most important equations in filter bank theory. It shows that the filter bank could be thought of as the cascade of two multi-input multi-output (MIMO) linear time-invariant systems that are inverses of each other. Notice that Equation (57) and Equation (58) hold independent of whether the h_i and g_i are causal, realizable, rational, etc. H_p and G_p represent MIMO systems corresponding to the analysis and synthesis filter bank, respectively. Moreover given an arbitrary MIMO system, there exists an analysis filter bank and a synthesis filter bank, of which the MIMO system could be considered to be the polyphase representation. Seen in the MIMO setting the basic results in filter bank theory become extremely simple. To appreciate the simplicity and elegance of the equation, an understanding of the history of filter bank theory is essential. Filter banks (actually two channel case) were first introduced in 1977 by Galand and Esteban [12]. For a long time, no non-trivial examples of *realizable* perfect reconstruction filter banks were known (for a history of filter banks see [43]). The first successes were in affecting alias cancellation (in which case $y(n) = x(n) * h(n)$, for some sequence $h(n)$ that is not $\delta(n)$), and in obtaining exact magnitude responses ($|Y(\omega)| = |X(\omega)|$) and phase responses. Finally, in 1984, Smith and Barnwell constructed exact reconstruction filter banks [41]. Later Vaidyanathan noticed the importance of a certain *unitariness* property in filter banks (the Smith-Barnwell examples satisfied these properties). Reinterpreting results from classical inverse scattering theory of two-port networks [1], the $\mathcal{Z}$-transform domain, Vaidyanathan et al., [44], transformed the factorization theorems for *lossless* scattering matrices to factorization theorems for rational matrix functions, unitary on the unit circle. Since then, *unitariness* has played a fundamental role in developing filter banks with desirable properties. In any case, from Equation (59), it is obvious that having perfect reconstruction filter banks is equivalent to the inversion of H_p and hence definitely feasible. From now on, by a filter bank we always mean a perfect reconstruction filter bank.

We now give the most fundamental result in filter bank theory. An $M \times M$ matrix function of the complex variable z, that has full rank for all z on the unit circle (*i.e.*, $z = e^{i\omega}$) is said to be in $GL^{\infty}_{M\times M}$. By the standard abuse of notation, for a $\mathcal{Z}$-transform $H(z)$, we will denote $H(e^{i\omega})$ by $H(\omega)$.

Lemma 1. *Let $h_i(n)$ denote the the analysis filter bank. Assume that for all i, $H_i(\omega) < \infty$. Then, there exists a synthesis filter bank with filters $g_i(n)$ iff, $H_p(\omega) \in GL^{\infty}_{M\times M}$, or equivalently if* $\det H_p(\omega) \neq 0$ *for all ω.*

Proof: If $H_p(\omega) \in GL^{\infty}_{M\times M}$, then it has full rank for all ω and is hence invertible; and therefore $H_p^{-1}(\omega)$ is well defined. Take $G_p(\omega) = [H_p^{-1}]^T(\omega)$. From $G_p(\omega)$, the numbers $g_{i,j}(n)$ and hence $g_i(n)$ can be obtained. ■

How does this result help in the construction of filter banks? Clearly for a given choice of $h_i(n)$, it is no easy task (in general) to check that $H_p(\omega)$ has full rank for all ω. So the trick is to start with an arbitrary $H_p(\omega) \in GL^{\infty}_{M\times M}$, and then construct $h_i(n)$ and $g_i(n)$ from it. One problem with this approach is that there is no explicit control over the filters $h_i(n)$ and $g_i(n)$, only over $h_{i,j}$ and $g_{i,j}$. A complete solution to this problem is known only in the special case of some (*unitary*) FIR filter banks. Nevertheless, this method gives rise to many filter banks and in some cases wavelet bases as we will shortly see. Another problem with this method is that the filters may not be realizable (causal). If the filters $h_{i,j}$ are causal then necessarily the filters h_i are causal. But if the filters $g_{i,j}$ are causal it does not mean that the filters g_i are as can be seen from Equation (49). However, the filters g_i are finitely non-causal and hence realizable. From now on, by causality of the filters we mean that $h_{i,j}$ and $g_{i,j}$ are causal.

We now proceed to show how realizability can be imposed as a constraint in the filter bank. We need some preliminary mathematical results. Let RH^{∞}, denote the set of real, causal (stable), rational functions. In other words, real rational functions, analytic in the exterior of the closed unit disc in the complex plane. Given a function $H(z) \in RH^{\infty}$, by reflecting all the zeros of the function outside the unit circle, inside the circle, it is well known that $H(z)$ has an *allpass/minimum-phase* factorization.

$$H(z) = H_{all}(z) H_{min}(z). \tag{60}$$

The function $H_{min}(z)$ has all its poles and zeros inside the unit circle and is called *minimum phase* because for all other functions in RH^∞ that have the same magnitude on the unit circle, $H_{min}(z)$ has the least phase. The function $H_{all}(z)$ is called allpass because, its magnitude on the unit circle is unity.

$$\left[H_{all}^T(z^{-1})H_{all}(z)\right]_{|z|=1} = 1. \tag{61}$$

The inverse of a minimum-phase function is also in RH^∞ since all its zeros are inside (or on) the unit circle. If the function has no zeros on the unit circle then it is in $GL_{1\times 1}^\infty$ and therefore invertible on the unit circle. This minimum phase rational functions with no zeros on the unit circle have causal (minimum-phase) rational inverses. Now let $RH_{M\times M}^\infty$ denote the set of $M \times M$ matrices with entries in RH^∞. A matrix $P \in RH_{M\times M}^\infty$, is said to be *unitary*, if it is unitary on the unit circle (such functions are sometimes called *inner* or *lossless*). We have,

$$\tilde{P}(z)P(z) = I_M = P(z)\tilde{P}(z) \tag{62}$$

where $\tilde{P}(z) = P^T(z^{-1})$. This generalizes the notion of a allpass function. A matrix $Q \in RH_{M\times M}^\infty$, is said to be *minimum phase*, if its inverse is also in $RH_{M\times M}^\infty$. This generalizes the notion of a minimum-phase function. Matrices in $RH_{M\times M}^\infty$ also have an *allpass/minimum-phase* factorization (see [14] and [23]).

Lemma 2. *Let $H \in RH_{M\times M}^\infty$. Then there exists an all pass $H_{all}(z)$ and minimum phase $H_{min}(z)$ such that*

$$H(z) = H_{all}(z)H_{min}(z). \tag{63}$$

Proof: See Francis [14]. ■

Let $U_{M\times M}$ denote the set of $M \times M$ unitary (allpass) functions. Clearly $U_{M\times M} \in GL_{M\times M}^\infty$. Therefore, given an arbitrary matrix $P \in U_{M\times M}$ one can construct a filter bank as follows. Take $H_p(z) = P(z)$ and $G_p(z) = P(z^{-1})$. Since $P(z)$ is analytic outside the unit disc, $P(z^{-1})$ is analytic inside the unit disc and therefore the analysis filters are causal and the synthesis filters are anti-causal. However, if the synthesis filters are *finitely non-causal*, then the filter bank is realizable. This occurs precisely when $P(z)$ is a polynomial in z^{-1}. We will shortly see how to construct all realizable unitary matrices $P(z)$. Also notice that given a minimum-phase $Q(z) \in RH_{M\times M}^\infty$. If $\det Q$ has no zeros on the unit circle $Q \in GL_{M\times M}^\infty$, by taking $H_p(z) = Q(z)$ and $G_p(z) = Q^{-1}(z)$ we have a filter bank. In this case since both Q and Q^{-1} are analytic outside the unit disc, we have a *causal*, rational and hence *realizable* filter bank.

From these observations we can construct a number of filter banks with the following recipe. Start with an arbitrary $H \in RH_{M\times M}^\infty \cap GL_{M\times M}^\infty$. Find its allpass and minimum-phase factors H_{all} and H_{min} and then construct two filter banks, one possibly non-realizable. In order to ensure that both are realizable we only have to start with a polynomial in z^{-1} for $H(z)$.

Another simpler way to impose realizability is to require that the analysis and synthesis filter are FIR. In that case,$H_p(z)$ and $G_p(z)$ are Laurent polynomials. We have the following result that characterized FIR filter banks.

Lemma 3. *h_i and g_i form an FIR filter bank iff one of the following is true*

1. *$H_p(z)$ or $G_p(z)$ is invertible over the Laurent polynomials*
2. *$\det H_p(z)$ or $\det G_p(z)$ is of the form Kz^k for some integer k and constant K.*

Proof: From Equation (59), $G_p^T(z) = H^{-1}(z)$ and hence

$$G_p^T(z) = \frac{1}{\det H_p(z)} \text{adj}(H_p(z)). \tag{64}$$

Since the adjoint of a Laurent polynomial is a Laurent polynomial, Equation (64) holds with $G_p^T(z)$, a Laurent polynomial, iff $\det H_p(z) = Kz^k$ for some integer k. ■

Lemma 64 is more useful than Lemma 1 in that given an arbitrary set of FIR filters one can easily check if there exits an FIR synthesis bank. Besides, one can start with an arbitrary polynomial matrix $P(z)$, whose determinant is of the form Kz^k. Set $H_p(z) = P(z)$ and $G_p^T(z) = P^{-1}(z)$ to construct an FIR filter bank. A special case occurs when $\det H_p(z) = K$, a constant. Such matrices are called *unimodular* matrices and can be constructed as the product of elementary polynomial matrices [25]. For such a $P(z)$, both $H_p(z)$ and $G_p(z)$ are causal.

Finally, notice that given an arbitrary Laurent polynomial matrix that is invertible on the unit circle, with appropriate multiplication by z^{-N} for large N one can also transform it into a polynomial matrix in z^{-1}, then do an allpass/minimum-phase factorization to obtain two other filter banks as before.

All of these approaches to construct filter banks suffer from one problem. It is not easy to impose design constraints on the analysis/synthesis filters. In the case of *allpass/minimum-phase* factorization, we have no control even over the polyphase matrix $H_p(z)$. This problem can be eliminated if we can construct arbitrary minimum phase $H(z)$ directly (without factorization) and can be achieved [20]. In the case of unitary filter banks a fairly complete answer to this problem is known from the elegant factorization theory for elements of $U_{M\times M}$ matrices. By the degree (or McMillan degree) of a rational matrix, we mean the smallest number of delays required to implement it as the transfer function of a linear system [25]. It turns out that for a unitary matrix the McMillan degree is simply $\deg(\det H_p(z))$. We now summarize the factorization results for rational unitary matrices of McMillan degree N.

Let $v \in \mathbf{C}^M$, with $\|v\| = 1$. Define for some $a \in \mathbf{C}$

$$P(z) = \left[I - v\tilde{v} + \frac{-a^\star + z^{-1}}{1 - az^{-1}} v\tilde{v} \right] \tag{65}$$

where $\tilde{v}$ denotes the transpose-conjugate of v. $P(z)$ is of degree 1 and it can be easily verified that

$$\tilde{P}(z)P(z) = I \tag{66}$$

where $\tilde{P}(z) = P^T(z^{-1})$, and furthermore if

$$P_i(z) = \left[I - v_i\tilde{v}_i + \frac{-a_i^\star + z^{-1}}{1 - az^{-1}} v_i\tilde{v}_i \right]. \tag{67}$$

Then the product $P_N P_{N-1} \ldots P_1$ is of McMillan degree N and is unitary. Up to a unitary constant matrix, it turns out that this is the most general rational, inner matrix of McMillan degree N

Lemma 4. *Let U_N be an arbitrary stable, rational, unitary matrix of McMillan degree N. Then, there exists a factorization of the form*

$$U_N(z) = P_N(z)P_{N-1}(z) \ldots P_1(z)U_0 \tag{68}$$

where U_0 is a constant unitary matrix where the P_i are of the form given in Equation (65). In particular, if $U_N(z)$ is a polynomial matrix in z^{-1}, the factors P_i can be shown to be of the form

$$P_i(z) = \left[I - v_i\tilde{v}_i + z^{-1} v_i\tilde{v}_i \right]. \tag{69}$$

Proof: See Vaidyanathan [46]. ■

Another result that we find useful when we discuss wavelet theory is the following representation theorem for *unitary* vectors on the unit circle. An $M \times 1$ vector $V(z)$ is said to be unitary on the unit circle if $\tilde{V}(z)V(z) = 1$. By McMillan degree of a unitary vector $V(z)$, we mean the degree of the least common multiple of the denominators of the entries in of $V(z)$. We have the following factorization theorem for unitary matrices of McMillan degree N analogous to the result for unitary matrices.

Lemma 5. *Let V_N be an arbitrary unitary vector of McMillan degree N. Then, there exists a factorization of the form*

$$V_N(z) = P_N(z)P_{N-1}(z) \ldots P_1(z)V_0 \tag{70}$$

where V_0 is a constant vector with $\|V_0\| = 1$ Moreover, if V_N is a polynomial vector in z^{-1} then the factors P_i can be shown to be of the form

$$P_i(z) = \left[I - v_i\tilde{v}_i + z^{-1} v_i\tilde{v}_i \right]. \tag{71}$$

Proof: See Vaidyanathan [44]. ■

Lemma 4 gives a complete characterization of all rational unitary filter banks. Designing such an filter bank is equivalent to choosing the parameters v_i and a_i, and clearly numerical optimization schemes for choosing these parameters for minimizing certain objective functions, that depend on the analysis filters h_i can be devised [33]. Another advantage of unitary filter banks is that, there is no need to solve for the synthesis bank by algebraic inversion of H_p since the synthesis filters are time-reverses of the analysis filters.

Lemma 6. *$(H_i(z), G_i(z))$ form the analysis and synthesis filters of an M-channel unitary filter bank iff (up to delays)*

$$G_i(z) = H_i(z^{-1}). \tag{72}$$

Proof: From Equation (49), Equation (48), and Equation (59),

$$G_i(z) = \sum_{k=0}^{M-1} z^k G_{i,k}(z) = \sum_{k=0}^{M-1} z^k H_{i,k}(z^{-1}) = H_i(z^{-1}). \tag{73}$$

■

This result implies that in the wavelet basis considered in Section 2, the scaling vector $h_0(n)$ and wavelet vector $h_1(n)$ form a unitary filter bank.

So far we have considered filter bank theory in the $\mathcal{Z}$-transform domain since it is much easier there than in terms of the sequences $h_i(n)$ and $g_i(n)$. Also recall that the fundamental result in filter bank theory, namely that $G_p^T = H_p^{-1}$ is of a purely algebraic character. It turns out that the algebraic result in the time domain looks relatively complex in that it gives no insight into what is happening. Nevertheless, it is the time domain equations that will be useful in dealing with wavelet theory. To give an example of the clarity of the $\mathcal{Z}$-domain approach, consider the following question. Can the analysis and synthesis filters be exchanged? It is not at all obvious that this is true from the time-domain. However, from the trivial fact that

$$G_p^T H_p = I = H_p^T G_p, \tag{74}$$

it follows that one could swap the analysis and synthesis filters. Since wavelet theory deals with sequences $h_i(n)$ in the filter bank, this *duality* comes out as a non-trivial result [3].

Lemma 7. *The pair $(H_i(z), G_i(z))$ forms analysis/synthesis filters of a unitary filter bank iff*

$$\sum_k h_i(k) h_j(k + Ml) = \sum_k g_i(k) g_j(k + Ml) = \delta(l)\delta(i - j). \tag{75}$$

Proof: We only need to show this result for the h_i since by Lemma 6 the same would be true for g_i. Also note that we have a unitary filter bank iff $G_p(z) = H^T(z^{-1})$, which is equivalent to

$$\sum_k H_{i,k}(z)H_{j,k}(z^{-1}) = \delta(i-j). \tag{76}$$

Now let $\alpha(l) = \sum_k h_i(k)h_j(k+Ml)$. Then from Equation (48) and Equation (76)

$$\begin{aligned}\mathcal{Z}(\alpha(l)) &= [\downarrow M]H_i(z)H_j(z^{-1}) \\ &= [\downarrow M]\sum_{k,l} z^{-k}H_{i,k}(z^M)z^l H_{j,l}(z^{-M}) \\ &= \sum_k H_{i,k}(z)H_{j,k}(z^{-1}) = \delta(i-j)\end{aligned} \tag{77}$$

■

This result has the following interpretation. In a unitary filter bank every filter $h_i(n)$ is orthogonal to the uniform translates of any other filter $h_j(n)$ (including itself) by multiples of M. Notice this result is exactly what we got for the scaling vector and wavelet vector in Section 2 and that it came from the orthonormality properties of the underlying scaling function and wavelet.

There is a corresponding result for general filter banks.

Lemma 8. *$(H_i(z), G_i(z))$ form analysis/synthesis filters of a filter bank iff*

$$\sum_k h_i(k)g_j(Ml-k) = \delta(l)\delta(i-j). \tag{78}$$

Proof: The fact that $G_p^T(z)H_p(z) = I$ implies that

$$\sum_k H_{i,k}(z)G_{j,k}(z) = \delta(i-j). \tag{79}$$

Define $\alpha(l) = \sum_k h_i(k)g_j(Ml-k)$. Then from Equation (48), Equation (49) and Equation (79) ,

$$\begin{aligned}\mathcal{Z}(a(l)) &= [\downarrow M]H_i(z)G_j(z) \\ &= [\downarrow M]\sum_{k,l} z^{-k}H_{i,k}(z^M)z^l G_{j,l}(z^M) \\ &= \sum_k H_{i,k}(z)G_{j,k}(z) = \delta(i-j)\end{aligned} \tag{80}$$

■

There is a *dual* time-domain characterization of both unitary and general filter banks which will be required in later sections.

Lemma 9. *$(H_i(z), G_i(z))$ forms analysis/synthesis filters of a unitary filter bank iff*

$$\sum_i \sum_k h_i(Mk+n_1)h_i(Mk+n_2) = \sum_i \sum_k g_i(Mk+n_1)g_i(Mk+n_2) = \delta(n_1-n_2). \tag{81}$$

Proof: Let $\sum_i \sum_k h_i(Mk+n_1)h_i(Mk+n_2) = f(n_1, n_2)$. Clearly, $f(n_1+M, n_2+M) = f(n_1, n_2)$ and therefore it is necessary only to consider $n_1 = 0, \ldots, M-1$. Let $n_2 = Mn + r$, where $r \in \{0, 1, \ldots, M-1\}$. Then, $f(n_1, n_2)$ is of the form

$$\sum_k \sum_i h_i(Mk + n_1)h_i(M(k + n) + r). \tag{82}$$

Taking the $\mathcal{Z}$-transform of this expression as a sequence in n, for fixed n_1 and n_2, we get

$$\sum_i H_{i,n_1}(z)H_{i,r}(z^{-1}) = \delta(n_1 - r) \tag{83}$$

and, therefore, is zero except when $n_1 = r$. When $n_1 = r$, the inverse $\mathcal{Z}$-transform is $\delta(n)$, and therefore the expression is zero except when $n = 0$. That is $n_2 = n_1$. ■

The dual result for general filter banks is the following.

Lemma 10. *$(H_i(z), G_i(z))$ form analysis/synthesis filters of a filter bank iff*

$$\sum_i \sum_k h_i(Mk + n_1)g_i(Mk + n_2) = \delta(n_1 - n_2). \tag{84}$$

Proof: The proof is identical to that in the previous result using the fact that $H_p G_p^T = I$. ■

Lemma 8 is dual to Lemma 10 in the sense that the former says that the rows of G_p are orthogonal to all but one of the columns of H_p, while the latter says that, the rows of H_p are orthogonal to all but one of the columns of G_p. In other words, one depends on the fact that $G_p^T H_p = I$, while the other depends on the fact that $H_p G_p^T = I$.

§4. Wavelet tight frames

Armed with filter bank theory, we are now ready to construct non-overlapping frequency decompositions of the DWT type. By a multiplicity M-DWT type frequency decomposition we mean the decomposition shown in Figure 5. Such a decomposition is naturally associated with an M-channel filter bank.

Therefore, unitary FIR filter bank theory can be used to describe all *compactly supported* multiplicity M orthonormal wavelet bases. Recall that in the 2-channel FIR *unitary* filter in Section 3 the analysis filters, and in particular h_0 satisfies (see Lemma 7)

$$\sum_k h_0(k)h_0(k+2l) = \delta(l). \tag{85}$$

For the non-overlapping DWT type frequency decomposition considered in Section 2, this is precisely the equation satisfied by the sequence $h_0(n)$. Furthermore, in that case $h_0(n)$ determined the frequency decomposition completely. For arbitrary h_0 satisfying Equation (85), there cannot be orthonormal wavelet bases since that would imply that the corresponding scaling function ϕ (see Section 2), would have to satisfy Equation (31),

$$\hat{\phi}(\omega) = \hat{\phi}(0)\prod_{i=1}^{\infty}[\frac{1}{\sqrt{2}}H_0(\omega/2^i)] \tag{86}$$

and if $H_0(0) \neq \sqrt{2}$, the infinite product diverges. But if $H_0(0) = \sqrt{2}$, it turns out that, the unitary filter bank does indeed give rise to a tight frame (a set of functions, that practically behave like an orthonormal basis as far as expansions and expansion coefficients are concerned).

In fact more is true. One need not start with $h_0(k)$ from a unitary filter bank. One can just start with a sequence h_0 that satisfies both Equation (32) and Equation (85). Recall that such sequences are called scaling vectors. Daubechies gives a parameterization of all scaling vectors [5]. An independent parameterization of scaling vectors has been obtained [35]. Using this parameterization, with some added constraints on the scaling vectors (referred to as regularity conditions) Daubechies constructed orthonormal bases for $L^2(\mathbb{R})$. Based on numerical evidence [29], Lawton realized and proved that all sequences $h_0(n)$ (independent of regularity assumptions) give rise to tight frames. More precisely,

Theorem 11. *Let h_0 be a scaling vector. Then there exists a $\phi(t) \in L^2(\mathbb{R})$ such that*

$$\phi(t) = \sum_k h_0(k)\phi(2t-k). \tag{87}$$

Define

$$h_1(k) = (-1)^k h_0(-k+1) \tag{88}$$

$$\psi(t) = \sum_k h_1(k)\phi(2t-k) \quad \text{and} \quad \psi_{j,k}(t) = 2^{j/2}\psi(2^j t-k). \tag{89}$$

Then $\psi_{j,k}$ forms a tight frame for $L^2(\mathbb{R})$. That is, for all $f(t) \in L^2(\mathbb{R})$ we have

$$f(t) = \sum_{j,k}\langle f\psi_{j,k}\rangle\psi_{j,k}(t). \tag{90}$$

Proof: See Lawton [29]. ■

Even though 90 holds, the functions $\psi_{j,k}$ need not be an orthonormal basis. This can be seen even in finite dimensions. The reason for this is that the functions $\psi_{j,k}$ may not be linearly independent. That is, one of them could be in the closure of the linear span of the others.

In his paper Lawton also proves that for almost all choices of h_0, $\psi_{j,k}$ forms an orthonormal basis. Necessary and sufficient conditions on the sequence h_0 such that $\phi_{j,k}$ give rise to an orthonormal basis have been independently obtained by Lawton and Cohen, but from a practical point of view they do not seem to be important since the numerical algorithms are the same.

We now try to extend the results to the case of multiplicity M DWT type frequency decompositions. We first obtain a parameterization of $h_0(n)$ that satisfy

$$\sum_k h_0(k)h_0(k+Ml) = \delta(l) \quad \text{and} \quad \sum_k h_0(k) = \sqrt{M}. \tag{91}$$

Such sequences are called *multiplicity M scaling vectors* or just *scaling vectors.*

Lemma 12. *Let h_0 be a multiplicity M scaling vector of length N. Write*

$$H_0(z) = \sum_k z^{-k} H_{0,j}(z^M)$$

and define

$$\mathbf{h}_0(z) = [H_{0,1}, H_{0,2}, \ldots, H_{0,M-1}]^T$$
$$\mathbf{e} = [\frac{1}{\sqrt{M}} \frac{1}{\sqrt{M}}, \ldots, \frac{1}{\sqrt{M}}]^T.$$

Then for every h_0 there exists an integer N, such that

$$\mathbf{h}_0(z) = P_N(z)P_{N-1}(z)\ldots P_1(z)\mathbf{e}. \tag{92}$$

From this it easily follows that,

$$\sum_k h_0(Mk+m) = \sqrt{M} \quad \text{for all} \quad m. \tag{93}$$

Proof: By the proof of Lemma 7,

$$\sum_j H_{0,j}(z^{-1})H_{0,j}(z) = 1 \tag{94}$$

and therefore $\mathbf{h}_0$ is a unitary vector. Hence by Lemma 5,

$$h_0(z) = P_N(z)P_{N-1}(z)\ldots P_1(z)V_0 \tag{95}$$

where P_i are 1st degree polynomial matrices as in Equation (69) and $V_0 \in R^M$ with $\|V_0\| = 1$. Now we are given that $\sum_k h_0(k) = \sqrt{M}$. Hence,

$$\sqrt{M} = \sum_k h_0(k) = H_0(z)|_{z=1} = [\sum_k z^{-k} H_{0,j}(z)]|_{z=1} = \sum_k H_{0,j}(z)|_{z=1}. \quad (96)$$

Now for all i,

$$P_i(z)|_{z=1} = [I - v\tilde{v}_i + z^{-1} v\tilde{v}_i]_{z=1} = I.$$

and, therefore, $\mathbf{h}_0(1) = V_0$, and hence

$$\sqrt{M} = \sum_k H_{0,j}(z)|_{z=1} = \sum_k V_{0,k} = 1.$$

The only solution to this equation for V_0, with $\|V_0\| = 1$ is $\mathbf{e}$. ■

We are now ready to give the Wavelet Tight Frames Theorem.

Theorem 13. *(Wavelet Tight Frames Theorem) Let h_0 be a scaling vector of length N. Then, there exists a function $\psi_0(t) \in L^2(\mathbb{R})$*

$$\psi_0(t) = \sqrt{M} \sum_k h_i(k) \psi_0(Mt - k) \quad (97)$$

and sequences $h_i, i = 1, 2, \ldots, M-1$, such that h_i form the analysis filters of a unitary filter bank. Define

$$\psi_i(t) = \sqrt{M} \sum_k h_i(k) \psi_0(Mt - k) \quad ; \quad \psi_{i,j,k}(t) = M^{j/2} \psi_i(M^j t - k) \quad (98)$$

then $\{\psi_{i,j,k}\}$ forms a tight frame for $L^2(\mathbb{R})$. That is,

$$f(t) = \sum_{i=1}^{M-1} \sum_{j,k} \langle f, \psi_{i,j,k}(t) \rangle \psi_{i,j,k}. \quad (99)$$

Also

$$f(t) = \sum_k \langle f, \psi_{0,0,k}(t) \rangle \psi_{0,0,k}(t) + \sum_{i=1}^{M-1} \sum_{j=1}^{\infty} \sum_k \langle f, \psi_{i,j,k}(t) \rangle \psi_{i,j,k}(t). \quad (100)$$

Proof: The proof is similar to the proof in the multiplicity 2 case in [29]. The difficulty in the multiplicity M case is the construction of wavelet sequences (*i.e.*, analogues of h_1). The outline of the proof is as follows. We first construct a $\psi_0(t)$, from $h_0(n)$, that is in $L^2(\mathbb{R})$ and satisfies Equation (97). We then construct the sequences h_i, and define functions $\psi_i(t) \in L^2(\mathbb{R})$ and $\psi_{i,j,k}$ as in Equation (98). We then establish Equation (90), and will arrive at the theorem.

Let $\psi_0^0(t)$ denote the characteristic function of $[0,1)$. Define

$$\psi_0^j(t) = \sqrt{M} \sum_k h_0(k) \psi_0^j(Mt - k). \tag{101}$$

This recursion converges weakly to a function in $L^2(\mathbb{R})$. To see this it suffices to know that for every j, $\|\psi_0^j\| = 1$ (*i.e.*, in the unit ball in $L^2(\mathbb{R})$). Then by the weak compactness of the unit ball in $L^2(\mathbb{R})$, there exists a g such that $\psi_0^j \rightharpoonup g$. Now for any j, $\langle \phi_0^j(t), \phi_0^j(t-k) \rangle = \delta_{0k}$ because $\langle \phi_0^0(t), \phi_0^0(t-k) \rangle = \delta(k)$ (recall $\phi_0^0(t)$ is the characteristic function of $[0,1)$), and by induction

$$\begin{aligned}
&\langle \psi_0^{j+1}(t), \psi_0^{j+1}(t-k) \rangle & (102)\\
&= \left\langle \sqrt{M} \sum_n h_0(n) \psi_0^j(Mt - n), \sqrt{M} \sum_l h_0(l) \psi_0^j(Mt - Mk - l) \right\rangle \\
&= \sum_{n,l} h_0(n) h_0(l) [M \int \psi_0^j(Mt - n) \psi_0^j(Mt - Mk - l)\, dt] \\
&= \sum_{n,l} h_0(n) h_0(l) \delta(n - Mk - l) \\
&= \sum_n h_0(n) h_0(n - Mk) = \delta(k). & (103)
\end{aligned}$$

Therefore for all j, $\|\psi_0^j\| = 1$.

We now need to show that g does indeed satisfy Equation (97). To see that first we show that ψ_0^j converges as a distribution to some function that satisfies Equation (97). Since the distribution topology is weaker than the weak topology [24] in $L^2(\mathbb{R})$, this would imply that $g = \psi_0^j$. The Fourier transform of Equation (101) gives

$$\hat{\psi}_0^{j+1}(\omega) = \hat{\psi}_0^j\left(\frac{\omega}{M}\right) \left[\frac{1}{\sqrt{M}} H_0\left(\frac{\omega}{M}\right) \right]. \tag{104}$$

Since $H_0(\omega)$ is a trigonometric polynomial and $H_0(0) = \sqrt{M}$, the sequence $\hat{\psi}_0^j(\omega)$ converges pointwise uniformly on compact subsets to the function

$$\hat{\psi}_0(\omega) = \prod_{j \geq 1} \left[\frac{1}{\sqrt{M}} H_0\left(\frac{\omega}{M^j}\right) \right]. \tag{105}$$

Let $C = \max \hat{\psi}_0^0(\omega)$ for $|\omega| \leq 1$. Let $A = \max [\frac{1}{\sqrt{M}} |H_0(\omega)|] \leq 1$ by Equation (91). For any $\omega, |\omega| \geq 1$,

$$M^{\lfloor log_M(\omega) \rfloor} \leq \omega \leq M^{\lceil log_M(\omega) \rceil} \tag{106}$$

and therefore from Equation (104)

$$\hat{\psi}_0^j(\omega) \leq C A^{\lceil \log_M(\omega) \rceil} \tag{107}$$

thus showing that $\hat{\psi}_0^j$ are bounded by polynomial growth for large ω. Hence in the time domain, ψ_0^j converges to a distribution ψ_0 whose Fourier Transform is $\hat{\psi}_0$ defined in Equation (105).

From Lemma 12, we know that $\mathbf{h}_0$ is of the form

$$\mathbf{h}_0(z) = P_L(z)P_{L-1}(z)\dots P_1(z)\mathbf{e} \tag{108}$$

for some L. Notice that for any scaling vector the corresponding $\mathbf{h}_0(z)$ will be of the above form and thus all scaling vectors could be generated from the above equation.

Now we need to construct a unitary FIR filter bank, such that h_0 is one of the analysis filters. That is we need to construct $h_i, i = 1, 2, \dots, M-1$. From Lemma 4, and Equation (108), it is clear that this problem is equivalent to constructing a unitary matrix U_0 whose first column is $\mathbf{e}$. This can be done by a Gram-Schmidt process in $\mathbb{R}^M$. A more efficient procedure using state-space techniques is given in [20]. This process is non-unique (unlike the multiplicity 2 case, where h_1 was determined by h_0), and therefore the sequences h_i are not uniquely determined.

We now need to show that $\psi_{i,j,k}$ as defined in Equation (98) form a tight frame. Firstly, since the sequences, h_i are all of finite length, say N and are supported in $[0, N-1)$ it can be shown using standard Paley-Weiner arguments that the functions $\psi_{i,j}(t)$ are compactly supported in $[0, M^{-j}(N-1)]$ [7]. Now define the operators

$$I_{i,J} = \sum_k \psi_{i,J,k} >< \psi_{i,J,k}. \tag{109}$$

The operators make sense for any f in $L^2(\mathbb{R})$. A simple check is to use the fact that $\{\psi_{i,j,(N-1)m+k}(t)/\|\psi_i\|\}$ is an orthonormal family for any fixed $k = 0, 1, 2, \dots, N-2$ (because $\psi_{i,j}$ are compactly supported) and hence by Bessel Inequality [40],

$$\|I_{i,j}f\| \leq (N-1)\|\psi_0\|^2\|f\|. \tag{110}$$

Notice also that the operator is uniformly bounded in j for all i. Now for any fixed j,

$$I_{0,j} = \sum_{i=0}^{M-1} I_{i,j-1} \tag{111}$$

$$\begin{aligned}
\sum_{i=0}^{M-1} I_{i,j-1} &= \sum_{i=0}^{M-1}\sum_k \psi_{i,j,k} >< \psi_{i,j,k} \\
&= \sum_{i=0}^{M-1}\sum_k\sum_l h_i(l)\psi_{0,j,Mk-l} >< \sum_m h_i(m)\psi_{0,j,Mk-m} \\
&= \sum_{i=0}^{M-1}\sum_k\sum_l h_i(Mk-l)\psi_{0,j,l} >< \sum_m h_i(Mk-m)\psi_{0,j,m}
\end{aligned}$$

$$= \sum_{l,m} \psi_{0,j,l} >< \psi_{0,j,m} \left[\sum_{i=0}^{M-1} \sum_k h_i(Mk-l) h_i(Mk-m) \right]$$

$$= \sum_{l,m} \psi_{0,j,l} >< \psi_{0,j,m} \left[\delta(l-m) \right]$$

$$= \sum_k \psi_{0,j,k} >< \psi_{0,j,k} = I_{0,j}. \tag{112}$$

Notice that we used Lemma 9 to obtain this result. By repeatedly substituting for $I_{0,j-1}$, $I_{0,j-2}$, etc. we conclude for a fixed J,

$$I_{0,J} = \sum_{i=1}^{M-1} \sum_{j=-\infty}^{J-1} I_{i,j}. \tag{113}$$

We will now show that $I_{0,j,k}$ approximates the identity operator in $L^2(\mathbb{R})$. That is for all $f \in L^2(\mathbb{R})$

$$\lim_{j \to \infty} I_{0,j} f = f. \tag{114}$$

This result is to be expected since $\psi_{0,j,0}$ approaches the Dirac measure at the origin as $j \to \infty$. From Equation (110) it suffices to show that $I_{0,j} f = f$, only for a dense subset of $L^2(\mathbb{R})$. For any continuous f with compact support, if j and k together approach infinity such that $M^{-j}k \to t$, then,

$$\lim_{j,k \to \infty} \langle \psi_{0,j,k}, f(x) \rangle = f(t) \tag{115}$$

uniformly in x. We will now show that $M^{-j/2}\psi_{0,j,k}(t)$ forms a partition of unity and this would imply that the result is true.

In the construction of $\psi_0(t)$, we started with $\psi_0^0(t)$, the characteristic function of $[0,1)$. Clearly, $\psi_0^0(t)$ forms a partition of unity since

$$\sum_k \psi_0^0(t+k) = 1. \tag{116}$$

Now by induction, we have for any j,

$$\sum_k \psi_0^j(t+k) = \sqrt{M} \sum_k \sum_l h_0(l) \psi_0^{j-1}(Mt - Mk - l). \tag{117}$$

Changing variables and interchanging the order of summation and invoking Equation (93) in Lemma 12 we get,

$$\begin{aligned} \sum_k \psi_0^j(t+k) &= \sqrt{M} \sum_k \sum_l h_0(Mk-l) \psi_0^{j-1}(Mt+l) \\ &= \sqrt{M} \sum_l [\sum_l h_0(Mk-l)] \psi_0^{j-1}(Mt+l) \\ &= \sum_l \psi_0^{j-1}(Mt+l) \\ &= 1 \end{aligned} \tag{118}$$

and, therefore, $M^{-j/2}\psi_{0,j,k}(t)$ forms a partition of unity. ■

The function $\psi_0(t)$ is called the scaling function and the $M-1$ functions, $\psi_i(t), i \neq 1$ are called wavelets. The corresponding sequences are called scaling vectors and wavelet vectors, respectively.

Multiplicity M wavelet tight frames have a natural multiresolution analysis structure associated with them. It essentially comes from the filter bank structure that lead to Equation (111). To see this, define the spaces,

$$V_{i,j} = \text{Span}\{\psi_{i,j,k}\}. \tag{119}$$

For any $f \in L^2(\mathbb{R})$, note that $I_{i,j}f \in V_{i,j}$. In particular, for $f \in V_{0,j}f$,

$$f = I_{0,j}f = \sum_{i=0}^{M-1} I_{i,j-1}f \tag{120}$$

leading to a decomposition

$$V_{0,j} = V_{0,j-1} \oplus V_{1,j-1} \ldots \oplus V_{M-1,j-1}. \tag{121}$$

Also from the fact that $I_{0,j}$ becomes I on $L^2(\mathbb{R})$ we have

$$\lim_{j\to\infty} V_{0,j} = L^2(\mathbb{R}). \tag{122}$$

It is also easily seen that

$$\lim_{j\to-\infty} V_{0,j} = \{0\}. \tag{123}$$

Thus we have a multiresolution analysis of $L^2(\mathbb{R})$, with a chain of closed subspaces

$$\{0\} \subset \ldots V_{0,-1} \subset V_{0,0} \subset V_{0,1} \ldots \subset L^2(\mathbb{R}). \tag{124}$$

In general the spaces $V_{i,j}$ for fixed j are not mutually orthogonal. But when we have an orthonormal basis, it is indeed true that for fixed j, and $i \neq l$, $V_{i_1,j} \perp V_{i_2,j}$

In Section 8 we will further explore these ideas and show that this type of multiresolution analysis has only to do with filter banks, and has little do with the fact that there exist scaling functions and wavelets.

In summary, given a unitary filter bank with a "DC" condition, there exist wavelet tight frames. In the multiplicity M case also necessary and sufficient conditions for orthonormality of the bases can be worked out as in [28]. An important point to note is that most of the tight frames would be orthonormal bases. Once we have an orthonormal basis, we can impose regularity on the functions $\psi_i(t)$, or equivalently on $\psi_0(t)$. Most of the work in [5] is in constructing wavelets with arbitrary large degree of differentiability. As is shown there, the regularity is related to the vanishing of the discrete moments of the sequences $h_0(n)$, or equivalently the flatness of $H_0(\omega)$ at $\omega = 0$. A number of researchers have stressed the importance of regularity of the analysis functions ([5], [47], [7], and [2]). In Section 8 we argue that that the regularity of the functions $\psi_i(t)$ may not be of any practical significance per se, but that the flatness of $H_0(\omega)$ may be required in some applications.

§5. Wavelet frames

From unitary FIR filter banks with the property that $H_0(\omega) = \sqrt{M}$, we can construct compactly supported wavelet tight frames and orthonormal bases. This raises the following interesting question: do all FIR filter banks modulo some linear conditions (like $H_0(0) = \sqrt{M}$) lead to expansions like Equation (90)? This section tries to answer this question. For the case $M = 2$ a reasonably complete answer to this question is given in [7]. Recall that in the construction of tight frames, only one sequence $h_0(n)$ was required. The filter bank and the wavelet tight frame was constructed from it. This was facilitated by the elegant factorization theory of unitary matrices over the Laurent polynomials. However, in the general case such an approach is not known. Yet given an FIR filter bank, with two "DC" constraints, $H_0(\omega) = G_0(\omega) = \sqrt{M}$, we can show that we have a wavelet frame (a dual set of functions that behaves like a biorthogonal basis and its dual as far as expansions and expansion coefficients are concerned).

Theorem 14. *(Wavelet Frames Theorem) Let h_i and g_i be the analysis and synthesis filters in an FIR FB such that,*

$$H_0(0) = G_0(0) = \sqrt{M} \tag{125}$$

and for all $i = 1, 2, \ldots, M-1$

$$H_i(0) = G_i(0) = 0. \tag{126}$$

Define the functions

$$\tilde{\psi}_0(\omega) = \prod_{j\geq 1} H_0\left(\frac{\omega}{M^j}\right) \tag{127}$$

$$\psi_0(\omega) = \prod_{j\geq 1} G_0\left(\frac{\omega}{M^j}\right). \tag{128}$$

These functions represent Fourier Transforms of distributions in general. Assume further that $\psi_0(t)$ and $\tilde{\psi}_0(t)$ are in $L^2(\mathbb{R})$ (notice that this implies restrictions on the h_i and g_i). Now define

$$\psi_i(t) = \sqrt{M}\sum_k h_i(k)\psi_0(Mt-k); \quad \psi_{i,j,k}(t) = M^{j/2}\psi_i(M^j t - k) \tag{129}$$

$$\tilde{\psi}_i(t) = \sqrt{M}\sum_k g_i(k)\tilde{\psi}_0(Mt-k); \quad \tilde{\psi}_{i,j,k}(t) = M^{j/2}\tilde{\psi}_i(M^j t - k) \tag{130}$$

then $\{\psi_{i,j,k}\}$ and $\{\tilde{\psi}_{i,j,k}\}$ form dual frames for $L^2(\mathbb{R})$. That is,

$$f(t) = \sum_{i=1}^{M-1}\sum_{j,k}\langle\tilde{\psi}_{i,j,k}(t)f\rangle\psi_{i,j,k} = \sum_{i=1}^{M-1}\sum_{j,k}\langle f\psi_{i,j,k}(t)\rangle\tilde{\psi}_{i,j,k}. \tag{131}$$

Also

$$f(t) = \sum_k \langle \tilde{\psi}_{0,0,k}(t), f \rangle \psi_{0,0,k}(t) + \sum_{i=1}^{M-1} \sum_{j=1}^{\infty} \sum_k \langle \tilde{\psi}_{i,j,k}(t), f \rangle \psi_{i,j,k}(t) \quad (132)$$

$$f(t) = \sum_k \langle f, \psi_{0,0,k}(t), f \rangle \tilde{\psi}_{0,0,k}(t) + \sum_{i=1}^{M-1} \sum_{j=1}^{\infty} \sum_k \langle f, \psi_{i,j,k}(t) \rangle \tilde{\psi}_{i,j,k}(t). (133)$$

Proof: The first part of the proof of this theorem parallels the wavelets tight frames theorem. The functions $\psi_i(t)$ and $\tilde{\psi}_i(t)$ are all compactly supported in an interval of length $N-1$ as in the tight frames case. Define two families of operators

$$\tilde{I}_{i,j} = \sum_k \tilde{\psi}_{i,j,k} >< \psi_{i,j,k} \quad (134)$$

$$I_{i,j} = \sum_k \psi_{i,j,k} >< \tilde{\psi}_{i,j,k}. \quad (135)$$

These operators make sense for all i and for an arbitrary $f \in L^2(\mathbb{R})$. To see this, one cannot use Bessel inequality directly as in the tight frames case, but notice that it suffices to show that given f,

$$\sum_k |\langle f, \psi_{i,j,k} |\rangle^2 \leq C_i \|f\|^2 \quad (136)$$

$$\sum_k |\langle f, \tilde{\psi}_{i,j,k} |\rangle^2 \leq \tilde{C}_i \|f\|^2 \quad (137)$$

for suitable non-negative constants. And this follows from Bessel inequality by noting that the functions $\{\psi_{i,j,(N-1)m+k}/\|\psi_i\|\}$ for fixed i, j and k, with $k = 0, 1, \ldots, N-2$, form an orthonormal family (since they don't overlap) and hence

$$\sum_k |\langle f \psi_{i,j,k} |\rangle^2 \leq (N-1) \|\psi_i\| \|f\|^2. \quad (138)$$

Similarly for the case of $\tilde{\psi}_{i,j,k}$. For any fixed j,

$$I_{0,j} = \sum_{i=0}^{M-1} I_{i,j-1} \text{ and } \tilde{I}_{0,j} = \sum_{i=0}^{M-1} \tilde{I}_{i,j-1}. \quad (139)$$

This result can be proved algebraically and only depends on the fact that the sequences h_i and g_i form a filter bank.

$$\sum_{i=0}^{M-1} I_{i,j-1} = \sum_{i=0}^{M-1} \sum_k \psi_{i,j,k} >< \tilde{\psi}_{i,j,k}$$

$$= \sum_{i=0}^{M-1} \sum_k \sum_l h_i(l)\psi_{0,j,Mk-l} >< \sum_m g_i(m)\tilde{\psi}_{0,j,Mk-m}$$

$$= \sum_{i=0}^{M-1} \sum_k \sum_l h_i(Mk-l)\psi_{0,j,l} >< \sum_m g_i(Mk-m)\tilde{\psi}_{0,j,m}$$

$$= \sum_{l,m} \psi_{0,j,l} >< \tilde{\psi}_{0,j,m} \left[\sum_{i=0}^{M-1} \sum_k h_i(Mk-l) g_i(Mk-m) \right]$$

$$= \sum_{l,m} \psi_{0,j,l} >< \tilde{\psi}_{0,j,m} \left[\delta(l-m) \right]$$

$$= \sum_k \psi_{0,j,k} >< \tilde{\psi}_{0,j,k} = I_{0,j} \tag{140}$$

Lemma 10 has been used in the proof. By repeatedly substituting for $I_{0,j-1}$, $I_{0,j-2}$, etc. we conclude that for a fixed J

$$I_{0,J} = \sum_{i=1}^{M-1} \sum_{j=-\infty}^{J-1} I_{i,j}. \tag{141}$$

Notice that $\psi_{i,j,k}(t)$ approaches the zero function for all i and k as, $j \to \infty$ and hence the infinite sum over j makes sense. We need to show that

$$\lim_{j\to\infty} I_{0,j} f = f \tag{142}$$

for all f in $L^2(\mathbb{R})$. Since $\psi_{0,0,0}(t)$ and $\tilde{\psi}_{0,0,0}(t)$ may *not* form a partition of unity, we cannot use the approach taken in the tight frames case.

But as in [7], we first show that there is weak convergence and then strengthen it to strong convergence in $L^2(\mathbb{R})$. For weak convergence we must have for any f, g in $L^2(\mathbb{R})$

$$\sum_{i=1}^{M-1} \sum_{j=-\infty}^{\infty} \langle g I_{i,j} f \rangle = \lim_{j\to\infty} \langle g, I_{0,j} f \rangle = \langle g, f \rangle. \tag{143}$$

Since both $\psi_{0,j,k}$ and $\tilde{\psi}_{0,j,k}$ approach the Dirac measure at $2^{-j}k$ for large j (recall that $H_0(0) = G_0(0) = \sqrt{M}$ and hence $\hat{\psi}_0(0) = \hat{\tilde{\psi}}_0(0) = 1$), this is expected, and in fact can be proved rigorously by noting that

$$\langle g, I_{0,j} f \rangle \tag{144}$$

$$= \sum_k \langle g, \psi_{0,j,k} \rangle \langle \tilde{\psi}_{0,j,k}, f \rangle$$

$$= M^j \sum_k \langle g, \psi_0(M^j t - k) \rangle \langle \tilde{\psi}_0(M^j t - k, f \rangle$$

$$=M^{-j}\sum_k[\frac{1}{2\pi}\int \hat{g}(\omega)\hat{\psi}_0^*(\frac{\omega}{M^j})e^{i\frac{\omega}{M^j}k}d\omega]\left[\frac{1}{2\pi}\int \hat{f}^*(\lambda)\hat{\tilde{\psi}}_0(\frac{\lambda}{M^j})e^{-i\frac{\lambda}{M^j}k}d\lambda\right]$$

$$=\frac{1}{2\pi}\int \hat{g}(\omega)\hat{\psi}_0(\frac{\omega}{M^j})\left[\hat{f}^*(\lambda)\hat{\psi}_0(\frac{\omega}{M^j})\frac{1}{2\pi M^j}\sum_k e^{i\frac{\omega-\lambda}{M^j}k}d\lambda\right]d\omega$$

$$=\frac{1}{2\pi}\int \hat{g}(\omega)\hat{\psi}_0(\frac{\omega}{M^j})\left[\hat{f}^*(\lambda)\hat{\psi}_0(\frac{\lambda}{M^j})\sum_l \delta(\frac{\omega-\lambda}{M^j}+2\pi l)d\lambda\right]d\omega$$

$$=\sum_l \frac{1}{2\pi}\int \hat{g}(\omega)\hat{\psi}_0(\frac{\omega}{M^j})\hat{f}^*(\omega+2\pi M^j l)\hat{\psi}_0(\frac{\omega}{M^j}+2\pi l)d\omega. \tag{145}$$

All terms in the right hand side are small for $l \neq 0$ and large j and the $l = 0$ term approaches the desired inner product. To strengthen this weak convergence to strong convergence we need the following result.

$$\sum_{i,j,k}|\langle f, \psi_{i,j,k}|\rangle^2 \leq C\|f\|^2 \tag{146}$$

for some $C \geq 0$. Notice that since the sum over i is finite it suffices to show that

$$\sum_{j,k}|\langle f, \psi_{i,j,k}|\rangle^2 \leq C_i\|f\|^2 \tag{147}$$

for some $C_i \geq 0$. This can be proved by elementary means using epsilon-delta arguments. The two important facts required for the proof are that for $i \neq 0$, firstly $\hat{\psi}_i(0) = 0$ and secondly, that $\hat{\psi}(\omega)$ decays faster than $|\omega|^{-\frac{1}{2}}$ for large ω. The first fact follows trivially from the hypothesis that $H_i(0) = G_i(0) = 0$, and the second from the fact that $\psi_i(t)$ is in $L^2(\mathbb{R})$. These two facts can be used to show that for j and k tending to ∞, the numbers $|\langle f, \psi_{i,j,k}|\rangle^2$ decay rapidly enough for the double sum to converge.

We have strong convergence because,

$$\left\|f - \sum_{|j|\leq L,|k|\leq L}\sum_i I_{i,j}f\right\| \tag{148}$$

$$= \sup_{\|g\|=1}\left|\langle g, f\rangle - \sum_{|j|\leq L,|k|\leq L}\sum_i\langle g, I_{i,j}f\rangle\right|$$

$$\leq \sup_{\|g\|=1}\sum_{i=1}^{M-1}\sum_{|j|>L,|K|>L}\left|\sum_{|j|>L,|k|>L}\sum_i\langle g, I_{i,j}f\rangle\right|$$

$$\leq \sup_{\|g\|=1}\left[\sum_{i=1}^{M-1}\sum_{j,k}|\langle g, \psi_{i,j,k}\rangle|\right]^{\frac{1}{2}}\left[\sum_{j>L,k>L}|\langle\tilde{\psi}_{i,j,k}, f\rangle|\right]^{\frac{1}{2}}$$

$$\leq C \left[\sum_{i=1}^{M-1} \sum_{j>L,k>L} |\langle \tilde{\psi}_{i,j,k}, f \rangle| \right]^{\frac{1}{2}}. \tag{149}$$

■

Wavelet Frames are also associated with a multiresolution analysis, but the structure of the multiresolution analysis is more complex than in the tight frames case. Define the spaces

$$V_{i,j} = \text{Span}\{\psi_{i,j,k}(t)\} \quad \text{and} \quad \tilde{V}_{i,j} = \text{Span}\{\tilde{\psi}_{i,j,k}(t)\}. \tag{150}$$

Then for any f in $L^2(\mathbb{R})$, $I_{i,j} f \in \tilde{V}_{i,j}$. For $f \in L^2(\mathbb{R})$, from Equation (139)

$$I_{0,j} f = \sum_{i=0}^{M-1} I_{i,j-1} f \quad \text{and} \quad \tilde{I}_{0,j} f = \sum_{i=0}^{M-1} \tilde{I}_{i,j-1} f \tag{151}$$

and from this we have a decompositions

$$V_{0,j} = V_{0,j-1} \oplus V_{1,j-1} \dots \oplus V_{M-1,j-1} \tag{152}$$

$$\tilde{V}_{0,j} = \tilde{V}_{0,j-1} \oplus \tilde{V}_{1,j-1} \dots \oplus \tilde{V}_{M-1,j-1}. \tag{153}$$

From the fact that we have a frame

$$\lim_{j \to \infty} V_{0,j} = L^2(\mathbb{R}) = \lim_{j \to \infty} \tilde{V}_{0,j} \tag{154}$$

and as before

$$\lim_{j \to -\infty} V_{0,j} = \{0\} = \lim_{j \to -\infty} \tilde{V}_{0,j}. \tag{155}$$

Thus we have a double multiresolution analysis of $L^2(\mathbb{R})$, with a chain of closed subspaces

$$\{0\} \subset \dots V_{0,-1} \subset V_{0,0} \subset V_{0,1} \dots \subset L^2(\mathbb{R}) \tag{156}$$

and

$$\{0\} \subset \dots \tilde{V}_{0,-1} \subset \tilde{V}_{0,0} \subset \tilde{V}_{0,1} \dots \subset L^2(\mathbb{R}). \tag{157}$$

What is the relationship between the spaces $V_{i,j}$ and $\tilde{V}_{i,j}$? When we have biorthogonal wavelet bases it turns out that have for all i, $V_{i,j} \perp \tilde{V}_{l,j}$ when $l \neq j$.

Necessary and sufficient conditions so that the frames become a biorthogonal basis can be derived exactly as in the multiplicity 2 case in [7]. Most of the wavelet frames will be biorthogonal bases. At this, point regularity conditions can also be imposed on the wavelets and scaling functions. The function $\tilde{\psi}_0(t)$ is called the dual scaling function and the functions $\tilde{\psi}_i(t)$ are called dual wavelets. Multiplicity 2 biorthogonal wavelet bases have been constructed by Vetterli [47], Chui [2] and Cohen [7]. But for Chui, the others have used filter bank theory as the starting point. Chui starts from a multiresolution analysis with spline functions and constructs spline wavelets. Since spline functions are as smooth as we require, these are regular wavelet bases. As mentioned earlier, the reasons to require the regularity of the functions are usually suspect.

§6. Computation in filter banks

6.1. Polyphase structure for FIR filter banks

For the general filter bank the most efficient way to implement the analysis and synthesis banks is to use polyphase filter structures [38]. The essential idea is to avoid computation of the filter outputs of the analysis bank at all time instants since only every Mth output is required. From Equation (57)

$$\mathbf{d}_p = H_p \mathbf{x}_p. \tag{158}$$

This can be represented in a block diagram as in Figure 7 and constitutes the polyphase implementation of the analysis filter bank.

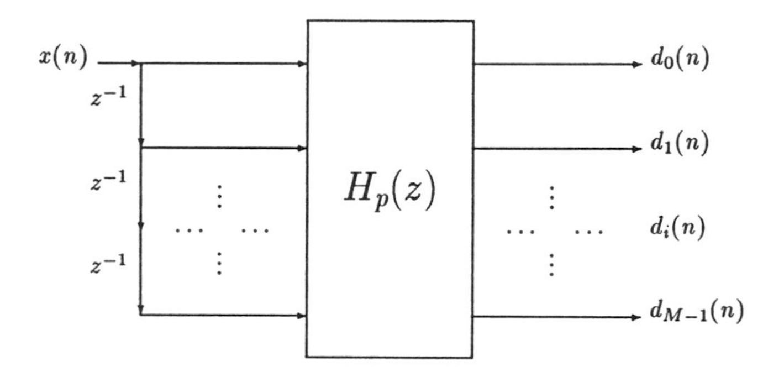

Figure 7. Polyphase structure for the analysis bank.

It is clear that the computational complexity of this implementation is less than that of the direct implementation by a factor of M, a substantial gain. And this is not surprising, since only every Mth output of the filters is being computed. Once again, this brings out the usefulness of considering the analysis as a MIMO linear shift-invariant system.

As for the synthesis bank, from Equation (58) we have

$$\mathbf{y}_p = G_p^T \mathbf{d}_p \tag{159}$$

and this can be implemented as shown in Figure 8. As in the analysis bank this implementation is more efficient by a factor of M.

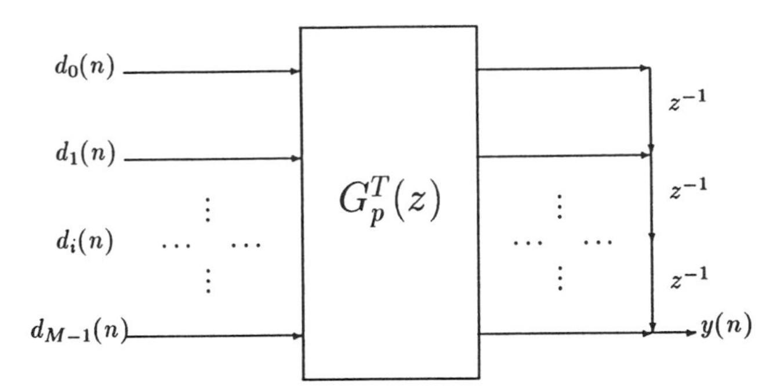

Figure 8. Polyphase structure for the synthesis bank.

6.2. The cascade structure for unitary filter banks

Given an FB, in general, the best possible implementation is the polyphase implementation. However, if the FBs are unitary, the factorization result in Lemma 4 can be used to further simplify the polyphase structure into smaller computational units. There are many other factorizations ([44] and [1]). It is useful to recall that for a constant unitary matrix, there are two well known factorizations, one based on Given's rotations, and the other based on Householder reflections [16]. Analogously, in the case of matrices unitary on the unit circle there is a factorization based on Given's rotations [44] and another based on Householder reflections. The factorization discussed in Section 3 is based on Householder reflections. From a computational perspective, in the two channel case, we find that the Given's rotation based factorizations gives a (moderately) more efficient structure to implement the filter bank. The structure is called the *lattice* structure [45]. In the general M channel case, the Householder reflection method is much more efficient. Factorization approaches are useful only for IIR and FIR rational, stable unitary matrices.

Consider an FIR filter bank with filters $h_i(n)$, all of length NM. For simplicity, let us further assume that they vanish for n less that 0 and n greater than $NM - 1$. Then the polyphase matrix $H_p(z)$ will be of (McMillan) degree $N-1$. Hence, by Lemma 4 H_p can be parameterized by $N-1$ unitary vectors $v_i, i = 1, 2, \ldots, N-1$ and a unitary matrix U_0.

$$H_p(z) = [I - v_{N-1}\tilde{v}_{N-1} + z^{-1}v_{N-1}\tilde{v}_{N-1}] \ldots [I - v_1\tilde{v}_1 + z^{-1}v_1\tilde{v}_1]U_0. \quad (160)$$

From this, it is clear that the analysis FB can be implemented as shown in Figure 9. The structure is regular and can be easy be coded. The thick

lines in Figure 9 correspond to vector data, and the thin lines to scalar data. Therefore the number of storage elements required to implement the entire filter bank is $N-1$, the McMillan degree of H_p and, hence, the realization

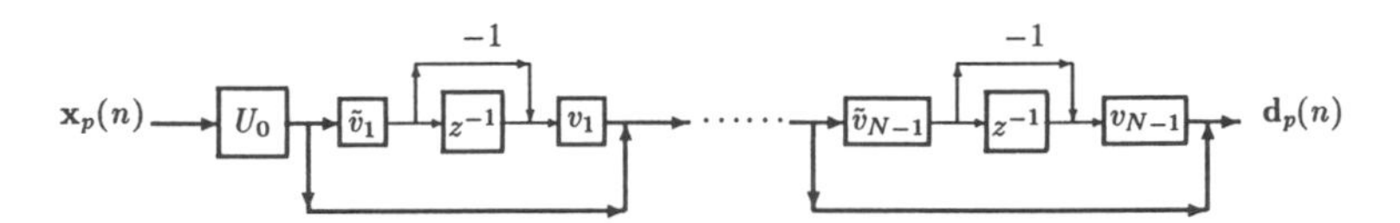

Figure 9. Cascade implementation of the analysis filter bank.

of the analysis FB is minimal. Let us estimate the computational complexity of the analysis FB. Recall that since each of the filters is of length NM, computation by direct convolution requires NM^3 multiplies and $(NM-1)M^2$ adds (for each vector output $\mathbf{d}_p$), while the polyphase structure requires NM^2 multiplies and $(N-1)M^2$ additions. As for the cascade implementation, there are $N-1$ blocks, followed by a unitary matrix multiplication. Since each block requires $2M$ multiplies and M adds, the entire cascade requires $[(N-1)2M+M^2]$ multiplies and $[(N-1)M+(M-1)M]$ additions. These counts could be moderately improved by taking advantage of the fact that the unitary matrix multiplication can be done more efficiently than a general matrix multiplication. As for storage requirements in a direct form implementation of $H_p(z)$, the obvious convolution implementation requires $(NM-1)M$ scalar storages (delays), the polyphase implementation $(N-1)M$ delays, and the cascade implementation requires just $N-1$ delays (the minimal possible since the McMillan degree of $H_p(z)$ is $N-1$).

The polyphase synthesis bank matrix is given by

$$G_p^T(z) = \tilde{H}_p(z) = \tilde{U}_0[I - v_1\tilde{v}_1 + zv_1\tilde{v}_1]\ldots[I - v_{N-2}\tilde{v}_{N-2} + zv_{N-2}\tilde{v}_{N-2}] \quad (161)$$

The corresponding structure for the synthesis bank is shown in Figure 10. The

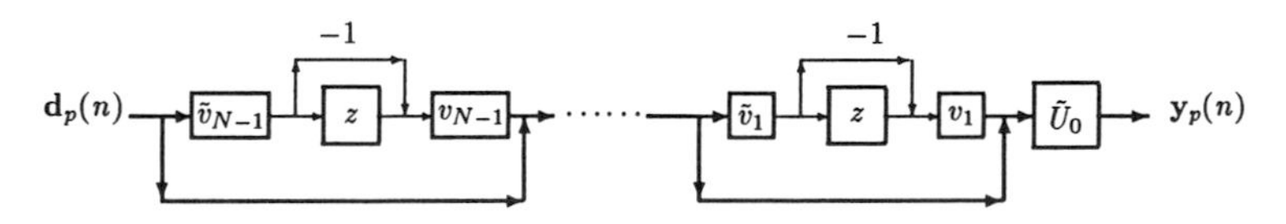

Figure 10. Cascade implementation of synthesis filter bank.

computational and storage complexity of this structure is identical to that of the analysis bank. However, because z-blocks are involved, the implementation is non-causal and cannot be realized. A fix is to reorganize the structure as shown in Figure 11. This implementation has the same complexity, but requires, $(M+1$ delays per block for a total of $(N-1)(M+1)$ storage elements.

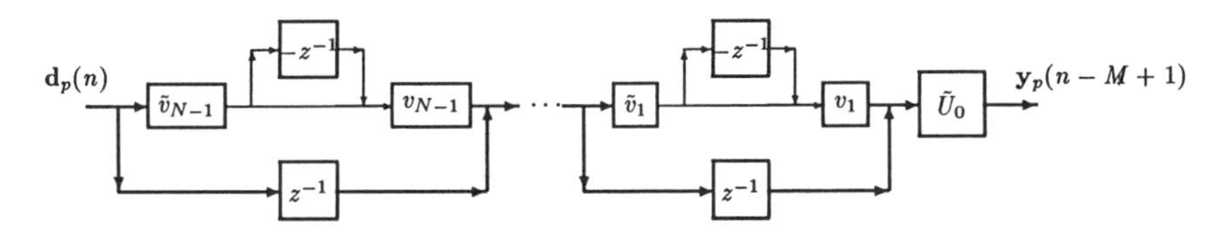

Figure 11. Cascade implementation of synthesis filter bank - causal.

We now show that the synthesis bank can also be implemented with the same number of delays and complexity as the analysis bank. Moreover, it turns out that the same program that was used for the analysis bank could be used for the synthesis bank. First note that $z^{-(N-1)}G^T(z)$, is a unitary polynomial matrix in z^{-1} and hence by Lemma 5, there exist unitary vectors w_i and a unitary matrix W_0 such that

$$z^{-(M-1)}G^T(z) = [I - w_{N-1}\tilde{w}_{N-1} + z^{-1}w_{N-1}\tilde{w}_{N-1}]\dots[I - w_1\tilde{w}_1 + z^{-1}w_1\tilde{v}_1]W_0. \tag{162}$$

Therefore,

$$\begin{aligned} z^{-(M-1)}\mathbf{y}_p(z) =& \{[I - w_{N-1}\tilde{w}_{N-1} + z^{-1}w_{N-1}\tilde{w}_{N-1}]\dots \\ & [I - w_1\tilde{w}_1 + z^{-1}w_1\tilde{w}_1]U_0\}\mathbf{d}_p(z) \end{aligned} \tag{163}$$

and this can be implemented exactly as in Figure 9. The computational and storage complexity is minimal. However, there is one disadvantage. Now, one has to compute the factorization parameters of both H_p and $z^{M-1}G_p(z)$ and store both of them.

§7. Computation in wavelet bases

7.1. Generation of ψ_i

Given a multiplicity M wavelet frame or tight frame, the first question for applications is to know how they look like. This section deals with the computation of the samples of ψ_i. It turns out that for a multiplicity M frame, computation of arbitrary samples of ψ_i is relatively difficult, while samples at the M-adic rationals (*i.e.*, numbers of the form $\frac{i}{M^J}$ for fixed J) can be computed easily. It turns out that M-adic samples of ψ_i can be computed using the filters h_i in the filter bank associated with the frame. We give two efficient methods to compute ψ_i at the M-adics. Notice that the samples do not really give a picture of ψ_i unless the J is large *and* the functions ψ_i are continuous, for they may behave wildly in between samples. Since it only makes sense to sample continuous functions, we will assume that ψ_i are continuous. We will also assume that ψ_i are compactly supported, or equivalently that $h_i(n)$ are FIR.

7.1.1. Interpolation method

In this method first $\psi_0(n)$ is obtained by solving the following eigenvalue problem.

$$\psi_0(n) = \sqrt{M} \sum_{k=0}^{N-1} h_i(k)\psi_0(Mn-k). \tag{164}$$

From this information $\psi_0(\frac{i}{M})$ is computed for all i. Using this information $\psi_0(\frac{i}{M^2})$ is computed. This recursive interpolation process is continued until the samples at the required accuracy, namely $\psi_0(\frac{i}{M^J})$ is obtained. Define the N-vector functions $\Psi_i(t), i = 0, 1, \ldots, M-1$ on $[0, 1)$, made up of the samples of $\psi_i(t)$ respectively at the integers.

$$[\Psi_i]_j(t) = \psi_i(t+j); \quad 0 \le j < N. \tag{165}$$

Without loss of generality we will assume that ψ_i is supported in $[0, N-1]$. This is equivalent to assuming that the $h_i(n)$ are finite sequences running from $n = 0$ to $n = N-1$. Since ψ_i is supported in $[0, N-1]$, it is clearly completely determined by $\Phi_i(t)$ for $t \in [0,1]$. Now define $N \times N$ matrices $M_{i,j}, i = 0, 1, \ldots, M-1, j = 0, \ldots, M-1$ by

$$(M_{i,j})(k,l) = \sqrt{M} h_i(Mk - l + i). \tag{166}$$

Then the recursive equations

$$\psi_i(t) = \sqrt{M} \sum_k h_i(k)\psi_0(Mt-k) \tag{167}$$

can be written in the compact form

$$\Phi_i(t) = M_{i,j}\Phi_0(Mt-j) \text{ for } \tfrac{j}{M} \le t \le \tfrac{j+1}{M},\ i = 0, 1, \ldots, M-1. \tag{168}$$

In particular when $t = 0$ and $i = 0$, we have

$$\Phi_0(0) = M_{0,0}\Phi_0(0) \tag{169}$$

an eigenvalue problem that can be solved for $\Phi_0(0)$, which give the values of $\psi_0(n)$. As an aside note that for any i, $\Phi_0(\frac{i}{M-1}) = M_{0,i}\Phi_0(\frac{i}{M-1})$. For continuous ψ_0, it is completely specified by its values at the M-adic rationals, that is, numbers of the form

$$\sum_{N_1}^{N_2} t_i M^i$$

with $t_i \in \{0, 1, \ldots, M-1\}$. Let t be an M-adic rational in $[0, 1)$. Then t has an M-ary expansion $.t_1t_2t_3\ldots t_K$. Then from Equation (168) by repeated substitution we obtain

$$\Psi_i(.t_1t_2\ldots t_K) = M_{i,t_1}M_{0,t_2}\cdots M_{0,t_K}\Psi_0(0). \tag{170}$$

Thus given any M-adic rational t, $\psi_i(t)$ can be obtained in a finite number of steps. Though the notation suggests that matrix products are involved in the computation, in reality, once $\phi_i(t)$ is obtained at the integers (by solving the system of linear equations in Equation (169) the rest of the computation can be carried out using the filter bank associated with the filters h_i [19]. The idea of representing Equation (129) as a matrix product was first used by H. Resnikoff to study the relationship between Weierstrass functions and compactly supported wavelets [39], and later used by I. Daubechies to study the differentiability properties of wavelets ([9] and [10]) both in the multiplicity 2 case.

7.1.2. Infinite product method

This method to compute the samples of ψ_i is approximate and works best for for M-adics of the form $\frac{i}{M^J}$ where J is approximately 8 to 10. This technique, essentially due to I. Daubechies [5], is based on the fact that ϕ_j^0 in the proof of Theorem 13, is always a staircase function, whose heights can be directly computed using the filter bank associated with the wavelet. These heights are taken to be the approximate samples of ψ_i. A very good approximation is obtained after about 8 or 9 iterations *i.e.*, the staircase approximation corresponding to ϕ_0^8 or ϕ_0^9.

In the first method the values of ϕ at the integers are first obtained exactly. Hence the values at the dyadic rationals are also obtained exactly within limits of numerical errors. In the second method since an infinite product is truncated only an approximation of ϕ is obtained. An advantage with the infinite product method is that there is no need to solve a system of linear equations. Both techniques are identical in their implementation in that the interpolation starts with the values at the integers and goes through a process while the infinite convolution method starts off with a delta and goes through the same process. Eventually they both converge to the same values.

7.2. Analysis/synthesis in wavelet bases

This section deals with the computational aspects related to analysis and synthesis of signals in wavelet bases. The *wavelet* expansion of a signal with respect to a wavelet frame $\psi_{i,j,k}$ is given by

$$f(t) = \sum_{i=1}^{M-1} \sum_{j,k} \langle \psi_{i,j,k}(t), f \rangle \tilde{\psi}_{i,j,k}. \tag{171}$$

The DWT of the function with respect to this frame is defined by

$$Wf(i,j,k) = \langle \psi_{i,j,k}(t), f \rangle \tag{172}$$

where $i = 1, 2, \ldots, M-1$. For computational purposes we define the Discrete Scaling Transform by

$$Wf(0,j,k) = \langle \psi_{0,j,k}(t), f \rangle. \tag{173}$$

Clearly for any fixed j the DST contains all the information in $V_{0,j}$.

Since only finite sequences can be stored and manipulated, both j and k have to be restricted to be finite. Since most signals are of finite duration it is natural to assume that $f(t)$ is compactly supported, and then if $\psi_i(t)$ are compactly supported, $Wf(i,j,k)$ vanishes outside a finite set.

Since $V_{0,j} \to L^2(\mathbb{R})$, any $f \in L^2(\mathbb{R})$ can be approximated to any desired accuracy by a function in $V_{0,j}$ for large enough j, say J_f (finest scale). Therefore, the DWT does not need to be computed for $j \le J_f$. Similarly some coarsest scale of interest J_c is also determined. Information in coarser scales is retained in the DST at that scale. That is,

$$f(t) = \sum_k Wf(0, J_c, k)\tilde{\psi}_{0,J_c,k}(t) + \sum_{i=1}^{M-1} \sum_{j=J_c}^{J_f} \sum_k Wf(i,j,k)\tilde{\psi}_{i,j,k}. \tag{174}$$

7.2.1. The projection onto the finest scale

In order to compute the DWT, the most efficient way is to start with $W(0, J_f, k)$ and then use the filter bank to compute the DWT. So the first step is the projection onto the finest scale V_{0,J_f}. In most applications $Wf(0, J_f, k)$ cannot be computed exactly, partly because the inner-products involved can seldom be computed analytically and partly because only the samples of $f(t)$ may be available.

From the samples, one can use polynomial, spline or any other interpolation technique and then compute $Wf(0, J_f, k)$ by numerical quadrature. This procedure is greatly simplified by the fact that the moments of the function $\psi_0(t)$ can be *exactly* computed and hence numerical quadrature reduces to a discrete convolution ([19] and [18]). Let $\mu(i,k)$ and $m(i,k)$ denote the kth moment of $\psi_i(t)$ and $h_i(n)$ respectively for all i.

$$\mu(i,k) = \int t^k \psi_i(t)\, dt \tag{175}$$

$$m(i,k) = \sum_n n^k h_i(n). \tag{176}$$

Then from Equation (129) and Equation (130), it can be shown that

$$\mu(i,k) = \frac{1}{M^{k+\frac{1}{2}}} \sum_{j=0}^{k} \binom{k}{j} m(i,j)\mu(0,k-j). \tag{177}$$

Thus the moments of all the functions $\psi_i(t)$ can be computed from the moments of the sequences $h_i(n)$ associated with them.

The fact that $\mu(0,k)$ can be exactly computed, and that $\psi_0(t)$ is compactly supported, can be combined to give a nice technique to compute $W(0, J_f, k)$ from the samples of f. For every k, if a local polynomial approximation of $f(t)$

around the point $M^{-J_f}k$ is taken in order to compute $W(0, J_f, k)$, then it can be shown that there exists a sequence $e(k)$, that depends on the order of the approximation, such that

$$W(0, J_f, k) = e(k) * f(M^{-J_f}k). \tag{178}$$

Perhaps the most common approach to assume that $f(M^{-J_f}k)$ is given and take $Wf(0, J_f, k) \approx f(M^{-J_f}k)$. The argument for this is that since $\psi_0(t)$ is compactly supported and $\int \psi(t)\,dt = 1$, for large j, $\psi_{0,j,k}$ approaches the Dirac delta measure at $M^{-j}k$. In practical cases, especially with wavelets that are reasonably regular (like the Daubechies' wavelets in [5]), this approximation gives very good results (see [26] and [18]) because of certain interesting relationships between the the moments $\mu(0, k)$.

7.2.2. Analysis

We are given $Wf(0, j, k)$ and we want to compute the numbers $Wf(i, j-1, k)$. From Equation (130) we can show that,

$$\tilde{\psi}_{i,j-1,n} = \sum_k h_i(k)\tilde{\psi}_{0,j,Mn+k} \tag{179}$$

and therefore

$$Wf(i, j-1, n) = \sum_k h_i(k)Wf(0, j, Mn+k) \tag{180}$$

and this equation can be implemented with an analysis filter bank where the analysis filters are the time reversals of $h_i(n)$. The input to the filter bank is $W(0, j, k)$ and the output is the desired numbers.

7.2.3. Synthesis

We are given $Wf(i, j-1, k)$ and we want $Wf(0, j, k)$. They can be shown to be related by the formula

$$Wf(0, j, n) = \sum_{i=0} M - 1 g_i(n - Mk)Wf(i, j-1, k) \tag{181}$$

and this is just the synthesis equations of a filter bank.

§8. Hilbert space decompositions

We have seen that wavelet frames and tight frames can be constructed from filter banks and consequently wavelet analysis can be done efficiently using discrete representations. In this section we show that filter banks not only induce decompositions of the spaces $V_{0,j}$ associated with the scaling function, but more generally all separable Hilbert spaces. Recall that when frequency decompositions were introduced in Section 2, it was mentioned that frequency decompositions can be thought of as a special case of the decomposition of a separable Hilbert space. Filter banks give one such structured, efficient decomposition. Furthermore, filter bank decompositions give a natural concept of *scale* to Hilbert spaces that are so decomposed.

Lemma 15. *(Tight Frames Decomposition Lemma) Let $H \in U_{M \times M}$. For a separable Hilbert space $\mathcal{H}$ equipped with a orthonormal basis (tight frame) $\{\phi_n\}$, H induces a natural tight frame decomposition*

$$\mathcal{H} = \mathcal{H}_0 \oplus \mathcal{H}_1 \ldots \oplus \mathcal{H}_{M-1}. \tag{182}$$

Proof: Take $H(z) = H_p(z)$ for some unitary filter bank. Let $h_i(n)$ denote the corresponding filters in the filter bank. Define

$$\psi_{i,k} = \sum_n h_i(Mk-n)\phi_n \tag{183}$$

and let $\mathcal{H}_i = \text{Span}\{\psi_{i,k}\}$. Then, $\mathcal{H}_i$ is the desired decomposition. Let I_i be an operator from $\mathcal{H}$ into $\mathcal{H}_i$ defined by

$$I_i = \sum_k \psi_{i,k} >< \psi_{i,k}. \tag{184}$$

Then by application of Lemma 9

$$\begin{aligned}
\sum_{i=0}^{M-1} I_i &= \sum_{i=0}^{M-1} \sum_k \psi_{i,k} >< \psi_{i,k} \\
&= \sum_{i=0}^{M-1} \sum_k \sum_l h_i(Mk-l)\phi_l >< \sum_m h_i(Mk-m)\phi_m \\
&= \sum_{l,m} \phi_l >< \phi_m \left[\delta(l-m)\right] \\
&= \sum_k \phi_k >< \phi_k = I,
\end{aligned} \tag{185}$$

the identity operator on $\mathcal{H}$. Furthermore, if $\{\phi_n\}$ forms an orthonormal basis for $\mathcal{H}$, then $\{\psi_{i,k}\}$ forms an orthonormal basis for $\mathcal{H}_i$. To see this note using Lemma 7 that,

$$\begin{aligned}
\langle \psi_{i,k} \psi_{i,l} \rangle &= \langle \sum_m h_i(Mk-m)\phi_m \sum_n h_i(Ml-n)\phi_n \rangle \\
&= \sum_m \sum_n h_i(Mk-m) h_i(Ml-n) \langle \phi_m, \phi_n \rangle \\
&= \sum_n h_i(Mk-n) h_i(Ml-n) = \delta(k-l).
\end{aligned} \tag{186}$$

■

The Lemma is extremely general and depends only on the fact that we have a unitary matrix H (there are no "DC" conditions that were required to construct wavelet tight frames). Let us assume that we have a multiplicity M wavelet tight frame. Then the space $V_{0,j}$ associated with it could be decomposed using any unitary matrix H, not necessarily the one from which $V_{0,j}$ was constructed. In particular, if the unitary matrix is the one from which $V_{0,j}$ is constructed we have a wavelet decomposition.

Lemma 16. *(Frames Decomposition Lemma) Let $H(z) \in GL^{\infty}_{M\times M}$. For a separable Hilbert space $\mathcal{H}$ with a biorthogonal basis (frame) $\{\phi_k\}$ and dual basis (frame) $\{\tilde{\phi}_k\}$, H induces two (dual) frame decompositions*

$$\mathcal{H} = \mathcal{H}_0 \oplus \mathcal{H}_1 \ldots \oplus \mathcal{H}_{M-1} \tag{187}$$

and

$$\mathcal{H} = \tilde{\mathcal{H}}_0 \oplus \tilde{\mathcal{H}}_1 \ldots \tilde{\mathcal{H}}_{M-1}. \tag{188}$$

Proof: By assumption, H is invertible on the unit circle. Let $G^T = H^{-1}$. Then, $G^T H = HG^T = I$. Consider a filter bank with $H_p = H$ and $G_p = G$. Define

$$\psi_{i,k} = \sum_n h_i(Mk-n)\phi_n; \tilde{\psi}_{i,k} = \sum_n g_i(Mk-n)\phi_n. \tag{189}$$

Let $\mathcal{H}_i = \text{Span}\{\psi_{i,k}\}$ and let $\tilde{\mathcal{H}}_i = \text{Span}\{\tilde{\psi}_{i,k}\}$. Then, $\mathcal{H}_i$ and $\tilde{\mathcal{H}}_i$ give the desired decompositions. For $f \in \mathcal{H}$ consider the following operator whose range is in $\mathcal{H}_i$

$$I_i = \sum_k \psi_{i,k} >< \tilde{\psi}_{i,k}. \tag{190}$$

By application of Lemma 10

$$\begin{aligned}\sum_{i=0}^{M-1} I_i &= \sum_{i=0}^{M-1}\sum_k \psi_{i,k} >< \tilde{\psi}_{i,k} \\ &= \sum_{i=0}^{M-1}\sum_k\sum_l h_i(Mk-l)\phi_l >< \sum_m g_i(Mk-m)\phi_m \\ &= \sum_{l,m} \phi_l >< \phi_m\,[\delta(l-m)] \\ &= \sum_k \phi_k >< \tilde{\phi}_k = I,\end{aligned} \tag{191}$$

the identity operator in $\mathcal{H}$ by assumption. If moreover, $\{\phi_n\}$ and $\{\tilde{\phi}_n\}$ form a biorthogonal basis for $\mathcal{H}$, then $\{\psi_i, k\}$ and $\{\tilde{\psi}_{i,k}\}$ form a biorthogonal basis for $\mathcal{H}_i$. This follows from Lemma 8, since

$$\langle \psi_{i,k}, \tilde{\psi}_{i,l}\rangle = \left\langle \sum_m h_i(Mk-m)\phi_m, \sum_n g_i(Ml-n)\tilde{\phi}_n \right\rangle \tag{192}$$

$$= \sum_m \sum_n h_i(Mk - m) g_i(Ml - n) \langle \phi_m, \tilde{\phi}_n \rangle$$
$$= \sum_n h_i(Mk - n) g_i(Ml - n) = \delta(k - l).$$

■

Notice how in both these proofs the dual lemmas, Lemma 8 and Lemma 10, play a role, one to show that we have a frame and the dual to show that if we have a basis to start with then the decomposition also has a basis.

We now discuss the corresponding recomposition lemmas for filter banks.

Lemma 17. *(Tight Frames Recomposition Lemma) Let $G \in U_{M \times M}$. Then, for a collection of M separable Hilbert spaces $\mathcal{H}_i$, equipped with an orthonormal bases (tight frames) $\{\psi_{i,k}\}$, respectively, G induces a natural orthonormal basis (tight frame) ϕ_k for $\mathcal{H}$ the direct sum of these separable Hilbert spaces. That is,*

$$\mathcal{H} = \mathcal{H}_0 \oplus \mathcal{H}_1 \ldots \oplus \mathcal{H}_{M-1} \tag{193}$$

and ϕ_k is an orthonormal basis (tight frame) for $\mathcal{H}$.

Proof: Take $G(z) = G_p(z)$ for some unitary filter bank. Let $g_i(n)$ denote the corresponding filters in the filter bank. Define,

$$\phi_k = \sum_i \sum_n g_i(k - Mn) \psi_{i,n}. \tag{194}$$

Then from Lemma 7

$$\begin{aligned}
\sum_k \phi_k >< \phi_k &= \sum_i \sum_l g_i(k - Ml) \psi_{i,l} >< \sum_j \sum_m g_j(k - Mm) \psi_{i,m} \\
&= \sum_{i,j,l,m} \psi_{i,l} >< \psi_{i,m} \left[\sum_k g_i(k - Ml) g_j(k - Mm) \right] \\
&= \sum_{i,j,l,m} \psi_{i,l} >< \psi_{i,m} \left[\sum_k g_i(k) g_j(k - M(l - m)) \right] \\
&= \sum_{i,j,l,m} \psi_{i,l} >< \psi_{i,m} [\delta(i - j) \delta(l - m)] \\
&= \sum_{i,l} \psi_{i,l} >< \psi_{i,l}.
\end{aligned} \tag{195}$$

pg Using Lemma 9 we can now show that, if, for all i, $\{\psi_{i,k}\}$ is an orthonormal basis for $\mathcal{H}_i$, then $\{\phi_n\}$ is an orthonormal basis for $\mathcal{H}$. ■

Lemma 18. *(Frames Recomposition Lemma) Let $G \in GL^{\infty}_{M\times M}$. Then, for a collection of M separable Hilbert spaces $\mathcal{H}_i$, equipped with a biorthogonal bases (frames) $\{\psi_{i,k}\}$ and dual bases (frames) $\{\tilde{\psi}_{i,k}\}$, respectively, G induces a natural biorthogonal basis (frame) ϕ_k with dual $\tilde{\phi}_k$ for $\mathcal{H}$ the (external) direct sum of these separable Hilbert spaces. That is, if*

$$\mathcal{H} = \mathcal{H}_0 \oplus \mathcal{H}_1 \ldots \oplus \mathcal{H}_{M-1} \tag{196}$$

then, ϕ_n and $\tilde{\phi}_n$ form dual frames $\mathcal{H}$.

Proof: Let $G(z) = G_p(z)$ for some unitary filter bank. Let $H^T = G^{-1}$, and $H_p = H$. Then (H_p, G_p) form a filter bank with filters h_i and g_i.

$$\phi_k = \sum_i \sum_n h_i(k - Mn)\psi_{i,n} \tag{197}$$

$$\tilde{\phi}_k = \sum_i \sum_n g_i(k - Mn)\psi_{i,n}. \tag{198}$$

Then from Lemma 8

$$\begin{aligned}
\sum_k \phi_k >< \tilde{\phi}_k &= \sum_i \sum_l h_i(k - Ml)\psi_{i,l} >< \sum_j \sum_m g_i(k - Mm)\tilde{\psi}_{i,m} \\
&= \sum_{i,j,l,m} \psi_{i,l} >< \tilde{\psi}_{i,m} \left[\sum_k h_i(k - Ml) g_j(k - Mm) \right] \\
&= \sum_{i,j,l,m} \psi_{i,l} >< \psi_{i,m} \left[\delta(i - j)\delta(l - m) \right] \\
&= \sum_{i,l} \psi_{i,l} >< \psi_{i,l}
\end{aligned} \tag{199}$$

Again, if for all i, $\{\psi_{i,k}\}$ forms a biorthogonal basis then $\{\phi_n\}$ forms a biorthogonal basis from Lemma 10. ■

Let us assume that we start with a Hilbert space and decompose it into M subspaces using a filter bank. Now each of the subspaces could themselves be decomposed into a different (not M) number of subspaces using some other filter bank and so on. Since the filter banks could all be different and of different multiplicities we have a rich class of decompositions for the Hilbert space. Moreover, some of the subspaces could also be recombined using filter banks again (as allowed by the recomposition lemmas). Notice also, that the filter banks used for the recomposition may have nothing to do with filter banks used for decomposition.

At this point, there is the need for the concept of a *scale* for these decompositions. To illustrate how it occurs, let us decompose $\mathcal{H}$ into 3 subspaces using a filter bank. Now if we recombine $\mathcal{H}_1$ and $\mathcal{H}_2$ to give $\mathcal{G}$ using some other filter bank, effectively, $\mathcal{H}$ has been decomposed into 2 subspaces. That is,

$$\mathcal{H} = \mathcal{G} \oplus \mathcal{H}_3. \tag{200}$$

Clearly this decomposition of $\mathcal{H}$ is quite different from any decomposition obtained with a 2 channel filter bank. In order to distinguish between these decompositions we have to introduce the concept of a scale. Given $\mathcal{H}$, all the subspaces obtained by decomposing it with an M channel filter bank are said to be at scale $\frac{1}{M}$. Notice that scale is also a measure of the number of basis elements in $\mathcal{H}_i$ relative to $\mathcal{H}$. And if we recombine them the scale of the resulting space is M times the scale of the constituent spaces. However, we have not been able to come up with a clean rigorous definition of *scale* for these decompositions. The full impact of these general decomposition/recomposition ideas is a subject of great interest. In any case, using the ideas described above we obtain the following result.

Theorem 19. *For every finite rational partition of the interval* $[0,1)$ *there is a decomposition of every separable Hilbert space into subspaces whose scales correspond to partitions of the interval.*

Proof: Obvious ■

We now use these results to discuss some of the issues concerning the regularity of wavelets in practical computations. It is often assumed, that wavelets have introduced an underlying set of continuous functions to filter bank analysis. Be that as it may, it is important to realize that digital signal processing had its roots in the Shannon sampling theorem that allows one to go from continuous time to discrete time. So underlying continuous functions for *analog* interpretation of filter banks are nothing new. Now consider the analysis of a signal wrt an orthonormal wavelet basis $\psi_{i,j,k}(t)$. Assume further that the samples (usually finite) of the signal f are given. Usually, the samples are *assumed* to be $Wf(0, J_f, k)$ and the numbers $Wf(i,j,k)$, for $J_c \leq j < J_f$ are then computed using a filter bank. Notice this is an approximate method to compute the DWT. We now try to give an *exact* interpretation of this process.

The idea is to interpret the samples $f(\frac{n}{T})$ as the the Nyquist rate samples of a bandlimited signal. These turn out to be the expansion coefficients with respect to the *sinc* basis. Notice that the sinc function is infinitely differentiable. There is no problem with the fact that there are only finitely many samples of f, since bandlimited signals could have finitely many Nyquist samples. Let $\mathcal{H}$ denote the space of bandlimited signals in $[0, \pi/T)$. Then $f \in \mathcal{H}$. By the decomposition lemmas discussed earlier, what we are doing is decomposing $\mathcal{H}$, recursively, and the numbers $W(i,j,k)$ obtained are precisely the expansion coefficients with respect to the bases in the decomposition subspaces. Since all the basis functions at any level are linear combinations (finite in the FIR case) of the sinc function, they are infinitely smooth, regardless, of whether the filters in the filter bank give rise to *regular* wavelets. Hence, even assuming we have a filter bank with tight frame wavelets, there is no reason to impose constraints on the sequences h_i so that they give smooth wavelets, because all practical computations can be interpreted as being done with some other smooth basis. The regularity of the filters are important *only* if we are going to have an infinite cascade of filter banks ($J_c \to \infty$), a practical impossibility.

The important implication of the above argument is that, for best frequency localization, the filters $h_i(n)$ need not be the ones that give rise to wavelet bases. Does this mean that it is unimportant to construct regular wavelet bases? The answer is no. In numerical analysis applications (like the approximation of differential operators), the vanishing moments property of the scaling vector (which causes smoothness of the corresponding $\psi_0(t)$) is important [21]. However, in signal processing applications, usually regularity is unimportant.

§9. Summary

The concept of frequency decompositions was used to introduce short-time Fourier analysis and wavelet analysis in a common framework. DWT type frequency decompositions were used to illustrate how naturally filter banks arise in such frequency decompositions. We then gave a concise account of filter bank theory as it relates to wavelet theory. Using results in filter bank theory (purely algebraic) and some analytical constraints on the filter banks, we showed that multiplicity M wavelet frames and wavelet tight frames can be obtained. We showed how multiresolution analyses arise naturally in the context of wavelet frames. Though precise conditions for wavelet frames to be bases are known, we did not discuss it since most wavelet frames are bases. Since all the computations in wavelet analysis are done in efficient structures for filter banks, we discussed the polyphase approach to general filter banks and the cascade approach to unitary filter banks. We then discussed computational aspects in basic analysis of signals in wavelet frames. Finally we explored the powerful algebraic aspects of filter banks, how they induce decompositions of abstract separable Hilbert spaces, and introduced a notion of *scale* in these abstract spaces. This was used to illustrate how most of the approximate practical computations in *wavelet* bases turn out to be *exact* calculations in some *smooth* bases when suitably interpreted. This meant that the regularity of the wavelets and scaling functions are probably not critical for practical applications, contrary to the popular belief. However, the regularity of the scaling vectors and the wavelet vectors may be important for some applications, and in those cases corresponding scaling function and wavelets turn out to be regular.

Acknowledgments

The authors would like to thank colleagues at Aware., Inc., Cambridge, MA, and members of the Computational Mathematics Laboratory, Rice University, for helpful discussions. Comments and suggestions from Howard Resnikoff and Wayne Lawton at Aware, and Raymond Wells and Jan Erik Odegard at CML are appreciated.

References

1. Belevitch, V., *Classical Network Synthesis*. Holden-Day, San Francisco, CA, 1968.
2. Chui, C. K., Wavelets and spline interpolation, in *Wavelets, Subdivisions, and Radial Functions*, W. Light (ed.), Oxford University Press, Oxford, 1991, to appear.
3. Chui, C. K. and J. Z. Wang. On compactly supported spline wavelets and a duality principle, *Trans. Amer. Math. Soc.*, 1991, to appear.
4. Cohen, L., Time-frequency distributions - a review, *Proc. IEEE* **77** (7) (1989), 941–981.
5. Daubechies, I., Orthonormal bases of compactly supported wavelets, *Comm. Pure and Appl. Math.* **41** (1988), 909–996.
6. Daubechies, I., The wavelet transform, time-frequency localization and signal analysis, *IEEE Trans. Inform. Theory* **36** (1990), 961–1005.
7. Daubechies, I., A. Cohen, and J. C. Feauveau, Biorthogonal bases of compactly supported wavelets, *Comm. Pure and Appl. Math.*, 1991, to appear.
8. Daubechies, I., A. Grossmann, and Y. Meyer, Painless non-orthogonal expansions. *J. Math. Phys.* **27** (5) (1986), 1271–1293.
9. Daubechies, I. and J. C. Lagarias, Two-scale difference equations: global regularity of solutions, *SIAM J. Appl. Math.*, to appear.
10. Daubechies, I. and J. C. Lagarias, Two-scale difference equations : local regularity and fractals, *SIAM J. Appl. Math.*, to appear.
11. Dunn, H. K., R. Koenig, and L. Y. Lacy, The sound spectrogram, *J. Acoustic Soc. Amer.* **18** (1946), 19–49.
12. Esteban D. and C. Galand, Applications of quadrature mirror filters to split band voice coding schemes, in *Proc. ICASSP* (1977).
13. Flanagan, J., *Speech Analysis and Synthesis Perception*, Springer-Verlag, New York, 1972.
14. Francis, B., *A Course in H^∞ Control Theory*, Springer-Verlag, Lecture Notes in Control and Information Sciences, 1987.
15. Friedlander B. and B. Porat, Detection of transient signals by the Gabor representation, *IEEE ASSP* **37** (2) (1989).
16. Froberg, Carl-Erik, *Numerical Mathematics*, Benjamin/Cummings Publishing Co., 1985.
17. Gabor, D., Theory of communications, *J. Inst. Elect. Eng.* **93** (1946), 429–457.
18. Gopinath, R. A., Moment theorem for the wavelet-Galerkin method, Technical report, Aware Inc., Cambridge, MA, 1990.
19. Gopinath, R. A., Wavelet transforms and time-scale analysis of signals, Master's thesis, Rice University, 1990.
20. Golpinath, R. A., State space approach to multiplcity m wavelets, Technical report, Computational Mathematics Laboratory, Rice University, 1991.
21. Gopinath, R. A., W. M. Lawton, and C. S. Burrus, Wavelet-Galerkin approximation of linear translation invariant operators, in *Proc. ICASSP* **3** (1991), 2021–2023.

22. Grossmann A. and J. Morlet. Decomposition of Hardy functions into square integrable wavelets of constant shape, *SIAM J. Math. Anal.* **15** (1984), 723–736.
23. Hoffman, K., *Banach Spaces of Analytic Functions*, Dover Publications, Inc., 1962.
24. Hormander, L., *Distribution Theory and Fourier Analysis*, Springer-Verlag, 1990.
25. Kailath, T., *Linear Systems*, Prentice Hall, Englewood Cliffs, NJ, 1980.
26. Lawton, W. M., Private communication.
27. Lawton, W. M., Lectures on wavelets, *CBMS Conference Series*, Lowell University, 1990.
28. Lawton, W. M., Necessary and sufficient conditions for existence of on wavelet bases, *J. Math. Phys.*, **32** (1991), 57–61. (1) (1991), 57–61.
29. Lawton, W..M., Tight frames of compactly supported wavelets, *J. Math. Phys.*, **31** (1990), 1898-1900.
30. Mallat, S., Multiresolution approximations and wavelets orthonormal bases of $L^2(\mathbb{R})$, *Trans. Amer. Math. Soc.* **315** (1989), 69–87.
31. Mallat, S., A theory for multiscale signal decomposition: the wavelet representation, *IEEE Pattern and Machine Intell.* **11** (7) (1989), 674–693.
32. Meyer, Y., Principe d'incértitude, bases Hilbertiennes et algèbres d'opérateurs, *Séminaire Bourbabi* **662**, 1985–1986.
33. Nguyen, T. Q and P. P. Vaidyanathan, Maximally decimated perfect-reconstruction FIR filter banks with pairwise mirror-image analysis and synthesis frequency responses, *IEEE ASSP* **36** (5) (1988), 693–706.
34. Oppenheim, A. and R. W. Schafer, *Discrete-Time Signal Processing*, Prentice Hall, 1989.
35. Pollen, $SU_I(F[z, 1/z])$ for F a subfield of $\mathbb{C}$, *J. Amer. Math. Soc.* **3** (3) (1990), 611–624.
36. Portnoff, M. R., Time frequency distributions of digital systems and signals based on short-time fourier analysis, *IEEE ASSP* **28** (1980), 55–69.
37. Quake, Steven, R. Coifman, Y. Meyer, and M. V. Wickerhause, Signal processing and compression with wave packets, preprint.
38. Rabiner, L. and D. Crochière, *Multirate Digital Signal Processing*, Prentice-Hall, 1983.
39. Resnikoff, H. L., Weierstrass functions and compactly supported wavelets, Technical Report AD900810, Aware Inc., Cambridge, MA, 1990.
40. Rudin, W., *Real and Complex Analysis*, McGraw-Hill, 3rd edition, 1987.
41. Smith, M. J. and T. P. Barnwell, Exact reconstruction techniques for tree-structured subband coders, *IEEE ASSP* **34** (1986), 434–441.
42. Parks, T. W. and C. S. Burrus, *Digital Filter Design*, John Wiley and Sons, Inc., 1987.
43. Vaidyanathan, P. P., Quadrature mirror filter banks m-band extensions and perfect-reconstruction techniques, *IEEE ASSP Magazine* **4** (1987), 4–20.

44. Vaidyanathan, P. P., Improved technique for the design of perfect reconstruction fir qmf banks with lossless polyphase matrices, *IEEE ASSP* **37** (7) (1989), 1042–1056.
45. Vaidyanathan, P. P. and Phuong-Quan Hoang, Lattice structures for optimal design and robust implementation of two-channel perfect reconstruction qmf banks, *IEEE ASSP* **36** (1) (1988), 81–93.
46. Vaidyanathan, P. P. and Z. Doganata, The role of lossless systems in modern digital signal processing, *IEEE Education* **32** (1989), 181–197.
47. Vetterli M. and C. Herley, Wavelets and filter banks: theory and design, *IEEE ASSP*, 1992, to appear.
48. Wornell, G., Karunhen-Loéve like expansion for $1/f$ noise processes, *IEEE Trans. Inform. Theory* **36** (1990), 859–861.

This work was supported by AFOSR under grant 90–0334 which was funded by DARPA.

R. A. Gopinath
Department of Electrical and Computing Engineering
Rice Unversity
Houston, TX 77251–1892
ramesh@dsp.rice.edu

C. S. Burrus
Department of Electrical and Computing Engineering
Rice Unversity
Houston, TX 77251–1892
csb@dsp.rice.edu

Second Generation Compact Image Coding with Wavelets

Jacques Froment and Stéphane Mallat

Abstract. We introduce a compact image coding algorithm that separates the edge from the texture information. It allows us to adapt the coding precision to the properties of the human visual perception. Multiscale edges are detected from the local maxima of the wavelet transform modulus. We reconstruct an image approximation from the important edges that are selected by the coding algorithm. An error image is computed by subtracting the reconstructed image from the original one. It carries mostly the texture information. The textures are coded within a wavelet orthonormal basis. We show image coding examples, with compression ratio of 40 and 100.

§1. Introduction

Compact image coding algorithms have been divided ([6] and [12]) into first and second generation methods. Among first generation coding algorithms, the most efficient techniques decompose the image in an orthogonal basis. Several orthogonal transforms have been used for this purpose: the Karhunen-Loeve transform [14], the discrete cosine transform [13], and different orthogonal sub-band decompositions ([3] and [16]) such as the wavelet transform [8] or wavelet packets [15]. The original image is decomposed in the corresponding orthogonal basis and the decomposition coefficients are quantized with various technics. For the discrete cosine transform as well as the wavelet transform, the orthogonal basis is chosen once for all. On the opposite, a Karhunen-Loeve transform or wavelet packets adapt the orthogonal basis to the statistical properties of each image. The basis is chosen in order to minimize the entropy of the decomposition coefficients and thus the number of bits required for the coding.

Second generation methods try to take advantage of the structural properties of images, and the way they are perceived by a human observer. The image information can be roughly divided in two categories. The first corresponds to edges, which are particularly important for image understanding. Degradations of the image contours generally affect severely the image quality.

Wavelets–A Tutorial in Theory and Applications
C. K. Chui (ed.), pp. 655–678.

ISBN 0-12-174590-2

The second category regroups the different textures. The notion of texture is ill-defined in computer vision, but seems to be very important for image perception. Textures correspond to an information that is less structured than edges, although this depends upon the reference scale. What we might label as edges when looking at an image at a given scale, can be considered as texture elements at a larger scale. In most cases, texture distortions are not highly visible. This suggests one to code the texture information to be less precisely than the edge information. Edge and texture coding algorithms have already been developed by Carlsson [2] and Kunt et al. [6]. These algorithms allow one to obtain high compressions, while introducing distortions that do not affect to much the image visualization. Right now, second generation coding procedures are mostly feasibility studies. The detection of edges and textures require non-linear transforms that are difficult to understand mathematically. The algorithms involved are complicated and require a lot of computations.

It has been shown by Zhong and one of us [9] that multiscale edges can be detected and characterized with a wavelet transform. For a particular class of wavelets, the local maxima of the wavelet transform modulus provide the locations of edges in images. A close approximation of the original image can be reconstructed from these multiscale edges [9]. The reconstructed image is visually identical to the original one. A compact coding algorithm has already been implemented [9] from multiscale edges. For images of 256 by 256 pixels, this algorithm compresses the image data by a factor of the order of 30 and restores an image without visible distortions along the edges. However, this compression algorithm removes all textures which is not acceptable when textures provide important information.

In this chapter, we introduce a double layer coding based on edges and textures. A first coding from multiscale edges is performed. We then compute the difference between the original image and the coded image. This error image includes all the textures. It is coded with an orthogonal wavelet transform. The orthogonal wavelet transform is well adapted to the statistical properties textures in images, in the sense that the entropy of the wavelet coefficients, on the average, is relatively small ([5], [8], and [16]). One can also adapt the quantization of the wavelet coefficients, to the human visual sensitivity [5]. The texture coding makes a coarse quantization of these wavelet coefficients, because texture distortions are not highly visible. This double coding algorithm is similar to the approach of Carlsson [2] but takes advantage of the multiscale properties of the wavelet transform. We give two examples of coding with a lady image of 256 by 256 pixels and 512 by 512 pixels. The compression ratios are respectively 40 and a 100.

§2. Multiscale edge detection from a wavelet transform

The edges of the different structures that appear in an image, are often the most important features for pattern recognition. This is well illustrated by our visual ability to recognize an object from a drawing that only outlines edges. Edge points are located where the image intensity has sharp transitions.

Hence, in computer vision, a large class of edge detectors look for points where the gradient of the image intensity has a modulus which is locally maximum. Canny's [1] edge detector is a multiscale version of this approach. The image is smoothed by a low-pass filter that is dilated by a scale factor. Edges are located by computing the gradient vector of the image intensity, at each pixel location. Such an edge detector can be reformalized within a wavelet framework, which allows us to understand more precisely the properties of multiscale edges. For a particular class of wavelets, edge points can be detected from the local maxima of the wavelet transform modulus. This section reviews the work of Zhong and one of us [9], on the detection of multiscale edges with a wavelet transform.

We call a smoothing function any function $\theta(x, y)$ whose double integral is equal to 1. It is the impulse response of a low-pass filter. Let us define two functions, respectively equal to the partial derivatives along x and y of a smoothing function $\theta(x, y)$:

$$\psi^1(x, y) = \frac{\partial\theta(x, y)}{\partial x} \text{ and } \psi^2(x, y) = \frac{\partial\theta(x, y)}{\partial y}. \tag{1}$$

These functions are called wavelets. Let us dilate each of these functions by a scale factor 2^j. We denote

$$\theta_{2^j}(x, y) = \frac{1}{4^j}\theta\Big(\frac{x}{2^j}, \frac{y}{2^j}\Big), \tag{2}$$

$$\psi^1_{2^j}(x, y) = \frac{1}{4^j}\psi^1\Big(\frac{x}{2^j}, \frac{y}{2^j}\Big) \text{ and } \psi^2_{2^j}(x, y) = \frac{1}{4^j}\psi^2\Big(\frac{x}{2^j}, \frac{y}{2^j}\Big). \tag{3}$$

For any function $f(x, y) \in \mathbf{L}^2(\mathrm{R}^2)$, the wavelet transform defined with respect to $\psi^1(x, y)$ and $\psi^2(x, y)$, has two components given by the convolution products

$$\mathcal{W}^1_{2^j} f(x, y) = f * \psi^1_{2^j}(x, y) \text{ and } \mathcal{W}^2_{2^j} f(x, y) = f * \psi^2_{2^j}(x, y). \tag{4}$$

We call a dyadic wavelet transform of $f(x, y)$ the sequence of functions

$$\big(\mathcal{W}^1_{2^j} f(x, y), \mathcal{W}^1_{2^j} f(x, y)\big)_{j \in \mathbb{Z}}. \tag{5}$$

Since the wavelet transform is defined by convolution products, the Fourier transform of each component is given by

$$\widehat{W}^1_{2^j} f(\omega, \xi) = \widehat{f}(\omega, \xi)\widehat{\psi}^1(2^j\omega, 2^j\xi) \text{ and } \widehat{W}^2_{2^j} f(\omega, \xi) = \widehat{f}(\omega, \xi)\widehat{\psi}^2(2^j\omega, 2^j\xi).$$

A dyadic wavelet transform is a complete signal representation, if and only if, the dilations of the Fourier transform of each wavelet cover completely the Fourier plane. This means that there exist two constants $A > 0$ and $B > 0$ such that

$$\forall(\omega, \xi) \in \mathrm{R}^2,\ A \le \sum_{j \in \mathbb{Z}} \mid \widehat{\psi}^1(2^j\omega, 2^j\xi) \mid^2 + \mid \widehat{\psi}^2(2^j\omega, 2^j\xi) \mid^2 \le B. \tag{6}$$

One can then prove that there exists a non-unique pair of reconstructing wavelets $\chi^1(x,y)$ and $\chi^2(x,y)$, whose Fourier transform satisfy

$$\sum_{j\geq 1} \widehat{\psi}^1(2^j\omega, 2^j\xi)\widehat{\chi}^1(2^j\omega, 2^j\xi) + \widehat{\psi}^2(2^j\omega, 2^j\xi)\widehat{\chi}^2(2^j\omega, 2^j\xi) = 1. \tag{7}$$

Any function $f(x,y)$ can be reconstructed from its wavelet transform with the following summation

$$f(x,y) = \sum_{j\in\mathbb{Z}} \mathcal{W}^1_{2^j} f * \chi^1_{2^j}(x,y) + \mathcal{W}^2_{2^j} f * \chi^2_{2^j}(x,y). \tag{8}$$

A dyadic wavelet transform is more than complete; it is a redundant signal representation. This means that any sequence of functions is not a priori the wavelet transform of some signal in $\mathbf{L}^2(\mathrm{R}^2)$. In order to be a dyadic wavelet transform, a sequence of function must satisfy a reproducing kernel equation that is given in [9].

Since we defined wavelets that are the partial derivatives of the smoothing function $\theta(x,y)$, one can derive from Equations (1) and (4), that the two components of the wavelet transform satisfy

$$\begin{pmatrix} \mathcal{W}^1_{2^j} f(x,y) \\ \mathcal{W}^2_{2^j} f(x,y) \end{pmatrix} = \begin{pmatrix} 2^j \frac{\partial}{\partial x}(f * \theta_{2^j})(x,y) \\ 2^j \frac{\partial}{\partial y}(f * \theta_{2^j})(x,y) \end{pmatrix} = 2^j \nabla (f * \theta_{2^j})(x,y). \tag{9}$$

The two components of the wavelet transform at the scale 2^j, define a vector that is proportional to the gradient vector of the image smoothed by $\theta_{2^j}(x,y)$. The modulus of the gradient vector is proportional to the modulus image

$$M_{2^j} f(x,y) = \sqrt{|\mathcal{W}^1_{2^j} f(x,y)|^2 + |\mathcal{W}^2_{2^j} f(x,y)|^2}. \tag{10}$$

At each point (x,y), the angle of the gradient vector with the horizontal is also given by

$$A_{2^j} f(x,y) = \arctan\left(\frac{\mathcal{W}^2_{2^j} f(x,y)}{\mathcal{W}^1_{2^j} f(x,y)}\right). \tag{11}$$

The gradient vector points in the direction along which the partial derivative of $f * \theta_{2^j}(x,y)$ has a maximum amplitude. Like in a Canny edge detection [1], we detect the points where the modulus of $\nabla(f * \theta_{2^j})(x,y)$ is locally maximum, in the direction where the gradient vector points to. At each scale 2^j, the modulus maxima of the wavelet transform are thus defined as points (x,y), where the modulus image $M_{2^j} f(x,y)$ is locally maximum along the gradient direction given by $A_{2^j} f(x,y)$. These modulus maxima are inflection points of $f * \theta_{2^j}(x,y)$. We record the position of each modulus maximum and the values of $M_{2^j} f(x,y)$ and $A_{2^j} f(x,y)$ at the corresponding location.

Images are measured at a finite resolution. We thus can not compute the wavelet transform below a given scale that we normalize to 1. In order to model this scale limitation, we introduce a real function $\phi(x,y)$ whose Fourier transform modulus satisfies

$$| \widehat{\phi}(\omega,\xi) |^2 = \sum_{j\geq 1} \widehat{\psi}^1(2^j\omega, 2^j\xi)\widehat{\chi}^1(2^j\omega, 2^j\xi) + \widehat{\psi}^2(2^j\omega, 2^j\xi)\widehat{\chi}^2(2^j\omega, 2^j\xi).$$

As a consequence of Equation (7), one can prove that $\phi(x,y)$ is a smoothing function and we define the smoothing operator

$$\mathcal{S}_{2^j} f(x,y) = f * \phi_{2^j}(x,y),$$

where $\phi_{2^j}(x,y)$ is a dilation of $\phi(x,y)$ by 2^j. Due to the finite resolution of measurements, we suppose that the input image is a smoothing at the scale 1 of some original image $f(x,y)$, and is thus equal to $\mathcal{S}_1 f(x,y)$. We can not compute the wavelet transform at very large scales because the signal is measured over a finite domain. The scale must remain smaller than the size of the signal support. Let us limit the scale parameter above a given scale 2^J. We call finite dyadic wavelet transform of $\mathcal{S}_1 f(x,y)$ the sequence

$$\left\{ \left(\mathcal{W}^1_{2^j} f(x,y), \mathcal{W}^2_{2^j} f(x,y)\right)_{1\leq j<J}, \mathcal{S}_{2^J} f(x,y) \right\}. \tag{12}$$

It gives the wavelet transform between the scale 1 and 2^J, and the remaining information is carried by the low-frequency image $\mathcal{S}_{2^J} f(x,y)$. One can prove that $\mathcal{S}_1 f(x,y)$ can be recovered from its finite dyadic wavelet transform. The discretization of this model is studied in more details in [9]. If the original image has N pixels, its finite dyadic wavelet transform can be computed with a fast algorithm that requires $O(N \log(N))$ computations. Figure 1 shows the finite dyadic wavelet transform of a lady image. The image $\mathcal{S}_1 f(x,y)$ is at the top left, and the low-frequency image $\mathcal{S}_{2^4} f(x,y)$ is at the top right. The first two column show respectively $\mathcal{W}^1_{2^j} f(x,y)$ and $\mathcal{W}^2_{2^j} f(x,y)$, and the scale increases from top to bottom. We can recognize the effect of the partial derivative along x and y, in each component of the wavelet transform. The wavelets used for this computations are defined from a smoothing function $\theta(x,y)$, that is a separable product along x and y of a one-dimensional cubic spline function of compact support. These wavelets are further defined in [9]. The modulus images $M_{2^j} f(x,y)$ and angle images $A_{2^j} f(x,y)$, are shown in the third and fourth columns. The fifth column gives the position of the edge points, defined as local maxima of the modulus image in the direction indicated by the angle image. When the scale 2^j is small, the smoothing introduced by $\theta_{2^j}(x,y)$ is negligible. Any small fluctuation, due to the image noise, creates an edge point. At coarser scales, the smoothing removes such variations and edges are located at the boundaries of more important image structures.

§3. Reconstruction of images from multiscale edges

For image coding, we need to reconstruct an image approximation from the multiscale edge information. Zhong and of us [9] have built an algorithm

Figure 1. The original image $\mathcal{S}_1 f(x,y)$ is at the top left. The top right is the low-frequency image $\mathcal{S}_{2^4} f(x,y)$. The first column from the left gives the images $\mathcal{W}^1_{2^j} f(x,y)$ and the scale increases from top to bottom. The second columns displays $\mathcal{W}^2_{2^j} f(x,y)$. Black, grey and white pixels indicate respectively negative, zero and positive sample values. The third column displays the modulus images $M_{2^j} f(x,y)$. Black pixels indicate zero values and white ones correspond to high amplitudes. The fourth column gives the angle images $A_{2^j} f(x,y)$. The angle value goes from 0 (black) to 2π (white). The fifth column displays in white the position of the points that are local maxima of $M_{2^j} f(x,y)$ in the direction given by the angle image $A_{2^j} f(x,y)$.

that reconstructs images from the multiscale edges obtained with a wavelet transform. When all the edge information is kept, the algorithm recovers an image that is visually identical to the original one. This section reviews this algorithm and adapts it for our coding purposes. At each scale, an edge point is defined by its position in the image plane, plus the values of $M_{2^j} f(x,y)$ and $A_{2^j} f(x,y)$, at the corresponding location. As it can be observed along the right column of Figure 1, the smoothing component at different scale modifies the position of edge points. Edge points that appear at a scale 2^{j+1}, can be found at a finer scale 2^j, but their position has often changed. To code different edge maps at different scales is expensive. We thus only detect edges at one scale of reference, from the local maxima of the wavelet transform modulus. At all other scales, we record the values of $M_{2^j} f(x,y)$ and $A_{2^j} f(x,y)$, at the same positions, as if the edge location did not move across scales. In our computations, we choose the reference scale equal to 2^2. Indeed, edges at the finest scale 2^1 are to much affected by the image noise, and at coarser scales, the detection of the edge locations is less precise. The right image of Figure 3 shows the edges obtained from the lady image, at the scale 2^2. Among these, one can select a sub-set of edges that are considered as "important" for the visualization of the image. The edges that appear in the right image of Figure 4 have been selected with an algorithm described in the next section.

Let $((u_n, v_n))_{n\in\mathbb{Z}}$ be the positions of the edge points that are selected at the scale 2^2. Let us suppose that we keep the values of the modulus and angle images at these locations:

$$(M_{2^j}f(u_n, v_n), A_{2^j}f(u_n, v_n))_{1\le j\le J, n\in\mathbb{Z}}\,.$$

This edge detection can be also interpreted as an adaptive sampling of the modulus and angle images. At the coarse scale 2^J, for coding purposes, we need to reduce the low-frequency image $\mathcal{S}_{2^J}f(x,y)$. One can show that the Fourier transform of this image has an energy mostly concentrated at frequencies smaller than $2^{-J}\pi$. We thus sample uniformly this low-frequency image, at the rate 2^{-J} along x and y. Given the values of the modulus $M_{2^j}f(x,y)$ and the angle $A_{2^j}f(x,y)$ at the edge locations $((u_n, v_n))_{n\in\mathbb{Z}}$, for $1\le j\le J$, plus the uniform sampling of $\mathcal{S}_{2^J}f(x,y)$, at the rate 2^{-J}, we want to recover the best possible approximation of $f(x,y)$. We expect to recover an image that contains the main edge features that have been selected, but where some details have disappeared.

To perform the reconstruction, we follow the same approach than in [9]. The goal is to recover an image $\mathcal{S}_1h(x,y)$ whose dyadic wavelet transform has the same values as the dyadic wavelet transform of $\mathcal{S}_1f(x,y)$, at the edge locations. At any edge location (u_n, v_n), from $M_{2^j}f(u_n,v_n)$ and $A_{2^j}f(u_n,v_n)$, we can compute $\mathcal{W}^1_{2^j}f(u_n,v_n)$ and $\mathcal{W}^1_{2^j}f(u_n,v_n)$, and vice-versa. The finite dyadic wavelet transform of $\mathcal{S}_1h(x,y)$ should therefore satisfy for any (u_n,v_n) and $1\le j\le J$,

$$\mathcal{W}^1_{2^j}h(u_n,v_n) = \mathcal{W}^1_{2^j}f(u_n,v_n) \text{ and } \mathcal{W}^2_{2^j}h(u_n,v_n) = \mathcal{W}^2_{2^j}f(u_n,v_n), \qquad (13)$$

and for any pair of integers (n,m)

$$\mathcal{S}_{2^J}h(n2^J, m2^J) = \mathcal{S}_{2^J}f(n2^J, m2^J). \qquad (14)$$

Outside the edge points (u_n,v_n), we do not know the values of the wavelet transform of $\mathcal{S}_1h(x,y)$. By default, we impose that $\mathcal{S}_1h(x,y)$ should be as smooth as possible. Equation (9) shows that the wavelet transform corresponds to the gradient of the image intensity. We thus try to recover a wavelet transform whose energy is as small as possible. To avoid any edge at other locations than the points (u_n,v_n), we also impose that each component of the wavelet transform should have as few oscillations as possible, so that we do not create local maxima in the modulus images $M_{2^j}f(x,y)$. If we suppose that both $\psi^1(x,y)$ and $\psi^2(x,y)$ are continuously differentiable, then the wavelet transform at each scale is also differentiable. Let us denote as $\|g\|$ the $\mathbf{L}^2(\mathrm{R}^2)$ norm of any function $g(x,y) \in \mathbf{L}^2(\mathrm{R}^2)$. We impose the smoothness constraints by minimizing the following Sobolev norm

$$\||\left((\mathcal{W}^1_{2^j}h, \mathcal{W}^2_{2^j}h)_{1\le j\le J}, \mathcal{S}_{2^J}h\right)\||^2 = \sum_{j=1}^{J}\left(\|\mathcal{W}^1_{2^j}h\|^2 + 2^{2j}\|\frac{\partial \mathcal{W}^1_{2^j}h}{\partial x}\|^2\right) \tag{15}$$

$$+\sum_{j=1}^{J}\left(\|\mathcal{W}^2_{2^j}h\|^2 + 2^{2j}\|\frac{\partial \mathcal{W}^2_{2^j}h}{\partial y}\|^2\right) + \|\mathcal{S}_{2^J}h\|^2.$$

The minimization of this norm yields a wavelet transform where the energy of $\mathcal{W}^1_{2^j}h$ and $\mathcal{W}^2_{2^j}h$ is small and which has as few oscillations as possible. The Sobolev norm uses a partial derivative along x for $\mathcal{W}^1_{2^j}h$, because this component is defined by a partial derivative with respect to x (Equation (9)), and thus oscillates mostly along the x variable. The same is true for $\mathcal{W}^2_{2^j}h$, along the y variable. The partial derivative terms are weighted by 2^j because the smoothness of $\mathcal{W}^1_{2^j}h$ and $\mathcal{W}^2_{2^j}h$ increases with the scale 2^j.

We recover the function $h(x,y)$ that minimizes the Sobolev norm (15), and satisfies the constraints (13) and (14), with the same alternative projection algorithm as described in [9]. Let us define the space $\mathbf{K}$ of all sequences of functions $\left((w^1_j(x,y), w^2_j(x,y))_{1\le j\le J}, s_J(x)\right)$ such that the Sobolev norm $\||\left((w^1_j(x,y), w^2_j(x,y))_1 \le j \le J, s_J(x)\right)\||$, defined in (15), is finite. The space $\mathbf{K}$ is a Hilbert space, with respect to this norm. Let $\mathbf{\Gamma}$ be the set of sequences $\left((w^1_j(x,y), w^2_j(x,y))_{1\le j\le J}, s_J(x)\right)$ that belong to $\mathbf{K}$ and such that at all edge points (u_n, v_n), and for $1 \le j \le J$,

$$w^1_j(u_n,v_n) = \mathcal{W}^1_{2^j}f(u_n,v_n) \text{ and } w^2_j(u_n,v_n) = \mathcal{W}^2_{2^j}f(u_n,v_n), \tag{16}$$

and for any pair of integers (n, m)

$$s_J(n2^J, m2^J) = \mathcal{S}_{2^J}f(n2^J, m2^J). \tag{17}$$

The set $\mathbf{\Gamma}$ is a closed affine space in $\mathbf{K}$. Let us also define the space $\mathbf{V}$ of all sequences of functions that are the finite dyadic wavelet transform of some function $\mathcal{S}_1 g(x,y)$, with $g(x,y) \in \mathbf{L}^2(\mathrm{R}^2)$. By definition, the set of dyadic wavelet transforms that satisfy the constraints (13) and (14) are the sequences of functions that belong to

$$\mathbf{\Lambda} = \mathbf{\Gamma} \cap \mathbf{V}. \tag{18}$$

The dyadic wavelet transform that satisfies these constraints and minimizes the Sobolev norm $\|| \ \||$ is thus the element of $\mathbf{\Lambda}$ whose norm is minimum. We recover this element through alternative projections on $\mathbf{\Gamma}$ and $\mathbf{V}$.

Let $\mathbf{P_\Gamma}$ be the orthogonal projector onto $\mathbf{\Gamma}$, with respect to the Sobolev norm $\|| \ \||$. This operator transforms any element of $\mathbf{K}$ into the closest element of $\mathbf{\Gamma}$, with respect to the norm $\|| \ \||$. The analytical expression of this operator is given in [9], as well as its numerical implementations. Let $\mathbf{P_V}$ be the orthogonal projector onto $\mathbf{V}$, with respect to the norm $\|\ \|$. This operator is also characterized in [9]. Let $\mathbf{P} = \mathbf{P}_{\mathbf{\Gamma}}.\mathbf{P}_{\mathbf{V}}$ be alternative projections on $\mathbf{\Gamma}$ and $\mathbf{V}$. Let $\mathbf{P}^n$ be n iterations over the operator $\mathbf{P}$. Since $\mathbf{\Gamma}$ is a closed affine space and $\mathbf{V}$ a Hilbert space, a classical result on alternative projections [17] proves that for any element of $\mathbf{K}$

$$X = \left(\left(w_j^1(x,y), w_j^2(x,y) \right)_{1 \le j \le J}, s_J(x) \right), \tag{19}$$

$$\lim_{n \to \infty} \mathbf{P}^n X = \mathbf{P}_{\mathbf{\Lambda}} X. \tag{20}$$

The alternative projections converge strongly to the orthogonal projection of X, on the affine space $\mathbf{\Lambda}$. If $X = 0$, which means that all the functions of the sequence (19) are equal to 0, then the limit (20) converges strongly to the element of $\mathbf{\Lambda}$ whose Sobolev norm $\|\ \|$ is minimum. This algorithm is illustrated by Figure 2. If the original image has N pixels, the implementation of $\mathbf{P}_{\mathbf{\Gamma}}$ and $\mathbf{P}_{\mathbf{V}}$, requires $O(N \log(N))$ operations [9]. The stability and convergence rate of this iterative algorithm is studied in more detail in [9]. We reconstruct an image by applying the inverse wavelet transform operator on the reconstructed dyadic wavelet transform. For image processing applications, 10 iterations are sufficient to recover an approximation that has no visible difference with the solution of our inverse problem. Let us emphasize that this solution is not equal to the original image $S_1 f(x,y)$. Meyer proved [10] that even when all the edge information is kept, we do not reconstruct exactly the original image. In our case, since we first select the edge information, we only recover an image approximation where the main image components are restored.

At the left of Figure 4 is an image reconstructed from the wavelet transform values along the edge map shown on the right. This image is obtained with 10 iterations on the operator $\mathbf{P}$. The wavelet transform was computed over 4 scales ($J = 4$). The edges that have been selected are well recovered, but the small edge details that have been removed from the edge map, have disappeared. The next section studies the application of this reconstruction algorithm to compact image coding.

§4. Compact coding of edges

A compact image coding based on edges involves two steps. We must first select from the "most important" edge points for the image visualization. Then we code efficiently this edge information. Instead of processing isolated edge points detected by the wavelet transform maxima, we chain together these edge points in order to build edge curves. At an edge location, the direction of the gradient vector is perpendicular to the edge curve that goes through this point. Hence, the chaining procedure links together two neighbor edge points, whose relative position is perpendicular to the direction given by the angle image. We also impose that the modulus value $M_{2^j} f(x,y)$ varies smoothly along an edge curve, so that the image intensity profile has no abrupt modification along this curve. The maxima chains that we obtain are generally located at

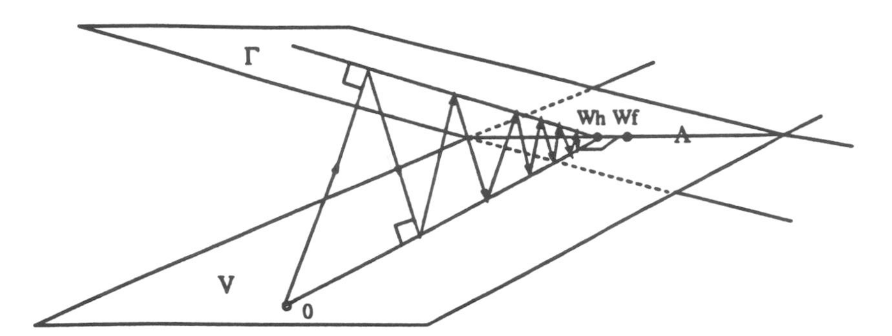

Figure 2. An approximation of the wavelet transform of $f(x,y)$ is reconstructed by alternating orthogonal projections on an affine space $\mathbf{\Gamma}$ and on the space $\mathbf{V}$ of all dyadic wavelet transforms. The projections begin from the zero element and converges to its orthogonal projection on $\mathbf{\Gamma} \cap \mathbf{V}$.

Figure 3. Left: original lady image of 256 by 256 pixels. Right: Edges obtained from the modulus maxima of the wavelet transform, at the scale 2^2.

the boundaries of the image structures. The selection of the most "important" edge curves requires sophisticated algorithms, if we take into account the image context. For example, in the lady image, it is important to introduce no distortions around the eyes because they are highly visible for a human observer. To take this factor into account, we need to find the location of the eyes, which

Figure 4. Left: image reconstructed from the wavelet transform values along the edge map shown on the right. This reconstruction is obtained with 10 iterations on the operator **P**. Right: edge curves selected from the edges shown at the right of Figure 3. The selection is done with a thresholding based on the length of the edge curves and their average modulus value.

is a difficult pattern recognition problem. In this chapter, we do not introduce such context information for the edge selection. The boundaries of important coherent structures often generate long edge curves. We thus remove any edge curve whose length is smaller than a given threshold. Among the remaining curves, we select the ones that correspond to the sharpest image variations. This is done by removing the edge curves, along which the average value of the wavelet transform modulus is smaller than an amplitude threshold. The edge map at the right of Figure 4 shows the edge curves, at the scale 2^2, that are selected by this simple algorithm. The parameters of the edge selection depend upon how many bits we can use for the image coding. The more edges after the selection, the more bits are needed. Figure 6 is the same lady image with 512 by 512 pixels, and Figure 7 shows the edge map at the scale 2^2. Figure 7 displays the edge curves selected by doubling the length threshold and the average modulus threshold, that were used for the edge curve selection in Figure 5. As a result, more edge curves are removed by this selection.

Once the edge selection is done, we must code efficiently the remaining information. This requires to code the position of the edge points that have been selected. At each scale 2^j, for $1 \leq j \leq J$, we must also code the modulus and angle values of the wavelet transform along these edge curves. Finally, we must store the sub-sampling of the the low frequency image $S_{2^J}f$. An algorithm for building a compact image code from edges, has already been developed by Zhong and one of us [9]. In this previous approach, the image was coded only from the edge information. In our case, the errors of the edge coding are then coded with an orthogonal wavelet transform. We can thus afford to quantize the edge information with very few bits, even-though the quantization errors can yield important visual distortions. These distortions are then reduced by the second coding layer.

The spatial geometry of the edge curves is coded with an algorithm described by Carlsson [2]. We record the position of the first point of each curve, and code the increment from one pixel to the next. Carlsson [2] showed that with an entropy coding, this requires on average, 10 bits for the first point and 1.3 bits for each point along an edge curve. To reduce the cost of this geometrical coding by a factor 2, we subsample by 2 the image grid where the edge curves are defined. From the edge curves on this sub-grid, we estimate the geometry of the edge curves on the full grid with linear interpolations.

Along an edge curve at the reference scale 2^2, we know that the angle value $A_{2^2}f(x,y)$ indicates the direction of the gradient vector, which is perpendicular to the tangent of the curve. We can estimate the value of $A_{2^2}f(x,y)$ at an edge point, by computing the direction of the tangent to the edge. We thus do not need to code these angle values. At the other scales 2^j, we approximate $A_{2^j}f(x,y)$ by the value of $A_{2^2}f(x,y)$ at the same location. Along each edge curve, the modulus values $M_{2^j}f(x,y)$ varies smoothly. The larger the scale, the smoother the evolution of $M_{2^j}f(x,y)$. We make a predictive coding of these values along the edges, that takes into account the increase of smoothness with the scale.

The image at the right of Figure 5, shows the interpolated edge map that is used for the lady image coding. At the left is the image reconstructed from the coded edge representation. This image coding requires 0.116 bits per pixel. Compared to the original lady image shown in Figure 3, that was coded with 8 bits per pixel, this corresponds to a compression factor of 72. The reconstructed images shows the main image features, but important distortions are visible and all the texture information has disappeared. Figure 9 is reconstructed from the coded wavelet transform values along the edge map shown in Figure 8 with 10 iterations on the operator **P**. This 512 by 512 image was coded on total of 0.043 bits per pixel. This corresponds to a compression ratio of 186, but the coded image is over-simplified. Few edge curves were selected for the coding, so we only recover the main image components. The reconstructed image is however sharp and does not include distortions along the edges such as Gibbs phenomena.

§5. Error image coding and final result

The edge coding algorithm restores the main image features but removes all the textures as well as fine image details. Like in Carlsson coding algorithm [2], we can compute an error image by subtracting the coded image from the original one. Figure 10 is obtained by subtracting the coded image in Figure 9, to the original one shown in Figure 6. The textures appear clearly in this error image, as well as some features such as the vertical bar on the left, whose edges have not been selected in Figure 8. The error image carries more information than the coded one, because we use a high compression rate in the edge coding.

Figure 5. Left: image reconstructed from the coded wavelet edge representation. It is reconstructed with 10 iterations on the operator **P**. This 256 by 256 image was coded on 0.116 bits per pixel. Right: edge curves selected and coded on a sub-grid, twice smaller.

To recover some of the texture information, we code the error image within a wavelet orthonormal basis [11].

Wavelet orthonormal bases provide one of the most effective transforms for compact image coding. These bases are well adapted to the statistical properties of images and the coding artifacts can be adapted to the human visual sensitivity ([5], [8], and [16]). At high compression rates, wavelet orthogonal bases produces Gibbs phenomena errors at the location of edges, which degrade considerably the image quality. Because we already coded most of the edge information, we can compress the error image with a high compression rate, without creating such errors. A wavelet orthogonal basis decomposes an image into "details" appearing at different resolutions, and within different spatial orientations. In two dimensions, one can build a wavelet orthonormal basis from three wavelets $\psi^1(x,y)$, $\psi^2(x,y)$ and $\psi^3(x,y)$. Each of these wavelets are dilated by scale factors 2^j, for $1 \leq k \leq 3$,

$$\psi^k_{2^j}(x,y) = \frac{1}{4^j}\psi^k\Big(\frac{x}{2^j}, \frac{y}{2^j}\Big) . \tag{21}$$

At each scale 2^j, these wavelets are translated over a uniform grid whose intervals are equal to 2^j. The three wavelets are chosen so that the resulting family of functions

$$\big(\psi^1_{2^j}(x-2^jn, y-2^jm), \psi^2_{2^j}(x-2^jn, y-2^jm), \psi^3_{2^j}(x-2^jn, y-2^jm)\big)_{(j,n,m)\in\mathbb{Z}^3} \tag{22}$$

is an orthonormal basis of $\mathbf{L}^2(\mathrm{R}^2)$. Any function is thus characterized by its inner products with each element of this basis.

Figure 6. Original lady image of 512 by 512 pixels.

An image is measured at a finite resolution that is normalized to 1. It is thus decomposed with wavelets dilated by scale factors larger than 1. Figure 11 shows the decomposition of the error image of Figure 10, with wavelets derived from a one-dimensional Daubechies wavelet [4] of compact support. At each scale 2^j, there are three images of size $2^{-j}N$, where N is the total number of pixels of the original image. These images correspond to the inner products of the original image with each of the three wavelets, dilated by 2^j and translated over a uniform grid of size $2^{-j}N$. White pixels indicate orthogonal coefficients whose absolute value have a large amplitude. As it can be observed, these coefficients are mostly concentrated in the texture regions of the error image. In this example, the error image is decomposed over 5 scales levels. The total number of samples of this orthogonal representation is equal to the number of sample of the original error image. A large portion of the orthogonal coefficients are nearly zero, and appear in black in Figure 10.

In order to reduce the number of wavelet coefficients to be coded, we set to zero as many coefficients as possible. This is done in a first step through

Figure 7. Edge detected from the local maxima of the wavelet transform modulus, at the scale 2^2.

a simple thresholding operation. At each scale 2^j, and for each orientation, we also measure the total energy of the wavelet coefficients in blocks of size 16.2^{-j} by 16.2^{-j} pixels. If this total energy is smaller than a given threshold, we put to zero all the coefficients of the block. It means that there is no texture in this image block and any non-negligible wavelet coefficient most probably corresponds to a less "important" image fluctuation. The quantization of the remaining non-zero coefficients is adapted to their probability distribution. One of us [10] showed that an exponential quantizer is well adapted to the statistical distributions of the wavelet orthogonal coefficients, in order to produce a small mean-square error. The error image in Figure 12 is coded in a wavelet orthonormal basis, with 0.38 bits per pixel. This coding restores the most irregular textures of the image. Figure 13 is obtained by adding the coded error image to the edge coded image of Figure 9. It requires a total of $0.043 + 0.038 = 0.081$ bits per pixel for its coding. This corresponds to a compression ratio of about a 100. There are clear visual differences with the

Figure 8. Edge curves selected with a thresholding on their length and on the average value of the wavelet transform modulus, at the scale 2^2. The length and modulus thresholds are twice larger than in Figure 5.

original image shown in Figure 6, however, given the very high compression ratio, the important edges and textures are well recovered. The upper left of Figure 14 is the error image for the 256 by 256 lady image, that is computed by subtracting the coded image in Figure 5 from the original image that is shown at the lower left of Figure 14. The upper right of Figure 14 is the result of coding the error image in a orthogonal basis, with 0.86 bits per pixel. The final coded image shown at the bottom right of Figure 14 is obtained by adding the coded error image with the edge coded lady image of Figure 5. It requires a total of $0.116 + 0.86 = 0.202$ bits per pixel. This represents a compression factor of about 40. Again, the main image features are well restored given the high compression ratio that is obtained.

§6. Conclusion

The double layer coding technique introduced in this chapter, enables us

Figure 9. Image reconstructed from the coded wavelet transform values, along the edge map shown in Figure 8. This coding requires 0.043 bits per pixel.

to compress images with factors between 40 and a 100. By coding specifically the edges and the textures of the image, we can take into account the fact that texture distortions are less visible than edge distortions. An important issue is to understand how to select optimally the image edges, given a final compression ratio. The edge selection algorithm that we described is quite simple and can be enhanced. We also need to understand precisely how to allocate bits between the the texture and the edge coding, in order to match the image information and it perception by human observers.

The wavelet transform is an important tool to detect edges in images. A precise characterization of edge types can be obtained from the behavior of the wavelet transform values across scales [9], which has applications in pattern recognition. Textures are more difficult to characterize and better statistical models are needed to optimize the texture coding.

Figure 10. This error image is obtained by subtracting the edge coded image of Figure 9 from the original image in Figure 6. Black, grey, and white pixels correspond respectively to negative, zero, and positive coefficients. Most of its energy is concentrated in texture regions.

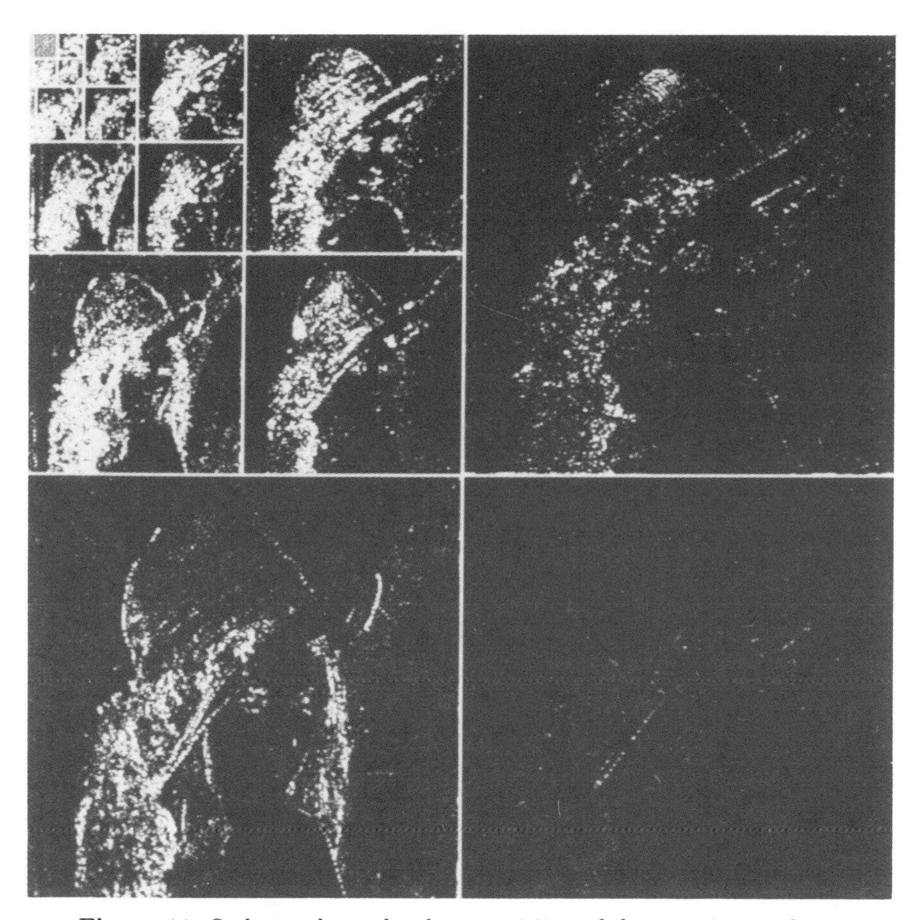

Figure 11. Orthogonal wavelet decompositions of the error image shown in Figure 10. We display the absolute value of the orthogonal coefficients. Black and white pixels correspond respectively to zero and high amplitude values. The high energy coefficients are mostly concentrated in the texture regions of the error image.

Figure 12. Error image coded on 0.038 bits per pixel, within a wavelet orthonormal basis. The main textures are restored.

Figure 13. Coded image obtained by adding the edge coded lady image of Figure 9 and the error coded image of Figure 12. This image is coded on a total of 0.081 bits per pixel.

Figure 14. Top left: error image obtained by subtracting the left image of Figure 5 to the original image which is at the lower left. Top right: error image coded in a wavelet orthogonal basis, with 0.086 bits per pixel. Lower left: original image. Lower right: image obtained by adding the coded error image at the top right with the coded edge image of Figure 5. This image was coded on 0.202 bits per pixel.

References

1. Canny, J., A computational approach to edge detection, *IEEE Trans. Pattern Anal. and Machine Intell.* **8** (1986), 679–698.
2. Carlsson, S., Sketch based coding of grey level images, *Signal Processing North-Holland* **15** (1) (1988), 57–83.
3. Crochiere, R. E., S. A. Webber, and J. L. Flanagan, Digital coding of speech in sub-bands, *Proc. Int. Conf. IEEE Acoust., Speech, Signal Processing*, Philadelphia, 1976.
4. Daubechies, I., Orthonormal bases of compactly supported wavelets, *Commun. Pure and Appl. Math.* **41** (1988), 909–996.
5. Froment, J., Traitement d'images et applications de la transformée en ondelettes, Thèse de l'Université Paris-IX, 1990, 131–180.
6. Kunt, M., A. Ikonomopoulos, and M. Kocher, Second-generation image-coding techniques, *Proc. IEEE* **73** (4) (1985), 549–574.
7. Kunt, M., M. Bénard, and R. Leonardi, Recent results in high-compression image coding, *IEEE Trans. Circuits and Systems* (1987), 1306–1336.
8. Mallat, S., A theory for multiresolution signal decomposition : the wavelet representation, *IEEE Trans. Patt. Anal. and Mach. Intell.* **11** (1989), 674–693.
9. Mallat, S. and S. Zhong, Characterization of signals from multiscale edges, New York University, Computer Science, Tech. Report, 1991; *IEEE Trans. Pattern Anal. and Machine Intell.*, to apper.
10. Meyer, Y., Un contre-exemple à la conjecture de Marr et à celle de S.Mallat, *CEREMADE*, University Paris-IX, 1991, preprint.
11. Meyer, Y., *Ondelettes et Opérateurs* **I**, Hermann, Paris, 1990.
12. Netravali, A. N. and J. O. Limb, Picture coding : a review, *Proc. of IEEE* **68** (3) (1980), 366–406.
13. Rao, K. R. and P. Yip, *Discrete Cosine Transform*, Academic Press, New York, 1990.
14. Rosenfeld, A. and A. C. Kak, *Digital Picture Processing*, Second Edition, Academic Press, Computer science and applied mathematics **1** (1982), 116–150.
15. Wickerhauser, M. V., Codage et compression du signal et de l'image par les ondelettes et les paquets d'ondelettes, *Problèmes Non Linéaires Appliqués, Ondelettes et Paquets d'ondes*, INRIA, 1991, 31–99.
16. Woods, J. W. and A. S. D. O'Neil, Subband coding of images, *IEEE Trans. Acoust. Speech and Signal Proc.* **34** (1986).
17. Youla, D. and A. H. Webb, Image restoration by the method of convex projections, *IEEE Trans. on Medical Imaging* **1** (1982), 81–101.

This research was supported by the NSF grant IRI-890331, AFOSR grant AFOSR-90-0040 and ONR grant N00014-91-J-1967.

Stéphane Mallat
Courant Institute of Mathematical Sciences
New York University
251 Mercer Street
New York, NY 10012, USA
mallat@robeye.nyu.edu

Jacques Froment
Université Paris V - René Descartes
UFR de Mathématiques, Statistique et Informatique
Sorbonne, 12 rue Cujas, 75005 Paris, FRANCE

Acoustic Signal Compression with Wavelet Packets

Mladen Victor Wickerhauser

Abstract. The wavelet transform is generalized to produce a library of orthonormal bases of modulated wavelet packets, where each basis comes with a fast transform. These bases are similar to adaptive windowed Fourier transforms, and hence give rise to the notion of a "best basis" for a signal subject to a given cost function. This paper discusses some early results in acoustic signal compression using a simple counting cost function.

§1. Introduction

By generalizing the method of multiresolution decomposition, it is possible to construct orthonormal wavelet packets which provide a family of orthonormal bases for $L^2(\mathbb{R})$. These wavelet packets are well localized in both space and frequency. Correlating digitally sampled signals with these wavelet packets and discarding small coefficients can substantially compress the signals. The wavelet packets can be adapted to optimize fast data transformation, or to facilitate signal recognition. This paper presents results of computer experiments with the author's signal compression algorithms on digitized speech signals.

Correlations with the entire family of orthonormal bases may be rapidly calculated for discrete functions of $\mathbb{R}^N$, where N is a positive integer power of 2. The algorithm used is based on the conjugate quadrature mirror filter method described by Mallat [4], with generalizations described in the Appendix. In the generalized case, it has the same complexity as the discrete "fast" Fourier transform. The complexity to give more than 2^N orthonormal bases all at once is $O(N \log_2 N)$. These bases are especially well suited for signal processing, and the availability of choices suggests adaptive algorithms which optimize for a given problem. Orthogonality insures that signal analysis and re-synthesis are computationally equivalent.

If the basis vectors of the family are arranged in an array whose dimensions are N columns by $n = \log_2 N$ rows, then any admissible subset of the entries will be an orthonormal basis of $\mathbb{R}^N$. Roughly speaking, an admissible subset is a choice of elements with the following three properties:

Wavelets–A Tutorial in Theory and Applications
C. K. Chui (ed.), pp. 679–700.

ISBN 0-12-174590-2

(i) Every column contains exactly one element;
(ii) Elements in a single row appear in contiguous blocks of 2^k elements, where k, with $0 \le k \le n$, is an integer; and
(iii) Row blocks begin an integer multiple of their length from the left edge of the array.

This notion will be made precise in the Appendix.

The redundancy of the bases has benefits for both numerical algorithms and signal processing. For example, it is possible to choose the basis (*i.e.*, the admissible subset of the array) which has the fewest large coefficients. The choice algorithm, as described in the appendix, can be implemented in $O(N \log_2 N)$ steps. The far smaller number of surviving coefficients (after thresholding) reduces the complexity of subsequent processing. Such an algorithm may also be used to compress audio signals. Applications to 2-dimensional signal compression may be found in [2].

Reconstructing the signal corresponding to an array of zeros with a single 1 gives a particular tone or tone-burst, corresponding to a note produced by a musical instrument. Graphs of individual wave packets bear a striking resemblance to bursts of sound, suggesting that these bases are well suited to representing acoustic signals. This provides a method for digital music synthesis.

Another example is the recognition of acoustic signatures. By displaying an array of grey scales proportional to the absolute values of the corresponding coefficients, one obtains a picture of the signal together with a family of Fourier-like transformations. Nicolas [5] has build an expert system which can read this picture and detect characteristic features of a particular signal with high probability.

Perhaps the most important application is the rapid transformation of digital data. The same wavelet packet bases that compress signals will conjugate a large class of (discretized) operators into sparse or even band-diagonal form. Thus data in compressed form is already prepared for fast and useful numerical algorithms. The wavelet transform methods have already produced fast numerical algorithms for matrix inversion and calculation of potentials [1]. Such a marriage of compression and transformation could have far reaching technical and economic consequences.

§2. Synthesis of discrete wavelets and wavelet packets

For a longer exposition on the construction of wavelets, see Daubechies [3]. The starting point is a summable sequence $h = \{h_k\} \in l^1(\mathbb{R})$ with the following three properties:

(i) $\sum_k h_{2k} = \sum_k h_{2k+1} = 2^{-1/2}$;
(ii) For all integers $l \neq 0$, $\sum_k h_k h_{k+2l} = 0$; and
(iii) There exists an $\epsilon > 0$, such that $\sum_k |h_k||k|^\epsilon < \infty$.

Finite sequences obviously satisfy (iii). Complete characterizations of the finite sequences $\{h_k\}_{-M\leq k<M}$ which satisfy conditions (i) and (ii) may be seen in [3]. We call such a finite sequence h a *summing filter*, and define the associated *differencing filter* $g = \{g_k\}$ by $g_k = (-1)^k h_{-k+1}$. The pair (h, g) is called a *conjugate mirror filter* of length $r = 2M$. These filters may be used to construct orthonormal bases in two ways.

First, there is a continuous and compactly supported real-valued function ϕ on $\mathbb{R}$ that solves the functional equation

$$\phi(x) = 2^{1/2} \sum_k h_k \phi(2x - k), \tag{1}$$

with $\hat{\phi}(0) = 1$. This function may be approximated very efficiently by iteration. Next, we define an associated function ψ by

$$\psi(x) = 2^{1/2} \sum_k g_k \phi(2x - k). \tag{2}$$

This function is also compactly supported—in fact, its support is contained in an interval of length r. Its dyadic dilates and integer translates generate a Hilbert basis for $L^2(\mathbb{R})$. Namely, writing ψ_{jk} for the function $\psi_{jk}(x) = 2^{-j/2}\psi(2^{-j}x - k)$, there is the following result of Daubechies and Meyer:

Theorem 1. *The collection $\{\psi_{jk}, j, k \in \mathbb{Z}\}$ is a complete orthonormal basis for $L^2(\mathbb{R})$.*

On the other hand, conjugate mirror filters provide a basis for $l^2(N)$, where N shall be shorthand for the finite set $\{1, 2, \ldots, 2^n\}$. To construct this, we introduce the summing and differencing operators H and G on $l^2(\mathbb{Z})$, respectively, as defined by

$$Hf(i) = \sum_k h_{k-2i} f(k), \qquad Gf(i) = \sum_k g_{k-2i} f(k). \tag{3}$$

These two operators have adjoints H^* and G^*, defined by

$$H^* f(k) = \sum_i h_{k-2i} f(i), \qquad G^* f(k) = \sum_i g_{k-2i} f(i). \tag{4}$$

The original three conditions imposed upon h guarantee that (i) $H^*H + G^*G = I$, (ii) $HG^* = 0$, and (iii) $l^2 = H^*l^2 \oplus G^*l^2$. Repeating this last decomposition n times yields

$$l^2 = \oplus \sum_{j=0}^{n-1} (H^*)^j G^* l^2 \oplus (H^*)^n l^2. \tag{5}$$

Now, it is possible to define $H, G : l^2(2^m) \to l^2(2^{m-1})$, and $H^*, G^* : l^2(2^{m-1}) \to l^2(2^m)$. First replace h, g with periodized filters $\tilde{h}_m, \tilde{g}_m$ defined by

$$\tilde{h}_m(k) = \sum_{l \equiv k \pmod{2^m}} h(l), \qquad \tilde{g}_m(k) = \sum_{l \equiv k \pmod{2^m}} g(l).$$

Then restrict the ranges of i and k in Equations (3) and (4) as follows:

$$Hf(i) = \sum_{k=1}^{2^m} \tilde{h}_m(k-2i)f(k), \quad Gf(i) = \sum_{k=1}^{2^m} \tilde{g}_m(k-2i)f(k); \tag{6}$$

$$H^*f(k) = \sum_{i=1}^{2^{m-1}} \tilde{h}_m(k-2i)f(i), \quad G^*f(k) = \sum_{i=1}^{2^{m-1}} \tilde{g}_m(k-2i)f(i). \tag{7}$$

For each $m > 0$, the periodized filters $\tilde{h}_m, \tilde{g}_m$ satisfy the same orthogonality conditions as h, g, even though they differ (in general) at different levels m. Hence, these operators H, G satisfy the relations

$$\begin{aligned} l^2(2^m) &= H^*l^2(2^{m-1}) \oplus G^*l^2(2^{m-1}), \\ HG^* = GH^* &= 0, \qquad \text{and} \qquad H^*H + G^*G = I_m, \end{aligned} \tag{8}$$

where I_m is shorthand for the identity operator on $l^2(2^m)$.

Counting dimensions in this finite-dimensional case shows that Equation (5) may be rewritten as

$$\begin{aligned} l^2(N) &= H^*l^2(N/2) \oplus G^*l^2(N/2) \\ &= \oplus \sum_{j=0}^{n-1} (H^*)^j G^* l^2(N/2^{j+1}) \oplus (H^*)^n l^2(1). \end{aligned} \tag{9}$$

Hence, taking the standard bases within each of the direct summands gives the so-called *wavelet basis* for the finite-dimensional space $l^2(N)$. Computation of the expansion of a vector in this basis may be done recursively in $O(Nr)$ operations, where r is the length of the filters h and g; *i.e.*, the cost of applying H or G.

Of course, however, each of the direct summands $l^2(N/2^j)$ may itself be expanded using the wavelet basis, and its summands recursively expanded still further, until all the summands are 1-dimensional. This yields the formula

$$l^2(N) = \oplus \sum_{j=0}^{N-1} F^*_{\epsilon_n(j)} \dots F^*_{\epsilon_1(j)} l^2(1), \quad \text{where} \quad F^*_\epsilon = \begin{cases} G^*, & \text{if } \epsilon = 0; \\ H^*, & \text{if } \epsilon = 1, \end{cases} \tag{10}$$

and $\epsilon_i(j)$ is the ith digit in the binary expansion of j. Taking one coefficient for each of the 1-dimensional summands in Equation (10) gives the lowest-level *wavelet packet basis* expansion of a vector. These functions are related to the classical Walsh functions, which are obtained just as above simply by taking filters of length 2. One obtains the bit-reversed and sequency-ordered Walsh function bases by respectively composing bit-reversal or Gray-encoding with the function ϵ. Longer filters give smooth generalizations of Walsh functions, which are totally new objects.

Individual wavelet packets may be constructed as follows. Fix $N = 2^n$ and define the jth wavelet packet $w_j^n \equiv w_j \in l^2(N)$ from Equation (10) above, by

$$w_j = F^*_{\epsilon_n(j)} \dots F^*_{\epsilon_1(j)} 1. \tag{11}$$

This is a unit vector in $l^2(N)$ by an easy induction argument.

Lemma 2. $\langle v, w_j \rangle = F_{\epsilon_1(j)} \dots F_{\epsilon_n(j)} v$, *where* F_ϵ *is the adjoint of* F^*_ϵ.

Proof: Starting from Equation (11), the operators F^*_ϵ may be transposed onto v, giving

$$\langle v, w_j \rangle = \langle v, F^*_{\epsilon_n(j)} \dots F^*_{\epsilon_1(j)} 1 \rangle = \langle F_{\epsilon_1(j)} \dots F_{\epsilon_n(j)} v, 1 \rangle.$$

However, the right-hand side is an inner product in $\mathbb{R}^1$. ■

§3. Other wavelet packets

The periodized lowest-level wavelet packet w_j^n is not localized. It stretches throughout the entire index interval $\{0, 1, \dots, 2^n - 1\}$, as is easily shown by induction on the steps of the reconstruction algorithm. Since the filter length is at least 2, the number of non-zero coefficients at least doubles with each application of F^*_ϵ. But there are n filter applications. This property distinguishes lowest-level wavelet packets from wavelets, which are supported in subintervals of length proportional to their scale. Since it contains only frequency but no position information, the inner product of a vector with $w_j^n, j = 0, 1, \dots, N-1$ is analogous to the Fourier transform. The lowest-level wavelet packets themselves are the analogs of pure tones.

The reconstruction algorithm may be generalized to provide wavelet packets supported in subintervals. For $m = 1, 2, \dots, n$, we define

$$w_j^m = F^*_{\epsilon_n(j)} \dots F^*_{\epsilon_{n-m+1}(j)} e_{\hat{j}}, \tag{12}$$

where $e_{\hat{j}}$ is the unit vector in $\mathbb{R}^{2^{n-m}}$ which has a 1 in the $\hat{j}$ position and zeros everywhere else, with $\hat{j} = j \bmod 2^{n-m}$. In other words, $\epsilon_k(j) = \epsilon_k(\hat{j})$ for $k = 1, 2, \dots, n-m$. Let w_j^0 be the standard basis vector in $\mathbb{R}^N$ which has a 1 in the jth position and zeros elsewhere. The component of $v \in \mathbb{R}^N$ in the w_j^m direction is given by the $\hat{j}$ coefficient in the vector $F_{\epsilon_{n-m+1}(j)} \dots F_{\epsilon_n(j)} v$. The proof is nearly identical with that of Lemma 2. That is, we have the following.

Lemma 3. $\langle v, w_j^m \rangle = \langle e_{\hat{j}}, F_{\epsilon_{n-m+1}(j)} \dots F_{\epsilon_n(j)} v \rangle$.

The width of the support of w_j^m is at most $2^m + (r - 2)(2^m - 1)$, where r is the length of the filters h, g. For small values of m and r, the width can be much smaller than 2^n. Thus, inner products with $w_j^m, j = 0, 1, \dots, N-1$, now contain some position information, just as the windowed Fourier transform. The narrower wavelet packets themselves correspond to tonebursts.

Write w^m for the orthonormal basis $\{w_j^m : j = 0, 1, \ldots, N-1\}$. Expanding a vector into w^n generates all the expansions into $w^1, \ldots, w^n$ simultaneously, and hence, suggests the following picture. Denote by w^0 the standard basis in $\mathbb{R}^N$. A vector in the w^0 basis may be expanded in the $\log_2 N$ wavelet packet bases $w^1, \ldots, w^n$ with the results placed row-by-row into a rectangular array. Each location in this array corresponds to a wavelet packet orthogonal to other wavelet packets in that row, that have support width increasing with the row number. The array thus contains $1 + \log_2 N$ representations of a vector, including analogs of the windowed Walsh (or Fourier) transforms at all dyadic window-widths. Likewise, if row m has a single 1 in column j with all its other entries 0, the rest of the array is determined, and the standard representation of w_j^m will appear in row 0.

Any wavelet packet can be extended to a smooth periodic function with the same degree of smoothness as the solution ϕ of the functional equation (1). (In the finite-dimensional case, *numerical smoothness of degree k* will mean rapid decrease of the first k successive differences.)

§4. Functions underlying the discrete algorithm

Wavelet packet coefficients can be interpreted as inner products of an underlying function of $\mathbb{R}$ with a family of smooth orthonormal elements of $L^2(\mathbb{R})$. Let ϕ, ψ be the continuous functions in $L^2(\mathbb{R})$ defined by Equations (1) and (2), and suppose that $S \in L^2(\mathbb{R})$ is a function to be approximated with one value in each subinterval of length δ. One possibility is to calculate an approximate value $s(k) = \delta^{-1} \int_{\mathbb{R}} S(x)\phi(\delta^{-1}x - k)\, dx$ at each integer k, relying on the good localization of the bump function ϕ. In fact, if we arrange that higher moments of ϕ vanish, this approximation is very good for smooth functions S; that is, we have the following.

Lemma 4. *If $\int_{\mathbb{R}} x^d \phi(x)\, dx = 0$ for all $0 < d < D$, and $S \in C^D(\mathbb{R})$, then for x near k, $S(x) = s(k) + O(\delta^D)$.*

Proof: Expand S in its Taylor series about k. The inner product ϕ evaluates S at $x = k$ and kills all lower order terms than D. ■

It is possible to arrange any fixed number of vanishing moments of ϕ. However, longer finite filters must be used, since the number of vanishing moments is at most half the filter length.

Now let us define a family of wave functions $W_n \in L^2(\mathbb{R})$, $n = 0, 1, \ldots$, recursively from ϕ and ψ as follows:

$$W_0(x) = \phi(x), \qquad W_1(x) = \psi(x), \qquad \text{and}$$

$$W_{2n}(x) = 2^{1/2} \sum_k h_k W_n(2x - k),$$
$$W_{2n+1}(x) = 2^{1/2} \sum_k g_k W_n(2x - k). \tag{13}$$

Since the resursive scheme is an orthogonal transformation in L^2, each W_n is a unit vector. From these definitions, we see that any sequence s that approximates a function S will have as its wave packet coefficients the inner products of S with the dyadic dilates and integer translates of W_n for various n. The following formula, noted by Coifman and Meyer, describes this correspondence:

Proposition 5. *For any integer k, and nonnegative integers m, n, the relationship between the wavelet packet coefficients and the underlying inner products is given by the formula:*

$$2^{-m/2} \int_{\mathbb{R}} S(x) W_n(2^{-m}x - k)\, dx = \langle e_k, F_{\epsilon_1(n)} \cdots F_{\epsilon_m(n)} s \rangle,$$

where $\epsilon_j(n)$ is the jth binary digit of n, padded with leading zeros if necessary, and e_k is the unit sequence in l^2 which has a single 1 in the kth position and zeros elsewhere.

Proof: The inner product with e_k merely picks out one coefficient. The rest of the formula may be demonstrated by considering the effect of a filter convolution on S. But if $\sigma = \sigma(k)$ is a partial result, we have:

$$\begin{aligned}
F_\epsilon \sigma(k) &= 2^{-q/2} \int_{\mathbb{R}} S(x) \Big(\sum_l f_\epsilon(l - 2k) W_p(2^{-q}x - l)\Big)\, dx \\
&= 2^{-q/2} \int_{\mathbb{R}} S(x) \Big(\sum_l f_\epsilon(l) W_p(2^{-q}x + 2k - l)\Big)\, dx \\
&= \begin{cases} 2^{-q/2} \int_{\mathbb{R}} S(x)\, 2^{-1/2} W_{2p}(2^{-q-1}x - k)\, dx, & \text{if } \epsilon = 0, \\ 2^{-q/2} \int_{\mathbb{R}} S(x)\, 2^{-1/2} W_{2p+1}(2^{-q-1}x - k)\, dx, & \text{if } \epsilon = 1. \end{cases}
\end{aligned}$$

Induction on q up to m completes the proof. ■

Wavelets correspond to $n \equiv 1$, $m = 0, 1, 2, \ldots$, and $k \in \mathbb{Z}$. Wavelet packets correspond to a fixed level $m \equiv l > 0$, $n = 0, 1, 2, \ldots$, and $k \in \mathbb{Z}$. The periodized case can be interpreted similarly. One can easily produce a series of graphs depicting approximations of these functions W_n. Filters with regular limits ϕ, ψ produce a library of wavelet packets with a remarkable resemblance to sound bursts or smoothly modulated notes. Expansion in the wavelet packet basis correlates a signal to these notes, and their appearance suggests that the library is suitable for acoustic signal analysis. Some examples of these graphs may be seen below:

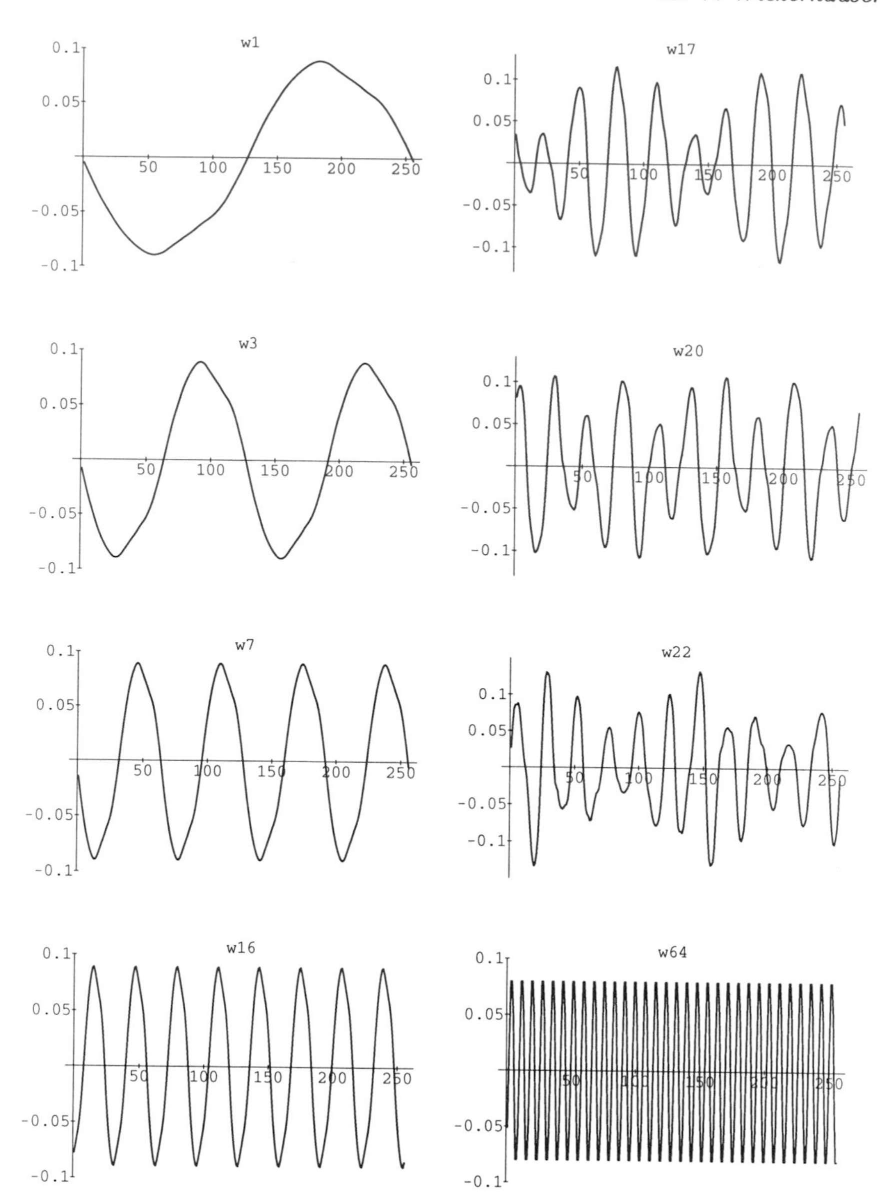

Figure 1. Periodized wavelet packets.

§5. Wavelet packets as musical tones and tonebursts

It is amusing to listen to wavelet packets. They approximate continuous periodic functions on $\mathbb{R}$, and can be played back as tones through a digital-to-analog converter and loudspeaker. The curious reader who wishes to hear these tones may obtain from the author a recording of a few examples.

The speed of the wavelet packet algorithm suggests that wavelet packet generators could be useful in sound synthesis. A single wavelet packet generator may replace a large number of oscillators. Through experimentation, a musician could determine combinations of wave packets that produce especially interesting sounds. Having an orthonormal basis offers the musician the widest possible range of sounds.

Also, consider approximating the sound of a musical instrument. A sample of the notes produced by the instrument may be decomposed into its wavelet packet coefficients, using a suitable filter length. Reproducing the note requires reloading those coefficients into a wavelet packet generator and playing back the result. Properties such as attack and decay may be controlled separately (*e.g.*, with envelope generators), or by using longer wavelet packets and encoding those properties as well into each note. Any of these processes may be controlled in real time, for example by a keyboard. Of course, the musical instrument could just as well be a human voice, and the notes, words or phonemes.

Conjugate mirror filters of length greater than 2 result in numerically smooth wavelet packets, which are presumably well suited to representing digitally-sampled smooth analog signals. Results with the two sample signals studied in the next section support this belief: longer filters result in less measurable and audible distortion.

Wavelet packet coefficients, like wavelet coefficients, are mostly very small for digital samples of smooth signals. The differencing filters remove correlations between adjacent, nearly identical samples. Discarding coefficients below some predetermined (or dynamically determined) cutoff introduces only small errors while effecting data compression for smooth signals. In effect, this means that a wave packet-based music synthesizer could store many complex sounds efficiently. Similarly, a wavelet packet-based speech synthesizer could be used to reconstruct highly compressed speech signals.

§6. Data compression with wavelet packets

Let F be a field, finite or infinite, and $v \in F^N$ be a vector, and consider the problem of representing v with fewer than N coefficients. If F is finite (*i.e.*, an alphabet), then there are numerous lossless data compression algorithms that can replace v with a substitution table and symbol list, removing some of the redundancy of information in the standard representation of v. Such algorithms rely on searching through v for repeated strings and tabulating some fraction of them. The best implementations (for small alphabets F) yield compression ratios between 2 and 10; *i.e.*, the compressed vector is 2 to 10 times shorter than the original.

Lossless data compression is also possible with wavelet packets. If more of the wavelet packet coefficients are small, then the entire expansion has lower entropy than the original signal. Fewer bits need to be transmitted for the small coefficients. As a simple test of this idea, one sample signal was compressed by Huffman coding, yielding a compression ratio of 2. For comparison, its complete best-level wave packet expansion was compressed by the same Huffman coding algorithm, yielding a compression ratio of 3. The reconstructed signal was identical to the original in both cases.

Such methods lose efficiency when F is very large or infinite, such as the case when it is taken to be $\mathbb{R}$. The alternative is a method that introduces some acceptable losses or distortions in exchange for greater efficiency.

Loss or distortion introduced in reconstruction shall be defined as the root-mean-square or l^2 norm of the difference between v and its approximation $\hat{v}$. For simplicity of comparison, this shall be normalized into the relative difference $\|v - \hat{v}\|/\|v\|$. Distortions of up to 5% are considered acceptable in some speech processing applications. An optimal compression method for a speech signal is one that yields no more than 5% distortion while maximizing the compression ratio. While it is too much to expect a single method to be optimal for all signals, it is not unreasonable to expect a method, which is close to optimal for one signal, to be close to optimal for similar signals—similar bandwidths, dynamic ranges, etc.

The class of algorithms from which the optimal compression method is to be chosen will be the following: A fixed signal of length $N = 2^n$ will be decomposed into the n wavelet packet bases $w^1, \ldots, w^n$. The entries in each row which are smaller than a predetermined cutoff (depending only on the norm of the signal) will be set to 0. Then the admissible subsets of the array will be searched to find that set which has the fewest nonzero coefficients. For filter length r, this algorithm takes $O(rN \log_2 N)$ steps: $rN \log_2 N$ operations to expand the signal, $N \log_2 N$ to cut off small coefficients, and $2N \log_2 N$ to find the optimal admissible subset. In addition, the compressed coefficients must have a header appended which defines their locations within the admissible subset. Enlarging the class of admissible subsets (which improves the compression ratio) causes this header to grow, forcing a tradeoff.

In the extreme case where the definition of admissibility is the set of conditions ((i), (ii), and (iii)) in the introduction, the number of admissible subsets is very large. Let A_n be the number of admissible subsets for a signal of length $N = 2^n$. Then $A_1 = 2$ (the standard basis, or { sum, difference }), and there is a recurrence relation deducible from the recursive construction of wavelet packets. Namely, the admissible subset is either w^0 or a choice of admissible subsets from the wavelet packet expansions of Gl^2 and Hl^2. This generalizes to provide the inductive step

$$A_{n+1} = 1 + A_n^2. \tag{14}$$

The constant 1 quickly becomes negligible as n increases, and an obvious simplification gives the estimate $A_{n+1} \geq 2^{2^n} = 2^N$. In this case the header attached to each coefficient must state which level it comes from, together with the position in that level. This requires at least $\log_2 N + \log_2 \log_2 N$ bits to transmit. In the appendix is a formula which evaluates the effect of this requirement upon the compression ratio. At the other extreme, always choosing the same admissible set—say, the lowest-level wavelet packets w^n—shortens the header to $\log_2 N$ bits per coefficient.

An intermediate case is to fix a segment length N and then allow within each segment only complete rows w^m, for $0 < m \leq n$, as the admissible subsets. This permits some adaptability to the signal while adding a header only $\log_2 \log_2 N$ bits long to an entire segment of coefficients, quite negligible compared to N as $N \to \infty$.

§7. Experiments with speech signal compression

A speaker recorded the phrase "Joe brought a young girl" into computer memory, using a 22kHz sampling rate and a resolution of 8 bits per sample. There was no pre-emphasis, and the analog-to-digital converter was assumed to be reasonably linear. This resulted in a signal v containing slightly fewer than 2^{15} samples, which was then padded at the end with dither to a length of 2^{15}. Hence $N = 32768$ and $n = 15$ for this example.

This signal was then transformed into a list of single-precision floating point numbers. For calculating the L^2 norm, the signal length was taken to be 1 with subinterval width $1/N$.

Three kinds of compression were tried on this list of floating-point numbers. Method 1 involved expanding v in the wavelet basis, using filters of length 2, 10, and 18, cutting off all terms less than ϵ, and then reconstructing the signal. Hardware limitations forced the signals to be divided into 2 segments of 16384 samples each before transforming to the wavelet basis. Table 1 shows the results obtained with this method.

For the next two methods, the signal was divided into 8 segments of $4096 = 2^{12}$ samples, and each segment expanded into the bases $w^1, \ldots, w^{12}$, using filters of length 2, 10, and 18. In method 2, each segment was reconstructed from its w^{12} representation after those coefficients less than ϵ were discarded. Results using this method are presented in Table 2. In method 3, each segment was reconstructed from its "best" representation w^b, $1 \leq b \leq 12$, where b was chosen for each segment to maximize the number of coefficients smaller than ϵ. This gave the best results, shown in Table 3.

In all the three methods, the values $\epsilon = 0.5, 2.0, 8.0$ were used, corresponding to approximately 0.6%, 2.5%, and 10% of the maximum value of the signal. This is expressed in dB below peak in Table 4, using the formula (dB below peak)$= -20 \log_{10}(\epsilon \text{ Peak})$.

The experiment was repeated with a male speaker and a female speaker. The dynamic range and rate of speech for the two samples were adjusted to be as similar as practicable, so that bandwidth was the principle difference between the signals.

While it is necessary to listen to the reconstructed signals in order to judge their intelligibility, some observations are possible from the estimates of errors introduced after reconstruction. First, the distortion seems to increase linearly with compression ratio. It becomes noticeable at a value of 5%, which occurs for compression ratios around 5 or 10. Second, at large cutoff values the longest filter gives the least distortion and the highest compression ratio. This is not surprising, given the smoothness of speech signals. Third, distortions become audible as superimposed hiss, while the speaker remains recognizable.

For contrast, Table 5 lists the distortions and compressions introduced by simply squelching the original signal at the same cutoff levels used in methods 1, 2, and 3. Even allowing large distortions, this method gives substantially smaller compression ratios.

Compression ratios are somewhat misleading out of context. In particular, if the sampling rate is more than twice the signal bandwidth, then oversampling contributes redundancy. Furthermore, silences between words or at the beginning and end of the phrase pad the signal length. Any compression scheme should remove these sources of redundancy. What remains is the Nyquist information content, and to judge a compression method it is important to measure the distortion introduced as the signal length is compressed below the Nyquist length. Estimating this length for the test signal above requires counting only those samples larger than some squelch level, and determining from a sonogram the essential bandwidth of the signal.

For the two speech samples studied, "silences" were considered to be any samples with values -1, 0, or 1 from the range $\{-128, \ldots, 127\}$. With the maximum amplitude of the male sample being 82, this represented a squelch level of about -38 dB. The following output is from a computer program that counted the squelched samples:

```
Please name the input file of bytes: jbayg.male
Set the squelch level (1,2,...,127): 1
Squelched: 7671 of 32768, or 23.4% of signal.
(True length 25097)
```

With the maximum amplitude of the female sample being 116, the squelch level represented -41 dB. Corresponding output for that signal is

```
Please name the input file of bytes: jbayg.female
Set the squelch level (1,2,...,127): 1
Squelched: 7653 of 32768, or 23.4% of signal.
(True length 25115)
```

The speech samples were produced by subjects intent on speaking at the same rate. Higher squelch levels result in diverging "True lengths," providing some justification for the rather arbitrary choice of 1.

Bandwidth determinations are less precise. One method is to find a frequency f such that $1-\epsilon$ of the signal energy, averaged over time, is at frequencies below f. It then remains to determine ϵ. Under the assumption that 5% rms errors are acceptable, we can set $\epsilon = 0.05$. Then the bandwidth for the male sample is 5.5 kHz, while for the female sample it is 7.5 kHz. Thus, the male's speech is oversampled by a factor of 2, while the female's is oversampled by a factor of 1.5.

Removing these redundancies would result in signals of Nyquist length. These are tabulated below:

Signal	**Bandwidth**	**Nyquist length**	**Compression**
Male	5.5kHz	12549 (38%)	2.6
Female	7.5kHz	16743 (51%)	2.0

Only compression ratios greater than the compression to the Nyquist length will be independent of the sampling method. The absolute compression ratio divided by the Nyquist compression will be called *compression below the Nyquist length* or the *invariant compression ratio*. For confirmation of the validity of this estimate, notice that long filters and negligible cutoffs result in compression to this Nyquist length, regardless of the method.

Whenever the invariant compression ratio exceeds 1, there will be at least some distortion. This distortion might be acceptable if it is small, or if the essential features of the signal are still intelligible. In fact, all of the reconstructed compressed signals retained recognizability as well as intelligibility: the speakers' friends could easily identify the speakers. Other compression methods, which exploit specific features of human speech to obtain much higher compression ratios, will be considered in a subsequent article.

Signal	Filter	Cutoff (dB)	Compression	Inv. Compression	Δ_{RMS} (%)
Male	2	44	1.4	0.5	0.0
		32	2.7	1.0	4.7
		20	7.9	3.0	14
	10	44	2.5	1.0	0.7
		32	6.3	2.4	4.0
		20	13.9	5.3	9.4
	18	44	2.6	1.0	0.7
		32	6.7	2.6	4.0
		20	14.3	5.5	9.1
Female	2	47	1.3	0.7	0.1
		35	2.1	1.1	3.2
		23	5.2	2.6	12
	10	47	2.1	1.1	0.4
		35	4.3	2.1	3.0
		23	8.7	4.3	8.0
	18	47	2.2	1.1	0.9
		35	4.6	2.3	2.9
		23	8.9	4.5	7.5

Table 1. Discard small wavelet components.

SIGNAL	FILTER	CUTOFF (dB)	COMPRESSION	INV. COMPRESSION	Δ_{RMS} (%)
Male	2	44	1.4	0.5	0.4
		32	2.8	1.1	5.0
		20	7.2	2.8	15
	10	44	2.3	0.9	0.7
		32	5.8	2.2	4.1
		20	14.0	5.4	10
	18	44	2.4	0.9	0.7
		32	6.1	2.3	4.0
		20	15.3	5.9	10
Female	2	47	1.3	0.7	0.3
		35	2.0	1.0	3.5
		23	4.6	2.3	12
	10	47	2.0	1.0	0.4
		35	4.0	2.0	3.0
		23	8.9	4.5	8.6
	18	47	2.1	1.1	0.5
		35	4.4	2.2	3.0
		23	9.7	4.9	8.2

Table 2. Discard small lowest-level wavelet packet components.

Signal	Filter	Cutoff (dB)	Compression	Inv. Compression	Δ_{RMS} (%)
Male	2	44	1.7	0.7	1.2
		32	3.3	1.3	4.9
		20	8.8	3.4	14
	10	44	2.6	1.0	0.8
		32	6.8	2.6	4.0
		20	16.6	6.4	9.7
	18	44	2.7	1.0	0.8
		32	7.3	2.8	4.0
		20	17.9	6.9	9.8
Female	2	47	1.5	0.7	0.7
		35	2.4	1.2	3.4
		23	5.6	2.8	11
	10	47	2.3	1.1	0.5
		35	5.0	2.5	3.0
		23	10.8	5.4	8.0
	18	47	2.4	1.2	0.5
		35	5.5	2.7	2.9
		23	12.4	6.2	7.7

Table 3. Choose best-level wavelet packets.

SIGNAL	PEAK	ϵ	dB BELOW PEAK
Male	82	0.5	44
		2.0	32
		8.0	20
Female	116	0.5	47
		2.0	35
		8.0	23

Table 4. Relative attenuation of certain cutoffs.

SIGNAL	CUTOFF (dB)	COMPRESSION	INV. COMPRESSION	Δ_{RMS} (%)
Male	44	1.1	0.4	0.0
	32	1.3	0.5	2.4
	20	2.4	0.9	17
Female	47	1.1	0.5	0.0
	35	1.3	0.7	1.7
	23	2.0	1.0	11

Table 5. Squelch small samples.

Appendix A: Fast algorithms for signal processing

All of the algorithms mentioned above and used to perform the experiments have been coded by the author in the C programming language. The experiments were performed on a Sun 3 minicomputer at the Yale University Department of Mathematics, using samples recorded on a Macintosh II personal computer.

Transformation to wavelet bases and back

The fast wavelet transformation and its inverse are described in a preprint of Beylkin, Coifman, and Rokhlin [1]. There they discuss fast multiplication by singular integral operators of the vectors thus encoded. A future paper will explore the applications of such operators to problems of noise suppression, speech acceleration, and speech recognition.

Note from Tables 1 and 3 that compression with wavelet packets is 25% to 38% tighter than compression with wavelets. This advantage may be offset by the greater computation time required.

Transformation to wavelet packet bases and back

Efficiently transforming to periodized wavelet packets requires organizing the applications of F_ϵ into a pyramid scheme. Suppose

$$v = v^0 = \sum_{j=0}^{N-1} v_j^0 w_j^0$$

is a vector in $l^2(N)$ written in the standard basis w^0. Then

$$v = \sum_{j=0}^{N-1} v_j^1 w_j^1 = \ldots = \sum_{j=0}^{N-1} v_j^n w_j^n$$

are the various representations of v in the n wavelet packet bases as $v^m \equiv (v_0^m, \ldots, v_{N-1}^m)$, $m = 1, 2, \ldots, n$. It will be shown that going from v^m to v^{m+1} requires Nr operations, where r is the filter length. Thus filling the entire rectangle $v^0 \mapsto v^1 \mapsto \cdots \mapsto v^n$ costs nrN operations.

However, $\{0, 1, \ldots, 2^n-1\}$ has a "dyadic decomposition" into subsets of 2^m consecutive indices, $m = 0, 1, \ldots, n$. Let us denote these subsets by

$$\Delta_i^m = \{i2^m, i2^m+1, \ldots, i2^m+2^m-1\}, \quad \text{for} \quad i = 0, 1, \ldots, 2^{n-m}-1,$$

and call m the *scale* of the interval. Then clearly $\Delta_i^m \cap \Delta_j^m = \emptyset$ if $i \neq j$, and each dyadic interval of scale m contains two disjoint (left and right) *daughters* of scale $m-1$, and is contained in a unique *parent* of scale $m+1$. In fact, $\Delta_{2i}^{m-1} \cup \Delta_{2i+1}^{m-1} \subset \Delta_i^m \subset \Delta_{\lfloor i/2 \rfloor}^{m+1}$, where as usual $\lfloor x \rfloor$ denotes the greatest integer less than or equal to x.

The convolutions H, G can be arranged so that their images are indexed by the daughters of the domain interval. Denote by V_i^m, $m = 0, 1, \ldots, n$, $i = 0, 1, \ldots, 2^{n-m}-1$, the 2^m-dimensional subspaces of $\mathbb{R}^N$ spanned by the standard basis vectors $\{e_j : j \in \Delta_i^m\}$. Then $\mathbb{R}^N = \oplus \sum_{i=0}^{2^{n-m}-1} V_i^m$, and composing with the appropriate injections gives, for each m,

$$H : V_i^m \to V_{2i}^{m-1}, \qquad \text{for } \; i = 0, 1, \ldots, 2^{n-m} - 1,$$
$$G : V_i^m \to V_{2i+1}^{m-1}, \qquad \text{for } \; i = 0, 1, \ldots, 2^{n-m}-1.$$

Applying H, G successively to $V_0^m, V_1^m, \ldots$ produces N new coordinates. Each operator requires $rN/2$ multiplications, giving a total of rN operations for this step. It remains to show that these steps successively develop the coefficients $v^1, v^2, \ldots, v^n$, but this is a straightforward induction. By Lemma 3, it follows that

$$v_j^{m+1} = \langle e_{\hat{\jmath}}, F_{\epsilon_{n-m}(j)} F_{\epsilon_{n-m+1}(j)} \cdots F_{\epsilon_n(j)}\, v \rangle = \langle e_{\hat{\jmath}}, F_{\epsilon_{n-m}(j)}\, v^m | \Delta_{\check{\jmath}}^m \rangle,$$

where the indices $\hat{\jmath}$ and $\check{\jmath}$ are determined uniquely by the bitwise decomposition $j = \check{\jmath} 2^{n-m} + \sum_{n-m}(j) 2^{n-m-1} + \hat{\jmath}$, and so

$$v_j^{m+1} = \begin{cases} \left(H\, v^m | \Delta_{\check{\jmath}}^m\right)_{\hat{\jmath}}, & \text{if } j \text{ is in the left daughter of } \Delta_{\check{\jmath}}^m; \\ \left(G\, v^m | \Delta_{\check{\jmath}}^m\right)_{\hat{\jmath}}, & \text{if } j \text{ is in the right daughter of } \Delta_{\check{\jmath}}^m. \end{cases}$$

Now let m range from 0 to $n-1$.

To reconstruct vectors from their wavelet packet representations, this pyramid scheme must be reversed. Composing with the appropriate injections gives

$$H^* : V_{2i}^{m-1} \to V_i^m, \qquad \text{for } i = 0, 1, \ldots, 2^{n-m}-1;$$
$$G^* : V_{2i+1}^{m-1} \to V_i^m, \qquad \text{for } i = 0, 1, \ldots, 2^{n-m}-1.$$

Since the ranges of H^* and G^* are orthogonal by Equation (8), as are the subspaces $V_i^m, i = 0, \ldots, 2^{n-m}-1$, the previous wavelet packet coefficients can be recursively computed, namely:

$$v_j^{m-1} = \left(H^* v^m | \Delta_{L(j)}^m + G^* v^m | \Delta_{R(j)}^m\right)_{\hat{\jmath}},$$

where $\hat{\jmath} \equiv j \bmod 2^{n-m+1}$, and $L(j), R(j)$ index the left and right daughters of the unique dyadic subset at level $m-1$ which contains j.

Each adjoint filter convolution requires $r2^m/2$ multiplications on each of the 2^{n-m} subspaces V_i^m, plus the results must be added at the end. The operation count is therefore $O(rN)$. Letting m range from m_0 to 0 rebuilds the signal from any complete level m_0 of coefficients.

It remains to show that any admissible subset of coefficients suffices to rebuild a complete level. The most general admissible subsets considered here are those of the next section, namely the disjoint dyadic covers. But if any member of the admissible subset is on the bottom level, it is there together with all the other members of its scale-n dyadic subset, and also with its twin sister on that level. The algorithm reconstructs each parent from its two daughters, which may then be discarded from the bottom, leaving another admissible subset with fewer levels. Clearly, the operation count to reconstruct some of the parents is at most the operation count to reconstruct all of them, which is $O(rN)$. By induction on the number of levels, the algorithm after at most n steps yields an admissible subset on level 0; *i.e.*, the original signal. The total operation count is $O(rN \log_2 N)$, since there are at most $n = \log_2 N$ levels.

Searching for optimal admissible subsets

Suppose that a vector $v \in \mathbb{R}^N$ has been expanded into all wave packet bases, with the coordinate vectors $v^0, \ldots, v^n$ forming rows in a rectangular array. An *admissible subset* Σ of this array, as defined in the introduction, corresponds to any disjoint cover of $\{0, 1, \ldots, N-1\}$ by the dyadic subsets $\Delta_j^m, m = 0, 1, \ldots, n, j = 0, 1, \ldots, 2^{n-m+1}$.

The correspondence is given by the following: w_j^m is part of the basis $\Longleftrightarrow$ $\Delta_j^m \in \Sigma$, where $j \in \Delta_j^m$ determines j uniquely.

An *optimal basis* for a given vector v and threshold ϵ is an admissible subset for which as many indexed coordinates are smaller than ϵ (in absolute value) as for any other. There are more than 2^N admissible subsets, but the optimal basis may be found in $O(N \log_2 N)$ operations via the following algorithm.

For simplicity of description, suppose that the array to be searched contains only "ones" (for large entries) and "zeros" (for entries smaller than the threshold). Finding an optimal basis means finding an admissible subset with the maximal number of zeros, or equivalently the minimal number of 1's. This search may be implemented recursively, and its properties proved by induction.

Consider a pair of dyadic subintervals Δ_{2i}^k and Δ_{2i+1}^k at level k, and suppose that each has been assigned the cost of "ones" present in the most efficient wavelet packet subbasis of itself. The sum of these costs may be compared to that of the joint parent Δ_i^{k-1}. If the parent is cheaper, it is marked "kept" and is assigned its own cost. If the two children sum to a lower cost, then this sum is assigned to the parent but the parent is not marked.

To start the induction, we mark as "kept" all the children at the bottom level n, and assign each of them their own cost in "ones." The induction on k procedes until $k = 0$. At each step, the chosen subset below each parent is the disjoint dyadic cover of that parent interval which indexes the maximal number of zeros. The union of these over all parents in that row is always an admissible subset. At the last step, when there is only one parent Δ_0^0, the chosen subset will be an optimal basis for the original signal. Of course, it is not necessarily unique.

Advancing from row m to $m-1$ requires counting zeros among N numbers for size, in groups of 2^m, then comparing these counts against adjacent sums from a table of $2N/2^m$ previous counts. Altogether, advancing one step takes $O(N)$ operations. There are $n = \log_2 N$ steps, resulting in the $O(N \log_2 N)$ complexity.

The topmost "kept" intervals in the marked tree constitute the best basis. They are exactly those intervals which are better than any combination of their descendents, but which have no better ancestors. These "best" intervals can be found by a depth-first search whose complexity is $O(N)$, the size of the tree of dyadic subintervals down to level $\log_2 N$.

To estimate the storage requirements of this algorithm, note that passing from step m to step $m-1$ requires a table of $2N/2^m$ integers from the set $0, 1, \ldots, 2^{m-1}$, which takes N bits of space. This may be reused. Recording the current best admissible subset at each row may be done with $2N$ bits. Order the family of $2N$ dyadic subintervals in any manner, labeling the result $\{\Delta(k), k = 1, 2, \ldots, 2N - 1\}$. Let D be the unique integer between 0 and $2^{2N} - 1$ with the property that $\epsilon_k(D) = 1 \iff \Delta(k)$ is a maximal subinterval of the admissible subset. Otherwise, put $\epsilon_k(D) = 0$. D is constructed during the search, and if the dyadic subsets are correctly ordered, the storage requirements for the fraction of D determined at step m are $N + N/2 + \ldots + N/2^m$. After the last step, D is equivalent to an optimal basis.

There are more efficient ways of coding the basis description, of course, for example by using standard lossless compression methods to eliminate repetition.

Transmission of compressed signals

After expansion into wavelet or wavelet packet coefficients and compression by discarding the small ones, it is necessary to transmit the surviving values as well as their positions in the dyadic cover. Suppose that the signal consists of N samples each b bits long, thus requiring Nb bits of storage. When transformed into the optimal basis, say that only N_ϵ of the coefficients exceed the cutoff ϵ. Since no cheating is allowed, the coefficients can only contain b bits each. Specifying their positions in an N by $\log_2 N$ array takes $\log_2 N + \log_2 \log_2 N$ bits, or equivalently the level and position within that level of the coefficient. This adds up to $N_\epsilon(b + \log_2 N + \log_2 \log_2 N)$ bits. The compression ratio of the transmission method is then

$$\rho = \frac{N_\epsilon}{N} \times \left(1 + \frac{\log_2 N + \log_2 \log_2 N}{b}\right).$$

The first factor is the compression ratio computed above. The second factor may be controlled by breaking up the signal into segments of approximately the same length as the dynamic range 2^b.

References

1. Beylkin, G., R. Coifman, and V. Rokhlin, Fast wavelet transforms and numerical algorithms I, *Comm. Pure and Appl. Math.* **44** (1991), 141–183.
2. Coifman, R., Y. Meyer, S. Quake, and M. V. Wickerhauser, Signal processing and compression with wave packets, *Proceedings of the Conference on Wavelets*, Marseilles, Spring 1989.
3. Daubechies, I., Orthonormal bases of compactly supported wavelets, *Comm. Pure and Appl. Math.* **41** (1988), 909–996.
4. Mallat, S., A theory for multiresolution signal decomposition: the wavelet reprsentation, *IEEE Pattern Anal. and Machine Intell.* **11** (1989), 674–693.
5. Nicolas, Acoustic signature recognition with arborescent wavelets, *Proceedings of the conference on wavelets*, Marseilles, Spring 1989.

This research was supported in part by ONR Grant N00014-88-K0020.

Mladen Victor Wickerhauser
Numerical Algorithms Research Group
Department of Mathematics
Yale University
New Haven, CT 06520
victor@math.yale.edu

Permanent address:
Mladen Victor Wickerhauser
Department of Mathematics
Washington University
St. Louis, Missouri 63130

Bibliography

1. Adelson, E. H., E. Simoncelli, and R. Hingorani, Orthogonal pyramid transforms for image coding, in *Proc. SPIE Conf. Visual Communication and Image Processing*, Cambridge, MA, 1987, 50–58.
2. Aldroubi, A. and M. Unser, Families of wavelet transforms in connection with Shannon's sampling theory and the Gabor transform, in *Wavelets: A Tutorial In Theory and Applications*, C. K. Chui (ed.), Academic Press, Cambridge, MA, 1992, 509–528.
3. Allen, J. B. and L. R. Rabiner, A unified theory of short-time spectrum analysis and synthesis, *Proc. IEEE* **65** (11) (1977), 1558–1564.
4. Alpert, B., Construction of simple multiscale bases for fast matrix operations, in *Wavelets and Their Applications*, M. B. Ruskai, G. Beylkin, R. Coifman, I. Daubechies, S. Mallat, Y. Meyer, and L. Raphael (eds.), Jones and Bartlett, Boston, 1992, 211–226.
5. Alpert, B., *Sparse Representation of Smooth Linear Operators*, Ph. D. Thesis, Yale University Department of Computer Science Report #814, August 1990.
6. Alpert, B., Wavelets and other bases for fast numerical linear algebra, in *Wavelets: A Tutorial In Theory and Applications*, C. K. Chui (ed.), Academic Press, Cambridge, MA, 1992, 181–216.
7. Alpert, B., G. Beylkin, R. Coifman, and V. Rokhlin, Wavelets for the fast solution of second-kind integral equations, *SIAM J. Sci. Statist. Comput.*, to appear.
8. Antoine, J. P., P. Carrette, R. Murenzi, and B. Piette, Image analysis with two-dimensional continuous wavelet transform, 1991, submitted to *IEEE Trans. Inform. Theory.*
9. Antonini, M., M. Barlaud, and P. Mathieu, Image coding using lattice vector quantization of wavelet coefficients, *Proc. IEEE Int. Conf. ASSP* 1991, 2273–2276.
10. Antonini, M., M. Barlaud, P. Mathieu, and I. Daubechies, Image coding using vector quantization in the wavelet domain, in *Proc. IEEE Int. Conf. ASSP*, Albuquerque, NM, 1990, 2297–2300.
11. Antonini, M., M. Barlaud, P. Mathieu, and I. Daubechies, Image coding using wavelet transforms, *IEEE Trans. ASSP* (1991), to appear.
12. Arneodo, A., F. Argoul, J. Elezgaray and G. Grasseau, Wavelet transform analysis of fractals: application to nonequilibrium phase transitions, in *Nonlinear Dynamics*, G. Turchetti (ed.), World Scientific, Singapore, 1988, 130.

13. Arneodo, A., F. Argoul, E. Freysz, J. F. Muzy, and B. Pouligny, The optical wavelet transform, in *Wavelets and Their Applications*, M. B. Ruskai, G. Beylkin, R. Coifman, I. Daubechies, S. Mallat, Y. Meyer, and L. Raphael (eds.), Jones and Bartlett, Boston, 1992, 241–273.
14. Arneodo, A., F. Argoul, E. Bacry, J. Elezgaray, E. Freysz, G. Grasseau, J. F. Muzy, and B. Pouligny, Wavelet transform of fractals: I, From the transition to chaos to fully developed turbulence, preprint.
15. Arneodo, A., F. Argoul, E. Bacry, J. Elezgaray, E. Freysz, G. Grasseau, J. F. Muzy, and B. Pouligny, Wavelet transform of fractals: II, Optical wavelet transform of fractal growth phenomena, preprint.
16. Argoul, F., A. Arneodo, J. Elezgaray, G. Grasseau, and R. Murenzi, Wavelet transform of two-dimensional fractal aggregates, *Phys. Lett. A.* **135** (1989), 327–336.
17. Argoul, F., A. Arneodo, G. Grasseau, Y. Gagne, E. J. Hopfinger, and U. Frisch, Wavelet analysis of turbulence reveals the multifractal nature of the Richardson cascade, *Nature* **338** (1989), 51–53.
18. Auscher, P., Ondelettes à support compact et conditions aux limites, preprint.
19. Auscher, P., Ondelettes fractales et applications, Thèse de Doctorat, Univ. Paris-Dauphine, 1989.
20. Auscher, P., Symmetry properties for Wilson bases and new examples with compact support, Université de Rennes, 1990, preprint.
21. Auscher, P., Wavelet bases for $L^2(\mathbb{R})$, with rational dilation factor, in *Wavelets and Their Applications*, M. B. Ruskai, G. Beylkin, R. Coifman, I. Daubechies, S. Mallat, Y. Meyer, and L. Raphael (eds.), Jones and Bartlett, Boston, 1992, 439–451.
22. Auscher, P., Wavelets with boundary conditions on the interval, in *Wavelets: A Tutorial In Theory and Applications*, C. K. Chui (ed.), Academic Press, Cambridge, MA, 1992, 217–236.
23. Auscher, P. and Ph. Tchamitchian, Bases d'ondelettes sur les courbes corde-arc, noyau de Cauchy et espaces de Hardy associés, *Rev. Mat. Iberoamericana* **5** cenk (1989), 139–170.
24. Auscher, P. and Ph. Tchamitchian, Ondelettes, pseudoacerétivité, noyan de Cauchy et espaces de Hardy, preprint.
25. Auscher, P., G. Weiss, and M. V. Wickerhauser, Local sine and cosine bases of Coifman and Meyer and the construction of smooth wavelets, in *Wavelets: A Tutorial In Theory and Applications*, C. K. Chui (ed.), Academic Press, Cambridge, MA, 1992, 237–256.
26. Auslander, L. and I. Gertner, Wide-band ambiguity function and $ax + b$ group, in *Signal Processing Part I: Signal Processing Theory*, L. Auslander, F. A. Grünbaum, J. W. Helton, T. Kailath, P. Khargonekar, S. Mitter (eds.), Springer-Verlag, New York, 1990, 1–12.
27. Auslander, L. and R. Tolimieri, Radar ambiguity functions and group theory, *SIAM J. Math. Anal.* **16** (1985), 577–601.

28. Bacry, H., A. Grossmann, and J. Zak, Proof of the completeness of lattice states in the kq-representation, *Phys. Rev. B* **12** (1975), 1118–1120.

29. Bacry, E., S. Mallat, G. Papanicolaou, Time and space wavelet adaptative scheme for partial differential equations, Communication at "Wavelets and Turbulence", Princeton, June 1991.

30. Bacry, E., J. F. Muzy, and A. Arneodo, Fractal dimensions from wavelet analysis: exact results, in preparation.

31. Bandt, C., Self-similar sets 5-Integer matrices and fractal tilings of $\mathbb{R}^n$, *Proc. Amer. Math. Soc.* **112** (1991), 549–562.

32. Basdevant, C., M. Holschneider, and V. Perrier, Méthode des ondelettes mobiles, *C. R. Acad. Sci. Paris*, Série I (1990), 647–652.

33. Bastiaans, M. J., A sampling theorem for the complex spectrogram, and Gabor's expansion of a signal in Gaussian elementary signals, *Opt. Eng.* **20** (1981), 594–598.

34. Bastiaans, M. J., Gabor's expansion of a signal into Gaussian elementary signals, *Proc. IEEE* **68** (1980), 538–539.

35. Bastiaans, M. J., On the sliding-window representation in digital signal processing, *IEEE Trans. ASSP* **33** (1985), 868–873.

36. Battle, G., A block spin construction of ondelettes. Part I: Lemarié functions, *Comm. Math. Phys.* **110** (1987), 601–615.

37. Battle, G., Heisenberg proof of the Balian-Low theorem, *Lett. Math. Phys.* **15** (1988), 175–177.

38. Battle, G., Phase space localization theorem for ondeletts, *J. Math. Phys.* **30** (1989), 2195–2196.

39. Battle, G., Wavelets: a renormalization group point of view, in *Wavelets and Their Applications*, M. B. Ruskai, G. Beylkin, R. Coifman, I. Daubechies, S. Mallat, Y. Meyer, and L. Raphael (eds.), Jones and Bartlett, Boston, 1992, 323–349.

40. Battle, G., Cardinal spline interpolation and the block spin construction of wavelets, in *Wavelets: A Tutorial In Theory and Applications*, C. K. Chui (ed.), Academic Press, Cambridge, MA, 1992, 73–90.

41. Benassi A., S. Jaffard, and D. Roux, Bases d'ondelettes orthonormées pour un opérateur différentiel, 1991, preprint.

42. Benedetto, J., Gabor representations and wavelets, in *Commutative Harmonic Analysis*, D. Colella (ed.), *Contemp. Math.* **19** Am. Math. Soc., Providence, 1989, 9–27.

43. Benedetto, J., Stationary frames and spectral estimation, in *Probabilistic and Stochastic Advances in Analysis, with Applications*, J. S. Byrnes and J. L. Byrnes (eds.), NATO-ASI Series C, Kluwer Academic Publishers, 1991, to appear.

44. Benedetto, J., Uncertainty principle inequalities and spectrum estimation, in *Recent Advances in Fourier Analysis and its Applications*, J. S. Byrnes and J. L. Byrnes (eds.), NATO-ASI Series C, Kluwer Academic Publishers, **315,** 1990, 143-182.

45. Benedetto, J., Irregular sampling frames, in *Wavelets: A Tutorial In Theory and Applications*, C. K. Chui (ed.), Academic Press, Cambridge, MA, 1992, 445–507.
46. Benedetto, J., C. Heil, and D. Walnut, Remarks on the proof of the Balian-Low theorem, *Canad. J. Math.*, to appear.
47. Benedetto, J. and W. Heller, Irregular sampling and the theory of frames, I, *Note Math.*, 1991.
48. Benedetto, J. and W. Heller, Irregular sampling and the theory of frames, II, preprint.
49. Berger, M. A., Wavelets as attractors of random dynamical systems, in *Stochastic Analysis: Liber Amicorum for Moshe Zakai*, E. Mayer-Wolf, M. Zakai, E. Merzbah, and A. Shwartz (eds.), Academic Press, NY, 1991.
50. Berger, M. A. and Y. Wang, Multi–scale dilation equations and iterated function systems, Report Math: 011491–001, Georgia Institute of Technology, preprint.
51. Berger, M. A. and Y. Wang, Multidimensional two-scale dilation equations, in *Wavelets: A Tutorial In Theory and Applications*, C. K. Chui (ed.), Academic Press, Cambridge, MA, 1992, 295–323.
52. Bertrand, J. and P. Bertrand, Time-frequency representations of broad-band signals, in *Wavelets*, J. M. Combes, A. Grossmann and Ph. Tchamitchian (eds.), 1989, 164–171.
53. Beylkin, G., On the representation of operators in compactly supported wavelet bases, *SIAM J. on Numerical Analysis*, to appear.
54. Beylkin, G., R. Coifman, and V. Rokhlin, Fast wavelet transforms and numerical algorithms I, *Comm. Pure and Appl. Math.* **44** (1991), 141–183.
55. Beylkin, G., R. Coifman, and V. Rokhlin, Fast wavelet transforms and numerical algorithms II, in progress.
56. Beylkin, G., R. R. Coifman, and V. Rokhlin, Wavelets in numerical analysis, in *Wavelets and Their Applications*, M. B. Ruskai, G. Beylkin, R. Coifman, I. Daubechies, S. Mallat, Y. Meyer, and L. Raphael (eds.), Jones and Bartlett, Boston, 1992, 181–210.
57. Boashash, B., Time-frequency signal analysis, in *Advances in Spectrum Analysis and Array Processing*, S. Haykin (ed.), Prentice Hall, Englewood Cliffs, NJ, 1990, 418–517.
58. de Boor, C., *A Practical Guide to Splines*, Springer-Verlag, NY, 1978.
59. de Boor, C, R. DeVore, and A. Ron, On the construction of multivariate (pre)wavelets, 1992, preprint.
60. Bourgain J., A remark on the uncertainty principle of Hilbertian basis, *J. of Funct. Anal.* **79** (1988), 136–143.
61. Bovik, A., T. Emmoth, and A. Restrepo, Localized measurement of emergent image frequencies by Gabor wavelets, *IEEE Trans. Inform. Theory* **38** (2) (1992), to appear.
62. Buhmann, M. and C. A. Micchelli, Spline prewavelets for non-uniform knots, IBM Research Report, 1990.

63. Burt, P. J., Fast algorithms for estimating local image properties, *Comput. Graphics and Image Processing* **21** (1983), 368–382.
64. Burt, P. J. and E. H. Adelson, The Laplacian pyramid as a compact image code, *IEEE Trans. Comm.* **31** (1983), 482–540.
65. Calderón, A. P., Intermediate spaces and interpolation, the complex method, *Studia Math.* **24** (1964), 113–190.
66. Calderón, A. P. and A. Zygmund, Singular integral operators and differential equations, *Amer. J. Math.* **79** (1957), 901–921.
67. Cavaretta, A. S., W. Dahmen, and C. A. Micchelli, Stationary subdivision, *Mem. Amer. Math. Soc.*, to appear.
68. Cenker, C., Master thesis, Kohärente reihendarstellungen von distributionen, Vienna, 1989.
69. Cenker, C., H. G. Feichtinger, and K. Gröchenig, Non-orthogonal expansions of signals and some of their applications, in *Image Acquisition and Real-Time Visualization, 14*, OEAGM Treffen, ÖCG, Bd. **56**, G. Bernroider and A. Pinz (eds.), May 1990, 129–138.
70. Chan, A. K. and C. K. Chui, Real-time signal analysis with quasi-interpolatory splines and wavelets, in *Curves and Surfaces*, P. J. Laurent, A. Le Méhauté, and L. L. Schumaker (eds.), Academic Press, 1991, 75–82.
71. Chan, A. K., C. K. Chui, J. Z. Wang, Q. Liu, and J. Zha, Introduction to B-wavelets and applications to signal processing, CAT Report #245, Texas A&M University, 1991.
72. Chui, C. K., Curve design and analysis using splines and wavelets, Trans. Eighth Army Conf. on Appl. Math. and Computing, 1990, 471–481.
73. Chui, C. K., An overview of wavelets, in *Approximation Theory and Functional Analysis*, C. K. Chui (ed.), Academic Press, Boston, MA, 1991, 47–72.
74. Chui, C. K., *An Introduction to Wavelets*, Academic Press, Boston, MA, 1992.
75. Chui, C. K., On cardinal spline-wavelets, in *Wavelets and Their Applications*, M. B. Ruskai, G. Beylkin, R. Coifman, I. Daubechies, S. Mallat, Y. Meyer, and L. Raphael (eds.), Jones and Bartlett, Boston, MA, 1992, 419–438.
76. Chui, C. K. (ed.), *Wavelets: A Tutorial in Theory and Applications*, Academic Press, Boston, 1992.
77. Chui, C. K., On wavelet analysis, in *U.S.-U.S.S.R. Workshop Vol.*, A. Gonchar and E. Saff (eds.), Springer-Verlag, 1992, to appear.
78. Chui, C. K., Wavelets–with emphasis on spline-wavelets and applications to signal analysis, in NATO Advanced Institute Studies, *Approximation Theory, Splines, and Applications*, S. Singh (ed.), to appear.
79. Chui, C. K., Wavelets: with Emphasis on Time-Frequency Analysis, under preparation.
80. Chui, C. K., Wavelets and spline interpolation, in *Advances in Numerical Analysis, II: Wavelets, Subdivision Algorithms, and Radial Basis Functions*, W. Light (ed.), Oxford University Press, Oxford, 1992, to appear.

81. Chui, C. K., Wavelets and signal analysis, in *Approximation Theory VII*, C. K. Chui, E. W. Cheney, and L. L. Schumaker (eds.), Academic Press, New York, 1992, to appear.
82. Chui, C. K. and C. Li, Non-orthogonal wavelet packets, CAT Report #261, Texas A&M University, 1991.
83. Chui, C. K. and C. Li, A general framework of multivariate wavelets with duals, CAT Report #268, Texas A&M University, 1992.
84. Chui, C. K. and H. N. Mhaskar, On trigonometric wavelets, CAT Report #254, Texas A&M University, 1991; *Constr. Approx.*, to appear.
85. Chui, C. K. and E. Quak, Wavelets on a bounded interval, in *Numerical Methods of Approximation Theory*, D. Braess and L. L. Schumaker (eds.), Birkhäuser Verlag, Basel, 1992, to appear.
86. Chui, C. K. and X. L. Shi, Inequalities of Littlewood-Paley type for frames and wavelets, CAT Report #249, Texas A&M University, 1991; *SIAM J. Math. Analysis*, to appear.
87. Chui, C. K. and X. L. Shi, On a Littlewood-Paley identity and characterization of wavelets, CAT Report #250, Texas A&M University, 1991; *J. Math. Anal. Appl.*, to appear.
88. Chui, C. K. and X. L. Shi, Characterization of scaling functions and wavelets, CAT Report #259, Texas A&M University, 1991.
89. Chui, C. K. and X. L. Shi, $N\times$ oversampling preserves any tight affine-frame for odd N, CAT Report #264, Texas A&M University, 1992.
90. Chui, C. K. and X. L. Shi, Bessel sequences and affine frames, CAT Report #267, Texas A&M University, 1992.
91. Chui, C. K. and X. L. Shi, Wavelets and multiscale interpolation, in *CAGD and Signal Analysis*, T. Lyche and L. L. Schumaker (eds.), Academic Press, 1992, 111–133.
92. Chui, C. K., J. Stöckler, and J. D. Ward, Compactly supported box-spline wavelets, CAT Report #230, Texas A&M University, 1990.
93. Chui, C. K., J. Stöckler, and J. D. Ward, Polynomial expansions for cardinal interpolants and orthonormal wavelets, in *Curves and Surfaces*, P. J. Laurent, A. Le Méhauté, and L. L. Schumaker (eds.), Academic Press, 1991, 83–90.
94. Chui, C. K. and J. Z. Wang, A cardinal spline approach to wavelets, *Proc. Amer. Math. Soc.* **113** (1991), 785–793.
95. Chui, C. K. and J. Z. Wang, A general framework of compactly supported splines and wavelets, *J. Approx. Theory*, to appear.
96. Chui, C. K. and J. Z. Wang, An analysis of cardinal-spline wavelets, *J. Approx. Theory*, to appear.
97. Chui, C. K. and J. Z. Wang, Computational and algorithmic aspects of cardinal-spline wavelets, CAT Report #235, Texas A&M University, 1990.
98. Chui, C. K. and J. Z. Wang, On compactly supported spline wavelets and a duality principle, *Trans. Amer. Math. Soc.*, 1992, to appear.
99. Cohen, A., Construction de bases d'ondelettes a-Hölderiennes, *Revista Matematica Iberoamericana*, (1991).

100. Cohen, A., Ondelettes, analyses multirésolutions et filtres miroir en quadrature, *Annales de l'IHP, analyse non linéaire* **7** (5) (1990), 439–459.
101. Cohen, A., Ondelettes, Analyses multirésolutions et traitement numérique du signal, Ph. D. Thesis, Univ. Paris-Dauphine, 1990.
102. Cohen, A., Wavelets and digital signal processing, in *Wavelets and Their Applications*, M. B. Ruskai, G. Beylkin, R. Coifman, I. Daubechies, S. Mallat, Y. Meyer, and L. Raphael (eds.), Jones and Bartlett, Boston, 1992, 105–121.
103. Cohen, A., Biorthogonal wavelets, in *Wavelets: A Tutorial In Theory and Applications*, C. K. Chui (ed.), Academic Press, Cambridge, MA, 1992, 123–152.
104. Cohen, A., Wavelet bases, approximation theory, and subdivision schemes, *Approximation Theory VII*, E. W. Cheney, C. K. Chui, and L. L. Schumaker (eds.), Academic Press, New York, 1992, xxx-xxx.
105. Cohen, A. and J. P. Conze, Régularité des bases d'ondelettes et mesures ergodiques, in *Comptes rendus de l'Académie des Sciences*, Paris, (1991), to appear.
106. Cohen, A. and I. Daubechies, Orthonormal bases of compactly supported wavelets III: Better frequency localization, AT&T Bell Laboratories, preprint, submitted to *SIAM J. Math. Anal.*
107. Cohen, A. and I. Daubechies, A stability criterion for biorthogonal wavelet bases and their related subband coding schemes, AT&T Bell Laboratories, 1991, preprint.
108. Cohen, A. and I. Daubechies, Nonseparable bidimensional wavelet bases, AT&T Bell Laboratories, 1991, preprint.
109. Cohen, A. and I. Daubechies, Private communication.
110. Cohen, A., I. Daubechies, and J. C. Feauveau, Biorthogonal bases of compactly supported wavelets, *Comm. Pure and Appl. Math.*, 1991, to appear.
111. Cohen, A., I. Daubechies, and P. Vial, Wavelets and fast wavelet transform on the interval, AT&T Bell Laboratories, 1991, preprint.
112. Cohen, A. and J. Johnston, Joint optimization of wavelet and impulse response constraints for biorthogonal filter pairs with exact reconstruction, AT&T Bell Laboratories, 1991, preprint.
113. Cohen, A. and J. M. Schlenker, Compactly supported wavelets with haxagonal symmetry, AT&T Bell Laboratories, 1991, preprint.
114. Cohen, L., Time-frequency distributions - a review, *Proc. IEEE* **77** (7) (1989), 941–981.
115. Coifman, R. R., Adaptive multiresolution analysis, computation, signal processing, and operator theory, ICM 90, Kyoto.
116. Coifman, R. R., Wavelet analysis and signal processing in *Signal Processing I: Signal Processing Theory*, L. Auslander, T. Kailath, and S. Mitter (eds.), Springer-Verlag, 1990.
117. Coifman, R. R. and Y. Meyer, Remarques sur l'analyse de Fourier à fenêtre, *C. R. Acad. Sci. Paris* Série I **312** (1991), 259–261.

118. Coifman, R., Y. Meyer, S. Quake, and M. V. Wickerhauser, Signal processing and compression with wave packets, in *Proc. Conf. on Wavelets*, Marseilles, Spring 1989.
119. Coifman, R., Y. Meyer, and M. V. Wickerhauser, Size properties of wavelet packets, in *Wavelets and Their Applications*, M. B. Ruskai, G. Beylkin, R. Coifman, I. Daubechies, S. Mallat, Y. Meyer, and L. Raphael (eds.), Jones and Bartlett, Boston, 1992, 453–470.
120. Coifman, R. R., Y. Meyer, and M. V. Wickerhauser, Wavelet analysis and signal processing, in *Wavelets and Their Applications*, M. B. Ruskai, G. Beylkin, R. Coifman, I. Daubechies, S. Mallat, Y. Meyer, and L. Raphael (eds.), Jones and Bartlett, Boston, 1992, 153–178.
121. Coifman, R. R. and R. Rochberg, Representation theorems for holomorphic and harmonic functions in L^p, *Astérisque* **77** (1980), 11–66.
122. Coifman, R. R. and M. V. Wickerhauser, Entropy-based algorithms for best basis selection, *IEEE Trans. Inform. Theory* **38** (2) (1992), to appear.
123. Colella, D. and C. Heil, Characterizations of scaling functions, I: continuous solutions, MITRE Corporations, McLean, Virginia, 1991.
124. Colella, D. and C. Heil, The characterization of continuous, four-coefficient scaling functions and wavelets, *IEEE Trans. Inform. Theory* **38** (2) (1992), to appear.
125. Combes, J. M., A. Grossmann, and Ph. Tchamitchian, *Wavelets, Time-Frequency Methods and Phase Space*, Lecture Notes on IPTI, Springer-Verlag, Berlin, New York, 1989.
126. Conze, J. P. and A. Raugi, Fonctions harmoniques pour un opérateur de transition et applications, Dept. de Math., Université de Rennes, France, 1990, preprint.
127. Dahmen, W., and A. Kunoth, Multilevel preconditioning, preprint.
128. Dahmen, W. and C. A. Micchelli, On stationary subdivision and the construction of compactly supported orthonormal wavelets, in *Multivariate Approximation and Interpolation*, W. Haußmann and K. Jetter (eds.), Birkhäuser Verlag, Basel, 1990, 69–89.
129. Dahmen, W. and C. A. Micchelli, Banded matrices with banded inverses II: locally finite decomposition of spline spaces, preprint.
130. Dahmen, W. and C. A. Micchelli, Using the refinement equations for evaluating integrals of wavelets, preprint.
131. Daubechies, I., Orthonormal bases of compactly supported wavelets, *Comm. Pure and Appl. Math.* **41** (1988), 909–996.
132. Daubechies, I., Orthonormal bases of compactly supported wavelets, II: variations on a theme, AT&T Bell Laboratories, 1989, preprint, *SIAM J. Math. Anal.*, to appear.
133. Daubechies, I., The wavelet transform, time-frequency localization and signal analysis, *IEEE Trans. Inform. Theory* **36** (1990), 961–1005.
134. Daubechies, I., Time-frequency localization operators: a geometric phase space approach, *IEEE Trans. Inform. Theory* **34** (1988), 605–612.

135. Daubechies, I., *Ten Lectures on Wavelets*, CBMS/NSF Series in Applied Mathematics #61, SIAM Publ., 1992.
136. Daubechies, I. and A. Grossmann, Frames in the Bargmann space of entire functions, *Comm. Pure and Appl. Math.* **41** (1988), 151–164.
137. Daubechies, I., A. Grossmann, and Y. Meyer, Painless nonorthogonal expansions, *J. Math. Phys.* **27** (1986), 1271–1283.
138. Daubechies, I., S. Jaffard, and J. L. Journé, A simple Wilson orthonormal basis with exponential decay, *SIAM J. Math. Anal.* **22** (2) (1991), 554–572.
139. Daubechies, I. and J. C. Lagarias, Sets of matrices all infinite products of which converge, *Linear Algebra Appl.*, to appear.
140. Daubechies, I. and J. C. Lagarias, Two-scale difference operations I: existence and global regularity of solutions, *SIAM J. Math. Anal.* **22** (1991), 1388–1410.
141. Daubechies, I. and J. C. Lagarias, Two-scale difference operations II: infinite matrix products, local regularity and fractals, *SIAM J. Math. Anal.*, to appear.
142. Daubechies, I., and J. C. Lagarias, Two-scale difference equations, Part I and II, *SIAM J. Math. Anal.*, (1990), to appear.
143. Daubechies, I. and T. Paul, Time-frequency localization operators - a geometric phase space approach II, *Inverse Problems* **4** (1988), 661–680.
144. Daubechies, I. and T. Paul, Wavelets–some applications, in *Proc. Inter. Conf. Mathematical Physics*, M. Mebkkout and R. Seneor (eds.), World Scientific, 1986, 675–686.
145. Daugman, J. G., Complete discrete 2-D Gabor transforms by neural networks for image analysis and compression, *IEEE ASSP* **36** (1988), 1169–1179.
146. DeVore, R., B. Jawerth, and B. Lucier, Surface compression, preprint.
147. DeVore, R., B. Jawerth, and B. L. Lucier, Image compression through wavelet transform coding, *IEEE Trans. Inform. Theory* **38** (2) (1992), to appear.
148. DeVore, R., B. Jawerth, and V. Popov, Compression of wavelet decompositions, preprint.
149. Delprat, N., et al., Asymptotic wavelet and Gabor analysis: extraction of instantaneous frequencies, *IEEE Trans. Inform. Theory* **38** (2) (1992), to appear.
150. Doganata, Z., P. P. Vaidyanathan, and T. Q. Nguyen, General synthesis procedures for FIR lossless transfer matrices, for perfect reconstruction multirate filter bank applications, *IEEE Trans. ASSP* **36** (1988), 1561–1574.
151. Dubuc, S., Interpolation through an iterative scheme, *J. Math. Anal. and Appl.* **114** (1986), 185–204.
152. Dufaux, F. and M. Kunt, Massively parallel implementation for real-time Gabor decompositions, in *SPIE Conf. in Visual Comm. and Image Processing*, Boston, November 1991,

153. Duffin, R. J. and A. C. Schaeffer, A class of nonharmonic Fourier series, *Trans. Amer. Math. Soc.* **72** (1952), 341–366.
154. Dutilleux, P., An implementation of the 'algorithme 'a trous' to compute the wavelet transform, in *Wavelets, Time-Frequency Methods and Phase Space*, J. M. Combes, A. Grossmann, and Ph. Tchamitchian (eds.), Springer-Verlag, Berlin, New York, 1989.
155. Duval-Destin, M., M. A. Muschietti, and B. Torrésani, From continuous wavelets to wavelet packets, in preparation.
156. Eirola, T., Sobolev characterization of solutions of dilation equations, Helsinki University of Technology, submitted to *SIAM J. Math. Anal.*, 1991.
157. Escudié, B., Wavelet analysis of asymptotic signals: A tentative model for bat sonar receiver, in *Wavelets and Applications*, Y. Meyer (ed.), Mason/Springer, to appear.
158. Escudié, B. and B. Torrésani, Wavelet analysis of asymptotic signals, Centre de Physique Théorique, CNRS-Luminy, 1989, preprint.
159. Escudié, B. and B. Torrésani, Wavelet representation and time-scaled matched receiver for asymptotic signals, *Proc. Conf. EUSIPCO V*, Barcelona, North Holland, 1990, 305–308.
160. Esteban, D. and C. Galand, Applications of quadrature mirror filters to split band voice coding schemes, in *Proc. ICASSP*, Hartford, Connecticut, 1977, 191–195.
161. Farge, M., The continuous wavelet transform of two-dimensional turbulent flows, in *Wavelets and Their Applications*, M. B. Ruskai, G. Beylkin, R. Coifman, I. Daubechies, S. Mallat, Y. Meyer, and L. Raphael (eds.), Jones and Bartlett, Boston, 1992, 275–302.
162. Farge, M., Transformée en ondelettes et application á la turbulence, J. Annuelle, Societé Mathématique de France, Mai 1990.
163. Farge, M. and D. Rabreau, Transformée en ondelettes pour détecter et analyser les structures cohérentes dans les écoulements turbulents bidimensionnels, *C. R. Acad. Sci.* **307** Série 2 (1988), 1479–1486.
164. Farge, M., Y. Guezennec, C. M. Ho, and C. Nineveau, Continuous wavelet analysis of coherent structures, Proceedings, Summer Program, Center for Turbulence Research, Stanford University and NASA-Ames, 1990.
165. Farge, M., M. Holschneider, and J. F. Colonna, Wavelet analysis of coherent structures in two-dimensional turbulent flows, in *Topological Fluid Mechanics*, Moffatt (ed.), Cambridge Univ. Press, 1990.
166. Feauveau, J. C., Analyse multirésolution par ondelettes non orthogonales et bancs de filtres numériques, Ph. D. Thesis, Univ. Paris-Sud, 1990.
167. Feauveau, J. C., Wavelets for Quincunx pyramid, in *Wavelets and Their Applications*, M. B. Ruskai, G. Beylkin, R. Coifman, I. Daubechies, S. Mallat, Y. Meyer, and L. Raphael (eds.), Jones and Bartlett, Boston, 1992, 53–66.
168. Feauveau, J. C., Nonorthogonal multiresolution analysis using wavelets, in *Wavelets: A Tutorial In Theory and Applications*, C. K. Chui (ed.),

Academic Press, Cambridge, MA, 1992, 153–178.

169. Feauveau, J. C., P. Mathieu, M. Barlaud, and M. Antonini, Recursive biorthogonal wavelet transform for image coding, in *Proc. Int. Conf. Acoust., Speech and Signal Processing*, Toronto, Canada, 1991, 2649–2652.
170. Federbush, P., Navier and Stokes meet the wavelet, 1991, preprint.
171. Feichtinger, H., Atomic characterizations of modulation spaces through Gabor-type representations, *Proc. Conf. in Constructive Function Theory*, Edmonton, July 1986, *Rocky Mount. J. Math.* **19** (1989), 113–126.
172. Feichtinger, H., Coherent frames and irregular sampling, *Proc. Conf. in Recent Advances in Fourier Analysis and Its Applications*, NATO conference, PISA, July 1989, J. S. Byrnes and J. L. Byrnes (eds.), Kluwer Acad. Publ., *NATO ASI Series C* **315** (1990), 427–440.
173. Feichtinger, H. and K. Gröchenig, A unified approach to atomic decompositions via integrable group representations. *Proc. Conf. in Function Spaces and Applications*, Lund, 1986, Lecture Notes in Math. 1302 (1988), 52–73.
174. Feichtinger, H. and K. Gröchenig, Banach spaces related to integrable group representations and their atomic decompositions I, *J. Funct. Anal.* **86** (1989), 307–340.
175. Feichtinger, H. and K. Gröchenig, Banach spaces related to integrable group representations and their atomic decompositions II, *Monatsh. f. Math.* **108** (1989), 129–148.
176. Feichtinger, H. and K. Gröchenig, Error analysis in regular and irregular sampling theory, Applicable analysis, to appear.
177. Feichtinger, H. and K. Gröchenig, Iterative reconstruction of multivariate band-limited functions from irregular sampling values, SIAM J. Math. Anal., to appear.
178. Feichtinger, H. and K. Gröchenig, Non-orthogonal wavelet and Gabor expansions, and group representations, in *Wavelets and Their Applications*, M. B. Ruskai, G. Beylkin, R. Coifman, I. Daubechies, S. Mallat, Y. Meyer, and L. Raphael (eds.), Jones and Bartlett, Boston, 1992, 353–375.
179. Feichtinger, H., K. Gröchenig, and D. Walnut, Wilson bases and modulation spaces, *Math. Nachrichten*, 1991, to appear.
180. Feichtinger, H. G. and K. Gröchenig, Gabor wavelets and the Heisenberg group: Gabor expansions and short time Fourier transform from the group theoretical point of view, in *Wavelets: A Tutorial In Theory and Applications*, C. K. Chui (ed.), Academic Press, Cambridge, MA, 1992, 359–397.
181. Flandrin, P., Some aspects of non-stationary signal processing with emphasis on time-frequency and time-scale methods, in *Wavelets: Time-frequency Methods and Phase Space*, J. M. Combes, A. Grossmann, and Ph. Tchamitchian (eds.), Springer-Verlag, New York, 1989, 68–98.
182. Flandrin, P., Wavelet analysis and synthesis of fractional Brownian motion, *IEEE Trans. Inform. Theory* **38** (2) (1992), to appear.
183. Folland, G. B., *Harmonic Analysis in Phase Space*, Princeton University Press, Princeton, NJ, 1989.

184. Frazier, M. and B. Jawerth, Decomposition of Besov spaces, *Indiana Univ. Math. J.* **34** (1985), 777–799.
185. Frazier, M. and B. Jawerth, A discrete transform and decomposition of distribution spaces, *J. Func. Anal.* **93** (1990), 34–170.
186. Frazier, M. and B. Jawerth, Applications of the ϕ and Wavelet Transforms to the theory of function spaces, in *Wavelets and Their Applications*, M. B. Ruskai, G. Beylkin, R. Coifman, I. Daubechies, S. Mallat, Y. Meyer, and L. Raphael (eds.), Jones and Bartlett, Boston, 1992, 377–417.
187. Frazier, M. and B. Jawerth, The ϕ-transform and applications to distribution spaces, in *Function Spaces and Application*, M. Cwikel et al. (eds.), Lecture Notes in Mathematics 1302, Springer, Berlin, 1988, 233–246.
188. Frazier, M., B. Jawerth, and G. Weiss, *Littlewood-Paley Theory and the Study of Function Spaces*, CBMS Regional Conference Series 79, AMS, Providence, RI, 1991.
189. Frisch, M. and H. Messer, The use of wavelet transform in the detection of an unknown transient signal, *IEEE Trans. Inform. Theory* **38** (2) (1992), to appear.
190. Froment, J. and S. Mallat, Second generation compact image coding with wavelets, in *Wavelets: A Tutorial In Theory and Applications*, C. K. Chui (ed.), Academic Press, Cambridge, MA, 1992, 655–678.
191. Gabor, D., Theory of communication, *J. IEE (London)* **93** (1946), 429–457.
192. Glowinski R., W. Lawton, M. Ravachol, and E. Tenenbaum, Wavelet solution of linear and nonlinear elliptic, parabolic, and hyperbolic problems in one space dimension, in *Proc. Ninth Inter. Conf. on Computing Methods in Appl. Sciences and Engineering*, R. Glowinski and A. Lichnewsky (eds.), SIAM Publ., Philadelphia, PA, 1990.
193. Gopinath, R. A., Moment theorem for the wavelet-Galerkin method, Technical Report, Aware, Inc., Cambridge, MA, 1990.
194. Gopinath, R. A., Wavelet transforms and time-scale analysis of signals, Master Thesis, Rice University, 1990.
195. Gopinath, R. A. and C. S. Burrus, Wavelet transforms and filter banks, in *Wavelets: A Tutorial In Theory and Applications*, C. K. Chui (ed.), Academic Press, Cambridge, MA, 1992, 603–654.
196. Gopinath, R. A., W. M. Lawton, and C. S. Burrus, Wavelet-Galerkin approximation of linear translation invariant operators, in *Proc. ICASSP, Toronto, Canada* **3** (1991) 2021–2023.
197. Goodman, T. N. T., S. L. Lee, and W. S. Tang, Wavelets in wandering subspaces, CAT Report #256 , Texas A&M University, 1991.
198. Gröchenig, K., Analyse multiéchelles et bases d'ondelettes, *CRAS Paris, Série 1* **305** (1987), 13–15.
199. Gröchenig, K., Describing functions: atomic decompositions versus frames, *Monatsh. Math.* **112** (1991), 1–42.
200. Gröchenig, K. and W. R. Madych, Multiresolution analyses, Haar bases, and self-similar tilings of R^n, *IEEE Trans. Inform. Theory* **38** (2) (1992),

to appear.

201. Gröchenig, K. and D. Walnut, A Riesz basis for Bargmann-Fock space related to sampling and interpolation, *Ark. f. Mat.*, to appear.
202. Grossmann, A., M. Holschneider, R. Kronland-Martinet, and J. Morlet, Detection of abrupt changes in sound signals with the help of wavelet transforms, in *Inverse problems: An interdisciplinary study; Advances in electronics and electron physics*, Supplement nr. 19, Academic Press, 1987.
203. Grossmann, A., R. Kronland-Martinet, and J. Morlet, Reading and understanding continuous wavelet transforms, in *Wavelets*, J. M. Combes, A. Grossmann, and Ph. Tchamitchian (eds.), Springer-Verlag, NY, 1989, 2–20.
204. Grossmann, A. and J. Morlet, Decomposition of Hardy functions into square integrable wavelets of constant shape, *SIAM J. Math. Anal.* **15** (1984), 723–736.
205. Grossmann, A. and J. Morlet, Decomposition of functions into wavelets of constant shape, and related transforms, in *Mathematics and Physics, Lectures on Recent Results*, L. Streit (ed.), World Scientific Publ., Singapore, 1985.
206. Grossmann, A., J. Morlet, and T. Paul, Transforms associated to square integrable group representations, I, *J. Math. Phys.* **26** (1985), 2473–2479
207. Grossmann, A., J. Morlet, and T. Paul, Transforms associated to square integrable group representations, II, *Annales l'IHP, Phys. Theor.* **45** (1986), 293–309.
208. Grossmann, A., G. Saracco, and Ph. Tchamitchian, Study of propagation of transient acoustic signals across a plane interface with the help of the wavelet transform, preprint.
209. Haar, A., Zur Theorie der orthogonalen Funktionen-systeme, *Math. Ann.* **69** (1910), 331–371.
210. Healy, D. M. and J. B. Weaver, Two applications of wavelet transforms in magnetic resonance imaging, *IEEE Trans. Inform. Theory* **38** (2) (1992), to appear.
211. Heil, C. and D. Walnut, Continuous and discrete wavelet transforms, *SIAM Review* **31** (1989), 628–666.
212. Heller, W., Frames of exponentials and applications, Ph. D. Thesis, University of Maryland, College Park, MD, 1991.
213. Heller, P. N., Higher multiplier Daubechies scaling coefficients, Technical Report, Aware, Inc., Cambridge, MA.
214. Heller, P., H. L. Resnikoff, and R. O. Wells, Jr., Wavelet matrices and the representation of discrete functions, in *Wavelets: A Tutorial In Theory and Applications*, C. K. Chui (ed.), Academic Press, Cambridge, MA, 1992, 15–50.
215. Herley, C. and M. Vetterli, Linear phase wavelets: theory and design, in *Proc. IEEE Int. Conf. ASSP*, Toronto, Canada, 1991, 2017–2020.
216. Herley, C. and M. Vetterli, Wavelets and recursive filter banks, Technical Report, Columbia University, 1992, submitted to *IEEE Trans. Signal*

Processing.
217. Hervé, L., Ph. D. Thesis, University of Rennes I, France, 1991.
218. Holschneider, M. R. Kronland-Marttinet, J. Morlet, and Ph. Tchamitchian, A real-time algorithm for signal analysis with the help of the wavelet transform, in *Wavelets, Time-Frequency Methods and Phase Space*, J. M. Combes, A. Grossmann, and Ph. Tchamitchian, eds., Lecture Notes on IPTI, Springer-Verlag, Berlin, New York, 1989, 286–297.
219. Holschneider, M. R. and Ph. Tchamitchian, On the wavelet analysis of the Riemann's function, Centre de Physique Théorique, CNRS-Luminy, 1988, preprint.
220. Hummel, R. and R. Moniot, Reconstruction from zero-crossing in space-space, in *IEEE Trans. ASSP* **37** (1989).
221. Jaffard, S., Analyse par ondelettes d'un problème elliptique singulier, *J. Math. Pures et Appl.*, to appear.
222. Jaffard, S., Construction et propriétés des bases d'ondelettes, Remarques sur la controlabilité exacte, Thèse de l'école Polytechnique, 1989.
223. Jaffard, S., Construction of wavelets on open sets, in *Wavelets*, J. M. Combes, A. Grossmann, and Ph. Tchamitchian (eds.), Springer-Verlag, Berlin, 1989, 247–252.
224. Jaffard, S., Exposants de Hölder en des points donnés et coefficients d'ondelettes, *C. R. Acad. Sci.* **308** Série 1 (1989), 79–81.
225. Jaffard, S., Local order of approximation by wavelets, 1991, preprint.
226. Jaffard, S., Pointwise smoothness, two-microlocalization and wavelet coefficients, *Publicacions Matematiques* **35** (1991), 155–168.
227. Jaffard, S., Propriétés des matrices "bien localisées" près de leur diagonale et quelques applications, *Annales de l'IHP*, série Problèmes non linéaires **7** (5) (1990), 461–476.
228. Jaffard, S., Sur la dimension de Hausdorff des points singuliers d'une fonction, 1991, preprint.
229. Jaffard, S., Wavelet methods for fast resolution of elliptic problems, *SIAM J. Numer. Anal.*, to appear.
230. Jaffard, S. and Ph. Laurençot, Orthonormal wavelets, analysis of operators, and applications to numerical analysis, in *Wavelets: A Tutorial In Theory and Applications*, C. K. Chui (ed.), Academic Press, Cambridge, MA, 1992, 543–601.
231. Jaffard,, S., P. G. Lemarié, S. Mallat, and Y. Meyer, Dyadic multiscale analysis of $L^2(R^n)$, manuscript.
232. Jaffard, J. and Y. Meyer, Bases d'ondelettes dans des ouverts de $\mathbb{R}^n$, *J. Math. Pures et Appl.* **68** (1989), 95–108.
233. Janssen, A. J. E. M., Gabor representation of generalized functions, *J. Math. Anal. Appl.* **83** (1981), 377–394.
234. Janssen, A. J. E. M., The Smith-Barnwell condition and non-negative scaling functions, *IEEE Trans. Inform. Theory* **38** (2) (1992), to appear.
235. Jensen, H. E., T. Hoholdt, and J. Justesen, Double series representation of bounded signals, *IEEE Trans. Inform. Theory* **34** (1988), 613–624.

236. Jia, R.-Q. and C. A. Micchelli, Using the refinement equations for the construction of pre-wavelets II: powers of two, in *Curves and Surfaces*, P.-J. Laurent, A. Le Méhauté, and L. L. Schumaker (eds.), Academic Press, San Diego, 1991, 209–246.
237. Jia, R.-Q. and C. A. Micchelli, Using the refinement equations for the construction of pre-wavelets V: extensibility of trigonometric polynomials, 1991, preprint.
238. Kadambe, S. and G. F. Boudreaux-Bartels, Application of the wavelet transform for pitch detection of speech signals, *IEEE Trans. Inform. Theory* **38** (2) (1992), to appear.
239. Kaiser, G., An algebraic theory of wavelets, Part I: Operational calculus and complex structure, *SIAM J. Math. Anal.***23** (1) (1992), to appear.
240. Kaiser, G., Generalized wavelet transforms, Part I: The windowed X–Ray transform, Technical Reports Series #18, University of Lowell, 1990.
241. Kaiser, G., Generalized wavelet transforms, Part II: The multivariate analytic–signal transform, Technical Reports Series #19, University of Lowell, 1990.
242. Kaiser, G., Phase–Space approach to relativistic quantum mechanics, Ph. D. Thesis, Mathematics Department, University of Toronto, 1977.
243. Kaiser, G. and R. R. Streater, Windowed radon transforms, analytic signals, and the wave equation, in *Wavelets: A Tutorial In Theory and Applications*, C. K. Chui (ed.), Academic Press, Cambridge, MA, 1992, 399–441.
244. Klauder, J. and K. Skagerstam (eds.), *Coherent States – Applications in Physics and Mathematical Physics*, Singapore, World Scientific Publ., 1985.
245. Kovaãević, J. and M. Vetterli, Nonseparable multidimensional perfect reconstruction filter banks and wavelet bases for $\mathbb{R}^n$, *IEEE Trans. Inform. Theory* **38** (2) (1992), to appear.
246. Kronland-Martinet, R., J. Morlet, and A. Grossmann, Analysis of sound patterns through wavelet transforms, *Int. J. of Pattern Recognition and Artificial Intelligence* **1** (1987), 273–301.
247. Laeng, E., Une base orthonormale de $L^2(\mathbb{R})$, dont les éléments sont bien localisés dans l'espace de phase et leurs supports adaptés à toute partition symétrique de l'espace des fréquences, *C. R. Acad. Sci. Paris* **311** (1990), 677–680.
248. Landau, H., An overview of time and frequency limiting, in *Fourier Techniques and Applications*, J. Price (ed.), Plenum Press, NY, 1985, 201-220.
249. Latto, A., and E. Tenenbaum, Compactly supported wavelets and the numerical solution of the Burgers' equation, *C. R. Acad. Sci. Paris* **311** Série I (1990), 903-909.
250. Laurençot, Ph., Résolution par ondelettes 1-D de l'équation de Burgers avec viscosité et de l'équation de Korteweg-De Vries avec conditions limites périodiques, Rapport de DEA, Université Paris XI, 1990.
251. Lawton, W. M., Lectures on wavelets, *CBMS Conference Series*, Lowell University, 1990.

252. Lawton, W. M., Necessary and sufficient conditions for constructing orthonormal wavelets bases, *J. Math. Phys.* **32** (1) (1991), 57–61.
253. Lawton, W. M., Tight frames of compactly supported affine wavelets, *J. Math. Phys.* **31** (8) (1990), 1898–1901.
254. Lawton, W. M. and H. L. Resnikoff, Multidimensional wavelet bases, Aware, Inc. Technical Report, submitted to *SIAM J. Math. Anal.*
255. Lemarié, P. G., Bases d'ondelettes sur les groupes de Lie stratifiés, *Bull. Soc. Math. France,* to appear.
256. Lemarié, P. G., Constructions d'ondelettes-splines, 1987, unpublished.
257. Lemarié, P. G., Ondelettes à localisation exponentielles, *J. Math. Pures et Appl.* **67** (1988), 227–236.
258. Lemarié, P. G., *Les ondelettes en 1989*, Lecture notes in Mathematics nr. 1438, Springer, Berlin, 1990.
259. Lemarié, P. G., Existence de fonction-père pour des ondelettes *à* support compact, *C. R. Acad. Sci. Paris*, to appear.
260. Lemarié, P. G. and G. Malgouyres, Support des fonctions de base dans une analyse multirésolution, *C. R. Acad. Sci. Paris*, to appear.
261. Lemarié, P. and Y. Meyer, Ondelettes et bases Hilbertiennes, *Rev. Mat. Iberoamericana* **2** (1986), 1–18.
262. Liandrat, J., V. Perrier, and Ph. Tchamitchian, Numerical resolution of the regularized Burgers equation using the wavelet transform, 1990, preprint.
263. Liandrat, J., V. Perrier, and Ph. Tchamitchian, Numerical resolution of nonlinear partial differential equations using the wavelet approach, in *Wavelets and Their Applications*, M. B. Ruskai, G. Beylkin, R. Coifman, I. Daubechies, S. Mallat, Y. Meyer, and L. Raphael (eds.), Jones and Bartlett, Boston, 1992, 227-238.
264. Liandrat, J. and Ph. Tchamitchian, Resolution of the 1D regularized Burgers equation using a spatial wavelet approximation, ICASE Report #90-83, NASA, 1990.
265. Light, W. (ed.), *Wavelets, Subdivisions, and Radial Functions*, Oxford University Press, Oxford, 1991.
266. Littlewood, J. E. and R. E. A. C. Paley, Theorems on Fourier series and power series, I, *J. London Math. Soc.* **6** (1931), 230–233.
267. Littlewood, J. E. and R. E. A. C. Paley, Theorems on Fourier series and power series, II, *Proc. London Math. Soc.* **42** (1936), 52–89.
268. Lorentz, R. A. H. and W. R. Madych, Spline wavelets for ordinary differential equations, GMD Technical Report.
269. Lorentz, R. A. H. and W. R. Madych, Wavelets and generalized box splines, GMD Technical Report, *Applicable Analysis*, to appear.
270. Lyubarskii, Y., Frames in the Bargmann Space of Entire Functions, *Proc. of Seminar on Complex Analysis*, Lecture Notes in Math. #15, Springer-Verlag, NY, to appear.
271. Madych, W. R., Multiresolution analyses, tiles, and scaling functions, in *Proc. of NATO ASI II Ciocco*, 1991, to appear.

272. Madych, W. R., Multiresolution analysis, wavelets, and homogeneous functions, BRC, preprint.
273. Madych, W. R., Polyharmonic splines, multiscale analysis, and entire functions, in *Multivariate Approximation and Interpolation*, W. Haussmann and K. Jetter (eds.), Birkhäuser Verlag, Basel, 1990, 205–216.
274. Madych, W. R., Translation invariant multiscale analysis, in *Recent Advances in Fourier Analysis and its Applications*, J. S. Byrnes and J. L. Byrnes (eds.), *NATO ASI Series C* **315**, Kluwer, Dordrecht, 1990, 455–462.
275. Madych, W. R., Some elementary properties of multiresolution analyses of $L^2(\mathbb{R}^n)$, in *Wavelets: A Tutorial In Theory and Applications*, C. K. Chui (ed.), Academic Press, Cambridge, MA, 1992, 259–294.
276. Mallat, S., A theory of multiresolution signal decomposition: the wavelet representation, *IEEE Trans. PAMI* **11** (1989), 674–693.
277. Mallat, S., Multifrequency channel decompositions of images and wavelet models , *IEEE Trans. ASSP* **37** (1989), 2091–2110.
278. Mallat, S., Multiresolution approximations and wavelet orthonormal bases of $L^2(\mathbf{R})$, *Trans. Amer. Math. Soc.* **315** (1989), 69–87.
279. Mallat, S., Multiresolution representation and wavelets, Ph. D. Thesis, Univ. of Pennsylvania, Philadelphia, 1988.
280. Mallat, S., Review of multifrequency channel decomposition of images and wavelet models, Technical Report 412, Robotics Report 178, NYU, 1988.
281. Mallat, S., Zero-crossings of a wavelet transform, *IEEE Trans. Inform. Theory* **37** (1991), 1019–1033.
282. Mallat, S. and W. L. Hwang, Singularity detection and processing with wavelets, *IEEE Trans. Inform. Theory* **38** (2) (1992), to appear.
283. Mallat, S. and S. Zhong, Complete signal representation with multiscale edges, Computer Science Technical Report #483, New York University, 1989, submitted to *IEEE Trans. PAMI.*
284. Mallat, S. and S. Zhong, Characterization of signals from multiscale edges, *IEEE Trans. PAMI* (1991), to appear.
285. Mallat, S. and S. Zhong, Wavelet transform maxima and multiscale edges, in *Wavelets and Their Applications*, M. B. Ruskai, G. Beylkin, R. Coifman, I. Daubechies, S. Mallat, Y. Meyer, and L. Raphael (eds.), Jones and Bartlett, Boston, 1992, 67–104.
286. Malvar, H., Lapped transforms for efficient transform/subband coding, *IEEE Trans. ASSP* **38** (1990), 969–978.
287. Marks, R., *Introduction to Shannon Sampling and Interpolation Theory*, Springer-Verlag, NY, 1991.
288. Marr, D. and E. Hildreth, Theory of edge detection, *Proc. Roy. Soc. London* **B 207** (1980), 187–217.
289. Meyer, Y., Ondelettes et calcul scientifique performant, INRIA, 1989.
290. Meyer, Y., Ondelettes et fonctions splines, Technical Report, *Séminaire EDP*, École Polytechnique, Paris, 1986.

291. Meyer, Y., *Ondelettes et Operateurs*, in two volumes, Hermann, Paris, 1990.
292. Meyer, Y., Ondelettes, fonctions splines et analyses graduées, Rapport CEREMADE **8703**, 1987.
293. Meyer, Y., Ondelettes sur l'intervalle, *Rev. Mat. Iberoamericana*, to appear.
294. Meyer, Y. (ed.), *Proc. Inter. Conf. on Wavelets*, May 1989, Marseille (France); to be published by Masson (Paris), (1991).
295. Meyer, Y., Principe d'incertitude, bases Hilbertiennes et algèbres d'opérateurs, *Séminaire Bourbaki* **662** (1985-1986).
296. Meyer, Y., *Séminaire Bourbaki* **38** (1985-86), 662.
297. Meyer, Y., The Franklin Wavelets, 1988, preprint.
298. Meyer, Y., (ed.), Wavelets and applications, Proceedings of the second wavelet conference, Marseille (1989), Masson/Springer, to appear.
299. Meyer, Y., Wavelets and applications, in *Proc. of the Int. Congress of Math.*, Kyoto, 1990.
300. Meyer, Y., Wavelets and operators, in *Analysis at Urbana*, Vol. 1, E. Berkson, N. T Peck, and J. Uhl (eds.), London Math Society, Lecture Notes Series 137, Cambridge Univ.Press, 1989, 256–364.
301. Meyer, Y. and R. R. Coifman, *Ondelettes et Opérateurs III–Opérateurs Multilinéaures*, Hermann, Paris, 1992.
302. Micchelli, C. A., Using the refinement equation for the construction of pre-wavelets, *Numerical Algorithms* **1** (1991), 75–116.
303. Micchelli, C. A. and H. Prautzsch, Uniform refinement of curves, *Lin. Alg. & Appl.* **114/115** (1989), 841–870.
304. Munch, J., Noise reduction in tight Weyl-Heisenberg frames, *IEEE Trans. Inform. Theory* **38** (2) (1992), to appear.
305. Murenzi, R., Ondelettes multidimensionelles et application *á* l'analyse d'images, Ph. D. Thesis, Université Catholique de Louvain, 1990.
306. Nawab, S. H. and T. Quatieri, Short-time fourier transform, in *Advanced Topics in Signal Processing*, J. S. Lim and A. V. Oppenheim (eds.), Prentice Hall Signal Processing Series, 1988.
307. Olsen, P. A. and K. Seip, A note on irregular discrete wavelet transforms, *IEEE Trans. Inform. Theory* **38** (2) (1992), to appear.
308. Paul, T., Ondelettes et mécanique quantique, Ph. D. Thesis, Université de Marseille, 1985.
309. Paul, T. and K. Seip, Wavelets and quantum mechanics, in *Wavelets and Their Applications*, M. B. Ruskai, G. Beylkin, R. Coifman, I. Daubechies, S. Mallat, Y. Meyer, and L. Raphael (eds.), Jones and Bartlett, Boston, 1992, 303–321.
310. Perrier, V., Ondelettes et simulation numériques, Thèse de l'Université Paris VI, February 1991.
311. Pollen, D., Parametrization of compactly supported wavelets, Aware Technical Report, Aware, Inc., Cambridge, MA.

312. Pollen, D., $SU_I(F[z, 1/z])$ for F a subfield of $\mathbb{C}$, *J. Amer. Math. Soc.* **3** (3) (1990), 611–624.
313. Pollen, D., Daubechies' scaling function on [0,3], in *Wavelets: A Tutorial In Theory and Applications*, C. K. Chui (ed.), Academic Press, Cambridge, MA, 1992, 3–13.
314. Porat, M. and Y. Y. Zeevi, The generalized Gabor scheme of image representation using elementary functions matched to human vision, in *Theory Foundations of Computer Graphics*, Springer Verlag, NY, 1197–1241.
315. Portnoff, M. R., Time frequency distributions of digital systems and signals based on short-time fourier analysis, *IEEE Trans. ASSP* **28** (1980), 55–69.
316. Quake, S., R. Coifman, Y. Meyer, and M. V. Wickerhauser, Signal processing and compression with wave packets, preprint.
317. Resnikoff, H. L., Weierstrass functions and compactly supported wavelets. Technical Report AD900810, Aware, Inc., Cambridge, MA, 1990.
318. Resnikoff, H. L. and R. O. Wells, Jr., *Wavelet Analysis*, in preparation.
319. de Rham, G., Sur une courbe plane, *J. Math. Pures et Appl.* **39** (1956), 25–42.
320. Riemenschneider, S. D. and Z. W. Shen, Wavelets and pre-wavelets in low dimensions, *J. Approx. Theory*, to appear.
321. Riemenschneider, S. D. and Z. W. Shen, Box splines, cardinal series, and wavelets, in *Approximation Theory and Functional Analysis*, C. K. Chui (ed.), Academic Press, NY, 1991, 133–149.
322. Rioul, O., A discrete-time multiresolution theory unifying octave band filter banks, Pyramid and wavelet transforms, *IEEE Trans. Signal Proccessing*, to appear.
323. Rioul, O., A unifying multiresolution theory for discrete wavelet transform, regular filter banks and pyramid transforms, IEEE Trans. ASSP, (1990), submitted.
324. Rioul, O. and P. Duhamel, Fast algorithms for discrete and continuous wavelet transforms, *IEEE Trans. Infor. Theory* **38** (2) (1992), to appear.
325. Ruskai, M. B., G. Beylkin, R. Coifman, I. Daubechies, S. Mallat, Y. Meyer, and L. Raphael (eds.), *Wavelets and Their Applications*, Jones and Bartlett, Boston, 1992.
326. Schoenberg, I. J., *Cardinal Spline Interpolation* CBMS/NSF Series in Applied Mathematics #12. SIAM Publ., 1973.
327. Schumaker, L. L., *Spline Functions: Basic Theory*, Wiley-Interscience, NY, 1981.
328. Schweinler, H. C. and E. P. Wigner, Orthogonalization methods, *J. Math. Phys.* **11** (1970), 1693–1694.
329. Seip, K., Mean value theorems and concentration operators in Bargmann and Bergman spaces, in *Wavelets*, J. M. Combes, A. Grossmann, and Ph. Tchamitchian (eds.), Springer-Verlag, NY, 1989, 209–215.
330. Seip, K., Reproducing formulas and double orthogonality in Bargmann and Bergman spaces, *SIAM J. Math. Anal.* **22** (1991), 856–876.

331. Seip, K., Wavelets in $H^2(\mathbb{R})$: Sampling, interpolation, and phase space density, in *Wavelets: A Tutorial In Theory and Applications*, C. K. Chui (ed.), Academic Press, Cambridge, MA, 1992, 529–540.
332. Smith, M. J. T. and T. P. Barnwell, III, Exact reconstruction techniques for tree-structured subband coders, *IEEE Trans. ASSP* **34** (1986), 434–441.
333. Stöckler, J., The construction of box-spline wavelets in arbitrary dimension, preprint.
334. Stöckler, J., Multivariate wavelets, in *Wavelets: A Tutorial In Theory and Applications*, C. K. Chui (ed.), Academic Press, Cambridge, MA, 1992, 325–355.
335. Strömberg, J.-O., A modified Franklin system and higher order spline systems on $\mathbf{R}^n$ as unconditional bases for Hardy spaces, in *Proc. Conf. in Honor of Antoni Zygmund*, Vol. II, W. Beckner, A. P. Calderón, R. Fefferman, and P. W. Jones (eds.), Wadsworth, NY, 1981, 475–493.
336. Strang, G., Wavelets and dilation equations: a brief introduction, *SIAM Review* **31** (1989), 614–627.
337. Strichartz, R. S., Wavelets and self-affine tilings, preprint.
338. Super, B. J. and A. C. Bovik, Shape from texture using Gabor wavelets, *SPIE Conf. in Visual Comm. and Image Processing*, Boston, November 1991.
339. Tchamitchian, Ph., Bases d'ondelettes et intégrales singulières: Analyse des fonctions et calcul sur les opérateurs, Habilitation, Université d'Aix-Marseille 2, 1989.
340. Tchamitchian, Ph., Biorthogonalité et théorie des opérateurss, *Rev. Math. Iberoamericana*, to appear.
341. Tchamitchian, Ph., Ondelettes et intégrale de Cauchy sur les Courbes Lipschitziennes, *Annals of Math.* **303** (1986), 215–218.
342. Tchamitchian, Ph., Calcul symbolique sur les opérateurs de Caldéron-Zygmund et bases inconditionnelles de $L^2(\mathbb{R}^n)$, *C. R. Acad. Sci. Paris* **129** (1989), 641–649.
343. Tchamitchian, Ph. and B. Torrésani, Ridge and skeleton extraction from the wavelet transform, in *Wavelets and Their Applications*, M. B. Ruskai, G. Beylkin, R. Coifman, I. Daubechies, S. Mallat, Y. Meyer, and L. Raphael (eds.), Jones and Bartlett, Boston, 1992, 123–151.
344. Tewfik, A. and M. Kim, Correlation structure of the discrete wavelet coefficients of fractional Brownian motion, *IEEE Trans. Inform. Theory* **38** (2) (1992), to appear.
345. Tewfik, A., D. Sinha, and P. Jorgensen, On the optimal choice of a wavelet for signal representation, *IEEE Trans. Inform. Theory* **38** (2) (1992), to appear.
346. Torrésani, B., Wavelet analysis of asymptotic signals: Ridge and Skeleton of the transform, *Wavelets and Applications*, Y. Meyer (ed.), Masson/Springer, to appear.

347. Torrésani, B., Wavelets associated with representations of the affine Weyl-Heisenberg group, *J. Math Phys.*, to appear.
348. Torrésani, B., Time-frequency representations: wavelet packets and optimal decompositions, *Ann. Inst. H. Poincaré*, to appear.
349. Unser, M., A. Aldroubi, and M. Eden, On the asymptotic convergence of B-spline wavelets to Gabor functions, *IEEE Trans. Inform. Theory* **38** (2) (1992), to appear.
350. Unser, M. and A. Aldroubi, Polynomial splines and wavelets–a signal processing perspective, in *Wavelets: A Tutorial In Theory and Applications*, C. K. Chui (ed.), Academic Press, Cambridge, MA, 1992, 91–122.
351. Vaidyanathan, P. P., *Multirate Systems and Filter Banks*, Prentice Hall, Englewood Cliffs, NJ, 1993, to appear.
352. Vaidyanathan, P. P., Quadrature mirror filter banks, M-band extensions, and perfect-reconstruction techniques, *IEEE ASSP Magazine* **4** (3) (1987), 4–20.
353. Vetterli, M., Wavelets and filter banks for discrete-time signal processing, in *Wavelets and Their Applications*, M. B. Ruskai, G. Beylkin, R. Coifman, I. Daubechies, S. Mallat, Y. Meyer, and L. Raphael (eds.), Jones and Bartlett, Boston, 1992, 17–52.
354. Vetterli, M. and C. Herley, Wavelets and filter banks: theory and design, *IEEE ASSP* (1992), to appear.
355. Vetterli, M. and K. Herley, Linear phase wavelets, *IEEE Trans. ASSP*, to appear.
356. Vetterli, M., J. Kovaăević and Le Gall, Perfect reconstruction filter banks for HDTV representation and coding, *Image Communication* **2** (1990), 349–364.
357. Villemoes, L., Energy moments in time and frequency for two scale difference equation solutions and wavelets, Math. Institute, Technical University of Denmark, DK2800 Lyngby, Denmark, 1991, preprint.
358. Volkmer, H., Distributional and square summable solutions of dilation equations, preprint.
359. Volkner, H., On the regularity of wavelets, *IEEE Trans. Inform. Theory* **38** (2) (1992), to appear.
360. Walnut, D. F., Continuity properties of the Gabor frame operator, JMAA, 1991, to appear.
361. Walnut, D. F., Lattice size estimates for Gabor decompositions, preprint.
362. Walter, G., A sampling theorem for wavelet subspaces, 1990, preprint.
363. Walter, G., Discrete wavelets, 1989, preprint.
364. Walter, G., Wavelets and generalized functions, in *Wavelets: A Tutorial In Theory and Applications*, C. K. Chui (ed.), Academic Press, Cambridge, MA, 1992, 51–70.
365. Wells, Jr., R. O., Parametrizing smooth compactly supported wavelets, *Trans. Amer. Math. Soc.* (1991), to appear.
366. Wexler, J. and S. Raz, Discrete Gabor expansions, *Signal Processing* **21** (1990), 207–220.

367. Wickerhauser, M. V., Acoustic signal processing with wavelet packets, Yale University, 1989, preprint.
368. Wickerhauser, M. V., Picture compression by best basis subband coding, Yale University, 1990, preprint.
369. Wickerhauser, M. V., Acoustic signal compression with wavelet packets, in *Wavelets: A Tutorial In Theory and Applications*, C. K. Chui (ed.), Academic Press, Cambridge, MA, 1992, 679–700.
370. Wilson, R., A. Calway, and E. Pearson, A generalized wavelet transform for Fourier analysis: the multiresolution Fourier transform and its application to image and audio signal analysis, *IEEE Trans. Inform. Theory* **38** (2) (1992), to appear.
371. Wornell, G. and A. Oppenheim, Wavelet-based representations for a class of self-similar signals with application to fractal modulation, *IEEE Trans. Inform. Theory* **38** (2) (1992), to appear.
372. Young, R. M., *An Introduction to Nonharmonic Fourier Series*, Academic Press, New York, 1980.
373. Zhong, S., *Edges representation from wavelet transform maxima*, Ph. D. Thesis, New York University, 1990.

Subject Index

D

E

WAVELET ANALYSIS AND ITS APPLICATIONS

CHARLES K. CHUI, SERIES EDITOR

1. Charles K. Chui, *An Introduction to Wavelets*
2. Charles K. Chui, ed., *Wavelets: A Tutorial in Theory and Applications*